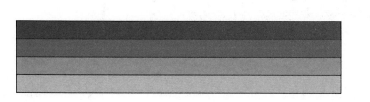

HUMAN
ANATOMY

Ira Fowler   *University of Kentucky Medical Center*

Wadsworth Publishing Company
Belmont, California
A Division of Wadsworth, Inc.

Biology Editor: Jack C. Carey

Production Editor: Hal Humphrey

Cover and Interior Designer: MaryEllen Podgorski

Copy Editor: Yvonne Howell

Cover Illustration: Leslie Laurien

Illustrators and photographers are listed on p. 617.

Printed in the United States of America

1 2 3 4 5 6 7 8 9 10—88 87 86 85 84

Library of Congress Cataloging in Publication Data

Fowler, Ira.
    Human anatomy.

    Includes bibliographies and index.
    1. Anatomy, Human.   I. Title.   [DNLM: 1. Anatomy.
QS 4 F786h]
QM23.2.F68 1984   611   83-16991
ISBN 0-534-02746-6

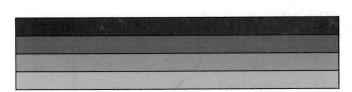

PREFACE

The human body is enormously complex in its arrangements of cells, tissues, and organs that together comprise the whole. This complexity can be reduced to a comprehensible level by considering the patterns of organization of the systems and how these patterns facilitate specific functions.

Structure responds to the specific demands of function. For example, ligaments that cross a movable joint support the joint because they resist tension, preventing the bones from being pulled apart. During growth and development of the individual, intermittent tension stimulates formation of the ligament; the greater the tension applied, the stronger the ligament becomes. This book presents the patterns of organization and their relations to function and thus reduces the complexity of the body to an understandable level for the beginning student.

Approach

We begin with a short history of anatomy and definitions of some of the more frequently used anatomical terms. The cells and tissues are described next, prior to considering each of the organ systems. Early embryonic development is described briefly in Chapter 3 as preparation for an account of the development of each of the systems in the appropriate chapters. Some anatomists consider a knowledge of development to be important in understanding the basic patterns of organization of the body. However, the sections on development may be omitted by instructors who feel that all the student's time should be devoted to the adult. The organ systems (skin, skeleton, and muscle) that are concerned with the somatic functions of support, protection, and movement are described prior to the systems that perform the visceral functions. The nervous and endocrine systems follow the muscles because these two systems are responsible for regulating and integrating all of the activities of the other organ systems. By having students acquire some knowledge of these most complex systems relatively early in the course, it not only becomes possible to discuss the regulation of each system, such as the cardiovascular system, in the appropriate chapter, but an opportunity is afforded for reinforcement of the student's understanding of the nervous and endocrine systems. Each chapter on the systems includes a section on the integration of the system by nerves and hormones. The chapter on body defenses is placed last because all systems contribute to defense mechanisms to some degree. There is no separate chapter on surface anatomy, but this important learning aid is incorporated into each of the chapters on the organ systems by brief descriptions of palpation points or of surface projections.

Accuracy and Readability

Accuracy and readability have been checked by more than fifty reviewers. The text and illustrations were carefully and thoroughly reviewed by scientists with various areas of expertise to insure accuracy, currency, and clarity of presentation. The reviewers included zoologists who teach anatomy and physiology, anatomists who teach in zoology departments, anatomists who teach in anatomy departments, neuroanatomists, electron microscopists, endocrinologists, an immunologist, and a pathophysiologist.

Illustrations

Anatomy is a visual science; the first drawings on the walls of caves were of human hands and their fingers. Some of the best means of recalling human structure are simple pictures built up from a diagrammatic skeletal background. Many of the illustrations in this book are single-concept diagrams designed to help the student remember one important anatomical fact. The verbal description and picture are aimed at helping the student retain a visual image of the anatomy involved and envisage the usefulness of its structure in performing its specific function.

The illustrations include several photographs of cadaver preparations that have proven useful in anatomy courses given for professional students at the University of Kentucky. In addition to cadaver dissections, these preparations include x rays of thick cadaver slices using soft x rays that enhance the contrast of specific structures. The University of Kentucky Medical Center was one of the first institutions in this country to obtain equipment for nuclear magnetic resonance (NMR) imaging. In this book, many regions of the body are shown by a special series of NMR imaging prepared using a healthy 25-year-old male subject. The NMR imaging can be compared to CAT scans of similar areas of the body. NMR is becoming an important diagnostic tool in the examination of soft tissues of the body.

Many of the photographs are accompanied by line drawings that accent the most important anatomical features shown in the illustration. This procedure makes it possible to identify the structures illustrated without the confusion that often results from adding lines to a photograph.

Throughout the text are photomicrographs, including scanning and transmission electron micrographs. There are also multiple-part figures to provide different perspectives and sequential illustrations.

Most of the chapters have tables that are placed within the text near the topics they summarize.

Clinical Applications

To motivate students and to stimulate their interest in learning normal anatomy prior to delving deeply into abnormal or pathological conditions, brief accounts of some clinical applications are presented in two forms:

1. Boxed inserts   (of a few sentences) are placed within the text near the discussion to which the insert is related. These inserts are not all discussions of diseases; many are special practical applications of principles of normal anatomy.

2. End-of-chapter clinical applications   are discussions of some of the major diseases of the system. These accounts are not extensive because it is important for students to have a basic knowledge of the normal structure and function of the human body before attempting to learn symptoms and treatments of diseases. Some knowledge of disorders can contribute to an understanding of how knowledge of structure and function is essential in the health field.

Study Aids

Each chapter begins with an *outline* that lists the major topics in the sequence in which they appear in the text. Each student should look over the outline prior to reading the text, to get a preview of the organization of the chapter. Next there is a list of *objectives* that can be used as a guide in determining what should be accomplished from the study of the chapter. The objectives should be noted before reading the text and again in review when the chapter has been completed. Each student should demonstrate the ability to perform the tasks called for by the objectives.

In the text, the most important words are printed in **bold face** the first time they are introduced. Words that are considered to be somewhat less significant but which should be learned are printed in *italics* the first time they are introduced. Most of the words printed in bold face type and some italicized words are listed at the end of the chapter as *words in review*. These words should be defined by each student upon completion of the chapter as one means of checking mastery of the content. The *glossary* at the end of the text provides definitions and phonetic pronunciation of these words in addition to many other words that are used in the text.

A *summary* appears at the end of each chapter, consisting of simple statements of the main points covered in the text. Usually the summary can be used by the student as a check on his or her understanding of the knowledge required by the objectives. If the statements in the summary are not clear, the text should be reviewed. There are *review questions* following the summary. The *references* at the end of each chapter may be used as sources of additional detail on the subject if desired.

Acknowledgments

This book has been produced by the cooperative efforts of many people. Jack Carey, science editor at Wadsworth, has worked on the project since its inception and has skillfully maneuvered me into developing a better book than I had intended. Many of his ideas are incorporated into the text and illustrations. Some of the many fellow anatomy teachers who reviewed the manuscript are listed on the following spread. Several of them read parts of the book three times and made numerous suggestions for each of the revisions. The book as finally printed shows the influence of their ideas, for which I am deeply indebted. I would also like to express my appreciation to my colleagues in the Anatomy Department, University of Kentucky. Some of them formally reviewed parts of the manuscript, but all were patient with my questions and cooperated by supplying information as to the current "state of the art" in their areas of expertise.

Illustrating an anatomy text requires the efforts of many talented people. The artists are listed on page 617.

The original photographs of gross anatomical material were taken by R. Ross, a University of Kentucky medical photographer. The models for the living (surface) anatomy photographs were Kathy and Ben Kelley and B. Kirol; T. Burke volunteered for the nuclear magnetic resonance imaging. I am especially indebted to my colleagues who supplied illustrations from their own work: D. H. Matulionis provided many electron photomicrographs and W. K. Elwood furnished the color photomicrographs of light microscopic preparations. Joyce R. Isbel at Alexandria Hospital, Alexandria, Virginia granted permission for the use of several x-ray films, and several black and white photomicrographs came from the Armed Forces Institute of Pathology. Special thanks are due to H. F. Parks, who carefully checked the illustrations for accuracy of construction and labeling.

Many people worked on the final revision of the manuscript and the conversion of the manuscript into a book. Robin Fox, a talented writer, served as development editor and did much to make the text more understandable. She and I were guided in our efforts by Mary Arborgast, a very efficient special projects editor. Many unnecessary words were eliminated by Yvonne Howell, a copy editor who never changed the meaning of a sentence. Hal Humphrey coordinated all aspects of production with the able assistance of the staff at Wadsworth. Peggy Mehan was very helpful with the permissions process.

My wife, Betsy, proofread the final manuscript and provided the necessary encouragement to keep me going through three revisions. My son, George, did most of the typing.

I am deeply grateful for all those who have contributed to this project. Any errors that remain are my responsibility.

Ira Fowler

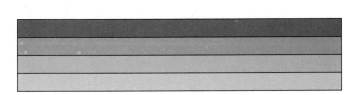

REVIEWERS

Donald M. Bertucci, Shasta College

Robert S. Benton, Medical Center, University of Kentucky

George W. Bond, Jr., Fitchburg State College

Gerald Collier, San Diego State University

Winifred Dickinson, University of Steubenville

R. M. Dom, Medical University of South Carolina

Edward Donovan, Avila College

Victor B. Eichler, Wichita State University

W. K. Elwood, Medical Center, University of Kentucky

Barbara Fidler, Salem College

Thomas G. Froiland, Northern Michigan University

Bruce Grayson, University of Miami at Coral Gables

Harlow Hadow, Coe College

B. J. Hirt, Wayne State College

Louise B. Katz, Sinclair Community College

Mary Kordisch, McNeese State University

Henry N. McCutcheon, Rhode Island College

Robert W. Maitlen, The Defiance College

D. H. Matulionis, Medical Center, University of Kentucky

H. E. Mayberry, University of North Carolina

Kathryn Podwall, Nassau Community College

Harry Reasor, Miami Dade Community College

Gary Resnick, Saddleback College

Thomas L. Roszman, University of Kentucky

Rose Leigh Vines, California State University, Sacramento

J. Vande Ven, Mesa College

Jocelyn Zika, Case Western Reserve University

BRIEF CONTENTS

CONTENTS

DETAILED CONTENTS

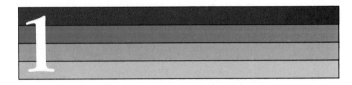

INTRODUCTION TO
THE STUDY OF THE CELL

After completing this chapter you should be able to:

1. List and define the branches of anatomy.

2. Contrast the study of systemic gross anatomy with that of regional gross anatomy, giving the advantages and disadvantages of each method.

3. Describe the levels of organization of the body and the techniques used to study them.

4. Describe the basic human body plan and name the body cavities.

5. Define or describe the anatomical position of the body.

6. List and define the common terms used to locate structures in relation to one another.

7. List and define the planes used to describe sections of the body.

8. Define or explain the most common general terms used in describing the movements of parts of the body.

Defining anatomy is considerably easier than knowing it. *Human anatomy* studies the structure of the human body, and that marvelously intricate and complex body provides the subject matter for an equally complex and intricate science. Anatomy is a basic science, the foundation for our understanding of how the body functions—which is physiology—and our attempt to maintain and repair it—which is medicine.

THE HISTORY AND BRANCHES OF ANATOMY

Curiosity about the human body is as old as humankind itself, but anatomy as a science, or systematic body of knowledge, probably began in ancient Egypt with observations of the structure of animals. However, it was the Greek Hippocrates (c. 400 B.C.), best known as the author of the medical oath still administered to graduating physicians, who wrote the first complete anatomical work, "On the Heart"; he is therefore considered the father of anatomy. Aristotle (384–322 B.C.), the quintessential scientist and philosopher of classical Greece, studied and wrote about the structure of many animals. In so doing he became the originator of **comparative anatomy,** the branch of anatomy that investigates the structural similarities and differences among organisms. Aristotle produced the first known system of classification of animals, based on his observations.

Rome, too, produced great anatomists. Rufus of Ephesus (c. 75 A.D.) wrote the first book on anatomical terminology. Galen of Pergamum (c. 150 A.D.), a gifted physician and surgeon, wrote voluminously and well on anatomical topics. Galen's work, considered authoritative through the Middle Ages, was based on studies of animals, since human dissection was forbidden.

In the Middle Ages science languished, and it was not until the Renaissance renewed interest in classical learning that anatomy was studied seriously again. The first dissections of human bodies were performed in Italy in the middle of the thirteenth century, and the practice became common in both Italy and France in the fourteenth century. Dissection gave anatomy its name, for the word *anatomy* is derived from the Greek "to cut up." Leonardo da Vinci studied anatomy because he wished to paint the human body accurately, and he made the subject part of the training for artists of his and later times. Anatomy advanced art, and art advanced anatomy. The work of Andreas Vesalius in the sixteenth century is particularly renowned for both scientific accuracy and artistic skill (see Fig. 1-1). It was the Renaissance scientists and artists who laid the foundations of the modern study of body parts, called **gross anatomy.**

The Renaissance anatomists were concerned with how the body is put together rather than with how it works. In the early seventeenth century, the English scientist William Harvey made the first study showing that the structure of the body provides clues to its function. Harvey's interest was the circulatory system, and his work, based on precise anatomical observations, showed the origin of blood flow in the heart and its pathway through the arteries and veins. The modern anatomist's concern with the functioning of organ systems had its beginning in Harvey's pioneering study.

Microscopes (known today as light microscopes to distinguish them from electron microscopes) were developed during the seventeenth century, and for the first time scientists could observe structures invisible to the naked eye. The term **cell** dates from this period, although the significance of cells was yet to be understood. By the eighteenth century microscopes were common, and the field of **microscopic anatomy** was unfolding. Anatomists studied unborn organisms to see how structure develops before birth, and they examined abnormal or malfunctioning structures as well as healthy ones. The first study grew into **embryology**, the second into **pathology.**

A concept crucial to our current understanding of the human body dates from the early nineteenth century, when Matthias J. Schleiden and Theodor Schwann, working independently, showed that all organisms are made up of cells. This finding led anatomists to study the various types of cells making up the body and ultimately led to the development of **cytology,** the study of cells, and of **histology,** the study of **tissues.** During the eighteenth and nineteenth centuries, a great deal of research was also devoted to comparative anatomy. Anatomical knowledge, greatly advanced since Aristotle's time, led to the modern scheme of classification and to assigning the human species to its proper place in the animal kingdom. The structure of our bodies dictates that we be classified with the *vertebrates*, animals with backbones.

The twentieth century has witnessed the greatest burst of anatomical knowledge ever. The electron microscope, developed in the 1930s, provides 100 times the magnification of the light microscope, revealing complexities never dreamed of 50 years ago. Improved experimental techniques in embryology, and other developments, have also contributed to the exponential expansion of our knowledge. This century has been one of discovery for anatomists, through it is still too early to say whether any of their findings will prove equal to Harvey's or to Schleiden and Schwann's. And much more remains to be learned; the greatest discoveries may well lie ahead.

ANATOMICAL ORGANIZATION

Contemporary anatomists examine the parts of the body at different levels of organization in an attempt to determine the structure of each part and to define the role of each part in the functioning of the body as a whole. The levels of anatomical organization, from simplest to most complex, are: **cells, tissues, organs,** and **organ systems.**

The fundamental structural and functional unit of the body is the *cell.* A cell, which consists of a complex material called *protoplasm* and is bounded by a membrane, is the smallest unit of organization that can carry on all the functions associated with life. Most cells are invisible to the naked eye and must be magnified for study. Cells are specialized in structure and function for specific roles in the body, but all have certain essential features, which we shall consider in Chapters 3 and 4.

Cells that are structurally and functionally related to each other are organized into aggregations called *tissues* (French, tissu = woven). There are four fundamental tissues of the body: *epithelium* (G., *epi* = upon, *thele* = nipple), or covering tissue, such as the outer layer of the skin; *connective tissues,* including bone and cartilage, which provide the principal support for the body; *muscle* (L., *musculus* = little mouse), which is tissue specialized for movement; and *nervous tissue,* which is specialized for the conduction of nerve impulses. Within each of these broad categories are various distinctive tissue types made up of different kinds of cells. We shall consider tissue types in detail in Chapter 4.

Figure 1-1 Andreas Vesalius (1514–1564), whose writings and paintings provided much of the basis for modern anatomy.

Human tissues are not found as isolated units. Two or more tissues form an anatomical unit called an *organ* (Gr., *orgnon* = a tool) that performs one or more specific functions. Each organ, such as the liver, has its own unique tissue pattern, with which a student of anatomy must become familiar. Each organ has a characteristic size, shape, and location within the body and is given a name.

The organs are components of and contribute to the overall functioning of larger organized units, *organ systems,* each of which is responsible for a bodily function. Organ systems include the integument that covers and protects the entire body (Fig. 1-2); the skeletal system, for support (Fig. 1-3); the muscular system, for movement (Fig. 1-4); the digestive system, for nutrition

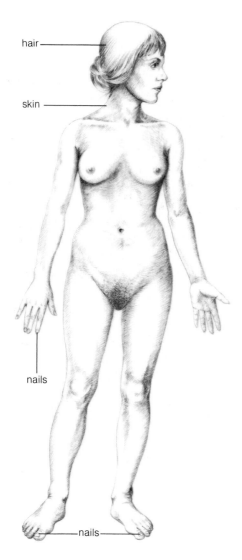

Figure 1-2 The integumentary system: the skin and its appendages, hair and nails, that may be seen at the surface.

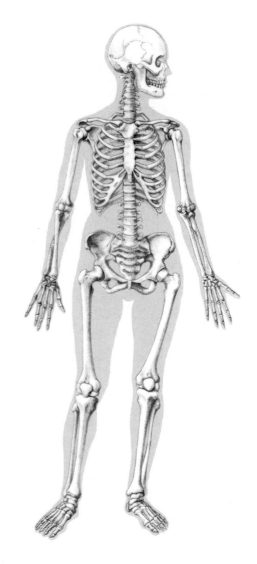

Figure 1-3 The skeletal system

(Fig. 1-5); the respiratory system, for respiration (Fig. 1-6); the circulatory system, for distribution (Fig. 1-7); the urinary system, for excretion (Fig. 1-8); the reproductive system, for continuing the species (Fig. 1-9); and the nervous and endocrine systems, for integration (Figs. 1–10 and 1–11). Each organ system contributes to the proper functioning of the body as a whole. Improper functioning of an organ or organ system results in illness. After considering cells and tissues, we shall consider each organ system and its component organs in turn.

Methods of anatomical investigation are determined by the

level of body organization being studied. The methods of gross anatomy, the study of organs and organ systems, are different from those of microscopic anatomy, which deals with cells, tissues, and the microscopic structure of organs.

Gross Anatomy

Gross anatomy, the primary subject of this book, is the study of structure as it can be seen with the unaided eye or with very low magnification such as with a hand lens. The usual method

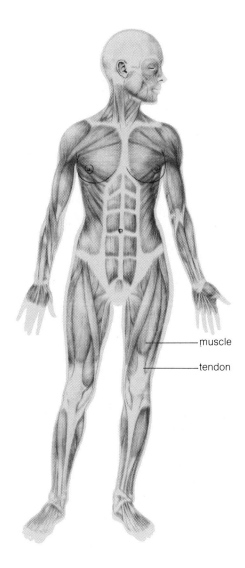

Figure 1-4 The muscular system

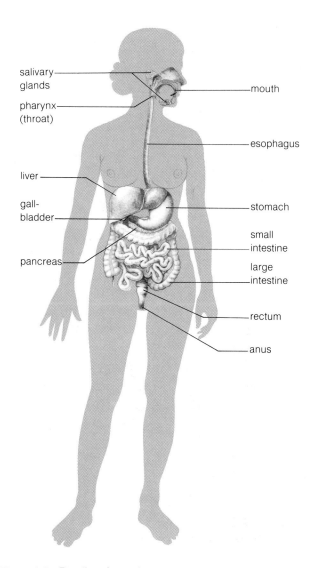

Figure 1-5 The digestive system

of study is dissection, with observations made during the separation or removal of parts of the body. However, gross observation is also done on living bodies. The study of external body structure by means of palpation (feeling) is called *surface anatomy*; the study of internal structure by means of x-ray films is called *radiological anatomy*. Courses in gross anatomy may be designed to study the organ systems one at a time, an approach called **systemic anatomy,** or they may be designed to study simultaneously all of the systems found in a specific area of the body, an approach called **regional anatomy.**

Systemic Anatomy In *systemic anatomy,* the approach used in this book, each organ system is examined in its entirety before another is considered; thus structure is correlated with function. Knowledge of structure without an understanding of function is of little value to students preparing for careers in which they strive to improve the physical well-being of the body.

In this book, we describe the skeletal system and the muscular system consecutively, because they are anatomically and functionally closely related. The relatively rigid skeletal system provides the principal support for the body, but it would be

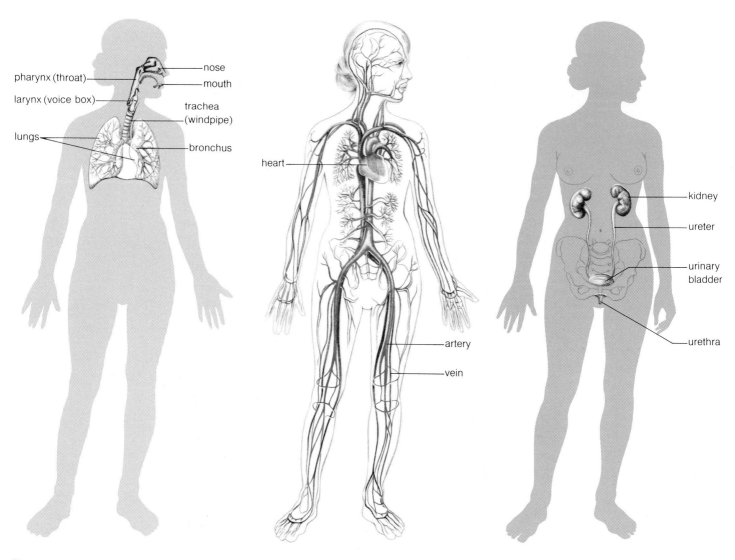

Figure 1-6 The respiratory system

pharynx (throat)
larynx (voice box)
lungs
nose
mouth
trachea (windpipe)
bronchus

Figure 1-7 The circulatory system

heart
artery
vein

Figure 1-8 The urinary system

kidney
ureter
urinary bladder
urethra

incapable of maintaining posture without the action of the muscles that extend from one bone to another. Similarly, the muscles, whose primary function is to move parts of the body and the body as a whole, would be ineffective without the bones upon which they pull. We also consider the nervous system and the endocrine system consecutively, because they frequently work together to integrate the activities of other organ systems. The other systems are described separately, but keep in mind that all the organ systems are dependent upon one another.

The weakness of systemic gross anatomy lies in its failure to show anatomical relations among organs of different systems.

This book describes the locations of the major body components in general terms only; more precise information on the spatial relations of anatomical structures may be obtained from anatomy books that use the regional approach.

Regional Anatomy The regional approach to the study of gross structure is used in courses that emphasize dissection. The cardiovascular system, for example, could not be dissected without disrupting many parts of the body; consequently, it is logical to examine all parts in an area as they become visible

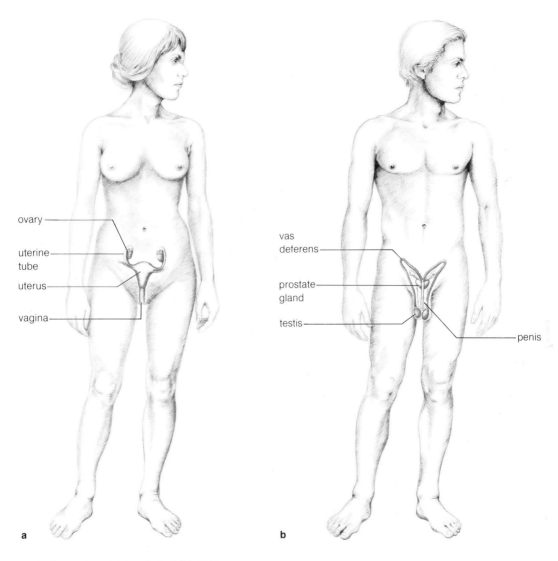

Figure 1-9 The reproductive system: (**a**) female and (**b**) male

during dissection. Regional gross anatomy is the method of choice for study of the location of structures in relation to one another and is especially useful for surgeons.

In *regional anatomy,* the body is divided into small areas, or regions, for dissection. These regions are usually the head, neck, thorax, abdomen, pelvis, upper extremities, and lower extremities. Within each region, structures are exposed at progressively deeper levels until all components have been examined. Anatomical relations of the parts of an organ system within a region are described precisely by using specific anatomical landmarks as reference points.

Microscopic Anatomy

Microscopic anatomy is the study of anatomical structure with the aid of magnification. Since it involves the study of cells and tissues, it embraces the fields of cytology and histology.

The methods of microscopic anatomy are *light microscopy* and **electron microscopy.** The term **microscopic structure** indicates the structure of cells, tissues, and organs as seen with the light microscope, which magnifies up to about 1500 times; the term *fine structure* or **ultrastructure** denotes the structure of cells and tissues as revealed by the electron microscope, which magnifies more than 150,000 times. Photography of

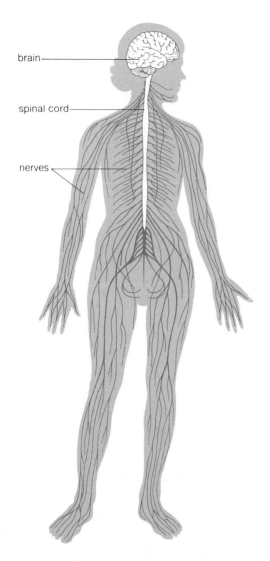

Figure 1-10 The nervous system

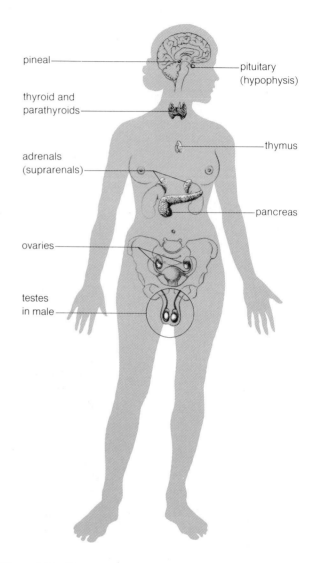

Figure 1-11 The endocrine system. Location of the major endocrine glands.

microscopic images is called *photomicrography*. Photomicrographs, which appear throughout this book, may be light micrographs showing microscopic structure or electron micrographs showing fine structure.

THE BASIC BODY PLAN

Humans have many features in common with other vertebrates. The vertebrate body (Fig. 1-12) consists of an outer tube and an inner tube—the body wall and the digestive tract, respectively. Between the two tubes is a cavity containing a small amount of fluid that prevents the tubes from sticking together.

The body wall is essentially a thick-walled tube composed of muscle and bone and covered on its outer surface by an impermeable epidermis (the outer layer of the skin). This wall provides support and protection for internal structures and is called the *soma* (Gr., body), or the **somatic** part of the body. In addition to supporting internal structures by means of membranes, the body wall provides a platform for the attachment of

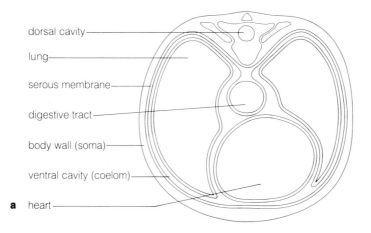

dorsal cavity

lung

serous membrane

digestive tract

body wall (soma)

ventral cavity (coelom)

a heart

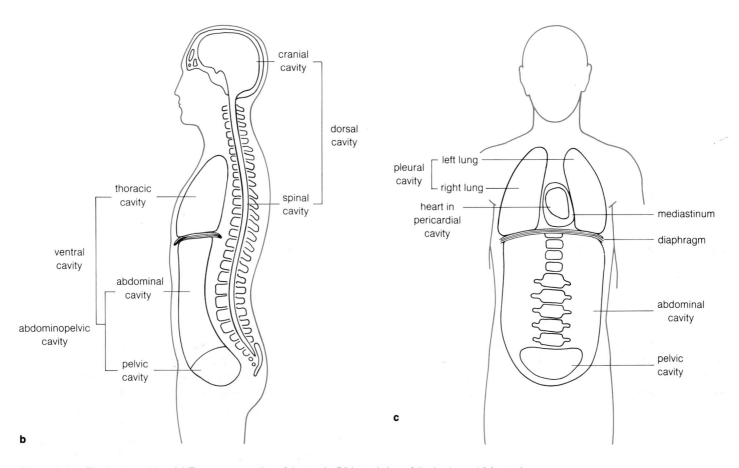

cranial
cavity

dorsal
cavity

spinal
cavity

thoracic
cavity

ventral
cavity

abdominal
cavity

abdominopelvic
cavity

pelvic
cavity

b

pleural
cavity

left lung

right lung

heart in
pericardial
cavity

mediastinum

diaphragm

abdominal
cavity

pelvic
cavity

c

Figure 1-12 The body cavities. (**a**) Transverse section of the trunk, (**b**) lateral view of the body, and (**c**) anterior view of the body.

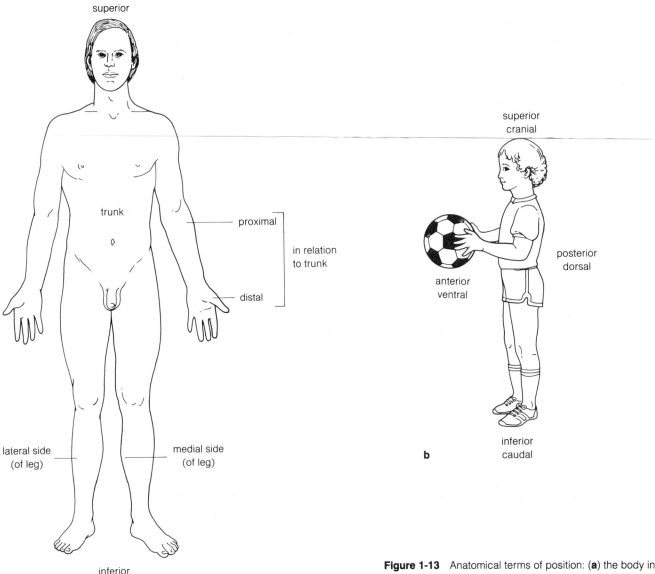

superior

trunk

proximal
distal
in relation
to trunk

lateral side
(of leg)

medial side
(of leg)

inferior

a

superior
cranial

posterior
dorsal

anterior
ventral

inferior
caudal

b

Figure 1-13 Anatomical terms of position: (**a**) the body in the anatomical position, anterior view, and (**b**) lateral view of the body

the head and limbs. These structures, also composed largely of skeletal muscle and bone, are considered part of the soma and are involved in the special somatic function of movement. Thus the three principal functions of somatic structures are *support, protection,* and *movement,* and the major components that provide these functions are the *bones* and skeletal *muscles.*

The *ventral body cavity,* or **coelom,** is the space between the digestive tract and the body wall. This cavity is completely lined with a membrane of a type called serous membrane. *Ventral* means "toward the front of the body," and the term is used

to distinguish the coelom from the much smaller dorsal body cavity—*dorsal* meaning "toward the back of the body." Sometimes the ventral body cavity is simply called *the body cavity.* The organs enclosed by the ventral cavity are called *viscera* (L., *viscera* = the soft parts). These include the digestive organs and the organs of the *circulatory, respiratory, excretory,* and *reproductive systems.* The ventral cavity and the organs it houses are the **visceral** (in contrast to the somatic) part of the body.

The ventral cavity is divided by the *diaphram* (Gr., *diaphragma* = a partition) into a **thoracic cavity** (L., *thorax* =

chest) above and an **abdominal cavity** (L., *abdomen* = belly) below (Fig. 1-12b). The thoracic cavity is further divided by serous membranes into the **pericardial cavity** (Gr., *peri* = around, *cardion* = heart), containing the heart, and the right and left **pleural cavities** (Gr., *pleura* = a rib), which house the lungs (Fig. 1-12c). All these cavities are enclosed by the body wall, and each is lined with serous membrane.

In addition to the cavities enclosed by the body wall, there is a cavity within the wall (Fig. 1-12b) that houses the *central nervous system* (the brain and spinal cord). This *dorsal body cavity* is enclosed by the bones of the skull and vertebral column (backbone); it is not lined with serous membrane. A tubular dorsal cavity containing the central nervous system is one of the fundamental characteristics of vertebrates.

ANATOMICAL TERMINOLOGY

In the study of gross anatomy, an understanding of certain anatomical directional terms is absolutely essential for communication. In describing the human, these directional terms refer to the body in the **anatomical position,** shown in Fig. 1-13. The body is upright with the hands down at the sides, the thumbs are turned outward, and the feet are close together. Regardless of the position of the body being observed, the description always refers to the anatomical position.

Terms of Relative Position

Structures are described in relation to one another by means of directional terms that indicate their relative position within the body. The following are some of the most commonly used terms (see Fig. 1-13):

Anterior (ventral): toward the front of the body (for example, the nose is anterior to the ears).

Posterior (dorsal): toward the back of the body (for example, the backbone is posterior to the breastbone).

Medial: toward the midline of the body (for example, the breastbone is medial to the arms).

Lateral: away from the midline (for example, the shoulder blades are lateral to the vertebral column).

Superior (cranial): upward or toward the head (for example, the nose is superior to the mouth).

Inferior (caudal): downward, or away from the head (for example, the chin is inferior to the mouth).

Deep (internal): away from the surface of the body (for example, the tongue is deep to the teeth).

Superficial (external): toward the surface of the body (for example, the skin is superficial to the breastbone).

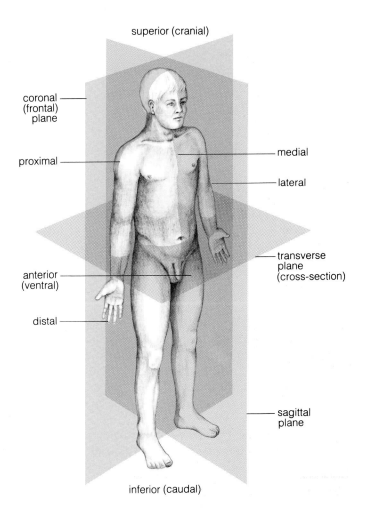

Figure 1-14 Planes of the body

Proximal (L., *proximalis* = next, nearest): closer to the point of origin of the part (for example, the elbow is proximal to the wrist).

Distal (L., *distalis* = at a distance): farther from the point of origin of the part (for example, the wrist is distal to the elbow).

Planes of the Body

In discussing internal structures, it is frequently useful to refer to a specific *plane* of the body. The planes represent imaginary sections through the body, at various levels and in various directions. The following planes are depicted in Fig. 1-14:

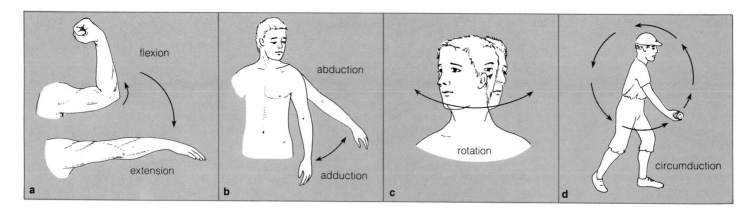

Figure 1-15 Terms of movement: (**a**) flexion and extension of the forearm, (**b**) abduction and adduction of the upper extremity, (**c**) rotation of the head, and (**d**) circumduction of the upper extremity

Median sagittal plane, or *midsagittal plane* (L., *sagitta* = arrow), a section passing vertically through the body dividing it into equal halves (right and left); it passes from the anterior midline to the posterior midline.

Sagittal plane, any section parallel to the median sagittal plane but not in the midline.

Coronal plane (L., *corona* = a garland, a crown), or *frontal plane,* a section passing vertically through the body at right angles to the sagittal planes; it divides the body into the anterior and posterior portions.

Transverse plane, or *cross section,* any section at a right angle to the longitudinal axis of the body. The transverse plane through the waist divides the body into upper (superior) and lower (inferior) portions.

Terms of Movement

Movements produce changes in the relative positions of the body. The following are the terms most commonly used in describing the movements that occur at joints (Fig. 1-15):

Flexion (L., *flecto* = to bend): movement that decreases the angle at a joint.

Extension (L., *extensio* = to stretch out): movement that increases the angle at a joint.

Abduction (L., *ab* = away from, *ductio* = to move): movement of a part, such as the arm, away from the body or away from the midline of the body.

Adduction (L., *ad* = toward): movement of a part toward the midline.

Rotation (L., *rotatio* = to revolve): any movement that involves the turning of a part on its axis without displacement of the axis.

Circumduction (L., *circum* = around, *ductio* = to move): movement or swinging of a part, such as the arm, in a circle; it involves displacement of the axis.

WORDS IN REVIEW

Abdominal cavity	Median sagittal plane
Abduction	Microscopic anatomy
Adduction	Microscopic structure
Anatomical position	Organ
Anterior	Organ system
Cell	Pathology
Circumduction	Pericardial cavity
Coelom	Pleural cavity
Comparative anatomy	Posterior
Coronal plane	Proximal
Cytology	Regional anatomy
Deep	Rotation
Distal	Sagittal plane
Electron microscopy	Somatic
Embryology	Superficial
Extension	Superior
Flexion	Systemic anatomy
Gross anatomy	Thoracic cavity
Histology	Tissue
Inferior	Transverse plane
Lateral	Ultrastructure
Medial	Visceral

SUMMARY OUTLINE

I. History and Branches of Anatomy.
   A. Hippocrates is known as the father of anatomy.
   B. Aristotle was the first to study comparative anatomy and to invent a system of animal classification.
   C. During the Renaissance dissection of the human body began, allowing the development of human gross anatomy.
   D. Harvey, in the seventeenth century, was the first to correlate structure with function when he demonstrated the circulation of blood.
   E. The light microscope was developed during the seventeenth century and by the eighteenth was in common use, allowing the growth of *microscopic anatomy* and leading to the growth of *embryology* and *pathology.*
   F. In the nineteenth century Schleiden and Schwann proposed the cell theory, which led to the development of *cytology* and *histology.*

G. The electron microscope and other advances have led to a great upsurge of anatomical discovery in the twentieth century.

II. Anatomical Organization
   A. The levels of anatomic organization are *cells, tissues, organs,* and *organ systems.*
   B. *Gross anatomy* is the study of structure visible with the unaided eye or with very low magnification. Its methods are dissection, palpation (surface anatomy), and x-ray imaging (radiological anatomy).
      1. *Systemic gross anatomy* is the study of each organ system separate from the others; it is useful for correlating structure with function.
      2. *Regional gross anatomy* is the study of all structures within a specific region of the body; it is useful for dissection.
   C. *Microscopic anatomy* is the study of structure with the aid of magnification. Its methods are light microscopy, which shows *microscopic structure,* and electron microscopy, which shows *fine structure,* or *ultrastructure.*

III. The Basic Body Plan
   A. The basic human body plan consists of a body wall surrounding a digestive tube.
   B. The body wall, including the limbs and head, is made up of muscle and bone and is called the *somatic* part of the body.
   C. The space between body wall and digestive tube is the ventral cavity, or *coelom,* which contains the viscera. Cavity and viscera constitute the *visceral* part of the body.
      1. The ventral cavity is divided by the diaphragm into a *thoracic* and an *abdominal* cavity.
      2. The thoracic cavity is divided into the *pleural cavities,* which contain the lungs, and the *pericardial cavity,* which contains the heart.
   D. The dorsal cavity houses the brain and spinal cord within the skull and vertebral column.

IV. Anatomical Terminology
   A. The body is always described in the *anatomical position:* erect, feet together, arms hanging at the sides, thumbs turned outward.
   B. Directional terms are used to give the relative position of the parts of the body.
      1. *Anterior,* or *ventral,* is toward the front of the body; *posterior,* or *dorsal,* is toward the back.
      2. *Medial* is toward the midline; *lateral* is away from the midline.
      3. *Superior* is toward the head; *inferior* is away from the head.

4. *Deep* is away from the body surface; *superficial* is closer to the surface.

5. *Proximal* is closer to the point of origin of a part; *distal* is farther away.

C. The planes of the body represent imaginary sections at various levels and in various directions.

1. The *median sagittal* plane passes through the body vertically, dividing it into right and left halves.

2. A *sagittal plane* is any section parallel to the median sagittal plane.

3. A *coronal,* or *frontal,* plane is a vertical section at a right angle to the sagittal planes.

4. A *transverse plane* is a section at a right angle to the longitudinal axis of the body.

D. Terms of movement describe changes in the position of parts of the body in relation to one another.

1. *Flexion* decreases the angle at a joint; *extension* increases it.

2. *Abduction* is movement of a part away from the midline of the body; *adduction* is movement toward the midline.

3. *Rotation* is turning of a part on its axis.

4. *Circumduction* is swinging a part in a circle.

REVIEW QUESTIONS

1. Describe the levels of organization of the human body, from the basic structural unit to the whole organism.

2. Describe the methods used to study the structure of the human body.

3. Distinguish between systemic and regional gross anatomy, giving the principal advantages and disadvantages of the two approaches to the study of human structure.

4. What are the approximate magnifications possible with the light and electron microscopes?

5. Describe the basic body plan. What are the principal characteristics that vertebrates have in common?

6. Describe the anatomical position.

7. List and define the planes of the body that are used to locate structures within the body.

8. List and define the directional terms that indicate the relative position of structures.

9. List and define the terms that are used to describe movement.

10. Anterior means toward the front (forward-facing) of the body and posterior means toward the rear; ventral means towards the belly side and dorsal toward the back (spine) side. In humans, anterior and ventral are the same, as are posterior and dorsal. What terms would be the same as dorsal and ventral in a four-footed animal? (Draw a diagram.)

SELECTED REFERENCES

Chaffee, E. E., and Lytle, I. M. *Basic Physiology and Anatomy.* 4th ed. Philadelphia: J. B. Lippincott Co., 1980.

Langebartel, D. A. *The Anatomical Primer.* Baltimore: University Park Press, 1977.

Montgomery, R. L. *Basic Anatomy for the Allied Health Professions.* Baltimore: Urban and Schwarzenberg, 1981.

Moore, K. L. *Clinically Oriented Anatomy.* Baltimore: Williams and Wilkins, 1980.

Snell, R. S. *Clinical Anatomy for Medical Students.* Boston: Little Brown and Co., 1981.

PROTOPLASM AND THE CELL

After completing this chapter you should be able to:

1. Name and describe the components of a cell as seen with the light microscope.

2. List the principal components of protoplasm and give their approximate relative amounts.

3. List and describe or define the principal physiological properties of protoplasm.

4. Describe the structure of the plasma membrane. Describe the methods by which substances enter and leave the cell.

5. Describe the structure of the nuclear envelope and its relations to the endoplasmic reticulum.

6. Correlate the structure of the endoplasmic reticulum with its functions in the synthesis of proteins and steroids.

7. Describe the structure of the Golgi complex and its role in the process of secretion.

8. Describe the structure of mitochondria and correlate their structure with their functions.

9. Describe the origin and function of secretory granules.

10. Describe the structure and function of lipid droplets of the cell.

11. Describe the structure of the centriole.

12. Describe the structure and functions of microfilaments and microtubules.

13. Describe ribosomes and polysomes, giving their chemical composition and functions in protein synthesis.

14. Give the structure and the general functions of lysosomes.

15. Describe the appearance, chemical composition, and functions of the nucleolus.

16. Describe the structures by which cells adhere to one another.

17. Describe the structure and general functions of cilia.

From the human anatomist's point of view, cells are the fundamental units of structure and function—the building blocks of the body. Yet a cell is itself an exceedingly complex structure. Many microscopic organisms are composed of but a single cell that contains within itself everything necessary for living and reproducing. These independently living cells vary greatly in form and physiology, but all display the most fundamental property of life, the ability to take matter from the environment and reorganize it into their own substance. In multicellular organisms like ourselves, each cell is specialized in function and depends on the others for its existence. Yet each of our cells displays this same fundamental ability: each is a living creature that uses nutrients delivered by the blood to maintain its own structure and function.

Cells vary in size from about 7 micrometers (a micrometer is one-thousandth of a millimeter) to more than 150 micrometers. (See Appendix A for a table of the units of measurement.) Cells obtain their nutrients from the fluid that surrounds them and are therefore limited in potential size by the distances the nutrients can be transported effectively. If cells were very large, then perhaps the nutrients entering a cell would be depleted before a sufficient quantity could reach the central region. And since each cell must eliminate its wastes, it cannot be so large that wastes accumulate excessively in the cell before being excreted. The small size of cells means that even a small organ of the body, such as a kidney, is composed of many billions of cells.

Early observers of cells used the term *protoplasm* to refer to the entire cell contents, and the term is still useful for discussing general properties of cells. Protoplasm has been called "the physical basis of life." But when we observe a cell with the light microscope (Fig. 2-1), it is apparent that protoplasm is not a homogeneous substance.

GENERAL STRUCTURE OF A CELL

Each cell of the body consists of a nucleus surrounded by **cytoplasm** and is enclosed by a **plasma membrane** that separates the cell from the environment. The *nucleus,* enclosed by the *nuclear envelope* that separates it from the cytoplasm, contains the genetic material that regulates the metabolic activities of the cell and is responsible for its reproduction. The genetic material, deoxyribonucleic acid, or DNA, is located in dark-staining masses called *chromatin granules* that are scattered throughout the nucleus. During cell division, the chromatin granules condense into rod-like structures termed *chromosomes* (also dark-staining), which will be described further in Chapter 3. Usually the chromatin granules and one or more masses called **nucleoli** (sing, nucleolus) are the only structures in the nucleus visible with the light microscope. The fluid component of the nucleus, the *karyoplasm* or **nucleoplasm,** contains substances that are too small for the light microscope to detect.

The cell *cytoplasm* contains organelles and inclusions floating in a ground substance. **Organelles** are cellular components with definite structural characteristics that perform specific functions within the cell. *Inclusions* are either chemical substances produced within the cell that accumulate before being secreted (such as the mucus produced by cells lining the nasal cavity), or nutrients obtained from the food we eat that are stored within cells (such as glycogen, a storage form of sugar that accumulates in cells of the liver). The *ground substance* is a soft gel that with the light microscope appears to be without form. However, the greater magnification of the electron microscope reveals that the ground substance also contains organelles and inclusions (which will be described later). The plasma membrane at the cell surface regulates the passage of material into and out of the cell. It is a selectively permeable membrane in that it permits the passage of certain substances and restricts the passage of others. In general terms, it allows water to pass freely but regulates most other substances.

CHEMICAL COMPOSITION OF PROTOPLASM

The chemical composition of living things ultimately determines their structure, so we must begin our study of cells with an analysis of the chemical composition of protoplasm. The composition of individual cells varies, but if we analyze the whole human body we find that protoplasm is approximately 75% water by weight, with about 1% inorganic (non-carbon-containing) substances in solution. The rest of its weight is made up of organic (carbon-containing) chemicals: protein (up to 20%); lipids, or fatty substances (about 3%); carbohydrates (about 1%); and nucleic acids (less than 1%). All these substances contain carbon, hydrogen, and oxygen; protein and nucleic acids also contain nitrogen; and nucleic acids contain phosphorus. Other organic chemicals present in minute amounts include adenosine triphosphate (ATP), a phosphorus-containing compound essential to synthesis in the cell. (We shall talk about ATP when we discuss cellular respiration.) Proteins, nucleic acids, and some carbohydrates are *macromolecules,* large molecules made up of smaller molecules linked together. The cell builds macromolecules from the smaller molecules it takes in from its environment.

Proteins

Of all the chemical constituents of the cell, *protein* is of the greatest interest to us, because the structure of a cell—and ultimately of higher levels of organization—depends mostly on the kinds of protein present. Furthermore, specific proteins determine not only the form of a specific cell but also its functions.

Proteins are made of smaller molecules called *amino acids,* of which there are 20 different kinds. Amino acids link together

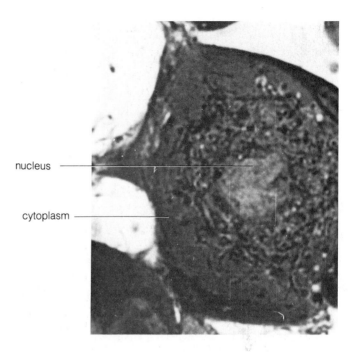

Figure 2-1   A nerve cell as seen with the light microscope, × 750 (courtesy of S. Scheff)

phe-try-pro-arg-leu-met-pro-met-ileu-phe-pro-cys-his-ala-try

Figure 2-2   Structure of a polypeptide. The sequence of amino acids in the chain.

into long macromolecules called *polypeptides* (peptide refers to the type of bond formed between amino acids). Each specific kind of polypeptide always has its amino acids uniquely arranged in a precise order (Fig. 2-2). A protein molecule may consist of a single polypeptide or of two or more polypeptides joined together at intervals. For example, hemoglobin, the oxygen-carrying pigment of red blood cells, is a protein made up of four polypeptides, two of one type and two of another. The length of polypeptides varies enormously, from perhaps one hundred amino acids in a chain to as many as several thousand.

Specific amino acids in one part of a polypeptide form chemical bonds with specific amino acids in other parts, causing the polypeptide to fold into a characteristic complex shape (Fig.

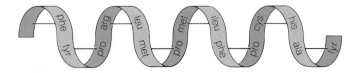

Figure 2-3   Coil formed by the amino acid chain

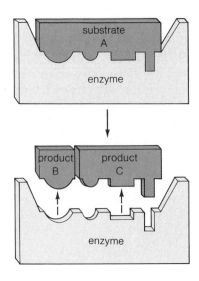

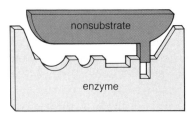

Figure 2-4   The surface of an enzyme has an active site on which its normal substrate fits. The reactants that fit the site are converted to the products of the specific reaction catalyzed by the enzyme. If another molecule fits the active site well enough to attach and prevent the substrate molecules from occupying the site, this nonsubstrate molecule may inhibit the action of the enzyme.

2-3). By this bonding the sequence of amino acids determines the structure and function of a protein. Substituting a single amino acid for another might—depending on the kind of amino acid and its location—completely distort the molecule's form and hence its function.

> Abnormal proteins can distort the entire cell in which they are located and lead to abnormal function. For example, the red blood cells of a person with sickle-cell anemia contain an abnormal form of the red pigment called hemoglobin. Because it is abnormal, the hemoglobin does not function properly as an oxygen carrier and causes the red blood cells to become crescent- or sickle-shaped.

Proteins play many roles in the life of the cell. Structural proteins are those that make up cellular structures and are responsible for the form of the cell; they include such proteins as the keratins, which give hair its shape and texture. Nonstructural proteins include the **enzymes,** which *catalyze,* or assist, biochemical reactions. Enzymes work by forming temporary bonds with other molecules, holding them in the correct position to react with each other (Fig. 2-4). Without this assistance, biochemical reactions would proceed far too slowly for life to exist. It is chiefly in their role as enzymes that proteins determine how cells function, for each kind of enzyme catalyzes a specific cellular reaction. There are tens of thousands of different kinds of enzymes in the body; some kinds are found in every cell, while others are found only in cells performing specialized functions.

Nonstructural proteins include some *hormones* (substances that travel via the bloodstream to reach specific target cells, upon which they exert a specific effect), insulin, for example. Other nonstructural proteins have a variety of functions (we have already mentioned hemoglobin). Nonstructural proteins are relatively small molecules that cannot be seen with the elec-

tron microscope except in aggregates. They are, however, of utmost importance in the physiology of the cell.

Nucleic Acids

The specific proteins that give a cell its characteristics are determined by the **nucleic acids** present in the cell. Nucleic acids are macromolecules made up of *nucleotides;* every link in the macromolecular chain is one of four kinds of nucleotides. There are two nucleic acids, DNA (deoxyribonucleic acid) and RNA (ribonucleic acid), which differ in the chemical composition of their nucleotides.

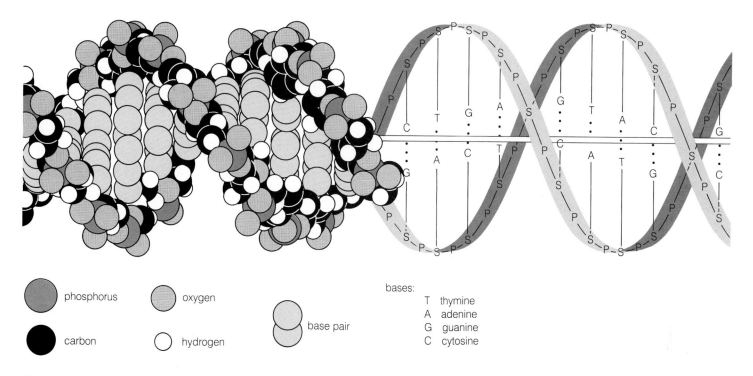

phosphorus

carbon

oxygen

hydrogen

base pair

bases:
T thymine
A adenine
G guanine
C cytosine

Figure 2-5   The Watson-Crick model of DNA as a double-stranded helix

The DNA molecule is made up of two helical strands of nucleotides, with each nucleotide bonded weakly to a nucleotide on the other strand (Fig. 2-5). DNA is found in the nucleus in structures called *chromosomes.* A chromosome is an exceedingly long DNA molecule wrapped around a protein core. The sequence of nucleotides in the DNA molecule constitutes the *genetic code,* the blueprint for the cell's proteins. Since these proteins include enzymes, which control the synthesis of all cell substances, the DNA blueprint for proteins is a blueprint for the entire structure and functioning of the cell. A short segment of the DNA molecule that codes for a specific polypeptide is called a *gene;* each DNA molecule (in a human chromosome) contains about a thousand genes.

The DNA molecule can reproduce itself, which it does between cell divisions. Its two strands of nucleotides unwind and separate; then each serves as a form, or *template,* for the synthesis of an identical strand from nucleotides present in the nucleus (Fig. 2-6). The result is two DNA molecules. During cell division, one of the DNA molecules goes to each new cell, so that both cells receive identical genetic information.

There are several forms of RNA in the cell (see Fig. 2-7). A molecule of messenger RNA, or mRNA, consists of a single hel-

ical strand of nucleotides. It moves about within the cell, carrying the instructions of DNA from the nucleus to the cytoplasm where protein is synthesized. The sequence of nucleotides in DNA dictates the sequence of nucleotides in messenger RNA, which in turn dictates the sequence of amino acids in polypeptides. We shall mention the other types of RNA later in this chapter.

Carbohydrates

Carbohydrates include *sugars* as well as macromolecules called **polysaccharides,** the units of which are sugar molecules. Examples of *sugars* are *glucose,* which consists of a single sugar molecule, and *sucrose* (table sugar), which is made up of two sugar molecules. Polysaccharides—carbohydrates made up of more than two sugar molecules—include glycogen, starch, and cellulose. *Glycogen* is found in the livers of humans and other animals and serves as a food reserve. When cells are short of fuel, glycogen is converted to glucose, which all cells of the body can use as an energy source, and the glucose is transported by the blood to wherever it is needed. *Starch* and *cellulose* are

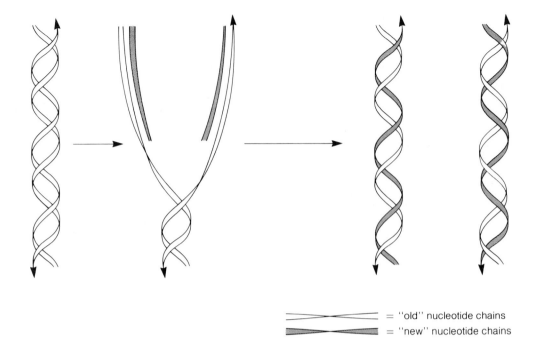

Figure 2-6 Arrangement of old and new chains of DNA in replication

= ''old'' nucleotide chains
= ''new'' nucleotide chains

plant products we ingest. Our digestive systems can break down starch to make glucose and other useful molecules, but we cannot digest cellulose, otherwise known as fiber or roughage. The carbohydrate present in the protoplasm of most cells is chiefly glucose; there is not very much of it because it is constantly being broken down to fuel the cell's activities.

Lipids

The term lipid applies to a variety of substances that are insoluble in water; it includes oils, fats, and waxy substances. Lipids, unlike proteins, nucleic acids, and polysaccharides, are not macromolecules. Lipids play many roles in the body; one of their best known functions is as a concentrated food reserve (fat). Some lipids are structural components of cells; for example, lipid molecules are part of the cell membrane.

PHYSIOLOGICAL PROPERTIES OF CELLS

All cells are composed of the same major elements organized into the same kinds of macromolecules. In consequence, all our body cells, no matter how different in structure and function, have basic physiological properties in common, which are summarized in Table 2-1. However, a property common to all cells

Table 2-1 Summary of Properties of Cells	
Property	Description
Metabolism	Process governing all changes in the chemicals of cells
Respiration	Breakdown of nutrients to produce energy, carbon dioxide, and water
Absorption	Intake of substances into the cell from the environment
Secretion	Process of extruding useful material from a cell
Excretion	Process of extruding waste material from a cell
Irritability	Capacity to respond to a stimulus
Conductivity	Capacity to transmit an impulse
Contractility	Capacity to shorten
Reproduction	Production of a new generation of cells or individuals
Growth	Increase in mass
Differentiation	Process of cellular change

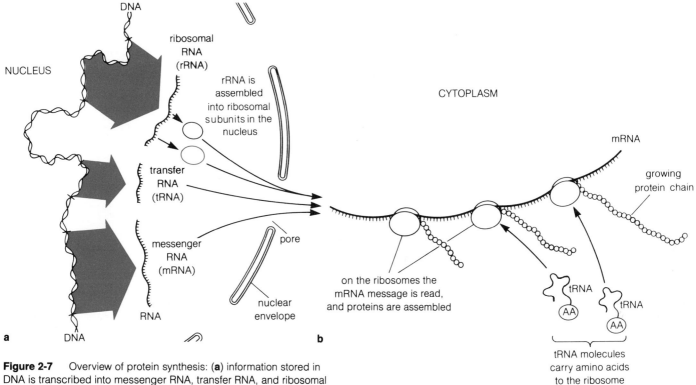

Figure 2-7 Overview of protein synthesis: (**a**) information stored in DNA is transcribed into messenger RNA, transfer RNA, and ribosomal RNA in the cell nucleus; (**b**) in the cytoplasm the information transcribed into RNA is used for assembling proteins.

may be more highly developed in some types of cells than in others.

Metabolism

Protoplasm exists in a state of constant change, building up molecules, tearing them down, continually destroying and renewing its own constituents. This self-renewal is what distinguishes living from nonliving things: when a cell ceases its activity, it is dead. All the physical, chemical, and energy changes that occur within each cell and within an organism are called *metabolism*.

Metabolic processes are of two general types. *Anabolism* includes all of the building-up aspects of metabolism—the synthesis of complex molecules from simpler ones. *Catabolism* includes all the tearing-down aspects—the breaking down of complex molecules into simpler ones. Both anabolism and catabolism are involved in the assimilation of nutrients and their conversion into living substances, but anabolic processes require energy, while catabolic processes usually liberate energy.

Respiration

In common usage, the term **respiration** is synonymous with breathing—the inhaling and exhaling of air by the lungs. But when we speak of *cellular respiration* we are referring to a catabolic process in which *nutrients*—the fuel of the cell—are broken down, liberating energy for use in anabolic processes. In the most important form of respiration carried on by human cells, nutrients such as glucose are broken down in stages to carbon dioxide and hydrogen, and the hydrogen combines with oxygen absorbed from the blood to produce water. The carbon dioxide and water leave the cell as waste products and are carried away by the blood.

The cell uses energy released by respiratory breakdown to manufacture ATP (adenosine triphosphate) from its constituents, ADP (adenosine diphosphate) and inorganic phosphate (Fig. 2-8). This means that some of the energy originally stored in the glucose, or other nutrient molecule, is captured by and stored in molecules of ATP. The energy-yielding (catabolic) reactions of respiration are *coupled* with the energy-using (anabolic) reactions of ATP formation (we do not know exactly how

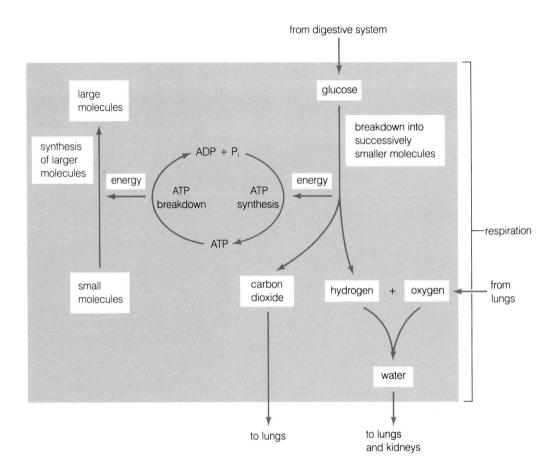

Figure 2-8 ATP synthesis and breakdown. ATP (adenosine triphosphate) is synthesized from ADP (adenosine diphosphate) and P_i (inorganic phosphate). The energy of this synthesis comes from respiration, as shown on the right. The breakdown of ATP into ADP and P_i provides the energy for synthetic reactions, as shown on the left.

this coupling works). ATP plays a critical role in the energy transactions of the cell, because it is broken down into its components quickly and easily, yielding large amounts of its stored energy for driving anabolic reactions. Just as the synthesis of ATP is coupled with the energy-yielding reactions of respiration, its breakdown is coupled with energy-using reactions, such as the synthesis of complex molecules as shown in the figure. The function of ATP is to transfer energy from catabolic reactions to anabolic ones, and for this reason it is sometimes called "the energy currency of the cell." *The function of respiration is to keep the cell supplied with ATP.*

Carbohydrates, proteins, and fats can all be broken down by digestion into smaller nutrient molecules that our cells can break down further by respiration. These three substances, which provide fuel as well as raw materials for the synthesis of cell components, are called *foods*. The respiration of a nutrient molecule involves a long series of breakdown reactions, each cat-

alyzed by a specific enzyme. The last step in the process of this *aerobic respiration* is the combining of hydrogen with free oxygen (oxygen that is not part of another molecule). That is why we must continuously supply our cells with oxygen by breathing. If no free oxygen is present in the cell, the final reaction of respiration cannot take place and the entire process grinds to a halt.

All cells respire, but not all do so in the same way. Our cells, and those of other higher animals and plants, depend primarily on aerobic respiration; even water-breathing animals such as fish use the atmospheric oxygen dissolved in water to respire aerobically. But some bacteria can live without free oxygen, because they use a different sequence of breakdown reactions that does not produce carbon dioxide and hydrogen as end products; this form of nutrient breakdown is called *anaerobic respiration*. Our own cells respire anaerobically as well as aerobically, although they cannot produce enough ATP by

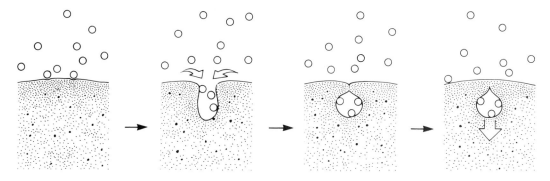

Figure 2-9 Pinocytosis. A macromolecular particle enters the cell via formation of a small vesicle from the plasma membrane.

anaerobic means to keep themselves alive for long. We shall discuss anaerobic respiration when we examine the functioning of muscles.

Absorption

Absorption (L., *ab* = to, in, *sorbeo* = to suck) is the process of taking substances into the cell from the cellular environment. Some substances cross the plasma membrane by *diffusion,* that is, by the random motion of molecules. Diffusion always results in a net movement of a specific kind of molecule from areas of greater concentration to areas of lesser concentration, as you know if you have ever smelled gas molecules spreading from a leak or perfume molecules from an open bottle. Thus, if a substance is found in higher concentration outside the cell than inside, it will move into the cell, provided it can get through the membrane. This diffusion of substance from regions of higher to regions of lower concentration is called "moving with the concentration gradient" (or "with the diffusion gradient") and requires no input of energy. Energy is required to move a substance from regions of lower to regions of higher concentration, or "against the gradient."

Water and small ions can enter and leave the cell by diffusion, or *passive diffusion,* as it is sometimes called. An ion is a charged particle—a molecule that has lost or gained one or more electrons and thus acquired a positive or negative electrical charge. Many chemical compounds *ionize,* or produce ions, in water; for example, when sodium chloride (table salt) is put in water, it dissociates (breaks up) into positive sodium ions and negative chloride ions. Sodium ions and chloride ions can both cross the plasma membrane by passive diffusion.

Osmosis is a special method of diffusion in which water passes through a membrane because a difference in the concentration of dissolved substances exists on opposite sides of the membrane, due to impermeability of the membrane to the dissolved substances (solute). Thus the solute cannot cross the membrane, but the water (solvent) will pass from the dilute solution to the more concentrated one to equalize the two sides. The force with which the more concentrated solution draws water into it is the osmotic pressure.

In contrast with passive diffusion, active processes of absorption require energy expenditure by the cell. Some molecules pass through the membrane by combining chemically with molecules of the membrane; the later release of the chemical bond leaves the transported molecules on the opposite side of the membrane. This is *active transport,* the principal method used by cells to move substances across a membrane against a concentration gradient.

Other active absorption processes include pinocytosis and phagocytosis, which involve the enclosing of particles by the cell membrane. Molecules that are dissolved in fluid may be taken into the cell by **pinocytosis,** (Fig. 2-9) in which the plasma membrane surrounds the molecule at the cell surface, enclosing it in a *vesicle* (L., *vesicula* = a blister). This vesicle pinches off from the plasma membrane and ruptures as it sinks into the cytoplasm, releasing the molecule. Most cells are capable of taking in water by pinocytosis. Any molecules that happen to be in solution in the water may be taken into the cell along with the water. **Phagocytosis** (Fig. 2-10) is a similar process but involves larger particles, large enough to be visible with the light microscope. During phagocytosis, the cell extends fingerlike projections, pseudopodia, that surround the particle being ingested. Once the particle is enclosed by plasma membrane,

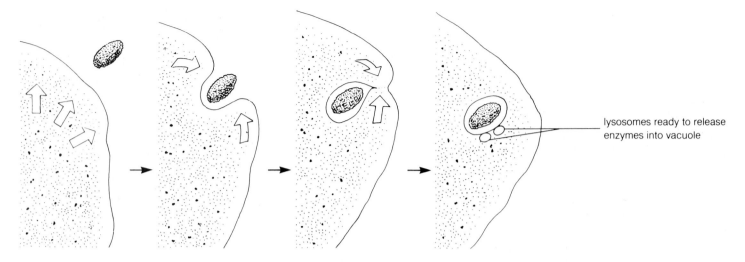

lysosomes ready to release enzymes into vacuole

Figure 2-10 Phagocytosis. Particulate matter is ingested by protoplasmic extensions called pseudopodia, forming a digestive vacuole within the cell.

the membrane pinches off from the cell surface, forming a vacuole containing the particle. Cells that take in material by phagocytosis are called **macrophages** (Gr., *macro* = large, *phagein* = to eat). Phagocytic cells include certain cells of the liver, spleen, and connective tissues as well as a type of white blood cell. These cells act as scavengers that rid the body of potentially harmful material. Collectively, pinocytosis and phagocytosis are known as **endocytosis,** which means "to bring into the cell."

Secretion

The release of substances through the cell membrane is called *emiocytosis* (L., *emitto* = to send forth) or **exocytosis.** *Secretion* (L., *se* = from, *cerno* = to separate) is the term applied to the extrusion of *useful* material from a cell. Secretion is highly developed in gland cells (Fig. 2-11), in which products are synthesized, wrapped in a membrane coat, and delivered to the cell surface. The membrane coat of the secretory granule fuses with the cell membrane, and the contents of the granule are released to the outside.

Excretion

Excretion (L., *ex* = out of, *cerno* = to separate) is the process of extruding waste or harmful material from a cell or from the body. Wastes are metabolic by-products of chemical reactions such as respiration. These potentially harmful substances pass from cells into the tissue fluid that surrounds them and enter the blood stream. They are eventually removed from the body, principally by the kidneys.

Irritability

Irritability (L., *irrito* = to excite) is the capacity of a cell or organism to respond to a stimulus. The stimulus may be a physical one, such as pressure, or a chemical one, such as a change in the acidity of the environment. The cell's response to a stimulus may be a change in any of the properties of protoplasm, including a change in its metabolism.

Conductivity

Conductivity (L., *con* = together, *duco* = to lead) is the capacity of cells to transmit an impulse from one part of the cell to another. This property is highly developed in nerve cells and is developed to a degree in muscle cells. It will be described in detail when we discuss nervous and muscular tissues.

Contractility

Contractility (L., *con* = together, *tractio* = to draw, pull) is the capacity of cells to shorten, or reduce their volume. This property is most highly developed in muscle cells, and the process of muscle contraction will be described in detail when muscle tissue is discussed.

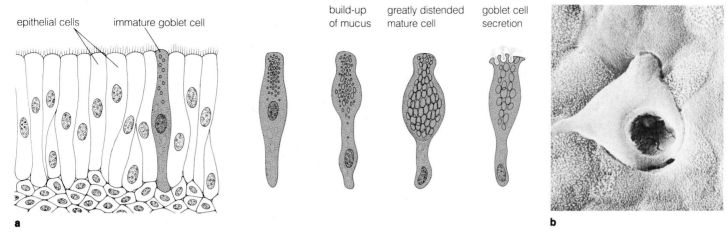

epithelial cells   immature goblet cell   build-up of mucus   greatly distended mature cell   goblet cell secretion

a   b

Figure 2-11   Secretion. (**a**) Secretory cells of the respiratory tract lining produce mucus in membrane-enclosed granules that are released through the cell surface. (**b**) Scanning electron micrograph of the surface of a goblet cell from which mucus is extruding (courtesy of M. Greenwood and P. Holland).

Reproduction

Perhaps the most striking property of living things is their ability to produce a new generation of organisms like themselves. **Reproduction** is a fundamental property of cells, although some cells lose their capacity as they become specialized. Cells reproduce by dividing in two, a process called **mitosis.** Before mitosis occurs the DNA of the *parent cell* is replicated (copied), and during mitosis the doubled DNA is divided equally between the two offspring, or *daughter cells.* As a result, the DNA content of each daughter cell is identical to that of the parent cell that divided.

Each of us began life as a single cell that multiplied by mitosis, producing the highly complex human body. That one cell, called a **zygote** or *fertilized egg,* was formed by the fusion of two *germ cells* (the egg and the sperm), each of which was produced by a special series of cell divisions called **meiosis.** Each germ cell contains precisely half the DNA content of other human cells; thus the fusion of two germ cells gives rise to a cell with the normal DNA content and the capacity to perpetuate the species.

Growth

After a cell divides by mitosis the daughter cells contain a normal amount of DNA, but each has received only half the cytoplasm of the parent cell. To attain normal size the daughter cells must grow, which they do by synthesizing new proteins and other cell components from the raw materials in their environment. **Growth** may be defined quite simply as an increase in mass. When a cell reaches a certain mass, it may divide again.

Multicellular organisms also grow, and their growth occurs in two ways: by increase in the number of their cells (by mitosis) and by an increase in the mass of individual cells. The former process is called *hyperplasia,* and the latter is called *hypertrophy.* Human beings stop growing after fifteen years or so, but many of their cells continue to grow and divide, replacing other cells that die or are worn away.

Differentiation

As the human embryo grows, it does more than simply increase in mass: it produces cells, tissues, and organs of diverse form and function. Although each of its cells contains an exact copy of the DNA of the zygote, each cell does not develop in the same way. **Differentiation** is the process of cellular change whereby new protoplasmic constituents are added to an unspecialized cell, making it capable of new functions. Differentiation usually leads to specialization of structure and function as specific properties become highly developed. For example, as muscle cells differentiate, their property of contractility is tremendously increased by their synthesis and accumulation of special contractile proteins. The genetic code, in ways only partially understood, can direct a given cell to develop kinds and amounts of proteins different from those of another cell.

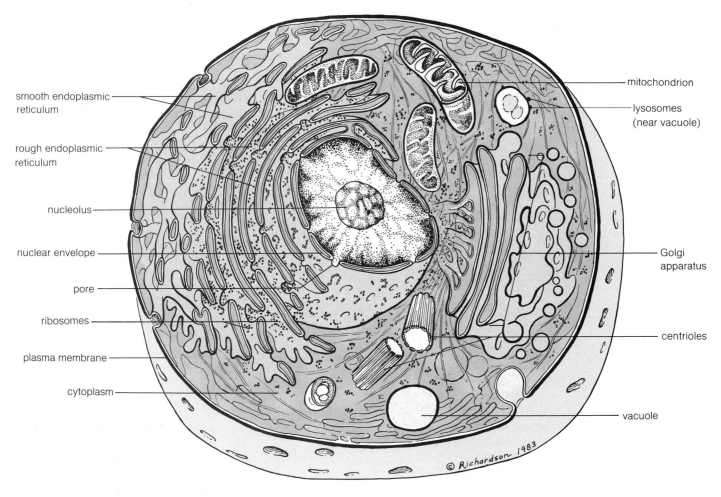

smooth endoplasmic reticulum

rough endoplasmic reticulum

nucleolus

nuclear envelope

pore

ribosomes

plasma membrane

cytoplasm

mitochondrion

lysosomes (near vacuole)

Golgi apparatus

centrioles

vacuole

© Richardson 1983

Figure 2-12 Generalized cell. Some structures can be seen with the light microscope; others are seen only with the electron microscope.

THE COMPONENTS OF THE CELL

When the cell is inspected with the light microscope (see Fig. 2-1), it appears to be largely empty, with widely scattered small *granules* (L., *granulum* = little grain) throughout the cytoplasm and a few larger clumps in the nucleus. Since most cells are about 75% water, it might be concluded that the empty spaces seen with the light microscope are spaces that are filled with water during the life of the cell. However, studies with the electron microscope have revealed that these apparently empty spaces are, in most cells, occupied by formed elements. The granules visible with the light microscope are only the largest of a vast array of cell organelles and inclusions. The most important ones are summarized in Table 2-2. Metabolic activities do not occur uniformly throughout the cytoplasm: each type of organelle plays a specific role in the metabolism of the cell. Fig. 2-12 is a composite drawing of cell features seen with the light and electron microscopes. The cell shown is a generalized one; that is, there is no cell type that looks exactly like it.

The Plasma Membrane

The cell is enclosed by a thin membrane, called the *plasma membrane,* or *cell membrane,* that controls the passage of substances into and out of the cell. When stained for viewing under a light microscope, the plasma membrane appears as a single fine line; but as seen with the electron microscope, the plasma

Table 2-2 Components of a Cell	
Name	Description
Plasma membrane	Surrounds cell. Selectively permeable. Unit membrane consisting of lipid bilayer with freely motile protein.
Nuclear envelope	Surrounds nucleus. Double layer of membrane with perforations closed by a diaphragm.
Endoplasmic reticulum (ER)	Membranous network of tubular cisternae. Rough ER, with ribosomes; smooth ER, without ribosomes. Connects to nuclear envelope, Golgi complex.
Golgi complex	Membranous organelle of stacked, disc-shaped saccules. Packages protein for export.
Ribosomes	Organelles made of RNA and protein. Sites of protein synthesis.
Mitochondria	Discrete membranous organelle. Contains respiratory enzymes.
Microtubules	Tubules composed of the protein tubulin. In centrioles and cilia, and free in cytoplasm.
Centriole	Cylindrical organelle consisting of nine triplets of microtubules. Functions in cell division.
Microfilaments	Elongated strands of protein that give support to cell. Some, such as actin and myosin, involved in contraction.
Lysosomes	Membrane-enclosed aggregations of hydrolytic enzymes.
Secretory granules	Membrane-enclosed aggregates of protein substances for export from the cell.
Lipid droplets	Oil (lipid) drop in a cell.
Nucleolus	Aggregation of RNA and protein in nucleus. Site of rRNA synthesis.

membrane is a *tri-laminar* (three-layered) structure called a *unit membrane.*

Chemical analysis shows that the plasma membrane is made up of proteins and *phospholipids.* A phospholipid is a lipid molecule with a negatively charged phosphate group at one end. Because the charged, or polar, end of the molecule is attracted to water, phospholipid molecules in a membrane are aligned in two rows, one with the polar ends pointing toward the watery interior of the cell and one with these ends pointing toward the watery surrounding environment. The nonpolar ends point inward, toward the inside of the membrane.

> Several diseases, such as rickets and gout, involve defects of transport across plasma membranes. In rickets, there is a failure of absorption of calcium from the intestine, resulting in defective growth of the skeleton. In gout, too much uric acid is returned to the bloodstream in the kidney, raising the blood levels of uric acid. This leads to the deposition of uric acid crystals in the lining of joints, causing swelling and pain.

Early models of the plasma membrane proposed that it consists of two rows of phospholipids with protein sandwiched between them. This model explained the three-layered appearance of the unit membrane; however, a current theory holds that the plasma membrane is actually a two-layered structure (Fig. 2-13). The protein molecules do not form a discrete layer, but move freely within a lipid *bilayer* that acts as a fluid. The lipid bilayer is formed by the two rows of phospholipid molecules.

> The properties of cell surfaces change in cancerous tissues. Contact between cell surfaces appears to restrict or regulate the division of normal cells, but tumor cells show unrestrained growth and the ability to spread to other locations by invasion and migration. Some types of cancer cells have surface proteins called antigens that are different than those of normal cells. These antigens are the subject of intensive research exploring their potential in the detection and treatment of cancer.

Many animal cells have *cell coats,* or *cell envelopes,* attached to the outside of the plasma membrane. The envelope, also called the *glycocalyx,* is made of a protein–carbohydrate com-

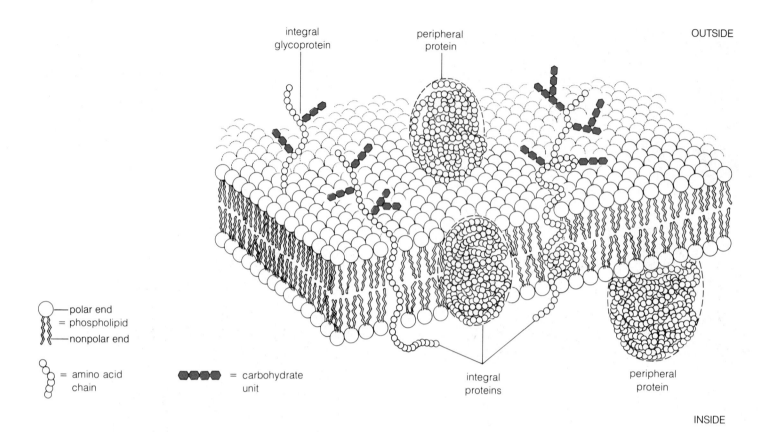

Figure 2-13 Model of the plasma membrane.

plex called *glycoprotein*. The glycoproteins are of various kinds and may act as specific receptors for substances such as hormones that react only with specific cells. They are also probably involved in selective transport across the membrane.

Nuclear Envelope

A double layer of membrane, perforated by irregularly spaced pores, separates the nucleus from the cytoplasm (Fig. 2-14). The two layers of this **nuclear envelope,** each of which is a unit membrane, are separated by a space 500 Å wide. Most exchanges of substances between the nucleus and cytoplasm occur through the pores. In most cells the pores are closed by a thin diaphragm that frequently has a granule in the center. The diaphragm is thought to regulate the passage of material through the envelope, and the granule may be material in transit.

The outer membrane of the nuclear envelope is directly continuous with a system of cytoplasmic membranes, the **endo-**

plasmic reticulum (Gk., *endo* = within; L., *recticulum* = small net). Both the outer nuclear membrane and the endoplasmic reticulum frequently have a granular appearance due to the presence of organelles called **ribosomes** on the membrane surface. Ribosomes are the sites of protein synthesis within the cell.

Endoplasmic Reticulum

The *endoplasmic reticulum,* or *ER,* is an organelle composed of membranes that extend throughout the cytoplasm (Fig. 2-15). In many places the membranes are arranged in parallel layers. The spaces enclosed by the membranes are called **cisternae** (sing., *cisterna*). The cisternae appear discontinuous in the two-dimensional slices of cells seen in electron micrographs, but they actually form a continuous, three-dimensional network of channels that serves as a pathway for the movement of substances within the cell. The ER, which cannot be seen with the

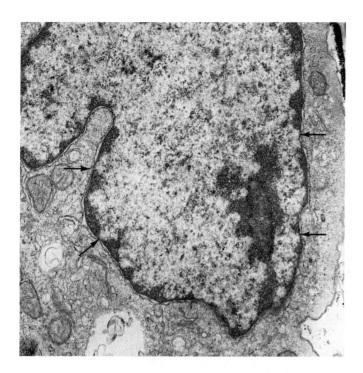

Figure 2-14 A portion of the nucleus of a cell of the mouse corpus callosum, showing the nuclear envelope. Nuclear pores are indicated by arrows. × 12,200 (courtesy of D. H. Matulionis).

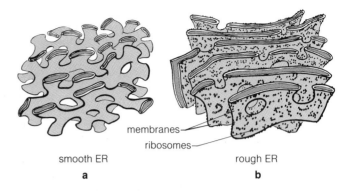

smooth ER   membranes—— ribosomes——   rough ER
a   **b**

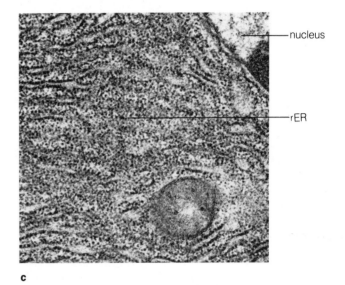

nucleus

rER

c

Figure 2-15 Endoplasmic reticulum: (**a**) tubular sacs of smooth ER; (**b**) layers of rough ER in section; (**c**) electron micrograph of rough ER in the mouse pancreas, × 30,000. (Micrographs courtesy D. H. Matulionis.)

light microscope, is connected in places to the outer nuclear envelope and to membranous organelles called **Golgi complexes.** There are two types of endoplasmic reticulum, rough and smooth.

Rough endoplasmic reticulum (rER) is studded with *ribosomes,* giving the membranes a beaded appearance (Fig. 2-15b,c). Rough endoplasmic reticulum is highly developed in secretory cells (such as the cells of the pancreas, which produce large quantities of digestive enzymes) and in other cells engaged in the synthesis of particularly large amounts of protein.

Smooth endoplasmic reticulum (sER) consists of membranes with no associated ribosomes (Fig. 2-15a). It is most highly developed in cells that secrete steroid hormones (such as those of the cortex of the suprarenal gland). Smooth endoplasmic reticulum is also prominent in the cells of stomach epithelium that produce hydrochloric acid.

The role of the ER in the synthesis of these substances will be described later.

The Golgi Complex

The Golgi complex (Fig. 2-16) stores protein and packages it for export outside the cell. Like the ER to which it is attached, it contains interconnected spaces enclosed by membranes. Its compartments, called *saccules,* are disc-shaped and arranged in a stack, like a stack of rather wavy pancakes.

Protein synthesized by the ribosomes associated with rough endoplasmic reticulum passes into the cisternae and migrates into the smooth ER. Here it becomes surrounded by membrane,

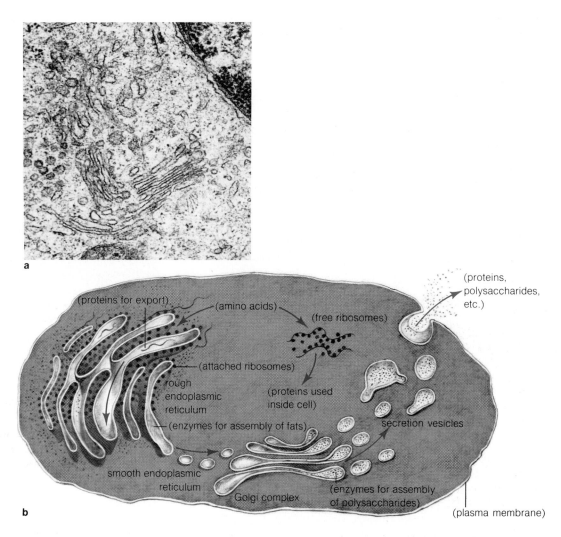

Figure 2-16 The Golgi complex: (**a**) electron micrograph of the Golgi apparatus of a pancreatic cell, × 62,500, showing the relations of the rER to the Golgi apparatus (courtesy of D. H. Matulionis); (**b**) roles of the rER and Golgi complex in the process of secretion. The dots in the cisternae represent concentrations of proteins.

producing small vesicles called *transfer vesicles*. The transfer vesicles detach from the smooth ER and migrate to a Golgi complex where they fuse with a saccule, releasing the protein inside the complex. Within the Golgi complex the proteins form larger, more condensed aggregates. These aggregates become surrounded by membrane to form *secretory vesicles,* which become **secretory granules** when they detach and move to the cell surface. Here they fuse with the plasma membrane and release the protein to the outside of the cell: this is the process of *secretion* (see Fig. 2-16b). Since packaging materials to be secreted is the principal function of the Golgi complex, it is not surprising that this organelle is most abundant in secretory cells.

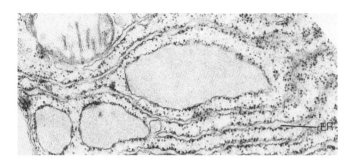

Figure 2-17 Electron micrograph of polysomes and ribosomes of mouse intestine, x 35,000 (courtesy of D. H. Matulionis).

Ribosomes

Ribosomes, the protein factories of the cell, are made of protein and a form of RNA called *ribosomal RNA* or *rRNA*. Aggregations of ribosomes (Fig. 2-17) are called *polysomes*. The synthesis of ribosomal RNA occurs in the nucleus in association with the *nucleolus*. All cells actively engaged in protein synthesis contain large numbers of ribosomes. Ribosomes involved in the synthesis of protein for export are attached to the endoplasmic reticulum; those involved in the synthesis of protein that remains in the cell occur as "free" polysomes, scattered throughout the cytoplasm. A ribosome acts as a surface on which amino acids from the cytoplasm are assembled into large protein molecules under the direction of *messenger RNA* (mRNA). Small RNA molecules called *transfer RNA* (tRNA) attach to amino acids in the cytoplasm and bring them to the ribosome, where they are assembled in the correct order by an mRNA molecule temporarily associated with the ribosome.

Mitochondria

Mitochondria (sing., mitochondrion) are membranous organelles in which respiration takes place (Fig. 2-18). They contain DNA; and since they have not been shown to be continuous with, or formed by, other membranes, they are considered to be largely self-reproducing. However, nuclear DNA has codes for some mitochondrial constituents, so mitochondria are not completely independent. They vary in shape from rod-like to essentially round. There may be several hundred mitochondria

scattered throughout the cytoplasm of a single cell, and they can be seen in the living cell with the phase microscope.

Electron micrographs (Fig. 2-17) show that a mitochondrion has an *outer membrane* and an *inner membrane,* the latter invaginated (folded inward) to form **cristae** (L., *crista* = crest). Between the cristae are some dark granules that may be aggregations of DNA.

Mitochondria contain the enzymes essential for respiration. Each step of the respiratory process described earlier is catalyzed by a specific enzyme located in the inner membrane. The cristae appear to serve as frameworks that hold enzymes in place in the sequence in which they take part in respiratory reactions. Since a mitochondrion supplies the cell with energy, in the form of ATP, it has been described as "the powerhouse of the cell."

Microtubules

Microtubules (Fig. 2-19) are thin, elongated tubules about 270 Å in diameter that run straight courses in the cytoplasm. They are composed of a protein called *tubulin* that can be fragmented rapidly into subunits or added to the ends of the tubules to

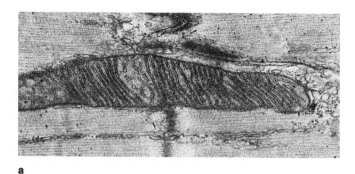

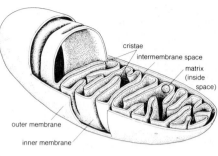

a

b

Figure 2-18 A mitochondrion (**a**) of mouse cardiac muscle, ×24,000 (courtesy of D. H. Matulionis) (**b**) diagram to show the typical arrangement of membranes.

Figure 2-19 Microtubules, x 45,000. (Courtesy D. H. Matulionis.)

increase their length. Microtubules are usually scattered throughout the cytoplasm and form an important part of the structure of **centrioles** and **cilia.**

The functions of the microtubules are not clear. It has been suggested that they provide a transport system within the cytoplasm for the movement of substances from one part of the cell to another. They may provide a skeletal framework that helps to maintain cell shape and may be involved in movement within the cell.

Centrioles

A *centriole* (Fig. 2-20) is a cytoplasmic organelle in the shape of a cylinder consisting of nine triplets of *microtubules*. Two centrioles, oriented at right angles to one another, are located near the nucleus. During cell division the centrioles move to opposite poles of the cell and are associated with the formation of *mitotic spindle*. The spindle fibers, microtubules that extend between the centrioles and the *chromosomes* (Gr., *chroma* = color, *soma* = body) are essential in the separation of the chromosomes, as will be discussed in Chapter 3.

Microfilaments

Microfilaments (Fig. 2-21) are elongated strands of protein that provide internal skeletal support for cells. They tend to be oriented toward specialized points of junction between cells where, for example, they appear to terminate in large numbers in the *desmosomes* (Gr., *desmos* = band, *soma* = body), structures which attach cells to one another. Not all microfilaments are the same. In muscle cells, microfilaments are of two types, actin and myosin, and are responsible for the intracellular movement occurring during muscle contraction. Movement of microfilaments may also be involved in the constriction of the cytoplasm by the plasma membrane during cell division.

Lysosomes

The **lysosomes** (Gr., *lysis* = dissolution), Fig. 2-22, are membrane-enclosed aggregations of hydrolytic enzymes. The ribosomes of rough endoplasmic reticulum produce hydrolytic

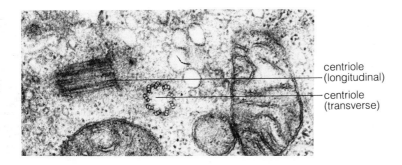

Figure 2-20 Centrioles. Longitudinal and transverse sectional views of the centrioles of a cell, ×20,000. (Courtesy D. H. Matulionis.)

enzymes that become wrapped in membrane within the Golgi complex. Instead of being discharged in the manner of secretory vesicles, these lysosomes remain within the cytoplasm and digest (ingest and breakdown) aged cellular components, such as mitochondria, and foreign material that may enter the cell, such as bacteria. The degraded molecules then diffuse from the cell. If lysosomes rupture, their digestive enzymes can destroy the entire cell. All cells except red blood cells contain lysosomes.

Lysosomes may become involved in disease by failing to digest nonbiological material. For example, coal miners inhale coal dust particles that are phagocytosed by lung macrophages. These particles cannot be digested by the lysosomes of the macrophages and consequently accumulate in the lungs, resulting in the tissue reactions known as black lung disease.

Cytoplasmic Inclusions

Secretory granules (Fig. 2-23a) are membrane-enclosed aggregates of substances that are ready to be discharged from the cell. Gland cells produce substances, usually enzymes, that aggregate into small globules, or droplets, that are surrounded by membrane. The process of aggregation and enclosure by membrane occurs in the Golgi complex. The membrane surrounding a secretory granule fuses with the plasma membrane in the process of releasing the contents of the granule outside the cell.

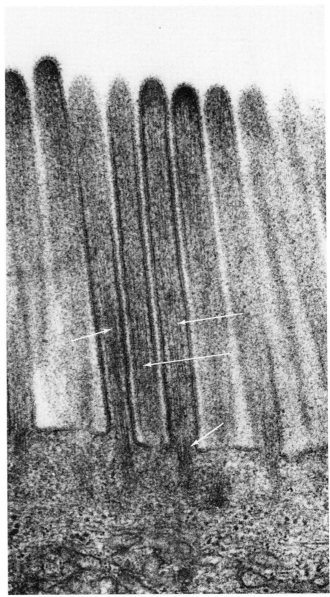

Figure 2-21 Microfilaments (arrows) in microvilli of a mouse small intestine and in the cytoplasm beneath the microvilli × 82,400. Microfilaments have been suspected of providing the force for the movement of microvilli. (Courtesy of D. H. Matulionis.)

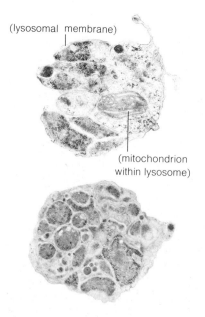

(lysosomal membrane)

(mitochondrion within lysosome)

Figure 2-22 Digestion of organelles as seen in a lysosome. × 12,000 (Gary W. Grimes)

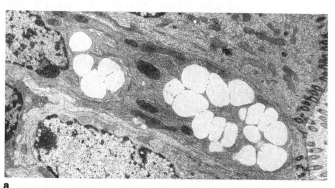

a

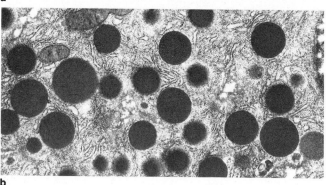

b

Figure 2-23 Cytoplasmic inclusions: **(a)** an electron micrograph of secretory granules in the cytoplasm, × 5500; **(b)** an electron micrograph of serous droplets in the cytoplasm (mouse pancreas), × 9600 (courtesy D.H. Matulionis)

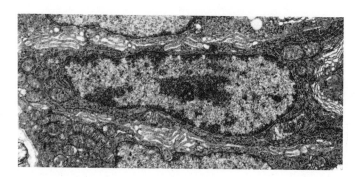

Figure 2-24 A nucleolus within the nucleus of a nondividing cell. The dark material of the nucleolus is largely RNA. The scattered chromatin granules are aggregations of DNA, x 6,000. (Courtesy of D. H. Matulionis.)

Lipid droplets (Fig. 4-11) function as storage vats in which lipids accumulate in the form of small drops of oil. As the drops become numerous they coalesce into larger and larger drops, until the cell is almost entirely filled with accumulated fat, with the nucleus displaced to one end and the cytoplasm present as a narrow rim beneath the plasma membrane. The cell then has the appearance of a signet ring and is called a *signet cell.* Large numbers of **fat cells** in certain areas of the body in obese individuals are designated as *adipose tissue* (L., *adeps* = fat).

Nucleus and Nucleolus

Each nucleus contains one or more aggregations of RNA called *nucleoli* (Fig. 2-24), which are centers of active RNA synthesis. The nucleolus produces ribosomal RNA molecules that pass through the nuclear pores into the cytoplasm where they form part of the structure of ribosomes.

The nucleoli vary in size and shape in different cell types and at different times within the same cell, having the capacity to fuse with one another or to split into two or more separate structures.

SURFACE SPECIALIZATIONS OF CELLS

The cell membranes of many cell types show structural modifications associated with special functions. The cell membrane not only isolates the cell from its environment and controls what enters and leaves the cell but also holds different cells together. Some surface specializations are designed to increase this adhesion. Others are designed to increase the surface area of the

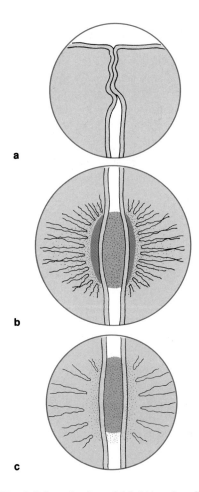

Figure 2-25 Cellular adhesions: (**a**) tight junction, (**b**) desmosome, and (**c**) intermediate junction.

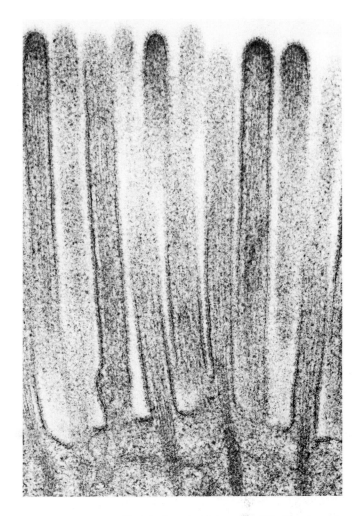

Figure 2-26 Microvilli of the intestinal lining, x 63,600. (Courtesy of D. H. Matulionis.)

cell by means of folds or projections, allowing faster transport of substances across the cell membrane. Still other surface specializations (cilia) provide a means of moving material along the surface of an epithelial sheet.

Cellular Adhesions

In general, epithelial cells adhere to one another by glycoprotein and by three specialized types of contact: tight junctions, desmosomes, and intermediate junctions.

The *tight junction* (Fig. 2-25c) is the adhesion between cells in which the outer layers of adjacent cell membranes are in direct contact with one another, obliterating the intercellular space. This type of adhesion prevents diffusion of substances through the intercellular space between cells in the area of tight junction.

The *desmosome* (Fig. 2-25b) is like a spot-weld between adjacent cells. Electron-dense material (material that looks dark in electron micrographs) condenses on the cell membrane at the spot where the cells are held together. Microfilaments (tonofibrils) usually radiate out from the desmosome, presumably providing support. This is a strong adhesion and joins cells in epithelial sheets, such as the epidermis, where forces such as friction might tend to separate them.

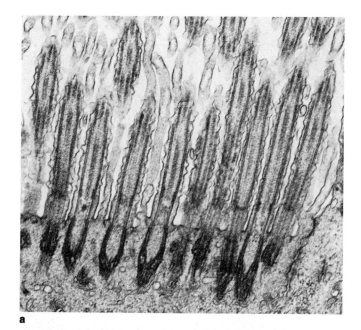

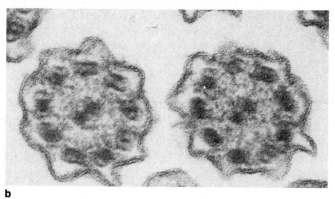

Figure 2-27   Cilia: (**a**) longitudinal section of cilia, × 80,000; (**b**) transverse section of cilia, × 153,000. (Courtesy of D. H. Matulionis.)

(Fig. 2-26). Microvilli occur at the free, or unattached, surface of a cell; similar infoldings of the cell membrane are found in the opposite surface of the same cell, next to the basement membrane. These membrane projections provide a tremendous increase in area of the cell surface, facilitating diffusion or active transport of substances across the cell membrane in either direction. Consequently, the epithelial sheets lining structures such as the digestive tract, where absorption and secretion are particularly important functions, bear microvilli on the free surface of the epithelial cells.

Cilia

Cilia (L., *cilium* = an eyelid) are slender processes projecting above the free surface of epithelial cells and are composed of microtubules capable of motion. Each cilium consists of *nine doublets* of microtubules arranged in a circle around two central microtubules (Fig. 2-27). The cilium contains a *basal body* that has the same structure as a centriole.

Smoking inhibits ciliary action, allowing particulate matter such as carbon to reach the lungs. Lung macrophages ingest the carbon particles, but since lysosomes cannot digest them they accumulate, blackening the lungs of heavy smokers. The carbon particles remain in the lung macrophages until released upon death of the macrophages. The carbon is released into the intercellular spaces of the lung and is quickly ingested by new macrophages that move into the area to clean up the debris of the dead macrophages. Thus the carbon remains in the lung tissue; there is no mechanism for clearing the lungs of material that is not degradable by the hydrolytic enzymes of lysosomes.

The *intermediate junction* (Fig. 2-25c) is a ribbon-like adhesion that frequently encircles the cell completely. It is the most extensive of the cell-to-cell adhesions but has less electron-dense material and fewer microfilaments at the junction than does the desmosome.

Microvilli

Microvilli (Gr., *mikros* = small; L., *vilus* = shaggy hair) are finger-like projections of the cell membrane at the cell surface

The mechanism of movement of cilia is not understood. The beat of the cilia of an epithelial sheet is in one direction, producing a current in the fluid at the surface that causes it to flow along the surface of the sheet. The currents created by the cilia on the epithelial sheet that lines the respiratory tract play an important role in removing particulate matter from the inspired air. Particles become trapped in mucus on the epithelial surface and float in the current created by the beating cilia toward the mouth, where they are swallowed. This prevents the passage of the particles into the lungs.

Clinical Applications: **Inborn Errors of Metabolism**

Many disorders are caused by mutations that alter the genetic constitution of an individual in a manner that disrupts normal function. These "inborn errors of metabolism" were first described by Archibald Garrod at the beginning of this century. He described four initially; now there are hundreds listed in standard pediatric textbooks.

An understanding of the role of deoxyribonucleic acid (DNA) in protein synthesis is essential in the understanding of these disorders. Genetic information is in the chromosomes of the nucleus encoded in DNA molecules. This information is transcribed to messenger ribonucleic acid (mRNA) that leaves the nucleus to regulate protein synthesis in the cytoplasm. The information carried by mRNA dictates the sequence of amino acids as protein molecules are synthesized from individual amino acids in the cytoplasm of the cells.

A change in the genetic code (mutation) may result in alteration of the structures of the cellular proteins either by changing the sequence of amino acids or by substituting one amino acid for another. If the protein that is altered happens to be an enzyme that is necessary for the synthesis of a specific substance in the body, that substance will not be present. Precursors of the substance are likely to accumulate excessively, adding to the problem.

One of the most common inborn errors of metabolism is phenylketonuria, or PKU. This condition results from the absence of the liver enzyme, phenylalanine hydrolase, which catalyzes the conversion of the amino acid phenylalanine to the amino acid tyrosine. Phenylalanine is necessary for protein synthesis, and it is present in excess in a normal diet. The dietary phenylalanine that is not needed for protein synthesis is used for energy in a metabolic pathway that requires its conversion into tyrosine. Owing to the enzymatic defect in this pathway, an individual with PKU accumulates excess amounts of phenylalanine that are converted into phenylpyruvic acid, which can then be converted to other metabolites. The accumulation of large quantities of phenylalanine and its metabolites causes severe mental retardation if untreated. Fortunately, people with PKU excrete large amounts of phenylpyruvic acid and its metabolites that can be detected in the urine soon after birth. The musty odor characteristic of untreated patients is thought to be due to the accumulation of phenylacetic acid. The treatment is to limit phenylalanine in the diet to just the amount required for protein synthesis. If the dietary intake of phenylalanine is controlled for at least the first five years of life, the infant with PKU can develop normally.

The number of recognized inborn errors of metabolism is constantly increasing and includes defects in carbohydrate and fat metabolism in addition to those of protein metabolism such as PKU. Many of these disorders can be detected early in life; large-scale screening tests for blood and urine are currently carried out. One of the current objectives of genetic research is to understand the genetic mechanisms involved so that it may become possible in the future to alter the genetic constitution of an individual and to overcome the inborn errors of metabolism.

WORDS IN REVIEW

Absorption	Endoplasmic reticulum	Microfilaments	Pinocytosis
Cellular adhesions	Enzyme	Microtubules	Plasma membrane
Centrioles	Excretion	Microvilli	Polysaccharides
Cilia	Exocytosis	Mitochondria	Reproduction
Cisternae	Fat cells	Mitosis	Respiration
Conductivity	Golgi complex	Nuclear envelope	Ribosomes
Contractility	Growth	Nucleic acids	Secretory granules
Cristae	Irritability	Nucleoli	Zygote
Cytoplasm	Lysosomes	Nucleoplasm	
Differentiation	Macrophages	Organelles	
Endocytosis	Meiosis	Phagocytosis	

SUMMARY OUTLINE

I. Composition of Protoplasm (approximate)
   A. Protein is about 20%, lipid 3%, carbohydrate 1%, water 75%, and salts 1% of the body as a whole.

II. Properties of Cells
   A. Irritability is the response to a stimulus.
   B. Conductivity is the transmission of an electrical impulse.
   C. Respiration is the process of the oxidation of nutrients.
   D. Absorption is the process of taking material into cells.
   E. Secretion is the discharge of useful material from the cell.
   F. Excretion is the discharge of wastes from the cell.
   G. Reproduction refers to cell division (mitosis) or to the production of a new individual.
   H. Differentiation means any change in structure and function of a cell.
   I. Growth refers to an increase in mass..

III. Components of the Cell
   A. The plasma membrane is a fluid-mosaic of lipid and protein enclosing the cell.
   B. The nuclear envelope is a double layer of unit membrane with pores.
   C. The endoplasmic reticulum is a cytoplasmic network of flattened membranous cisternae or tubules involved in the synthesis of cellular products.
   D. The Golgi complex is a series of flattened, elongated vesicles involved in packaging secretory products in vacuoles or granules for discharge from the cell.
   E. The mitochondria are membranous cytoplasmic organelles that contain oxidative enzymes.
   F. Secretory granules are membrane-enclosed aggregates of enzymes.
   G. Lipid droplets are a storage form of fat.
   H. Centrioles are cylinders of nine triplets of microtubules that produce the mitotic spindle.
   I. Microtubules are thin, elongated tubules composed of the protein tubulin; they possibly provide an internal transport system or skeletal support for cells.
   J. Microfilaments are elongated strands of protein within cells that provide for a mechanism of contraction and also for skeletal support.
   K. Ribosomes are accumulations of RNA and protein that serve as substrates in the synthesis of protein.
   L. Lysosomes are membrane-enclosed aggregations of hydrolytic enzymes that function in the digestion of proteins within the cell.
   M. The nucleolus is an aggregation of RNA within the nucleus; it is responsible for the production of ribosomal and transfer RNA.

IV. Surface Specializations of Cells
   A. Cellular adhesions are specializations of the plasma membrane that enable cells to stick together.
      1. A tight junction is the adhesion in which adjacent outer layers of cell membranes are in direct contact.
      2. A desmosome consists of electron-dense material at the spot of the adhesion.
      3. Intermediate junctions are ribbon-like adhesions that frequently encircle the cell.
   B. Microvilli are finger-like projections of the plasma membrane.
   C. Cilia are slender, motile processes extending from the cell surface; they are composed of nine doublets of microtubules arranged in a circle around two central microtubules.

REVIEW QUESTIONS

1. What are the principal constituents of protoplasm?

2. List and define the physiological properties of protoplasm.

3. Diagram a cell showing the components that are visible with the light microscope. Give the principal function of each component.

4. Diagram a cell showing the components that are visible only with the electron microscope. Give the principal function of each component.

5. Describe the three types of cell adhesions.

6. Describe the morphology of cilia and relate their structure to their general function.

SELECTED REFERENCES

Bloom, W., and Fawcett, D. W. *A Textbook of Histology*. 10th ed. Philadelphia: W. B. Saunders, 1975.

Chayen, J.; Bitensky, B. T.; and Butcher, R. G. *Practical Histochemistry*. New York: John Wiley and Sons, 1973.

DeRobertis, E. D.; Nowinski, W. W.; and Saez, F. A. *Cell Biology*. 5th ed. Philadelphia: W. B. Saunders, 5th ed. 1970.

Fawcett, D. W. *An Atlas of Fine Structure: The Cell*. Philadelphia: W. B. Saunders, 1966.

Lash, J. W., and Burger, M. M., eds. *Cell and Tissue Interactions*. New York: Raven Press, 1977.

Weiss, L., and Greep, R. O. *Histology*. 4th ed. New York: McGraw-Hill, 1977.

Yunis, J. J. *New Chromosomal Syndromes*. New York: Academic Press, 1977.

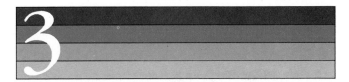

REPRODUCTION OF THE CELL AND THE ORGANISM

After completing this chapter you should be able to:

1. Describe the physiological events of the phases of the cell cycle and relate these events to morphological changes within the cell.

2. Define mitosis and describe its stages.

3. Define meiosis and explain its importance in the life cycle of the individual.

4. Describe the morphological changes during spermatogenesis and oogenesis.

5. Define a karyotype and give the normal chromosomal complement of the human.

6. Describe development of the early embryo from fertilization through the origin of body form.

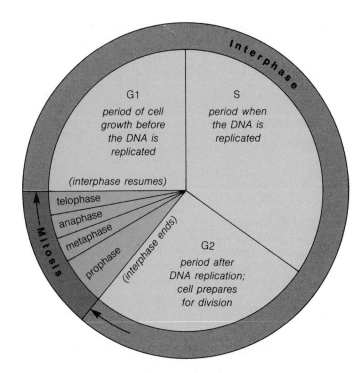

Figure 3-1 The cell cycle. Interphase includes G1, S, and G2 of the cycle. The time spent in phase G1 is variable, but the other stages are of relatively constant length for a given type of cell.

Every somatic cell goes through a life cycle, or **cell cycle,** comparable to the life of the individual. It is "born" at the moment of its formation as the offspring of the parent cell and "dies" when it reproduces, giving rise to two identical offspring. Its life is divided into two main stages: **interphase,** the nondividing stage of the cycle, and **mitosis,** the dividing stage. After mitosis the cycle begins again for each daughter cell. Synthesis of new cytoplasmic components occurs only during the interphase. During interphase a cell may produce proteins or other substances that were not present in the parent cell; if these new substances form distinctive structural components, the cell is said to have differentiated. Some types of cells, such as nerve cells, become so highly differentiated, or specialized, that they lose the ability to divide.

During interphase a cell precisely duplicates each of its **chromosomes,** copying the DNA exactly. When the cell undergoes mitosis, each of its offspring receives one of the "offspring" of each chromosome. Thus each new cell has exactly the same number of chromosomes as the parent cell and an exact copy of the parental DNA.

In **meiosis,** the special cell division necessary for sexual reproduction, each parent cell produces germ cells, or sex cells, that have only half the DNA content of somatic, or nonreproductive, cells. In higher animals, including humans, meiosis occurs only in cells producing germ cells. The purpose of a germ cell is to unite with another germ cell, producing a cell with the normal DNA content. This zygote, through repeated mitoses and

cellular **differentiation**, eventually produces an individual composed of trillions of cells of varied forms and functions.

Meiosis, mitosis, and cellular differentiation are the essential mechanisms by which new individuals are produced to perpetuate the human species. This chapter describes the cell cycle, including the principal events of mitosis, discusses meiosis and the formation of eggs and sperm, and gives a brief account of tissue differentiation and the early embryonic development of new individuals.

THE CELL CYCLE

A cycle can be described beginning with any phase, but the cell cycle (Fig. 3-1 and Table 3-1) is usually considered to begin at the end of mitosis. This is the start of the *interphase* stage, during which the cell is not dividing. Interphase is divided into three phases: G1, S, and G2; G1 and G2 refer to the generative or growth phases of the cycle and S to the period of DNA synthesis. Mitosis is called the M phase of the cycle. The time required by

Table 3-1 The Cell Cycle and Mitosis	
Stage	Principal Events
Interphase	Cell is not dividing. Period of synthesis.
G1 phase	Synthesis of cytoplasmic components
S phase	Synthesis (replication) of DNA, creating doubled (two-chromatid) chromosomes
G2 phase	Synthesis of cytoplasmic components essential for spindle formation
M phase	Mitosis (cell division)
Prophase	Doubled chromosomes begin to condense and become visible. Each consists of two chromatids, joined at the centromere. Nuclear envelope disappears. Spindle is produced.
Metaphase	Doubled chromosomes line up in equatorial plane, and spindle fibers attach to their centromeres. Chromatids are still joined.
Anaphase	Centromeres double, and chromatids become separate chromosomes. As formerly joined chromosomes separate, they move toward opposite poles.
Telophase	Plasma membrane constricts, dividing the cell into two daughter cells. The nuclear envelope appears, and chromosomes uncoil. Nucleolus appears.

different types of cells to complete the cycle varies greatly; however, rapidly growing cells such as those that give rise to certain blood cells complete a cycle in about 17 hours.

G1 Phase

During the *G1 phase* of interphase the cell rapidly synthesizes cytoplasmic components. When the parent cell divides by mitosis, each daughter cell receives a full complement of DNA; but the original cytoplasm, without being duplicated, is divided roughly equally between the two cells. Thus the cytoplasm of a new cell is not adequate for all the work it must do. Protein synthesis is activated, and the new cell produces more *cytoplasmic organelles* such as ribosomes, the Golgi complex, and endoplasmic reticulum. This process restores the volume and composition of the cytoplasm to approximately that of the parent cell.

S Phase

Replication (or synthesis) of the chromosomes in preparation for the next mitosis occurs during the *S phase* of interphase. At the end of this phase, a cell contains twice as much DNA as it does at the end of mitosis. Interphase chromosomes are largely uncoiled, forming a tangled mass within the nucleus. Small segments of the chromosomes remain condensed and are visible with staining as *chromatin granules.* During the S phase each long, drawn-out chromosome is duplicated, and the duplicates, called **chromatids,** remain connected to each other until they move apart during mitosis. While DNA is being synthesized, most other metabolic processes cease; few, if any, proteins are added to the cytoplasm during the S phase.

Chemotherapeutic drugs and radiation used in the treatment of cancer often interfere with DNA synthesis during the S phase of the cell cycle. These agents can kill cancer cells but are nonspecific and will also kill dividing cells, including those that give rise to blood cells. Thus it is difficult to kill all cancer cells in a patient without reducing the number of blood cells below that essential for life.

G2 Phase

In the *G2 phase* of interphase the cell synthesizes cytoplasmic constituents that play an important role in mitosis. The DNA doubled during the preceding S phase; but before the cell can divide, it must assemble the materials necessary for rapid formation of the *spindle,* a structure involved in separating the chromosomes. The spindle, you will recall, is made of microtubules composed of tubulin subunits. Tubulin is synthesized during G2, and it remains as disassociated subunits in the cytoplasm until it is needed during mitosis, when rapid assembly of the subunits into microtubules results in the sudden appearance of the spindle.

M Phase, or Mitosis

Mitosis, the *M phase* of the cell cycle, is the physical process that produces two daughter cells, each containing a chromosome complement identical to that of the parent cell. Every human cell nucleus (except in the germ cells) contains forty-six chromosomes. During the S phase of the cell cycle each of these chromosomes replicates, so that the parent cell contains forty-six doubled chromosomes, each consisting of two identical chromatids attached to one another. During mitosis the chromatids are pulled away from each other, and each daughter cell

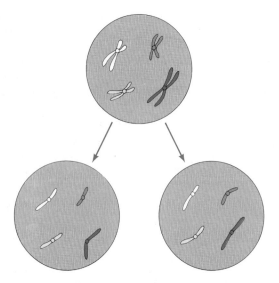

Figure 3-2 The result of mitosis in a cell with four chromosomes. The parent cell contains four two-chromatid chromosomes, the result of DNA replication during interphase. Each daughter cell contains four one-chromatid chromosomes, which will replicate during the next interphase.

receives one chromatid of each chromosome. Once pulled apart, the chromatids are considered chromosomes in their own right, so each daughter cell has forty-six chromosomes at the end of mitosis. During the discussion of the details of mitosis, keep the essence of the process in mind: In mitosis, each doubled chromosome separates into two new chromosomes, and one of the new chromosomes goes to each new cell.

STAGES OF MITOSIS

During division the cell undergoes progressive changes that may be described as four stages: **prophase, metaphase, anaphase,** and **telophase** (Fig. 3-2). Keep in mind, however, that mitosis is really a continuous process with no break between one stage and the next.

Prophase

During late interphase the chromosomes begin to coil and condense. Prophase begins when the chromosomes first become visible with the light microscope and lasts until they are attached to the spindle, the apparatus that will separate the chromatids.

As they continue to condense during prophase, the chromosomes become thicker and shorter until each can be seen to consist of two chromatids attached to each other at a point called the **centromere**. During late prophase the chromosomes move toward the midline, or equator, of the cell.

As the chromosomes take form, changes are occurring in the cytoplasm. The nucleoli shrink and soon disappear as the nucleolar RNA disperses. At the same time the *centrioles* undergo duplication, and the two pairs of centrioles slowly migrate toward opposite poles of the cell. The *nuclear envelope* breaks into fragments and disappears completely by the end of prophase. By this time the two pairs of centrioles have reached the opposite poles of the cell, and the spindle begins to organize around them.

Metaphase

Metaphase begins when the two-chromatid chromosomes are aligned at the equator of the cell and can be seen to be attached to *spindle fibers*. The spindle arises during prophase from the joining of tubulin subunits to form microtubules, which become organized around the centrioles. The centrioles are essential for spindle formation and form its two poles. The individual microtubules of the spindle are the spindle fibers, some of which extend across the cell from one centriole to the other, while others extend only from the centriole to the centromeres of the chromosomes.

Anaphase

Anaphase begins with the *doubling* of the *centromere* and the separation of the chromatids into two independent sets of chromosomes. It may be considered that the parent cell dies at the end of metaphase, and the two new cells are born at the beginning of anaphase. After the chromatids separate, the duplicate chromosomes migrate toward the opposite poles of the spindle. The forces moving the chromosomes apart have not been ascertained, but they are probably pulls and pushes exerted by the spindle. Pulling would result from the shortening of the microtubules attached to the centromeres, pushing from the lengthening of the microtubules extending between the centrioles. At the close of anaphase the two sets of chromosomes are clustered near the centrioles. Anaphase ends with the onset of the *cytoplasmic constriction* that will pinch the cell in two.

Telophase

Telophase starts with the constriction of the cytoplasm by the plasma membrane, beginning the separation of the cell into two daughter cells. The mechanism whereby the plasma membrane cuts the cell into two more or less equal parts is unknown, but it obviously involves intracellular movement, perhaps the result of microtubule movements. Splitting the cytoplasm takes only a few minutes in most cell types. While the plasma membrane is constricting between the two future cells, fragments of nuclear envelope begin to appear around each set of chromosomes and soon coalesce to form complete nuclear envelopes. As the nuclear envelopes form, the chromosomes uncoil and fade from view, and a nucleolus appears in each new nucleus. Restoration of the nuclear envelopes marks the end of telophase. The two daughter cells, identical in DNA content to the parent cell, are now in the G1 phase of the next cell cycle.

A fairly common abnormality of mitosis is nondisjunction of the chromosomes, which produces daughter cells that do not have the same DNA content. In nondisjunction the two *chromatids* of a chromosome do not separate at anaphase. Instead, both chromatids enter the same cell, giving one cell an extra chromosome and leaving the other cell one chromosome short. In the adult this usually does not cause much of a problem since not many cells are involved. But if nondisjunction occurs during embryonic development, the abnormal cells give rise to a large proportion of the body of the new individual who becomes a mixture of normal and abnormal tissues.

MEIOSIS

Meiosis (Gr., *meiosis* = a lessening) occurs only in cells of the gonads (testes or ovaries) which give rise to *sex cells.* The result of meiosis is a reduction of the number of chromosomes in the daughter cells to half that of the parent cell. This reduction in the chromosome complement is an essential part of the production of both male and female germ cells; if it did not occur, the union of egg and sperm would produce a zygote with twice the normal chromosome number.

The forty-six chromosomes of our somatic cells come in pairs: For every one of the twenty-three chromosomes derived from one parent, there is a corresponding, or *homologous,* chromosome derived from the other parent. Homologous chromosomes are similar in appearance, and they carry genes for the same traits (that is, for the same cellular proteins) in the same positions. However, there may be slight differences in the molecular structure of a gene derived from one parent and the

corresponding gene derived from the other parent, leading to variations in the traits for which they code. For example, a gene on a chromosome derived from your father may code for the production of a keratin protein that causes hair to curl, whereas the corresponding gene on the chromosome derived from your mother may give rise to a slightly different keratin that results in straight hair. These differences arise from occasional changes in DNA called mutations, which are faithfully copied whenever the DNA replicates. The science of genetics is, in part, the study of how the different genes derived from two parents interact to produce an individual's characteristics. Sexual reproduction, by combining two sets of chromosomes in one individual, gives rise to virtually infinite variation in the genetic makeup of humans.

The number of chromosomes found in somatic cells is called the *diploid number;* for humans this number is forty-six. Diploid means having two sets of chromosomes. Sex cells contain the *haploid number* of chromosomes, meaning they have a single set of twenty-three chromosomes. The diploid number is represented as $2n$ and the haploid number as n. The union of two haploid sex cells results in a diploid zygote.

The production of male sex cells is called *spermatogenesis* and the production of female sex cells is *oogenesis.* The two processes differ considerably in detail because they are designed to produce cells of very different forms. However, both involve meiosis, so we will discuss the important features of meiosis before going on to the production of eggs and sperm.

Reducing the Chromosome Number

The chromosomes replicate in an interphase cell that is going to undergo meiosis which, like mitosis, begins with forty-six doubled chromosomes. However, meiosis involves not one, but two cell divisions (Fig. 3-3). In the first division the chromatids are not pulled apart. Instead, one member of each homologous chromosome pair goes to one daughter cell while the other member of the pair goes to the other. Thus each daughter cell ends up with twenty-three chromosomes, each of them still doubled. This is called a *reduction division.* The second division is very much like mitosis: the chromatids comprising each double chromosome are pulled apart, one going to one daughter cell and the other to the other. The result of the second division is four daughter cells, each with twenty-three one-chromatid chromosomes. The second division is necessary, because without it each chromosome would have twice the normal amount of DNA.

A crucial difference between meiosis and mitosis is the pairing of homologous chromosomes in the first division of meiosis. In mitosis the doubled chromosomes line up in random order along the cell's equator (see Fig. 3-2); but in the first meiotic division every chromosome is paired with its homolo-

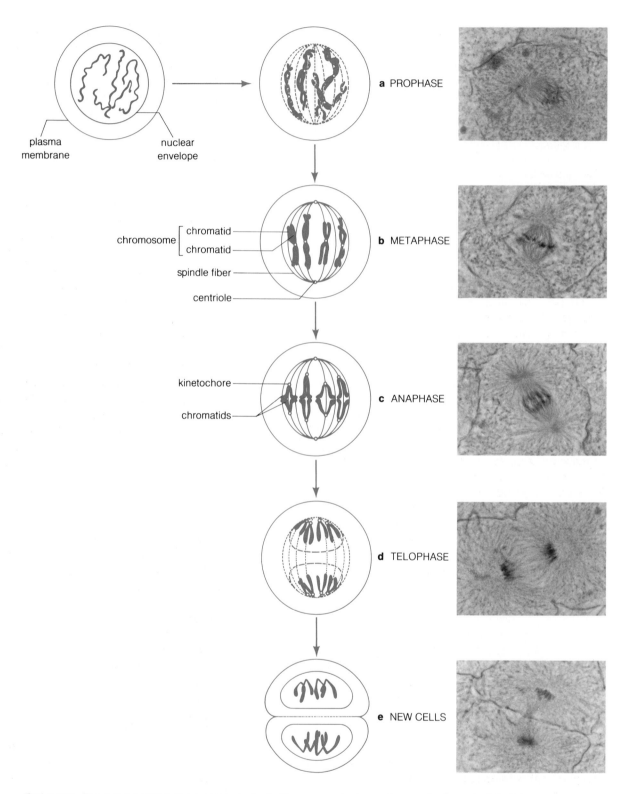

Figure 3-3 The major events in the process of mitosis. The process is shown diagramatically on the left and as it appears in the whitefish embryo on the right, x 1000. (Photographs courtesy of Joan Creager.)

gous chromosome, or homolog (Fig. 3-4). The intertwined homologs form a structure called a **tetrad,** so named because it includes four chromatids. During anaphase the homologs are pulled in opposite directions, so one member of each pair ends up in each daughter cell.

Increasing Genetic Diversity

When the tetrads line up at the cell's equator, the orientation of the homologs is random; that is, it is a matter of chance whether the chromosome derived from the mother or the chromosome derived from the father faces a given pole of the dividing cell. Thus it is also a matter of chance how many maternally derived chromosomes and how many paternally derived ones end up in each new cell. This chance distribution, called *random assortment* of chromosomes, helps meiotic divisions produce a vast array of genetically different sex cells in a single individual. With twenty-three tetrads, each with two possible orientations, there are 2^{23} possible arrangements at the cell's equator and 2^{23} possible chromosome combinations for every sex cell. Meiosis not only reduces the chromosome number in each generation of cells, but also guarantees enormous variation in the genetic makeup of sex cells.

There is another feature of meiosis that increases genetic variability. As the paired homologous chromosomes intertwine during the first meiotic division, chromatids from one chromosome exchange varying amounts of material with chromatids from the other. This exchange of genes is called *crossing over.* Since the chromatids are aligned exactly, they exchange segments containing corresponding genes. However, as we have already pointed out, corresponding genes are often not identical, so exchanging chromosome segments changes the genetic makeup of each chromosome. Crossing over combined with random assortment produces an almost infinite variety of sex cells in a single individual.

Spermatogenesis

Spermatogenesis (Gr., *sperm* = seed, *genesis* = production) is the process of maturation of the male germ cells (Fig. 3-5). It involves reduction of chromosomes to the haploid number and cellular differentiation that changes the immature germ cells to mature **spermatozoa,** or sperm. This process occurs in the male gonad, the *testis,* beginning at *puberty* and continuing throughout life, although the number of spermatozoa produced decreases with advancing age. During differentiation of spermatozoa the cells undergo morphological changes that, for descriptive purposes, may be classified in stages.

The **spermatogonia** are diploid immature germ cells of the testis. They divide by mitosis to produce more spermatogonia, each containing the diploid number of chromosomes. After several mitotic divisions some of them enlarge to become **primary spermatocytes,** which undergo the first meiotic division—the reduction division—producing **secondary spermatocytes.** A secondary spermatocyte contains one set of chromosomes, but each chromosome still consists of two chromatids. The second meiotic division produces four haploid **spermatids** that begin the process of differentiation into mature spermatozoa with no further cell divisions. The spermatid is the intermediate stage between the secondary spermatocyte and the mature spermatozoan.

The mature spermatozoan (Fig. 3-6) consists of a *head,* a *middle piece,* and a motile *tail.* The head contains the nucleus, with its single set of chromosomes, and the *acrosome,* a cap over the nucleus formed from the Golgi complex of the spermatid. The acrosome has enzymes that probably help the sperm penetrate the ovum during fertilization. The middle piece, which is involved in the movements of the tail, has a core of longitudinal microtubules derived from the centriole. This core is surrounded by a sheath of mitochondria that supply the energy needed for the rapid tail movements. The tail, or *flagellum,* is similar to a cilium, having two central and nine peripheral doublets of microtubules (Fig. 2-27). Movements of the tail propel the sperm towards the ovum. The entire spermatozoan is about 50 micrometers long, but the head is only about 3 micrometers in diameter; its minute size facilitates its penetration of the ovum.

Oogenesis

Oogenesis (Gr., *oon* = egg), the development of the female germ cells, or **ova** (Fig. 3-7), occurs in the female gonad, called the *ovary.* The **oogonia,** or diploid immature germ cells, multiply by mitosis in the ovary of the fetus during its development. At birth each ovary contains all the oogonia the female will have in her lifetime—no further mitoses occur. Each oogonium becomes surrounded by a covering composed of ovarian cells, called the *follicle.*

At puberty the oogonia begin growing into large cells called **primary oocytes.** A primary oocyte, containing the diploid number of chromosomes, undergoes the reduction division of meiosis, but in this case the division does not produce two equal cells. Rather, all of the cytoplasm is retained by one daughter cell, the **secondary oocyte.** The other product of the division is a tiny structure—not much more than a nucleus—called a *polar body,* that contains half the chromosomes of the parent cell; but having no cytoplasm it cannot carry on normal cell functions and eventually degenerates.

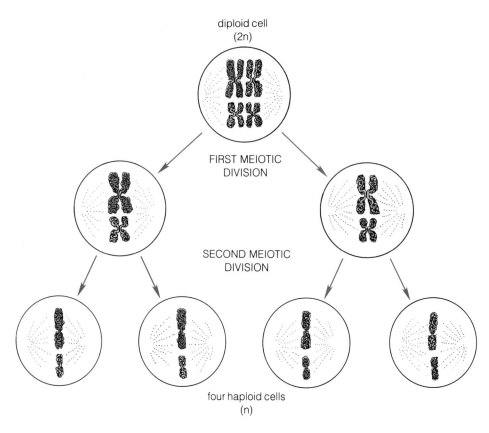

diploid cell
(2n)

FIRST MEIOTIC
DIVISION

SECOND MEIOTIC
DIVISION

four haploid cells
(n)

Figure 3-4   The result of meiosis in a cell with four chromosomes.

The secondary oocyte contains a single set of chromosomes, each still consisting of two chromatids. It accumulates additional cytoplasm while in the follicle, becoming an **ootid.** When discharged from the ovary (ovulation), the ootid, which is now capable of being fertilized, still contains twenty-three doubled chromosomes. When penetrated by the sperm, it begins the second meiotic division: the chromatids separate into chromosomes, and one set is discharged from the ootid as part of a *second polar body.* The ootid retains the other set and becomes the *ovum,* a haploid cell. Both the ootid and the ovum are commonly called the *egg.* Sometimes the first polar body undergoes a second meiotic division, producing two more polar bodies. Thus oogenesis may produce four haploid products, but only one of them is a functional cell. The polar bodies disintegrate shortly, leaving the ovum with one complete set of chromosomes and all the cytoplasm of the parent cell.

The two unequal divisions of oogenesis result in a single cell with an enormous amount of cytoplasm; about 120 microm-eters in diameter, it contains enough cytoplasmic organelles and food reserves to carry on the metabolism of the developing embryo through several cell divisions. However, the ovum is short-lived since its nucleus soon merges with the nucleus of the sperm to produce a zygote.

THE KARYOTYPE

The **karyotype** is the complete chromosomal complement of a diploid nucleus as it exists at metaphase of mitosis, showing both the number and the appearance of the chromosomes. A human karyotype varies according to the sex of the individual and whether the chromosome complement is normal or abnormal. Each pair of chromosomes has a characteristic size and shape when fully condensed at metaphase (Fig. 3-8). The chromosomes have been classified and numbered according to size,

SPERMATOGENESIS

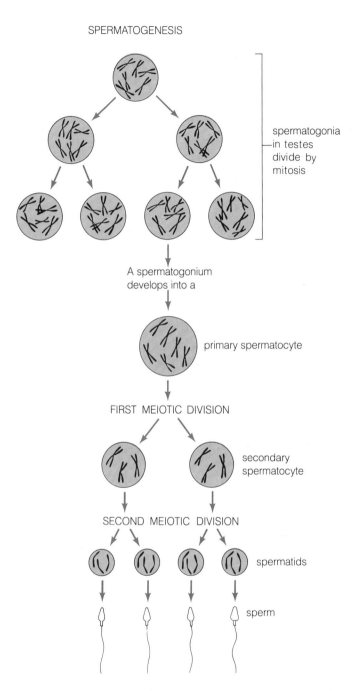

A spermatogonium
develops into a

primary spermatocyte

FIRST MEIOTIC DIVISION

secondary
spermatocyte

SECOND MEIOTIC DIVISION

spermatids

sperm

Figure 3-5 The process of spermatogenesis represented in cells having a diploid number of six chromosomes.

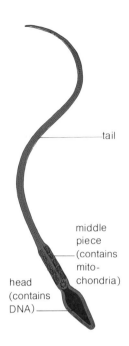

tail

middle
piece
(contains
mito-
chondria)

head
(contains
DNA)

Figure 3-6 A mature human spermatozoon.

Errors in meiosis can lead to the production of germ cells with missing or added chromosomes. If either the egg or the sperm involved in a fertilization has a missing chromosome, the zygote will have only one of that pair (monosomy). If the egg or sperm has an extra chromosome, the zygote will have three of that kind (trisomy). An autosomal monosomy usually results in the death of the embryo. Embryos with trisomies of certain chromosomes may survive but will have multiple abnormalities, including mental retardation. *Down's syndrome,* for example, is trisomy of chromosome 21; persons with Down's syndrome are mentally retarded and highly susceptible to infectious diseases. Karyotyping the cells of such a person reveals the extra chromosome.

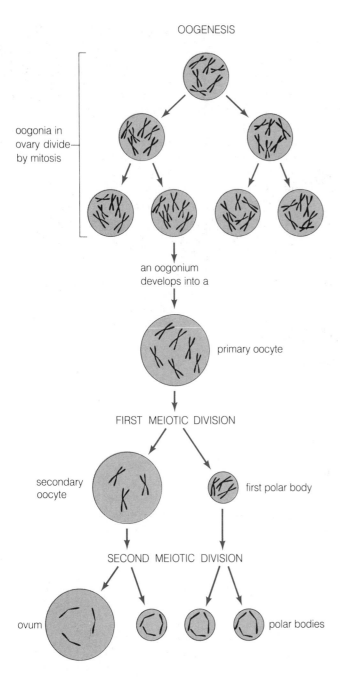

OOGENESIS

oogonia in ovary divide by mitosis

an oogonium develops into a

primary oocyte

FIRST MEIOTIC DIVISION

secondary oocyte

first polar body

SECOND MEIOTIC DIVISION

ovum

polar bodies

Figure 3-7 The process of oogenesis represented in cells having a diploid number of six chromosomes.

shape, and position of the centromere—the constricted region that attaches to the spindle. Twenty-two of the human chromosome pairs consist of two very similar homologs; these forty-four chromosomes are called *autosomes* and are the same in both sexes. The twenty-third pair consists of similar chromosomes (called X chromosomes) in the female and dissimilar chromosomes (called X and Y chromosomes) in the male. These are the *sex chromosomes*.

Karyotyping (determining the karyotype) may be accomplished with any cell type that will undergo mitosis, but it is usually done with lymphocytes, a kind of white blood cell that divides rapidly. Lymphocytes may be obtained from peripheral blood and cultured in a medium that stimulates them to divide. After enough lymphocytes have been produced, the drug colchicine is added to the medium to block mitosis at metaphase. (Colchicine blocks the assimilation of tubulin into microtubules, preventing the formation of the mitotic spindle; in the absence of the spindle, the chromosomes cannot separate, so they remain at metaphase in a highly condensed state.) The lymphocytes are fixed and spread on microscope slides, and a cell that shows all the chromosomes clearly is chosen to be photographed. The pictures of the chromosomes are cut out and lined up according to size and shape, depicting the karyotype.

Karyotyping is an important laboratory procedure in genetic counseling of parents who have given birth to an abnormal baby. Sometimes knowing the karyotype of the parents provides information needed to predict the odds that potential parents will have a normal child.

EARLY EMBRYONIC DEVELOPMENT

People have long sought immortality in various guises, but they have usually failed to recognize the true immortality of the germ cells. Each of us arose from a line of cells unbroken since the beginning of human life on earth, and each of us has the power to perpetuate this same cell line. A heart attack or a smashed auto may initiate your return to ashes before you reach the age of forty. However, if you have paused along the way to reproduce, one of your cells will contribute to the development of the little one who emerges kicking and screaming from the womb to repeat the quest for a long and happy life. Though the body has a limited span of life, the *germ cell line* will continue without interruption until the human species extinguishes itself.

The child who emerges from the womb has already undergone nine months of more rapid growth and more extensive

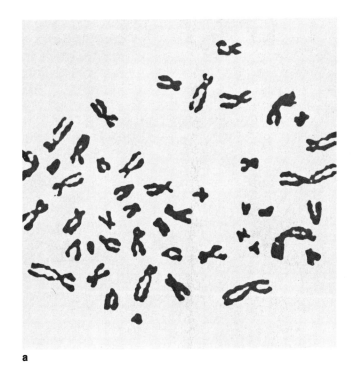

a

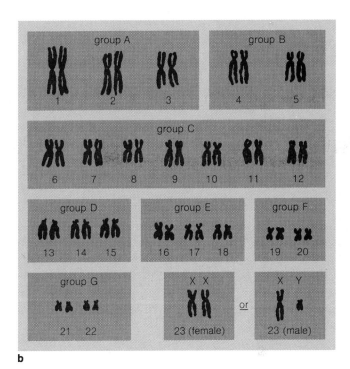

b

Figure 3-8 Human chromosomes. (**a**) Chromosomes of a normal human cell, × 2300. This is the historic photograph by which Drs. Joe Hin Tjio and Albert Levin demonstrated that human cells contain forty-six chromosomes. (Courtesy of Dr. Tjio and the National Institute of General Medical Studies, National Institute of Health.) (**b**) Karyotype for human chromosomes. (Courtesy of Starr and Taggart.) See text for description.

differentiation than it will ever experience again. All of the adult structures are present in recognizable form at birth, though they are not fully developed. The rest of this chapter will give a brief account of early human development, as an introduction to the study of mature tissues and organs in the rest of the book. An understanding of embryonic development contributes greatly to an understanding of adult structure.

Differentiation of Tissues

All our cells arise from the same cell, so barring occasional errors of mitosis, all must carry the same genes. How then can they produce different proteins, giving rise to specialization of structure and function? A great deal of research is being devoted to answering this question, but at present our understanding of differentiation is very limited. We do know that some genes have a regulatory function—they are involved in turning other genes off and on in specific cells. We also know that the environment of a cell can affect its development. Substances produced by neighboring cells, or even distant ones, may encourage or inhibit the functioning of specific genes.

One of the great mysteries of embryonic development is the aggregation of cells that are destined to differentiate into a specific tissue or structure. For example, embryonic cells that will differentiate into a muscle move from their original site to the region where the muscle will be located in the adult. Here they proliferate, forming a compact mass of undifferentiated

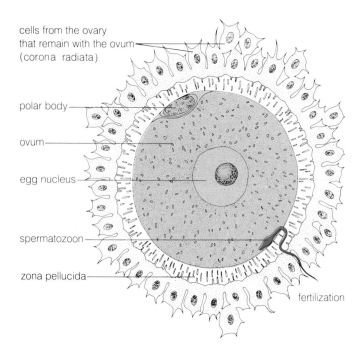

cells from the ovary
that remain with the ovum
(corona radiata)

polar body

ovum

egg nucleus

spermatozoon

zona pellucida

fertilization

Figure 3-9   The process of fertilization.

cells that has the general shape of the future muscle. The mechanism that regulates this aggregation is not understood, When tissue-specific proteins are first synthesized, only a few cells may be involved, but soon more and more of the cells join in producing the protein until most of them are involved. The structure or tissue begins its special functions as soon as enough specific protein accumulates.

Beginnings of the Embryo

This section traces the development of the single cell into a multicellular embryo with its protective coverings, called *fetal membranes*. (A human developing in the uterus is called an embryo for the first eight weeks and a fetus thereafter.) It also deals briefly with the establishment of the *placenta,* a temporary organ composed of both fetal and maternal tissue that provides a close association between the fetal and maternal vascular systems. We take up the story of human beginnings where we left off, with the fertilization of the ovum.

Fertilization   After it is discharged from the ovary, the ovum enters the *uterine tube,* the normal site of **fertilization.** You will recall that the ovum has undergone the first meiotic division in the ovary. When it is penetrated by the head and middle piece of the sperm (Fig. 3-9), it undergoes the second meiotic division, giving off a second polar body containing one set of chromosomes. The chromosomes of the ovum join those of the sperm near the center of the cell, fusing into a single nucleus. The fertilized ovum, or zygote, now contains two sets of chromosomes; the sex of the future person has been determined by the presence of either two X chromosomes or an X and Y, and the cell is ready to get down to the business of multiplying.

Cleavage   Following fertilization the zygote begins its six-day voyage down the uterine tube to the *uterus.* As it travels, the zygote begins mitotic cell division, a process termed **cleavage.**

Unlike many animals that have bizarre cleavage patterns, humans have a rather uninspired sort of cleavage. The cells simply divide almost equally until about 64 cells have accumulated into a solid ball, called a morula (Fig. 3-10b). As the cells multiply, there is a reduction in the size of the daughter cells; so for the time being there is little or no increase in total mass.

Formation of the Inner Cell Mass (Blastocyst Stage)   Even at this early stage there is some morphological evidence of cellular differentiation, for not all of the cells are precisely identical in size. As development proceeds, the disparity in cell size increases and an off-center fluid-filled cavity appears within the oval mass (Figs. 3-10c, d, e). The result is an **inner cell mass,** which will give rise to the embryo proper, and an outer cell mass called the *trophoblast,* which will form the chorion. The chorion will protect the embryo and provide the necessary connection between the embryonic and maternal tissues, through which the embryo will obtain oxygen and nutrients from the mother and discharge its wastes into her bloodstream. The cavity between the inner cell mass and the trophoblast is called the *blastocoel,* and the entire structure is called the blastocyst.

Implantation   About six days after fertilization the blastocyst reaches the lumen (cavity) of the uterus and implants in the uterine wall—usually in the rear wall. **Implantation** begins when trophoblast cells above the inner cell mass attach to the *endometrium,* or uterine lining. Endometrium in contact with the embryo is destroyed, probably by enzymes produced by the embryo, and cells proliferating from the trophoblast begin to invade the endometrium (Fig. 3-11). These cells will give rise to the fetal components of the placenta. The cell membranes

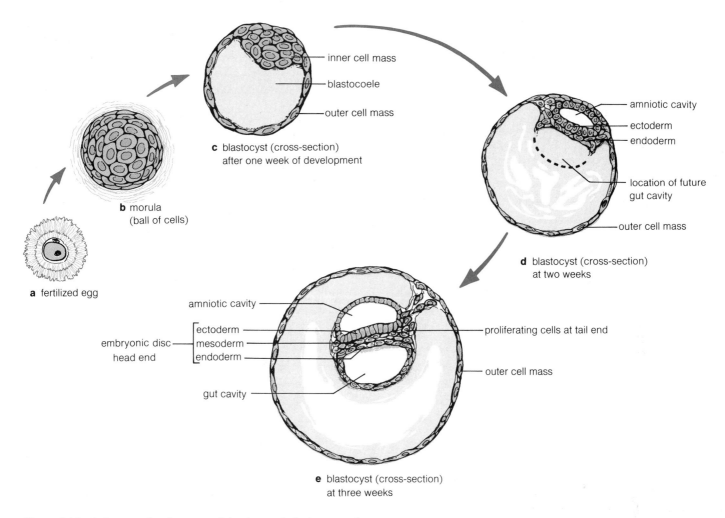

Figure 3-10 A diagram of early stages of development of a human embryo.

separating the nuclei in the outer trophoblast break down, producing a multinucleated mass of cytoplasm that expands extremely rapidly (Fig. 3-12). The innermost layer of trophoblast cells retains its cell membranes. Soon the invading mass of the trophoblast reaches the blood vessels of the uterine wall, and these, too, are eroded, resulting in a pool of maternal blood around the embryo. As the embryo continues to burrow into the endometrium, the uterine epithelium grows over the wound, safely enclosing the embryo in a layer of maternal tissue.

Since the seal at the point of implantation may at first be imperfect, there is frequently some bleeding or spotting, which occurs at or about the time of onset of menstruation and may lead the prospective mother to assume that she has begun menstruation. This phenomenon is responsible for many errors in the determination of the date of conception. The woman may have become pregnant a month earlier than she thinks.

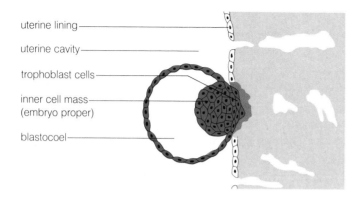

Figure 3-11 Initiation of implantation. Trophoblast cells overlying the embryo invade the uterine lining, or endometrium.

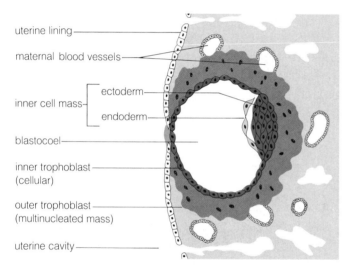

Figure 3-12 Expansion of the trophoblast and embedding of the embryo within the endometrium. Two germ layers, ectoderm and endoderm, can be distinguished in the inner cell mass, or embryo proper.

As the embryo implants in the uterine wall, two major processes begin. One is the formation of the placenta from the developing extraembryonic membranes and the surrounding maternal tissue, a process called **placentation.** The other is the formation of three *germ layers,* embryonic tissue that will give rise to specific adult tissues and organs. We shall first describe placentation and then consider how the germ layers—ectoderm, mesoderm, and endoderm—are formed and how they contribute to the adult body plan. However, since the two

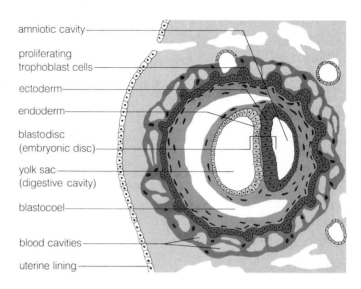

Figure 3-13 Early stage in the formation of the fetal membranes and placenta. Localized disintegration of the outer trophoblast cells allows maternal blood to pass closer to the embryo. The amniotic cavity, yolk sac, and embryonic disc have formed. Extraembryonic mesodermal cells have begun to differentiate from the inner trophoblast and proliferate in the blastocoel.

processes go on simultaneously, the discussion of placentation will inevitably include some references to the germ layers. The illustrations of embryonic development show both processes in the same picture.

Placentation The pool of maternal blood in which the embryo lies, from which it obtains nutrients and oxygen and into which it exudes wastes, begins a slow ebb and flow of circulation. This unevenly circulating blood is brought closer to the embryo proper by the development of cavities in the trophoblast, into which maternal blood flows (Fig. 3-13). While this is happening in the extraembryonic membranes, *extraembryonic mesoderm* begins to differentiate from the inner trophoblast. This tissue, which will eventually connect to the *mesoderm* (middle germ layer) of the embryo proper, forms a layer associated with the inner trophoblast (Fig. 3-14a). With the addition of mesoderm to the inner trophoblast, this layer becomes the *chorion,* the outermost fetal membrane. The chorion completely encloses the embryo, and its mesodermal component is continuous with the mesoderm of the embryo via the *body stalk,* the future umbilical cord.

The next step in placentation is the development of *chorionic villi,* fingerlike extensions of the chorion that grow into

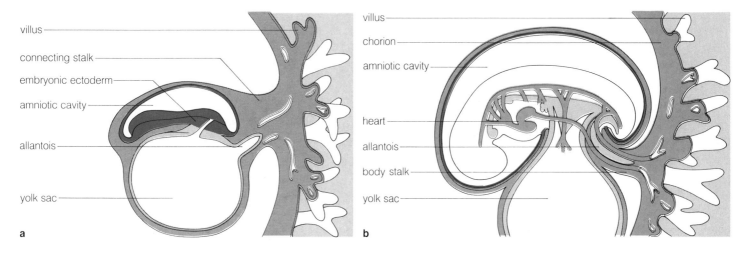

Figure 3-14 Further development of the fetal membranes and placenta. (**a**) Chorionic villi extend into the trophoblast in the region of the forming placenta. A cavity has formed in the extraembryonic mesoderm that filled the blastocoel, so this mesoderm now lines the embryo and trophoblast. (**b**) An older embryo with its mesoderm lining that forms the chorion, the outer fetal membrane. Extraembryonic mesoderm forms a body stalk (umbilical cord) connecting the placenta to the intraembryonic mesoderm, which has arisen from the embryonic disc.

the trophoblast and penetrate the areas between the pools of maternal blood. These villi (sing., *villus*) remain covered by the trophoblast, which separates them from maternal blood. Mesoderm penetrates the villi, setting the stage for the development of fetal blood vessels—both veins and arteries—in each villus (Fig. 3-14b). By now the embryo's heart has begun to beat, forcing blood not only through the vessels of the embryo proper but also through vessels in the body stalk. The trophoblast around the villi becomes very thin, allowing the passage of dissolved substances between fetal and maternal blood, and the volume of maternal blood in the cavities increases greatly. Fetal blood is separated from maternal blood by the lining of the fetal vessels, the inner trophoblast or chorion, and the outer trophoblast, which are all fetal tissues. No maternal tissues separate fetal and maternal blood. The endometrium underlying the developing embryo gives rise to the maternal part of the placenta (Fig. 3-15). At birth not only the baby emerges from the birth canal; the placenta, the rest of the fetal membranes, and the endometrium are discharged as the "afterbirth." This leaves the basal portion of the endometrial glands to restore the uterine lining to its former state.

The Primary Germ Layers

As the embryo develops, its cells form three layers, the **primary germ layers,** from which all of the tissues of the body differentiate. Complicated folds in the layers and irregular growth

patterns result in the origin of organs, such as the liver. The formation of these early germ layers will be described prior to the consideration of the adult tissues.

Formation of the Germ Layers At the time of implantation the inner cell mass, or embryo proper, is a hemisphere of cells enclosed within the trophoblast (Fig. 3-10c). Once implantation is established, the embryo proper increases in mass as the cells continue to divide and rearrange themselves in two morphologically different layers, *ectoderm* and *endoderm* (Fig. 3-12). With further growth a cavity appears within each of the two primary germ layers (Fig. 3-13): an *amniotic cavity* surrounded by a layer of ectoderm and a *yolk sac* surrounded by endoderm. The yolk sac is so called because it is analogous to the yolk sac in the embryos of birds and reptiles, which is an important food-storage organ. The fluid-filled amniotic cavity expands to sur-

Amniotic fluid can be aspirated through the mother's abdominal wall in order to obtain fetal skin cells that have been shed into the fluid. This procedure, termed *amniocentesis,* can be used to examine the fetal karyotype for suspected fetal abnormalities, such as Down's syndrome which is caused by having an extra chromosome. The sex of the fetus can be determined by this procedure

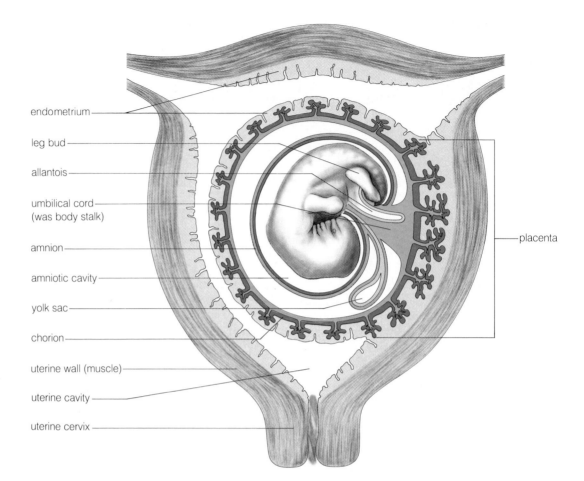

endometrium

leg bud

allantois

umbilical cord
(was body stalk)

amnion

amniotic cavity

yolk sac

chorion

uterine wall (muscle)

uterine cavity

uterine cervix

placenta

Figure 3-15 Relations of the fetal membranes and endometrium. The placenta is made up of both fetal membranes and endometrium, with fetal blood vessels penetrating the maternal tissue. As the fetus enlarges, the endometrium surrounding it fuses with the endometrium of the uterine wall, obliterating the uterine cavity.

round the entire embryo (Fig. 3-16a3) and helps create a digestive cavity from part of the yolk sac. The pinched-off yolk sac remains connected to the digestive tract at first, but in humans and other mammals that are nourished through a placenta it has no nutritive function and eventually disappears.

The outer layer of ectoderm forms the amniotic membrane, or *amnion,* which encloses the embryo in a protective, fluid-filled sac (the "bag of waters"), shown in Figs. 3-16a1, a2, a3. The inner layer of the surface of the embryo proper is destined to become the epidermis, the outer layer of skin. The amnion also forms the outer covering of the umbilical cord. The amnion usually ruptures during birth, releasing its fluid into the birth canal.

While the amniotic cavity and yolk sac are forming (Fig. 3-16), cells arising from the inner trophoblast layer proliferate in the blastocoel, forming the extraembryonic mesoderm. Eventually (Figs. 3-15 and 3-17), the extraembryonic mesoderm lines

the trophoblast and yolk sac, forms the body stalk that connects the embryo proper with the chorion, and begins to penetrate the chorionic villi where it will form blood vessels.

In the case of abnormal development during these very early stages the embryo usually dies, and a miscarriage (spontaneous abortion) occurs. The mother does not weep with the embryo's passing; she merely thinks she was a little late with her menstrual period and never knows she was pregnant for a short time.

After the amniotic and digestive cavities have been formed, the embryo proper consists of the *embryonic disc* (Figs. 3-11, 3-12, and 3-17), which is an oval plate of uniform thickness

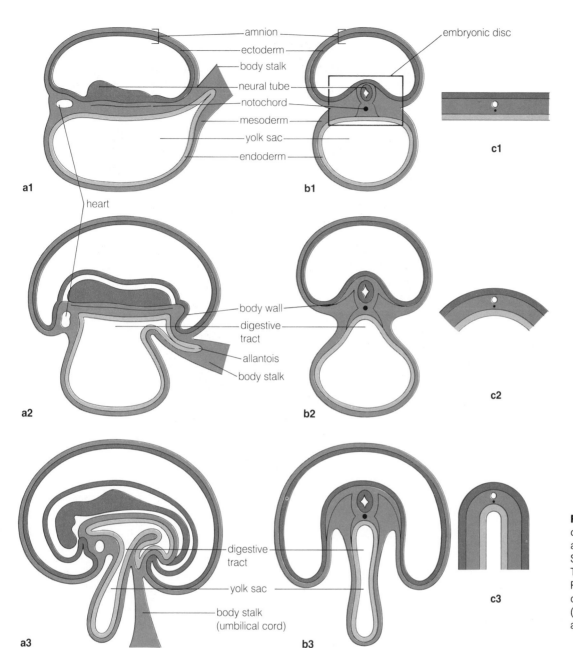

Figure 3-16 Three stages in development of the body wall and digestive tract. (a1, a2, a3) Sagittal sections. (b1, b2, b3) Transverse sections. (c1, c2, c3) Representation of the embryonic disc as shown in b as a flat plate (seen on edge) that is rolling into a tube.

consisting of two layers of cells, ectoderm and endoderm. As growth proceeds, the disc elongates, and a thickened streak, called the *primitive streak*, appears near one end. The primitive streak is the result of rapid localized cell proliferation in the ectoderm. Cells arising from this thickened area of the ectoderm break loose and migrate between the ectoderm and the endoderm, giving rise to the third germ layer, the *mesoderm*

(intraembryonic mesoderm). The ectoderm in the region of the primitive streak functions as a growth center, producing cells that go out on their own and continue proliferating rapidly.

The embryo is now essentially a flat plate of three cell layers. These are the *primary germ layers,* ectoderm, endoderm, and mesoderm, which will give rise to all the structures of the body.

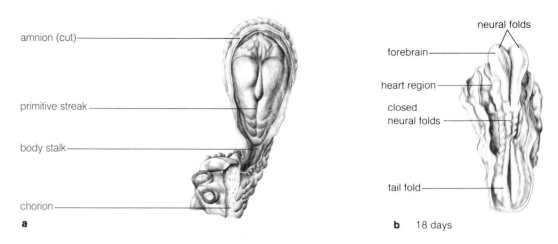

amnion (cut)

primitive streak

body stalk

chorion

a

neural folds

forebrain

heart region

closed
neural folds

tail fold

b   18 days

Figure 3-17   The primitive streak. (**a**) Dorsal surface of the primitive streak, with membranes cut away. (**b**) Dorsal view of an 18-day embryo.

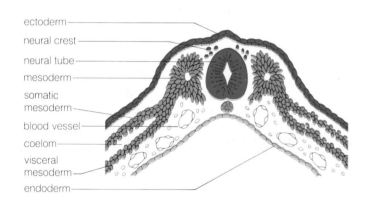

ectoderm
neural crest
neural tube
mesoderm
somatic
mesoderm
blood vessel
coelom
visceral
mesoderm
endoderm

Figure 3-18   Origin of the coelom by splitting of the mesoderm, seen in a transverse section of the trunk region of the embryo. The neural tube was formed earlier by an infolding of the ectoderm.

Derivatives of the Germ Layers   The *ectoderm* covers the outer surface of the embryo and gives rise to the *epidermis* and its *derivatives* such as hair, nails, and sweat glands. Epidermis is composed of a type of epithelium, a tissue that covers surfaces and lines cavities. In addition the ectoderm gives rise to *nervous tissue* and to certain *endocrine glands*.

The *endoderm* gives rise to the linings of the *digestive tract* and the *respiratory tract* and to the secretory elements of all those *glands* that develop as outgrowths of the lining of the digestive tract. These endodermally derived tissues are also epithelium, although not identical to that of the epidermis.

The *mesoderm* gives rise to some representatives of all four of the fundamental tissues. Two types of epithelium are mesodermal in origin: (1) the simple squamous epithelial lining of the *blood vessels* and the epithelium of the *serous membranes* of the body and (2) the epithelium of the *kidneys* and of the *urinary tract*. All the *connective tissues* and *muscle* are mesodermal in origin. One type of *supporting cell* (microglia), a morphological and functional part of *nervous tissue,* also derives from the mesoderm.

Development of Somatic and Visceral Components

For descriptive purposes the body may be divided into a framework, or somatic portion, and a softer, or visceral, portion. The *somatic* portion constitutes the body wall and the limbs, which support and protect the internal *viscera*.

Origin of the Body Wall   We can picture the human body form as a three-layered flat plate that has been rolled into a tube (Fig. 3-16). The resulting structure is lined on the inside with endoderm and covered on the outside with ectoderm, with mesoderm initially filling the space in between these two layers.

When the nervous system and the cardiovascular system begin to form, the embryo is still essentially a three-layered structure (Fig. 3-17). As development proceeds, the three layers roll into a tube (digestive tract) within a tube (the body wall).

Origin of the Coelom   The mesoderm splits into two layers (Fig. 3-18), an outer somatic layer adjacent to the ectoderm and an inner visceral layer adjacent to the endoderm. The space between the two layers of mesoderm is the beginning of the *coelom,* or ventral body cavity.

Origin of the Digestive Tract   The inner tube, or digestive tract, is composed of endoderm on the inside and the *splanchnic* layer of mesoderm on the outside. As the flat plate begins to roll into a tube, the large yolk sac stands in the way of progress and is cut in two by pressure from the body wall, closing the digestive tract. Ventral growth of the body wall occurs from all directions: cranially, caudally, and from the sides. This process is illustrated in Fig. 3-18a.

Completion of the formation of the digestive tract finishes development of the body form. The origin of the systems will be discussed with the adult anatomy of each organ system.

Anaphase

Cell cycle

Centromere

Chromatid

Chromosome

Cleavage

Differentiation

Fertilization

Implantation

Inner cell mass

Interphase

Karyotype

Meiosis

Metaphase

Mitosis

Oogenesis

Oogonia

Ootid

Ova

Placentation

Primary germ layers

Primary oocyte

Primary spermatocyte

Prophase

Secondary oocyte

Secondary spermatocyte

Spermatid

Spermatogenesis

Spermatogonia

Spermatozoa

Telophase

Tetrad

SUMMARY OUTLINE

I. The Cell Cycle
   A. During the G1 phase cytoplasmic constituents are synthesized.
   B. During the S phase DNA is replicated.
   C. The G2 phase is the period during which there is synthesis of cytoplasmic constituents in preparation for mitosis.
   D. The M phase is the period of cell division.

II. Stages of Mitosis
   A. Prophase is the stage during which chromosomes shorten and the nuclear envelope disappears.
   B. During metaphase the spindle appears, and duplicate pairs of chromosomes attach to the spindle in the center of the cell.
   C. During anaphase homologous pairs of chromosomes separate and move toward opposite poles of the cell.
   D. During telophase cytoplasmic constriction divides the cell into two equal cells. The nuclear membrane forms around the chromosomes as they lengthen and lose their identity. Each cell contains the diploid number of chromosomes.

III. Meiosis
   A. Spermatogenesis is the process of the production of spermatozoa. Four haploid cells are produced as a result of two divisions of the primary germ cell.
   B. Oogenesis is the process of the production of ova. A single haploid cell is produced by two divisions of the primary germ cell.

IV. The Karyotype
V. Early Embryonic Development
   A. Fertilization is the union of spermatozoan with ovum to form a zygote.
   B. Cleavage is the division of the zygote by mitosis.
   C. Inner cell mass is an aggregation of cells that will become the embryo proper, surrounded by cells that will become the extraembryonic membranes.
   D. Implantation is the process by which the embryo becomes embedded in the wall of the uterus.
   E. Placentation is the process by which extraembryonic (fetal) membranes develop a means of exchanging material with the maternal tissues.
   F. Germ layer formation is the process by which the inner cell mass initially differentiates into ectoderm and endoderm. Later the middle layer (mesoderm) is derived from ectoderm.

VI. Derivatives of Germ Layers
   A. Ectoderm gives rise to the epidermis and its appendages, nervous tissue, and some endocrine glands.
   B. Endoderm gives rise to the lining of the digestive tract, the respiratory tract, and associated glands of both tracts.
   C. Mesoderm differentiates into the epithelium of blood vessels and the urogenital system, the connective tissues, muscle, and the microglia of the central nervous system.

VII. Development of Somatic Components
   A. The body wall is formed by the three-layered flat plate rolling into a three-layered tube.
   B. The coelom is the cavity in the mesoderm of the three-layered tube.
   C. The digestive tract develops from the inner lining of the three-layered tube.

REVIEW QUESTIONS

1. Define cell cycle. Name the phases of the cell cycle and state the principal physiological activities that occur in each phase.

2. Define mitosis. Name and describe its stages.

3. When in mitosis does the duplication of the chromosomes become visible?

4. Describe the differences between mitosis and meiosis.

5. What is the function of meiosis? of fertilization?

6. Describe the similarities and differences between spermatogenesis and oogenesis.

7. Define a karyotype and give its significance.

8. Describe the processes of implantation and placentation.

9. Name the primary germ layers and list the principal derivatives of each layer.

SELECTED REFERENCES

Bevelander, G., and Ramaley, J. A. *Essentials of Histology.* 8th ed. St. Louis: C. V. Mosby, 1979.

Bloom, W., and Fawcett, D. W., *A Textbook of Histology.* 10th ed. Philadelphia: W. B. Saunders, 1975.

Copenhaver, W. M.; Kelly, D. E.; and Woods, R. W. *Bailey's Textbook of Histology.* 17th ed. Baltimore: Williams and Wilkins, 1978.

Fawcett, D. W., and Bedgord, J. M., eds. *The Spermatozoon.* Baltimore: Urban and Schwarzenberg, 1979.

Flickinger, C. J. *Medical Cell Biology.* Philadelphia: W. B. Saunders, 1979.

Moore, K. L. *The Developing Human.* Philadelphia: W. B. Saunders, 1973.

Rodin, J. A. G. *Histology.* New York: Oxford University Press, 1974.

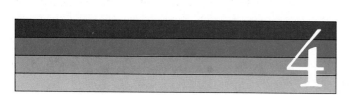

THE FUNDAMENTAL TISSUES

After completing this chapter you should be able to:

1. Classify membranous epithelium on the basis of the shape of the cells and the thickness of the sheet.

2. Give the morphological characteristics of each type of membranous epithelium and show how these characteristics are correlated with the functions performed by the epithelial types.

3. Give specific examples of places in the body where each type of epithelium is located.

4. Describe glandular epithelium. Classify multicellular glands as simple or compound and contrast the morphology of the two types.

5. List the major functions of the connective tissues.

6. List the cell types that are commonly present in loose connective tissue. Describe the morphological features of each cell type and give its major function(s).

7. List and describe the three types of connective tissue fibers and say in what type of connective tissue each is found.

8. Describe the morphological features of dense connective tissue and give examples of its location in the body.

9. Describe the three types of cartilage and give examples of the location of each type.

10. Describe the structure of bone. Compare the structure of compact bone with that of spongy bone. Contrast the structure of bone with that of cartilage.

11. List and describe each of the constituents of blood. Give the general function(s) of each constituent.

12. Name and describe the structure of the three types of muscle tissue.

13. Describe the structure and function of a neuron. Define the terms central nervous system, peripheral nerve, ganglion, and supporting cells.

Cells of similar structure tend to aggregate and work together to perform specific tasks. Such aggregations, together with the intercellular substance surrounding the cells, are called *tissues*. Each tissue is characterized by chemical and morphological features that contribute to one or more of its specific functions. As you learned in Chapter 1, tissues of different types combine to form organs such as the heart that function as parts of larger systems such as the circulatory system. The study of tissue structure and function is called histology.

TISSUE TYPES

There are four fundamental tissue types, each with a wide distribution in the body: **epithelium, connective tissue, muscle tissue,** and **nervous tissue** (Fig. 4-1). Although their structure varies at different locations due to functional requirements, certain characteristics are fundamental to each type of tissue.

Epithelium lines cavities, covers surfaces, and forms the secretory tissue of glands. Its cells are packed closely together in sheets with little intercellular substance between them. The outer layer of the skin is an impermeable epithelial tissue that protects underlying tissue. Epithelium that lines cavities or covers internal structures is often permeable, allowing transport of substances between the blood and the underlying tissues. *Connective tissue* serves many functions including support of the body and the binding together of organs; tendons, ligaments, bone, blood, and fat are connective tissues. Connective tissue cells are relatively far apart, separated by nonliving intercellular fibers produced and maintained by the cells. *Muscle tissue* moves parts of the body by contracting its cells. Thus, muscle tissue is characterized by its extreme contractility. *Nervous tissue* coordinates the activities of other tissues by conducting electrical impulses along its cells and by releasing chemical substances that initiate responses in other cells. Nerve tissues are characterized by extreme irritability and conductivity.

In this chapter we shall discuss epithelial and connective tissues in detail. We shall describe muscle and nervous tissue only briefly, for we will consider them again in the chapters on the muscular and nervous systems.

EPITHELIUM

All epithelium rests upon a **basement membrane** important in holding the epithelial sheet in place. As seen with the electron microscope, the basement membrane is composed of two layers: that nearer the epithelial cells is the **basal lamina,** elaborated by the epithelium itself; the other is the reticular lamina, produced by the underlying connective tissue. Specialized intercellular adhesions hold the tightly packed epithelial cells together. In other types of tissue, dissolved substances may move through the intercellular spaces; but the intercellular adhesions of epithelial cells prevent this. Epithelium is usually *avascular;* that is, it is usually not penetrated by blood vessels. Substances

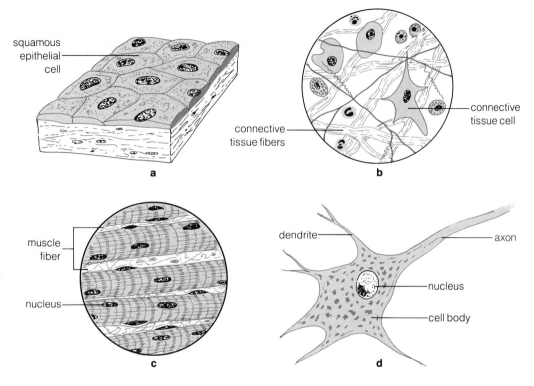

Figure 4-1 Diagram of a representative of each type of tissue. (**a**) Epithelium of the simple squamous type. (**b**) Loose connective tissue. (**c**) Skeletal muscle fibers. (**d**) A neuron (nerve cell).

Table 4-1   Classification of Epithelium

I. Membranous
   A. Simple
      1. Simple squamous
      2. Simple cuboidal
      3. Simple columnar
      4. Pseudostratified (pseudostratified columnar)
   B. Stratified
      1. Stratified squamous
         a. Keratinized
         b. Nonkeratinized
      2. Transitional
      3. Stratified cuboidal
      4. Stratified columnar
II. Glandular
   A. Unicellular
   B. Multicellular
      1. Simple
         a. Simple tubular
         b. Simple alveolar
         c. Simple tubulo-alveolar
      2. Compound
         a. Compound tubular
         b. Compound alveolar
         c. Compound tubulo-alveolar

move by diffusion between epithelial cells and the underlying connective tissue, which does have a vascular supply.

Epithelium is divided into two major categories. That which covers surfaces and lines cavities is called **membranous epithelium,** and that which forms the secretory cells of glands is **glandular epithelium.** These categories are subdivided as shown in Table 4-1.

Membranous Epithelium

The classification of membranous epithelium depends upon the thickness of its cellular sheet and the shape of its cells (Fig. 4-2). If the sheet consists of a single cell layer it is called *simple;* if it is made up of two or more cell layers it is called *stratified.* The morphology of the epithelial sheet indicates its function, shown in Table 4-2.

In general, epithelium occurring as sheets provides mechanical protection for underlying structures and as membranes controls the passage of chemical substances. The following description of membranous epithelia is restricted to those morphological features that may be readily correlated with function. The types described are the most common of the membranous epithelia of the body.

Simple Epithelium   Since simple epithelium consists of a single cell layer, every cell in the sheet is attached to the basement membrane. **Simple squamous epithelium** (Fig. 4-2a) is a single layer of thin, flattened cells. The flat surface of the sheet is smooth and moist to reduce friction when it is in contact with other structures; it covers the outer surfaces of structures such as the digestive tract and the spleen (Fig. 4-3) that constantly rub against adjacent structures; in these locations it is called *mesothelium*. Simple squamous epithelium also lines blood vessels, whose walls are frequently bumped by rapidly moving blood cells; here it is known as *endothelium*. Squamous cells are small with relatively little cytoplasm. Because they have limited numbers of mitochondria they are capable only of metabolic processes that require little energy. Therefore substances pass through their membranes by passive diffusion without expending energy. A squamous cell has no obvious polarity, that is, no detectable morphological difference between its free and its attached surface, and diffusion across the cell membrane may occur in either direction depending upon the concentration gradient of the substances involved. Thus the simple squamous epithelial lining of the smallest blood vessels, or capillaries, facilitates exchange of materials between the tissues and the blood.

Simple cuboidal epithelium (Fig. 4-2b) is a single layer of **cuboidal,** or cube-shaped, cells that are larger and have considerably more cytoplasm than squamous cells. Like squamous cells, they display no visible evidence of polarity. Cells of this type use energy to move substances in either direction across the cell membrane; that is, they engage in active transport, which is the use of energy by cells to move substances against a concentration gradient. This feature is illustrated by the thyroid gland in which cuboidal epithelium plays an essential role in secreting thyroid hormone into the follicle for storage, and then in returning and transporting it across the epithelium into the blood stream. Cuboidal epithelium also covers the external surface of the ovary (Fig. 4-4).

Simple columnar epithelium (Fig. 4-2c) is a single layer of tall, rectangular columnar cells. Since these are big cells, they contain a very large volume of cytoplasm, which suggests that they are well-equipped for producing energy for active transport and protein synthesis. Because the nuclei are usually located near the attached ends, the cells have an obvious polarity. Simple columnar epithelium lines the parts of the digestive tract in which digestion and absorption of food take place, and the uterus where substances are secreted into the uterine cavity (Fig. 4-5). Digestion is the result of enzymatic action that breaks large food molecules into smaller ones that can be absorbed into the blood stream. Many of the enzymes, synthesized in the columnar epithelial cells of the digestive glands, are secreted through the free surfaces of these cells into the ducts leading into the digestive tract. The products of digestion (glucose, amino

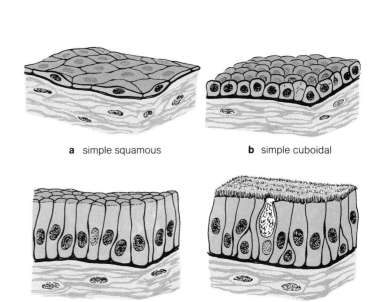

a simple squamous

b simple cuboidal

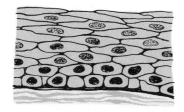

c simple columnar (nonciliated)

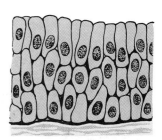

d pseudostratified columnar (ciliated)

e stratified squamous

f stratified columnar

g stratified cuboidal

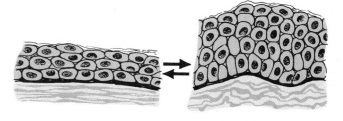

h transitional (distended)

i transitional (undistended)

Figure 4-2   Classification of epithelial tissues.

Table 4-2 Membranous Epithelium		
Type	Description	Function
Simple squamous	Single layer of thin, flattened cells	Passive transport across sheet in either direction
Simple cuboidal	Single layer of cube-shaped cells	Active transport in either direction
Simple columnar	Single layer of tall, rectangular cells	Active transport and secretion
Pseudostratified columnar	Single layer of cells with nuclei at more than one level	Usually has cilia and mucus to facilitate removal of particles from air
Stratified squamous	Several layers of squamous cells	Protection against friction Impermeable (keratinized) or nearly so (nonkeratinized)
Transitional	Several layers of cube-shaped cells	Impermeable, capable of stretch

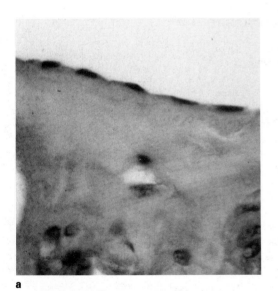

a

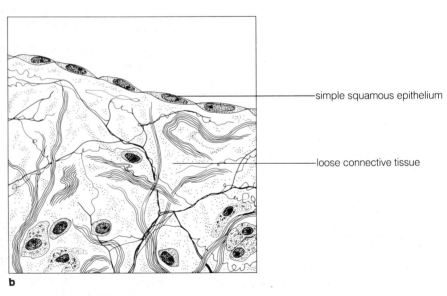

b

— simple squamous epithelium

— loose connective tissue

Figure 4-3   Simple squamous epithelium. (**a**) Mesothelium of the serous membrane on the vermiform appendix of the human, ×3600. (Courtesy of W. K. Elwood.) (**b**) Sketch of the photomicrograph.

acids, and fatty acids) are taken into the epithelial lining of the gut. They enter the free surface of a cell by active transport and are discharged through its attached surface into the underlying tissue where they enter vessels of the circulatory system for distribution. You can see that the morphological polarity of simple columnar epithelial cells is associated with a functional polarity: substances may pass into and out of these cells in one direction only. In those portions of the gut where the most absorption of nutrients takes place, finger-like projections of the plasma membrane, called **microvilli,** greatly increase the surface area of the epithelial cells and their capacity to transport substances across the membrane (Fig. 2-26).

structures, such as in small blood vessels and around the secretory elements of some glands. This type of muscle undergoes slow contractions characteristic of muscle cells that are unstriated. The *autonomic nervous system,* which operates without our awareness, regulates the contraction of smooth muscle. The autonomic nervous system continuously adjusts the rate and strength of contractions to meet the physiological needs of organs according to data received constantly by the nervous system. Unlike skeletal muscle, which cannot contract unless stimulated by nerve impulses, smooth muscle also contracts when stimulated physically, for example, by being stretched.

Skeletal Muscle

Skeletal muscle tissue (Fig. 4-24) is voluntary muscle consisting of elongated, multinucleated *muscle fibers,* whose myofibrils have wide dark bands and narrow light ones, giving it a

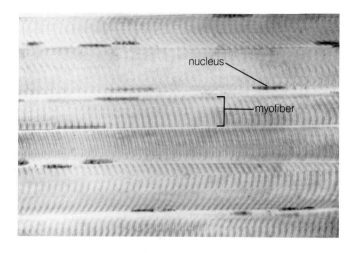

Figure 4-24 Skeletal muscle. Longitudinal section of human skeletal muscle, × 800. (Courtesy of W. K. Elwood.)

Clinical Applications: Regeneration and Repair of Tissues

Human cells differ greatly in their life spans and in their ability to replace themselves. Nerve cells normally live for almost the entire life of the individual and have a very limited capacity for regeneration or repair. The same is true of muscle cells. Much of the apparent regeneration of muscles that occurs following injury is due to the activity of connective tissue rather than of muscle tissue; in effect, any muscle tissue destroyed is replaced by scar tissue. Although they cannot multiply, individual skeletal muscle fibers can hypertrophy (increase in size) as a result of use, such as in a planned exercise program. The fibers increase in diameter as well as in length through the addition of myofibrils, without increase in the number of nuclei. Complete disuse causes individual muscle fibers to undergo atrophy (severe reduction in size) due to reduction in the myofibril content.

On the other hand, many epidermal cells, such as those in the outer layer of the skin, are constantly being rubbed away at the surface and are being replaced by new cells produced in the basal epithelial layer. The cells of many adult organs, such as the liver, normally do not divide but will do so if cells are lost due to trauma or disease. In such organs loss of cells appears to be the stimulus for initiating mitosis in other cells. Replacement or regeneration of an entire structure, such as a limb, does not occur in humans, but repair of an injured part, such as a broken bone, occurs readily.

In some tissues, such as blood, certain cell types have definite life spans, so complete replacement (turnover) of all cells occurs within a definite time period. For example, red blood cells live for 128 days, requiring constant cell production at a rate adequate to replace red cells as they die and are removed from the circulation. Other connective tissues contain mixed cell populations, which include cells that have lost their capacity to divide as well as undifferentiated cells that may actively produce any of the specialized cell types. For example, bone cells (osteocytes) embedded in bone matrix cannot divide, but cells that are capable of producing bone (osteoblasts) can undergo mitosis until they differentiate into osteocytes.

In cases of trauma, such as a knife wound through the skin, the open cut is usually closed initially by a blood clot. Blood cells migrate into the clot to combat infection, forming a mixture of cells and fibrin termed granulation tissue. Connective tissue cells (fibroblasts) are stimulated to produce connective tissue fibers, and macrophages begin to ingest the cellular debris of the granulation of tissue. At the edges of the wound the basal epithelial cells proliferate and gradually spread into the area of the wound, closing the gap. If the gap is not too extensive, the skin is soon completely repaired, with no visible evidence of the wound. If the gap in the skin is too large to be filled in readily by cell proliferation at the edges of the wound, the gap is closed by excessive production of collagenous connective tissue fibers. The network of fibers becomes so dense that a solid, dry sheet of collagen is exposed at the surface. It appears white and is commonly called a scar, or cicatricial tissue.

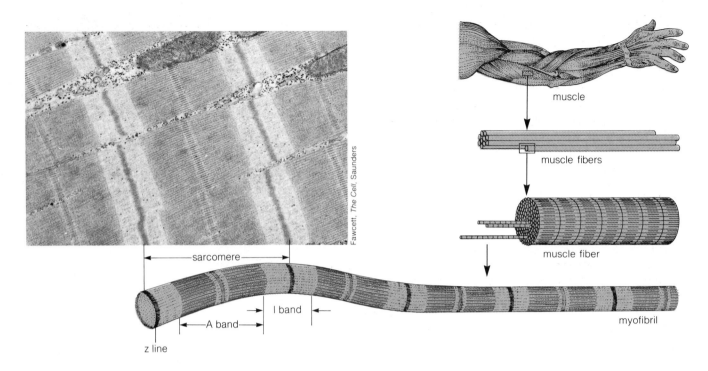

Figure 4-25 Structure of skeletal muscle.

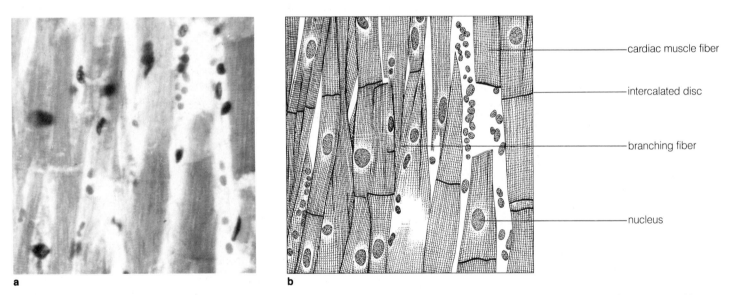

Figure 4-26 Cardiac muscle. (**a**) Longitudinal section of human cardiac muscle, × 3000. (Courtesy of W. K. Elwood.) (**b**) Sketch of the photomicrograph.

striated appearance. The dark bands are due to the presence of myosin myofilaments, and the light bands contain thin **myofilaments** of actin. Skeletal muscle fibers occur in a bundle, termed a *muscle,* that works as a unit under the conscious control of the central nervous system. An individual muscle fiber may extend the entire length of the muscle and contain many nuclei located at the periphery of the fiber. Skeletal muscle fibers (Fig. 4-25) do not branch and are enclosed individually by a thin covering of connective tissue termed *endomysium.* Small bundles of muscle fibers, bound together by a connective-tissue covering called the *perimysium,* form a larger unit of muscle tissue termed a *muscle fascicle.* The entire muscle, composed of many muscle fascicles, is enclosed by a connective tissue sheath termed the *epimysium.* At the junction of the muscle with its tendon the connective tissue fibers of the endomysium are continuous, not only with the fibers of the perimysium and epimysium but also with the collagenous fibers of the tendon. Thus when the muscle fibers contract, tension is exerted on the tendon which exerts a pull on a bone. Note that when we speak of muscle "fibers" we are refering to living cells, but when we speak of the "fibers" of the perimysium, epimysium, and tendon, all of which are connective tissues, we mean nonliving intercellular elements.

Cardiac Muscle

Cardiac muscle tissue (Fig. 4-26) is involuntary, striated muscle, consisting of uninucleated, freely branching muscle fibers that form a network. Cardiac muscle is found only in the heart. Its rate of contraction is regulated by the autonomic nervous system. The myofibrils within the branching fibers have alternating light and dark bands similar to the myofibrils of skeletal muscle but do not extend the entire length of the muscle fibers; they terminate at the plasma membrane, termed *intercalated disc,* which separates the nuclei. Thus a cardiac muscle fiber consists of many uninucleated cells arranged end-to-end. The fibers form elongated strands that branch and anastomose (reconnect) repeatedly, creating a network of muscle tissue.

Cardiac muscle has an intrinsic capacity to contract rhythmically in the absence of stimulation. The rate of its contraction is regulated by extrinsic factors, but it does not require stimulation to initiate a contraction. This is unlike skeletal and smooth muscle which require stimulation by extrinsic factors to initiate contraction.

NERVOUS TISSUE

Nervous tissue consists of *nerve cells,* or **neurons,** modified to conduct electrical impulses, and their *protective* or *supporting* cells.

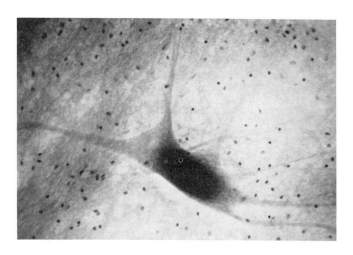

Figure 4-27 Nerve tissue, central nervous system. The large cell is a neuron and the small cells are neuroglia, × 2400. (Courtesy of W. K. Elwood.)

Neurons

The neuron (Fig. 4-27) is the structural unit of the nervous system. Each neuron has a *cell body,* containing the *nucleus,* and a variable number of cytoplasmic extensions called *processes* or *fibers.* (Note that "fiber" has yet another meaning here.) The nerve cell body contains the nucleus and the cytoplasmic organelles responsible for the essential metabolic processes of the cell. The processes are of two types, named according to the direction of impulse flow. Those that conduct the nerve impulse toward the cell body are *dendrites,* and the single process conveying the impulse away from the cell body is the *axon.* The axon may send out many branches (collateral fibers) before it terminates.

The ends of an axon lie very close to the ends of dendrites of other nerve cells, but there is always a small gap at the *synapse* between the ends of the two processes. A *neuron* can respond to a stimulus by generating an electrical impulse. When the impulse reaches the terminal ends of the axon, the ends release a *neurontransmitter* substance, a chemical that can transmit the nerve impulse across a synapse to another neuron or stimulate a muscle to contract or a gland cell to secrete.

Most neurons are located entirely within the central nervous system (brain and spinal cord), but some have cell bodies within the central nervous system and send processes out into the periphery. Any processes that occur in bundles outside the central nervous system are called *peripheral nerves.* Nerve cell bodies may also be located outside the central nervous system, occurring as aggregations termed *ganglia.*

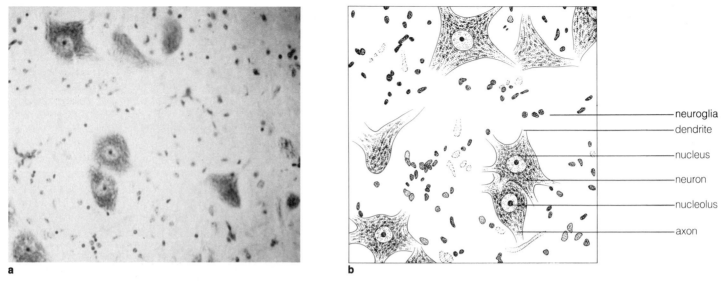

neuroglia
dendrite
nucleus
neuron
nucleolus
axon

Figure 4-28 Nerve tissue. (**a**) Motor neurons, × 1200. (Courtesy of K. K. Elwood.) (**b**) Sketch of the photomicrograph.

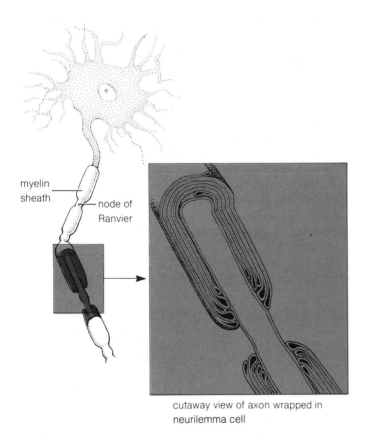

myelin sheath

node of Ranvier

cutaway view of axon wrapped in neurilemma cell

Figure 4-29 Myelinated axon of a neuron.

Supporting Cells

Supporting cells form protective coverings for nerve cell bodies and nerve fibers and also deposit layers of membrane, called *myelin,* on the surface of nerve fibers. Myelin is composed of multiple layers of plasma membrane derived from the supporting cells and serves as an insulation material that enables nerve fibers to conduct nerve impulses without stimulating adjacent nerve fibers. The supporting cells within the central nervous system differ from those of the peripheral nerves.

The supporting cells of the central nervous system, collectively called the **neuroglia,** are of several morphological and functional types. In general terms, they are highly branched cells whose cytoplasmic processes form protective covers for the nerve fibers and blood vessels (Fig. 4-28). The supporting cells of peripheral nerves are called **neurilemma cells,** or Schwann cells (Fig. 4-29). A single neurilemma cell encircles a segment of a single myelinated nerve fiber, having produced the myelin around that segment. The gaps in the myelin at each end of the segment enclosed by a single neurilemma cell are termed *nodes of Ranvier.* In addition to the protective covering provided by the neurilemma cells, peripheral nerve fibers are encased in loose connective tissue.

WORDS IN REVIEW

Actin
Agranulocyte
Amorphous ground
substance
Basal lamina
Basement membrane
Basophil
Bone
Cardiac muscle
Cartilage
Collagenous fiber
Compact bone
Connective tissue
Cuboidal
Elastic cartilage
Elastic fiber
Endocrine gland
Fat cell
Fibroblast
Fibrocartilage
Glandular epithelium
Granulocyte
Hemocytoblast
Heparin
Histamine
Hyaline cartilage
Keratin
Leukocyte
Ligament
Lymphocyte
Macrophage

Mast cell
Megakaryocyte
Membranous epithelium
Microvilli
Monocyte
Muscle tissue
Myeloid tissue
Myofibril
Myofilament
Myosin
Nervous tissue
Neurilemma cell
Neuroglia
Neuron
Plasma cell
Platelet
Primitive reticular cell
Pseudostratified epithelium
Reticular fiber
Simple columnar
epithelium
Simple epithelium
Simple squamous
epithelium
Skeletal muscle
Smooth muscle
Spongy bone
Stratified epithelium
Tendon
Thromboplastin

SUMMARY OUTLINE

I.   Epithelium lines cavities, covers surfaces, and forms the parenchyma of the glands. It functions as a protective covering in transport of substances to and from underlying tissue and in secretion.

   A. Membranous epithelium is arranged as sheets of cells and is classified according to the shape of the cells and the thickness of the sheet.

   1. Simple squamous epithelium is a single layer of thin, flattened cells that is capable of transport across epithelium in either direction.
   2. Simple cuboidal epithelium is a single layer of cube-shaped cells involved in active transport across epithelium in either direction.
   3. Simple columnar epithelium is a single layer of tall, rectangular cells involved in protein synthesis and active transport.
   4. Pseudostratified columnar epithelium is a single layer of cells with nuclei at more than one level; it usually bears cilia and secretes mucus.
   5. Stratified squamous epithelium consists of several layers of squamous cells. If it is keratinized, it has a hard, dry surface that thickens with wear; if nonkeratinized, it has a moist surface that resists friction.
   6. Transitional epithelium consists of several layers of cube-shaped cells, is relatively impermeable, and is capable of stretching.

   B. Glandular epithelium forms the parenchyma of glands.
   1. Simple glands have an unbranched duct with secretory cells at the end of the duct.
   2. Compound glands have a branching duct system with secretory cells at the end of the ducts.

II.   The connective tissues bind structures together, serve as a supporting framework, store fat, distribute substances throughout the body, provide for defense against disease, and provide pathways for nerves and blood vessels.

   A. Loose connective tissue consists of cells, fibers, and amorphous ground substance.
   1. Cells.
      a. A fibroblast is a stellate cell that produces and maintains fibers.
      b. A macrophage is a phagocytic cell.
      c. A plasma cell produces antibodies.
      d. A mast cell contains heparin and histamine.
      e. A fat cell stores fat.
      f. A primitive reticular cell forms reticular fibers.
      g. A leukocyte is a white blood cell that may leave the blood stream and enter the connective tissue.
   2. Connective tissue fibers are nonliving strands of protein found between the cells of connective tissue.
      a. Collagenous fibers are composed of strands of collagen having a characteristic repeating periodicity.
      b. Reticular fibers are composed of branching strands of collagen.
      c. Elastic fibers are composed of elastin and are capable of stretch and recoil.
   3. Amorphous ground substance varies from fluid to stiff gel and has no visible form with a light microscope.

B. Dense connective tissue has great fiber content with fewer cells. Its fibers may be arranged in an irregular pattern (network) or regular pattern (parallel to one another).
   1. Tendons connect muscle to bone.
   2. Ligaments connect bone to bone.
C. Cartilage has cells embedded in a matrix that contains chondromucoid.
   1. Hyaline cartilage has a homogeneous matrix.
   2. Elastic cartilage has a matrix that contains visible elastic fibers.
D. Bone is a rigid tissue with a matrix that contains collagen (organic) and calcium (inorganic).
   1. Compact bone is a solid mass that contains Haversian systems.
   2. Spongy bone is arranged as irregular plates of bone separated by marrow spaces.
E. Blood functions in distribution.
   1. Erythrocytes contain hemoglobin and have no nucleus.
   2. Leukocytes are white blood cells that combat disease.
   3. Hemopoiesis is the formation of blood cells.

III. Muscle tissue consists of contractile cells or fibers.
  A. Smooth muscle consists of uninucleated tapered cells with unbanded myofibrils.
  B. Skeletal muscle consists of multinucleated fibers that do not branch and that contain banded myofibrils.
  C. Cardiac muscle consists of uninucleated cells arranged end-to-end; the fibers branch and contain banded myofibrils.

IV. Nervous tissue conducts electrical impulses.
  A. A neuron consists of a cell body containing a nucleus with extensions from the cell body that are either axons or dendrites.
  B. The supporting cells are termed neuroglia in the central nervous system and neurilemma in the peripheral nerves.

REVIEW QUESTIONS

1. Define a tissue. List the four types of tissue found in the body and give the principal functions of each.

2. Describe the general functions of membranous epithelium. Compare the structure and functions of simple and stratified epithelia.

3. Classify the types of membranous epithelium according to the shape of the cells and the thickness of the sheet. Explain how the structure of each type is correlated with its functions. Give examples of the location of each type of epithelium.

4. Define a gland. Explain how multicellular exocrine glands are classified.

5. Describe the general characteristics of connective tissues and list their principal functions.

6. List the types of connective tissue cells and give the principal function(s) of each. List the three types of connective tissue fibers and describe the chemical and physical characteristics of each.

7. Describe the characteristic features of cartilage.

8. Describe the characteristics of bone.

9. List and describe the various types of blood cells. What is the origin of the blood platelet?

10. Describe the structural and functional characteristics of the three types of muscle tissue.

11. Define a neuron. Describe the components of a neuron and its protective coverings.

SELECTED REFERENCES

Bevelander, G., and Ramaley, J. A. *Essentials of Histology.* 7th ed. St. Louis: C. V. Mosby, 1974.

Bloom, W. B., and Fawcett, D. W. *A Textbook of Histology.* 10th ed. Philadelphia: W. B. Saunders, 1975.

Copenhaver, W. M.; Kelly, D. E.; and Wood, R. W. *Bailey's Textbook of Histology.* 17th ed. Baltimore: Williams and Wilkins, 1978.

Difore, M. S. M. *An Atlas of Histology.* 3rd ed. Philadelphia: Lea and Febiger, 1967.

THE SKIN AND ITS APPENDAGES

OBJECTIVES

After completing this chapter you should be able to:

1. Name the two layers of skin and give the tissue classification of each layer.

2. List the principal functions of the skin.

3. Describe each of the five layers of the epidermis that are normally visible in thick skin.

4. Relate the morphological features of each of the five layers of the epidermis to the metabolic events occurring in the layer.

5. Relate the morphological features of the epidermis to each of the major functions of skin.

6. Describe the process of replacement of cells of the epidermis.

7. Describe the dermis and relate its structure to the general functions of the skin.

8. List the pigments that are responsible for skin color and describe the role of melanocytes in pigmentation of the skin.

9. Describe the growth of hair and relate the hair follicles to the sebaceous glands.

10. Describe the sebaceous gland and discuss the mechanism of its secretion.

11. Describe the structure and growth of nails.

12. Describe the sweat gland and relate its structure to functions of the skin.

13. Describe the structure of the breast and relate its structure to the function of milk secretion.

The skin, or **integument,** covers the surface of the body and is its largest single organ, making up about 16% of its weight. It functions as a covering and protects against abrasion and against invasion by bacteria and other organisms. The skin also prevents excessive water loss, helps regulate body temperature, and shields the body from ultraviolet radiation. It is responsible for development of its appendages, which include *hair, nails, sweat glands, sebaceous glands,* and *mammary glands.* This chapter will consider the structure of the skin and then give a brief account of the skin appendages.

THE SKIN

The skin (Fig. 5-1) consists of two main layers: an epithelial layer, the **epidermis,** at the surface and an underlying connective tissue layer, the **dermis,** or *corium.* The surface of the epidermis is marked by irregularities, or grooves, that form patterns unique to each person. These patterns are particularly distinct on the fingers, providing the fingerprints used for identification. The deep surface of the epidermis is also irregular, indented by connective-tissue projections termed **dermal papillae** (sing., *papilla*) separated by *epidermal ridges.* The boundary between the epidermis and dermis is sharp, but the fibrous elements of the dermis interlace with the underlying loose connective tissue, with no sharp boundary between them. This loose connective tissue is the **hypodermis,** also called *superficial fascia* and *subcutaneous layer,* which in many regions contains aggregations of fat cells.

The two layers of skin vary in thickness in different regions of the body, from about 0.5 mm on the back to 4.0 mm on the soles of the feet and palmar surface of the hand. The terms thick or thin skin usually refer to the thickness of the epidermis, although the dermis also varies in thickness.

The Epidermis

The epidermis is *stratified squamous epithelium.* Thick skin such as that of the palm (Fig. 5-2) has five layers, representing epithelial cells in various stages of differentiation (thin skin has only four). The basal layer adjacent to the dermis is the **stratum germinativum**, a single layer of columnar or cuboidal cells that undergo mitosis, producing cells that move outward into the second layer, the **stratum spinosum,** or squamous cell layer. The name spinosum reflects the spiny appearance of the layer due to numerous *desmosomes* joining the cells. These cells are rich in RNA and other cytoplasmic components essential in protein synthesis. The cytoplasm contains numerous microfilaments termed *tonofibrils,* and as the cells mature they produce a proteinaceous cement called **keratohyaline** that binds the tonofibrils together, forming granules. Eventually the cells differentiate into the third layer, the **stratum granulosum,** composed of two or three layers of cells containing prominent keratohyaline granules. This layer is also characterized by the presence of *lysosomal enzymes* that eventually cause the dissolution of the nucleus and other cell organelles as the cells become packed with keratohyaline.

The fourth layer, found only in thick skin, is the **stratum lucidum.** The translucent cells of this layer are without cytoplasmic granules due to the accumulation of *soft keratin,* or eleidin (in contrast with the *hard keratin* of hair and nails). The

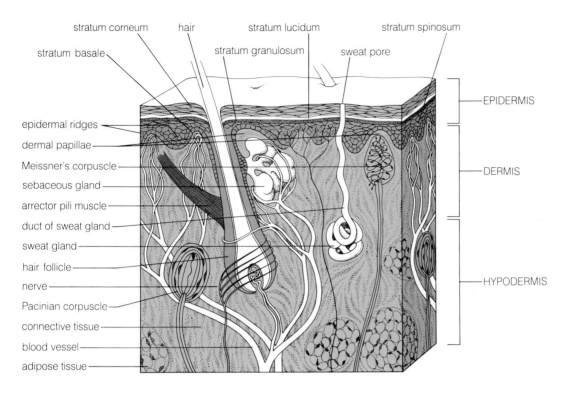

Figure 5-1   Major constituents of skin.

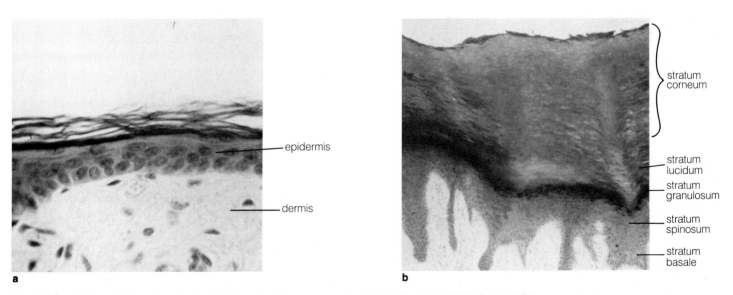

Figure 5-2   Histological structure of skin. (**a**) Thin skin of the human axilla, ×200. (Courtesy of W. K. Elwood.) (**b**) Thick skin of the human palm, ×200. (Courtesy of W. K. Elwood.)

stratum lucidum is rich in *protein-bound phospholipids* that probably act as a barrier to water penetration. Only a few degenerating nuclei are present in this layer. In thin skin, which lacks a stratum lucidum, cells pass directly from the stratum granulosum to the external epidermal layer.

The external layer of the epidermis, the **stratum corneum,** or horny layer containing soft keratin, is variable in thickness and consists of flattened, dead, enucleated cells that constantly slough off at the surface. They are replaced by the proliferation of the stratum germinativum, thus keeping the epidermis intact. In areas such as the trunk, covered by thin skin, the stratum corneum is much thinner than in the thick skin of the palm and soles.

Vitamin D, essential for the growth and maintenance of bone, is synthesized in the epidermis in the presence of ultraviolet rays from sunlight. Vitamin D deficiency leads to the condition known as rickets in which bones develop abnormal curvatures due to withdrawal of calcium.

The **keratin** that accumulates in the cells of the stratum corneum protects the underlying cells from friction or abrasion; it also forms a barrier that is difficult for bacteria to penetrate and is impermeable to water. Although the skin of the fetus is not subjected to friction, it begins to thicken on the palms and soles prior to birth. After birth, friction stimulates cellular proliferation and synthesis of keratin in the stratum germinativum, producing a thicker stratum corneum in the areas subjected to the friction.

All epithelia seem to have an extreme distaste for discontinuity, and the epidermis is no exception. A wound in the epidermis triggers a rapid migration of epithelial cells to the injured area from its periphery, to cover the area. The cells remaining at the periphery undergo a burst of mitosis and continue dividing until the original thickness of the epidermis is restored. Frequently, there is an overcompensation and the thickness of the epithelium in the wounded area eventually exceeds the original thickness. The entire depth of the epidermis does not have to be injured to initiate proliferation. The surface layers may be stripped away with adhesive, for example, which will cause rapid cellular proliferation to replace the lost cells.

In the normal human epidermis, mitosis usually occurs only in the stratum germinativum. The columnar basal cells are oriented with their long axes perpendicular to the epidermal–dermal junction. As each cell divides, it usually splits at right angles to the long axis, so that the lower daughter cell remains in the basal layer and the upper one is pushed by the growth of the underlying cells into the stratum spinosum. Each basal cell divides approximately every 19 days, and a new cell takes from 26 to 42 days to work its way into the stratum granulosum. It then takes an average of about 14 days for the cell to reach the surface of the stratum corneum and slough off. Thus the total renewal time for the epidermis is 59 to 75 days in a normal person. In certain skin diseases such as *psoriasis* renewal time may be only 8 to 10 days.

The Dermis

The dermis serves for attachment of the epidermis above and is attached to the superficial fascia at its deep surface. Except for its outer margin it consists of dense irregular connective tissue. The margin of the dermis adjacent to the epidermis is irregular, with fingerlike extensions, the *dermal papillae,* projecting into concavities in the epidermis. This portion of the dermis, frequently termed the **papillary layer,** consists of loose connective tissue. It has a dense network of capillaries that provides the epidermis with blood, since blood vessels do not penetrate epithelium. In hot weather or when the body temperature is elevated by fever or physical exercise the small arteries supplying the skin dilate, increasing blood flow to the capillaries and making the skin appear pink. The heat brought near the surface of the body by the blood is thus lost to the environment. This is part of the mechanism of *temperature regulation* that keeps body temperature fairly constant.

Overstretching the skin, as may occur during pregnancy, may break the elastic fibers and permanently stretch the collagen, resulting in the linear markings known as striae, or stretch marks.

The deeper part of the dermis, termed the **reticular layer,** consists of dense connective tissue with collagenous fiber bundles oriented mostly parallel to the surface. A dense network of elastic fibers anchors the collagenous fibers in place and allows the skin to stretch under tension and to return to its former configuration when tension is released. Loss of this elasticity is

Clinical Applications: **Skin Grafts**

A skin graft is a segment of skin that has been excised from a *donor site* and transplanted to another part of the body, termed the *recipient site,* or *graft bed.* Such a transplant is much less likely to be rejected by the body's immune system than tissue donated by another individual. Three types of grafts are used to aid in the healing of skin wounds resulting from trauma or burns.

The *pinch graft* is a small cone-shaped piece of skin that includes the entire thickness of the epidermis and dermis. The area of epidermis is broader than the dermis, which is the tip of the cone.

The *full thickness graft* is a large single strip of skin, including the epidermis and the full thickness of the dermis. The thickness of the graft depends upon the thickness of the skin.

The *split thickness graft* is also a single strip of skin but does not contain all of the dermis. An instrument termed a *dermatome* is used to split the skin so that the deeper part of the

dermis left at the dono[...] the epidermis can rege[...] the recipient site, cont[...] maintains the epidermis[...]

The successful trans[...] as the graft "take." Durin[...] by imbibing plasma that [...] plantation site. The plasm[...] ing the graft in place. Th[...] the bed and its weight i[...] during the first forty-eight hours. Vascularization begins in eighteen to seventy-two hours by delicate capillaries that arise from capillary loops in the bed and invade the graft in all directions. Some of these vessels connect with the severed vessels of the graft, restoring circulation and bringing about a rapid change in color of the graft from white to pink. Blood flow is usually well established by the third or fourth day if the graft has taken.

---

normal in the aging process and results in wrinkles. One test for the physiological age of the skin is to pinch the back of the hand and time the flattening of the fold of skin.

> Lines of tension develop in the skin due to the orientation of the connective tissue fibers in the dermis. These lines, known as *Langer's lines,* are taken into consideration by surgeons in planning incisions. An incision at right angles to Langer's lines will retract more extensively and be more difficult to suture than an incision parallel to these tension lines.

In most areas of the body, the dermis is continuous with the subcutaneous connective tissue, or superficial fascia. This underlying tissue is loose and slides easily when the skin rubs against a firm object, reducing friction and the possibility of abrasion. However, on the palms of the hands and soles of the feet the dermis is firmly attached to the fascia, which prevents the skin from sliding. Loose skin would interfere with grasping by the hands and with gripping the ground by the feet. In these areas, protection against friction is provided by the increased thickness of the epidermis.

The skin is richly supplied with sensory nerve fibers located in the dermis and terminating in specialized endings near the epidermis. These fibers make us aware of touch, pressure, pain, and temperature. *Itch,* a common symptom of diseased skin, seems to be closely related to the sensation of pain. If a weak acid causes damage to the skin that is below the pain threshold, the resulting sensation may be felt as an itch.

Skin Color

The color of the skin is due to three pigments, of which only one is produced by cells of the skin. The yellow tinge of healthy skin is due to **carotene** obtained from the diet, the same yellow pigment found in carrots. The pink color is imparted by *oxyhemoglobin,* the pigment in the red blood cells that circulate through the capillaries of the dermal papillary layer. The pigment produced within the skin is **melanin,** responsible for the brown to black color. Special pigment-producing cells, **melanocytes,** are present in the basal layers of the epidermis and at the dermal–epidermal interface. They produce melanin which is transferred to the epidermal cells, giving color to the skin, masking the other colors to a greater or lesser degree, and protecting against the ultraviolet rays of sunlight. Production of melanin is enhanced by sunlight.

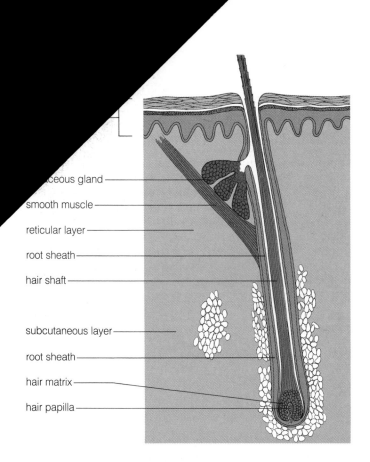

ceous gland
smooth muscle
reticular layer
root sheath
hair shaft

subcutaneous layer
root sheath
hair matrix
hair papilla

Figure 5-3 A hair follicle and related structures.

The color of a person's skin depends upon the relative amounts of carotene and melanin stored in the epidermis and the visibility of the red color of the blood in the dermis. The number of melanocytes per unit area of skin is approximately the same in all individuals. The increased pigmentation in dark skin is due to a higher rate of melanin synthesis than in light skin.

APPENDAGES OF THE SKIN

The principal appendages of the skin are hair, sebaceous glands, nails, sweat glands, and mammary glands; all are derivatives of the epidermis.

Hair

Hairs are the products of proliferative centers of the epidermis called **hair follicles,** which are extensions of the stratum germinativum that project into the dermis (Fig. 5-3). They are cup-shaped aggregations of cells that develop from the epidermis and accumulate hard keratin within their cytoplasm as they proliferate and differentiate. As the keratin accumulates, the cells die, forming a *hair shaft* of hard, dead material. The shaft is

Clinical Applications: **Diseases of the Skin**

The skin is taken for granted until something goes wrong with it. The unfortunate victim of a simple rash becomes obnoxious to family and friends, whose only advice is, "Don't scratch!" In Biblical days, people with skin diseases were all classified as unclean and were frequently isolated. The word leprosy was so dreaded that it was compared to the evil of sin.

At present, the skin is recognized as the largest organ of the body. Dermatology has kept pace with the other specialties of medicine. A thorough physical examination begins with the skin, for it pictures the general health of a patient perhaps better than any other organ.

A recent survey of more than 20,000 Americans ages 1 to 74 who were examined by dermatologists revealed that one-third had one or more skin diseases. Estimates are that between 1 and 3 million people have *psoriasis,* a condition whose cause is unknown but which involves an increase in cellular proliferation of the stratum germinativum of the epidermis. Individuals with psoriasis have epidermal lesions that are characterized by nonmalignant but uncontrolled skin growth. There are also changes in the blood vessels at the sites of the lesions which may stimulate proliferation.

The skin of the psoriatic patient shows a doubling of the rate of DNA synthesis when compared to normal human skin. The cause of this misregulated cell growth can only be suggested; it involves as-yet-undefined hereditary and environmental factors. Since it was first described as a clinical entity by Robert Willan (1757–1812), psoriasis has been the subject of extensive research by scientists from several disciplines, such as genetics, cellular biology, biochemistry, and immunology. The development of basic and clinical research data may be translated into improved patient care, or psoriasis may remain an extremely discomforting condition. Mark Twain may have been right when he stated, "Science is a wonderful field. For a very small investment of fact you get such a tremendous return of theory."

surrounded below the skin surface by a *sheath* of living cells that is continuous with the stratum germinativum of the epidermis.

For centuries man has been literally scratching his head and searching for means of correcting or retarding the condition called male pattern baldness. Thousands of cures have been devised and gullible men have applied them liberally in the hope that hair would reappear upon thinning scalps. The treatment of baldness with lotion or salve began over 5,000 years ago in Egypt and continues even to this day. In the early 1960s, hair transplantation became a popular means of restoring hair to the bare scalp.

Several techniques have been devised by which skin-bearing hair is removed from one site on the individual to the bald scalp area. The most commonly used procedure is the punch graft. A small plug of skin-containing hairs is removed from a donor site and inserted into a hole on the skin in the area receiving the transplant. Usually the donor plug of skin is obtained using a punch that is approximately 4 mm in diameter. The hole into which the graft is to be inserted is made by a punch that is about 2.5 mm in diameter. The transplant must fit snugly, but not too tight. About 75 of these transplants are made at a session; after 3 or 4 sessions a haircut might be in order.

Sebaceous Glands

The epithelial cells on one side of the *hair sheath* are continuous with the **sebaceous gland,** a solid gland (Figs. 5-3, 5-4) that produces an oily product called *sebum* and discharges it onto the surface of the hair shaft. Sebaceous glands are *holocrine glands,* which discharge their products by rupturing their cells; the secretion includes all components of the cells. Discharge of the secretions is accelerated by contraction of the **arrector pili,** a smooth muscle that extends from the hair follicle to its superficial attachment (attachment at the upper surface) in the dermis. The smooth muscle passes beneath the sebaceous gland

Sebaceous glands are stimulated by androgenic ("male") hormones produced by the testes in males and, in much smaller quantities, by the ovaries and suprarenal glands in females. Glandular activity is low until puberty, at which time the increased production of sebum, especially in the male, tends to block the sebaceous gland ducts. This results in acne, which is particularly prevalent in adolescent males.

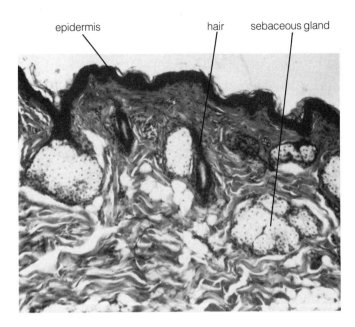

epidermis   hair   sebaceous gland

Figure 5-4 A light photomicrograph of a human sebaceous gland, × 500. (Courtesy of W. K. Elwood.)

so that its contraction not only straightens the hair shaft, causing the hair to stand up, but also exerts pressure on the sebaceous gland, forcing secretions.

Nails

The **nails** are flat plates of hard keratin covering the dorsal surface of the distal ends of the fingers and toes (Fig. 5-5). Nails are formed by the growth and differentiation of epidermal cells of the *nail bed,* on which the *nail plate* rests.

The most common problem that occurs with nails is the condition termed ingrown nails. These occur regularly in toenails but are rarely seen in fingernails. An ingrown nail is the result of a curvature in the nail plate that causes the plate to pierce or put abnormal pressure upon the lateral fold epithelium. This may be caused by uneven cutting or tearing of the nail plate. The earliest signs of ingrown nails are pain and slight swelling in the area. The solution to the problem is obviously the extraction of the intruding nail plate from the lateral nail fold. This is frequently more difficult to do than would be expected. A better solution is the prevention of ingrown nails by keeping the nails properly trimmed.

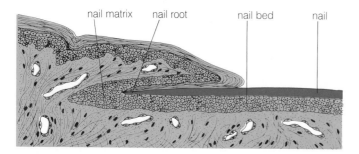

Figure 5-5 Sagittal section of the nail, showing its histological structure.

nail matrix   nail root   nail bed   nail

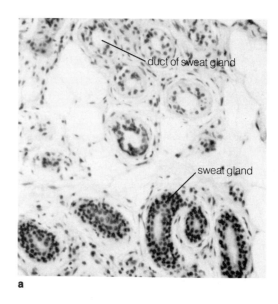

duct of sweat gland

sweat gland

a

The nail bed is a layer of stratified squamous epithelium that is continuous proximally, distally, and laterally with the *stratum basale* of the epidermis. Most of the growth of the nail occurs through cell proliferation at the proximal end of the nail bed; as these cells accumulate keratin, becoming a part of the nail plate, the nail lengthens as it slides distally on the nail bed.

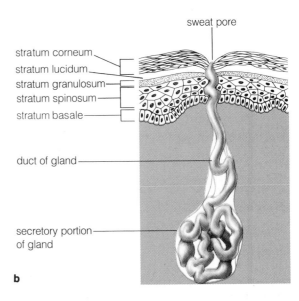

sweat pore

stratum corneum
stratum lucidum
stratum granulosum
stratum spinosum
stratum basale

duct of gland

secretory portion of gland

b

Figure 5-6 A sweat gland. (**a**) Histological structure of sweat glands and ducts (courtesy of W. K. Elwood). (**b**) Diagram of a sweat gland and duct.

Toenails grow more slowly than fingernails. It takes from 12 to 18 months for toenails to be replaced completely while fingernails are replaced in about 6 months. This time difference must be taken into consideration in treating fungus infection of the nail plate since fungus buried in the nail bed of the toenails will be protected for a longer period of time than will fungus of the fingernails.

Sweat Glands

The **sweat glands** (Fig. 5-6), also derivatives of the epidermis, develop as simple tubules extending into the dermis. Each gland consists of a straight portion that opens through a pore in the surface of the epidermis and a coiled portion deep within the dermis, containing secretory cells at its blind end. The fluid called sweat, elaborated by the secretory cells, is propelled toward the surface by contraction of special contractile cells, called myoepithelial cells, around the blind end of the sweat gland.

Sweating is important in regulating body temperature, especially in warm climates. Sweating patterns differ with different stimuli. Heat produces sweat over the entire body, but very little on the palms and soles. Fear or anxiety produces sweat mainly on the palms and soles and in the axillae (armpits).

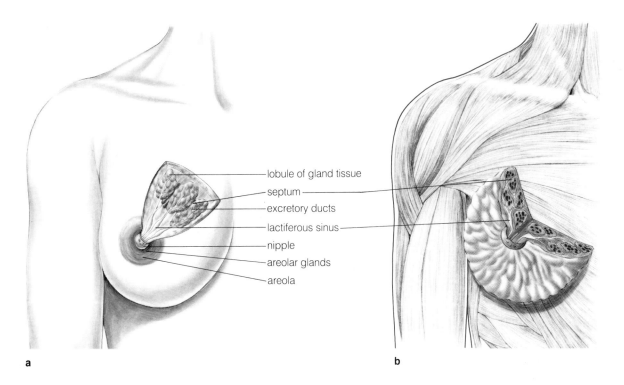

lobule of gland tissue
septum
excretory ducts
lactiferous sinus
nipple
areolar glands
areola

a

b

Figure 5-7 The mammary gland. (**a**) Partially dissected to show gland lobules. (**b**) A section has been removed to show relations of the glands to underlying muscles.

The sweat glands eliminate excess water and salts and also help the body maintain a constant temperature by promoting heat loss through evaporation of water on the body surface.

Mammary Glands

The **mammary glands** (Fig. 5-7) are accessory reproductive organs and develop as modified sweat glands. They arise in the fetus as 15 to 20 solid *epithelial cords* growing into the underlying mesoderm. Each solid cord differentiates into a separate exocrine gland and becomes a component of the female breast, consisting of a **lactiferous duct** dilated into an *ampulla,* or lactiferous sinus, near its opening to the surface at the nipple. The nipple is usually located below and lateral to the center of the breast; it is surrounded by a disc-shaped **areola** that is darker in color than the adjacent skin. In females the nipple and areola enlarge as a result of hormonal secretions during puberty.

Each lactiferous duct branches as it passes into the breast, and in the lactating breast the branches terminate in sac-like secretory portions called alveoli. All of the branches and alveoli drained by a single lactiferous duct constitute a *lobe* of the breast. Each lobe is subdivided by connective tissue into smaller units termed *lobules.* Heavy strands of fibrous connective tissue form septa (partitions) between the lobes and extend from the skin of the breast to the deep fascia covering the pectoralis major muscle. These fibrous strands constitute the *suspensory ligaments* (of Cooper), the principal support of the breast. During puberty adipose tissue accumulates in the connective tissue separating the lobules. Individual variation in the amount of adipose tissue is largely responsible for differences in breast size; the amount of glandular tissue is much the same in all women.

> The period of lactation is a time of decreased fertility. Ovulation is often suppressed due to inhibition of the production of certain hormones. But because this effect is erratic, conception can occur during lactation; in such instances lactation continues during pregnancy although the quantity of milk secreted is reduced.

In the nonlactating breast (Fig. 5-8a) each mammary gland consists only of a system of branched ducts. Under the influence of the hormones of pregnancy the ducts grow in length and the

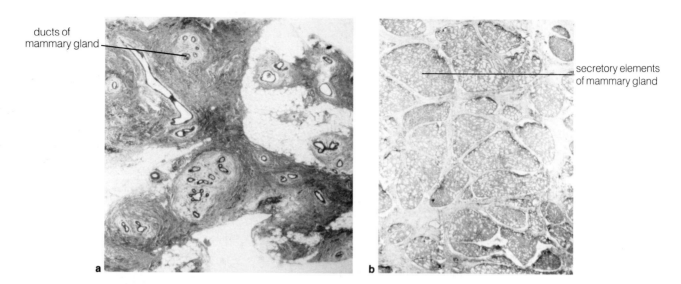

ducts of mammary gland

secretory elements of mammary gland

Figure 5-8 Histological structure of the mammary gland. (**a**) Nonlactating human mammary gland, × 1000. (Courtesy of W. K. Elwood.) (**b**) Lactating human mammary gland, × 600. (Courtesy of W. K. Elwood.)

alveoli differentiate (Fig. 5-8b). Following childbirth the stimulus of suckling is responsible for hormonal maintenance of the lactating breast.

The process of secretion is unusual in that the proteins of milk are not enclosed in membrane vesicles. Milk droplets collect in the cytoplasm near the apical end of the cell and form a drop that is secreted by pinching off the end of the cell. Each discharged drop of milk is enclosed by the pinched-off piece of cell membrane. Since a small amount of cytoplasm is lost, the process is similar to the *apocrine type* of secretion; but the amount of cytoplasm involved is so small that the mammary gland is usually considered a *merocrine gland*.

WORDS IN REVIEW

Areola	Mammary gland
Arrector pili	Melanin
Carotene	Melanocyte
Dermal papillae	Nail
Dermis	Papillary layer
Epidermis	Reticular layer
Hair	Sebaceous gland
Hair follicle	Stratum corneum
Hypodermis	Stratum germinativum
Integument	Stratum granulosum
Keratin	Stratum lucidum
Keratohyaline	Stratum spinosum
Lactiferous duct	Sweat gland

SUMMARY OUTLINE

I. Integument (skin) functions as protection against friction, water loss, and penetration by harmful organisms; it is important in temperature regulation and for protection from ultraviolet radiation.

  A. The epidermis is keratinized stratified squamous epithelium.

  1. Stratum germinativum is the basal layer and has cells that divide to replace cells lost at the surface.

  2. Stratum spinosum is the squamous cell layer; its cells contain cytoplasmic components for the synthesis of keratohyaline.

  3. Stratum granulosum contains keratohyaline granules.

  4. Stratum lucidum is the layer whose cells contain soft keratin, which is not visible.

  5. Stratum corneum consists of flattened dead cells that slough off at the surface.

B. The dermis is connective tissue that is continuous with the superficial fascia.
   1. The papillary layer consists of loose connective tissue that contains capillaries supplying blood to the epidermis; it is involved in temperature regulation.
   2. The reticular layer is the dense irregular connective tissue that attaches the skin to deeper structures.

II. Skin color is due to the balance between the pigments carotene and oxyhemoglobin, which are not produced in the skin, and melanin, which is produced by melanocytes in the skin and transferred to epidermal cells.

III. Appendages of the skin are all derived from the epidermis.
   A. Hair consists of rods of hard keratin produced by follicles that are continuous with the stratum germinativum.
   B. Sebaceous glands are solid glands continuous with the hair sheath; the entire cell is discharged as the secretion.
   C. Nails are flat plates of hard keratin produced by the cells of the nail bed, which is continuous with the stratum germinativum.
   D. A sweat gland has a straight duct that opens through the epidermis at the surface; the duct drains the coiled secretory portion of the gland located in the dermis.
   E. Mammary glands have 15 to 20 branched lactiferous ducts that drain the secretory alveoli; the milk droplet at the apical end of the cell is pinched off in the discharge of secretions.

REVIEW QUESTIONS

1. List the functions of the integument and state the component of skin that performs each of the functions.

2. Name the layers of keratinized (thick) epidermis and describe the morphological characteristics of each layer.

3. Describe the dermis; name the layers of the dermis and relate the structure of each layer to its functions.

4. Explain the basis for variation in skin color among individuals of different races.

5. Describe the structure and development of hair.

6. Describe the location of sebaceous glands in relation to hair follicles and relate their structure to their function.

7. Describe the structure of the nails. Describe the role of the nail bed in the growth of the nail.

8. Describe the structure and function of sweat glands.

9. Describe the structure and functions of the mammary gland. Relate the structure of the mammary gland to the structure of the female breast.

SELECTED REFERENCES

Epstein, E., and Epstein, Jr., E., eds. *Skin Surgery*. 4th ed. Springfield, Ill.: Charles C. Thomas, 1977.

Montagna, W., and Parakkal, P. F. *The Structure and Function of Skin*. New York: Academic Press, 1974.

Rhodes, E. L. *Dermatology for the Physician*. London: Bailliere Tindall, 1979.

Seiji, M., and Bernstein, I. A., eds. *Biochemistry of Cutaneous Epidermal Differentiation*. Baltimore: University Park Press, 1977.

THE SUPPORTING
FRAMEWORK

6

OBJECTIVES

After completing this chapter you should be able to:

1. Compare and contrast connective tissues proper, cartilage, and bone, with regard to the type of cells and the nature of the intercellular substance.

2. Compare and contrast connective tissues proper and cartilage with bone, with regard to the methods by which these tissues grow.

3. Classify connective tissues proper according to their location within the body.

4. List and give the distinguishing characteristics of the types of cartilage found in the body.

5. Describe the principal physical and chemical characteristics of bone. List and give the characteristics of the major components of the organic and inorganic matrix of bone.

6. List the three ways in which bone may be classified and name the types of bones according to each method of classification.

7. Describe the process of intramembranous ossification.

8. Describe the process of endochondral ossification.

The *supporting tissues* of the body are connective tissues that provide a framework for the attachment of muscles as well as protective wrappings for muscles and other structures. The three types of supporting tissues are connective tissue proper, cartilage, and bone. The principal function of these tissues is mechanical and is based on the strength of the fibers embedded in the intercellular substance. The metabolic activity of the cells of these tissues is limited to maintenance and is relatively unimportant in comparison with the role of the fibers.

This relative insignificance of cellular activity in the supporting tissues is in striking contrast with the roles of the cells in organs such as the liver, where cells are closely packed with little intercellular material. In the liver, cells are actively engaged in the synthesis of proteins, which are useful not only to the organ involved but also to the body as a whole.

The fourth type of connective tissue, blood, has no fibers except when clotted and does not function in support of the body. It is classified as a connective tissue because many kinds of blood cells move freely from the circulating blood into other connective tissues where they function as a normal component.

CONNECTIVE TISSUES PROPER

The classification of connective tissues proper according to their microscopic characteristics was described in Chapter 4. This classification into loose connective tissue (see Fig. 4-10) and dense connective tissue (see Fig. 4-15) is based upon the relative amounts and arrangement of the fibers in relation to the cells. In gross anatomy the connective tissues proper that do not have specific names (such as *tendon* or *ligament*) are given the general name *fascia* (L., *fascia* = a band) and are classified as *superficial* or *deep fascia* according to their position in the body (Fig. 6-1).

Superficial Fascia

In all regions of the body, the skin rests on a layer of fascia, termed **superficial fascia,** which contains a variable amount of fat. The superficial fascia is loose connective tissue that serves not only as padding and insulation but also as the substrate, or supporting medium, for the nerves and blood vessels that supply the skin.

In some parts of the body the superficial fascia contains muscles, such as the voluntary skeletal muscles of facial expression or the involuntary smooth muscle fibers of the nipple and the scrotum. The deeper part of the superficial fascia in the lower, less distensible part of the abdomen has a membranous appearance owing to its high collagenous fiber content. It assists the abdominal wall muscles in resisting intra-abdominal pressure and prevents the abdominal wall from sagging under the weight of the viscera.

There are sex differences in the amount and distribution of fat in the superficial fascia. The rounded contours of the female are due largely to the accumulation of fat in specific areas such as the breasts, buttocks, and thighs. In the male, for reasons that are not understood, fat tends to accumulate more in the upper part of the abdominal wall than in the lower part.

> The main constituent of the superficial fascia is fat, which insulates the body against the loss of heat. It has been shown that obese people survive longer in cold weather than do people with little subcutaneous fat. Whales have a very thick layer of fat (blubber) under their skins. Some animals, such as beaver, have thick hair for insulation and have little fat in their superficial fascia.

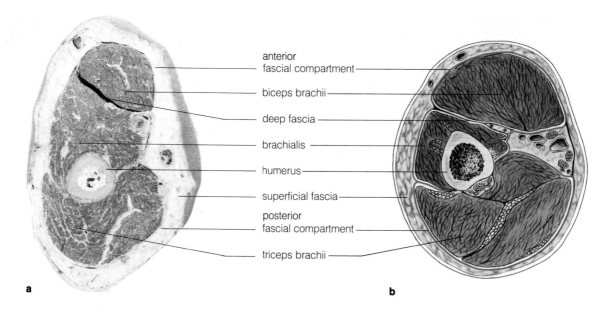

anterior
fascial compartment
biceps brachii
deep fascia
brachialis
humerus
superficial fascia
posterior
fascial compartment
triceps brachii

a   b

Figure 6-1 Transverse section of the human arm. (**a**) Section approximately at mid-humerus (photograph by R. Ross). (**b**) Diagram of a section near the distal end of the humerus. Note attachments of the fasciae to the humerus.

Deep Fascia

Beneath the skin and superficial fascia, the skeletal muscles of the body are enclosed by sheaths of connective tissues collectively designated the **deep fascia,** which has a greater fiber content than the superficial fascia. The deep fascia contains appreciable amounts of fat only in special areas such as the buccal fat pad in the cheek and around the eyeball in the orbit. In most areas the deep fascia *encases* structures and has extensions, or *septa,* that penetrate the structures and hold their components together. For example, muscles are wrapped in connective tissue that not only covers the surface of the muscle with a layer called **epimysium,** but also penetrates the muscle, enclosing small bundles of muscle fibers (the fascicles) with **perimysium** and forming a thin sheath, called **endomysium,** around individual muscle fibers (see Fig. 4-25). In this way the deep fascia provides support and protective covering for muscle, and a means of attachment of muscles to bones by way of the tendons. Collagenous fibers of the *periosteum,* the connective tissue wrapping of bone, become entrapped within the bone matrix and provide a firm attachment of the periosteum to the bone; thus muscle contraction moves the bone instead of pulling the periosteum away from the bone. The periosteum is special in that it contains cells capable of forming bone for the repair of fractures and for the normal, constant reorganization of bone tissue.

Infection of a gland, such as the parotid (salivary) gland, is usually confined within the capsule of the gland. One result of the infection is the accumulation of excess fluid, but the gland cannot enlarge because its fibrous capsule resists stretching. The resulting pressure upon sensory nerve endings within the gland causes the pain that occurs in a disease like mumps.

In regions of the body that are not distensible, such as the extremities, deep fascia forms a tight enclosure that does not stretch appreciably and holds muscles firmly in place. Muscles that tend to function as a group are usually bound together, forming a **fascial compartment** that separates the entire group from other muscles in the vicinity (Fig. 6-1). In distensible parts of the body such as the abdomen, the deep fascia covering the muscles is of necessity relatively thin, allowing considerable flexibility. If the deep fascia enclosing the abdomen formed a rigid covering, eating a big meal would lead to severe discomfort, and pregnancy would create problems during its later stages.

The deep fascia encloses all the organs of the body and sends septa into them to such an extent that the removal of all tissues except the fascia would leave an organ unaltered in size and shape. Without the fascia for support the body would become

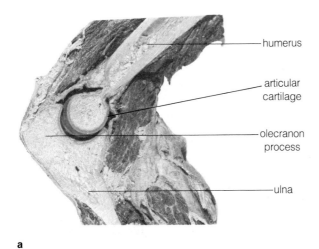

- humerus
- articular cartilage
- olecranon process
- ulna

a

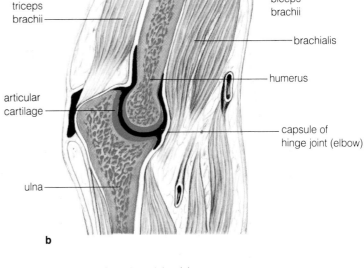

- triceps brachii
- biceps brachii
- brachialis
- humerus
- articular cartilage
- capsule of hinge joint (elbow)
- ulna

b

Figure 6-2 The elbow joint. (**a**) Cadaver preparation of a sagittal section of the human elbow joint (photograph by R. Ross). (**b**) Diagram of a similar sagittal section. (**c**) X ray of a lateral view of the elbow joint (from the teaching collection, Department of Anatomy, University of Kentucky).

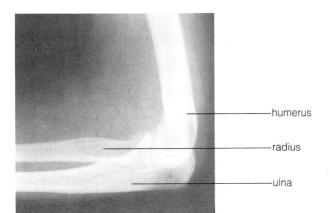

- humerus
- radius
- ulna

c

a shapeless mass, with only hard structures such as cartilage and bone retaining their normal morphology.

Much of the deep fascia occurs in the form of sheets or membranes that are not easily penetrated by fluid. One result of this is that infections tend to spread along **fascial planes** and may be confined within fascial spaces or compartments.

CARTILAGE

Cartilage, you will recall, consists of cells called chondrocytes and intercellular material containing chondromucoid that takes

the form of a *firm gel*. In this section we discuss the histological characteristics of the three types of cartilage (hyaline, elastic, and fibrocartilage) shown in Fig. 6-2, their distribution, and their roles in the supporting framework of the body.

Hyaline Cartilage

Hyaline cartilage (Fig. 6-2a), the most common type of cartilage in the body, is glassy white or slightly bluish with a firm, homogeneous consistency. On examination with the light microscope the intercellular material of hyaline cartilage appears homogeneous, because the collagenous fibers have the same refractive

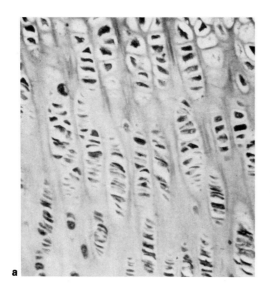

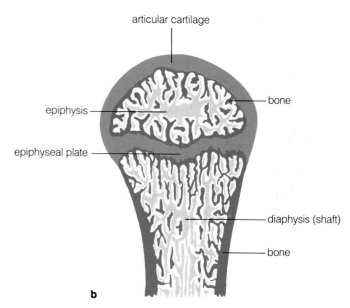

Figure 6-3 The epiphyseal plate. (**a**) Light micrograph of hyaline cartilage of an epiphyseal plate of a human long bone (courtesy of W. K. Elwood). (**b**) Diagram of an epiphyseal plate and related bone of the head and shaft of a long bone.

index (bend light at the same angle) as the matrix. However, this tissue does contain numerous fibers, which can be seen with the electron microscope; these increase the resilience and tensile strength of the cartilage. The chondrocytes synthesize and secrete the collagenous fibers as well as the chondromucoid components of the matrix during growth and differentiation of the tissue.

Hyaline cartilage differentiates from embryonic mesenchymal cells and forms a cartilaginous skeleton that provides the supporting framework for the embryo before the bony skeleton develops. This cartilaginous skeleton serves as a model on which most of the bones form during later embryonic development. As bone differentiates, it replaces the cartilage except at the ends of bones that are in contact with one another. This remaining hyaline cartilage, the *articular cartilage* (Fig. 6-3), provides a smooth, compressible covering of the articular (adjoining) surfaces of bones at movable joints. Thus hyaline cartilage has a wide distribution in the skeletal system, with an essential friction-reducing and shock-absorbing role in all movable joints.

Hyaline cartilage also plays an important role in the growth of the skeleton, and thus of the individual, through its capacity to increase in mass by both internal and surface growth. Cells within a cartilage mass retain their ability to multiply by mitosis,

thereby increasing the number of chondrocytes available for synthesis of intercellular material. As this synthesis proceeds, the elaboration of additional fibers and amorphous ground substance pushes the chondrocytes farther apart, increasing the size of the mass. Increase in mass as a result of internal growth is termed **interstitial growth,** in contrast to growth that occurs through the addition of cells and intercellular substance at the surface of the mass, termed **appositional growth.** The latter is the only method by which bone can grow, although cartilage grows in both ways. A long bone, such as a bone of the arm, can increase in diameter by depositing new bone at its surface, but it cannot increase in length through multiplication of bone cells. During embryonic life, infancy, and childhood a long bone grows in length by interstitial growth of hyaline cartilage plates called **epiphyseal plates** (Fig. 6-3). At the same time cartilage is constantly being replaced by bone in a process described later in this chapter. When all the hyaline cartilage of the epiphyseal plates has been replaced by bone tissue, the bone can no longer grow in length and adult height has been reached.

Articular cartilage tends to accumulate minerals, or calcify, and becomes harder and more brittle as a normal part of the aging process. As a result of this calcification old people have less flexible joints than young people and eventually have predictable problems with certain movements that are not essential for life. Hyaline cartilage is not found in locations where its calcification would be life-threatening.

In addition to covering the surfaces of bones at the joints, hyaline cartilage also serves as the internal skeleton of much of the respiratory tract, where its function is to maintain an open passage. Its increasing brittleness with advancing age does not interfere with this supporting function which depends on rigidity.

Fibrocartilage

Fibrocartilage (Fig. 6-4) is much like hyaline cartilage except that *collagenous fibers* predominate in its intercellular substance. This tissue consists of nests of hyaline cartilage scattered between large bundles of fibers, which provide it with far greater tensile strength than could hyaline cartilage alone. Strength is increased still more by the orientation of most of the fibers parallel to the direction of applied force. The small clumps of cells between the rows of fibers do not reduce the strength of the cartilage. Fibrocartilage is found in sites where stress is great and little movement is necessary; for example, it forms an important part of the *intervertebral discs* between the bodies of the vertebrae, which bear most of the weight of the body.

Elastic Cartilage

Elastic cartilage (Fig. 6-5) contains large numbers of *elastic fibers* in its intercellular substance and is therefore more flexible than hyaline cartilage. This flexibility is retained throughout life because elastic cartilage does not tend to calcify. Consequently, structures whose flexibility is essential for life contain elastic cartilage for support. For example, the *epiglottis* closes the air passage during swallowing, thereby preventing food from entering the larynx. The action of elevating the larynx and pushing the food back with the tongue forces the flexible epiglottis into position over the laryngeal opening. If the epiglottis were to become rigid, it would be unable to deflect food away from the opening and we would run the risk of strangling when swallowing. Fortunately the elastic cartilage of the epiglottis remains flexible even in old age. On the other hand, elastic cartilage is also found in the auditory tube and external ear, where flexibility can hardly be considered essential for life.

BONE

Bone tissue consists of osteocytes, or bone cells, embedded in a mineralized matrix. It is further distinguished from cartilage in having a lamellar structure with layers of matrix and rows of cells organized around blood capillaries.

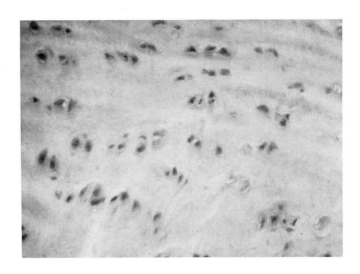

Figure 6-4   Photomicrograph of fibrocartilage (courtesy of W. K. Elwood).

Bone Matrix

Bone matrix has two components: an **organic matrix** composed largely of collagenous fibers and an **inorganic matrix** containing crystalline calcium and phosphate.

Organic Matrix   The organic matrix contributes about 25% of the weight of adult bones. Approximately 95% of the organic matrix is *collagen,* in the form of fibers that are essentially the same as those of connective tissue proper. These fibers are composed of bundles of microfibrils that have a periodicity due to alternating light and dark bands running across them in a characteristic fashion; thus they may readily be identified with the electron microscope.

The remaining 5% of the organic matrix consists of an amorphous ground substance made of protein–carbohydrate complexes called *mucopolysaccharides.* These complexes contain *hyaluronic acid* and *chondroitin sulfate,* which are also present in the chondromucoid matrix of cartilage. Since they have a high capacity to bind ions, mucopolysaccharides may be an important factor in the deposition of minerals in bone.

Inorganic Matrix   The inorganic matrix of bone materials includes calcium, magnesium, and sodium, that are reacted with phosphate, carbonate, and citrate. Most of the mineral present is in the form of crystals of **hydroxyapatite,** which contains calcium, phosphate, and hydroxyl (OH^-) ions.

Crystal formation (mineralization) proceeds rapidly during bone formation. Each crystal is hydrated, that is, it binds a layer

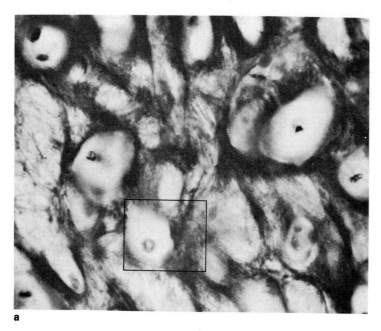

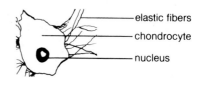

elastic fibers
chondrocyte
nucleus

Figure 6-5 (**a**) Elastic cartilage from the larynx × 412 (Armed Forces Institute of Pathology, negative number 71-9216). (**b**) Diagram of a single chondrocyte and associated elastic cartilage.

of water molecules at the surface of the crystal. This water attracts ions and thus is important in the exchange of ions with the blood and other body fluids. Like all the tissues of the body, bone is in a constant state of change; it serves as a storage place for calcium, essential for the metabolism of all cells, so its calcium content fluctuates continuously. Hormones from the thyroid and parathyroid affect the exchange of calcium between bone and blood.

Types of Bone Tissue

Unlike cartilage, which is classified on the basis of its fibers, bone tissue is classified as compact or spongy according to the organization of its cellular and intercellular elements.

Compact Bone   Compact bone (Fig. 6-6) is made up of *Haversian systems* that consist of concentric layers of bone matrix called concentric lamellae and rows of lacunae containing cells, oriented around a blood vessel. Spaces between Haversian systems are filled with irregular layers of bone termed *interstitial lamellae*. At the periphery of the shaft of the bone *circumferential lamellae* provide a relatively smooth surface that is covered by the periosteum. This type of bone is found in the shaft of a long bone, forming the solid wall of a hollow tube. The center of the tube is lined by *endosteum* and is filled with *bone*

marrow composed, in some bones, of blood-forming cells. Blood vessels from the bone marrow or from surrounding tissues penetrate the bone and occupy the canals in the centers of the Haversian systems.

Spongy Bone   Spongy bone (Fig. 6-7) consists of thin plates of bone, called trabeculae, separated by narrow spaces filled with marrow containing the blood vessels that nourish the osteocytes (Fig. 4-14). This structure gives the appearance of being perforated by empty spaces, hence its name. Spongy bone is found at each end of long bones, with the plates oriented along lines of greatest stress, an arrangement that provides the greatest strength with the least weight.

Kinds of bones

Bone tissue occurs in the body as discrete skeletal elements called bones that form the major part of the supporting framework for muscle attachment and the protective housing for viscera such as the brain and the heart. Each bone of the body has a characteristic size and shape, and should be considered to be an organ since it consists of more than one type of tissue with all tissues contributing to its functions. Bones may be classified according to their shape as *short, long, flat,* or *irregular.*

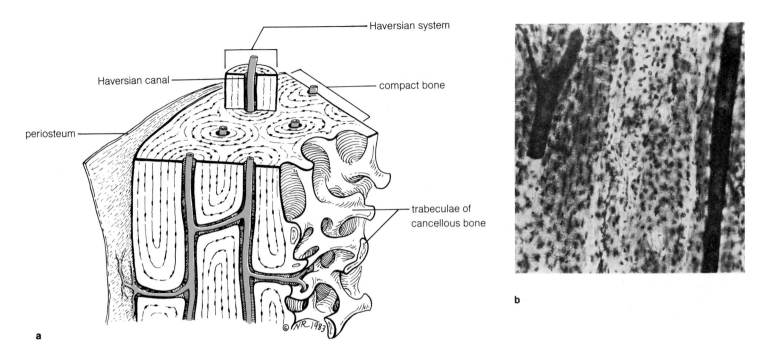

Figure 6-6 Section through a long bone showing the Haversian systems of compact bone and the trabeculae of cancellous bone. (**a**) Diagram. (**b**) Photomicrograph (courtesy of Joan Creager).

Short Bones Short bones (Fig. 6-8) such as those of the wrist and the ankle have a rim of compact bone surrounding spongy bone and a centrally placed marrow cavity. Blood-forming cells fill the *marrow cavity,* which is lined by a layer of bone-forming cells termed *endosteum* that separates the marrow from the bone. The *trabeculae,* or beams, of the spongy bone form an interlacing network attached to the more solid compact bone. At the outer surface of the bone hyaline cartilage covers the points of articulation with adjacent bones. In areas not covered by hyaline cartilage bone is ensheathed by the *periosteum.* This sheet of dense irregular connective tissue contains, in addition to the usual connective tissue cell types, bone-forming cells essential in the repair of fractures, as well as nerves and blood vessels. Short bones develop from a *single center of ossification* in an embryonic cartilage model; that is, ossification begins in the center of the model and spreads toward the periphery.

Long Bones A long bone (Fig. 6-9) consists of a *shaft,* or **diaphysis** (Gr., *diaphysis* = between or to grow apart), with a *head,* or **epiphysis** (Gr., *epiphysis* = to grow on top of), at each end. The diaphysis of a long bone such as the humerus (the bone of the arm) is its longest part, tubular in form with *compact bone* forming most of the wall of the tube. A longitudinal section of a long bone reveals that toward each end of the diaphysis

spongy bone fills much of the *marrow space,* with its *bony trabeculae* oriented along lines of stress. The marrow found within the shaft of a long bone is *yellow marrow,* which contains fat and is not involved in the formation of blood cells. The diaphysis is not absolutely straight; therefore it does not receive the full force of a blow or shock but absorbs some of the force by bending slightly. The spongy bone adds strength to long bones, increasing their capacity to withstand the stresses and strains associated with weight-bearing. The hollow-tube construction is well adapted to provide the strength and flexibility needed by the extremities.

The epiphysis at each end of a long bone consists of *spongy bone* covered at the articular surface by a thin layer of compact bone, which is covered by hyaline cartilage. Thin plates of the spongy bone partially separate marrow spaces, which contain the blood vessels essential for the maintenance of the bone. A thin layer of compact bone, termed the **epiphyseal line,** separates the epiphysis from the diaphysis. The epiphyseal line is the remnant of the epiphyseal plate (Fig. 6-3).

A long bone develops from at least three separate centers of ossification: the *primary center* in the diaphysis and a *secondary center* in each of the epiphyses. During the development of the bone, the epiphyseal plate separates the primary and secondary ossification centers. This plate of hyaline cartilage grows continuously by cell division, increasing the length

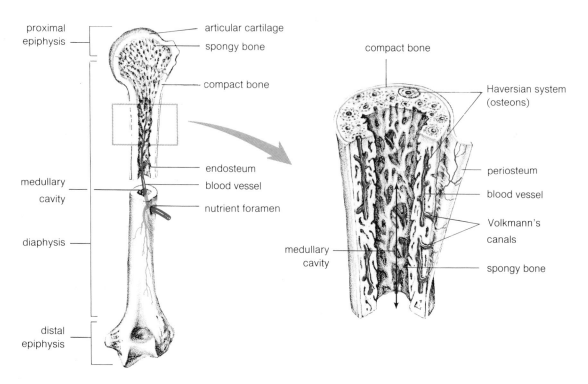

Figure 6-7 Diagrams illustrating the structure of cancellous (spongy) bone and its location within a long bone.

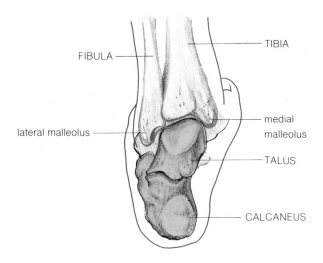

Figure 6-8 Short bones; posterior view of articulation of ankle with leg bone.

of the bone. Eventually bone proliferating from the ossification centers completely replaces the plate, leaving only the epiphyseal lines as visible evidence of its previous position.

Flat Bones Flat bones (Fig. 6-10) consist of *two plates* of *compact bone* separated by spongy bone, which provides cross connections between the plates. Marrow fills the spaces between the trabeculae of the spongy bone. As with all bones periosteum covers the surface, and endosteum lines the marrow spaces. The sternum, ribs, and the bones forming much of the brain case are flat bones. The flat bones of the skull develop directly in connective tissue and have no cartilaginous models in the embryo; the ribs and sternum are formed as cartilaginous models before ossification.

Irregular Bones Any bone that is neither short, long, nor flat is classified as an irregular bone. *Irregular bones,* such as the *vertebrae* (Fig. 6-11), has extensions called processes, or spines, projecting from the main bony element. The main element, as well as the processes, has compact bone at the periphery and spongy bone and marrow filling the central portion. Irregular bones include some of the complex bones of the *cranial cavity,* the *face,* and the *pelvic girdle.* In some of the irregular skull bones the marrow spaces are filled not with marrow, but with air sinuses that are connected with the nasal cavity.

Surface Features of Bone

Although adult bone is a hard, rigid tissue, it is remarkably responsive to external tension and pressure, altering its internal

Figure 6-9 A long bone; the ulna.

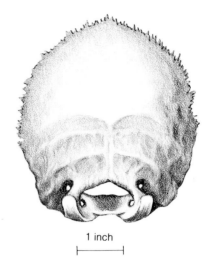

Figure 6-10 A flat bone; the occipital.

somewhat less below the anatomical neck, and the bone is more easily fractured in this narrower region, which is therefore called the *surgical neck.*

> The relationship between the attachment of the joint capsule and the epiphyseal line is of clinical importance. If the epiphyseal line is within the area enclosed by the joint capsule, an infection in the bone shaft may spread into the joint cavity. However, if the epiphyseal line is outside the joint capsule, the infection may spread along the shaft but will not penetrate the joint cavity.

organization and external contours in response to mechanical factors. Each bone has a genetically determined size and shape, but forces applied by nonskeletal elements during the growth and development of the bones determine many of the surface irregularities. Every long bone has an elongated diaphysis with a rounded, smooth-surfaced epiphysis at each end, but there are many irregularities and variations in the surface of the bone. At the point of junction of the head with the shaft there is a rough area encircling the bone caused by tension applied by the attachment of the joint capsule. In a long bone such as the humerus or femur this rough area marks the *anatomical neck of the bone* (Fig. 6-9). Usually the diameter of the long bone is

Irregularities on the surface of the diaphysis are given names that are usually descriptive of their form. A **condyle** (Gr., *kondylos* = knuckle) is a rounded articular surface at the end of a bone, whereas a **ramus** (L., *ramus* = a branch) is a broad projection from the main part of a bone. Other projections are named more or less according to their size. A **trochanter** (Gr., *trochanter* = a runner) is a larger bony prominence than a **tuberosity** (L., *tuber* = a knob), which is larger than a **tubercle** (L., *tubercle* = little knob), which in turn is larger than a **spine** (L., *spina* = a short, sharp process). An elongated projection or prominence is called a **ridge,** a **crest,** or a **line.** Almost all the bony prominences to which these names are

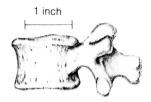

1 inch

Figure 6-11 An irregular bone; a lumbar vertebra.

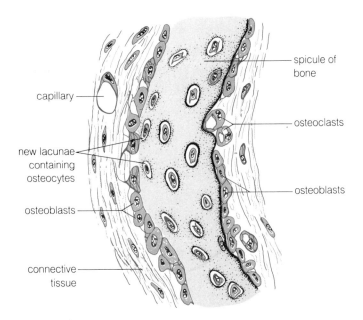

Figure 6-12 Diagram of intramembranous ossification.

applied have muscles attached to them, and many of them will not form without muscle pull during development. An elongated depression is called a **sulcus,** or **groove,** and is usually occupied by a soft structure such as a nerve or a blood vessel. Less elongated depressions are called **fossae** (L., *fossa* = a trench) or **foveae** (L., *fovea* = a pit). A large cavity in a bone is termed a **sinus** (L., *sinus* = cavity) or **antrum** (Gr., *antron* = a cave). An oval or round opening in bone is termed a **foramen** (L., *foro* = to pierce); a slit is called a **fissure** (L., *fissura* = a deep furrow). If the opening has length, it may be called a **canal,** a **meatus** (L., *meatus* = a passage), a **hiatus** (L., *hiatus* = an aperture), or an **aqueduct** (L., *aqua* = water, *ductus* = a leading). Most openings in bones transmit nerves and/or blood vessels. The use of most of these terms is not restricted to the skeletal system.

Bone Formation (Ossification)

Bones may develop in the embryo from the mesoderm (middle germ layer), either by differentiation in embryonic connective tissue membranes, a process called **intramembranous ossification,** or by replacement of hyaline cartilage in cartilaginous models of future bone, a process called **endochondral ossification.** In either process the cells responsible for the production of bone matrix (ossification) are embryonic mesenchymal cells that differentiate into bone-forming cells, the *osteoblasts;* adult bone is the same in structure and function regardless of the method of development. Intramembranous ossification is the simpler method of bone differentiation, and will be described first.

Intramembranous Ossification   Intramembranous ossification is the process by which bone is formed in a connective tissue sheet without the prior formation of cartilage. During embryonic development mesenchyme condenses in areas where bone will form and deposits fine collagenous fibers between the stellate mesenchymal cells, forming a membrane. Some of the cells differentiate into osteoblasts, and clumps of osteoblasts

form isolated *foci,* or centers of ossification, depositing intercellular substance on the fibers already within the matrix. The material deposited forms bars of dense matrix, called *osteoid,* which is not calcified but constitutes the *organic component* of bone matrix. The cells within the osteoid cause minerals to be deposited within the organic matrix, a process called *calcification,* producing a matrix with the chemical nature of mature bone. The bars of calcified matrix become the trabeculae (Fig. 6-12). Cells that are trapped within the matrix have become *osteocytes* which, unlike osteoblasts, cannot divide. The foci expand by appositional growth until they fuse, forming a single bone.

Development of the skull is not completed at the time of birth, so the skull of a newborn has soft areas, called fontanels, that are covered by membrane rather than bone. This makes the skull less rigid than if it were completely ossified, which allows it to be compressed as it passes through the birth canal.

The osteocytes are not arranged in rows at first, and the matrix does not form layers, or *lamellae,* but rather assumes the structure of a loose, interlacing network. This loose organization of immature, nonlamellar bone is termed **woven bone.** The reor-

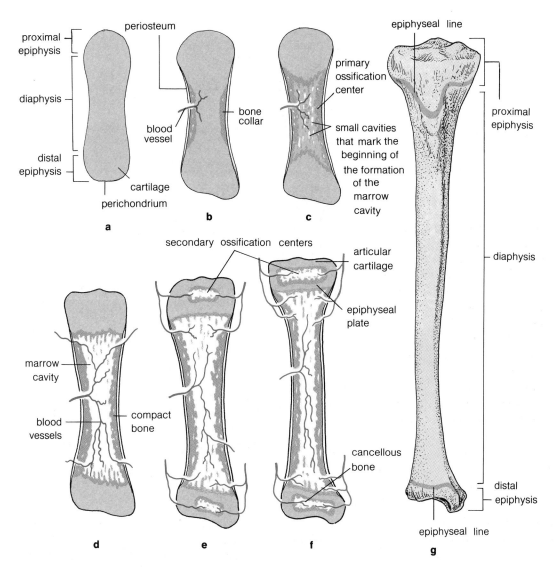

Figure 6-13 The formation of endochondral bone: (**a**) a cartilage model, (**b**) formation of collar, (**c**) beginning of primary ossification, (**d**) formation of marrow cavity and entry of blood vessels into it, (**e**) formation of secondary ossification centers, (**f**) cartilage remaining at articular surfaces and epiphyseal plates, (**g**) fully formed bone with remaining epiphyseal lines.

ganization of woven bone into **lamellar bone** begins almost as soon as calcification. At the surface of the bony trabeculae multinucleated cells called *osteoclasts* appear, and *bone resorption* begins under their influence. The resorption of woven bone is accompanied by the deposition of new bone by the osteoblasts in the vicinity. Some of the new bone is woven bone, and some is lamellar bone with osteocytes in rows; the lamellae of bone matrix separate the rows of cells. The bone continues to grow by the addition of new bone at the periphery and reorganization of the bone internally, contributing to the adult size and shape. The connective tissue that always remains at the surface is the periosteum, which contains cells that have bone-forming capability.

Endochondral Ossification   In bones that differentiate by endochrondal ossification, embryonic mesenchyme differentiates initially into a hyaline cartilage model that is gradually replaced by bone as follows. The hyaline cartilage model forms in the approximate shape of the future bone (Fig. 6-13). This model is surrounded by a connective tissue sheath, the *peri-*

Clinical Applications: **Bone Diseases**

Since bone is not a static tissue, anything that upsets the balance between the deposition and resorption of bone will inevitably lead to disease. Several of these conditions fall into the general category of metabolic bone diseases; others are due to infection.

Osteoporosis

Osteoporosis is the most common bone disease because it is part of the aging process. It is a condition in which there is a reduction in bone mass, a reduction associated with decreased production of the organic component of bone matrix. The chemical composition of bone matrix remains normal, but there is less of it than is needed to bear the weight of the body. The weakened bones are liable to fracture. The most common fractures associated with osteoporosis are fractures of the neck of the femur and the upper end of the humerus. The most common complaint is back pain. Usually, x rays of the back show a generalized decrease in bone density and slightly expanded intervertebral discs. The vertebral bodies are susceptible to fracture and tend to become wedge-shaped as they collapse. There is no effective treatment for the disease, but the disease process can usually be slowed by good nutrition and plenty of exercise.

Osteomalacia

Osteomalacia is a condition in which organic bone matrix is produced normally but mineralization is defective, due to a deficiency of calcium or phosphate in the extracellular fluid. The deficiency may be caused by lack of calcium, phosphate, or vitamin D in the diet. (Vitamin D is essential for absorption of calcium from the digestive tract.) Osteomalacia may also be caused by renal diseases in which calcium is excreted excessively, diminishing the availability of calcium.

The principal symptom is bone pain, particularly in weight-bearing bones. There may be difficulty in walking because of pain and because of muscle weakness. The condition may be alleviated by supplements of vitamin D and calcium, but the underlying cause should be determined and treated.

Rickets is a similar condition that occurs in children. The most obvious manifestation is at the epiphyseal plate, where the newly formed cartilage and osteoid fail to calcify. Bone growth is inhibited, and the child's adult height may be affected.

Osteomyelitis

Osteomyelitis may be caused by a number of bacteria, but usually the offending agent is staphylococcus. The bacteria enter the blood stream at the site of infection, such as a skin abrasion, and settle on the marrow cavity of a bone. Pus forms in the confined space of the marrow cavity and gradually forces its way along the Volkmann canals to the surface of the bone. The pus then spreads between the periosteum and bone along the surface of the bone and may break out into the surrounding soft tissues. As it spreads along the bone, the pus disrupts the blood supply of the area and some of the bone tissue dies. After the infection is cleared by the use of antibiotics, new bone is formed to replace the dead tissue and the bone returns to normal.

Bone Growth Disturbances

There are many known genetic disturbances of bone growth. *Achondroplasia* is the most common form of *short-limbed dwarfism* and has been recognized since ancient times. It is an example of autosomal dominant inheritance, for the gene is consistently expressed when present. The trunk, head, and neck of the individual are normal in size, but the limbs are disproportionately short and legs are frequently bowed. Limb shortening usually occurs more on the proximal than in the distal skeletal elements. Achondroplasts usually live a normal life span and have good health generally. There are about 5000 achondroplasts in the United States.

Another type of disorder resulting in short stature is *pituitary dwarfism*. Perhaps the best known pituitary dwarf was Tom Thumb, the famous performer in P. T. Barnum's circus. All types of pituitary dwarfism have a genetic basis. It has been known for years that failure of the ephiphyseal plates to grow normally is due to lack of growth hormone or to lack of ability of the plate to respond to the hormone.

Hyperparathyroid Bone Disease

Increased secretion of parathyroid hormone such as may be caused by a parathyroid gland tumor results in bone disease. The major effect of excess parathyroid hormone is to increase resorption of bone and decrease activity of osteoblasts. The number of osteoclasts is increased and these cells invade the Haversian canals. Trabeculae are thinned because of increased bone resorption. There is frequently chronic renal failure associated with hyperthyroidism. Afflicted persons usually complain of bone pain, vague musculoskeletal pains, and bone tenderness.

Paget's Disease of Bone

Paget's disease is characterized by high rates of bone deposition and bone resorption. Both osteoblastic and osteoclastic activity are increased. Bone resorption usually exceeds deposition in the early phases of the disease. This is followed by a period of increased osteoblastic activity during which bone is deposited irregularly at a rate that exceeds bone resorption. The cause of Paget's disease is unknown, but it may result from a virus infection of bone. The disease is common in the United States and usually occurs in about 3% of persons older than age thirty.

chondrium, that contains cells capable of forming either cartilage or bone, depending upon the conditions. During the early development of the cartilage model, perichondrial cells add to its size by depositing cartilage at the surface (appositional growth). Growth also occurs within the model through cell proliferation and the elaboration of additional cartilage matrix (interstitial growth). The process of endochondral ossification is perhaps best described in the development of a long bone.

In a long bone, bone formation is initiated in the perichondrium surrounding the center of the future diaphysis. Cells of the perichondrium *hypertrophy* (enlarge), becoming *osteoblasts,* and begin to deposit bone on the surface of the cartilage until bone surrounds the middle of the future diaphysis. This bone is the *periosteal bone collar,* and the perichondrium around this region becomes the periosteum.

As the bone collar is forming, the cartilage within the model undergoes visible changes. The cartilage cells beneath the bone collar begin to hypertrophy, and the associated matrix undergoes *calcification.* The enlarged cartilage cells die and disintegrate, leaving empty spaces separated by rods of calcified cartilage matrix. From the periosteum, blood vessels and elements of connective tissue grow through openings in the bone collar into the lacunae that were formerly occupied by cartilage cells. These *lacunae* represent the beginnings of the marrow spaces. The *periosteal buds* consist of blood vessels and cells that form the marrow components.

> In a long bone that has an epiphyseal plate at either end, one plate may contribute more to growth in length of the bone than the other because the cartilage cells proliferate more rapidly for a longer period. Premature cessation of cellular proliferation may result from damage to the epiphyseal plate.

Cells of the periosteal buds, osteoblasts, begin to deposit bone on the surface of the calcified cartilage matrix producing an **ossification center**. As bone deposition continues, this *primary ossification center* expands, extending toward both ends. It forms the diaphysis of the long bone and grows in length by the gradual replacement of cartilage toward each end.

In a long bone, development of the primary ossification center in the diaphysis is followed by the formation of a secondary ossification center at each end of the cartilage model. The secondary ossification centers become the epiphyses and develop without bone collars. Otherwise the process of ossification is similar to that described for the primary center. Cartilage remains at the surface of the ends of the cartilage model and serves as the articular cartilage of the mature bone.

The primary and secondary ossification centers do not immediately fuse, but remain separated by a layer of hyaline cartilage, the epiphyseal plate. Cartilage cells of the epiphyseal plate multiply by mitosis and continue to lay down cartilage matrix, which is gradually replaced by bone. Thus the bone grows in length until the individual reaches maturity. At maturity, bone replaces the cartilage of the epiphyseal plate, which becomes the epiphyseal line of the mature bone. After elimination of the epiphyseal plate, called closure of the epiphysis, the individual no longer grows in height. Closure of the epiphyses in the long bones of the body is normally complete by about twenty-five years of age.

Growth in diameter occurs by deposition of bone at the periphery by the activity of periosteal cells. At first the bone matrix is deposited on the surface of the calcified cartilage that remains following the death of the cartilage cells. Later, in the process of internal reorganization, all of this bone and the enclosed cartilage matrix are resorbed through the activity of osteoclasts and replaced by new, lamellar bone without the cartilage matrix. Initially, the bone produced is spongy bone, but later remodeling converts the spongy bone into the compact bone of the shaft.

WORDS IN REVIEW

Antrum	Diaphysis	Fascial compartment	Hiatus
Appositional growth	Endochondral ossification	Fascial plane	Hydroxyapatite
Aqueduct	Endomysium	Fissure	Inorganic matrix
Canal	Epimysium	Foramen	Interstitial growth
Condyle	Epiphyseal line	Fossae	Intramembranous
Crest	Epiphyseal plate	Foveae	ossification
Deep fascia	Epiphysis	Groove	Lamellar bone

Line	Ridge
Meatus	Spine
Organic matrix	Superficial fascia
Ossification center	Trochanter
Perimysium	Tuberosity
Ramus	Woven bone

SUMMARY OUTLINE

I. Connective Tissues Proper
   A. Superficial fascia serves as padding and insulation material beneath the skin; it contains a varying amount of fat.
   B. Deep fascia has a higher fiber content than does superficial fascia, usually has no fat, and may occur as fascial sheets or membranes. It encloses organs and provides an internal framework for soft structures.

II. Cartilage
   A. Hyaline cartilage provides skeletal support for the embryo and is largely replaced by bone during early life. Cartilage remains at the articular surfaces of bones in synovial joints. It provides for growth in the length of long bones.
   B. Fibrocartilage has an increased fiber content in the matrix and thus provides greater tensile strength than does hyaline cartilage.
   C. Elastic cartilage is flexible due to elastic fibers in the matrix.

III. Characteristics of Bone
   A. The organic matrix consists of 95% collagenous fibers plus 5% amorphous ground substance containing mucopolysaccharides.
   B. The inorganic matrix consists of hydroxyapatite crystals composed largely of calcium, phosphate, and hydroxyl ions.
   C. A short bone has a rim of compact bone surrounding a core of spongy bone with marrow spaces.
   D. A long bone is composed of a shaft, or diaphysis, and an epiphysis at each end.
   E. A flat bone is composed of two flat plates of compact bone connected by spongy bone.
   F. An irregular bone is composed of a main element of bone with extensions, or processes, adding to its complexity. The bone has a shell of compact bone enclosing spongy bone with marrow spaces.

IV. Bone Markings
   Surface markings of bone consist of elevations caused by tension, or pull, such as that exerted by muscle attachment. Grooves or depressions are caused by pressure and may suggest the location of a nerve, blood vessel, or tendon.

V. Bone Formation
   A. Intramembranous ossification is the development of bone directly in a connective tissue membrane.
   B. Endochondral ossification is the development of bone in a cartilage model as the cartilage is replaced by bone.

REVIEW QUESTIONS

1. What are the function and location of the superficial fascia?

2. How does deep fascia differ from superficial fascia in structure, location, and function?

3. What is the role of hyaline cartilage in joints?

4. What is the structure of fibrocartilage that increases its tensile strength?

5. What are the functional characteristics of elastic cartilage that differ from those of hyaline cartilage?

6. What are the characteristics of organic and inorganic bone matrix?

7. Contrast compact bone with spongy bone.

8. Name and describe the types of bones as classified according to shape.

9. What are the principal causes of markings on the surface of bones?

10. What are the principal features of intramembranous ossification?

11. List the major morphological events that occur in endochondral ossification.

SELECTED REFERENCES

Bloom, W., and Fawcett, D. W. *A Textbook of Histology.* 10th ed. Philadelphia: W. B. Saunders, 1975.

Crafts, R. C. *A Textbook of Human Anatomy.* 2nd ed. New York: John Wiley & Sons, 1979.

Hollinshead, W. H. *Textbook of Anatomy.* 3rd ed. New York: Harper & Row, 1974.

Romanes, G. J., ed. *Cunningham's Textbook of Anatomy.* New York: Oxford University Press, 1981.

THE AXIAL SKELETON

After completing this chapter you should be able to:

1. List the parts of the axial skeleton and give the general functions of each part.

2. Define the skull and list its most important functions.

3. Define the cranium and list its most important functions. Name the bones that form the cranium.

4. Describe the frontal bone. Relate its principal morphological features to the orbits and the forehead.

5. Describe and locate the paired parietal and unpaired occipital bones.

6. Name the parts of the ethmoid bone and relate these parts to the nasal cavity and the orbits.

7. Describe the sphenoid bone. Name its processes and relate these parts to the body of the sphenoid and to other bones of the skull.

8. Describe the temporal bone. Name the parts of the temporal bone and relate the parts to one another.

9. Define the face and list its main functions. Name the bones that form the face.

10. Describe the principal morphological features of the mandible. Relate the most prominent parts of the mandible to the skull.

11. Describe the maxilla. Name its four processes and relate the processes to other bones of the skull.

12. Describe the zygomatic bone and locate it in relation to other bones of the skull.

13. Locate and describe the nasal bones.

14. Describe and locate the lacrimal, palatine, vomer, and inferior nasal conchae.

15. Describe the anterior, middle, and posterior cranial fossae and give the boundaries of each fossa.

16. Describe the orbit. List the bones in each of its walls.

17. Describe the development of the skull, face, and palate.

18. Diagram or describe a typical vertebra.

19. Name the regions of the vertebral column, giving the number of vertebrae in each of the regions. Describe the principal regional differences in the vertebral column.

20. Describe the development of the vertebral column.

21. Describe the ribcage. Relate the ribs to the vertebrae, the costal cartilages, and the sternum.

22. Describe the formation of the ribs and the sternum.

The *skeleton* is a framework of bones that articulate with one another to form the supporting structure of the body. With the help of the muscles that are attached to the bones, the skeleton enables us to stand erect and to move with extraordinary speed and grace. It may be divided into two parts for descriptive purposes (Fig. 7-1). The **axial skeleton,** which forms the long axis of the body, consists of the *skull,* the *vertebral column,* the *ribs,* and the *sternum.* The **appendicular skeleton,** which includes the skeletal appendages, consists of the *pectoral* and *pelvic girdles* and the *bones* of the *extremities.*

The axial skeleton provides *support* and *protection* for the head, neck, and trunk. With its associated muscles, it also serves as a platform for the *attachment* and *support* of the *limbs.*

THE SKULL

The *skull* (Fig. 7-2) protects the brain and special sense organs and provides support and protection for the air and food entrances. "Skull" is the general term for the skeletal support of the head. It includes the **cranium,** whose bones directly enclose the brain, and the **face,** whose bones surround the openings of the respiratory and digestive tracts. Bones of the cranium are, for the most part, classified as flat bones; most of the bones of the face are classified as irregular bones.

The Cranium

The eight bones of the *cranium* (summarized in Table 7-1) include four unpaired bones, frontal, occipital, sphenoid, and ethmoid, and two paired bones, parietal and temporal. These bones articulate with one another at fibrous joints called **sutures** and provide a solid protective covering for the brain. Openings (foramina) in the bones provide for passage of nerves and blood vessels.

Frontal Bone   The **frontal bone** forms the *vertical plate* of the forehead and has two *orbital plates* that contribute to the roofs of the **orbits.** The ridges above the orbit, the *superciliary ridges,* make the eyebrows more prominent. The elevation between the two superciliary ridges is the *glabella,* and the depression at the root of the nose is the *nasion.* The frontal bone contains the *frontal sinuses,* which communicate with the *middle meatuses* of the nasal cavity. The *supraorbital foramen* (or notch), a small opening at the superior rim of each orbit, transmits nerves and blood vessels to the forehead and upper eyelids.

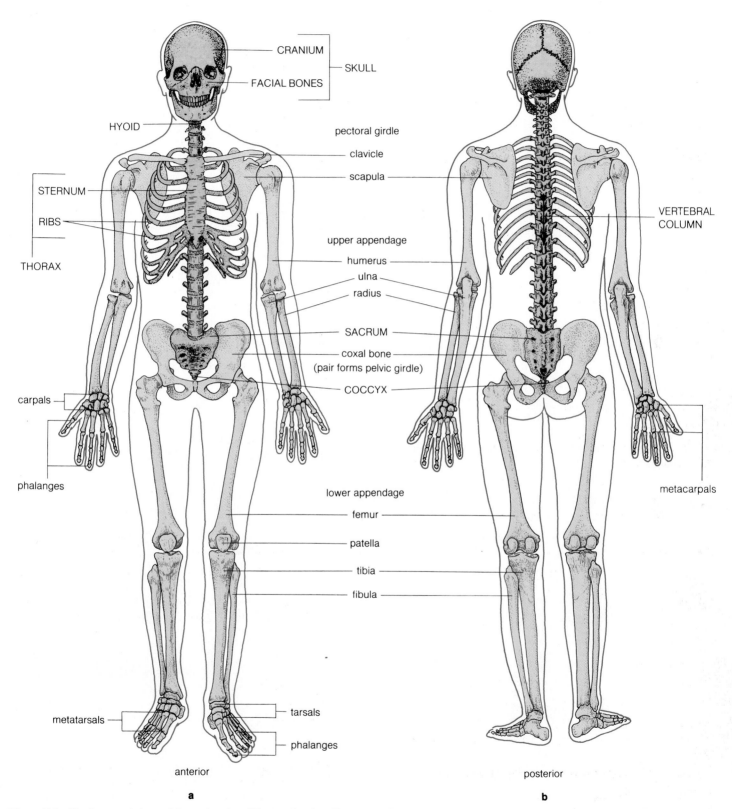

Figure 7-1 The human skeleton: (**a**) anterior view, (**b**) posterior view. The bones of the axial skeleton are shaded; those of the appendicular skeleton are shown in outline.

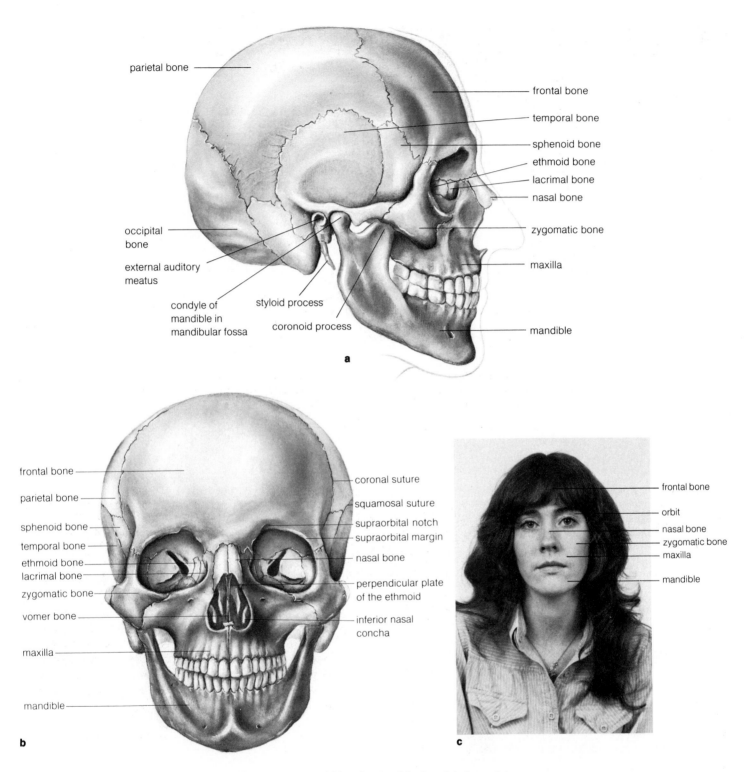

parietal bone

frontal bone

temporal bone

sphenoid bone

ethmoid bone

lacrimal bone

nasal bone

zygomatic bone

maxilla

occipital bone

external auditory meatus

condyle of mandible in mandibular fossa

styloid process

coronoid process

mandible

a

frontal bone

parietal bone

sphenoid bone

temporal bone

ethmoid bone

lacrimal bone

zygomatic bone

vomer bone

maxilla

mandible

coronal suture

squamosal suture

supraorbital notch

supraorbital margin

nasal bone

perpendicular plate of the ethmoid

inferior nasal concha

b

frontal bone

orbit

nasal bone

zygomatic bone

maxilla

mandible

c

Figure 7-2 The adult skull: (**a**) lateral aspect, (**b**) anterior aspect. (**c**) Landmarks of the face (photograph by R. Ross).

Table 7-1 Bones of the Cranium

Name and Number	Description	Special Characteristics
Ethmoid, 1	Perpendicular plate contributes to nasal septum; lateral plates form part of lateral walls of nasal cavity.	Contains air cells. Cribriform plate contains openings for nerve filaments.
Frontal, 1	Vertical plate forms forehead; orbital plates form roofs of orbits.	Contains frontal sinuses and supra-orbital foramina.
Occipital, 1	Single flat bone forms posterior portion of cranium and part of floor of cranial cavity.	Contains foramina (foramen magnum, and medial walls of jugular foramen and foramen lacerum) and processes (occipital condyles).
Parietal, 2	Paired flat bones contribute to sides and roof of cranial cavity.	Articulate at sagittal, coronal, lambdoidal, and squamosal sutures.
Sphenoid, 1	Complex bone positioned in midline of floor of skull. Body contains sella turcica.	Contains sphenoid sinus, processes (greater wings, lesser wings, medial pterygoid plates, lateral pterygoid plates), foramina (optic, ovale, rotundum), and the superior orbital fissure.
Temporal, 2	Paired flat bones contribute to sides and floor of cranial cavity; houses middle and inner ears.	Each has four parts (squamous, petrous, tympanic, mastoid) and has processes (mastoid, styloid, zygomatic) and mastoid air cells.

Parietal Bones The **parietal bones** are paired flat bones that meet each other in the midline at the *sagittal suture* to form most of the roof of the skull. They articulate posteriorly with the *occipital bone* at the *lambdoidal suture* and anteriorly with the *frontal bone* at the *coronal suture*.

Occipital Bone The **occipital bone** (Fig. 7-3) forms the posterior portion of the cranium and part of the floor of the cranial cavity. It contains the large *foramen magnum,* through which the spinal cord enters the skull. On each side of the foramen magnum are the *occipital condyles,* which articulate with the first cervical vertebra. The occipital bone forms the medial walls of the two *jugular foramina.* An internal *jugular vein* and three cranial nerves pass through each jugular foramen. The occipital bone also forms the medial walls of the *foramina lacerum,* at each of which an internal *carotid artery* enters the cranial cavity. The *hypoglossal canal* is an opening superior to the occipital condyles, through which the *hypoglossal nerve* passes.

In *tomography,* the x-ray film and the x-ray tube move in synchrony and in opposite directions in order to blur structures above and below the desired level. Thus, only structures in a narrow plane are in focus. In *computerized axial tomography (CAT Scan),* all structures except those in a narrow plane are completely eliminated mathematically. The CAT Scan is a computerized reconstruction of a series of x rays of a specific plane of the body. The resulting image is displayed as a gray scale with light areas indicating the most dense structures and darker shades of gray indicating structures that are progressively less dense. Black areas are the least dense and indicate fat or air. CAT Scans may be enhanced by the injection of radio opaque material prior to the scan.

Ethmoid Bone The **ethmoid bone** (Fig. 7-4) consists of three vertical plates: a *perpendicular plate* forms much of the

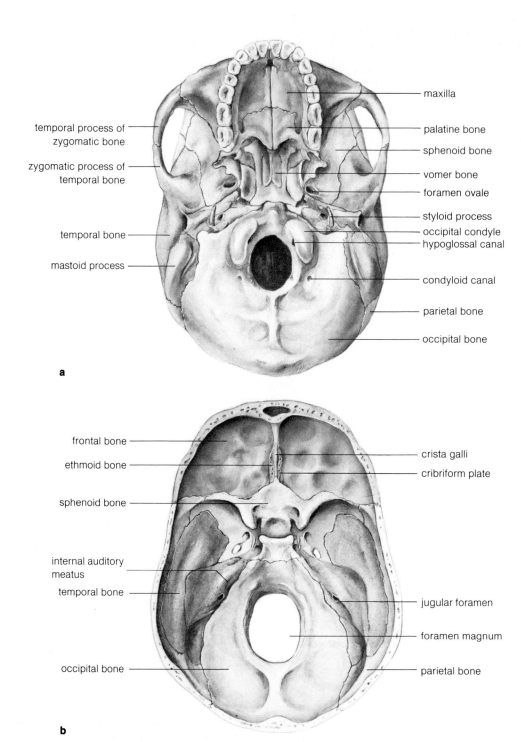

temporal process of
zygomatic bone

zygomatic process of
temporal bone

temporal bone

mastoid process

maxilla

palatine bone

sphenoid bone

vomer bone

foramen ovale

styloid process

occipital condyle

hypoglossal canal

condyloid canal

parietal bone

occipital bone

a

frontal bone

ethmoid bone

sphenoid bone

internal auditory
meatus

temporal bone

occipital bone

crista galli

cribriform plate

jugular foramen

foramen magnum

parietal bone

b

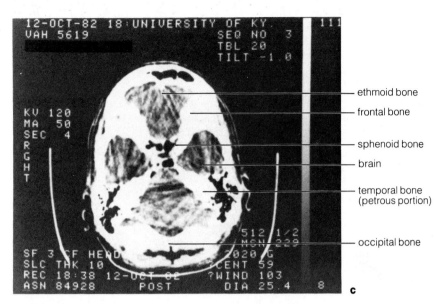

Figure 7-3 *(Facing page and above.)* The adult skull. (**a**) Inferior view of the base of the skull. (**b**) Internal aspect of the base of the skull. (**c**) CAT scan of the base of the skull (courtesy of S. J. Goldstein).

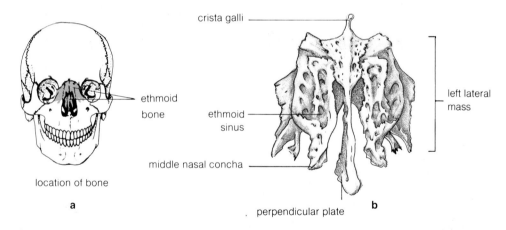

Figure 7-4 The ethmoid bone. (**a**) Anterior view of the skull showing the location of the ethmoid bone. (**b**) Enlarged anterior view of the ethmoid bone.

nasal septum and two *lateral plates* form part of the *lateral walls* of the *nasal cavity* and part of the *medial walls* of the *orbits.* The lateral plates contain numerous *ethmoid air cells,* which communicate with the nasal cavity through several small openings. Superiorly, the ethmoid forms part of the anterior cranial fossa (the anterior part of the cranial floor). This part of the ethmoid consists of the *cribriform plate,* which has numerous openings (cribriform means "sievelike") that transmit filaments of the olfactory nerves, and the *crista galli,* a projection to which the *falx cerebri* of the *cranial dura mater* attaches.

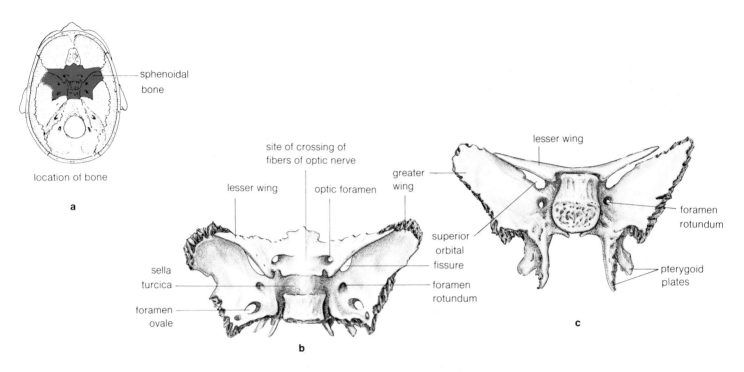

Figure 7-5 The sphenoid bone. (**a**) Superior view of the internal floor of the skull showing the location of the sphenoid bone. (**b**) Enlarged superior view of the sphenoid bone. (**c**) Enlarged posterior view of the sphenoid bone.

Sphenoid Bone The **sphenoid** (Fig. 7-5) is a complex bone positioned in the midline of the floor of the skull. The *body* of the sphenoid, which contains a depression on its superior surface called the *sella turcica,* extends anteriorly to the cribriform plate of the ethmoid. The *sphenoid sinus,* which communicates with the nasal cavity, is also within the body of the sphenoid.

Projecting from the body of the sphenoid are four paired processes: the lesser wings, the greater wings, the medial pterygoid plates, and the lateral pterygoid plates. The *lesser wings* project anteriorly to form part of the roof of the orbit; each contains an *optic foramen,* through which the optic nerve and ophthalmic artery pass to enter the orbit. The foramen *ovale* and foramen *rotundum,* through which nerves pass, and a *foramen spinosum,* which transmits the middle meningeal artery, are in the greater wings of the sphenoid. On the inside of the cranial cavity, the lesser wings form the posterior boundary of the *anterior cranial fossa.* Inferior and posterior to the lesser wing, the *greater wing* forms part of the lateral wall of the orbit and meets the parietal bone on the lateral surface of the cranium. The *superior orbital fissure,* through which several cranial nerves enter the orbit, separates the greater and lesser wings. The *medial pterygoid plates,* which support the soft palate, and the *lateral*

pterygoid plates, which serve for the attachment of the medial and lateral pterygoid muscles (for chewing), project inferiorly at the bases of the greater wings.

> The study of the sella turcica in x-ray films is important because the pituitary gland, located in the sella turcica, can give rise to tumors that can cause erosion of bone in the area. Moreover, the internal carotid artery passes nearby, and a dilation (aneurysm) of this vessel can also erode the bone. Bone erosion or resorption can usually be detected on x-ray films.

Posterior to the ethmoid and the frontal bones the lesser wing of the sphenoid forms a ledge extending laterally from its most posterior projection, the *anterior clinoid processes.* The depressed area posterior to the anterior clinoid processes is the *hypophyseal fossa,* which houses the pituitary gland. The little shelf anterior to the hypophyseal fossa and between the two

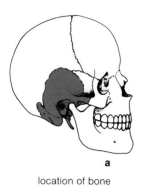

a

location of bone

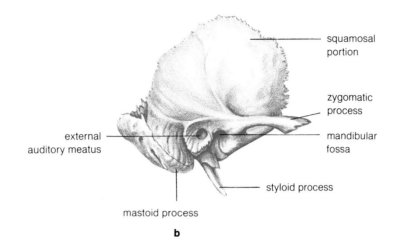

squamosal
portion

zygomatic
process

mandibular
fossa

styloid process

external
auditory meatus

mastoid process

b

Figure 7-6   The temporal bone. (**a**) Location of bone in skull. (**b**) Lateral view of the external surface of the left temporal bone.

anterior clinoid processes is the *tuberculum sellae;* the projection behind the hypophyseal fossa that resembles the back of a saddle is the *dorsum sellae,* from which the *posterior clinoid processes* project superiorly on each side. These parts, which resemble the front, seat, and back of a saddle, make up the *sella turcica,* which means "Turkish saddle."

Temporal Bones   The temporal bones (Fig. 7-6) are paired bones, each consisting of four parts fused into a single bone. The *squamous portion,* with its *zygomatic process,* is thin and flat and contributes to the sides of the brain case inferior to the parietals; inferiorly, it forms the *mandibular fossa* that articulates with the mandibular condyle. The *petrous portion* of the temporal bone forms a prominent ridge in the floor of the cranial cavity; it contains the *carotid canal,* through which the internal carotid artery passes, and the *internal auditory meatus,* which leads into the middle and inner ears. The external opening of this passage is the *external auditory meatus.* The *mastoid process,* a bony prominence posterior to the external auditory meatus, serves as an attachment for the sternomastoid muscle and contains the *mastoid air cells,* which communicate with the middle ear cavity. The *tympanic portion* of the temporal bone

forms the boundaries of the external auditory meatus and bears the *styloid process,* which projects ventrally. The *stylomastoid foramen,* which transmits the facial nerve, is located between the *styloid* and *mastoid processes.* The tympanic portion of the temporal bone forms the posterior boundary of the mandibular fossa but does not articulate with the mandible.

Due to the continuity of the middle ear cavity and mastoid air cells, infections of the middle ear cavity may spread to the mastoid air cells. Such infections usually originate in the pharynx as tonsillitis and spread to the middle ear via the auditory (Eustachian) tube. Infection of the mastoid air cells, called mastoiditis, was frequently fatal before the discovery of antibiotics. It is still a dangerous condition when the offending bacteria are resistant to antibiotics.

The bones of the skull include the three *ear ossicles* (malleus, incus, and stapes) found in the middle ear cavity. These will be described in the chapter on the sense organs.

Table 7-2 Bones of the Face		
Name and Number	Description	Special Characteristics
Inferior nasal conchae, 2	Form curved ledges on inferior portion of lateral walls of nasal cavity.	Articulate firmly with maxilla.
Lacrimals, 2	Small rectangular bones in medial wall of the orbits.	Form part of groove housing lacrimal sac of the nasolacrimal duct.
Mandible, 1	Horseshoe-shaped body that forms the lower jaw and bears teeth; has two rami with processes that articulate with cranium.	Has processes (condyloid and coronoid), foramina (mandibular and mental), and ridge (mylohyoid line).
Maxillae, 2	Fused in anterior midline to form upper jaw. Bear upper teeth.	Has processes (alveolar, frontal, palatine, zygomatic), foramen (infraorbital), and sinus (maxillary).
Nasals, 2	Narrow, flat, rectangular bones that form bridge of nose.	
Palatines, 2	Horizontal plates form posterior part of hard palate. Vertical plates contribute to lateral wall of nasal cavity.	Have processes (orbital, pyramidal, sphenoid).
Vomer, 1	Has flared superior portion, tapered inferior portion. Forms posterior part of nasal septum.	
Zygomatic, 2	Forms prominences of cheeks in lateral face. Temporal processes articulate with zygomatic processes of the temporal bones, forming the zygomatic arches.	Have processes (frontal, maxillary, orbital, temporal) and foramina (zygomaticofacial, zygomaticotemporal).

The Face

The face is formed by six paired and two unpaired bones (Table 7-2). The six paired bones are the **maxillae, zygomatic bones, lacrimal bones, inferior nasal conchae, nasals,** and **palatine** bones; the two unpaired bones are the **mandible** and the **vomer.** The facial bones provide support and protection for the openings of the respiratory and digestive tracts and house the special sense organs (olfactory epithelium, eyeballs, taste buds). These bones also serve as points of muscle attachments for most of the muscles of the head.

Mandible   The mandible (Fig. 7-7), which contains the lower teeth, is the lower skeletal support for the mouth and is the only movable bone of the skull. It consists of a horseshoe-shaped horizontal portion called the *body* and two vertical extensions from each posterior surface of the body, termed *rami.* Each ramus has two processes: the coronoid and the condyloid. The *coronoid process,* to which the temporalis muscle is attached, projects superiorly from the anterior border of the ramus; the *condyloid process,* which articulates with the temporal bone, extends superiorly from the posterior border of the ramus. Movement of the condyloid process as the mouth opens may be palpated by placing a finger in front of the ear and below the zygomatic arch.

The area of junction between the ramus and the body is termed the *angle* of the mandible, which may be palpated on your jaw. There is a faint ridge in the midline of the external surface of the body, called the *symphysis,* which is wider at the inferior border of the mandible where it forms the *mental protuberance.* A small *mental foramen* located on either side in the external surface of the body of the mandible below the premolar tooth transmits nerves and blood vessels to the lower lip and adjacent gingiva (gums).

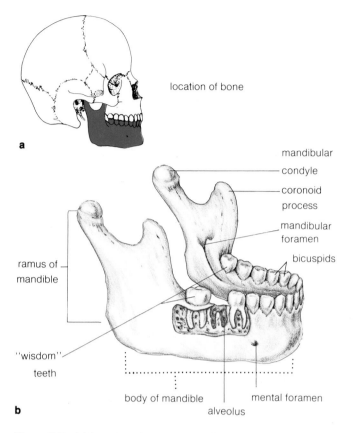

Figure 7-7 (**a**) Location of the mandible. (**b**) Lateral view of the mandible.

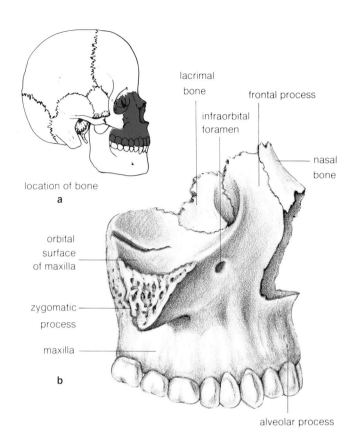

Figure 7-8 (**a**) Shaded area of skull shows location of maxilla, lacrimal, and nasal bones. (**b**) Lateral view of these bones.

During its development the mandible may not grow properly in relation to the maxillae; it may project too far anteriorly or not far enough. In either case there may be not only a cosmetic problem but also a problem of malocclusion (the teeth do not come together properly while chewing). Such problems can be corrected surgically by transecting the rami of the mandible and shifting the body of the mandible either forward or backward to establish good occlusion and improve facial features.

On the internal surface of the body of the mandible an oblique ridge, the *mylohyoid line,* extends from the lower portion of the symphysis backward and upward to the ramus. The *mandibular foramen,* which transmits the nerves and blood vessels supplying the lower teeth, is located on the inner surface of the ramus near its center. The mandible articulates with the squamous portion of the temporal bones by means of the rounded head of the condyloid processes.

Maxillae   The maxillae (Fig. 7-8) are joined by a suture to form the upper jaw which bears the upper teeth. Each maxilla consists of a body with four processes (frontal, alveolar, zygomatic, and palatine). The *frontal process* articulates with the frontal bone and contributes to the medial wall of the orbit and the lateral wall of the nasal cavity. The *alveolar process* provides the bony sockets for the upper teeth. The *zygomatic process* articulates with the zygomatic bone of the cheek and contributes to the contours of the cheek. The *palatine processes* fuse with one another in the midline to form most of the hard palate of the roof of the oral cavity.

The *body* of the maxilla contains the large *maxillary sinus* which communicates with the middle meatus of the nasal cavity.

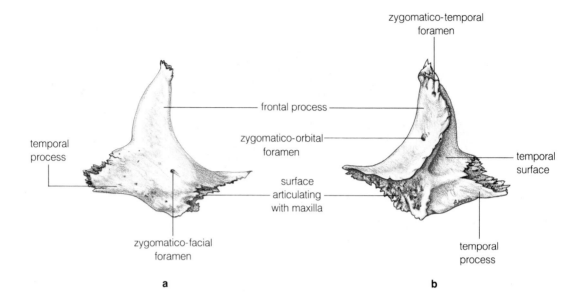

Figure 7-9 Right zygomatic bone. (**a**) Lateral aspect. (**b**) Medial aspect.

zygomatico-temporal foramen

frontal process

temporal process

zygomatico-orbital foramen

surface articulating with maxilla

temporal surface

zygomatico-facial foramen

temporal process

a

b

The maxilla forms the floor of the nasal cavity and part of the medial-anterior wall of the orbit. Just below the orbit, the *infraorbital foramen,* which transmits nerves and blood vessels to the face in the region of the upper lip, penetrates the body of the maxilla.

> When the teeth are extracted, the bone of the alveolar processes of the maxillae and mandible are resorbed. As a result these bones become smaller, changing the shape of the face. The height of the lower face is reduced, producing wrinkles. Dentures, if properly made, can compensate for these changes, minimizing the changes in the contours of the face.

Zygomatic Bones   The zygomatic bones (Fig. 7-9) form the prominence of the cheeks in the upper lateral face. Each zygomatic bone articulates with four other bones by means of processes named according to the bones with which they articulate: the *temporal,* the *maxillary,* the *frontal,* and the *orbital.* The zygomatic process of the temporal bone and the *temporal process* of the zygomatic bone form the zygomatic arch. The

zygomatic bone contributes about one-third of the rim of the orbit along the ventro-lateral border (see Fig. 7-10). This bone contains two small foramina, the *zygomaticofacial* and the *zygomaticotemporal,* which transmit nerves and blood vessels to the area.

Nasal Bones   The nasal bones (Fig. 7-11) are narrow, flat, almost rectangular bones that form much of the bridge of the nose. They articulate with one another in the midline (Fig. 7-10), with the frontal bone and ethmoid posteriorly, with the maxilla laterally, and with the nasal cartilage anteriorly.

Lacrimal Bones   The **lacrimal bones** (Fig. 7-12) are small rectangular bones in the medial walls of the orbits (Fig. 7-10), named for the lacrimal glands. The central portion of each lacrimal bone contains a vertical ridge, the *posterior lacrimal crest,* that forms the posterior boundary of a groove containing the lacrimal sac, an expanded portion of the nasolacrimal duct. This groove, called the *lacrimal sulcus,* is bounded anteriorly by the maxillary bone; it opens inferiorly, providing a pathway for the *nasolacrimal duct* through which tears drain into the inferior meatus of the nasal cavity. The lower anterior portion of the lacrimal bone ends as the *lacrimal hamulus,* a small, hooklike projection.

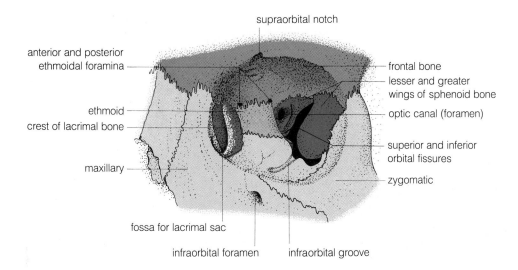

anterior and posterior ethmoidal foramina

supraorbital notch

frontal bone

lesser and greater wings of sphenoid bone

ethmoid

crest of lacrimal bone

optic canal (foramen)

maxillary

superior and inferior orbital fissures

zygomatic

fossa for lacrimal sac

infraorbital foramen

infraorbital groove

Figure 7-10 Anterior view of the right orbit.

Palatine Bones The palatine bones (Fig. 7-13) lie posterior to the maxillae and contribute to the wall of the nasal cavity, the floors of the orbits, and the roof of the mouth. Each palatine bone consists of a *horizontal* and a vertical *plate*. The horizontal plate forms the posterior portion of the hard palate (Fig. 7-14) and contains the *greater* and *lesser palatine foramina* that transmit nerves and blood vessels; the *vertical plate* contributes to the posterior part of the lateral wall of the nasal cavity. In addition to the vertical and horizontal plates three processes add to the irregularity of the bone. The *orbital process* is directed superiorly and laterally and forms part of the floor of the orbit. The *sphenoid process* extends superiorly and medially to articulate with the sphenoid bone. The *pyramidal process* projects laterally and posteriorly from the point of junction with the vertical and horizontal plates.

Vomer The vomer (Fig. 7-15), meaning plowshare, is an unpaired bone that forms the posterior, inferior portion of the nasal septum. It is an irregular bone with a flared superior portion that articulates with the sphenoid (Fig. 7-14) and ethmoid bones and a narrow, tapered part that contributes to the posterior part of the nasal septum and joins the superior surface of the hard palate.

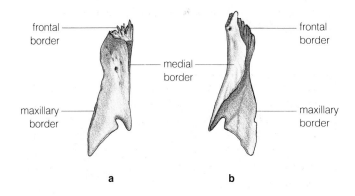

frontal border

medial border

frontal border

maxillary border

maxillary border

a

b

Figure 7-11 Right nasal bone. (**a**) External aspect. (**b**) Internal aspect.

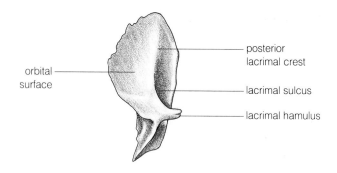

Figure 7-12 Right lacrimal bone. External aspect.

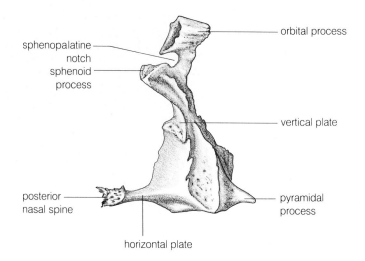

Figure 7-13 Right palatine bone. Posterior aspect.

Inferior Nasal Conchae The inferior nasal conchae form the inferior portion of the lateral walls of the nasal cavity, each projecting as a curved ledge into the cavity. Each inferior nasal concha articulates with a maxilla and appears to be a process of that bone. In addition to the maxilla, the inferior nasal concha also articulates with the ethmoid, lacrimal, and palatine bones.

The Cranial Fossae

The floor of the cranial cavity (Fig. 7-15) has three depressions, the **cranial fossae,** known as the anterior, middle, and posterior fossae. These house three major divisions of the brain: the frontal lobes, the temporal lobes, and the cerebellum and brainstem.

The large frontal bone forms the forehead and has an orbital plate that turns inward, forming most of the roof of the orbit as well as part of the floor of the cranial cavity. This portion of the floor of the cranial cavity, composed of the frontal and parts of the ethmoid and sphenoid bones, is the *anterior cranial fossa,* whose posterior boundary is the anterior clinoid process of the sphenoid and the posterior border of the lesser wing of the sphenoid. The *middle cranial fossa* is the area of the cranial floor between the anterior clinoid process and the petrous portion of the temporal bone. The portion of the middle cranial

fossa formed by the sphenoid bone resembles a butterfly. The body of the butterfly represents the body of the sphenoid and its wings represent the greater wings of the bone. The *posterior fossa* lies posterior to the petrous portion of the temporal bone and the dorsum sellae of the sphenoid bone.

The Orbit

The orbit (Fig. 7-10) is a large fossa formed by portions of several bones that provides an incomplete protective housing for the eyeball. Each of three bones contributes approximately one-third of the rim of the orbit. The *frontal bone* forms the rim superiorly and contributes to the rim medially and laterally where it joins the maxilla and the zygomatic bone, respectively. The *zygomatic bone* and the *maxilla* complete the formation of the rim of the orbit, uniting in the inferior rim. A slit, called the *superior orbital fissure,* in the posterior wall of the orbit between the two portions of the sphenoid bone, transmits several nerves to the orbit. Like the ethmoid, the sphenoid is a very complicated, irregular bone with several important flat plates extending out from the main body. Within the orbit, the *greater wing* of the *sphenoid* is lateral and the *lesser wing* of the *sphenoid* is medial to the superior orbital fissure. The *optic foramen,* through which the optic nerve (responsible for vision) passes,

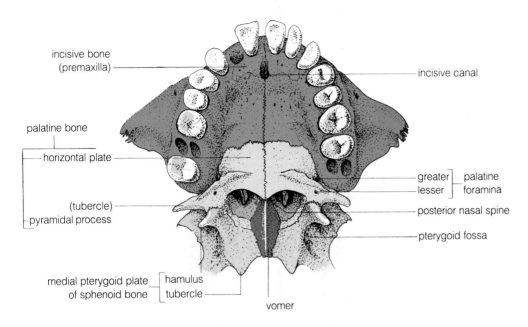

Figure 7-14 The hard palate. Inferior aspect.

incisive bone (premaxilla)

incisive canal

palatine bone

horizontal plate

greater / lesser — palatine foramina

posterior nasal spine

(tubercle)

pyramidal process

pterygoid fossa

medial pterygoid plate of sphenoid bone — hamulus / tubercle

vomer

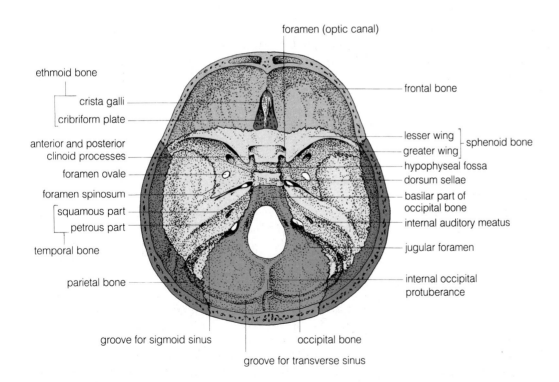

Figure 7-15 The cranial fossae. Internal aspect of the floor of the cranial cavity.

foramen (optic canal)

ethmoid bone

crista galli

cribriform plate

frontal bone

lesser wing / greater wing — sphenoid bone

hypophyseal fossa

anterior and posterior clinoid processes

foramen ovale

dorsum sellae

foramen spinosum

basilar part of occipital bone

squamous part

petrous part

internal auditory meatus

temporal bone

jugular foramen

parietal bone

internal occipital protuberance

groove for sigmoid sinus

occipital bone

groove for transverse sinus

Table 7-3 Summary of Openings of the Skull

Opening	Location	Structures Transmitted
Carotid canal	Petrous temporal	Internal carotid artery
Greater palatine foramen	Horizontal plate of palatine	Greater (anterior) palatine nerve and descending palatine vessels
Hypoglossal canal	Superior to occipital condyle	Hypoglossal nerve
Inferior orbital fissure	Between maxilla and greater wing of sphenoid	Maxillary nerve and infraorbital vessels
Infraorbital foramen	Maxilla below rim of orbit	Infraorbital nerve and vessels
Jugular foramen	Between petrous temporal and occipital, lateral to hypoglossal canal	Internal jugular vein; glossopharyngeal, vagus, and spinal accessory nerves
Foramen lacerum	Between sphenoid, petrous temporal, and occipital	Internal carotid artery
Lesser palatine	Horizontal plate of palatine	Lesser (posterior) palatine nerves and branches of descending palatine vessels
Foramen magnum	Occipital	Spinal cord, vertebral arteries, spinal accessory nerve
Mandibular foramen	Medial surface of ramus of mandible	Inferior alveolar nerve and vessels
Mental foramen	Lateral surface of mandible below second premolar	Mental nerve and vessels
Nasolacrimal canal	Between lacrimal, frontal, and maxilla	Nasolacrimal duct
Optic foramen or canal	Lesser wing of sphenoid	Optic nerve and ophthalmic artery
Foramen ovale	Posterior edge of greater wing of sphenoid	Mandibular nerve
Foramen rotundum	Medial edge of greater wing of sphenoid	Maxillary nerve
Foramen spinosum	Posterior edge of greater wing of sphenoid	Middle meningeal artery
Stylomastoid foramen	Between styloid and mastoid processes of temporal	Facial nerve
Superior orbital fissure	Between greater and lesser wings of sphenoid	Nerves: oculomotor, trochlear, abducens, frontal, lacrimal, and nasociliary; ophthalmic vein
Supraorbital foramen (or notch)	Frontal bone above rim of orbit	Supraorbital nerve and vessels
Zygomaticofacial foramen	Zygomatic bone, superficial surface	Zygomaticofacial nerve and vessels
Zygomaticotemporal foramen	Zygomatic bone, medial to frontal process	Zygomaticotemporal nerve and vessels

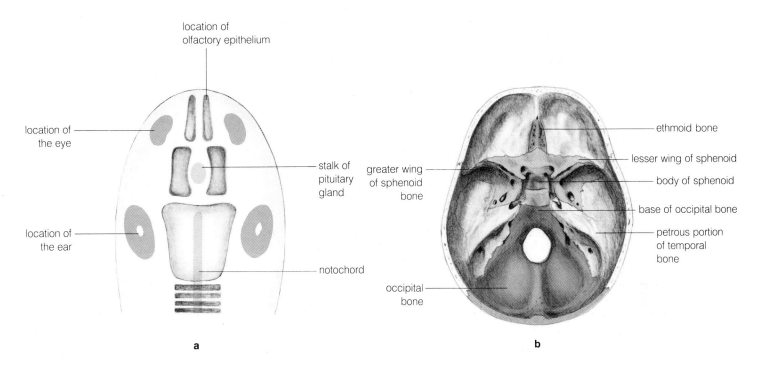

location of
olfactory epithelium

location of
the eye

stalk of
pituitary
gland

location of
the ear

notochord

greater wing
of sphenoid
bone

ethmoid bone

lesser wing of sphenoid

body of sphenoid

base of occipital bone

petrous portion
of temporal
bone

occipital
bone

a

b

Figure 7-16 Diagram illustrating the primordia of the bones of the base of the skull. (**a**) Cartilages that contribute to the development of the skull. (**b**) Derivatives of components shown in a.

is in the lesser wing. The *inferior orbital fissure,* which transmits nerves and blood vessels, lies between the bone of the floor of the orbit (maxilla) and that of the lateral wall (greater wing of the sphenoid). Table 7-3 contains a summary of the openings in the skull.

> The medial and inferior walls of the orbit are extremely thin and fragile. The sudden pressure of a blow striking the eyeball can be transmitted to the walls of the orbit, causing a "blowout" fracture. If the fracture is of the medial wall, the ethmoid air cells will be penetrated. A fracture of the inferior wall will usually involve the maxillary sinus.

Development of the Skull

As noted in the preceding chapter, the bones of the cranial floor are produced by ossification in a cartilage model (endochondral ossification), whereas those of the face and upper cranium are laid down directly in embryonic membranes (intramembranous ossification). The floor of the skull develops initially around the notochord, pituitary gland, and paired sense organs (nose, eyes, and ears). The earliest cartilages form as separate structures, expand, and ultimately fuse into a single plate (Fig. 7-16). Ossification centers within this cartilage plate give rise to the individual bones of the base of the skull, with most adult bones resulting from the fusion of two or more ossification centers. The membranous bones covering the dorsal and lateral brain surfaces develop later.

Development of the Face   The development of the bones of the face depends upon differentiation of mesenchyme around the nose and the embryonic pharynx. A lateral view of the embryo (Fig. 7-17) reveals a series of bulges in the ventrolateral aspect of the pharyngeal region that are due to rodlike accumulations of mesenchyme between the ectoderm and endoderm. These accumulations are referred to as **visceral arches** (or branchial arches), and the spaces they enclose, lined with endodermal extensions of the lining of the pharynx, are called **visceral**

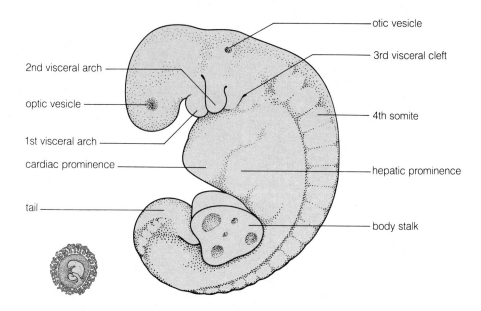

Figure 7-17 Lateral view of a human embryo at about four weeks of development (courtesy of Joan Creager). Note the visceral (branchial) arches in the future neck region.

Labels: otic vesicle, 3rd visceral cleft, 4th somite, hepatic prominence, body stalk, 2nd visceral arch, optic vesicle, 1st visceral arch, cardiac prominence, tail

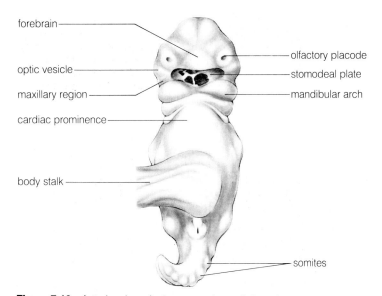

Figure 7-18 Anterior view of a human embryo of about four weeks of age.

Labels: forebrain, olfactory placode, optic vesicle, stomodeal plate, maxillary region, mandibular arch, cardiac prominence, body stalk, somites

pouches. The arches extend ventrally toward the midline (Fig. 7-18), giving rise to skeletal muscles associated with head and neck structures and contributing to the skeletal elements of the face and palate. Ventrally, at the cranial end of the gut tube, the endoderm lining the gut lies next to the surface ectoderm with no intervening mesoderm, forming a special plate termed the *oral membrane* that marks the future mouth opening. The lower (mandibular) portion of the first visceral arch, called the *mandibular process,* fuses with the corresponding structure on the other side to form the lower jaw. The upper (maxillary) portion of the first arch forms a maxillary process lateral to the oral opening and is destined to become the lateral wall of the future oral cavity.

Dorsally the roof of the mouth is thickened by mesenchyme, the frontal prominence (Fig. 7-19a). Ectodermal thickenings called olfactory placodes, which contain cells that will differentiate into the neurons of the olfactory epithelium, appear bilaterally above the maxillary processes. The *olfactory placodes* soon invaginate, forming *olfactory pits* that open inferiorly into the oral cavity (Fig. 7-19b). The accumulations of mesenchyme around the pits are called *nasal processes,* medial and lateral relative to the olfactory pits. The medial nasal process grows inferiorly and laterally to meet the maxillary process. When these two processes fuse (Fig. 7-19d), the olfactory pits become tubes that open cranially to the outside and caudally into the oral cavity.

The extension of the medial nasal processes beneath the olfactory pits (Fig. 7-19c) produces a small part of the upper jaw, called the *premaxilla,* in which the incisors will form. Externally the maxillary process fuses with the lateral nasal process; it also fuses with the medial nasal process at the edge of the median groove of the upper lip (Fig. 7-19d). Failure of the maxillary processes to fuse with the median nasal process leads to *cleft lip,* which may be bilateral or unilateral and is frequently associated with *cleft palate.*

Formation of the Palate Separation of the nasal and oral cavities is completed by the formation of the palate, which begins with the growth of *palatal processes* from the maxillary processes (Fig. 7-20). At first the palatal processes project downward toward

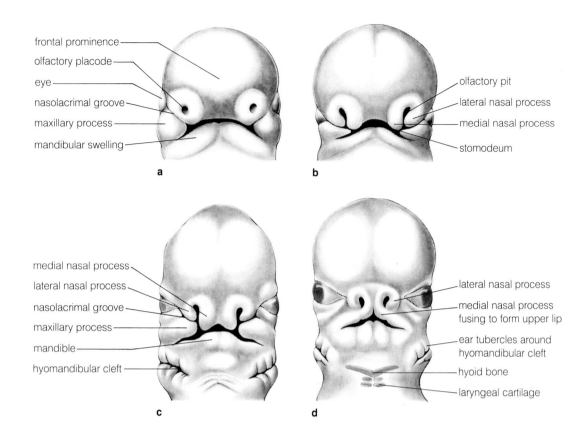

frontal prominence
olfactory placode
eye
nasolacrimal groove
maxillary process
mandibular swelling

a

olfactory pit
lateral nasal process
medial nasal process
stomodeum

b

medial nasal process
lateral nasal process
nasolacrimal groove
maxillary process
mandible
hyomandibular cleft

c

lateral nasal process
medial nasal process
fusing to form upper lip
ear tubercles around
hyomandibular cleft
hyoid bone
laryngeal cartilage

d

Figure 7-19 Anterior view of four stages in the development of the face: (**a**) 5-week embryo, (**b**) 5½-week embryo, (**c**) 6-week embryo, (**d**) 7-week embryo.

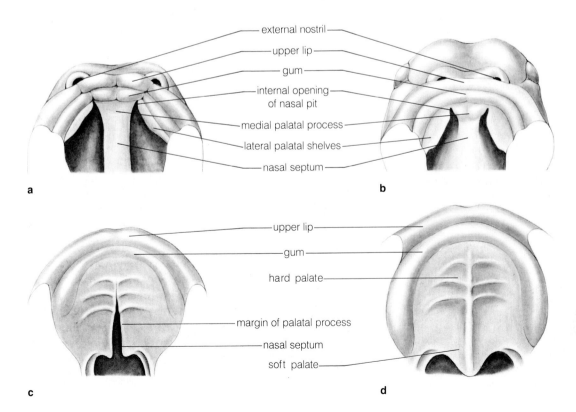

external nostril
upper lip
gum
internal opening
of nasal pit
medial palatal process
lateral palatal shelves
nasal septum

a

b

upper lip
gum
hard palate
margin of palatal process
nasal septum
soft palate

c

d

Figure 7-20 Inferior view of four stages in the development of the palate. (**a**) Early development of the lateral palatal shelves. (**b**) Early development of the nasal septum. (**c**) Fusion of the lateral palatal shelves in the midline, beginning anteriorly. (**d**) Fusion of lateral palatal shelves in the midline is complete.

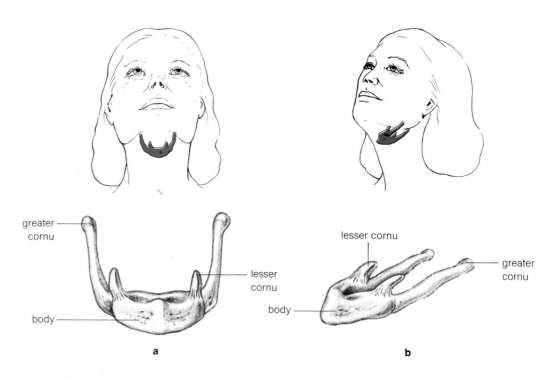

Figure 7-21 The hyoid bone: (**a**) anterior view, (**b**) oblique left lateral view.

the floor of the mouth (Fig. 7-20a), perhaps because of the oversized tongue that is present during early development. Later as the tongue assumes a more reasonable relative size, the palatal shelves project medially in a horizontal plane (Fig. 7-20b) and eventually meet in the midline to fuse with each other and with the septal process, which extends down from the frontal process. Fusion of the processes begins anteriorly (Fig. 7-20c) with the fusion of the maxillary process and frontonasal process (primitive palate) and spreads caudally until the nasal and oral cavities are completely separated (Fig. 7-20d).

THE HYOID BONE

The **hyoid bone** (Fig. 7-21) is a small, U-shaped, isolated bone that does not articulate directly with other bones. It is located in the upper anterior part of the neck where it serves as the point of attachment for muscles of the floor of the mouth and the tongue and for muscles (strap muscles of the neck) that extend inferior to this bone. It is suspended from the styloid

processes of the temporal bone by the stylohyoid ligaments. Some anatomists believe the hyoid bone articulates with the skull at the styloid process, but most list the hyoid as the only bone that does not articulate with another bone.

The hyoid bone consists of a *body,* directed anteriorly, from which project the *greater* and *lesser cornua* (sing., *cornu*), or horns. The greater cornu extends posteriorly from the lateral edge of the body and serves for the attachment of muscles. The lesser cornu, to which the stylohyoid ligament is attached, extends obliguely as it projects superiorly and posteriorly from the upper lateral border of the body.

THE VERTEBRAL COLUMN

The **vertebral column** (Fig. 7-22) consists of a series of irregularly shaped bones called vertebrae arranged one above another in a column extending from the skull to the lower end of the pelvis. It provides flexible support for the trunk and protection for the spinal cord. Each vertebra (Fig. 7-23a) has two parts: a

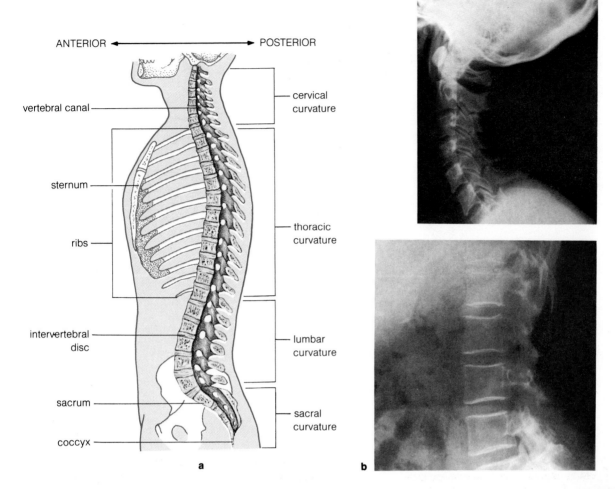

ANTERIOR ◀————————▶ POSTERIOR

vertebral canal

cervical
curvature

sternum

ribs

thoracic
curvature

intervertebral
disc

lumbar
curvature

sacrum

sacral
curvature

coccyx

a

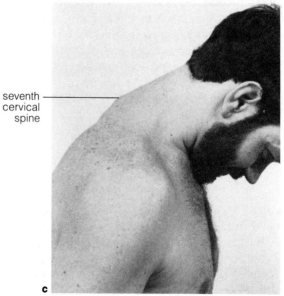

b

seventh
cervical
spine

c

Figure 7-22 The ribs, sternum, and vertebral column. (**a**) Lateral view showing individual bones and natural curvature of the vertebral column. (**b**) Lateral x-ray view of cervical (top) and lumber (bottom) vertebrae. (X-rays courtesy of Alexandria Hospital, Alexandria, Virginia, Joyce R. Isbel, R.T.) (**c**) Surface anatomy of the neck. The spine of the seventh cervical vertebra is usually the most prominent (photograph by R. Ross).

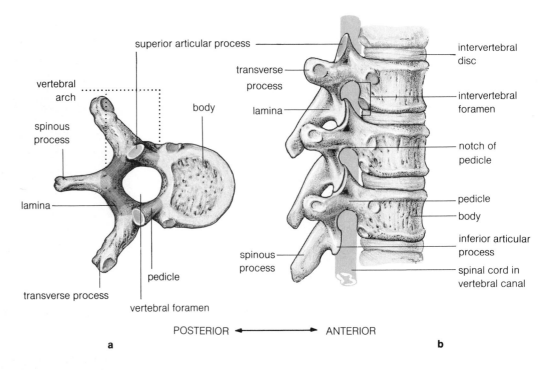

Figure 7-23 A typical vertebra: (**a**) superior view, (**b**) right lateral view.

vertebral body that bears most of the weight and a *neural arch* that forms part of the wall of the *vertebral canal* housing the spinal cord. The vertebrae articulate with one another by means of *facets* on the processes of the neural arches and by means of fibrocartilage **intervertebral discs** between the vertebral bodies (Fig. 7-23b). The intervertebral discs function as shock absorbers and allow slight movement so that the column is flexible and resilient. When the body is upright, its weight is distributed uniformly through the disc; but when the body bends, one part of the disc is compressed while the opposite side is under tension. Pressure upon the intervertebral disc is relieved only when the body is lying down. The vertebral column can support the pressure of maintaining posture not only because of the fibrous nature of the intervertebral disc but also because strong *anterior* and *posterior longitudinal ligaments* extend for the length of the column along the anterior and posterior surfaces of the vertebral bodies (Fig. 7-24). These collagenous ligaments attach to each vertebra and prevent excessive movement.

The neural, or vertebral, arch attaches on each side of the posterior aspect of the vertebral body, creating an opening called **the vertebral foramen.** When aligned within the vertebral column these foramina comprise the vertebral canal that is occupied by the spinal cord. Each neural arch bears a pair of

transverse processes and an unpaired posterior *spinous process* to which muscles are attached. In the thoracic (chest) region the transverse processes articulate with ribs. Each neural arch also bears a superior and an inferior pair of articular processes having facets by which the arch articulates with vertebrae above and below. These joints prevent excessive movement between adjacent vertebrae.

Nuclear magnetic resonance (NMR) imaging comes from the ability to make certain atomic nuclei resonate or oscillate in a magnetic field. When the nuclei are stimulated by signals from radiofrequency coils, they become "excited." As they relax, they release energy in the form of radio frequency signals, which are collected and fed to a computer for processing. The image is displayed electronically and the operator can look at any slice at any angle. NMR is perhaps the most exciting and most promising diagnostic tool thus far developed. It employs powerful magnetic and radiofrequency waves to produce details of the human body with no known risk to the patient or staff. It is superior to the CAT scan in its analysis of soft tissues.

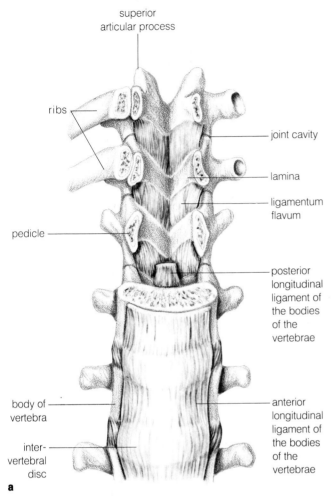

superior
articular process

ribs

joint cavity

lamina

ligamentum
flavum

pedicle

posterior
longitudinal
ligament of
the bodies
of the
vertebrae

body of
vertebra

anterior
longitudinal
ligament of
the bodies
of the
vertebrae

inter-
vertebral
disc

a

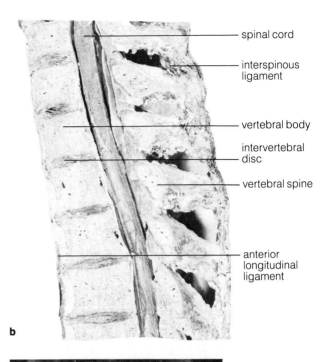

spinal cord

interspinous
ligament

vertebral body

intervertebral
disc

vertebral spine

anterior
longitudinal
ligament

b

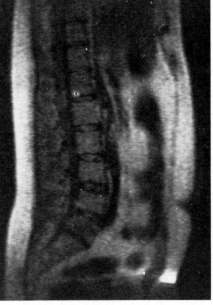

c

Figure 7-24 The vertebral column. (**a**) Anterior view of a partially dissected segment of the vertebral column, illustrating ligaments of the vertebral column. (**b**) Photograph of a sagittal section of a segment of vertebral column with the spinal cord in the vertebral canal (photograph by R. Ross). (**c**) Nuclear magnetic resonance (NMR) image of a sagittal section of a normal 35-year-old male (courtesy of Paul Wang). See adjacent box for explanation of technique.

Regions of the Vertebral Column

Individual vertebrae differ in size and shape, depending upon their location within the vertebral column; thus an isolated vertebra can be assigned to its correct region even though all vertebrae have the same general structure (Fig. 7-25). There are usually thirty-three bones in the vertebral column: seven **cerv-**ical, twelve **thoracic,** and five **lumbar vertebrae** that are separate bones, five vertebrae that are fused to form the **sacrum,** and four small bones that form the **coccyx** at the lower end of the vertebral column.

The vertebrae of the three upper regions are labeled C1 through C7 (cervical), T1 through T12 (thoracic), and L1

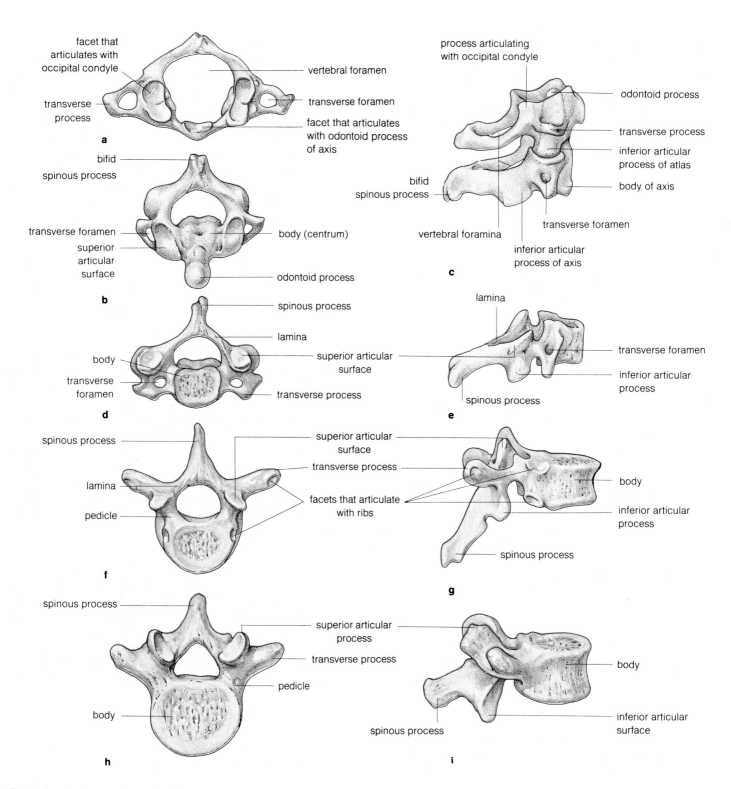

Figure 7-25 Characteristics of certain vertebrae: (**a**) atlas, (**b**) axis, (**c**) articulation of atlas and axis, (**d**) superior view of cervical vertebra, (**e**) lateral view of cervical vertebra, (**f**) superior view of thoracic vertebra, (**g**) lateral view of thoracic vertebra, (**h**) superior view of lumbar vertebra, (**i**) lateral view of lumbar vertebra.

through L-5 (lumbar). The vertebral column, when viewed from the side, normally has four curves that alternate in direction (Fig. 7-22); the cervical and lumbar regions are convex in relation to their anterior surfaces, and the thorax and sacrum are convex in relation to their posterior surfaces. The curvatures of the vertebral column are important in maintaining posture and in absorbing the shocks that result from activities such as walking or running.

The thoracic and sacral curvatures of the vertebral column originate during fetal development. Cervical curvature, related to holding the head erect, begins at about three months of age. Lumbar curvature appears as the infant begins to walk, at about twelve months.

The weight borne by vertebrae increases progressively from the upper cervical level to the sacrum; consequently the vertebral bodies and the intervertebral discs increase in size progressively from the cervical region through the lumbar region. The discs also vary in thickness in different regions of the vertebral column. The larger discs of the lumbar region permit flexion, extension, and side movements to occur relatively freely, but the articulations at the facets prevent rotation. In the thoracic region the intervertebral discs are relatively thin, so very little movement is possible between any two vertebrae; however, all types of movement including rotation are possible over the entire length of the region. Thus the thorax is more flexible than casual observation might suggest. The cervical region has thick intervertebral discs that permit more movement in all planes, including rotation, than is possible for the thorax.

Some individuals may have thirty-two or thirty-four vertebrae due to the development of an extra vertebra or the failure of one to form. This usually does not cause functional disability and is considered to be an anatomical variation rather than an abnormality. Such variation occurs most frequently in the coccyx where either three or five bones may develop instead of the usual four. In other regions of the column a vertebra that normally develops into one type may develop into another; for example an individual whose twelfth ribs fail to develop will have eleven thoracic vertebrae and six lumbar vertebrae, with no change in the total number.

Cervical Vertebrae The cervical vertebrae (Figs. 7-25a, b, d) are the smallest. The first and second vertebrae are modified for articulation with the skull: the first, the *atlas* (Fig. 7-25a), has no body or spine; the second, the *axis* (Fig. 7-25b), has an oval processs, the *dens,* that projects superiorly into the ring of the atlas enabling the head to rotate by movement at the joint between the atlas and the axis. Each of the short transverse processes of the cervical vertebrae (Fig. 7-25d) contains a foramen (the foramen transversarium); usually all of these foramina, except that of the seventh vertebra, transmit the vertebral artery as it passes on its way to the cranial cavity. The spines of the second through the sixth vertebrae are small bifid (forked), but the spine of the seventh is relatively long and projects posteriorly, causing a palpable prominence at the junction of the neck and thorax (Fig. 7-22). The most prominent spine, either C-7 or T-1, provides a useful landmark for determining vertebral levels for some procedures such as the anesthetization of the nerves that emerge in the root of the neck and supply limb structures.

Thoracic Vertebrae The thoracic vertebrae (Figs. 7-25f and g), intermediate in size between the cervical and lumbar, are easily distinguished by the facets on the sides of the vertebral bodies which articulate with the heads of the ribs. Also two transverse processes arise from the vertebral arch, and each articulates with a rib by means of a small concave facet near its lateral edge. The spine of the first thoracic vertebra is longer than that of other vertebrae of this region and projects almost horizontally. The spines become progressively shorter and thicker inferiorly so that the spine of the twelfth vertebra is a short, rectangular process resembling the spines of the lumbar vertebrae.

Lumbar Vertebrae The lumbar vertebrae (Figs. 7-25h and i), the largest of the flexible part of the vertebral column, may be recognized by the absence of facets for articulation with the ribs. The bodies are wider transversely than anteroposteriorly and are thicker anteriorly than posteriorly. The articular processes are well developed; the superior two project from the pedicles and the inferior two are attached to the vertebral arches at the laminae. The transverse processes and spines are short and thick for attachment of the muscles and ligaments that assist these vertebrae in supporting much of the weight of the body.

Sacrum The sacrum (Fig. 7-26) is a single bone formed by the fusion of five vertebrae during its development. It is concave anteriorly, increasing the size of the pelvic cavity, and convex posteriorly, where it serves for the attachment of muscles and ligaments. There are four transverse lines on the anterior surface of the sacrum that mark the divisions between the bodies

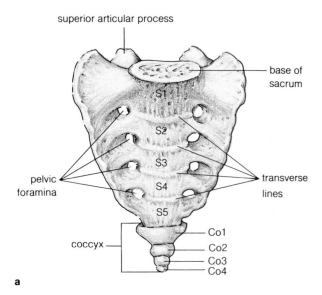

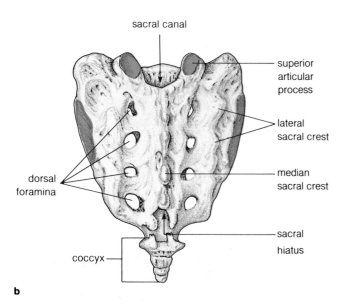

Figure 7-26 The sacrum and coccyx: (**a**) anterior view, (**b**) posterior view.

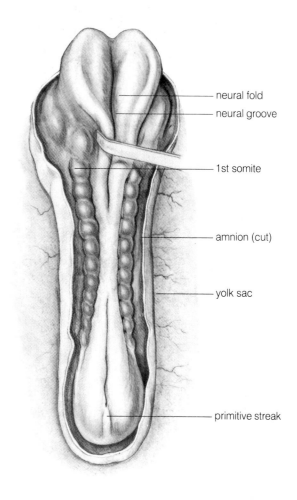

Figure 7-27 Dorsal view of an embryo, showing closure of the neural tube.

of the vertebrae. Posteriorly the tubercles on the *median sacral crest* represent rudimentary spines and indicate the middle portions of the vertebral arches dorsally. The *intermediate crests* lateral to the median crest are derived from the fusion of the articular processes; the lateral crests represent fused transverse processes. Inferiorly the last one or two vertebral arches fail to form dorsally, leaving the vertebral canal open at the *sacral hiatus,* which is used by anesthesiologists to inject drugs to block nerve impulses in nerve roots within the canal.

Coccyx The coccyx (Fig. 7-26) usually consists of four rudimentary bones but may have three or five. There is morphological evidence that each bone represents the body and the articular and transverse processes of a vertebra whose vertebral arch and spine have failed to develop. Though coccygeal bones usually fuse, the first bone may remain as a separate piece; the last three taper inferiorly and are usually fused into one piece. The coccyx is mainly a vestigial structure, but it does serve to anchor muscles of the pelvic floor.

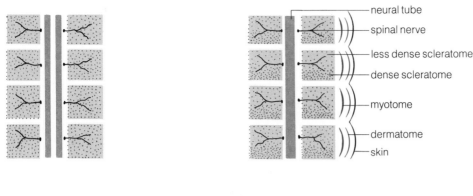

neural tube
spinal nerve
less dense scleratome
dense scleratome
myotome
dermatome
skin

a

b

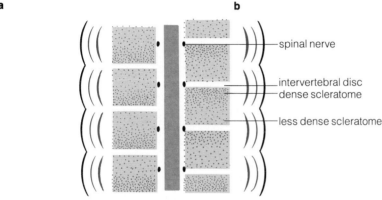

spinal nerve

intervertebral disc
dense scleratome

less dense scleratome

c

Figure 7-28 Differentiation of the sclerotome in the formation of the verte-brae. (**a**) Early somite stage. (**b**) Relations of parts of the somite. (**c**) Development of the vertebral body. Left side is earlier than the right side. (See text for description.)

Development of the Vertebral Column

Upon completion of the body wall by fusion ventrally the embryo has a dorsal tubular nervous system supported by a stiff rod of mesodermal cells, the notochord. At this stage the notochord is the embryo's only skeletal support.

Somites (Fig. 7-27), blocklike condensations of meso-derm, develop on each side of the neural tube. Formation of the somites begins first in the cervical region and spreads cra-nially and caudally as the embryo grows and as closure of the neural tube proceeds. Approximately thirty-eight somites are formed, with about four migrating into the head to contribute to the head mesenchyme.

The somites differentiate quickly into three separate com-ponents: a dorso-lateral portion, the *dermatome,* which gives rise to the dermis; a medial portion, the *myotome,* which forms most of the skeletal muscle of the body; and a ventro-medial portion, the *sclerotome,* which forms the supporting tissues,

including the skeletal system. The sclerotomes are the primor-dia of most of the bones of the axial skeleton.

The relations of the somites and the neural tube (Fig. 7-28a) may be visualized in a frontal section through the embryo in the somite stage of development. The sclerotome differen-tiates by the segregation of its cells into a dense caudal half and a less dense cranial half (Fig. 7-28b). The dense portion of each sclerotome fuses with the less dense portion of the sclerotome caudal to it, and a cleft appears between the newly created masses that establishes the position of the definitive vertebra (Fig. 7-28c). Intervertebral discs develop in the region of the clefts from mesenchyme of the dense portion of the sclerotome, and the notochord degenerates except in the middle of the intervertebral disc where it persists as the *nucleus pulposus.* Anterior fusion in the midline of the sclerotomes of the two sides converts the bilateral primordia into a single vertebral body.

Clinical Applications: **Abnormalities of the Vertebral Column**

Abnormalities of the vertebral column are important because any deformity, such as an abnormal curvature of the spine, may reduce the volume of the thorax or abdomen, causing respiratory, circulatory, or digestive problems. There is risk of pressure on the spinal cord or nerves that could cause pain or paralysis of muscles. The three main types of abnormal spinal curvature are *kyphosis, lordosis,* and *scoliosis.* Other abnormalities, such as *spina bifida,* may directly involve the nervous system.

Kyphosis

Kyphosis (Gr.) means hill or mountain, implying a hump on the back. This reduces the height of the person and may lead to pressure on the spinal cord or peripheral nerves that affects the functioning of the muscles of the limbs. The condition may be congenital, due to disturbances in growth or even absence of vertebral bodies during fetal development. Frequently, congenital abnormalities of the vertebral column are associated with defects of the spinal cord. In adolescent kyphosis, the vertebral bodies fail to grow, especially in the lower thoracic region. This is an inherited condition and it varies in its severity. In old age, osteoporosis may lead to severe kyphosis. Other conditions such as infections or tumors of the spine may produce similar effects.

Lordosis

The word lordosis is derived from the description of a person leaning backwards in a "lordly" fashion. In such a position the spine may become rigid and cause problems. If there is a thoracic kyphosis, there will be a compensatory lumbar and cervical lordosis, but this type of abnormal curvature is rare.

Scoliosis

Scoliosis (Gr.) means "worm-like" and suggests that the spine is bent from side to side. This defect is a combination of lordosis, lateral flexion, and rotation. It may be congenital, due to abnormal development of the vertebrae, or it may be the result of muscle imbalance or of nervous system disturbances. In many cases the cause of scoliosis cannot be determined. The abnormal spine curvature reduces the height of the person and puts abnormal pressure on soft organs of the thorax and abdomen. The muscle balance is disturbed so that one group of muscles must work harder than others and becomes fatigued more quickly. This fatigue may lead to local pain or even to exhaustion of the whole body. Thus, work that can be done by such an individual is much less than could be done by a normal person.

Spina Bifida

Spina bifida is a congenital defect of the vertebral column in which the neural arches either fail to develop or fail to fuse dorsally. This occurs most commonly in the lower lumbar region and may be visible at birth as a dimple or small tuft of hair at the site of the defect. If the spinal cord and its protective coverings, the meninges, remain in place, there is usually no loss

---

THE RIB CAGE AND STERNUM

The rib cage (Fig. 7-29) consists of the twelve pairs of ribs that articulate with the thoracic vertebrae posteriorly and with the sternum anteriorly.

Ribs

Of the twelve pairs of ribs only seven, the *true ribs,* articulate directly with the sternum by means of *costal cartilages.* The costal cartilages are rods of hyaline cartilage that serve to strengthen the ribs anteriorly and to increase their flexibility. The eighth, ninth, and tenth ribs, called false ribs, are attached to one another and to the seventh rib by way of their costal cartilages. The eleventh and twelfth ribs, also false ribs, are unattached anteriorly and are often called "floating ribs." These two ribs have short, conical costal cartilages at their free ends.

> Separation of a rib is the expression used to denote dislocation of the junction of the rib and its costal cartilage. A slipping rib means that a costal cartilage (eighth, ninth, or tenth) has separated from the costal cartilage above it and moved upward, overriding the cartilage above it. This can be very painful.

of function. A more serious condition occurs when the meninges of the spinal cord protrude as a fluid-filled sac through the bony defect, often covered only by very thin skin. This sac may leak fluid and lead to severe infection. There may be neurological signs of function changes in such cases. Even more serious is when the spinal cord herniates out of the vertebral canal into the fluid-filled sac. There may be abnormal pressure on the nerves arising from the involved part of the spinal cord, resulting in paralysis of muscles of the lower limbs and sensory loss on the skin. Frequently, there is loss of control of bladder and bowel function.

Back Pain

Back pain is one of the most common conditions requiring medical treatment; about 80% of all persons in the United States will experience back pain at some time during their lives. Most of the time there is recovery without specific treatment within a month or two. However, recurrences are frequent and inability to work makes these episodes very expensive. Perhaps as many persons lose time from work because of back problems as they do with cardiac problems. Unlike cardiac problems, chronic back disability cannot be diagnosed by definitive tests. Most of the back difficulties seem localized in the lumbar region of the spine and some may involve the intervertebral discs, most frequently between the fourth and fifth lumbar vertebrae. If the pain is caused by herniation of the intervertebral disc, the nucleus pulposus usually pushes through the posterolateral aspect of

the disc and puts pressure on the spinal nerves that exit from the vertebral canal at that level. Such a condition usually requires surgery to correct.

Most low back problems, however, appear to involve only soft tissues and show no abnormalities on x rays of the vertebral column. Soft tissue strain, such as that involving muscles or tendons and ligaments, is assumed to be the cause of any chronic back pain that cannot be explained by clinical or laboratory tests. Soft tissue strain usually brings on pain suddenly as the result of some sort of trauma. The trauma may be as simple an event as bending over to pick up a pencil. The pain is aggravated by activity and is usually lessened by rest. Any action, such as coughing or sneezing, that inceases stress on the back tends to intensify the pain. Soft tissue strain usually responds well to rest, and the symptoms may disappear in a week or two.

There are a number of factors that appear to predispose a person to soft tissue strain. Obesity and poor muscular tone increase the likelihood of low back problems by increasing the lumbar lordosis. This adds to the tension on the discs, ligaments, and muscles as an upright position is maintained. A daily program of postural exercises and the maintenance of proper weight will do much to prevent the recurrence of low back pain.

A typical rib (Fig. 7-30) consists of a *head, neck, tubercle,* and *body,* or *shaft.* The head of most ribs articulates with the bodies of two vertebrae by means of two facets separated by a horizontal crest. The neck is the flattened portion, about 2.5 cm. long, that extends between the head and the tubercle and by which the rib articulates with the transverse process of a vertebra. Each tubercle has an articular surface and a nonarticular portion to which a ligament is attached. The body of a typical rib is curved so that it slopes inferiorly before twisting to ascend anteriorly toward the sternum. The point of greatest curvature, called the *angle of the rib,* is marked by a line for the attachment of muscle (the iliocostalis part of the sacrospinalis). The internal surface of the body is marked by a *costal groove* for the nerves

and blood vessels supplying each intercostal space. The anterior or sternal end of each rib is flattened, presenting a porous surface for union with the costal cartilage.

Sternum

The sternum (Fig. 7-29) is an elongated, flattened bone located in the anterior midline of the wall of the thorax. It is composed of three parts: the manubrium, the body (gladiolus), and the xiphoid process. The manubrium, the superior portion, articulates with the body inferiorly and with the clavicles and first ribs laterally near its cranial border. The middle portion of its cranial border is concave and forms a depression called the

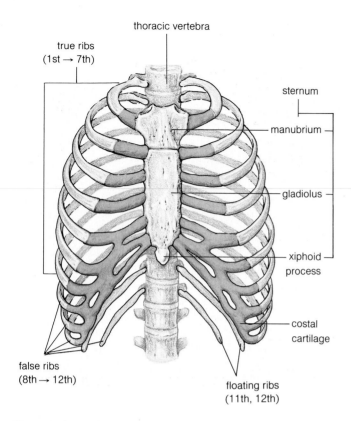

true ribs
(1st → 7th)

thoracic vertebra

sternum

manubrium

gladiolus

xiphoid
process

costal
cartilage

false ribs
(8th → 12th)

floating ribs
(11th, 12th)

Figure 7-29 The ribs and sternum.

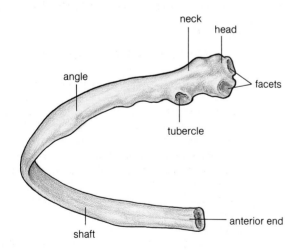

neck

head

angle

facets

tubercle

anterior end

shaft

Figure 7-30 Detailed view of a rib.

jugular notch. On either side there are oval surfaces, or notches, for articulation with the clavicles and first costal cartilages, with smaller articular facets on the sides inferiorly for articulation with the second costal cartilages as they join the sternum at the *sternomanubrial junction* (sternal angle). The body extends inferiorly from the sternal angle as a flat rectangular plate of bone that is thinner than the manubrium. Facets for articulation with the costal cartilages of the second through the seventh ribs are found along the lateral borders of the body as it narrows to join the xiphoid process at its inferior tip. The xiphoid process, the smallest of the three pieces of the sternum, is made of hyaline cartilage in youth, usually becoming somewhat ossified in older people.

> The second rib articulates with the sternum at the sternomanubrial junction (the sternal angle). This is an important landmark because the first rib lies underneath the clavicle and cannot be palpated. In order to count a patient's ribs, necessary in many procedures such as locating the proper intercostal space for listening for heart sounds, the second rib must be located by finding the sternal angle and then counting from that point.

Formation of the Ribs and Sternum

The ribs begin as simple extensions of the sclerotomes and grow laterally and ventrally in the form of dense mesenchymal rods that follow the curvature of the body wall. With the differentiation of cartilage in the rods, joint cavities (articulations) appear between the vertebrae and the ribs. A single endochondral ossification center occurs at the angle of the ribs, and bone formation spreads from this center toward each end of the rib. Bone formation does not reach the anterior end of the rib, so a costal cartilage is left between the rib and the sternum.

The sternum is unusual in that it matures into a single midline structure with a segmented appearance, but begins as a paired structure with no suggestion of segmentation. The earliest indication of a sternum (Fig. 7-31) is the appearance of a pair of *mesenchymal bands* in the ventrolateral body wall that gradually migrate together and fuse as cartilage develops. The segmented appearance of the sternum is acquired secondarily as a result of ossification centers that appear within the cartilaginous sternum between the pairs of ribs.

Figure 7-31 Diagrammatic representation of the development of the sternum. (**a**) Mesenchymal (sternal) bands. (**b**) Fusion of bands into a single cartilaginous bar. (**c**) Formation of ossification centers adjacent to rib articulations.

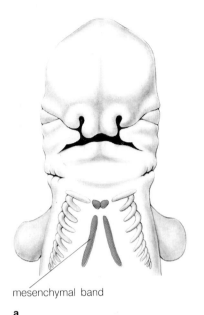

mesenchymal band

a

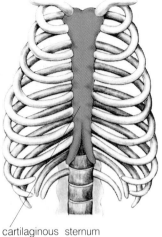

cartilaginous sternum

b

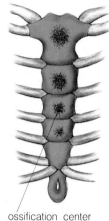

ossification center

c

WORDS IN REVIEW

Appendicular skeleton	Orbit
Axial skeleton	Palatine bone
Cervical vertebrae	Parietal bone
Coccyx	Rib
Cranial fossae	Sacrum
Cranium	Skull
Ethmoid bone	Somite
Frontal bone	Sphenoid bone
Hyoid bone	Sternum
Inferior nasal concha	Suture
Intervertebral disc	Thoracic vertebrae
Lacrimal bone	Vertebral column
Lumbar vertebrae	Vertebral foramen
Mandible	Visceral arch
Maxilla	Visceral pouch
Nasal bone	Vomer
Occipital bone	Zygomatic bone

SUMMARY OUTLINE

I. The Axial Skeleton

The axial skeleton provides support and protection for head, neck, and trunk; it consists of skull, vertebral column, hyoid bone, ribs, and sternum.

II. The Skull

The skull provides support and protection for head structures and consists of cranium and face.

    A. The cranium protects brain and inner ear and consists of eight bones that form the boundaries of the cranial cavity.

        1. The frontal is the bone of the forehead and superior rim of the orbit; it forms most of the roofs of the orbits.

        2. The parietal bones form part of the sides and roof of the cranial cavity; they articulate with each other, the frontal, and the occipital at the sagittal, coronal, and lambdoidal sutures.

        3. The occipital bone forms the back of the skull and part of the floor of the cranial cavity; it contains the foramen magnum, articulates with the first cervical vertebra via occipital condyles, contains the hypoglossal canals, and forms the medial boundary of the foramina lacerum and the jugular foramina.

4. The ethmoid has a median perpendicular plate that contributes to the nasal septum and two lateral plates that form much of the lateral walls of the nasal cavity and contain ethmoid air cells; its cribriform plate transmits olfactory nerve filaments.

5. The sphenoid bone has a body in the midline containing the sella turcica and bearing four paired processes: the greater and lesser wings and the medial and lateral pterygoid plates. The sphenoid contains the foramina rotundum, ovale, optic, and spinosum.

6. The temporal bone consists of four parts: squamous, petrous, tympanic, and mastoid process; the tympanic part bears the styloid process and the stylomastoid foramen between the styloid and mastoid processes.

B. The face consists of 14 bones that provide support and protection for entrances of the respiratory and digestive tracts, house special sense organs, and serve for muscle attachment.

1. The mandible is the lower jaw; it has two rami bearing condyloid and coronoid processes and two foramina (mandibular and mental). It is the only movable bone of the skull.

2. The maxillae are paired bones that are fused to form the upper jaw; its processes are the frontal, alveolar, zygomatic, and palatine; and it contains the maxillary sinus and infraorbital foramen.

3. The zygomatic is the cheek bone; its processes are the maxillary, frontal, and orbital; its foramina are the zygomaticofacial and zygomaticotemporal.

4. The nasal bones are small, paired, rectangular bones forming the bridge of the nose.

5. The lacrimal bones are paired, rectangular bones in the medial wall of the orbit; they help provide support for the system that drains tears into the nasal cavity.

6. The palatines are paired, L-shaped bones with horizontal plates that contribute to the hard palate and vertical plates that contribute to the lateral walls of the nasal cavity.

7. The vomer is located between the sphenoid bone and the hard palate; it forms the posterior part of the nasal septum.

8. The inferior nasal concha is firmly attached to the maxilla; it projects as a ledge from the lateral wall into the nasal cavity.

C. The cranial fossae are depressions in the floor of the cranial cavity; they are named anterior, middle, and posterior, according to their relative position.

D. The orbit is a large, deep fossa that provides a protective housing for the eyeball.

III. Development of the Skull

A. The skull begins as a cartilage plate that forms around notochord, pituitary gland, and paired sense organs.

B. The face originates from mesenchymal condensations of the visceral arches and nasal processes.

C. The palate originates from medial nasal processes and palatal processes of maxillae.

IV. The Hyoid Bone
The hyoid bone is a small U-shaped bone in the upper anterior neck region; it is the only bone that does not articulate with another bone.

V. The Vertebral Column
The vertebral column is composed of irregular bones arranged in sequence that vary morphologically in regions. There are 7 cervicals, 12 thoracics, 5 lumbars, 5 fused to form the sacrum, and 4 coccygeals; intervertebral discs between vertebral bodies and articular facets of articular processes between the laminae permit movement at all levels except the sacrum.

VI. Formation of Vertebrae
A single vertebra originates from paired caudal and cranial halves of adjacent sclerotomes.

VII. The Rib Cage
The rib cage consists of 12 pairs of ribs that articulate posteriorly with thoracic vertebrae. Seven pairs articulate directly with sternum anteriorly; ribs 8, 9, and 10 have costal cartilages that articulate with cartilage of the next higher rib; ribs 11 and 12 are unattached anteriorly. Ribs develop as extensions of sclerotomes with a single ossification center.

VIII. The Sternum
The sternum consists of the manubrium, body, and xiphoid process. The sternal angle is the point of junction of the manubrium with the body; it can be palpated to locate the second rib, which articulates at this point. The sternum originates from paired mesenchymal bands that fuse anteriorly; ossification centers form in relation to the ribs.

REVIEW QUESTIONS

1. What are the components (in general terms) of the axial skeleton? In what region of the body is each of these components located?

2. What are the two major parts of the skull? How many bones are found in each part?

3. What are the major functions of the cranium? What bone forms the vertical plate of the forehead? Name the bones that form the sides and roof of the cranial cavity. What bone forms the posterior part of the floor of the cranial cavity? What two bones form the middle portion of the floor of the cranial cavity? The cribriform plate is part of what bone? In what bone is the foramen magnum located? What are the four parts of the temporal bone? In which part is the inner ear located?

4. What are the major functions of the facial bones? What bone forms the lower jaw? Which facial bone is movable? What processes project from the body of the maxilla? Name the four processes of the cheek bone. What bones form the bridge of the nose? Where is the lacrimal bone located and what is its major function? What part of the palatine bone contributes to the lateral wall of the nasal cavity?

5. What are the boundaries that separate the anterior, middle, and posterior cranial fossae?

6. What three bones form the rim of the orbit? What bone that contributes to the medial wall of the orbit contains air cells? What bone of the floor of the orbit contains an air sinus? What bone of the roof of the orbit contains an air sinus?

7. The floor of the skull initially develops as a cartilage plate that forms in relation to what structures?

8. What are the parts of the hyoid bone? How is the hyoid bone related anatomically to other bones?

9. What are the anatomical regions of the vertebral column? How many vertebrae are usually found in each region? What structures permit movement of the vertebral column?

10. From what embryological structures does the vertebral column develop?

11. What are the components of the rib cage? How are these parts related to one another?

12. What are the parts of the sternum? What is the significance of the sternal angle?

SELECTED REFERENCES

Anderson, J. E. *Grant's Atlas of Anatomy*. 8th ed. Baltimore: Williams & Wilkins, 1983.

Langman, J. *Medical Embryology*. 4th ed. Baltimore: Williams & Wilkins, 1981.

Moore, K. L. *Clinically Oriented Anatomy*. Baltimore: Williams & Wilkins, 1980.

Romanes, G. J. *Cunningham's Textbook of Anatomy*. New York: Oxford University Press, 1981.

8

THE APPENDICULAR
SKELETON

After completing this chapter you should be able to:

1. List the major components of the superior and inferior extremities. Discuss the structural and functional similarities and differences between these two limbs.

2. Describe the scapula and its location relative to the ribs, the clavicle, and the humerus. Name its processes and fossae.

3. Describe the clavicle and its location relative to the ribs, the sternum, and the scapula.

4. Describe the humerus. Name and locate its prominent anatomical features.

5. Describe the ulna. Give its relations to the bones with which it articulates. Name and locate its chief anatomical features.

6. Describe the radius, including its anatomical features and major relations.

7. Name the components of the hand, giving the numbers of bones in each component.

8. Name the bones of the wrist and describe their arrangement.

9. Describe the metacarpals. List the parts that are visible or can be palpated on the back of the hand.

10. Describe the phalanges, giving the numbers found in each of the digits.

11. Name and describe the components of the os coxae. Relate these components to the sacrum, the femur, and to one another.

12. Describe the femur and its position relative to other bones of the lower limb. Name and locate its prominent anatomical features.

13. Locate the patella in relation to the femur, tibia, and fibula.

14. Describe the tibia and fibula, naming their prominent anatomical features.

15. Name the parts of the foot. Name and give the numbers of bones in each of the parts.

16. Describe the development of the limbs in general terms. Discuss the effects of opposite rotation of the limbs during development upon movements of the limb.

The appendicular skeleton consists of the bones of the upper and lower limbs and the girdles with which they articulate. The muscles attached to the bones of the appendicular skeleton not only assist in supporting the limbs, but also provide for a wide range of movements. The arrangement of the skeletal elements and muscles of the upper limbs allows great freedom of movement but sacrifices stability. The lower limbs have less freedom of movement but the arrangement of their skeletal elements provides the stability essential in supporting and moving the body. In both the upper and lower limbs, the problem of providing stability at the joints where movements occur is partially solved by having the muscles cross the joints, so that the joints are supported by the same muscles whose contractions also cause movement.

For the upright human the upper limb functions as an extension of the trunk, enabling the individual to reach out to grasp an object and to draw it closer to the trunk. In such a maneuver movement occurs at all the joints of the upper limb, and many muscles act together by alternately contracting and relaxing to accomplish the objective with a minimum of effort. In the study of the anatomical features that make such a complex process possible, one should remember that the limb functions as a structural unit; the arrangement of muscles and bones at the joints is designed to facilitate movement while at the same time maintaining the integrity of the joint by keeping the articular surfaces of the bones in their proper relations to one another.

Any decrease in the efficiency of our upper extremities affects our capacity to work. In our mechanical age, injuries to the upper limb, particularly to the hand, are common. In treating such injuries it is important to attempt to preserve or restore function.

The same principles apply to the lower extremity although it functions to provide support and locomotion rather than reaching and grasping. The bones and muscles of the lower extremity are arranged to support the body while allowing movement at all joints in such activities as walking or turning. The bones articulating at the joints of the lower extremity also function as a unit with the muscles crossing the joint which provide support for the limb and produce movements by contracting.

BONES OF THE PECTORAL GIRDLE AND UPPER LIMB

The bones of the **pectoral girdle** and upper limb are illustrated in Fig. 8-1 and summarized in Table 8-1. The shoulder region is supported by the two bones of the *pectoral girdle,* the

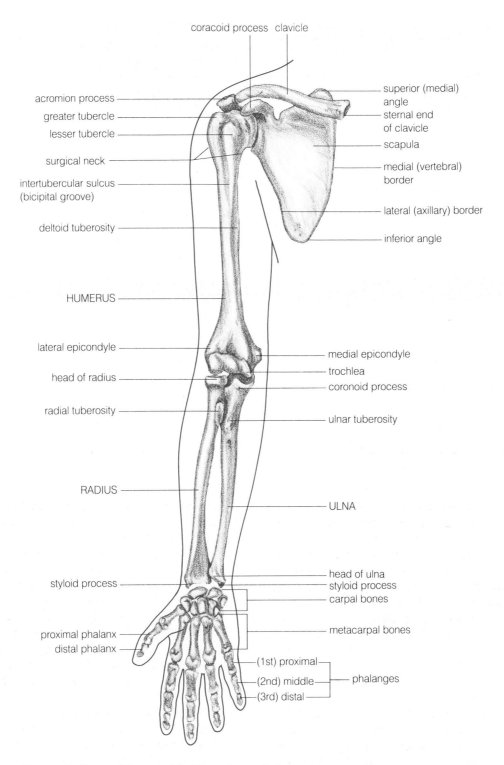

coracoid process   clavicle

acromion process

greater tubercle

lesser tubercle

surgical neck

intertubercular sulcus
(bicipital groove)

deltoid tuberosity

HUMERUS

lateral epicondyle

head of radius

radial tuberosity

RADIUS

styloid process

proximal phalanx
distal phalanx

superior (medial)
angle

sternal end
of clavicle

scapula

medial (vertebral)
border

lateral (axillary) border

inferior angle

medial epicondyle

trochlea

coronoid process

ulnar tuberosity

ULNA

head of ulna
styloid process
carpal bones

metacarpal bones

(1st) proximal
(2nd) middle
(3rd) distal

phalanges

Figure 8-1   Bones of the pectoral girdle and upper limb.

Table 8-1 Bones of the Pectoral Girdle and Upper Limbs

Name and Number	Location	Gross Anatomical Features
Scapula, 2	Posterior aspect of thorax, overlapping upper seven ribs	Spine, acromion process, coracoid process, supraspinous and infraspinous fossae, subscapular fossa, glenoid fossa
Clavicle, 2	Anterior-superior aspect of thorax (base of neck)	Curvatures: medial two-thirds concave anteriorly, lateral one-third convex anteriorly
Humerus, 2	Arm; between shoulder and elbow	Head, tubercles (greater and lesser), intertubercular groove, epicondyles (medial and lateral), fossae (coronoid and olecranon)
Ulna, 2	Medial side of forearm	Olecranon process, trochlear notch, styloid process
Radius, 2	Lateral side of forearm	Head, radial tuberosity, styloid process
Carpal, 16	Wrist of manus (hand)	Two rows of four bones in each wrist
Metacarpal, 10	Palm of the hand	A single row of bones in line with the fingers
Phalanx (28)	Fingers (digits)	Two bones in the thumb; three bones in each finger

scapula (shoulder blade), and the **clavicle** (collar bone). The **humerus** is the bone of the **brachium** (arm), and the **radius** and **ulna** are the bones of the **antebrachium** (forearm). The **manus** (hand) is composed of three parts: the **carpus** (wrist), composed of eight small bones in two rows; the *hand proper,* including the palm and the back of the hand with five *metacarpal* bones; and the five *digits,* numbered from the lateral side, supported by small bones called **phalanges.**

Pectoral Girdle

Scapula The scapula (Fig. 8-2) is a triangular bone overlying the second to the seventh ribs on the posterior aspect of the thorax (see Fig. 7-22). It articulates with the clavicle and humerus but is connected to the axial skeleton only by muscles. Its *body* is concave on the costal (anterior) surface, which faces the ribs, and convex posteriorly. The body is divided posteriorly into two parts by a triangular *spine,* continuous laterally with the **acromion process,** that projects above the shoulder joint

as it curves anteriorly to meet the clavicle; the posterior surface above this spine is the *supraspinous fossa,* and that below it is the *infraspinous fossa.*

The scapular body has three borders, which meet at three angles. The *medial border* of the body meets the *lateral border* at the *inferior angle* and the *superior border* at the *superior angle.* The medial border is also called the vertebral border because it faces the vertebral column; the lateral border is also called the axillary border because it faces the axilla, or armpit. The superior border is horizontal and is marked by the deep *scapular notch* through which nerves pass to the *supraspinous fossa.* The anterior or costal surface is called the *subscapular fossa.*

The thickened lateral angle called the *head* of the scapula is concave, forming a shallow depression called the *glenoid fossa* for articulation with the head of the humerus. The fibrocartilaginous rim of the glenoid fossa increases the depth of the fossa and serves for the attachment of the joint capsule. Above the cavity is the *supraglenoid tubercle,* a rough area that provides for the attachment of the long head of the biceps muscle

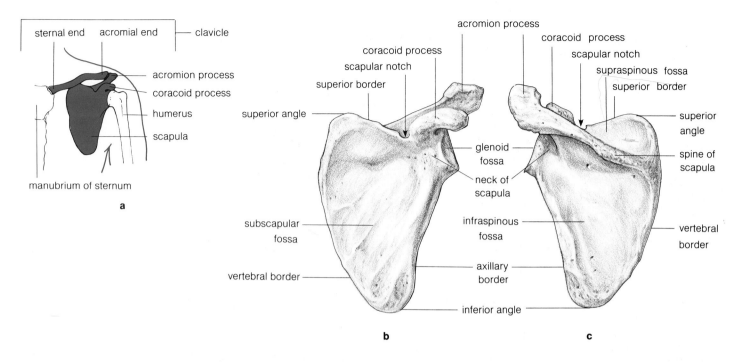

Figure 8-2 The scapula: (**a**) anterior view of the left part of the pectoral girdle (ribs have been removed); (**b**) anterior view of the scapula; (**c**) posterior view of the scapula.

of the arm. The head of the scapula is separated from the body by a narrow constriction called the *neck*. From the upper part of the head and neck the curved **coracoid process** projects anteriorly below the clavicle. It is shaped like a bent finger, with its lower part directed superiorly and inferiorly (its tip may be palpated inferior to the lateral third of the clavicle). The coracoid process is the point of attachment for *three muscles* (the coracobrachialis, the short head of the biceps brachii, and the pectoralis minor) and *two ligaments* (the coracoacromial and coracoclavicular).

be felt along their entire length between the attached muscles. The acromion and the spine serve for the attachment of muscles (the deltoid and trapezius, respectively). Rough areas on the lateral and medial borders and on the *subscapular fossa* of the costal surface also serve for muscle attachment.

Clavicle   The clavicle (Fig. 8-3) is a curved bone lying horizontally across the superior border of the thorax; it articulates medially with the sternal manubrium and laterally with the acromion process of the scapula.

> The acromion process develops from an ossification center that is independent of the developing scapular spine. Sometimes the union of the two structures is incomplete, remaining as a fibrous union instead of fusion of the bony components. On an x-ray film this may appear as a break and lead to an incorrect diagnosis.

> The clavicle is one of the most frequently broken bones in the body. It begins ossification at about the fifth week of life, earlier than any other bone. Consequently when the baby is born, the clavicle is rather rigid and may be fractured during birth.

The acromion process may be palpated above the shoulder joint where it articulates with the clavicle, and both the acromion process and the posterior border of the scapular spine may

The medial two-thirds of the clavicle is convex forward and the lateral one-third is concave forward. The inferior surface is marked medially by a depression for the attachment of the *costoclavi-*

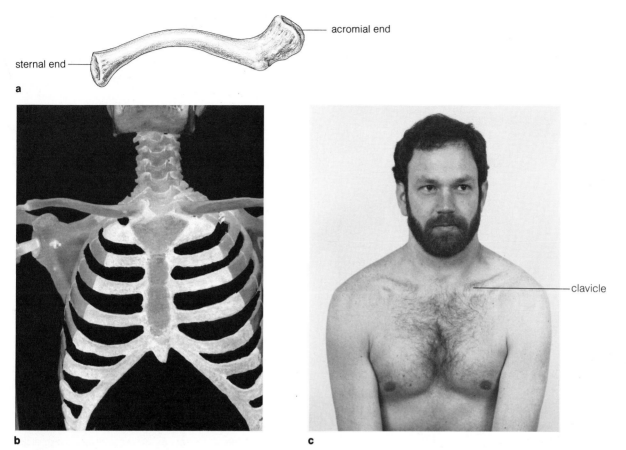

sternal end

acromial end

clavicle

a

b

c

Figure 8-3 The clavicle: (**a**) inferior view; (**b**) relations of clavicle to rib cage, anterior view (photograph by R. Ross); (**c**) surface anatomy (photograph by R. Ross).

cular ligament that binds the clavicle to the first rib. Laterally the *conoid tubercle* and the *trapezoid line* are rough areas for attachment of the strong *coracoclavicular ligament* that binds the clavicle and scapula together. The lateral (acromial) end of the clavicle has an oval facet for articulation with the acromion process. The medial (sternal) end is enlarged and flattened into an articular surface that moves on the articular disc of the sternoclavicular joint and on the first costal cartilage (see Fig. 7-22).

> The clavicle is usually shorter, thinner, and straighter in the female than in the male. In those individuals who exercise regularly or perform manual labor, the clavicle becomes thicker and more curved, and the ridges to which muscles attach become more prominent.

Arm

The upper limb consists of three segments: proximal, middle, and distal. The proximal segment, known in anatomy as the *arm,* extends from the shoulder to the elbow and contains the humerus. The middle segment is the *forearm* and contains the radius and ulna. The distal segment, the *hand,* is subdivided for descriptive purposes into wrist and hand proper. The wrist contains the carpals; the hand proper consists of the palm, supported by the metacarpals, and the digits, supported by the phalanges.

Humerus   The *humerus* (Fig. 8-4) is a long bone with an *epiphysis,* or head, at each end of the *diaphysis,* or shaft. It extends from the scapula to the elbow joint and is almost completely covered by muscles, but it can be palpated through the muscles along its entire length. The round head of the humerus is joined

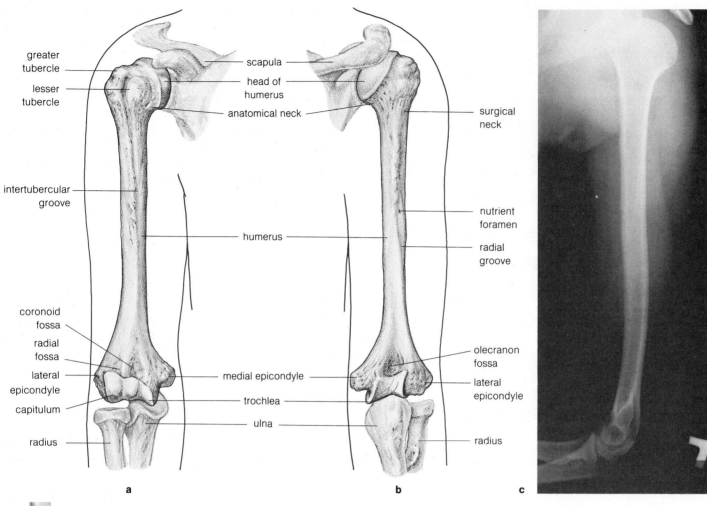

greater tubercle

lesser tubercle

intertubercular groove

coronoid fossa

radial fossa

lateral epicondyle

capitulum

radius

scapula

head of humerus

anatomical neck

humerus

medial epicondyle

trochlea

ulna

surgical neck

nutrient foramen

radial groove

olecranon fossa

lateral epicondyle

radius

a

b

c

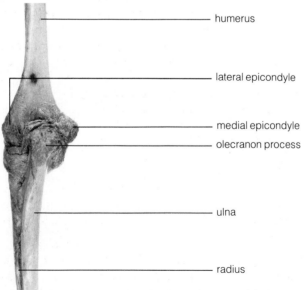

humerus

lateral epicondyle

medial epicondyle

olecranon process

ulna

radius

d

Figure 8-4 The right humerus: (**a**) anterior view; (**b**) posterior view; (**c**) x ray of posterior view (rotated), showing articulation with radius and ulna (courtesy of Alexandria Hospital, Alexandria, Virginia, Joyce R. Isbel, R. T.); (**d**) posterior view of a cadaver preparation (photograph by R. Ross).

to the superior end of the shaft by the *anatomical neck,* a slightly constricted narrow strip encircling the bone at the edge of the articular surface, and by the *surgical neck,* the region inferior to the head and tubercles, where fractures often occur. The *greater tubercle,* the prominence on the lateral surface that can be palpated just below the acromion process, is the point of insertion of the muscles that rotate the arm. The *lesser tubercle,* serving for the attachment of the *subscapularis muscle,* is on the anterior surface and can be felt through the *deltoid muscle* just inferior and lateral to the coracoid process. Extending below each tubercle are crests forming the sides of the *intertubercular groove* occupied by the tendon of the long head of the biceps. The large muscle of the back inserts into the floor of the groove and the large anterior chest muscle into the crest of the lesser tubercle. At about the middle of the shaft a rough area on the lateral surface, the deltoid tuberosity, marks the location of the insertion of the deltoid; an oblique groove indicates the course of the radial nerve.

At the distal end the shaft is wide and flattened anteroposteriorly, forming two depressions, the *coronoid fossa* in front and the *olecranon fossa* in back, for articulation with the ulna. Between the two fossae is the distal end of the humerus, termed the *trochlea* (pulley), which articulates with the ulna. Lateral to the trochlea is a rounded prominence, the *capitulum* (little head), that serves for articulation with the radius. Posterior to the capitulum the *lateral epicondyle* projects laterally and serves for the attachment of forearm muscles. On the medial side the prominent *medial epicondyle* also serves for the attachment of forearm muscles.

> The lower end of the humerus bends forward 45 degrees on the shaft, an angulation easily seen on an x-ray film. Any decrease in this angulation indicates a fracture above the condyles that displaces the distal end of the humerus backwards.

Forearm

Ulna The medial bone of the forearrm, the ulna (Fig. 8-5), is a long bone with a large proximal end identifiable by two projecting processes enclosing a concavity. The **olecranon process,** the larger of the two, is in line with the shaft and forms the point of the elbow, which can be felt at the back of the joint. Posteriorly the olecranon process is smooth and subcutaneous, whereas anteriorly it forms part of the articular surface of the *trochlear notch,* which is completed by the *coronoid process* as it projects anteriorly from the shaft. Just distal to the coronoid process the small *ulnar tuberosity* serves for insertion of a muscle (brachialis) of the arm. On the lateral side of the coronoid process the *radial notch* provides for articulation with the head of the radius. The lateral surface of the shaft is marked by the interosseous border, a distinct ridge for the attachment of the *interosseous membrane* joining the ulna to the radius. Posteriorly the ulna may be palpated throughout its length. At the lower end the ulna enlarges into a distally flattened head that articulates with the articular disc between the head and the triquetrum of the carpus. The posteromedial aspect of the head bears the *styloid process,* which projects distally and serves for the attachment of the *ulnar collateral* ligament of the wrist.

Radius The lateral bone of the forearm, the radius (Fig. 8-5), is shorter than the ulna and does not overlap the humerus. The *head* is a thick disc with a concave superior surface for articulation with the capitulum of the humerus and a smooth edge encircled by the radial notch of the ulna and the annular ligament that holds it there. Most of the *shaft* of the radius is triangular in shape with a slight lateral convexity. It bears the *radial tuberosity,* a prominence near its proximal end for the insertion of the tendon of the biceps brachii. A prominent medial edge, or interosseous border, extends the length of the shaft for the attachment of the interosseous membrane. The remaining surfaces of the radius are marked indistinctly by muscle attachments.

The lateral surface of the distal end of the bone is prolonged inferiorly as the thick *styloid process* to which the radial collateral ligament of the wrist is attached. The medial surface has a concave articular surface, the *ulnar notch,* for articulation with the head of the ulna. The distal, or carpal, articular surface is concave to fit against the scaphoid and lunate bones of the wrist.

Hand

The hand (Fig. 8-6) consists of the *wrist* (containing the carpals), the *hand proper* (containing the metacarpals), and the *digits* (made up of phalanges).

Carpals The bones of the wrist are eight carpal bones in two rows; the proximal row articulates with the radius and the distal row with the metacarpals. From the medial to the lateral side the bones of the proximal row are named *pisiform, triquetral, lunate,* and *scaphoid.* The bones of the distal row, from the medial to the lateral side, are the *hamate, capitate, trapezoid,* and *trapezium.* The pisiform and the tubercle of the scaphoid may be palpated at the distal skin crease on the anterior surface of the wrist. The bones in each row form a transverse arch whose concavity faces forward with the marginal bones projected farther forward than the other carpals. Each of the four marginal bones bears a bony prominence to which is attached a strong, broad ligament, the *flexor retinaculum,* which forms a protective covering for the slender tendons of the flexors of the fingers. The posterior aspect of the wrist is smoothly convex with no bony prominences.

Metacarpals The bones of the palm, or hand proper, are five metacarpals numbered I through V from the medial to the lateral side. These are small, long bones, each having a flat *base*

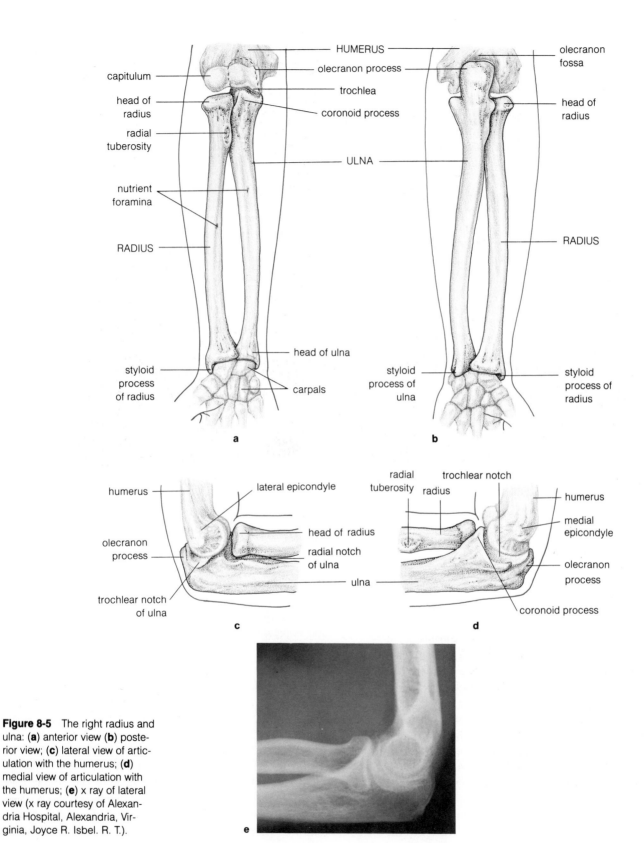

capitulum

head of radius

radial tuberosity

nutrient foramina

RADIUS

styloid process of radius

HUMERUS

olecranon process

trochlea

coronoid process

ULNA

head of ulna

carpals

a

olecranon fossa

head of radius

RADIUS

styloid process of ulna

styloid process of radius

b

humerus

olecranon process

trochlear notch of ulna

lateral epicondyle

head of radius

radial notch of ulna

ulna

c

radial tuberosity

radius

trochlear notch

humerus

medial epicondyle

olecranon process

coronoid process

ulna

d

e

Figure 8-5 The right radius and ulna: (**a**) anterior view (**b**) posterior view; (**c**) lateral view of articulation with the humerus; (**d**) medial view of articulation with the humerus; (**e**) x ray of lateral view (x ray courtesy of Alexandria Hospital, Alexandria, Virginia, Joyce R. Isbel. R. T.).

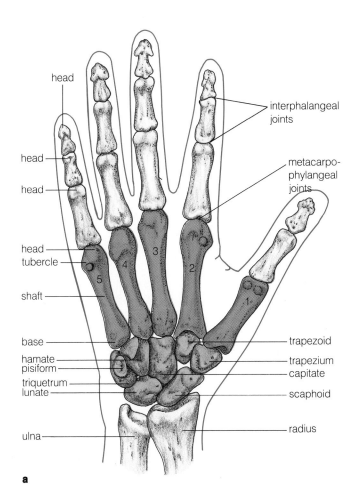

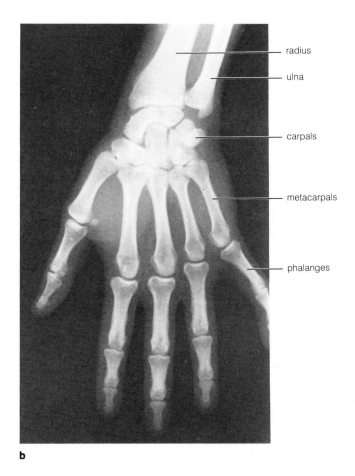

Figure 8-6 The hand. (**a**) Palmar aspect of the bones of the right hand. (**b**) X ray of the hand (from the teaching collection, Department of Anatomy, University of Kentucky).

and an expanded round *head* connected by a thick round *shaft*. Adjacent to the wrist the bones of the metacarpals are close together, but they diverge distally so that the skeleton of the palm is fan-shaped. The round head of each metacarpal is flattened anteriorly, facilitating flexion at the metacarpophalangeal joints. The shafts may be palpated at the back of the hand where they are not covered by muscle. The extent of movement possible may be investigated by palpation. When the fists are clenched, the heads of the metacarpals (knuckles) are readily visible, and the metacarpophalangeal joint may be felt near the distal edge of the knuckle on either side of the fingers. Note the greater mobility of the metacarpal to which the thumb is attached.

Phalanges The bones of the digits are the three phalanges (sing., phalanx) of each of the four fingers and the two phalanges of the thumb. Like the metacarpals the phalanges are miniature long bones expanded at their articular surfaces at each end of the shaft. Within each finger the proximal phalanx is the largest; the distal is the smallest and has a tapered end flattened posteriorly to serve as the *nail bed*. The phalanges of the thumb are similar in structure but are thicker and wider to accommodate the muscles associated with its greater mobility. The phalanges of all the digits are slightly concave anteriorly to provide space for the tendons that flex the fingers. In most individuals these round tendons may be palpated with the fingers flexed against resistance. On the posterior surface the pha-

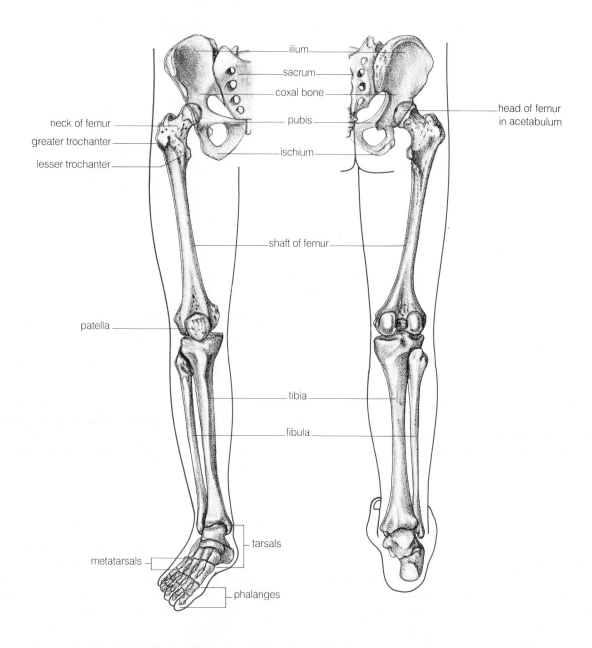

ilium

sacrum

coxal bone

neck of femur

greater trochanter

lesser trochanter

head of femur
in acetabulum

pubis

ischium

shaft of femur

patella

tibia

fibula

tarsals

metatarsals

phalanges

Figure 8-7   Bones of the pelvic girdle and lower limbs.

Table 8-2 Bones of the Pelvic Girdle and Lower Limbs

Name and Number	Location	Gross Anatomical Features
Hip bone, 2	Hip region; form the pelvic girdle. With sacrum form a complete circle called the pelvis	Composed of ilium, ischium, and pubis; articulates with femur at acetabulum and with other hip bone at pubic symphysis
Femur, 2	Bone of the thigh between hip and knee	Head, neck, trochanters (greater and lesser), condyles (medial and lateral)
Patella, 2	Anterior surface of knee	A sesamoid bone in tendon of anterior thigh muscle
Tibia, 2	Medial side of leg between knee and ankle	Tibial tuberosity, condyles (medial and lateral), medial malleolus
Fibula, 2	Lateral side of leg	Head, lateral malleolus, styloid process
Tarsal, 14	Ankle of pes (foot)	7 bones bound tightly together
Metatarsal, 10	Instep of sole of foot	A single row of bones in line with the toes
Phalanx, 28	Toes (digits)	Two bones in great toe, three in each of the other toes

langes are just beneath the skin and may be palpated readily; the extensor tendons lie on each side of the bones and, being flattened, are not palpable. The interphalangeal joints are not as large as the knuckles but are palpable, especially along the sides of the fingers.

BONES OF THE LOWER LIMB

The skeleton of the lower limb (Fig. 8-7) is joined to the trunk by the **pelvic girdle,** which consists of two coxal bones (**os coxae,** hip bones). The coxal bones articulate with the sacrum posteriorly and meet each other inferiorly and anteriorly in the midline at the **pubic symphysis.** The coxal bones and sacrum together form the skeleton of the *pelvis,* the lowest part of the trunk.

The freely movable lower limb that articulates with the hip bone at the **acetabulum** consists of the **thigh** (including the knee), the **crus** (leg), and the **pes** (foot). The bones of the *thigh* are the **femur** and the **patella** (kneecap); those of the *crus* are the large **tibia** medially and the slender **fibula** on the lateral

side. The bones of the *pes* includes the **tarsus** (ankle), consisting of seven **tarsal bones**; the **metatarsus**, made up of five *metatarsals;* and five **digits** (toes). Four of the digits have three *phalanges* each and one has two. These bones are summarized in Table 8-2.

Pelvic Girdle

The *pelvis,* from the Latin for *basin* (Fig. 8-8) is stronger than the wall of the cranial or thoracic cavities. It is divided into the *greater* and *lesser pelvis* by a line interconnecting the *pelvic brims* of the two sides. The pelvic brim is a ridge consisting of the *prominence* of the *sacrum,* the *arcuate* and *pectineal lines,* and the superior margin of the *pubic symphysis.* The greater pelvis is above the pelvic brim and the lesser pelvis is below, or distal, to the brim.

There are differences between the male and female pelvis. The female pelvis is wider in all diameters, including the diameter of the inferior aperture, and the pelvic cavity is shorter and less funnel-shaped. The *superior* and *inferior apertures* of her lesser pelvis are larger and the bones are less massive.

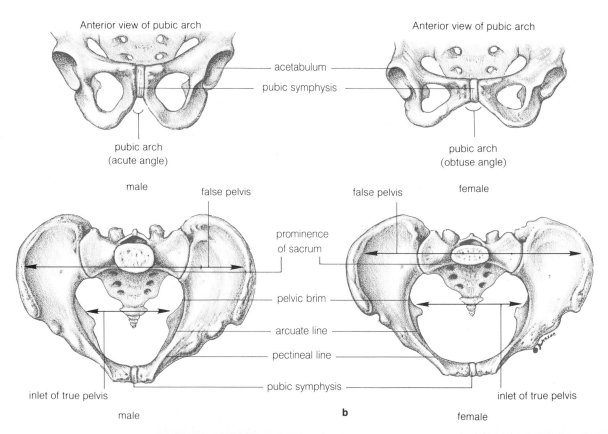

Anterior view of pubic arch

Anterior view of pubic arch

acetabulum

pubic symphysis

pubic arch
(acute angle)

pubic arch
(obtuse angle)

male

false pelvis

false pelvis

female

prominence
of sacrum

pelvic brim

arcuate line

pectineal line

pubic symphysis

inlet of true pelvis

inlet of true pelvis

a

male

b

female

Figure 8-8 Differences in the structure of the pelvis related to sex: (**a**) anterior view of a male pubic arch and superior view of male pelvic girdle; (**b**) anterior view of a female pelvic arch and superior view of female pelvic girdle; (**c**)x ray. Is this a male or female pelvis? (x ray courtesy of Alexandria Hospital, Alexandria, Virginia, Joyce R. Isbel, R. T.).

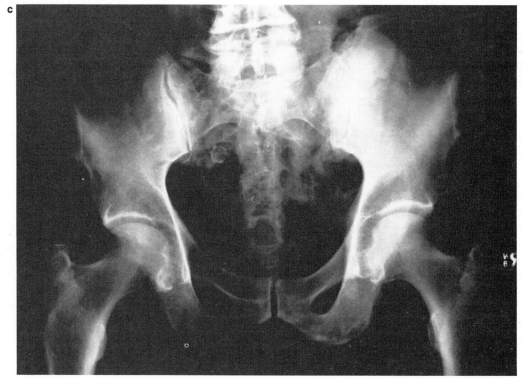

c

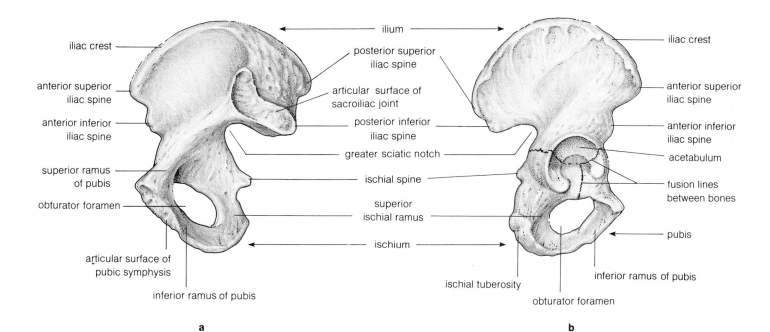

a

b

Each os coxae (Fig. 8-9) is a large, irregularly shaped bone formed by three bones that meet and fuse in the *acetabulum,* the socket for articulation with the head of the femur. The three components of the coxal bone are the **ilium, ischium,** and **pubis.**

Ilium The ilium (Fig. 8-9), the largest portion of the hip bone, extends superiorly and ends with a convex border called the *iliac crest* that can be felt from its junction with the sacrum to its termination anteriorly at the *anterior superior iliac spine.* The iliac crest serves principally for the attachment of muscles of the anterior and posterior abdominal wall. The anterior superior iliac spine is the lateral point of attachment of the *inguinal ligament,* which passes medially to the pubic tubercle. Below the anterior superior iliac spine the anterior border of the ilium extends downward to reach the acetabulum. Above the margin of the acetabulum is a rounded projection, the *anterior inferior iliac spine,* for the attachment of the rectus femoris muscle.

The posterior border of the ilium extends downward from the *posterior superior iliac spine,* to which the deep fascia of the back is attached, and follows the curvature of the posterior boundary of the sacroiliac joint to reach its point of fusion with the ischium at the acetabulum. As it descends, the posterior margin is deeply indented by the *greater sciatic notch,* which

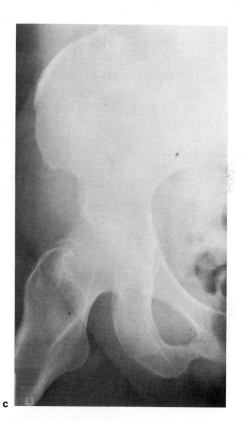

c

Figure 8-9 The coxal bone: (**a**) view of the internal surface; (**b**) view of the external surface; (**c**) x ray view showing head of femur in acetabulum (x ray courtesy of Alexandria Hospital, Alexandria, Virginia, Joyce R. Isbel, R. T.).

is filled by the piriformis muscle that arises from the anterior surface of the sacrum and goes to the femur.

The lateral aspect of the *ilium* (gluteal surface) is marked by three lines that curve superiorly and anteriorly from the greater sciatic notch and mark the boundaries between the three gluteal muscles. On the medial side the concave anterior portion, the *iliac fossa,* is occupied by the iliacus muscle, and most of the medial-posterior surface serves for articulation with the sacrum and is roughened for the attachment of ligaments of the sacroiliac joint.

Ischium   The ischium (Fig. 8-9) may be divided into a *body* that fuses with the ilium and forms part of the acetabulum and a ramus that fuses with the pubis. As the body of the ischium passes inferiorly its posterior margin bears two projections: the *ischial spine,* to which the *sacrospinous ligament* and a muscle are attached, and the *ischial tuberosity,* to which the sacrotuberous ligament and the hamstring muscles are attached. The ischial tuberosity marks the lower boundary of the ischium; it is the bone on which the trunk rests when one is sitting down. From this point the ramus passes anteriorly to fuse with the pubis.

Pubis   The pubis (Fig. 8-9) has a *body* with a flat surface for articulation with the pubis of the other side. From the body two rami project: a *superior ramus,* which passes upward and laterally to contribute to the acetabulum, and an *inferior ramus,* which passes downward and laterally to fuse with the ramus of the ischium. The anterior border of the body is thickened to form a *pubic crest* for the attachment of the rectus abdominis muscle. At the lateral end of the crest the *pubic tubercle* receives the inguinal ligament, and extending laterally from the pubic tubercle the *pectineal line* serves for the attachment of the *conjoint tendon* and *lacunar ligaments* that reinforce the inguinal canal.

The rami of the pubis and the ischium with the acetabulum above surround a large opening, the *obturator foramen,* which is closed by a fascial membrane that serves for the attachment of the obturator muscles.

Thigh

Femur   The femur (Fig. 8-10) extends from a rounded *head* that fits into the acetabulum to the knee joint. Its lower end is flattened anteroposteriorly with the large medial and lateral *condyles* on either side and an extensive articular surface for the tibia. About two-thirds of the ball-shaped head is covered with its articular surface, which contains a pit where an *intraarticular ligament* attaches. The neck runs obliquely inferiorly and laterally to join the shaft, of which it is functionally a part. At the junction of the *neck* with the *shaft* the **greater trochanter** projects upward to about the level of the center of the hip joint and serves for the attachment of gluteal muscles. The *intertrochanteric crest* is a ridge extending posteriorly between the greater and **lesser trochanters;** it serves for the attachment of the *quadratus femoris muscle.* The *lesser trochanter,* located at the inferior end of the crest in the angle between the junction of the neck and shaft, serves for attachment of the *psoas major muscle.*

A broken hip usually means a fracture of the neck of the femur. Such injuries are common in individuals older than 60 years and are more common in women than in men. One common feature of the aging process in postmenopausal women is that resorption of bone is faster than its formation. This weakens the bones, so that weight-bearing bones such as the femur are easily fractured.

The *shaft* of the femur is convex forward, adding to the fullness of the front of the thigh. On its posterior surface the shaft narrows in its middle portion into a rough ridge, the *linea aspera,* which is the point of attachment for several large thigh muscles. Below the linea aspera the posterior surface widens to form the flat triangular *popliteal surface,* which is bounded on each side by the *medial* and *lateral supracondylar lines,* each of which becomes continuous inferiorly with the epicondyle on its side. The *lateral* and *medial epicondyles* are rounded elevations to which leg muscles attach. Between the epicondyles are articular surfaces borne on the two large condyles and separated by the *intercondylar notch.*

In the erect position, the femurs slope medially so that the knees are closer together than the hip joints. This slope varies in different individuals and is usually greater in the female than in the male because of the greater width of the female pelvis.

Patella   The patella (Fig. 8-11) is a sesamoid bone (a bone formed in a tendon) in the tendon of the *quadriceps femoris muscle.* It is oval in outline but tapers inferiorly at its apex for the attachment of the patellar ligament. The anterior surface is rough and slightly convex. The posterior surface is an oval articular surface divided by a vertical ridge into medial and lateral parts that fit corresponding surfaces on the femur. The patella protects the front of the knee and increases the leverage of the large muscle of the front of the thigh by making it act at a greater angle.

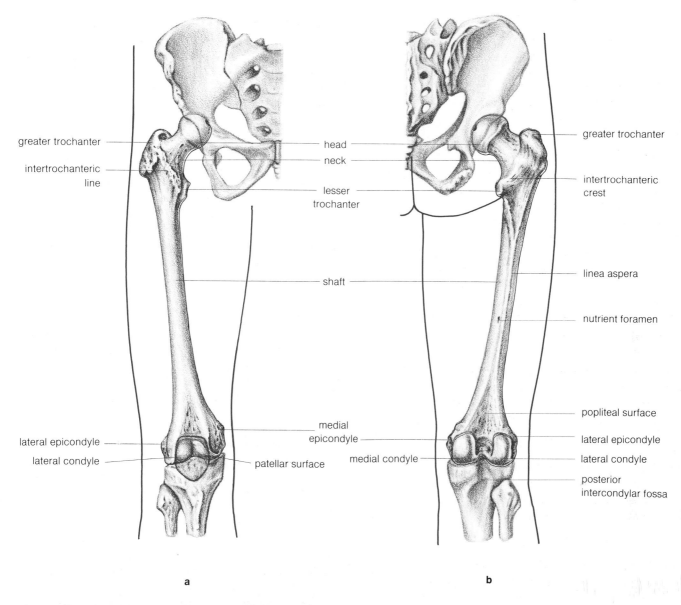

Figure 8-10 The right femur: (**a**) anterior view; (**b**) posterior view.

Leg

Tibia The tibia (Fig. 8-11), the medial bone of the leg, is palpable beneath the skin from the knee to the ankle. The proximal end is expanded to form the *medial* and *lateral condyles,* which articulate with the condyles of the femur. The proximal surface is slightly concave, with the two halves of the articular surface separated by an elevation called the *intercondylar eminence* that projects upward for the attachment of the *menisci,* the articular discs of the knee joint. The condyles are roughened on each side laterally by the attachment of fascia and medially by the attachment of a ligament of the leg as well as fascia. The lateral condyle bears a facet for articulation with the fibula. At the proximal end of the shaft a prominent eminence, the *tibial tuberosity,* serves for the attachment of the patellar tendon (tendon of the quadriceps femoris muscle) and marks the proximal end of the sharp anterior border of the shaft. The lateral surface is marked posteriorly by the sharp interosseous border to which the interosseous membrane connecting tibia and fibula is attached. The posterior surface is roughened in its proximal part for the attachment of muscles and has an oblique line for the attachment of the soleus, which extends from the articular

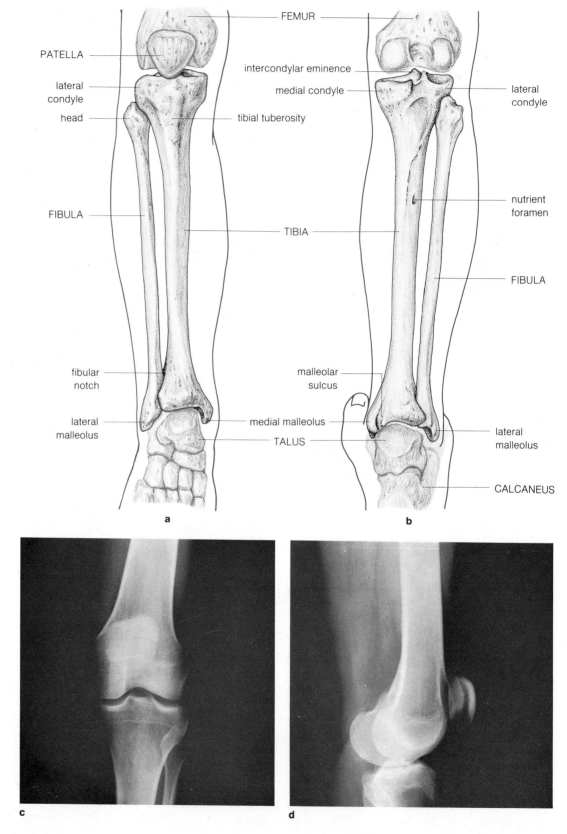

Figure 8-11 The right tibia and fibula: (**a**) anterior view; (**b**) posterior view; (**c**) x ray of posterior view; (**d**) x ray of lateral view (note patella to left and above joint) (x rays courtesy of Alexandria Hospital, Alexandria, Virginia, Joyce R. Isbel, R. T.).

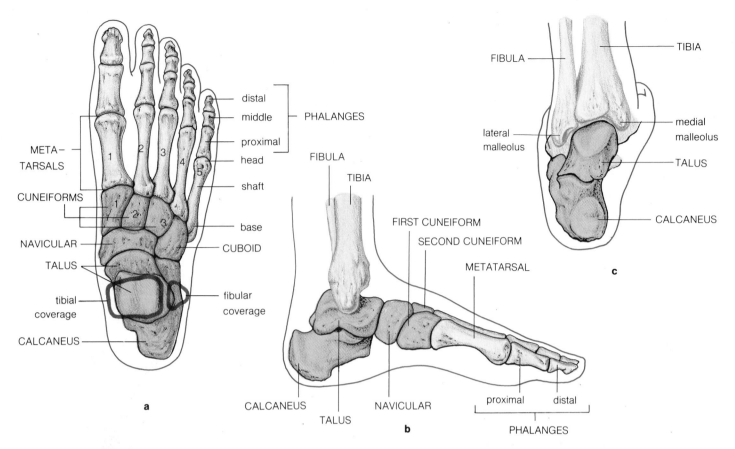

Figure 8-12 (a) The right foot: superior view; (b) medial view of left foot; (c) posterior view of articulation of left foot with leg bones. Shaded areas indicate tarsals.

facet for the fibula about one-third of the way down the shaft.

The distal end of the tibia is enlarged and prolonged medially by the *medial malleolus,* a projection that can be palpated on the medial surface of the ankle where the strong deltoid ligament of the ankle joint attaches. The articular surface has an anteroposterior ridge that fits the groove on the articular surface of the talus (the uppermost tarsal, or ankle bone). The lateral surface of the tibia has a concavity called the fibular notch for articulation with the fibula.

Fibula   The fibula (Fig. 8-11) is a slender bone with two enlarged ends on the lateral side of the leg. The *head* can be felt as a knob laterally distal to the knee, where the tendon of the biceps femoris attaches to the lateral aspect of the head. Medially the head bears an oval facet for articulation with the tibia. The surface of the shaft has an anterior and a posterior border in the form of ridges for the attachment of muscles.

The distal end of the fibula is prolonged laterally as the *lateral malleolus,* a prominence that can be felt on the lateral surface of the ankle joint. The medial aspect of the lateral malleolus has an articular facet that fits the lateral facet on the talus. Its lateral side provides for the attachment of ligaments of the ankle joint.

Foot

The foot (Fig. 8-12) is that part of the lower limb distal to the leg; it begins at the articulation of the tibia and fibula with the tarsal bone called the talus. The foot is homologous with the hand, that is, it resembles it in structure and development even though its functions are support and locomotion rather than grasping. The bones of the foot are the **tarsals** (constituting the *tarsus*), the **metatarsals** (constituting the *metatarsus*), and the

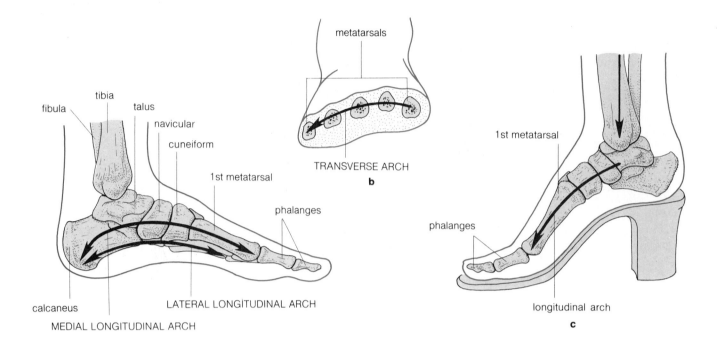

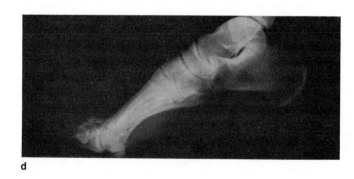

Figure 8-13 Arches of the right foot: (**a**) the longitudinal arch; (**b**) the transverse arch; (**c**) alteration of weight distribution during the wearing of high-heeled shoes; (**d**) x ray showing medial view (x ray courtesy of Alexandria Hospital, Alexandria, Virginia, Joyce R. Isbel, R. T.).

phalanges. There are seven tarsal bones in the ankle, which bears the weight of the body and functions as a shock absorber for the forces resulting from such activities as walking, running, and jumping. The most superior tarsal bone, the *talus,* unites inferiorly with the *calcaneus,* the largest of the tarsals, and anteriorly with the *navicular,* which is interposed between the talus and the three *cuneiform* bones. The calcaneus projects posteriorly as the palpable heel and anteriorly beneath the talus to join the *cuboid* bone, which also articulates with the two lateral metatarsals. The three cuneiform bones articulate with the three medial metatarsals.

Each of the five metatarsals, like the metacarpals, has a base, shaft, and head and articulates distally with the proximal pha-

lanx of a digit. As the tarsals and metatarsals are pulled tightly together, they form the longitudinal and transverse arches of the foot (Fig. 8-13). In standing, the normal foot contacts the ground at three points: at the calcaneus, at the head of the fifth metatarsal, and at the head of the first metatarsal (usually supported by two underlying sesamoid bones). The longitudinal arch functions as a shock-absorbing spring and is higher medially than laterally. The less prominent transverse arch is seen most readily at the heads of the metatarsals where only the first and fifth normally make contact with the surface.

The transverse arch extends posteriorly to the bases of the metatarsals. The toes touch the ground but do not support much of the weight of the body in normal activity. The bones of the

  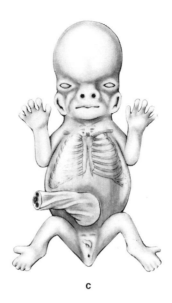

a   b   c

Figure 8-14 Ventral view of embryos, illustrating limb development: (**a**) limb bud stage; (**b**) prior to rotation of the limbs; (**c**) early phase of limb rotation. (See text for description.)

toes, like those of the fingers, are called phalanges and are similar to those of the fingers. The first (big) toe has two phalanges, and all the others have three.

If the transverse arch fails, the centrally placed metatarsals, not normally weight-bearing, may be unable to withstand the strain of supporting the body and may fracture during walking. A fracture of this kind is called a march fracture.

DEVELOPMENT OF THE LIMBS

For some time after the closure of the body wall, as the embryo attains its general mammalian form, it has no limbs. The first indication that limbs are about to appear is the formation of short, paired ridges that project laterally from the cranial and caudal ends of the trunk region, at about the end of the fourth week of gestation. These ridges are the *limb buds,* developing as thickenings in the *somatic mesoderm* of the trunk wall. As the limb buds lengthen (Fig. 8-14), they initially project laterally at right angles to the trunk, each having a dorsal and ventral surface. A slight constriction appears in the elongated bud, marking the region of the future elbow or knee. The limbs then bend ventrally on the trunk so that the former ventral surfaces become medial. Next the upper and lower limbs undergo a 90 degree rotation in opposite directions. The upper limb rotates clockwise so that its medial surface (palm of hand) faces ventrally (anteriorly), with the thumb on the lateral side. The lower limb rotates counterclockwise so that its medial surface (plantar surface of the foot) faces posteriorly with the big toe on the medial side.

Within the limb buds the skeletal elements differentiate in the mesenchyme along the proximal-distal axis of the central portion of the rodlike limb. The *pre-muscle masses* that appear *ventral* to the bones differentiate into *flexors,* and those that appear *dorsal* to the bones become the *extensors* of the limbs. Due to the opposite rotation of the upper and lower limbs the extensors of the upper limb are on the posterior surface of the arm, whereas those of the lower limb are on the anterior surface of the thigh. The pre-muscle masses appear in the limb mesenchyme with no connection to the somites from which the skeletal muscles of the trunk are derived. However, the pattern of innervation suggests a myotomal origin of limb muscles, perhaps resulting from migration of myotome cells into the early limb buds.

In the embryo the development of the upper limb precedes that of the lower limb, so at birth the upper limb is longer than the lower limb. After birth the lower limb grows more rapidly than the upper and attains approximately equal length in about two years. After puberty the lower limb exceeds the length of the upper limb by about one-sixth.

Clinical Applications: **Bone Fractures and Repair**

A bone may be broken, or fractured, as a result of disease, but most fractures (Fig. 8-15) are due to trauma (injury). A fracture is usually accompanied by soft-tissue damage such as bruised muscles or ruptured blood vessels. Traumatic fractures are termed *incomplete* (greenstick) if the bone is only partially severed, or *complete* if there is discontinuity of the bone. A third type of fracture, termed *comminuted,* occurs when a bone is splintered by more than one fracture line, so that bone fragments occur at the fracture site. Fractures are also classified as *simple* if the skin covering the area remains intact, or *compound* if the broken bone protrudes through the skin. A compound fracture is the more dangerous because of the greater likelihood of infection at the site of the injury.

Fractures may be described according to the location or direction of the fracture line in bone as *transverse, oblique, longitudinal,* or *spiral.* These descriptive terms may be used in addition to the terms complete or incomplete and simple or compound.

The repair of a fracture is facilitated by approximating (bringing together) the fracture ends and stabilizing them by means of a splint or cast, or by inserting a bone pin. Proper alignment of fracture ends is essential for optimal repair.

When a bone is broken, blood vessels within the bone and adjacent periosteum rupture, and a blood clot soon forms in the fracture site. This clot is essential for the normal healing process, for it provides the substrate (physical base) upon which the cells involved in the healing process are organized. Phagocytic cells from the blood (granulocytes and monocytes) migrate into the clot and begin ingesting the red blood cells. This produces a granulation tissue that is more extensive, usually, than the gap in the bone, extending into the marrow cavity and protruding beyond the external surface of the bone. Fibroblasts and osteoblasts migrate in from the periosteum and endosteum, organizing the clot into a *procallus* (tissue that precedes callus), which rapidly becomes fibrous with the elaboration of collagenous fibers. Organization of the procallus usually occurs within a week.

The repair process varies after the procallus stage, depending on the severity of the fracture and the adequacy of the blood supply in the area. The procallus may be transformed into fibrocartilage prior to bone deposition. The formation of fibrocartilage, termed *temporary callus,* affords some support but is inadequate to support the weight of the body. Bone replaces the temporary callus by endochondral ossification, quite similar to ossification in the development of a long bone. The cartilage matrix undergoes calcification, and osteoblasts deposit bone matrix on the calcified cartilage rods, producing spongy bone that bridges the gap at the fracture site. This bone, termed the *callus,* bridges the marrow cavity and extends beyond the external surface of the bone, causing a palpable bump in regions where the fracture site is not covered by muscle. Over a period of several months the spongy bone of the callus undergoes reorganization into compact bone, and excess bone is eliminated, leaving a smooth surface with little or no evidence of the fracture.

PATTERNS OF OSSIFICATION OF THE BONES OF THE EXTREMITIES

The long bones of the extremities develop from a primary ossification center in the shaft and one or more secondary ossification centers at each end. There is a characteristic time of appearance of the ossification centers for each bone. Primary centers appear at different times in different bones, but virtually all appear before birth; most between the seventh and twelfth weeks of intrauterine life.

The times at which the secondary centers of ossification appear and at which they fuse with the primary centers may be useful in determining the ages of individuals. The time of appearance of a secondary center in any one epiphysis and of its fusion with the primary center varies in different children, but the sequence is quite constant and the time intervals are much the same in different individuals. In other words, if fusion takes place early in one bone, it will occur early in all bones of that person.

It is a general rule that a secondary ossification center that appears early in a long bone will fuse late and *vice versa.* For example, the secondary center of the head of the humerus appears at about the time of birth, but it does not fuse with the primary center until the person is 18 to 21 years old. The secondary centers of the distal end of the humerus appear between 2 and 14 years of age and fuse relatively quickly at 14 to 18 years of age. The end of the humerus with the epiphysis, which fuses (closes) late, grows for a longer period. Thus the proximal end of the humerus is "the growing end" of the long bone.

The presence of secondary ossification centers may be used to determine if a baby is full term. For example, if x-ray films of a newborn infant show ossification centers in the proximal ends of the humerus, femur, and tibia, then it may be assumed that the baby was full term when born.

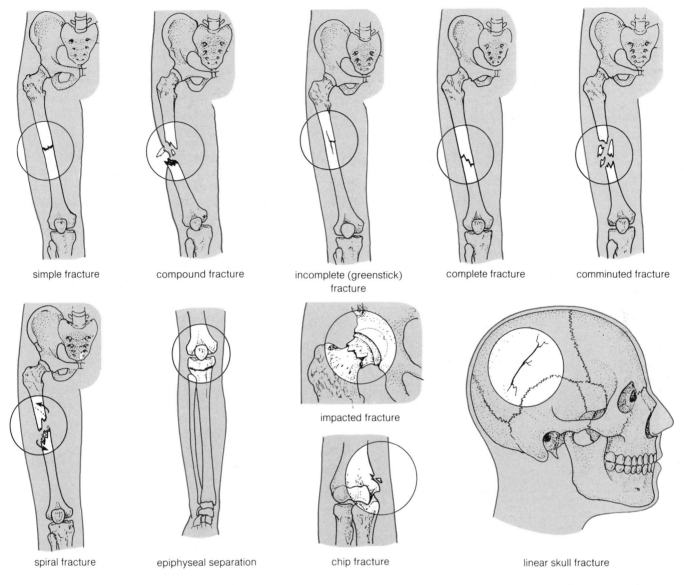

simple fracture

compound fracture

incomplete (greenstick) fracture

complete fracture

comminuted fracture

spiral fracture

epiphyseal separation

impacted fracture

chip fracture

linear skull fracture

Figure 8-15 Some different kinds of fractures.

WORDS IN REVIEW

Acetabulum	Coracoid process	Manus	Phalanges	Thigh
Acromion process	Femur	Olecranon process	Pubic symphysis	Trochanter (greater and lesser)
Antebrachium	Fibula	Os coxae	Pubis	Ulna
Brachium	Humerus	Pectoral girdle	Radius	
Carpus	Ilium	Pelvic girdle	Scapula	
Clavicle	Ischium	Pes	Tarsus	

SUMMARY OUTLINE

I. The Appendicular Skeleton.

The appendicular skeleton consists of bones of the pectoral and pelvic girdles and the attached limbs.

II. The Bones of the Upper Limb

A. The pectoral girdle, consisting of the scapula and the clavicle, provides support for the upper limb.

1. The scapula is a triangular bone on the posterior aspect of the thorax; it articulates with the clavicle at the acromion process and with the humerus at the glenoid fossa.

2. The clavicle is a curved bone on the anterior thorax; it articulates medially with the sternum and first costal cartilage.

B. The arm is that part of the upper extremity containing the humerus; it extends from the shoulder to the elbow.

1. The humerus is a long bone with an epiphysis at each end of the diaphysis.

2. The humerus articulates proximally with the scapula and distally with the ulna and radius.

C. The forearm is that part of the upper extremity containing the ulna and radius; it extends from the elbow to the wrist.

1. The ulna is the medial bone of the forearm; it articulates proximally with the humerus and radius and distally with the radius and a carpal bone, the triquetrum.

2. The radius is the lateral bone of the forearm; it articulates proximally with the humerus and ulna and distally with the ulna and two carpals, the scaphoid and lunate bones.

D. The hand is composed of carpals, which form the wrist; metacarpals, which form the hand proper; and phalanges, which form the digits.

1. The carpals consist of two rows of four bones each; they articulate proximally with the ulna and radius and distally with the metacarpals.

2. The metacarpals consist of five bones in a single row; each is in line with a digit.

3. The phalanges number 14 bones; three are arranged in series in each finger and two in the thumb.

III. The Bones of the Lower Limb

A. The pelvic girdle consists of two os coxae that articulate with each other in the anterior midline at the pubic symphysis and with the sacrum posteriorly.

1. The os coxae are composed of the ilium, ischium, and pubis, which meet and fuse at the acetabulum, the socket for the head of the femur.

2. The ilium is a broad, flared bone that articulates posteriorly with the sacrum; its prominent features include the iliac crest, the anterior superior and anterior inferior iliac spines, and the greater sciatic notch.

3. The ischium consists of a body fused with the ilium and a ramus fused with the pubis; prominent features include the ischial spine and the ischial tuberosity.

4. The pubis consists of a body that articulates with the pubis of the opposite side and two rami; prominent features include the pubic crest, pubic tubercle, and the pectineal line.

B. The thigh is the part of the lower extremity that contains the femur and patella; it extends from the hip to the knee.

1. The femur is a long bone with a proximal head that fits into the acetabulum and a neck between the head and the shaft; the shaft bears the greater and lesser trochanters proximally and the lateral and medial epicondyles distally; two condyles articulate distally with the tibia.

2. The patella (kneecap) is a sesamoid bone in the tendon of the quadriceps femoris muscle.

C. The leg is the part of the lower extremity containing the tibia and fibula; it extends from the knee to the ankle.

1. The tibia is the medial bone of the leg; it bears two condyles proximally for articulation with the femur. Distally the medial malleolus projects inferior to the articular surface. The tibia articulates distally with the fibula and the talus.

2. The fibula is the lateral bone of the leg. It is a slender bone that articulates proximally with the tibia and distally with the tibia and the talus.

D. The foot is composed of tarsals, which form the ankle, metatarsals, which form the instep and sole; and phalanges, which form the digits.

1. The tarsals consist of seven bones that bear the weight of the body. They articulate proximally with the tibia and fibula and distally with the metatarsals.

2. The metatarsals consist of five bones in a single row; each bone is in line with a toe.

3. The phalanges include 14 bones; the great toe has two bones and each of the other toes has three.

IV. Limb Development

The limbs develop from thickenings in somatic mesoderm that grow distally, differentiating along a proximal-distal axis; in development the upper and lower limbs rotate in opposite directions so that the anterior surfaces of the upper limbs and posterior surfaces of the lower limbs are the flexor surfaces.

REVIEW QUESTIONS

1. The scapula overlies the posterior aspect of what specific part of the rib cage?

2. With what part of the scapula does the clavicle articulate?

3. With what part of the scapula does the humerus articulate? What are the protuberences to which muscles attach near the proximal end of the humerus?

4. What process of the ulna can be palpated on the back of the elbow?

5. What process of the radius can be palpated on the lateral aspect of the radius?

6. Contrast the carpal bones with the tarsals in regard to numbers of bones, arrangement of the bones, connections to bones of the forearm or leg, and connections to bones of the hand or foot proper.

7. Contrast the pectoral and pelvic girdles with respect to numbers of bones, arrangement of the bones, and connections to bones of the axial skeleton.

8. With what bones does the femur articulate?

9. In what way is the patella different from other bones of the lower extremity?

10. What is the significance of the rotation of the limb buds during the development of the limbs?

SELECTED REFERENCES

Langebartel, D. A. *The Anatomical Primer. An Embryological Explanation of Human Gross Morphology.* Baltimore: University Park Press, 1977.

Hollinshead, W. H. *Anatomy for Surgeons. The Limbs and Back.* Vol. 3. 2nd ed. New York: Harper & Row, 1971.

Snell, R. S. *Clinical Anatomy for Medical Students.* Boston: Little, Brown & Co., 1981.

9

JOINTS AND OTHER FRICTION-REDUCING DEVICES

OBJECTIVES

After completing this chapter you should be able to:

1. Define or describe the following friction-reducing devices.
   A. A fascial cleft
   B. A bursa
   C. A tendon sheath
   D. A synovial cavity

2. Describe the following types of joints.
   A. Synarthrosis
      1. Syndesmosis
      2. Synostosis
      3. Synchondrosis
   B. Synovial (diarthrosis)

3. Classify the ligaments of synovial joints according to their relations to the joint capsules.

4. Describe the mechanisms that contribute to the stability of synovial joints.

5. Classify and describe the various diarthrodial joints and give examples of each type.

Movement is a continuous and essential characteristic of life in all the tissues and organs of the body. Within cells molecules and organelles constantly stream from one part of the cell to another during metabolic processes such as protein synthesis and secretion. Friction between cellular components is not a problem because of the fluid nature of cytoplasm. On a larger scale, however, the sliding of multicellular masses against one another does create problems because of their weight and the way connective tissues bind structures together. The body has special devices for reducing friction, the complexity of which depends upon such factors as the amount of movement that occurs between adjacent structures and the amount of weight borne at the contact points. Because most movements that cause friction result from the activity of the muscles, the friction-reducing devices will be described before considering the specific components of the muscular system involved in the production of movement.

STRUCTURE OF FRICTION-REDUCING DEVICES

To function satisfactorily, friction-reducing devices must permit an appropriate range of movement while stabilizing structures

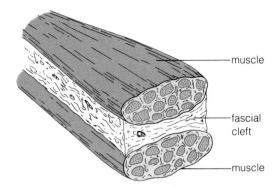

Figure 9-1 The fascial cleft, loose connective tissue that separates adjacent muscles.

in proper relation to each other. Some body parts have far greater freedom of movement than others; for example, the almost unlimited motion possible at the shoulder joint greatly exceeds that permitted at the elbow joint, where movement is virtually restricted to one plane. Stabilizing the shoulder joint therefore requires a different type of anatomical construction from that needed to stabilize the elbow joint. Heavy *collagenous fibers* support joints like the elbow, but they are inappropriate for the shoulder joint because they would restrict movement. The shoulder joint is therefore supported principally by the tendons of *muscles* that cross the joint.

Bones are not the only structures that move relative to one another, and joints are not the only friction-reducing devices in the body. Movement can occur between two structures—adjacent muscles, for example, or a bone and a tendon—that do not require special support. Fascial clefts, bursae, and tendon sheaths prevent friction between such structures. Bursae and tendon sheaths have synovial membranes, which are also involved in the structure of many joints.

Fascial Cleft

A **fascial cleft** (Fig. 9-1) is an area of loose connective tissue separating two compact structures, such as adjacent muscles, that exert little tension or pressure on each other. Adjacent muscles that contract independently must slide against each other with little resistance. Such muscles are encased in deep fascia whose connective tissue fibers are oriented in the same direction as the muscle fibers. The connective tissue that fills the fascial cleft, or plane, is less dense than the fascial covering of the muscles, and the orientation of its fibers has little relation to the orientation of the muscle fibers. This fascia allows one muscle to move without moving the other by separating the muscle surfaces. The abundant intercellular fluid in the loose areolar connective tissue of the fascial cleft reduces the possibility of the two muscles adhering to one another.

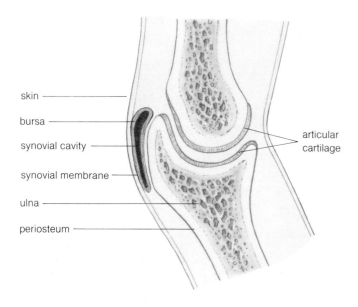

skin

bursa

synovial cavity

synovial membrane

ulna

periosteum

articular cartilage

Figure 9-2 A bursa.

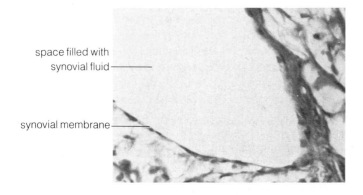

space filled with synovial fluid

synovial membrane

Figure 9-3 Histological structure of the subdeltoid bursa of the rat, × 450. (Courtesy of B.F. Sisken.)

Bursa

A **bursa** (Fig. 9-2) is a fluid-filled sac enclosed by a **synovial membrane** (Gr., *syn* = with, *oon* = egg). A synovial membrane consists of a thin sheet of fibrous connective tissue, usually with an incomplete layer of flattened connective tissue cells that resemble epithelium at the inner surface of the sheet. Bursae develop from fascial clefts between structures whose movements apply tension or pressure as they slide on one another. Movement stimulates the accumulation of intercellur fluid within the fascial cleft; and as the fluid collects, a synovial membrane (Fig. 9-3) differentiates, forming a sac that assumes a shape dic-

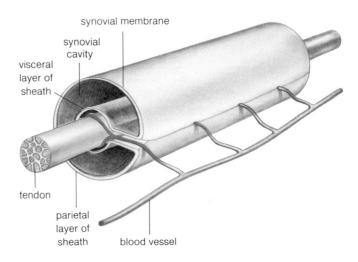

synovial membrane

synovial cavity

visceral layer of sheath

tendon

parietal layer of sheath

blood vessel

Figure 9-4 A tendon sheath. A double layer of synovial membrane enclosing the tendon.

tated by the pressure of the adjacent structures. Numerous blood capillaries invade the synovial membrane and become the source of the fluid, now called **synovial,** that fills the bursa. Bursae between adjacent muscles, between muscles and bone, between tendon and bone, or between bone and skin permit much greater movement with less friction than do fascial clefts.

Tendon Sheath

A **tendon sheath** (Fig. 9-4) is a double-layered tubular sleeve of synovial membrane that forms a fluid-filled **synovial cavity** enclosing a tendon. It is more complex than a bursa and, unlike a bursa, completely encloses the structure it protects. The inner layer of synovial membrane fuses anatomically with the surface of the tendon. The other layer forms the external boundary of the cavity, which contains the synovial fluid that lubricates the membranes as they slide on one another when the tendon moves.

Disruption of the synovial membrane from infection or injury may lead to abnormal fusion of the outer and inner layers of the tendon sheath. Then any muscle contraction that applies tension to the tendon also pulls on the outer synovial membrane, which initiates impulses in nearby nerves, causing pain. Abnormal fusion of structures is generally referred to as *adhesion,* and most adhesions are painful.

Synovial membranes enclose tendons wherever the tendons are pulled down against bone by muscle contractions. Hyaline cartilage covers the bone surface at the points of contact. At a finger joint, for example, the tendon slides across the joint capsule covering the hyaline cartilage of the articular surface of the bones as the fingers bend; the synovial membrane and the smooth surface of the hyaline cartilage reduce friction.

A tendon sheath can enclose several tendons, as at the wrist. See Fig. 9-5 for a diagram of the complex arrangement of tendon sheaths in the wrist and hand. The sheath that encloses several tendons in the wrist and palm is continuous with the sheaths of the thumb and little finger, but the other three fingers have independent sheaths. Infection of the tendon sheath in any part of the complex usually spreads rapidly throughout the sheath but does not spread to the independent sheaths.

Tendon strains consist of irritative lesions or inflammation of the tendon attachments, the sites of greatest stress and the poorest blood supply. Many of these conditions are due to chronic overstrain, and some are specifically associated with a certain sport. For example, *lateral epicondylitis* is most frequently associated with the game of tennis and is known as "tennis elbow." In a backstroke, the wrist and forearms are pronated, placing the extensors of the wrist and hand under maximum tension. These extensors take origin from the lateral epicondyles of the humerus. Thus, constant use of the extensors may cause irritation in the area of the lateral epicondyle, causing tenderness in the area and pain on extending the elbow on pronation. If there are no complications, treatment for three to six weeks will usually be enough to allow the individual to resume the sport.

It is interesting to note that *medial epicondylitis,* or "golfer's elbow," is essentially the opposite of tennis elbow. The flexors of the wrist and hand arise mostly from the medial epicondyle. They are placed under tension as the forearm flexes with the supination that occurs when a golf ball is driven. The symptoms and treatments are the same as for tennis elbow.

The tendon sheath is an effective friction-reducing device but is not designed to bear weight; excessive pressure would quickly erode the synovial membrane. Specialized synovial cavities are specifically modified for reducing friction while bearing weight.

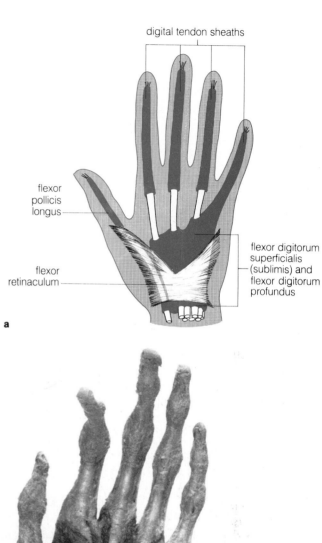

a

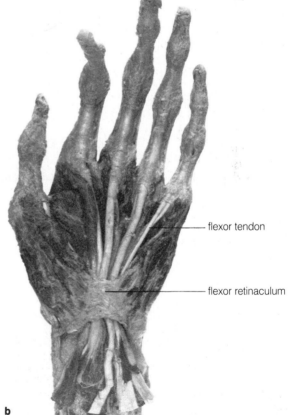

b

Figure 9-5 Wrist and hand. (**a**) Tendon sheaths, anterior aspect. (**b**) Flexor tendons of cadaver preparation (photograph by R. Ross).

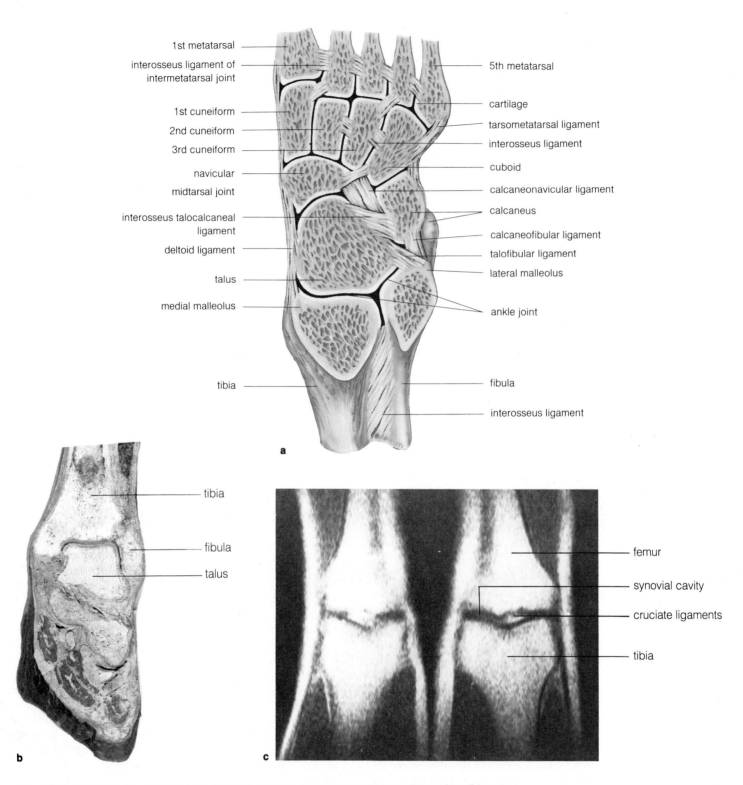

1st metatarsal

interosseus ligament of intermetatarsal joint

1st cuneiform

2nd cuneiform

3rd cuneiform

navicular

midtarsal joint

interosseus talocalcaneal ligament

deltoid ligament

talus

medial malleolus

tibia

5th metatarsal

cartilage

tarsometatarsal ligament

interosseus ligament

cuboid

calcaneonavicular ligament

calcaneus

calcaneofibular ligament

talofibular ligament

lateral malleolus

ankle joint

fibula

interosseus ligament

a

tibia

fibula

talus

b

femur

synovial cavity

cruciate ligaments

tibia

c

Figure 9-6   Synovial joints: (**a**) right ankle and foot, shown in a section through the ankle and foot; (**b**) vertical section of a cadaver preparation (photograph by R. Ross); (**c**) nuclear magnetic resonance (NMR) image of a frontal section of the knee joints of a 25-year-old normal male (courtesy of Paul Wang); see box adjacent to Fig. 7-24 for explanation of NMR.

Synovial Cavity

A synovial cavity, or joint cavity (Fig. 9-6), is the fluid-filled space between the articular surfaces of bones at freely *movable joints*. The synovial cavity of a joint differs from a bursa or tendon sheath in that the synovial membrane does not completely line the cavity; it lines the *joint capsule* without covering the articular cartilages at the surfaces of the bones. The *joint cavity* is enclosed by the joint capsule, which is lined with synovial membrane, and by the articular surfaces of the bones, which are lined with hyaline cartilage. The cavity is filled with synovial fluid, which lubricates and cushions the joints, such as the knee, that must bear the weight of the body while moving.

CLASSIFICATION OF JOINTS

A joint is the point where bones articulate, and the classification of joints is based upon the type and arrangement of the tissues that connect them. The material that binds bones together may be *fibers* (ligaments), *cartilage*, or even *bone* itself (in which case the continuity between bones allows little or no movement). In a *movable joint* a synovial cavity separates the contiguous articular surfaces.

Synarthrosis

Synarthrosis is the general classification of all joints that have no synovial cavity. These joints acquire stability at the sacrifice of flexibility. Synarthrodial joints include syndesmoses (slightly movable fibrous joints), sutures (immovable fibrous joints), and synchondroses (cartilaginous joints) which include symphyses and intervertebral discs.

Syndesmosis   A **syndesmosis** (Gr., *syn* = together, *desis* = a binding, *osis* = condition) is a joint in which bones are bound by ligaments that may be long enough to permit some movement by stretching. These ligaments cross the joint and attach to the bones by interlacing with the periosteum that covers them. The *distal tibiofibular joint* permits some flexibility in this fashion, a highly desirable feature because a rigid joint would lead to frequent fractures of the slender fibula (Fig. 9-7).

Sutures

A suture (L., *sutura* = a seam) is a joint in which adjacent bones are united by collagenous fibers that cross the narrow space separating the two bones, binding them together tightly so that no movement is possible. (Fig. 9-8). This type of union occurs between the flat bones of the skull where rigidity is essential

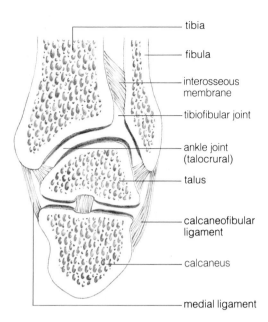

Figure 9-7   A syndesmosis: diagram of the ankle joints, vertical section. Note the distal tibiofibular joint in which the bones are connected by collagenous fibers of the interosseous membrane.

for protecting the brain. It is characteristic of sutures that the aging process frequently includes the deposition of bone within the joint, replacing the fibrous connective tissue; when bone completes the union, the ossified joint is classified as a **synostosis** (Gr., *syn* = together, *osteon* = bone).

Synchondrosis   In a **synchondrosis** such as a **symphysis** (Gr., *symphysis* = a growing together) the articular surfaces of adjacent bones are covered with *hyaline cartilage* and are connected by *fibrocartilage*. In the *symphysis pubis*, (Fig. 9-9), which joins the two pubic bones of the two halves of the pelvic girdle in the anterior midline, the union is tight and prevents movement. In the *intervertebral discs* the center of the fibrocartilage is occupied by a soft, gelatinous mass called the *nucleus pulposus* which facilitates some movement (Fig. 9-10). The movement between any two vertebrae is slight, but the collective movement of all twenty-four intervertebral discs between the vertebrae of the neck, thorax, and lumbar regions allows considerable flexibility. Besides providing for flexibility, the nucleus

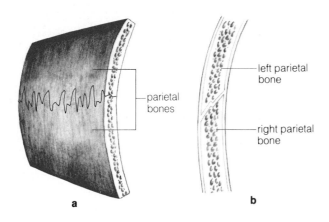

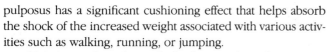

Figure 9-8 Diagram of a suture: (**a**) surface view, (**b**) transverse section.

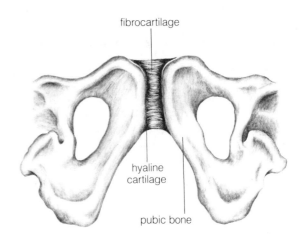

Figure 9-9 The pubic symphysis, anterior aspect.

pulposus has a significant cushioning effect that helps absorb the shock of the increased weight associated with various activities such as walking, running, or jumping.

The union of two bones by hyaline cartilage alone occurs between the primary and secondary ossification centers joined by the epiphyseal plate and constitutes a temporary synchondrosis. Normally movement does not occur at the epiphyseal plate.

Diarthrosis

A *synovial joint* (Fig. 9-11) contains a cavity filled with synovial fluid and is lined in part by a synovial membrane and in part by the articular cartilages of the opposing bones. The synovial membrane lining the fibrous joint capsule elaborates the synovial fluid. Hyaline cartilage covers the articular surfaces of adjacent bones, providing a surface that is both smooth and slightly compressible, and assists in absorbing the shock associated with movement and weight-bearing. The fibrous capsule encircles the synovial cavity and connects the two bones by becoming continuous with their periosteum at the edges of the articular cartilages. A tight capsule restricts movement and may allow it in one plane only.

As is generally true of connective tissue, the strength of the fibrous capsule depends chiefly upon the intermittent stress to which it is subjected during its development. The intermittent stress of exercise strengthens joint capsules by stimulating the

production of collagenous fibers, thickening the fibrous capsule. There is a great deal of variation in the strength of the capsules of different joints in the body and in the shapes of the articular surfaces of bone.

There are also differences in complexity within synovial cavities that depend upon the presence or absence of a complete or incomplete *articular disc* or a *bursa*. Most of the movable joints of the body have a single synovial cavity that is uncomplicated by extensions or partitions (Fig. 9-11a). In the fingers, for example, this simple type of joint has a capsule reinforced on the lateral and medial sides of the joint by ligaments that restrict movement to flexion and extension in a single plane. Simple synovial joints include all the joints by which the ribs articulate with the vertebrae and all the joints of the extremities except the knee.

The articular disc of the knee may be injured in violent sports, such as football. When the capsule is tight, as when the leg is fully extended, a sudden, unexpected blow from the side—say a football block—may stretch the capsule, tearing the attached cartilaginous disc. Because cartilage repairs slowly, the injury may be serious, taking a long time for the knee joint to return to normal function. In some cases the meniscus must be removed to facilitate recovery of function.

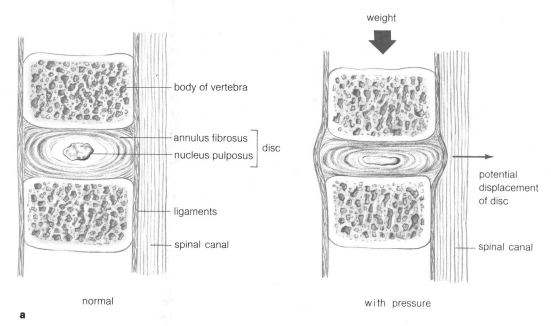

body of vertebra

annulus fibrosus
nucleus pulposus — disc

ligaments

spinal canal

normal

weight

potential displacement of disc

spinal canal

with pressure

a

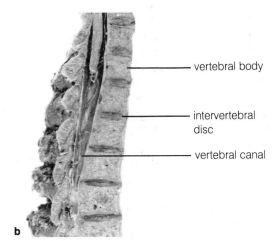

vertebral body

intervertebral disc

vertebral canal

b

Figure 9-10 A symphysis. (**a**) An intervertebral joint. (**b**) A sagittal section of the vertebral column, a cadaver preparation (photograph by R. Ross).

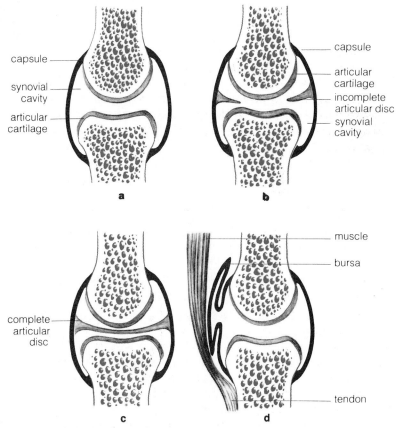

capsule

synovial cavity

articular cartilage

a

capsule

articular cartilage

incomplete articular disc

synovial cavity

b

complete articular disc

c

muscle

bursa

tendon

d

Figure 9-11 Types of synovial joints: (**a**) no articular disc; (**b**) incomplete articular disc; (**c**) complete articular disc; (**d**) synovial joint cavity with connecting bursa.

The synovial cavity of the knee joint is the largest in the body and is partially separated into upper and lower cavities by a crescentic *medial* and *lateral articular disc* (**or meniscus**) of fibrocartilage. The free edge of this incomplete articular disc projects into the cavity (Fig. 9-11b) and attaches peripherally on the medial side to the articular capsule, which is tight because of the presence of a heavy ligament.

In the sternoclavicular and the temporomandibular joints a complete articular disc of fibrocartilage divides the synovial

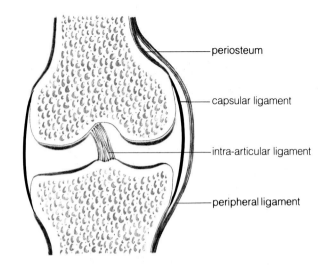

peripheral ligament

intra-articular ligament

capsular ligament

periosteum

Figure 9-12 Position of ligaments in relation to synovial joints.

cavity into two cavities (Fig. 9-11c). Movement occurs between the articular disc and one of the bones, with the other bone remaining fixed at the articulation. For example, in the temporomandibular joint the articular surface of the mandible (mandibular condyle) and the articular disc slide forward as the mouth opens, with no movement of the temporal bone. Each synovial cavity of this type of joint is lined by synovial membrane on the inner surface of the fibrous capsule and by the fibrous tissue or fibrocartilage of the articular disc, in addition to the hyaline cartilage of the articular surfaces of the bone involved.

In other joints the synovial cavity is expanded by communication with a bursa (Fig. 9-11d). The synovial cavity of the knee joint extends superiorly, anterior to the femur and communicates with the suprapatellar bursa between the extensor muscles on the anterior surface of the thigh and the femur. This bursa reduces friction between the muscles and the bone during extension and flexion of the leg.

CLASSIFICATION OF LIGAMENTS OF DIARTHROSES

Ligaments are bands of collagenous fibers that connect bones. Because intermittent stress stimulates the production of collagenous fibers, tension at a joint leads to the formation of a ligament whose fibers are oriented to resist the stress. The ligament's strength develops in direct proportion to the tension exerted upon it. Thus the ligaments of weight-bearing joints such as the knee, where traction is frequently great, are heavier than those of the shoulder joint, which is not constantly carrying a heavy load.

Ligaments may develop in three places relative to a joint and are classified accordingly (Fig. 9-12). **Capsular ligaments** are simply localized thickenings in the fibrous joint capsule and are usually given names that include the word *capsular*. The thickness of capsular ligaments varies, being generally heavier in muscular individuals than in those whose muscles are small. This thickness is probably the result of weight rather than of muscle activity. **Intra-articular ligaments** are fibrous bands that cross the synovial cavity from one articular cartilage to the other. In the knee joint, for example, intra-articular ligaments (called cruciate ligaments) may function as "check" ligaments, preventing hyperextension of the leg, but they appear to provide little support for the joint, functioning mainly as pathways for the blood vessels that supply joint structures. **Peripheral ligaments** are fibrous bands that connect bones by crossing joints external to the joint capsule. The names of these ligaments usually include the word *collateral*; for example, the **lateral collateral ligament** of the knee crosses the lateral aspect of the joint to connect the femur and the fibula.

MECHANISMS PROVIDING FOR THE STABILITY OF DIARTHROSES

Since a certain amount of movement is possible at all synovial joints, the articulation cannot be absolutely rigid. The stability of most joints depends in varying degrees on three anatomical features (Fig. 9-13): the way the bones fit at their articular surfaces (bony configuration), the strength of the ligaments crossing the joint, and the tone (constant contraction) of the muscles crossing the joint.

Bony Configuration

In ball-and-socket joints the rounded articular surface of one bone fits into the cuplike socket of the other, providing bony support that helps prevent dislocation of the joint. The effectiveness of this bony configuration depends largely upon the depth and strength of the wall of the socket. In the hip joint (Fig. 9-13a), for example, the heavy rim of the socket (the acetabulum) almost completely encloses the head of the femur, making it difficult to dislocate the joint without withdrawing the head from the socket. This construction of the hip joint restricts movement: less posterior movement (extension) is possible than anterior movement (flexion).

Of course, the hip joint receives some support from the joint capsule and from the ligaments that cross the joint, as well as from the muscles acting on the joint. All these are important in keeping the head of the femur in the acetabulum, but the ability of the hip joint to bear the weight of the body depends primarily on the bony configuration. In some ball-and-socket joints such as the shoulder joint, the socket is shallow, allowing great freedom of movement but requiring support by some other means.

> The hip joint is most easily dislocated with the hip in a flexed position. With the femur in the adducted position the head of the femur may be dislodged from the socket without fracture of the acetabulum. With the femur abducted, dislocation by shoving the femur posteriorly is accompanied by fracture of the lip of the acetabulum.

Ligaments

The strength of collagenous ligaments provides stability for joints, but their lack of elasticity reduces the flexibility of the joint and restricts movement to that allowed by the orientation of the fibers of the ligaments. A tight, fibrous capsule tends to restrict movement at any synovial joint. For example, in the sacroiliac joint (Fig. 9-13b) the principal structures holding the bones together are the strong *intra-articular (interosseous) ligaments* that bind the articular surfaces together in the posterior half of the joint. So many intra-articular ligaments cross these articular surfaces that the synovial cavity is reduced to a small cavity in the anterior part of the joint. Collateral ligaments crossing the synovial capsule externally provide additional support to maintain the integrity of the sacroiliac joint. The interosseous ligaments, which are short, cross the narrow gap between adjacent bones and stabilize the joint for the purpose of supporting the weight of the body; but in so doing they virtually eliminate the possibility of movement.

Muscles

The greatest freedom of movement occurs at joints where stability is provided chiefly by muscles; muscle contractions here not only produce movement but also keep the articular surfaces of the adjacent bones in their proper relations with one another. The best example of such a joint is the shoulder joint (Fig. 9-13c), a ball-and-socket joint with a shallow socket. The *glenoid fossa* of the scapula is somewhat concave with cartilage built up around its rim, increasing the depth of the socket. Despite the cartilage rim the fossa is still too shallow to surround the head of the humerus, and the capsule of the shoulder joint is thin and loose without heavy capsular ligaments that would limit movement. Short muscles that arise from the scapula and insert on the humerus near its head provide the principal support and function in the rotation of the arm and in keeping the head of the humerus in the glenoid fossa.

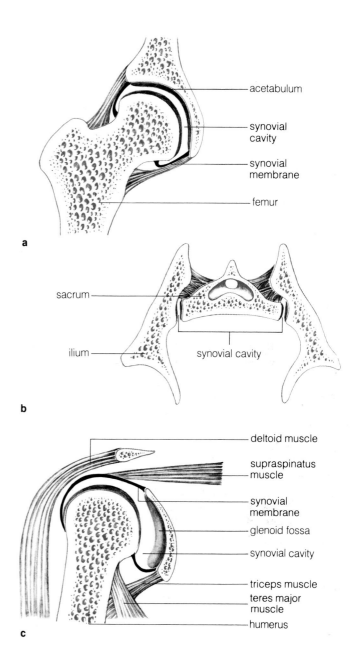

Figure 9-13 Diagrams illustrating mechanisms providing stability for synovial joints: (**a**) hip joint, stability provided by bony configuration; (**b**) sacroiliac joints, stability provided by ligaments; (**c**) shoulder joint, stability provided by muscles.

> The shoulder joint is the most commonly dislocated major joint. The joint is least protected inferiorly, and the head of the humerus is easily pulled out of the glenoid fossa when the arm is abducted.

Table 9-1 Classification of Diarthroses

Type of Joint	Structural Characteristics	Movements Permitted	Examples
Plane (articulatio plano)	Articular surfaces are flat	Sliding	Wrist, ankle
Hinge (ginglymus)	Bone conformation restricts extension	Flexion, extension	Elbow, interphalangeal
Ball-and-socket (articulatio spheroidea)	Cup-like articular fossa into which head of opposing bone fits	All movements	Shoulder, hip
Pivotal (articulatio trochoidea)	Rounded head fits into notch of another bone	Rotation	Proximal ulna-radial, atlas articulation with axis
Ellipsoid (articulatio ellipsoidea)	Rounded condyle fits into elliptical depression of another bone	Flexion, extension, abduction, adduction	Metacarpophalangeal
Saddle (articulatio sellaris)	Rounded projection fits into saddle-shaped depression of another bone	Flexion, side to side	Trapezium with metacarpal of thumb

CLASSIFICATION OF DIARTHROSES

The shapes of the articulating surfaces determine to a large extent the motion possible at a joint and provide a means of classifying diarthodial joints. This classification is summarized in Table 9-1.

Plane Joints

Plane joints (articulatio plano), such as most of the *wrist* (Fig. 9-14d) and *ankle joints* (Fig. 9-14a,e), are the simplest type of synovial joint, in that the articular surfaces of opposing bones are virtually flat so that only a *sliding movement* is possible. Movement is greatly restricted by the fibrous capsule and by collateral ligaments.

Hinge Joints

Hinge joints (ginglymus) allow movement in only *one plane,* usually flexion and extension, with lateral movement prevented by strong collateral ligaments. The conformation of the bones usually restricts extension to about 180 degrees or a straight line between the bones at the joint. This type of joint is found in the *interphalangeal joints* of the fingers and in the *elbow* (Fig. 9-14b).

Ball-and-Socket Joints

The **ball-and-socket joint** (articulatio spheroidea) has a rounded, ball-like articular surface on one bone that fits into a concave, cup-like articular fossa on the opposing bone. This joint permits a *full range of movement,* including flexion, extension, abduction, adduction, rotation, and circumduction. Movement is restricted ultimately by the bone contours and the arrangement of muscles in relation to the joint. The *hip joint* (Fig. 9-14c) and *shoulder joint* are examples of ball-and-socket joints.

Pivotal Joints

In **pivotal joints** (articulatio trochoidea) the manner in which the bones fit together and the way in which the ligaments are arranged permit only *rotation* of one bone on the other. This type of joint is found in the forearm at the *proximal articulation* of the *radius* with the *ulna*. During pronation and supination the head of the radius rotates in a ring formed by the *radial notch* of the ulna. The *anular ligament* supports and maintains the integrity of the joint (Fig. 9-14b). Another pivotal joint is seen at the articulation of the skull with the first (atlas) and second (axis) vertebrae, where rotation of the head occurs

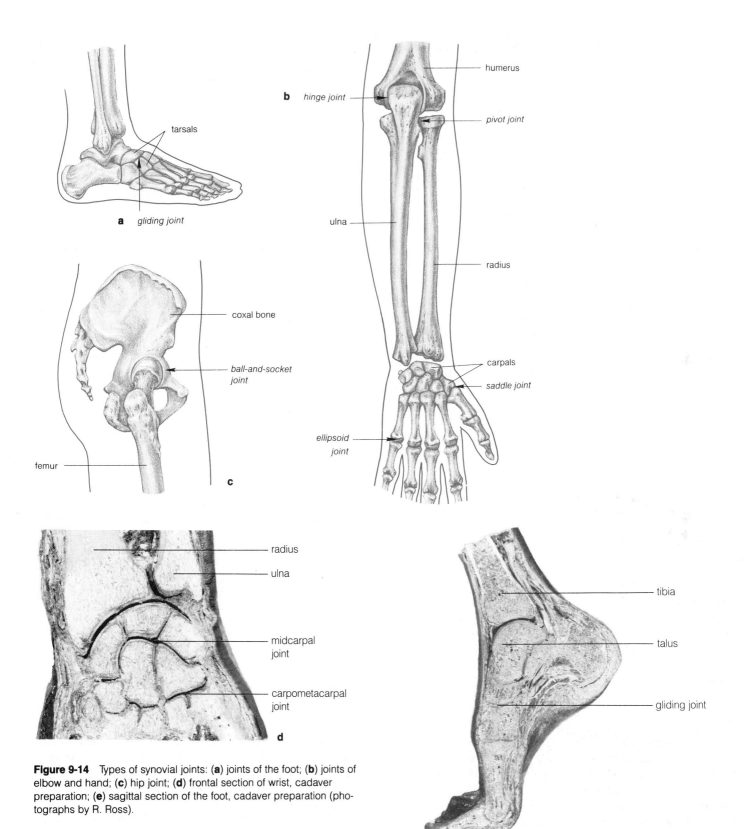

Figure 9-14 Types of synovial joints: (**a**) joints of the foot; (**b**) joints of elbow and hand; (**c**) hip joint; (**d**) frontal section of wrist, cadaver preparation; (**e**) sagittal section of the foot, cadaver preparation (photographs by R. Ross).

between the atlas and the dens (odontoid process) of the axis. Flexion and extension occur at the articulation of the occipital condyles and the articular facets of the atlas.

Ellipsoid Joint

The **ellipsoid joint** (articulatio ellipsoidea) resembles a ball-and-socket joint whose articular surfaces are elliptical, which prevents rotation at the joint while allowing all other types of movement. The articular capsule encloses a single pair of articulating surfaces. An example of this joint is seen in the *metacarpophalangeal joints* of the hand.

Condylar Joints

In **condylar joints** (articulatio condylaris) a single bone articulates with another bone at two separate articular surfaces called *condyles.* The condyles may be widely separated and enclosed by two separate articular capsules, as in the *temporomandibular joints,* or close together and enclosed by a single joint capsule, as in the *knee joint.* In either case an articular disc allows a sliding motion in addition to the hinge movement. In the temporomandibular joint the joint capsule is loose in front and behind, and a sliding movement is possible. Most of the support of the joint is due to strong collateral ligaments.

Saddle Joints

An example of a **saddle joint** (articulatio sellaris) is the articulation of a carpal, the *trapezium,* with the metacarpal of the thumb (Fig. 9-14b). The articular surface of the trapezium, which is shaped like a saddle, fits into the troughlike articular surface of the metacarpal so that movement is possible in a forward direction as well as toward the sides. This allows the thumb to be turned against the palm (opposed).

Ball-and-socket joint

Bursa

Capsular ligament

Collateral ligament

Condylar joint

Ellipsoid joint

Fascial cleft

Hinge joint

Intra-articular ligament

Meniscus

Pivotal joint

Plane joint

Saddle joint

Symphysis

Synarthrosis

Synchondrosis

Syndesmosis

Synostosis

Synovial cavity

Synovial fluid

Synovial membrane

Tendon sheath

SUMMARY OUTLINE

I. Friction-Reducing Devices

   A. A fascial cleft, or space, is a loose area between denser structures.

   B. A bursa is a membrane-enclosed, fluid-filled sac between structures.

   C. A tendon sheath is a double-layered tubular sleeve of synovial membrane forming a fluid-filled cavity that encloses a tendon.

   D. A synovial cavity is a fluid-filled space between the articular surfaces of adjacent bones at movable joints.

II. Classification of Joints

   A. Synarthroses are immovable or slightly movable joints that have no synovial cavity.

      1. A syndesmosis is joined by collagenous fibers that may allow slight movement; it includes sutures, which are immovable.

      2. A synchondrosis is joined by cartilage, usually a combination of hyaline on the surface of bones and fibrocartilage joining the hyaline articular cartilages.

      3. A synostosis is joined by bone that eliminates the joint.

   B. Synovial joints, diarthroses, have synovial cavities between articular surfaces of adjacent bones. The synovial cavity is enclosed by the hyaline cartilage covering the articular surfaces of the bones and by the synovial membrane on the inner surface of the joint capsule. The joint capsule may be strengthened by capsular ligaments which are a part of the capsule. The joint may also be stabilized by peripheral ligaments external to the capsule and by intra-articular ligaments between the articular surfaces. An articular disc of hyaline, fibrocartilage, or dense fibrous connective tissue may partition the synovial cavity.

III. Mechanisms for Stability of Synovial Joints

   A. Bony configuration, the way the bones fit together at their articular surfaces, tends to prevent dislocation of the joint.

   B. Ligaments that do not stretch stabilize joints but also restrict movement.

   C. Muscles crossing the joint help maintain the integrity of the joint and allow great freedom of movement.

IV. Classification of Diarthroses

   A. Plane joints have flat articular surfaces that permit sliding movements; examples are the wrist and ankle joints.

   B. Hinge joints have strong collateral ligaments that restrict movements to one plane; examples are interphalangeal joints and the elbow.

   C. Ball-and-socket joints have a round head of one bone that fits into a cuplike socket of another bone; a full range of movement is permitted; examples are the shoulder and hip joints.

   D. Pivotal joints are characterized by ligaments that restrict movement to rotation of one bone on the other; an example is the proximal articulation of the radius and the ulna.

   E. Ellipsoid joints have elliptical articular surfaces that permit all movements except rotation; examples include metacarpophalangeal joints.

   F. Condylar joints have a single bone that articulates with another bone by two articular surfaces called condyles; examples are the temporomandibular joint and the knee.

   G. Saddle joints have one articular surface that is shaped like a saddle and fits the troughlike articular surface of the other bone; an example is the articulation of the trapezium with the metacarpal of the thumb that allows the thumb to oppose the fingers.

REVIEW QUESTIONS

1. Why is it possible for a muscle to contract without moving adjacent muscles with which it appears to adhere closely?

2. What is the principal difference between a bursa and a fascial cleft? A bursa and a tendon sheath?

3. What are the components of a synovial joint that are present in all synovial joints?

4. What morphological features of a synovial joint permit movement of articulating bones without noticeable friction as the two bones move against one another?

5. What morphological feature(s) prevent movement at sutures?

6. Why is the intervertebral disc classified as a synchondrosis?

7. What is the relationship of a peripheral ligament to the capsule of a diarthrodial joint?

8. What features of the hip joint are largely responsible for its stability?

9. What prevents dislocation of the shoulder joint when a person falls on outstretched hands?

10. Why is movement in only one plane possible at a hinge joint?

11. What type of synovial joint allows the greatest freedom of movement?

12. What type of joint permits the thumb to oppose the fingers? Between what two bones is this joint located?

SELECTED REFERENCES

Barnett, C. H., *et al. Synovial Joints.* London: Longmans, 1961.

Bloom, W., and Fawcett, D. W., *A Textbook of Histology.* 10th ed. Philadelphia: W. B. Saunders, 1975.

Freeman, M. A. R., ed. *Adult Articular Cartilage.* London: Pitman Medical, 1973.

Hamerman, D., and Rosenberg, L. C. Diarthrodial Joints Revisited. *Journal of Bone and Joint Surgery* 52A: 725-ff, 1970.

Snell, R. S. *Clinical Anatomy for Medical Students.* 2nd ed. Boston: Little, Brown & Co., 1981.

Wilson, F. C. ed. *The Musculoskeletal System. Basic Processes and Disorders.* 2nd ed. Philadelphia: J. B. Lippincott Company, 1983.

10

THE STRUCTURE AND FUNCTION OF SKELETAL MUSCLE

OBJECTIVES

After completing this chapter you should be able to:

1. Define or describe the contractile unit (sarcomere) of a skeletal muscle fiber.

2. Give the morphological characteristics of a muscle fiber.

3. Describe a muscle fascicle.

4. Describe the organization of a skeletal muscle that suggests that each muscle should be considered an organ.

5. Compare and contrast the four types of muscle attachments.

6. Explain the typical relation of a muscle to the synovial joint crossed by the muscle.

7. List and demonstrate an understanding of the methods by which muscles are named.

8. Name and diagram the types of muscle fiber arrangements found in skeletal muscles.

9. Explain the relationship of muscle fiber arrangement to the range of movement and the force exerted by contraction of the muscle.

10. Describe the influence of the point of insertion of a muscle upon the range of movement and the force exerted by the contraction of the muscle.

11. Define or explain the following types of muscle actions.
   A. Prime movers
   B. Antagonists
   C. Fixators
   D. Synergists

12. Define the following terms denoting the direction of movement.
   A. Flexion
   B. Extension
   C. Abduction
   D. Adduction
   E. Pronation
   F. Supination
   G. Rotation
   H. Circumduction

13. Describe in general terms the fascial relations of skeletal muscles.

Skeletal muscles are organs highly specialized for the maintenance of posture and the production of voluntary movements. These functions are possible only because skeletal muscles have the capacity to contract, or shorten, by as much as 43% of their length. Some skeletal muscles serve as padding and protect underlying viscera, but their chief importance is their ability to contract in a controlled fashion.

Each muscle is considered an *organ* because it is composed of more than one tissue and functions as a unit. Like all organs, each muscle is named and has a specific location within the body. A muscle is held in place by its specific attachments.

CELL STRUCTURE AND FUNCTIONING OF SKELETAL MUSCLES

Skeletal muscle consists of elongated multinucleated cells called *muscle fibers,* or *myofibers*. Myofibers are bound together in bundles called fascicles, which in turn make up the larger bundles called muscles. A muscle usually attaches at both ends to bones.

Myofibers (Fig. 10-1) are bound by a cell membrane called the **sarcolemma** and contain cytoplasm (**sarcoplasm**) within which are longitudinally oriented contractile protein strands called *myofibrils*. The endoplasmic reticulum, called the sarcoplasmic reticulum, and other organelles are found around and between the myofibrils; the nuclei are located between the myofibrils and the sarcolemma. The myofibrils are composed of *microfilaments* (myofilaments) made of the contractile proteins actin and myosin; the arrangement of the microfilaments produces light and dark bands in the myofibrils and myofibers, giving skeletal muscle its striated appearance.

To understand the mechanism of muscle contraction one must understand the relationship between the **actin** and **myosin** microfilaments and the shifts that occur in this relationship during contraction. When a muscle fiber contracts, the actual shortening occurs within segments of the myofibrils called **sarcomeres.** Within a single muscle fiber all sarcomeres shorten at the same time, reducing the length of the muscle fiber.

The Sarcomere

The structural and functional unit of skeletal muscle is a sarcomere (Fig. 10-2). It is composed of the thick microfilaments of myosin and the thin microfilaments of actin that are located within a myofibril between adjacent *Z-bands,* also called Z-lines or Z-discs. The myofibrils in a skeletal muscle fiber show alternating dark and light bands (A-band and I-band, respectively) along their length. The most obvious of the other bands visible in special preparations of muscle is the Z-band, located in the middle of the I-band; the band in the middle of the A-band is the H-zone. Thick myosin microfilaments are largely responsi-

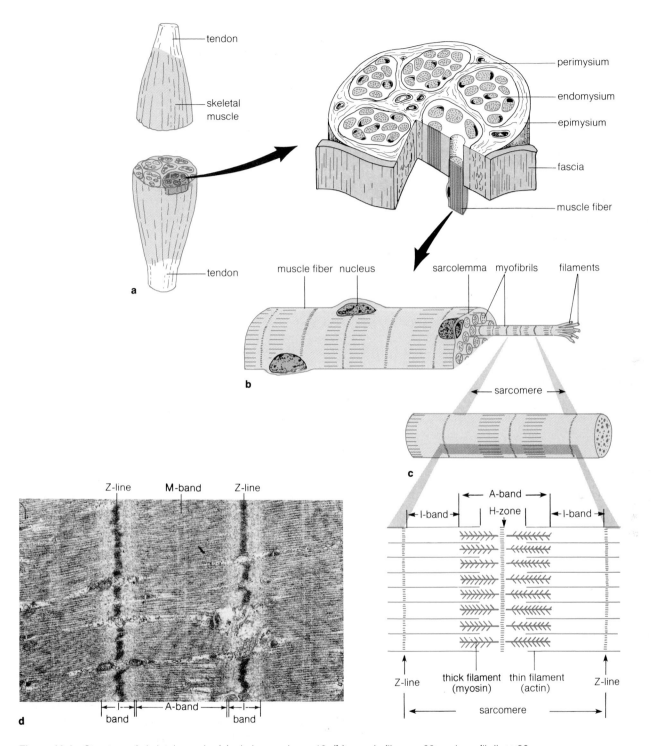

Figure 10-1 Structure of skeletal muscle: (**a**) whole muscle, × 10; (**b**) muscle fiber, × 30, and myofibril, × 90; (**c**) diagram of a sarcomere; (**d**) electron photomicrograph of skeletal muscle of the rat, × 17,000. The relatively narrow I-band is the result of contraction of the muscle fiber (courtesy of D. H. Matulionis).

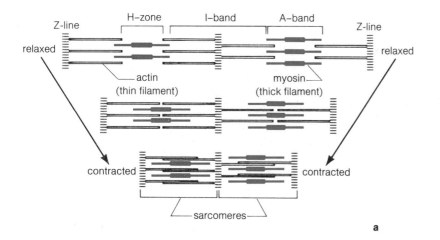

a

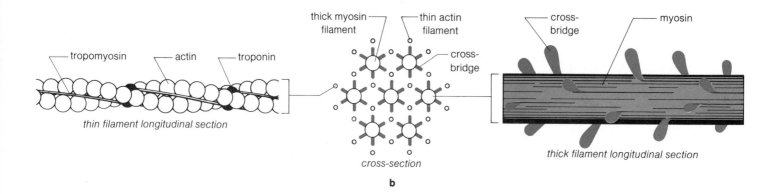

b

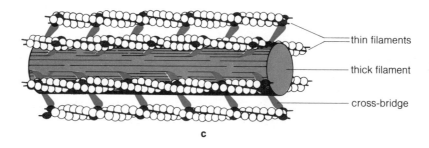

c

Figure 10-2 The arrangement of filaments in a sarcomere: (**a**) diagram of the change in the arrangement of filaments from the relaxed to the contracted state and a photomicrograph of filaments (photomicrograph courtesy of Patricia Schulz); (**b**) structures and arrangement of filaments; (**c**) interrelation of actin and myosin.

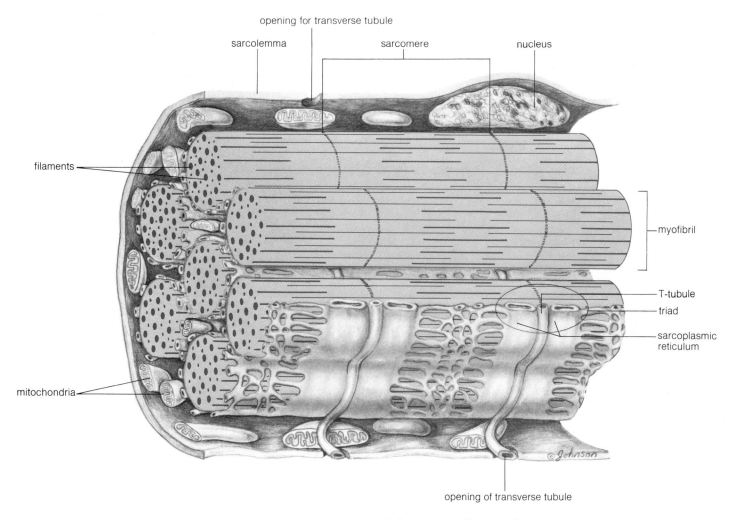

Figure 10-3 Arrangement of T-tubules and sarcoplasmic reticulum around myofibrils in a muscle fiber, or cell.

ble for the A-band; thin actin microfilaments are responsible for the I-band and extend into the A-band as far as the H-zone, overlapping the thick filaments. The thin filaments begin at the Z-band where they are attached to the protein molecules of *tropomyosin* that make up the Z-band. Individual myofibrils, therefore, are strands consisting of thick myofilaments in the center of the sarcomere and thin myofilaments at the ends, with the two types overlapping at the center of the sarcomere.

Near the junction of the A- and I-bands the sarcolemma sends tubelike extensions, called *T-tubules* (transverse tubules), into the skeletal muscle fiber (Fig. 10-3), bringing the extracellular environment close to the terminal cisternae of the sarcoplasmic reticulum. Neural stimulation of the sarcolemma causes an influx of *calcium* into the fiber from the T-tubules, activating the microfilaments and resulting in the contraction of the muscle fiber. The enzyme ATPase, found in the microfilaments, is essential in the production of the energy required for muscle contraction.

The Process of Muscle Contraction

During muscle contraction (Fig. 10-3) the thin filaments slide past the thick ones, obliterating the H-band and shortening the distance between the Z-bands. Thus neither the thin nor the thick strands decrease in length; they simply change their relative positions.

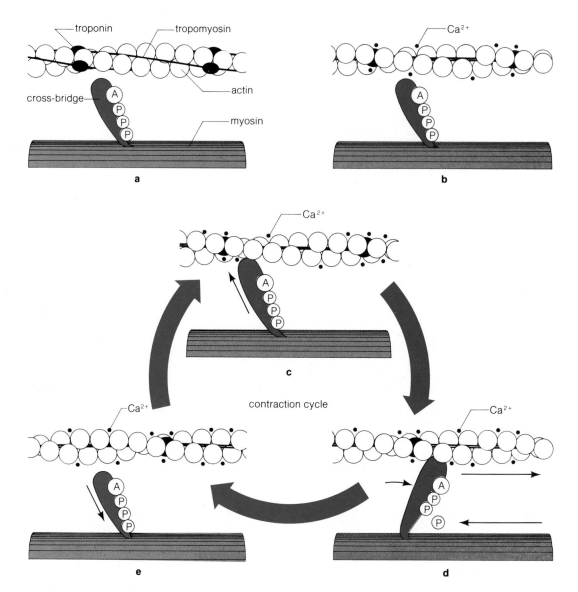

Figure 10-4 Steps in the contraction process: (**a**) ATP attaches to myosin; (**b**) calcium ions cause the troponin–tropomyosin complex to move away from the reactive sites on actin; (**c**) energized cross-bridge binds to actin; (**d**) the energized myosin head swivels, causing the filaments to slide past one another; (**e**) new ATP attaches to myosin, and the head returns to its earlier shape, ready to form another cross-bridge. Steps c, d, and e constitute the contraction cycle and occur over and over again, 50 to 100 times per second, as long as calcium ions are available in adequate concentration.

Each thick filament is a strand composed of many myosin molecules. Each molecule has a long tail region and a head region that can bind to actin, creating *cross-bridges* (Fig. 10-4) between the thick and thin filaments. In addition to actin the thin filaments contain the protein **tropomyosin** which, when bound to actin, prevents the binding of actin and myosin. A

molecule of **troponin** is associated with each tropomyosin molecule. Troponin binds Ca^{++} during the initiation of muscle contraction, which causes a structural change in tropomyosin that rotates it out of the way of the myosin-binding site on actin. At the same time, the ATP bound to the myosin molecule is split by myosin ATPase, activating the myosin so that it binds to an actin molecule and forms the myosin cross-bridge essential to contraction. The combination of myosin with actin causes the myosin molecules to change shape and promotes the swiveling of the myosin cross-bridges, so that the actin filaments slide on the myosin filaments toward the center of the sarcomere. The splitting of ATP provides the energy for contraction, but the level of intracellular Ca^{++} regulates the process. The influx of Ca^{++} is brought about by neural stimulation.

Prolonged contraction of a muscle leads to muscle fatigue, due to the inability of the contractile and metabolic processes of the muscle fibers to continue doing the same amount of work. The contraction becomes weaker because of the decrease in the energy supply (ATP) in the muscle fibers.

Adenosine triphosphate (ATP), the major energy source in muscle fibers, is needed not only for the contractile process but also for the relaxation of muscle fibers. Muscle cramp, in which a muscle contracts spasmodically and cannot relax completely, is probably due to lack of an adequate supply of ATP. The muscle rigidity that follows death, known as *rigor mortis,* is due to the decrease in the level of the intracellular ATP.

The Motor Unit

The nervous system controls the contraction of skeletal muscle by means of nerve fibers (processes from nerve cells) that terminate on individual muscle fibers in special nerve endings called motor end plates (Fig. 10-5). A single nerve cell may supply only one or two muscle fibers as in the innervation of some of the eyeball muscles, or it may supply more than a hundred muscle fibers as in some of the large thigh muscles. When a neuron (nerve cell) transmits a nerve impulse, all of the muscle fibers supplied by the neuron contract maximally; there is no such thing as partial contraction of an individual myofiber. The neuron and all the muscle fibers it supplies act as a unit, termed a **motor unit.**

The force of the contraction of any muscle depends upon the number of motor units stimulated; the theoretical maximum force involves all the motor units of an entire muscle. The delicacy of control of the force of contraction depends upon the size of the motor unit; that is, a motor unit composed of one nerve cell controlling a single muscle fiber achieves the finest control of movement. The small size of the motor units in the eyeball muscles enables us to follow visually a rapidly moving object such as a tennis ball, which would be impossible with large motor units. Only gross movements such as weight lifting are possible with large motor units; the skilled movements used in writing or playing the piano require small motor units for fine control. From this it may be concluded that the motor units of the muscles that move the fingers are smaller than those of the arms and thighs. It may also be inferred that not all muscle actions are alike except that their principal function is to contract.

There are two types of skeletal muscle fibers that differ in their rates of contraction and in appearance. *Red,* or *slow, fibers* have a high *myoglobin* content, which is responsible for the red color of the fibers. Myoglobin is a sarcoplasmic protein capable of binding the oxygen used in cellular respiration; it makes it possible for the energy requirements of the slow contracting muscle fibers to be met by oxidative metabolism (aerobic respiration) based on oxygen supplied by the circulation. *White,* or *fast, fibers* do not contain appreciable amounts of myoglobin and therefore cannot obtain the energy required for contraction by oxidative metabolism. Their energy is derived instead from the breakdown of glycogen without oxygen, a process called anaerobic respiration. This process is inefficient and cannot provide energy for long periods, but it is very fast and can supply the energy required for fast contraction for a short period of time.

All muscle fibers are red at birth; white fibers differentiate from the red fibers as the individual matures. All the fibers within a motor unit are of the same type; thus the nerve of the motor unit appears to influence the fiber type. Exercise also has an influence on muscle development. Short-duration exercise, such as push-ups, causes hypertrophy of white fibers due to the synthesis of more myofibrils. This produces muscle fibers with more myofibrils contracting in unison, which can develop more force. In contrast, long-duration exercise, such as jogging, increases the capacity of red fibers for aerobic respiration with no increase in fiber size. This increased efficiency of the utilization of oxygen is correlated with an increase in the blood vessels within the muscles being used, so that a greater supply of oxygen is available. Because the latter type of exercise uses only aerobic respiration, it is called aerobic exercise.

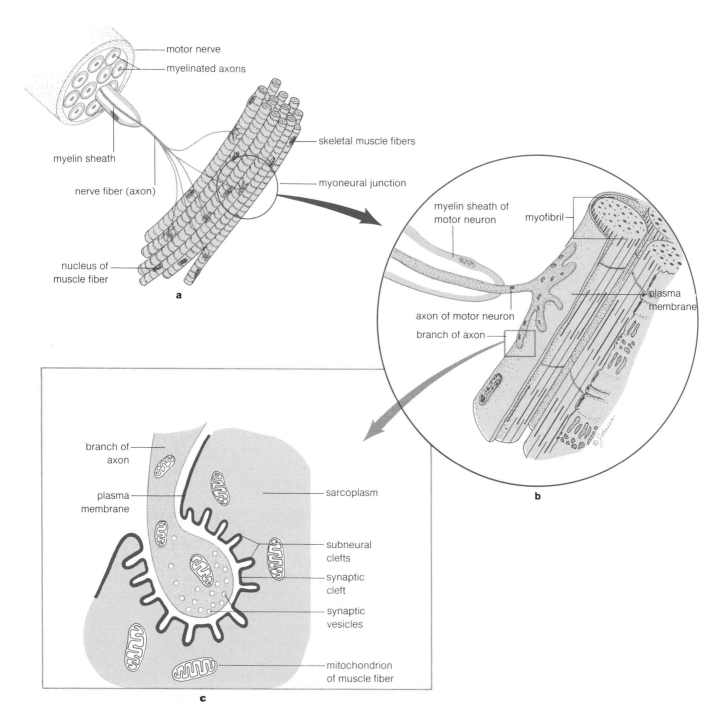

Figure 10-5 How an axon innervates a muscle fiber. (**a**) In a motor unit an axon of a motor neuron branches to several muscle fibers. (**b**) Each branch of the axon forms a myoneural junction; (**c**) The branch of the axon rests in a groove in the muscle fiber, the motor end plate.

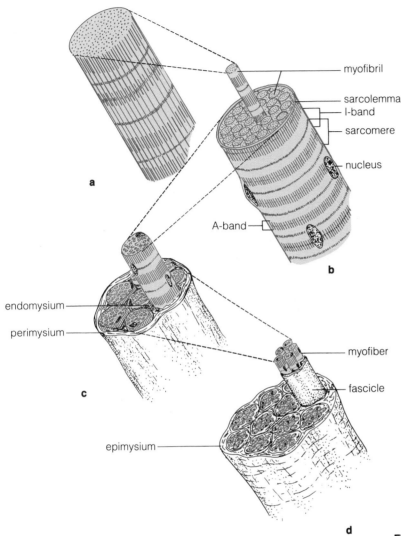

myofibril

sarcolemma
I-band
sarcomere
nucleus

A-band

a

b

endomysium
perimysium

myofiber

fascicle

c

epimysium

d

Figure 10-6 Components of a muscle: (**a**) myofibril; (**b**) muscle fiber; (**c**) muscle fascicle; (**d**) muscle.

THE ARRANGEMENT OF SKELETAL MUSCLE FIBERS

The arrangement of muscle fibers within a skeletal muscle is shown in Fig. 10-6. Individual myofibers are enclosed by the *sarcolemma,* which is structurally similar to the plasma membrane of other cells. A network of fine reticular fibers forms the supporting *stroma* (binding connective tissue) of the muscle as a whole and encloses each muscle fiber external to the sarcolemma and its cell coat. The supporting stroma includes the fascial investments for individual fibers, groups of fibers, and the entire muscle, which are continuous with one another; they are individually named only for convenience. Fine loose connective tissue, the **endomysium,** ensheaths each muscle fiber and the reticular fiber network on the surface of the sarcolemma. The reticular fibers intermingle with the thin collagenous fibers, forming a loose sheath that facilitates sliding of muscle fibers during contraction. Muscle fibers, too small to be seen with the unaided eye, are wrapped by connective tissue into small bundles, muscle *fascicles,* that are visible grossly. Fascicles

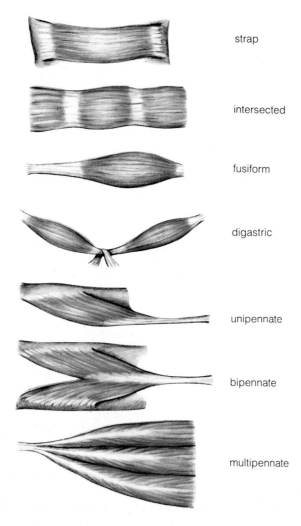

strap

intersected

fusiform

digastric

unipennate

bipennate

multipennate

Figure 10-7 Diagrams of the types of muscle fiber arrangements

nective tissue sheaths into the interior of the muscle. The nerve fibers terminate in the *motor end plates* on the muscle fibers, and the blood vessels form a rich capillary network that closely invests each muscle fiber. Muscle fibers and fascicles have a definite arrangement relative to the fascial coverings and attachments. The arrangement of muscle fiber is constant in any given muscle in a specific individual but varies greatly in the different muscles of the body (Fig. 10-7).

Strap

In the **strap muscles** of the neck and in other muscles such as the *sartorius* of the thigh, muscle fascicles are *parallel* to one another and to the long axis of the muscle and may extend for the entire length of the muscle.

Intersected

In the *rectus abdominis,* the vertical muscles of the abdomen, the muscle fascicles are parallel to one another but do not extend the entire length of the muscle. Instead, the muscle is divided into segments by connective tissue (tendinous) septa. These transverse partitions are usually termed **intersections.**

Fusiform

The *fleshy belly* of many muscles is much larger than the tendon to which it is attached. Thus the muscle fascicles taper near the muscle-tendon junction, giving a **fusiform** (tapering) shape to the muscle as a whole. In this type of fiber arrangement not all of the individual muscle fibers reach the grossly visible tendon. The *biceps brachii* of the arm, which boys take pride in demonstrating, is a muscle whose belly is larger than either end.

Digastric

If a *central tendon* separates the muscle into parts, the muscle is said to be **digastric,** meaning "two-bellied." Usually each belly is fusiform in fiber arrangement, as illustrated by the *digastric* muscle below the lower jaw and by the *omohyoid,* an infrahyoid muscle of the neck.

Unipennate

In the **unipennate** pattern of fiber arrangement the fascicles are parallel to one another, but they join the tendon from only one side as in the *semimembranosus* muscle of the thigh.

are ensheathed by fascia called **perimysium,** which contains considerably more collagenous fibers than the endomysium. Unlike the endomysium, which is so thin it is not visible to the unaided eye, the perimysium can be seen grossly. The larger blood vessels and nerves supplying the muscle fascicles run through the perimysium in position to send branches into several adjacent fascicles. Muscle fascicles, in turn, are wrapped in progressively larger bundles, until finally the entire muscle is encased in a sheath of deep fascia called the **epimysium.**

The blood vessels and nerves that supply muscles pass through the *fascial clefts* between muscles. Branches of the blood vessels and nerves penetrate the muscles and follow the con-

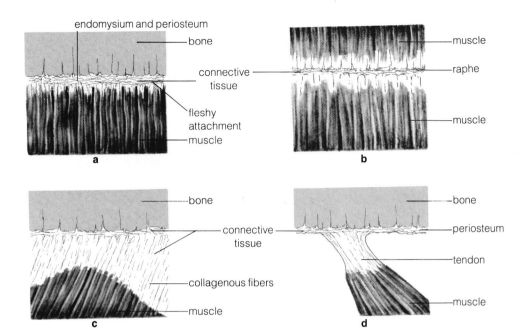

Figure 10-8 Diagrams of the types of muscle attachments: (**a**) fleshy attachment; (**b**) raphe; (**c**) aponeurosis; (**d**) tendon.

Bipennate

The **bipennate** fiber pattern is very much like the pattern of a down feather. The fascicles approach the centrally located tendon from two sides as in the *rectus femoris* of the thigh.

Multipennate

Some muscles such as the *deltoid* muscle of the shoulder are highly complex in their fiber arrangements and may contain many anastomosing (branching) tendons, with muscle fibers arranged chiefly in a bipennate fashion around each tendon.

MUSCLE ATTACHMENTS

To produce movement of a body part by contracting, a muscle usually attaches to two bones and crosses at least one movable joint. The endomysium, which forms sleevelike coverings of the muscle fibers, provides for muscle attachment by becoming continuous with the periosteum covering the bones to which the muscle is attached; connective tissue fibers of the endomysium interlace with those of the periosteum. Contraction of the muscle results in tension, or pull, upon the bone because

periosteal fibers (Sharpey's fibers) trapped in the bone matrix anchor the periosteum firmly to the bone.

All muscles have at least two points of attachment, one of which moves more than the other when the muscle contracts. The point of attachment that is either immovable or relatively so is called the *origin* and is usually nearer the midline of the body; the more movable attachment, the *insertion,* is usually located away from the midline. It may be difficult to determine which point of attachment is more movable if some movement occurs at both points. The origins and insertions of some muscles may be reversed if one fixes the point usually regarded as the insertion and moves the point usually regarded as the origin. Such a reversal can be demonstrated for the muscles that usually flex the forearms, for example, by grasping and pulling on a rigid object so that the entire body moves, as occurs in the chinning exercise.

The attachments of individual muscles vary greatly, each being unique in morphology. Attachments are classified according to the amount and gross appearance of the connective tissue involved (Fig. 10-8) and range from a *fleshy attachment,* or *raphe* (pr. *ra*-fay), in which little connective tissue is visible in the gross specimen, to attachment by collagenous fibers that form either a broad sheet called an **aponeurosis** or a slender cord called a *tendon.*

Fleshy Attachment

In the *fleshy attachment* (Fig. 10-8a) the connective tissue fibers of the endomysium surrounding the individual muscle fibers interlace with those of the periosteum, with little connective tissue between the muscle and the bone to which it is attached. Usually the fleshy attachment is approximately the width of the rest of the muscle; for example, the intercostal muscles bridge the intercostal spaces with short fibers that form a sheet of muscle having a fleshy attachment along most of the length of the two adjacent ribs. This arrangement has the advantage of bridging the intercostal space with contractile fibers, with little of the space going to noncontractile elements.

Raphe

Raphe (Gr., *raphe* = a seam) is a term used in anatomy to designate a line of junction between two structures whose parts interdigitate at the line. In some cases bilateral muscles attach to one another in the midline by interlacing the connective tissue of their fascial coverings. This interdigitation creates a fibrous band in a groove that runs at right angles to the direction of the muscle fibers (Fig. 10-8b). For example, the *pharyngeal raphe* is a fibrous band that extends downward from the base of the skull along the posterior wall of the pharynx in the median plane and serves for the attachment of the pharyngeal constrictor muscles. These muscles encircle the pharynx to insert anteriorly on skeletal elements. Their contraction reduces the lumen (cavity) of the pharynx during swallowing.

Aponeurosis

An **aponeurosis** (Fig. 10-8c) is a broad, flat sheet of dense regular connective tissue that is composed principally of collagenous fibers. Connective tissue fibers of the endomysium interlace with and are continuous with the fibers of the aponeurosis, which are oriented for the most part in the same direction as the muscle fibers. Because the collagenous fibers do not stretch, contraction of the muscle produces tension directly upon the structure to which the aponeurosis is attached. Usually the aponeurosis is approximately the width of the muscle, which is also generally a broad, sheetlike structure. A good example of an aponeurosis is seen in the abdominal wall, where each of the three muscles (external and internal obliques and transversus abdominis) inserts into an aponeurosis that does not attach to bones, but instead interdigitates in the anterior midline with the aponeurosis of the same muscle of the opposite side. This arrangement provides the anterior abdominal wall with strong fascial support that can be tightened by muscle contraction, which compresses the abdominal viscera and raises the intra-abdominal pressure.

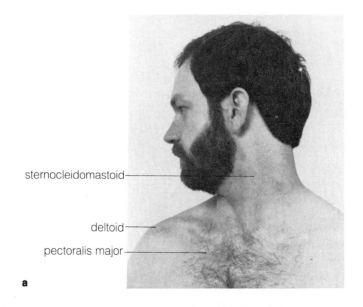

sternocleidomastoid

deltoid

pectoralis major

a

Figure 10-9 (**a**) Surface anatomy, anterior view (photograph by R. Ross). (**b**) Some muscles of the human body (anterior view).

Tendon

A tendon (Fig. 10-8d) is a cordlike structure of dense regular connective tissue, mostly in the form of thick bundles of collagenous fibers. The size and shape of tendons vary greatly, from the long slender cords of the muscles of the forearm, which insert by slender tendons onto bones of the fingers, to the broad, thick, and relatively short tendons such as the Achilles tendon at the back of the ankle. A tendon is always smaller in diameter than the muscle to which it is attached, but it is much stronger than the muscle; thus excessive tension usually injures the muscle rather than the tendon. The cordlike tendons are the means by which a muscle can pull upon a bone located at a distance from the muscle belly. For example, muscles of the forearm move the fingers by means of long tendons that extend from the forearm across the wrist and palm or back of the hand to attach to the bones of the fingers.

THE GROSS ANATOMY OF MUSCLES

Skeletal muscles (Figs. 10-9 and 10-10) comprise the bulk of the flesh of the body, contributing about 36% of the total body weight in women and about 42% in men; these are average figures for the total population, the range of variation being as

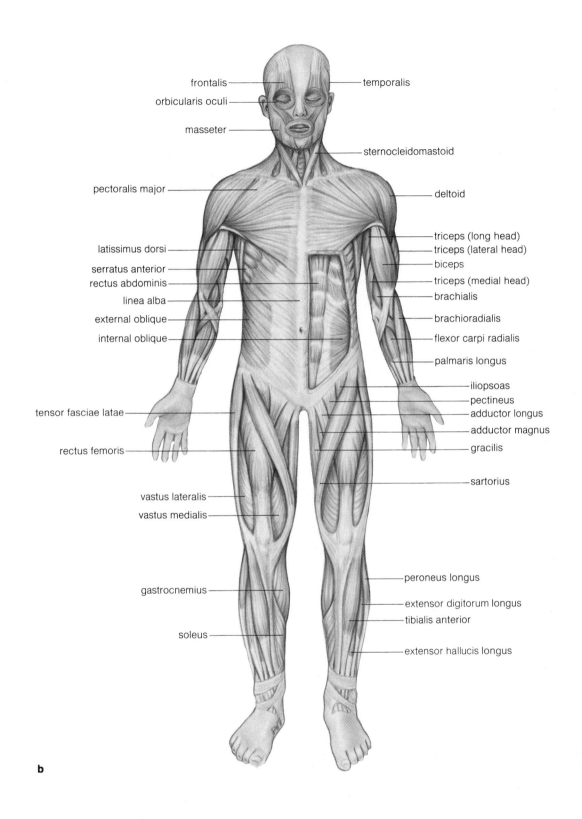

frontalis
temporalis
orbicularis oculi
masseter
sternocleidomastoid
pectoralis major
deltoid
triceps (long head)
latissimus dorsi
triceps (lateral head)
serratus anterior
biceps
rectus abdominis
triceps (medial head)
linea alba
brachialis
external oblique
brachioradialis
internal oblique
flexor carpi radialis
palmaris longus
iliopsoas
pectineus
adductor longus
tensor fasciae latae
adductor magnus
rectus femoris
gracilis
sartorius
vastus lateralis
vastus medialis
peroneus longus
gastrocnemius
extensor digitorum longus
tibialis anterior
soleus
extensor hallucis longus

b

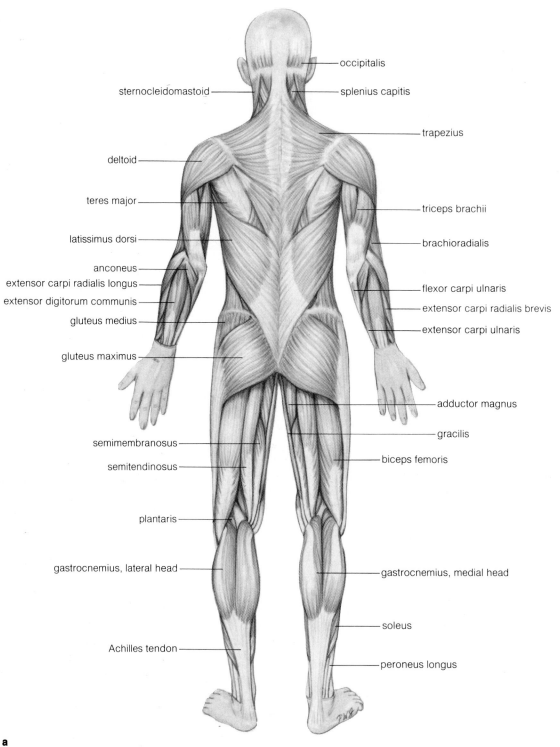

occipitalis

sternocleidomastoid

splenius capitis

trapezius

deltoid

teres major

triceps brachii

latissimus dorsi

brachioradialis

anconeus

extensor carpi radialis longus

extensor digitorum communis

flexor carpi ulnaris

extensor carpi radialis brevis

gluteus medius

extensor carpi ulnaris

gluteus maximus

adductor magnus

gracilis

semimembranosus

biceps femoris

semitendinosus

plantaris

gastrocnemius, lateral head

gastrocnemius, medial head

soleus

Achilles tendon

peroneus longus

a

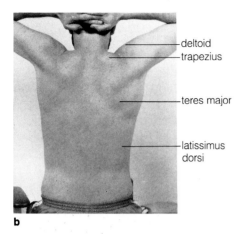

b

Figure 10-10 (**a**) Some muscles of the human body (posterior view). (**b**) Surface anatomy (posterior view). (Photograph by R. Ross)

deltoid
trapezius
teres major
latissimus dorsi

wide as the variation in body weight. The total weight of skeletal muscle is about three times the total weight of bone in both males and females. The close correlation between the weight of muscle and bone is because bulky muscles are usually attached to a heavy skeletal framework.

The study of the skeletal muscles includes learning their names, their relations to other structures, their attachments, action, blood supply, and innervation.

The Naming of Skeletal Muscles

Muscles are named in a variety of ways. Usually the name of a muscle contains considerable information about it, so it will be helpful to understand the meanings of the words as you learn the names.

Shape Some muscles are named for their shape, as illustrated by the large shoulder muscle, the *deltoid,* so named because it is *triangular* and because delta in Latin means triangle.

Size The name *latissimus dorsi* describes the size as well as the location of the muscle. *Dorsi* refers to the *back* and *latissimus* means "widest" or "broadest"; therefore this is the broadest muscle of the back. It is a large, fan-shaped muscle that originates on the back of the trunk and inserts on the humerus.

Location Some muscles are named to indicate their position in the body. For example, the *dorsal interossei* are muscles that are located between bones; *inter* means "between" and *osseus* means "bone."

Direction of Fibers The names of some muscles indicate the direction of their fibers relative to the structures to which they are attached. Examples are *transversus* (across), *obliquus* (oblique, or slanted), and *rectus* (straight).

Action Frequently part of the name of the muscle indicates its action. For example, the *adductor magnus* muscle adducts the thigh.

Organization Muscles that have multiple heads of origin may have the number of heads as part of their names. For example, the *biceps brachii* has two heads, *triceps brachii* has three, and *quadriceps femoris* has four.

Attachments Muscles are also named simply by combining the names of their points of origin and insertion. For example, the *sternohyoid* designates a muscle that originates on the *sternum* and inserts on the *hyoid* bone.

Combination Most muscles take their names from a combination of the preceding methods. For example, *flexor digitorum profundus* literally means the *deep* (profundus) *flexor* of the *fingers* (digitorum). Its name tells its location and action.

Anatomical Relations of Muscles

The skeletal muscles and their fascial coverings have a close anatomical relationship to the skeletal elements to which they attach. This close relationship usually exists between a muscle and the two bones of its origin and insertion. Movement results from contraction of muscles attached to adjacent bones that articulate with one another at a synovial, or movable, joint. A typical arrangement of a muscle in relation to a synovial joint is diagrammed in Fig. 10-11. The muscle takes origin proximal to the joint and inserts distal to the joint; thus the bone distal to the joint is the one usually moved by contraction of the muscle. The biceps brachii, for example, originates on the scapula, and its tendon crosses the elbow joint to insert upon the radius of the forearm, with no attachment to the capsule of the elbow joint. Contraction of the biceps flexes the forearm while leaving the joint capsule loose, so that the fibrous capsule does not restrict movement.

Not all muscles insert on bone. Most of the muscles controlling facial expression originate from bone and insert in the skin. Therefore they cross the superficial fascia of the head and neck and have different fascial relations from other muscles. Almost all other skeletal muscles have the anatomical arrangement shown in Fig. 10-12.

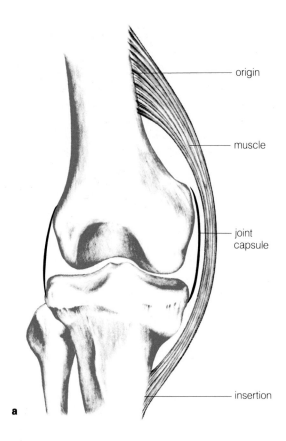

origin

muscle

joint
capsule

insertion

a

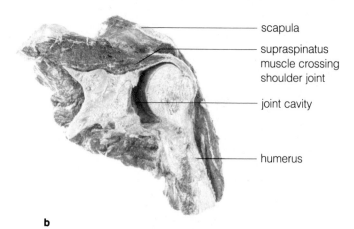

scapula

supraspinatus
muscle crossing
shoulder joint

joint cavity

humerus

b

Figure 10-11 Relations of a skeletal muscle to a synovial joint. (**a**) Diagram; (**b**) sagittal section of the shoulder joint, cadaver preparation (photograph by R. Ross).

The skin, which covers the external surface of the body, is not sharply separated from the subcutaneous connective tissue, called *superficial fascia* in gross anatomy. Histologically the superficial fascia is loose areolar connective tissue that varies in thickness and in fat content in various parts of the body. The *deep fascia,* which is in direct contact with the muscle, is denser than the superficial fascia and contains very little fat. It is tough, strong connective tissue, in contrast to the superficial fascia which serves principally as a padding and insulation material. When two muscles lie adjacent to one another, as in the figure, a sheet of deep fascia covers each of them. A looser type of deep fascia fills the space between the muscles, providing a pathway for the large nerves and blood vessels that supply the muscles. This loose fascia is the *fascial plane,* or *cleft,* described in Chapter 9, which permits the muscles to slide back and forth on one another.

When two or more muscles are enclosed in a fascial sheet that more or less isolates them from adjacent structures, a *fascial compartment* is the result. Figure 10-13 illustrates the fascial relations in the arm, where each of the two principal muscles anterior to the humerus, the biceps brachii and the brachialis,

has its own fascial covering and occupies the anterior fascial compartment of the arm. The muscles are enclosed by deep fascia that attaches to the humerus, dividing the arm into anterior and posterior compartments.

MUSCLE ACTION

Types of Muscle Action

The voluntary control of muscle activity "wills" whole movements rather than the contractions of specific muscles. Most movements require the cooperative action of several muscles functioning as a group. Muscles may be categorized according to the effects produced by their contraction.

Prime Movers A muscle that contracts to produce a particular movement is a **prime mover.** Most movements are the result of the contraction of more than one muscle, and frequently a single muscle contributes to the production of two or more movements. A familiar example is the action of the biceps bra-

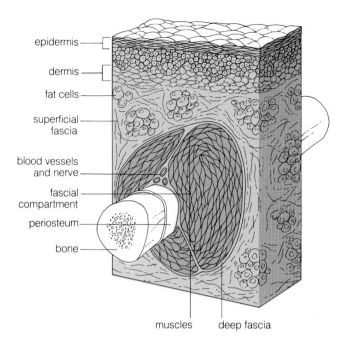

Figure 10-12   The fascial relations of muscles.

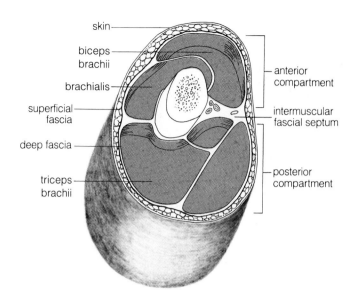

Figure 10-13   The two fascial compartments of the arm.

chii of the arm. When it contracts, the biceps *flexes* the forearm; but it is also a *supinator* because it can turn the palm of the hand upward. This dual function can easily be demonstrated. Place your hand under a table or other heavy object and attempt to lift (flex) it; note that the biceps tightens. Next place your hand palm downward on top of the table and observe the relaxed state of the biceps. Finally, palpate the biceps as you turn the palm upward (supination) and note the contraction of the biceps, which confirms the involvement of the biceps as a prime mover in both flexion and supination of the forearm.

Antagonists   Muscles that oppose one another upon contraction are **antagonists.** For example, the *biceps flexes* the forearm, but the *triceps brachii,* on the posterior aspect of the arm, *extends* the forearm. If the biceps is to produce movement, the triceps must relax. If both contract at the same time with equal force, there will be no movement. This *isometric contraction,* in which the muscle maintains a constant length, is the basis for the isometric exercises that oppose one muscle group against another with no movement. This type of contraction contrasts with *isotonic contraction,* in which the muscle shortens while maintaining a constant force.

Synergists   Most of the muscles are **synergists,** working together to increase their effectiveness. One example is the action of the *extensors* of the *hand* in conjunction with *flexion* of the *fingers.* Try flexing your hand at the wrist and then clench-

ing your fist; note the difficulty in exerting much force with the fingers. Much greater force is possible if you extend the hand at the wrist and then flex the fingers. The extensors on the back of the hand cooperate with the flexors of the fingers in making a fist or in grasping an object. This is accomplished without having to think about it.

Fixators   The principal function of many muscles is the simple fixation of a part so that other muscles can exert a greater effect; these are called **fixators.**

Perhaps you have seen runners at the end of a race hang onto a pole or other structure above their heads to fix their shoulders, which enables them to take deeper breaths. Several muscles that elevate the rib cage during forced inspiration are attached to the shoulder girdle. For these muscles to exert an effect upon the ribs the scapula must be fixed in place. This can be accomplished by rigidly fixing the arms by holding onto something, particularly something above the head. With the scapula fixed in this manner the contraction of muscles attached to both the scapula and the rib cage elevates the ribs, resulting in deep inspiration.

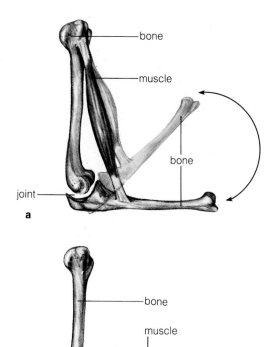

Figure 10-14 Diagram illustrating the influence of the point of insertion of muscles upon the movement produced.

Types of Movements

Movement, which may be defined as a change in the relationship of two parts of the body, is determined by the shapes of the articular surfaces of bones at the joints and by the direction of muscle pull. Some of the terms used to describe movements produced by muscle contraction are part of our everyday language. The following terms describe most of the possible movements, and most are illustrated in Fig. 1-5.

Flexion is the reduction of the angle between two bones at a joint.

Extension is an increase in the angle between two bones at a joint.

Abduction is the movement of the body away from the midline.

Adduction is the movement of a part of the body toward the midline.

Pronation is rotation, or turning, of the forearm so that the hand is turned downward or backward. In this movement the radius is rotated so that it crosses the ulna.

Supination is rotation of the forearm so that the palm is turned upward or forward with the thumb lateral in position. The radius and ulna are parallel to one another.

Rotation involves turning a bone on its axis at a joint without displacement of the axis.

Circumduction is a movement that swings the moving part in a circle. The proximal end remains fixed while the distal end describes a cone.

Structural Influences on Muscle Action

The arrangement of the muscle fibers and the location of the muscle insertion affect the action of a muscle.

Muscle Fiber Arrangement   The arrangement of fibers within a muscle influences its range of movement and the force it can exert in contracting. The *range of movement* depends on the average length of the muscle fibers whereas the *strength* (force) of a muscle contraction is a function of the number of fibers contracting.

A muscle fiber can shorten only to about 40% of its relaxed (unstretched) length, and the greatest range of muscle movement is produced by a parallel arrangement of fibers, which allows the entire muscle to contract the same distance as the individual fibers. Such an arrangement is found in the strap muscles. However, this type of muscle is not as strong—that is, it cannot exert as much force—as the same volume of muscle with a pennate fiber arrangement because the strap muscle has fewer fibers in the same muscle volume.

Point of Insertion   The point of insertion affects the strength of contraction and the amount of movement produced. The ratio of the distance between a muscle origin and the joint to the distance between the joint and the muscle insertion is directly proportional to the extent of movement produced and is inversely proportional to the force that can be exerted.

This rule is illustrated in Fig. 10-14. The bones and joints form a system of levers in which the joints serve as fulcra and the bones bear the weight of the part of the body that is to be moved. Attachment of the muscle close to the joint gives less mechanical advantage to the muscle than attachment closer to the weight (farther from the joint). Therefore the closer the muscle attachment is to the joint, the weaker it will be, but the greater will be the movement at the distal end of the bone.

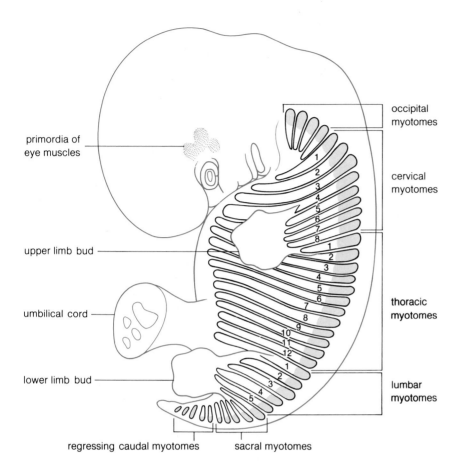

primordia of
eye muscles

upper limb bud

umbilical cord

lower limb bud

regressing caudal myotomes   sacral myotomes

occipital
myotomes

cervical
myotomes

thoracic
myotomes

lumbar
myotomes

Figure 10-15   Diagram of
embryonic myotomes at about
four weeks development. Stip-
pled areas show approximate
size and location of myotomes;
unstippled areas indicate the
area into which the myotome
extends as it develops.

DIFFERENTIATION OF SKELETAL MUSCLE

All skeletal muscles are derived from *embryonic mesoderm.*
Some of the muscles of the head and neck are derived from
mesoderm of the *visceral arches,* but the others come from the
myotome portion (Fig. 10-15) of the *segmented mesoderm*
(somites). All evidence of segmentation is usually lost during
proliferation of the mesodermal cells, and changes in the orientation of the fibers occur as differentiation proceeds.

The myotome cells **(myoblasts)** are initially spindle-shaped
uninucleated cells. Those that form a single muscle aggregate
into a recognizable muscle primordium, where they multiply
rapidly and become arranged in rows packed closely together.
Soon the plasma membranes between the cells begin to break
down, giving rise to multinucleated **myotubes.**

The nuclei of the myotubes do not undergo mitosis, and
the myotubes grow in length through addition of myoblasts at
the ends of the developing fiber. Not all of the potential myoblasts become part of the myotubes; some remain in the con-

nective tissue of the muscle as undifferentiated cells capable of
differentiating into muscle. Such cells remain throughout life
and are capable of contributing to the repair of muscle in cases
of mild injury.

Initially the nuclei are centrally located in the myotubes,
but as actin, myosin, and tropomyosin are synthesized within
the myotubes and become arranged in the myofilaments, the
nuclei are gradually pushed toward the periphery of the fiber.
The cytoplasm (sarcoplasm) containing the organelles, such as
mitochondria, is also found principally at the edge of the fiber.

The myofilaments formed within the myotube have cross-
striations as soon as they are produced. The sarcomeres of the
embryo have the same length as those of the adult; muscle fibers
grow in length by addition of new sarcomeres at each end of
the muscle fiber. As the muscle fiber increases in length, the
nuclei move farther apart.

While the muscle fibers are differentiating, the connective
tissue elements are being elaborated. A basement membrane is
deposited on the surface of the muscle fiber, around which the

Clinical Applications: **Muscle Diseases**

The diagnosis of a muscle disease depends upon the symptoms, rather than upon laboratory tests. However, laboratory studies are useful in determining that muscle weakness is due to muscle wasting and not caused by deficiency of innervation. There are many types of muscle diseases that are genetically determined.

Muscular Dystrophies

The diagnosis of *muscular dystrophy* in cases of progressive muscle weakness may be very difficult unless other members of the family have had the disease. Diagnosis is usually confirmed by muscle biopsy that reveals histological changes in the muscle fibers.

Duchenne muscular dystrophy is the most common of the dystrophies and is the most severe. The disease is present at birth and can be diagnosed at this time by the presence of high levels of muscle enzymes in the blood stream of the infant. Clinical symptoms of muscle wasting are usually not very obvious until about three years of age. By this time, the child may have difficulty in playing games that require running. It is characteristic of the disease that proximal muscles of the limbs are more severely affected than are the distal. Some of the limb muscles may appear to be enlarged but are actually very weak. The disease progresses rapidly and the child may be confined to a wheelchair before ten years of age and usually dies because of respiratory or cardiac insufficiency by the age of twenty.

Duchenne muscular distrophy is transmitted as a sex-linked recessive disease that is carried on the x-chromosome. Females are carriers but do not show manifestations of the disease; males who have the x-chromosome with the gene all are severely incapacitated. There is no known effective treatment for the disease.

Myotonic muscular dystrophy is a muscle-wasting disease that is different from the other dystrophies. The disease gets the name *myotonia* from the inability of the muscles to relax following contraction. The patient is unable to make a fist and then open it quickly. The fingers open slowly. Myotonia may be painful, may be more severe in the cold, and may be relieved by heat or exercise. There is considerable variation in the progress of the disease even among members of the same family. Many patients can continue walking for many years, but the course of the disease is progressive and many organ systems are affected. Death may be due to cardiac dysrythmias.

Myotonic muscular dystrophy is transmitted as an autosomal dominant trait. There is no cure at present, but the myotonia symptoms may be relieved by medication.

There are several other dystrophies, but it is beyond the scope of this book to describe them.

Myositis

Any inflammatory disorder of muscle is known as *myositis;* it may be of known cause, such as *trichinosis* (caused by a parasitic infection), or of unknown etiology, such as *polymyositis.* Polymyositis is characterized by inflammation of several muscles, with muscle fiber destruction. There is usually weakness in the proximal muscles of both limbs, accompanied by pain, tenderness, and swelling of these muscles. Neck muscles may be involved, including muscles of mastication and the pharynx. Involvement of respiratory muscles may be life-threatening. In *dermatomyositis,* there is a characteristic skin rash around the eyelids. The disease progresses rapidly during the first few weeks unless treated. Myositis can be diagnosed by muscle biopsy; the most striking feature is the infiltration of mononuclear cells that accumulate initially around the blood vessels but may spread throughout the entire muscle.

---

endomysial sleeve of fine connective tissue is formed. After differentiation of the tendons and other muscle attachments from muscle tissue, the differentiation is complete.

TYPES OF MUSCLE CONTRACTION

According to the "all or none" principle of muscle contraction, when a muscle fiber is stimulated, it shortens as much as possible, which depends on the relationship of the contraction force to the resistance. If the force exceeds the resistance, the contracting fibers shorten; if the force is less than the resistance, then stimulation of the muscle fibers will cause tension without shortening and without movement. The following terms are sometimes used in describing muscle contraction.

Isotonic Contraction

Isotonic contraction is the production of movement by maximal shortening of the muscle when there is no resistance against the movement. These circumstances produce movement without tension, which usually occurs only when movement is produced by gravitational pull or some other external force. This type of contraction can be demonstrated by extending an upper extremity above the shoulder, and then allowing the hand to drop backwards toward the shoulder by relaxing the extensor (triceps brachii) on the posterior aspect of the arm. Observe that there is flexion of the forearm without the development of tension and that the biceps shortens while remaining flaccid. This type of contraction is rare in daily activity; most of our activities involve the development of some tension during the movement of the point of attachment. The tension that develops represents the force of the muscle contraction and must exceed the resistance for movement to occur.

Isometric Contraction

Isometric contraction, muscle contraction without the production of movement, occurs when the resistance is equal to or greater than the force exerted—such as when attempting to lift a weight heavier than one can handle. However, isometric contraction usually implies that muscles of the body are working against one another. Many isometric exercises have been designed that call for simultaneous contraction of antagonists, or pit the force of the contraction of similar groups of muscles in the right and left extremities against one another. For example, interlace fingers with the palms together in front of the chest and push the hands against one another. The same muscles in both shoulders and arms are being contracted with identical force and without the production of movement. Now, still keeping the palms in contact, attempt to pull the hands apart. In this exercise many muscles antagonistic to those used in the first exercise are called into action, with a more important role for some of the muscles of the fingers. Such exercises are especially useful for individuals who need to build up a specific group of muscles.

Tetanus

Tetanus is the contraction of muscle during repeated stimulation. When a muscle is repeatedly stimulated, the tension in the muscle rises slowly due to the gradual involvement of additional motor units and then is maintained at a constant level until the stimulation ceases. After stimulation ceases the tension rapidly returns to zero as the muscle relaxes. Tetanus is a steady, progressive contraction of muscle, without twitching; it should not be confused with *tetany,* a condition of abnormal muscle twitching resulting from abnormal calcium metabolism or hypofunction of the parathyroid glands. There is a disease called tetanus, or "lockjaw," in which *muscle spasm* (inability to relax), especially of the muscles of mastication, is caused by bacterial toxins.

WORDS IN REVIEW

Abduction	Myoblast
Actin	Myosin
Adduction	Myotube
Antagonist	Perimysium
Aponeurosis	Prime mover
Bipennate	Pronation
Circumduction	Raphe
Digastric	Rotation
Endomysium	Sarcolemma
Epimysium	Sarcomere
Extension	Sarcoplasm
Fixator	Strap muscle
Flexion	Supination
Fusiform	Synergist
Intersection	Tropomyosin
Motor unit	Troponin
Multipennate	Unipennate

SUMMARY OUTLINE

I. Internal Structure of Skeletal Muscles
    A. A muscle fiber (cell), or myofiber, contains longitudinally oriented myofibrils composed of thick myofilaments of myosin and thin myofilaments of actin.
    B. A sarcomere is that part of a myofiber between adjacent Z-bands; T-tubules penetrate near the junction of A- and I-bands.
    C. The process of muscle contraction is initiated by neural stimulation, which causes an influx of Ca^{++} that binds to troponin, causing tropomyosin to rotate out of the way of the myosin-binding site on actin; the myosin head binds to actin, forming cross-bridges between the thick and thin filaments; the myosin head rotates, drawing thin filaments toward the center of the sarcomere and reducing the distance between Z-bands.

II. Regulation of Muscle Contraction
    Muscle requires neural stimulation for the regulation of contraction. A motor unit consists of a neuron and all the muscle fibers it supplies.

III. Muscle Fiber Arrangements

Muscle fiber arrangements differ in the internal orientation of their fibers in relation to their tendons. Strap muscles have parallel fibers extending the length of the muscle belly. In the intersected arrangement the parallel fibers are interrupted by connective tissue partitions. Fusiform muscles contain many fibers that do not extend the length of the belly. Some muscles have fibers arranged much like a feather and may be unipennate, bipennate, or multipennate.

IV. Muscle Attachments

In muscle attachments the connective tissue (endomysium) around individual muscle fibers provides for attachment of muscle to bone by becoming continuous with the periosteum of the bone. The morphological types of attachments may be a fleshy attachment with little grossly visible connective tissue, an aponeurosis in the form of a fibrous sheet, or a tendon composed of collagenous fibers forming a cordlike structure. Muscles may also attach to one another through interdigitating fibers, forming a raphe.

V. Muscle Names

Muscles are usually named with descriptive terms that give information on the shape, size, location, direction of fibers, action, internal organization, or attachments.

VI. Fascia

Each muscle is encased by fascia that provides for the attachment of the muscle and separates the muscle from adjacent structures. Fascial clefts (spaces) between muscles enable muscles to contract independently.

VII. Types of Muscle Action

A. Prime movers are muscles that produce a particular movement upon contraction.

B. Antagonists are muscles that oppose one another upon contraction.

C. Fixators are muscles whose principal action is to stabilize a part of the body so that other muscles may exert an effect.

D. Synergists are muscles that work together to increase their effectiveness.

VIII. Types of movements

Types of movements include flexion, extension, abduction, adduction, pronation, supination, rotation, and circumduction.

IX. Muscle Differentiation

Skeletal muscle differentiates from embryonic mesoderm cells that aggregate into premuscle masses; single nucleated cells, myoblasts, fuse into multinucleated myotubes in which myofilaments are formed, producing multinucleated myofibers.

REVIEW QUESTIONS

1. What are the molecules responsible for the A- and I-bands seen with the light microscope in skeletal muscle fibers?

2. How are the thick and thin microfilaments arranged in the sarcomere?

3. List the sequence of events that occurs within a muscle fiber following neural stimulation of muscle contraction.

4. What is a motor unit? What is the influence of the size of the motor unit upon the fineness of control of muscle action?

5. In what type of muscle fiber arrangement do most of the muscle fibers extend the entire length of the muscle?

6. Name the four general types of muscle attachments and describe the connective tissue components of each type of attachment.

7. List the principal methods by which muscles are named.

8. What morphological features of the fascial arrangements around muscles allows muscles to contract independently?

9. What are prime movers? How do antagonists cooperate in the production of movements?

10. Define the following types of movement: flexion, extension, abduction, adduction, pronation, supination, rotation, and circumduction.

11. List the sequence of events essential in the differentiation of uninucleated myoblasts into multinucleated skeletal muscle fibers.

SELECTED REFERENCES

Bevelander, G., and Ramaley, J. A. *Essentials of Histology*. 8th ed. St. Louis: C. V. Mosby, 1979.

Copenhauer, W. M., *et al. Bailey's Textbook of Histology*. 17th ed. Baltimore: Williams and Wilkins, 1978.

Flickinger, C. J. *Medical Cell Biology*. Philadelphia: W. B. Saunders, 1979.

Moore, K. L. *Clinically Oriented Anatomy*. Baltimore: Williams and Wilkins, 1980.

THE AXIAL MUSCLES

OBJECTIVES

After completing this chapter you should be able to:

1. Name the principal muscles of facial expression associated with the mouth, nose, eyes, and ears.

2. Name and give the origins, insertions, and actions of the muscles of mastication.

3. List and give the functions of the muscles of the soft palate.

4. Name and give the attachments of the suprahyoid muscles that lie in the floor of the mouth.

5. List and locate the extrinsic muscles of the tongue.

6. List and give the principal actions of the muscles of the orbit.

7. Name the major flexors of the trunk and give the direction of the muscle fibers of each.

8. List and give the relative positions of the layers of the anterior abdominal wall.

9. List and describe the muscles of the wall of the thorax and the diaphragm.

10. Name the two major groups of deep back muscles and describe their general arrangement.

11. Name and describe the general pattern of arrangement of the principal muscles that close the pelvic outlet.

12. Name and give the general attachments of the superficial neck muscles, the infrahyoid muscles, the prevertebral muscles, the lateral vertebral muscles, and the muscles of the pharynx.

13. Describe the development of the axial muscles.

The axial muscles are those muscles that have their origins and insertions upon components of the axial skeleton. These muscles are essential to many body functions, such as maintaining posture, chewing, swallowing, breathing, speaking, and controlling eye movements. In this chapter the axial muscles will be described in functional groups in the various regions of the body, beginning with the muscles of the head.

MUSCLES OF THE HEAD

In addition to supporting and protecting head structures such as the brain, the skull provides for the attachment of skeletal muscles in the head and upper neck region.

The *intrinsic* muscles of the head are those muscles that have their origin and insertion on bones of the skull or on structures housed within or on the surface of the skull. These include the muscles of *facial expression* which act principally upon the skin, the *eyeball* muscles, the muscles of *mastication,* the muscles of the *soft palate* and *pharynx* for swallowing, the muscles of the *floor* of the *mouth* (suprahyoid muscles) for supporting the tongue and opening the mouth, and the muscles of the *tongue* which manipulate the tongue during speech, chewing, and swallowing.

Extrinsic muscles of the head originate on the cervical vertebrae and insert on the head. They will be discussed as part of the neck.

Muscles of Facial Expression

The muscles of facial expression (Fig. 11-1) are delicate strands of skeletal muscle located in the superficial fascia of the face and neck. They attach to the deep aspect of the skin at at least one point and usually to bone either directly or indirectly at one point. These muscles, summarized in Table 11-1, are anatomically and functionally related to the facial orifices: mouth, nose, eyes, and ears; they function as *sphincters*, constricting these openings, or as *dilators,* increasing them.

Because the facial muscles are attached to the skin, an incision at right angles to the direction of pull of the muscle fibers causes a gaping wound that will result in a scar unless the wound is closed carefully with stitches. The superficial fascia of the face is loose, so bleeding beneath the skin, as in a bruise, tends to spread widely through this fascia.

The **orbicularis oris** (Fig. 11-2) is a circular muscle in the lips that surrounds the mouth and draws the lips together. When tightly contracted it puckers the lips, as in kissing or whistling. The lips are pulled apart by several muscles, such as the **levator labii superioris,** which approaches the mouth obliquely and interdigitates with the orbicularis oris. These muscles not only open the mouth but also contribute to expression of emotion: smiling or grimacing. A special muscle of the group that interdigitates with the orbicularis oris, the **buccinator,** located in the cheek, has the essential role of keeping food between the teeth during chewing. Paralysis of the buccinator results in the accumulation of food in the vestibule between the cheek and teeth while eating.

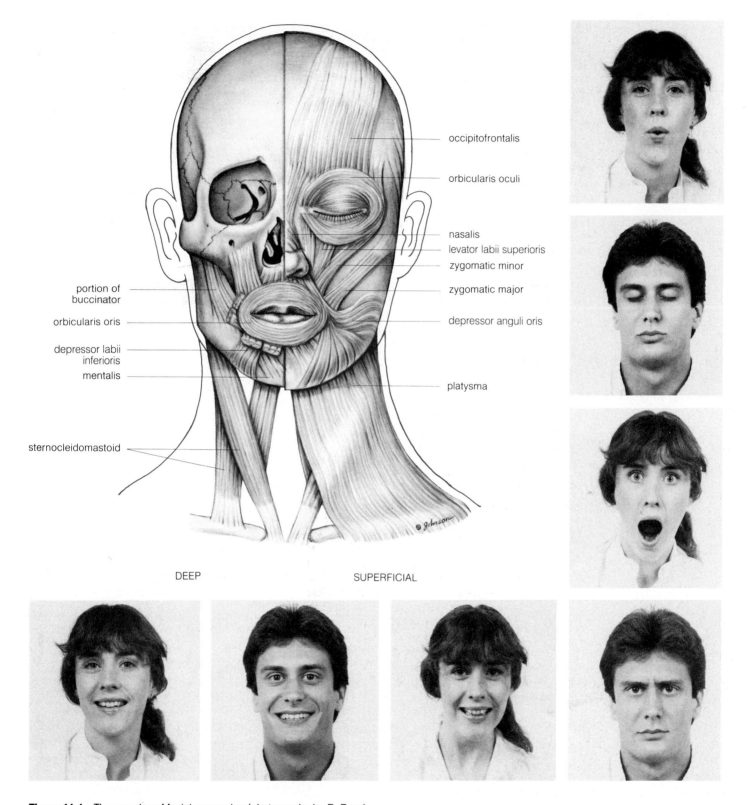

occipitofrontalis

orbicularis oculi

nasalis

levator labii superioris

zygomatic minor

zygomatic major

depressor anguli oris

platysma

portion of buccinator

orbicularis oris

depressor labii inferioris

mentalis

sternocleidomastoid

DEEP

SUPERFICIAL

Figure 11-1 The muscles of facial expression (photographs by R. Ross).

Table 11-1 Muscles of Facial Expression			Innervation: Facial nerve
Muscle	Origin	Insertion	Action
Muscles Related to the Eyelids			
Corrugator supercilii	Superciliary arch	Skin above orbital arch	Causes vertical wrinkles
Occipitofrontalis	Galea aponeurotica	Occipital part, superior nuchal line, frontalis part blends with orbicularis oculi	Tightens scalp, eyebrows
Orbicularis oculi	Nasal part of frontal, frontal process of maxilla, medial palpebral ligament	Lateral palpebral raphe	Sphincter of the eyelids
Muscles Related to the Mouth			
Buccinator	Pterygomandibular raphe	Blends with muscles of angle of mouth	Compresses cheeks
Depressor anguli oris	Oblique line of mandible	Angle of mouth	Pulls angle downward
Depressor labii inferioris	Oblique line of mandible	Skin, lower lip	Draws lower lip downward
Levator anguli oris	Canine fossa, maxilla	Angle of mouth	Deepens furrow at angle of mouth
Levator labii superioris	Lower margin of orbit	Upper lip	Elevates upper lip
Levator labii superioris alaeque nasi	Frontal process, maxilla	Alar cartilage of nose and upper lip	Dilates nares, deepens nasolabial furrow
Mentalis	Incisive fossa, mandible	Skin of chin	Raises and protrudes lower lip
Orbicularis oris	Skin and mucous membrane, lips	Skin and mucous membrane, lips; maxilla, nasal septum	Closes lips
Rizorius	Fascia over masseter	Angle of mouth	Draws angle posteriorly
Zygomaticus major	Zygomatic bone	Angle of mouth	Draws angle upward and back
Zygomaticus minor	Zygomatic bone	Upper lip	Deepens furrow at angle

Table 11-1	Muscles of Facial Expression	(cont.)	Innervation: Facial nerve
Muscle	Origin	Insertion	Action
Muscles Related to the Nose			
Depressor septi	Incisive fossa, maxilla	Septum and ala, nose	Constricts the nose
Nasalis	Skin over nose and maxilla	Alar cartilage, nose	Dilates the nares
Procerus	Fascia covering lower part of nasal bone	Skin over forehead	Causes transverse wrinkles over nose
Muscles Related to the Ear			
Auricularis anterior	Fascia of temporal area	Anterior surface of helix	Draws auricle forward and upward
Auricularis posterior	Mastoid portion, temporal bone	Cranial surface, concha	Draws auricle posteriorly
Auricularis superior	Fascia over temporalis	Cranial surface, auricle	Draws auricle upward
Muscle of the Neck			
Platysma	Mandible	Skin inferior to clavicle	Tightens skin of neck

The arrangement of muscles around the eyes (Fig. 11-3) is similar but less complex. The sphincter of the eyelids, the **orbicularis oculi,** draws the eyelids together, affording protection for the eyes.

ator palpebrae superioris, a muscle of the orbit that does not belong to the group of muscles of facial expression, opens the eye.

The frequent contraction of the orbicularis oculi, blinking, helps regulate the flow of tears across the eyeball, which keeps the cornea moist. Tears produced by the *lacrimal gland,* located in the upper lateral corner of the orbit, drain into the nasal cavity through the *nasolacrimal canal.* A dry cornea is very painful due to the irritation of nerve endings that results from friction as the eyelids rub against the dry surface.

The functions of the facial muscles are best appreciated by observing the results when these muscles are paralyzed on one side. The mouth is pulled to the unaffected side by the unopposed contraction of the intact muscles. The affected eyelids cannot be closed tightly as the lower eyelid drops down. This allows tears to overflow onto the cheek. Chewing is disrupted as food tends to accumulate outside the teeth and may escape between the lips on the affected side.

When the eyes are opened, the lower eyelids are not pulled downward because no muscles enter the lower eyelid from below. Elevation of the upper eyelids by contraction of the *lev-*

Muscles around the openings into the nose dilate or compress the nostrils, but their action is too slight to be effective in closing or significantly increasing the openings. The external

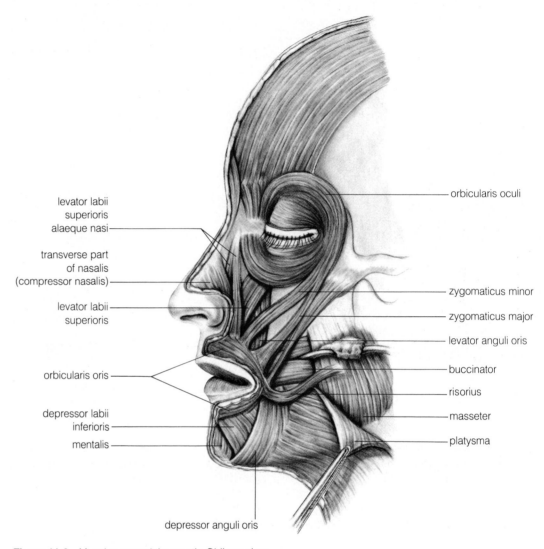

levator labii
superioris
alaeque nasi

transverse part
of nasalis
(compressor nasalis)

levator labii
superioris

orbicularis oris

depressor labii
inferioris

mentalis

orbicularis oculi

zygomaticus minor

zygomaticus major

levator anguli oris

buccinator

risorius

masseter

platysma

depressor anguli oris

Figure 11-2   Muscles around the mouth. Oblique view.

ear (Fig. 11-4) also has muscles that are too small to have any real significance in humans. In some animals, such as dogs, with large external ears, these muscles have an important role in raising the auricles to expose the external auditory meatus to sound waves, increasing the capacity to hear.

The muscles of facial expression also include muscles of the scalp and neck. The *scalp* (Fig. 11-3) has a musculoaponeurotic sheet, the *galea aponeurotica,* composed of the *occipitalis* muscle at the back of the head, the *frontalis* of the forehead, and the *aponeurosis* that joins them. The frontalis causes horizontal creases in the skin of the forehead and elevates the eyebrows in the expression of surprise. Another muscle not related

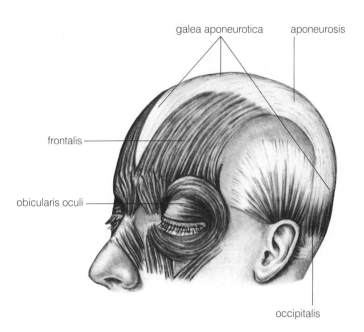

Figure 11-3 Muscles around the eyes. Lateral view.

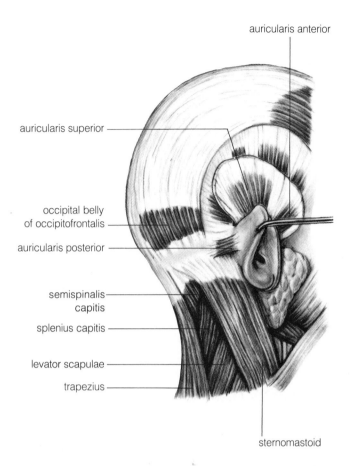

Figure 11-4 Muscles of the external ear and back of the neck. The external ear muscles are intrinsic head muscles; the neck muscles are extrinsic. Lateral view.

to the openings is the *platysma* (Fig. 11-5), a broad sheet of muscle whose fibers pass vertically from the mandible within the superficial fascia of the neck to reach the upper chest wall. The platysma is thus an extrinsic muscle of the head as well as a neck muscle: it tenses the skin of the neck. It gradually weakens with age, which together with the loss of skin elasticity causes wrinkles in the skin of the neck.

Muscles of Mastication

The four *muscles* of *mastication* (Table 11-2) are the principal muscles involved in the process of biting and chewing; all produce movement of the mandible upon contraction. These muscles are the masseter, temporalis, medial pterygoid, and lateral pterygoid.

The **masseter** (Fig. 11-6) is the most lateral, or superficial, of the muscles of mastication. It originates on the *zygomatic arch* and inserts into the bone and the fascial sling around the lower border of the *angle* of the *mandible*. The masseter primarily *elevates* and slightly *protrudes* the mandible. Alternate contraction of the right and left masseter muscles results in a

rocking, grinding type of motion that facilitates the breakdown of food. The outline of the masseter may be traced by palpating the muscle as it contracts while clenching the teeth.

The **temporalis** (Fig. 11-7) has a broad origin from the *temporal lines* on the side of the skull and from the entire *temporal fossa;* it passes deep to the zygomatic arch to insert on the *coronoid process* of the *mandible* and the *anterior border* of the *mandibular ramus.* It is a powerful elevator of the mandible, and its contraction may be felt by placing the fingers on the side of the head while clenching the teeth.

The **pterygoid muscles** are shown in Fig. 11-8. The *medial pterygoid* arises by two heads: the *deep head* originates from the *medial surface* of the *lateral pterygoid plate* of the *sphenoid,*

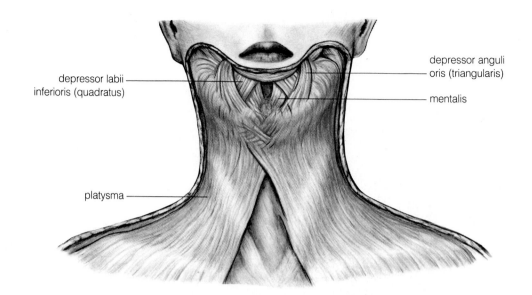

depressor labii
inferioris (quadratus)

depressor anguli
oris (triangularis)

mentalis

platysma

Figure 11-5 The platysma.
Anterior view.

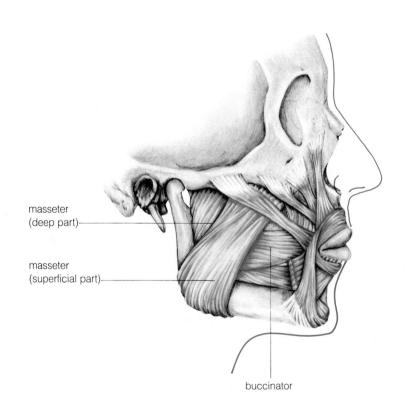

masseter
(deep part)

masseter
(superficial part)

buccinator

Figure 11-6 The masseter
muscle. Lateral aspect.

Table 11-2	Muscles of Mastication		Innervation: Trigeminal nerve
Muscle	Origin	Insertion	Action
Lateral pterygoid	Greater wing of sphenoid and infratemporal crest, and lateral surface of lateral pterygoid plate	Neck of condyle, capsule and articular disk of temporomandibular joint	Opens jaws, pulls condyle and articular disc forward
Masseter	Zygomatic arch and process	Angle of mandible	Closes jaws
Medial pterygoid	Medial surface of lateral pterygoid plate	Angle of mandible; joins insertion of masseter by means of fascial sling	Closes jaws
Temporalis	Temporal fossa and deep temporal fascia	Coronoid process and anterior border of ramus of mandible	Closes jaws

and the *lateral head,* or *superficial head,* arises from the maxilla. Fibers of the medial pterygoid descend along the deep surface of the mandibular ramus toward its angle and are parallel to the fibers of the masseter; consequently this muscle acts in conjunction with the masseter. It elevates and slightly protracts the mandible. It is the deepest of the muscles of mastication, forming the medial boundary of the pterygomandibular space through which the nerves pass that supply the lower teeth and the tongue.

The *lateral pterygoid* originates from two heads: an *upper head* from the *greater wing* of the *sphenoid* and *lower head* from the *lateral surface* of the *lateral pterygoid plate*. It inserts into the *capsule* and *articular disc* of the *temporomandibular joint* and the *neck* of the *mandible.* The fibers of this muscle pass at a right angle to those of the other three muscles of mastication. Its fibers lie in a horizontal plane and pass posteriorly and laterally. The lateral pterygoids move the head of the mandible forward while muscles in the floor of the mouth lower the mandible.

Muscles of the Soft Palate

The soft palate (Fig. 11-9) is a ledge of soft tissue attached to the posterior surface of the hard palate; it forms the posterior part of the partition between the oral cavity and the nasopharynx. It is a muscular sheet covered with mucous membrane and

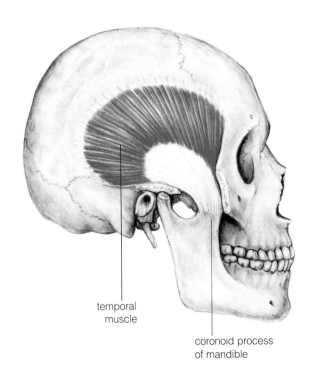

temporal muscle

coronoid process of mandible

Figure 11-7 The temporalis muscle. Lateral aspect.

Figure 11-8 The medial and lateral pterygoid muscles. Lateral view.

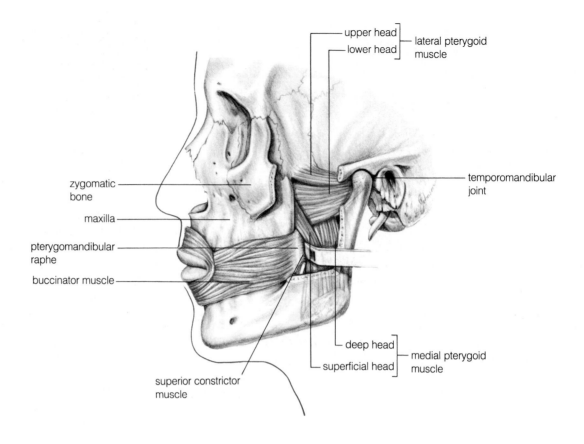

upper head | lateral pterygoid muscle
lower head

temporomandibular joint

zygomatic bone

maxilla

pterygomandibular raphe

buccinator muscle

deep head | medial pterygoid muscle
superficial head

superior constrictor muscle

Table 11-3   Muscles of the Soft Palate

Muscle	Origin	Insertion	Action	Innervation
Levator veli palatini	Petrous portion of temporal bone and medial wall of auditory tube	Blends with muscle of other side	Raises soft palate during swallowing	Vagus
Palatoglossus	Blends with muscle of other side	Blends with tongue musculature	Pulls tongue backward and upward	Vagus
Palatopharyngeus	Palatine aponeurosis	Interlaces with middle pharyngeal constrictor	Assists in raising the pharynx during swallowing	Vagus
Tensor veli palatini	Base of medial pterygoid plate and wall of auditory tube	Tendon hooks around pterygoid hamulus, passes medially to palatine aponeurosis	Tightens soft palate during swallowing	Trigeminal

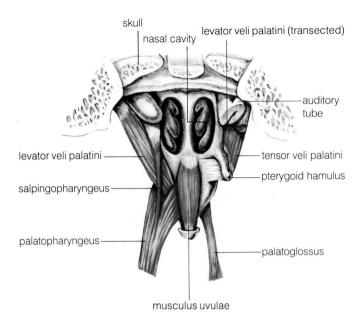

skull

nasal cavity

levator veli palatini (transected)

auditory tube

levator veli palatini

tensor veli palatini

salpingopharyngeus

pterygoid hamulus

palatopharyngeus

palatoglossus

musculus uvulae

Figure 11-9 Muscles of the soft palate. Posterosuperior view. Muscles on the left are posterior to those on the right.

supported by a central fibrous band, the palatine aponeurosis. The muscles of the soft palate (Table 11-3) tighten and elevate the soft palate during swallowing, preventing food from entering the nasal cavities. These muscles attach to either the upper or lower surface of the palatine aponeurosis. The aponeurosis attaches to the posterior border of the hard palate and extends as a sheet toward the posterior tip of the *uvula,* the small extension at the free margin of the soft palate.

The **tensor veli palatini** arises from the edge of the auditory tube and the adjacent area of the temporal bone. Its muscle fibers end in a tendon that hooks around the pterygoid *hamulus* (of the *medial pterygoid plate*) and flares out, forming the central aponeurosis of the soft palate. Upon contraction the tensor veli palatini tightens the soft palate, resisting the pressure of the bolus (the mass of chewed food) against the undersurface of the palate during swallowing.

The **levator veli palatini** also arises from the edge of the *auditory tube* and descends to insert into the upper surface of the *palatal aponeurosis.* It elevates the soft palate during swallowing, preventing food from entering the nasal cavities.

When the palatal muscles are paralyzed, the soft palate is not elevated properly during swallowing. Food or liquid may enter the nasal cavity and run out of the nose. Paralysis of the tensor veli palatini may allow the entire soft palate to prolapse into the nasal cavity.

The **palatoglossus** takes origin from the *inferior surface* of the *aponeurosis* and descends to enter the lateral aspect of the tongue. It creates a fold called the *anterior pillar* of the *fauces,* or palatoglossal fold, that marks the posterior boundary of the oral cavity. It pulls the tongue posteriorly and superiorly. The pull of this muscle is resisted by the actions of the tensor veli palatini and levator veli palatini muscles during swallowing.

The **palatopharyngeus** also arises from the lower surface of the aponeurosis and descends to blend with the muscles of the pharynx. In its course it creates a fold in the mucous membrane called the *posterior pillar* of the *fauces.* The fossae between the anterior and posterior pillars are occupied by the *palatine tonsils.* Contraction of the palatopharyngeus assists other muscles in the elevation of the pharynx and larynx during swallowing.

The *uvula* contains a few longitudinal fibers termed the *musculus uvulae.* Its contraction causes the soft palate to become somewhat shorter and thicker.

The Suprahyoid Muscles

The **suprahyoid muscles** (Fig. 11-10) extend from the hyoid bone to the mandible and the tongue and are the major structures of the floor of the oral cavity. Several of these muscles are attached to the *hyoid bone,* which is not part of the skull, and are thus extrinsic head muscles. However, they have a close functional relationship with the oral cavity. The suprahyoid muscles act to support the tongue and to elevate the hyoid bone during swallowing or to depress the mandible during chewing. When they depress the mandible, the hyoid bone is fixed by contraction of the infrahyoid muscles (which will be discussed with the neck muscles). The muscles discussed here are listed in Table 11-4.

The most superficial muscle of the upper neck and the floor of the mouth is the *digastric* muscle; its *posterior belly* arises from the *mastoid process* of the *temporal bone,* and its *anterior belly* attaches to the *mandible* near the midline. The two bellies are connected by an *intermediate tendon* that attaches to the hyoid bone by means of a fascial sling. During swallowing, the digastric elevates the hyoid bone and thus helps elevate the larynx, which is attached to the hyoid bone by the *thyrohyoid membrane.* When the hyoid bone is fixed by the infrahyoid muscles, the anterior belly of the digastric assists in lowering

Table 11-4	The Suprahyoid Muscles			
Muscle	Origin	Insertion	Action	Innervation
Digastricus	Medial aspect of mastoid process; intermediate tendon attached to hyoid bone by means of fascial sling	Anterior belly attached to inferior border of mandible	Elevates hyoid bone and depresses mandible	Trigeminal (anterior belly) Facial (posterior belly)
Geniohyoideus	Internal surface of symphysis menti	Body of hyoid bone	Draws hyoid bone and tongue anteriorly; can depress mandible with hyoid bone fixed	Spinal nerve, C1
Mylohyoid	Hyoid	Medial surface of mandible	Depresses mandible	Trigeminal
Stylohyoideus	Styloid process	Greater cornu of hyoid bone	Elevates hyoid bone	Facial

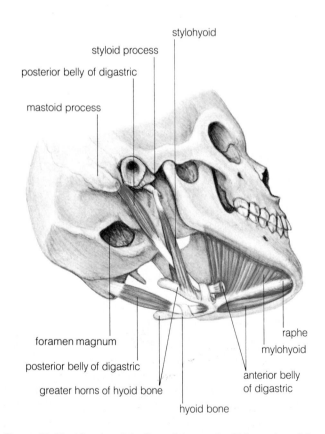

stylohyoid

styloid process

posterior belly of digastric

mastoid process

foramen magnum

posterior belly of digastric

greater horns of hyoid bone

raphe

mylohyoid

anterior belly of digastric

hyoid bone

Figure 11-10 Muscles of the floor of the mouth. Oblique view of the inferior aspect.

the mandible as the mouth opens. The *mylohyoid muscle* originates on the *hyoid* bone and inserts on the inner surface (mylohyoid line) of the *mandible*. The *geniohyoid* muscle is deep to the mylohyoid and passes to its attachment on the inner surface of the *mandible* near the anterior midline. All three of the muscles may either *depress* the *mandible* or *elevate* the *hyoid* bone. By elevating or stabilizing the hyoid bone, to which the base (the root) of the tongue is attached, these muscles enable the tongue to move freely from a stable base.

The Extrinsic Tongue Muscles

The tongue (Fig. 11-11), a muscular structure located in the floor of the mouth, is divided into two lateral halves by a medial fibrous septum extending posteriorly from the tip of the tongue to the hyoid bone. In each half of the tongue there are extrinsic muscles whose origins are outside the tongue and intrinsic muscles whose origins are entirely within it (Table 11-5). The fibers of the two sets of muscles interlace and function together in tongue movements.

The three pairs of *extrinsic muscles* of the *tongue* arise from skeletal elements external to the tongue. These muscles are arranged to protrude, retract, and move the flexible tongue in any direction. The *genioglossus* arises from the *genial tubercles* of the inner surface of the mandible and passes posteriorly to enter the root of the tongue, from which its fibers radiate into the body of the tongue. The *hyoglossus* arises from the

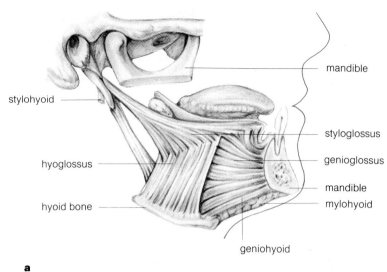

a

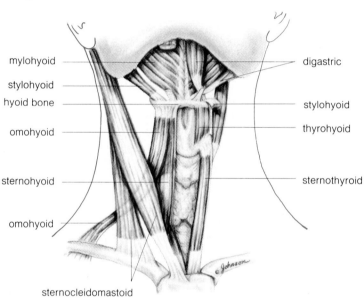

b

Figure 11-11 Muscles of the floor of the mouth and anterior neck. (**a**) Extrinsic muscles of the tongue, lateral aspect. (**b**) Neck muscles, anterior aspect.

hyoid bone and passes superiorly to intermingle with the other tongue muscles at the root. The *styloglossus* originates from the *styloid process* of the *temporal* bone, from which it passes anteriorly to the root. The *palatoglossus,* which sends fibers into the tongue from the palate, is considered a palatal muscle rather than a muscle of the tongue because its innervation is by the nerve that supplies palatal muscles.

If the tongue muscles are paralyzed on one side, the tongue will deviate to the affected side as it is protruded. Paralysis of the genioglossus muscle brings the risk of obstruction of the airway, because the tongue tends to drop back over the opening to the larynx when it cannot be pulled forward by muscular tone of the genioglossus.

Table 11-5 Muscles of the Tongue			Innervation: Hypoglossal nerve
Muscle	Origin	Insertion	Action
Genioglossus	Superior genial tubercle of mandible	Passes posteriorly to blend with other tongue muscles	Pulls tongue forward
Hyoglossus	Body and greater cornu of hyoid bone	Passes superiorly to blend with other tongue muscles	Depresses tongue
Styloglossus	Anterior and lateral surfaces of styloid process	Passes anteriorly to blend with other tongue muscles	Draws tongue up and back
Intrinsic tongue muscles	Interlace with other tongue muscles	Interlace with other tongue muscles	Alter shape of tongue

Muscles of the Orbit

Six skeletal muscles attached to the eyeball are responsible for eye movement (Fig. 11-12, Table 11-6). They arise from fibrous elements attached to the bony orbit and insert into the sclera, the outer connective tissue coat (or "white") of the eyeball.

The four **rectus muscles** (straight muscles)—the superior rectus, the inferior rectus, the medial rectus, and the lateral rectus—take origin from a tough *fibrous ring,* a thickening of the periosteum of the area, that surrounds the optic foramen at the back of the orbit. Each rectus muscle passes anteriorly to insert into the *sclera* in the anterior half of the eyeball. The contraction of each of these muscles moves the eyeball in the direction suggested by the name.

The **superior oblique muscle** arises from the bone of the *posterior aspect* of the *orbit* above the optic foramen and passes to the upper anteromedial corner of the orbit, where its tendon passes through a pulleylike arrangement of fascia, the *trochlea,* prior to its insertion into the *sclera* of the *upper posterior half* of the eyeball. Contraction of the muscle tilts the front of the eyeball downward and laterally.

The **inferior oblique muscle** arises from the *floor* of the *orbit* near its *anterior border* and inserts into the *posterior half* of the *eyeball* on the lower lateral quadrant. Its contraction elevates the front of the eyeball and turns it laterally. The extrinsic eye muscles of the two eyes function as a group, moving the eyes together to follow a rapidly moving object, such as a baseball thrown at 90 miles per hour. A similar cooperative effort occurs in reading, as the eyeballs move at a relatively uniform rate along lines of print, bringing the words into focus during the pause between each movement.

One muscle of the orbit, the *levator palpebrae superioris,* does not attach to the eyeball; it arises from the posterior aspect of the roof of the orbit and passes into the upper eyelid where its fibers intermingle with the fibers of the orbicularis oculi. Contraction of the levator palpebrae superioris raises the upper eyelid, opening the eye. Injury to this muscle or to its nerve means that the orbicularis oculi is unopposed and the eye remains closed.

> Because the muscles of the eyeballs work together to align the right and left eyes in focusing upon an object, paralysis of one or more muscles unilaterally results in misalignment. The usual result is double vision, or diplopia.

THE TRUNK MUSCLES

Most of the muscles of the trunk attach directly or indirectly to the vertebral column and may be grouped into those located anterior and those located posterior to the transverse processes of the vertebrae. The anterior muscles are flexors and rotators of the trunk whereas the posterior muscles are extensors and rotators. Both sets of muscles assist in maintaining posture, and the anterior muscles also compress the viscera when they contract, increasing the intra-abdominal pressure. The anterior muscles are thus concerned with visceral functions such as breathing, urination, and defecation, which are controlled largely on a reflex basis.

Table 11-6 Muscles of the Orbit

Muscle	Origin	Insertion	Action	Innervation
Inferior oblique	Maxilla near rim of orbit	Sclera of eyeball posterior to equator	Draws eyeball upward and laterally	Oculomotor
Inferior rectus	Below optic foramen	Sclera of inferior surface of eyeball	Pulls eyeball downward and medially	Oculomotor
Lateral rectus	Lateral to optic foramen	Sclera of lateral surface of eyeball	Turns eyeball laterally	Abducens
Levator palpebrae superioris	Inferior surface of lesser wing of sphenoid	Upper eyelid	Raises upper eyelid	Oculomotor
Medial rectus	Medial to optic foramen	Sclera of medial surface of eyeball	Turns eyeball medially	Oculomotor
Superior oblique	Above optic foramen, above and medial to superior rectus	Tendon passes through fibrocartilaginous ring attached to frontal bone to insert posterior to equator of eyeball	Tilts eyeball downward and laterally	Trochlear
Superior rectus	Superior to optic foramen	Superior surface of eyeball	Pulls eyeball upward and medially	Oculomotor

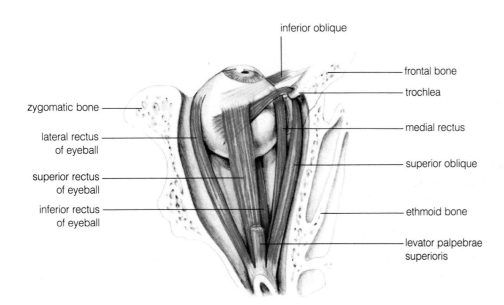

Figure 11-12 Muscles of the orbit. Superior view of the left orbit.

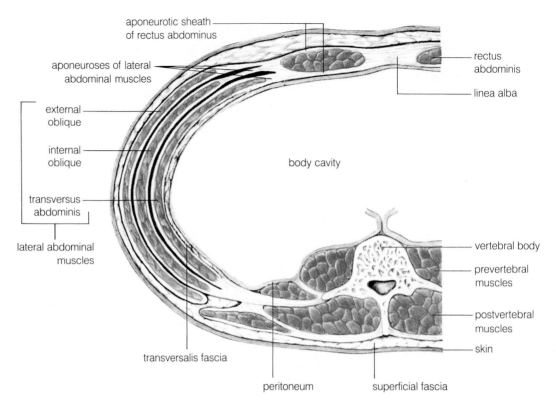

Figure 11-13 Layers of the abdominal wall. Diagram of a transverse section of the abdominal wall.

Table 11-7	Abdominal Wall Muscles			
Muscle	Origin	Insertion	Action	Innervation
External oblique	Lower eight ribs	Iliac crest and aponeurosis	Compresses abdominal viscera, flexes trunk	Spinal nerves T8 to L1
Internal oblique	Inguinal ligament, iliac crest, and lumbodorsal fascia	Last three or four ribs and aponeurosis	Compresses abdominal viscera, flexes trunk	Spinal nerves T8 to L1
Rectus abdominis	Pubic bone	Cartilages of ribs 5, 6, and 7	Compresses abdominal viscera, flexes trunk	Spinal nerves T7 to T12
Transversus abdominis	Inguinal ligament, iliac crest, lumbodorsal fascia, and last six ribs	Aponeurosis	Compresses abdominal viscera	Spinal nerves T7 to L1

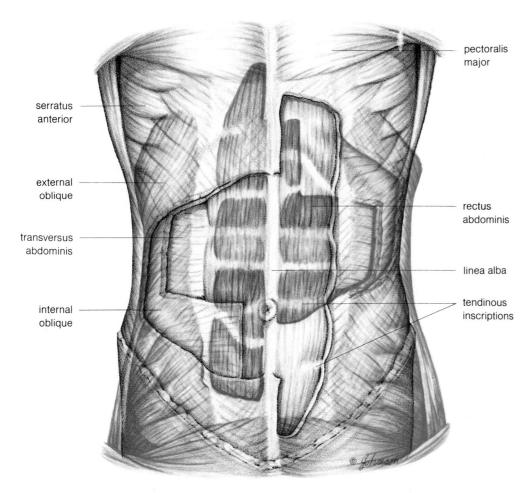

Figure 11-14 Muscles of the anterior and lateral abdominal wall. Anterior view.

The trunk muscles include the muscles of the anterior abdominal wall and thorax, and the posterior deep muscles of the back.

The Abdominal Wall

The abdominal wall (Fig. 11-13) surrounds the abdominal (or peritoneal) cavity, forming a flexible protective covering for the abdominal viscera. It is covered externally by skin and superficial fascia, and it is lined internally by a serous membrane called the *peritoneum* and subserous fascia called *transversalis fascia*. Each half of the wall consists of an anterior muscle (rectus abdominis) with fibers running vertically and three lateral muscle layers (**external oblique, internal oblique,** and **transversus abdominus**) with their fibers oriented in different directions. All of the lateral muscles are attached to the anterior midline of the abdominal wall by means of aponeuroses (Table 11-7).

Skin is quite elastic, and the superficial fascia, composed of loose connective tissue, usually contains a considerable quantity of fat and offers little resistance to expansion and contraction of the abdominal wall. The lateral abdominal muscles are covered with deep fascia that is thin and weak, allowing expansion of the cavity and its contents upon relaxation of the muscles (as after eating a large meal). The transversalis fascia is thin in most areas of the wall and serves principally as the passageway for the many small blood vessels that supply fluid for moistening the peritoneum. The smooth, moist surface of the peritoneum prevents the abdominal viscera, such as the gut, from sticking to the body wall.

The Rectus Abdominis The paired **rectus abdominis** muscles (Figs. 11-13, 11-14, 11-15) are straplike muscles extend-

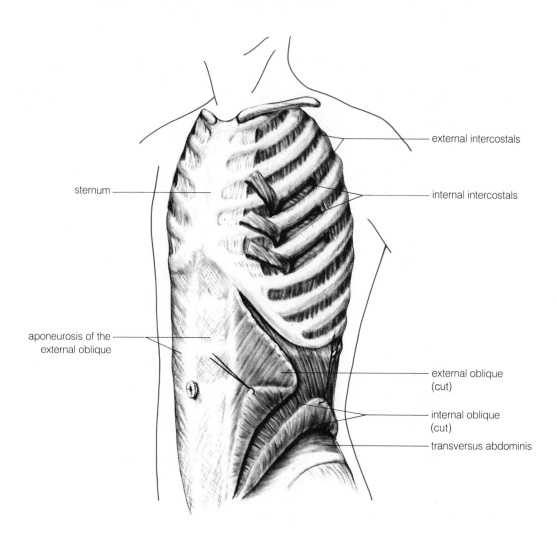

external intercostals

sternum

internal intercostals

aponeurosis of the
external oblique

external oblique
(cut)

internal oblique
(cut)

transversus abdominis

Figure 11-15 Muscles of the
trunk. Oblique view.

ing vertically on each side of the anterior midline of the abdominal wall from the ribs and sternum to the pubis. An aponeurotic sheath, an extension of the aponeuroses of the lateral abdominal muscles, encloses each rectus abdominis muscle (Fig. 11-13). This fascial sheath (rectus sheath) adds considerable strength to the anterior abdominal wall while allowing a certain degree of flexibility. The externally visible linea alba (white line) in the abdominal midline (Fig. 11-13) is the line of fusion of the rectus sheaths of the two sides and serves as the insertion of the lateral abdominal wall muscles. The parallel fibers of the rectus abdominis are interrupted by transverse collagenous tendinous intersections, which fuse with the anterior surface of the rectus sheath and prevent the muscle from sliding freely within the sheath. The rectus abdominis flexes the trunk and is one of the principal muscles contracting when one does sit-ups.

During the later stages of pregnancy the rectus abdominis undergoes tremendous elongation, and the two muscles tend to separate as the rectus sheath stretches. After parturition the rectus abdominis returns to its former length slowly and only if properly exercised. Doing sit-ups with a twisting motion for a few weeks following pregnancy gradually brings the two muscles back toward the midline and eliminates the tendency of the anterior abdominal wall to sag.

The Lateral Abdominal Muscles The lateral abdominal wall consists of three muscles (Fig. 11-14) arranged as sheets, with the muscle fibers of each sheet oriented in a different direction.

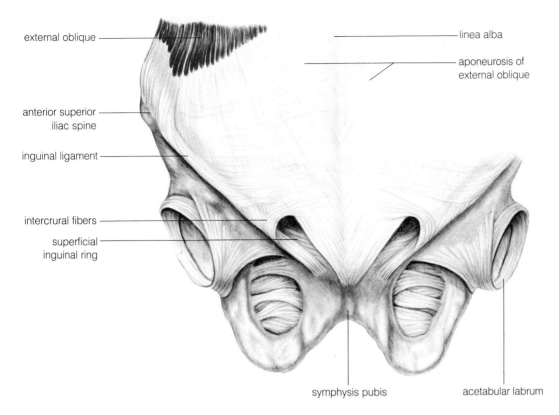

external oblique

linea alba

aponeurosis of
external oblique

anterior superior
iliac spine

inguinal ligament

intercrural fibers

superficial
inguinal ring

symphysis pubis

acetabular labrum

Figure 11-16 The inguinal region. Anterior view.

The fibers of the outer muscle, the **external oblique** (Fig. 11-15), arise posteriorly from the lower ribs and lumbodorsal fascia; they course anteriorly and inferiorly to insert into the aponeurosis that passes anterior to the rectus abdominis and into the *inguinal ligament,* a fibrous band extending from the anterior superior iliac spine to the pubic tubercle. The inguinal ligament is the inferior, thickened border of the aponeurosis of the external oblique. Just above the inguinal ligament is a triangular gap in the external oblique, known as the *superficial inguinal ring.* In males this opening serves for the passage of the spermatic cord from the body cavity to the scrotum (Fig. 11-16); in females it is the passageway for the round ligament that anchors to the uterus.

The inguinal ligament also serves as part of the origin of the second lateral abdominal muscle, the **internal oblique.**

These fibers arise from the ligaments, the iliac crest, and the lumbodorsal fascia; they pass anteriorly and superiorly, almost at right angles to the fibers of the external oblique, to insert into the aponeurosis of the internal oblique and into the lower ribs. Unlike the aponeurosis of the external oblique, which passes anterior to the rectus abdominis muscle, the aponeurosis of the internal oblique splits into two layers in the upper part. The anterior layer joins the aponeurosis of the external oblique, whereas the posterior sheet passes posterior to the rectus abdominis where it is joined by the aponeurosis of the third muscle, the **transversus abdominis.** The transversus abdominis consists of fibers that run horizontally from the lumbodorsal fascia, inferior six costal cartilages, iliac crest, and inguinal ligament to the aponeurosis which passes posterior to the rectus abdominis.

Table 11-8   Muscles of the Thorax

Muscle	Origin	Insertion	Action	Innervation
Diaphragm	Lower border, rib cage, xiphoid process, and vertebrae, L1 to L3	Central tendon	Draws tendon downward	Phrenic (C3 to C5)
External intercostals	Caudal border, ribs	Cranial border, rib below	Draws ribs together, raises rib cage	Intercostals
Internal intercostals	Caudal border, ribs and costal cartilages	Cranial border, rib below	Draws ribs together, lowers rib cage	Intercostals
Levatores costarum	Transverse processes, vertebrae	Ribs, near angle	Raises ribs	Intercostals
Innermost intercostals	Inner surface, ribs and costal cartilages	Cranial border, rib below		Intercostals
Subcostals	Inner surface, ribs, near angle	Inner surface, ribs 2 or 3 below	Lowers rib cage	Intercostals
Transversus thoracis	Inner surface, sternum	Inner surface, costal cartilages	Lowers rib cage	Intercostals

The point at which the spermatic cord passes through the inguinal ligament may be explored with a finger. A large or loose superficial inguinal ring suggests the possibility of a segment of gut working its way through the body wall into the scrotum, a condition termed an *inguinal hernia*.

The abdominal cavity contains viscera that are not protected by bone. Muscles of the anterior abdominal wall afford some protection, especially if they are in good condition. During most of human evolution the flexible abdominal wall was an adequate defense, for people who were attacked could simply bend forward and defend themselves with their fists. At present the flexible body wall may prove to be inadequate if the object coming toward you happens to be a truck.

As a primary function the abdominal wall musculature provides the body wall with flexibility and strength. The rectus sheath provides a great deal of strength and is flexible but not very distensible. Much of the capacity of the abdominal wall to expand comes from the obliques and the transversus abdominis muscles of the lateral abdominal wall. These muscles function as a unit, relaxing and contracting at the same time. For example, as the diaphragm contracts and pushes downward to allow air to enter the lungs during inspiration, the lateral abdominal muscles normally relax so that pressure is not exerted upon the abdominal viscera. Following inspiration, contraction of the abdominal wall muscles may be involved in forced expiration, a situation that may be demonstrated by observing your own breathing. During quiet breathing the abdominal muscles remain relaxed.

The Thorax

Muscles of the thorax (Table 11-8) include the intercostal muscles whose contractions elevate the ribs and the diaphragm whose contraction increases the superior–inferior extent of the thorax. The intercostal muscles occupy the intercostal spaces, contributing to the lateral and anterior walls of the thorax. The diaphragm is attached to the inferior margin of the rib cage and separates the thorax and abdomen.

The Intercostal Muscles   The *intercostal muscles* of the thorax (Fig. 11-17) bridge the spaces between adjacent ribs. They occur

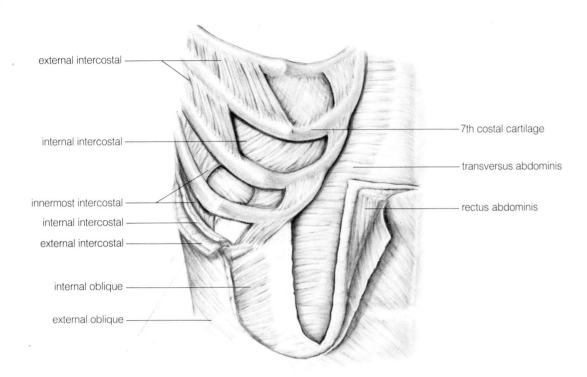

external intercostal

internal intercostal

innermost intercostal

internal intercostal

external intercostal

internal oblique

external oblique

7th costal cartilage

transversus abdominis

rectus abdominis

Figure 11-17 The intercostal muscles. Oblique view of a segment of the trunk wall.

in three layers that correspond to those of the lateral abdominal wall (Figs. 11-15, 11-17). The **external intercostals** form the outer muscle sheets on each side of the thorax, with fibers coursing in the same direction as those of the external oblique of the same side. The external intercostals extend only to about the costochondral (bone–cartilage) junction. Anterior to the costochondral junctions fascial membranes replace the muscles. The **internal intercostals** have fibers coursing in the same direction as those of the internal obliques; they extend from the costal angle of the ribs to the sternum. Posterior to the costal angle this layer of muscle is replaced by a fascial membrane. The **innermost intercostals** occupy only the middle portion of the thorax, forming an inner complex of muscles that includes the *transversus thoracis* anteriorly and the *subcostals* posteriorly; this mass corresponds to the transversus abdominis although the fiber directions do not correspond.

The external intercostals elevate the rib cage during inspiration, thus expanding the transverse and anterior–posterior diameters of the thorax and allowing air to enter the lungs. In normal, quiet breathing the elastic recoil of the lungs is ade-

quate for expiration and requires no muscle contraction, but in forced expiration the internal intercostals assist by depressing the rib cage, reducing the volume of the thorax. Radiographs taken during various phases of breathing show that the intercostal spaces are wider at inspiration than at expiration. The role of the intercostal muscles in this process has not been determined.

The Diaphragm The **diaphragm** is a dome-shaped musculofibrous partition that separates the thoracic from the abdominal cavity. The peripheral muscular part (Fig. 11-18) originates from the xiphoid process of the sternum, the inferior margin of the ribcage (the last six ribs on either side), and the bodies of the first two or three lumbar vertebrae. The muscle fibers extend inward from their circular origin to insert into a central tendon, a thin but strong aponeurosis of collagenous fibers.

The diaphragm has three large openings through which structures pass between the thorax and abdomen. The aortic hiatus (opening) is the most dorsal of the apertures. The esoph-

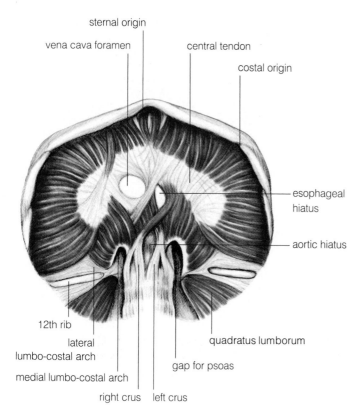

sternal origin
vena cava foramen
central tendon
costal origin
esophageal hiatus
aortic hiatus
12th rib
lateral lumbo-costal arch
quadratus lumborum
medial lumbo-costal arch
gap for psoas
right crus   left crus

Figure 11-18 The diaphragm. Inferior aspect.

They function as extensors and rotators of the head, neck, and trunk and are important in the maintenance of posture. These muscles may be divided into two groups: the large sacrospinalis group and the smaller and deeper transversospinalis complex (Table 11-9).

The **sacrospinalis,** frequently called the *erector spinae,* is a large muscle mass that originates from the lumbodorsal fascia, the sacrum, and the iliac crest; it is divided rather artificially into three groups of muscles. As the main muscle passes superiorly, it splits at intervals into smaller masses, each of which joins one of the three divisions: the lateral division, or *iliocostalis,* which inserts on the ribs; the intermediate division, or *longissimus,* which inserts on the ribs and on the transverse processes of the vertebrae; and the medial division, or *spinalis,* which originates from the spines of the vertebrae and passes a considerable distance superiorly to insert onto the spines at higher levels of the vertebral column. All of the muscles in this group function as extensors of the trunk and contribute to the maintenance of posture; the balance between the contraction of the extensors and flexors maintains the body in an upright position, resisting gravitational pull.

The **transversospinalis** (Fig. 11-20) group lies deep to the sacrospinalis and runs obliquely from its multiple origins at the transverse processes of the vertebrae to insert as slips of muscle at higher levels on the vertebral spines. These muscles are shorter than the sacrospinalis, crossing no more than eight vertebrae each. Unilateral contraction of these muscles rotates the trunk; bilateral contraction assists in extension of the trunk.

The upright posture of the human is a relatively recent evolutionary development, and the back muscles have a task that requires considerable effort. This in part explains the frequency with which individuals complain of pain in the back region. The most common sites of aches and pains are in the lower back and the neck, the two most mobile parts of the vertebral column and the areas in which muscle control is most essential.

---

ageal hiatus is in the muscular part of the diaphragm; the vena cava foramen is in the central tendon.

Contraction of the diaphragm expands the thoracic cavity at the expense of the abdominal cavity. The diaphragm pushes inferiorly against the viscera, causing the anterior abdominal wall to push outward.

The Deep Back Muscles

The muscles of the back include superficial muscles (Fig. 11-19) such as the trapezius and latissimus dorsi that belong to the limbs more than to the trunk. These muscles are involved in limb movements and are discussed with the appendicular muscles in Chapter 12. The *deep back muscles,* or *genuine back muscles* (Fig. 11-20), are located posterior to the transverse processes of the vertebrae, with fibers coursing longitudinally or obliquely in relation to the long axis of the vertebral column.

Two additional groups of neck muscles are also listed in Table 11-9. The suboccipital muscles are deep muscles placed between the first and second vertebrae and the skull. They function chiefly in rotation of the head. The splenius muscles are the most superficial of the deep back muscles of the neck. They wrap around all of the other muscles of this region, extending between the spines of the upper thoracic and lower cervical vertebrae to the skull or to the transverse processes of the cervical vertebrae.

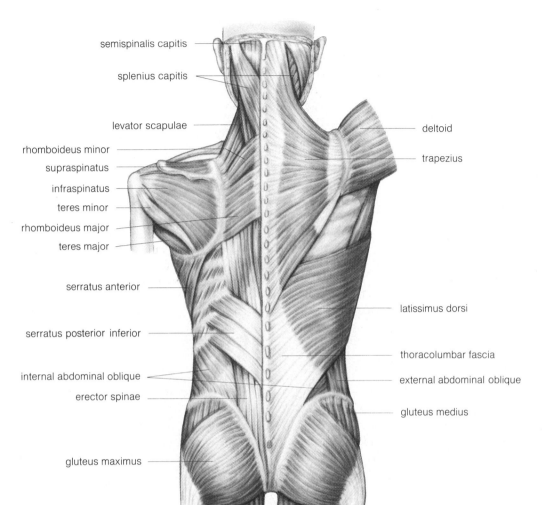

semispinalis capitis

splenius capitis

levator scapulae

rhomboideus minor

supraspinatus

infraspinatus

teres minor

rhomboideus major

teres major

serratus anterior

serratus posterior inferior

internal abdominal oblique

erector spinae

gluteus maximus

deltoid

trapezius

latissimus dorsi

thoracolumbar fascia

external abdominal oblique

gluteus medius

a

Figure 11-19 Back muscles. (**a**) Superficial muscles of the back and trunk, with most superficial muscles shown on the right and the next deeper layer of muscles shown on the left. (**b**) Photograph of the surface of a young adult male (photograph by R. Ross).

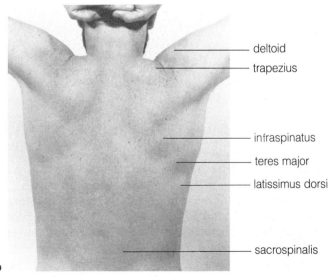

deltoid

trapezius

infraspinatus

teres major

latissimus dorsi

sacrospinalis

b

Figure 11-20 The deep back muscles. Posterior aspect. The right shows the sacrospinalis muscles; the left side shows the deeper transversospinalis group.

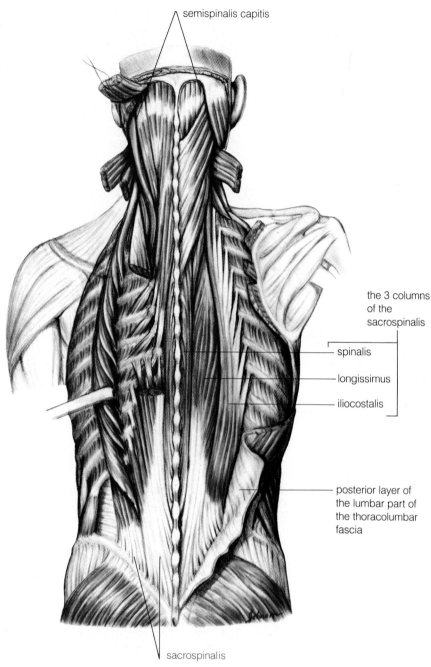

semispinalis capitis

the 3 columns of the sacrospinalis

spinalis

longissimus

iliocostalis

posterior layer of the lumbar part of the thoracolumbar fascia

sacrospinalis

The Muscles of the Neck

The muscles that are anatomically part of the neck include muscles that attach to the skull or thorax and those that originate or insert on neck structures. The platsyma, a neck muscle within the superficial fascia, was described with the muscles of facial expression, to which it belongs by virtue of its location, function, and innervation. The suboccipital muscles on the back of the neck are functionally related to the sacrospinalis and transversospinalis complex. The trapezius, which lies in part on the posterior aspect of the neck, is functionally a muscle of the pectoral girdle and will be considered in Chapter 12. The remaining neck muscles are the sternomastoid muscles, the infrahyoid muscles, the prevertebral muscles, the lateral vertebral, or scalene, muscles, and the muscles of the pharynx (Table 11-10).

Table 11-9 The Deep Back Muscles

Muscle	Origin	Insertion	Action	Innervation
The Splenius Muscles				
Splenius capitis	Vertebral spines of cervical vertebrae	Occipital bone and mastoid process	Pulls head back and rotates it laterally	Dorsal primary rami, cervical spinal nerves
Splenius cervicis	Spines of vertebrae T3 to T6	Transverse processes of first three cervical vertebrae	Rotates head and pulls it backward	Dorsal primary rami, cervical spinal nerves
The Sacrospinalis Group				
Sacrospinalis (erector spinae)	Sacrum, iliac crests, lumbodorsal fascia	Ribs, transverse processes, and spines of vertebrae	Extends vertebral column and bends it to one side	Dorsal primary rami, spinal nerves
The Transversospinalis Group				
Transversospinalis	Transverse processes of vertebrae	Spines of vertebrae, one or more above	Extends and rotates vertebral column	Dorsal primary rami, spinal nerves
The Suboccipital Muscles				
Rectus capitis posterior major	Vertebral spine of axis	Inferior nuchal	Extends and rotates head	Dorsal primary ramus, spinal nerve C1
Rectus capitis posterior minor	Arch of atlas	Inferior nuchal line	Extends the head	Dorsal primary ramus, spinal nerve C1
Obliquus capitis inferior	Vertebral spine of axis	Transverse process of atlas	Rotates atlas	Dorsal primary ramus, spinal nerve C1
Obliquus capitis superior	Transverse process of atlas	Occipital bone	Extends head and bends it laterally	Dorsal primary ramus, spinal nerve C1

The Sternomastoid The *sternomastoid* muscle (Fig. 11-21) crosses the lateral aspect of the neck obliquely from its origin on the sternum and clavicle to its insertion on the mastoid process of the temporal bone. Its function, flexing the neck or pulling the head forward, may best be demonstrated by contracting the muscles against resistance. Place your right elbow on a table, and with your chin in the palm of your hand press your head forward. The entire length of the taut sternomastoid may be palpated along its course through the neck. The sternomastoid and trapezius are considered *superficial neck muscles*.

The Infrahyoid Muscles The **infrahyoid muscles** (Fig. 11-22), also known as the "strap" muscles, are a group of ribbonlike muscles with a parallel muscle fiber arrangement that are located in the neck inferior to the hyoid bone. The *omohyoid* takes origin from the *scapula* and has two bellies separated by an intermediate tendon anchored to the clavicle by means of a fascial sling. The *superior belly* of the *omohyoid* inserts upon the *hyoid bone*. The *sternohyoid* lies in the same plane as the omohyoid and extends upward from the *sternum* to insert on the *hyoid bone* medial to the insertion of the omohyoid. Deep to the sternohyoid the *sternothyroid* arises from the *sternum* and *clavicle* and inserts obliquely on the lateral surface of the *thyroid cartilage* of the larynx. The *thyrohyoid* originates from the *oblique line* of insertion of the sternothyroid and extends upward to attach to the *hyoid bone* deep to the attachment of

Table 11-10 Muscles of the Neck

Muscle	Origin	Insertion	Action	Innervation
Sternomastoid	Manubrium and medial third of clavicle	Mastoid process	Rotates head and bends it forward	Spinal accessory
Infrahyoid Muscles				
Omohyoid	Cranial border of scapula and intermediate tendon	Body of hyoid bone	Depresses hyoid bone	Ansa cervicalis (C1 to C3)
Sternohyoid	Clavicle and manubrium	Body of hyoid bone	Depresses hyoid bone	Ansa cervicalis (C1 to C3)
Sternothyroid	Manubrium and first rib	Oblique line of thyroid cartilage	Draws thyroid cartilage obliquely downward	Ansa cervicalis (C1 to C3)
Thyrohyoid	Oblique line of thyroid cartilage	Greater cornu of hyoid bone	Pulls hyoid bone inferiorly	Spinal nerve C1
Prevertebral Muscles				
Longus capitis	Transverse processes of cervical vertebrae 3-6	Occipital bone	Flexes head	Spinal nerves C1 to C3
Longus colli	Transverse processes of cervical vertebrae 3-5	Atlas	Flexes neck	Spinal nerves C2 to C7
Rectus capitis anterior	Atlas	Occipital bone	Flexes head	Spinal nerves C1, C2
Rectus capitis lateralis	Transverse process of C1	Occipital bone	Bends head laterally	Spinal nerves C1, C2
Lateral Vertebral (Scalene) Muscles				
Scalenus anterior	Transverse processes of cervical vertebrae 3-6	Scalene tubercle of first rib	Raises first rib and can flex neck	Lower cervical spinal nerves
Scalenus medius	Transverse processes of last six cervical vertebrae	First rib, lateral to scalene tubercle	Raises first rib and can rotate neck	Lower cervical spinal nerves
Scalenus posterior	Transverse processes of last three cervical vertebrae	Second rib	Raises second rib and can rotate neck	Spinal nerves C6 to C8

Table 11-10   Muscles of the Neck (cont.)

Muscle	Origin	Insertion	Action	Innervation
Muscles of Pharynx				
Superior pharyngeal constrictor	Pterygomandibular raphe	Pharyngeal raphe	Constricts pharynx	Vagus
Middle pharyngeal constrictor	Hyoid bone	Pharyngeal raphe	Constricts pharynx	Vagus
Inferior pharyngeal constrictor	Thyroid and cricoid cartilage	Pharyngeal raphe	Constricts pharynx	Vagus
Stylopharyngeus	Styloid process	Thyroid cartilage with middle constrictor	Elevates pharynx	Glossopharyngeal
Salpingopharyngeus	Auditory tube	Thyroid cartilage with middle constrictor	Elevates pharynx	Vagus
Palatopharyngeus	Soft palate	Thyroid cartilage with middle constrictor	Elevates pharynx	Vagus

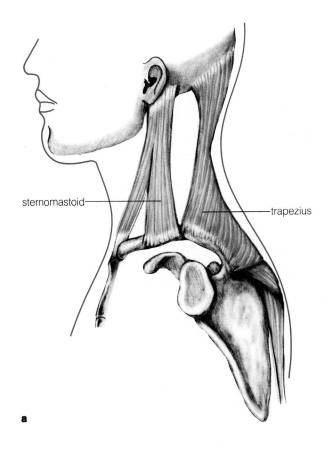

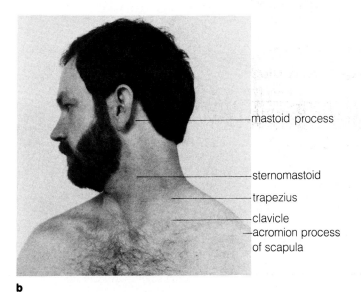

Figure 11-21   The sternomastoid and trapezius muscles. (**a**) Diagram of lateral aspect. (**b**) Photograph of surface anatomy of the neck (photo by R. Ross).

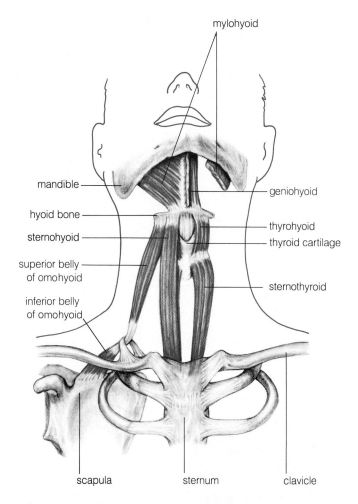

Figure 11-22 The infrahyoid muscles. Anterior aspect.

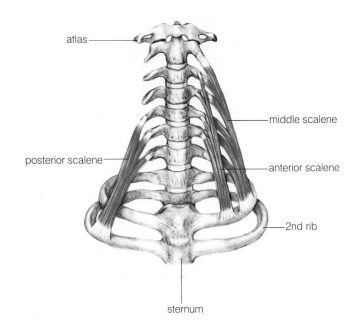

Figure 11-23 The scalene muscles. Anterior view.

the sternohyoid. As a group these muscles depress the hyoid bone and are important in providing stability for the floor of the mouth and the tongue by anchoring the hyoid bone to the sternum.

The Lateral Vertebral Muscles The *lateral vertebral muscles* (Fig. 11-23), or **scalene muscles,** are a group of three functionally related muscles that arise from the *transverse processes* of *cervical vertebrae* and descend in the neck anterior to the vertebral column to insert on the *ribs*. The *anterior* and *middle scalenes* insert on the *first rib;* the *posterior scalene* passes superficial to the first rib to insert on the *second rib*. The sca-

lenes may be used to flex the neck but are less powerful than the sternomastoid. Their most important role appears to be the maintenance of the normal level of the rib cage and the elevation of the rib cage during forced inspiration.

Muscles of the Pharynx The pharynx is the upper part of the digestive and respiratory tracts, located between the mouth and the esophagus. The muscles of the pharynx (Fig. 11-24) consist of *constrictor muscles* whose contractions create a peristaltic wave, moving food down the pharynx, and *longitudinally oriented muscles* that elevate the pharynx during swallowing. All of these muscles are skeletal (striated) but are not really under voluntary control, for once the food enters the pharynx from the mouth, it is difficult to prevent it from being swallowed. Each pair of constrictor muscles (superior, middle, and inferior) originates from fibrous or skeletal structures anteriorly and encircles the pharynx; the members of each pair meet at the posterior midline along the median raphe into which they insert. The *superior constrictor* originates from the *pterygomandibular raphe,* a fibrous cord in the posterior aspect of the cheek that extends from the pterygoid hamulus to the mandible. The *middle constrictor* arises principally from the *hyoid bone,* and the *inferior constrictor* originates from the *thyroid* and *cricoid* cartilages of the larynx. Contractions of these muscles provide the principal force that moves food down the pharynx into

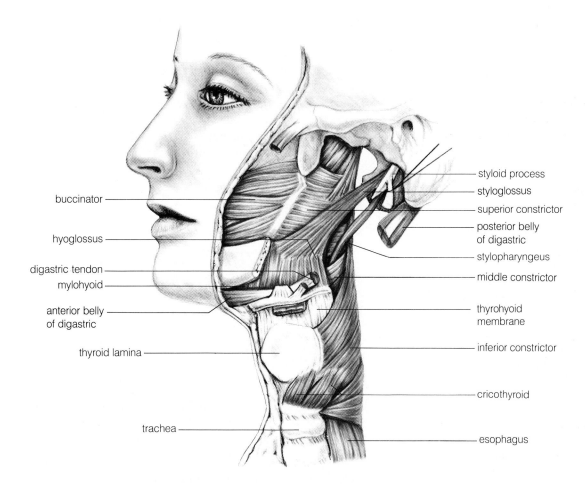

buccinator

hyoglossus

digastric tendon

mylohyoid

anterior belly
of digastric

thyroid lamina

trachea

styloid process

styloglossus

superior constrictor

posterior belly
of digastric

stylopharyngeus

middle constrictor

thyrohyoid
membrane

inferior constrictor

cricothyroid

esophagus

Figure 11-24   Muscles of the pharynx. Lateral aspect.

the esophagus. Three longitudinal muscles of the pharynx that assist in elevating the pharynx during swallowing insert by intermingling with the fibers of the middle constrictor. The *stylopharyngeus,* the most powerful of the three, takes origin from the *styloid process* of the *temporal bone* and passes inferiorly external to the superior constrictor to interlace with the middle constrictor. Both the *salpingopharyngeus,* originating from the *auditory tube,* and the *palatopharyngeus,* arising from the *aponeurosis* of the *soft palate,* pass internal to the superior constrictor to interlace with the middle constrictor.

Muscles of the Pelvic Outlet

The pelvic outlet (Figs. 11-25, 11-26) is closed by true pelvic muscles and muscles of the lower extremity that originate within the pelvis. The latter muscles are discussed with the appendicular muscles. The true pelvic muscles (Table 11-11) and their associated fascia may be considered to form two membranes closing the pelvic outlet: the pelvic diaphragm and the urogenital diaphragm. These muscles not only close the area between the bones of the pelvic girdle but also serve as muscle sphincters

Clinical Applications: **Injuries of Muscles and Tendons**

Athletic Injuries

There is no agreement on the terminology that should be used to categorize athletic injuries to muscle. The least severe injuries (*first-degree strains*) are said to be due to "overstretching," and the expression "pulled muscle" is frequently used to describe the condition. In this injury, the limit of muscle elasticity is reached but not exceeded. The individual muscle fibers remain intact and this is a reversible functional muscle injury. There is sudden pain when the injury occurs, followed by pain on active movement, but there is no swelling.

In a partial muscle tear (*second-degree strain*), some muscle fibers are ruptured or torn but the muscle as a whole remains intact. In a second-degree strain the limits of stability and elasticity of the muscle fibers have been exceeded, tearing some of the muscle fibers and associated connective tissue. This disrupts the blood supply, and swelling follows quickly. There is retraction of the muscle fibers at the injury site, leaving a small depression that soon fills with blood. Application of ice to the area tends to inhibit swelling. Massage is contraindicated, as it might cause further damage to the muscle fibers and prolong bleeding in the area. Repair may take two months or longer.

In a complete tear or rupture (*third-degree strain*), the entire muscle ruptures, leaving a palpable and visible depression at the injury site. The injury is a sudden one and is associated with sharp pain and unwillingness to use the muscle. This type of injury usually requires surgical repair.

There are several factors that increase the chances of athletes receiving muscle injuries. Exercising in the cold may be a factor, for muscle tension increases when the body is cold. Overfatigue of muscles not only leads to a decline in muscle performance, but also to a decline in muscle elasticity. Insufficient warming-up and limbering-up prevent muscles from being adequately warmed and elastic. Finally, infectious diseases in which muscles show signs of inflammation reduce muscle function.

Skeletal Muscle Repair or Regeneration

In skeletal muscle injured by an incision wound such as might be made by a sharp knife, repair proceeds by sprouting from the stumps of the severed muscle fibers. These sprouts may not make proper connection with their original counterparts, but they soon bridge the gap in the area of the wound. In large wounds, sprouting may be inadequate to close the gap. In this event, undifferentiated satellite cells within the muscle begin to divide and are converted into myoblasts. The myoblasts fuse into myotubes that, if innervated, will continue to grow and differentiate into skeletal muscle fibers filling in the wound area.

Atrophy

Any muscle that is not used will undergo a reduction in size and in the number of muscle fibers. This atrophy may be the result of an acute disease or a progressive muscular disorder. Acute atrophy occurs in poliomyelitis, in which rapid destruction of the skeletal-muscle neurons leads to rapid atrophy of the muscles that are no longer innervated. In the progressive types of muscle atrophy, such as occur in children as a result of congenital spinal cord defects, the muscles atrophy to such an extent that the disease is fatal within about five years.

Myoneural Junction Disorders

Normal function of muscle depends upon the transmission of the neural stimulus from the nerve fiber to the muscle fiber. Some muscle diseases appear to result from defects in transmission at the motor end plates.

Myasthenia gravis is considered to be due to a failure of neuromuscular transmission at the motor end plates. The exact causes of this failure are unknown.

In the disease called the *myotonias,* contraction of a muscle continues after stimulation ceases. The result is lack of adequate control of movements. The causes are unknown but may involve defects of the myoneural junction.

---

for the tubes of the urinary tract (urethra), the female reproductive tract (vagina), and the digestive tract (anal canal), which open externally in this region.

The *pelvic diaphragm* (Figs. 11-25, 11-26) consists of the **coccygeus** muscle, the **levator ani** muscle, and fascia; the levator ani includes some muscle fibers that encircle the anal canal adjacent to the external anal sphincter. The urogenital diaphragm is composed of fascia and the **deep transverse perinei** muscle, which functions in part as the external sphincter of the urethra.

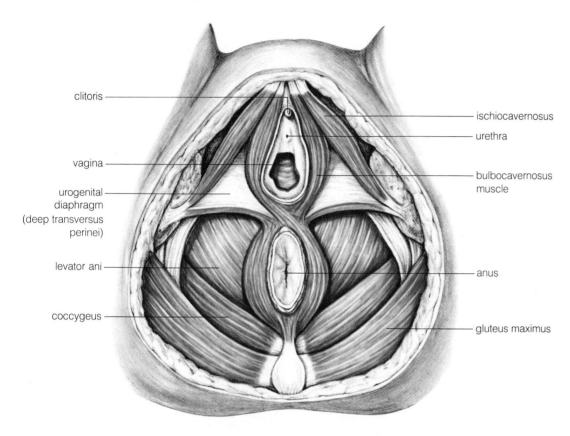

clitoris

ischiocavernosus

urethra

vagina

bulbocavernosus
muscle

urogenital
diaphragm
(deep transversus
perinei)

levator ani

anus

coccygeus

gluteus maximus

Figure 11-25 Muscles that close the pelvic outlet in the female. Inferior view.

DEVELOPMENT OF THE AXIAL MUSCLES

Most of the skeletal muscles of the body differentiate from the *myotome* portion of the somites. With rapid growth by cell proliferation the myotomes extend dorsally in between the spinal cord and superficial ectoderm and also ventrally into the somatic mesoderm of the body wall (Fig. 11-27). The myotomes segregate into a dorsal muscle mass (epimere) and a ventral muscle mass (hypomere), separated by an intermuscular fibrous septum (Fig. 11-28). The *epimere* differentiates into the deep back muscles; the *hypomere* gives rise to the muscles of the lateral and anterior body walls.

The myotomes retain their segmented characteristics as they extend into the various regions of the body. A single myotome may be the sole contributor to a muscle or group of muscles, as in the case of the intercostal muscles of a single intercostal space, or it may join with several other myotomes to form a single muscle, such as the external oblique. In the case of the external oblique, the muscle fibers that develop from a single myotome are aggregated into a bundle that appears separate from adjacent bundles (Fig. 11-29a). During development this gives the muscle a segmented appearance; but as development proceeds and muscle fibers are added to the mass, the muscle fuses into a single sheet, without evidence of segmentation (Fig. 11-29b).

Some muscles such as the strap (infrahyoid) muscles of the neck are derived from more than one myotome but show no evidence of segmentation during their development.

Table 11-11 Muscles of the Pelvic Outlet

Muscle	Origin	Insertion	Action	Innervation
Coccygeus	Spine of ischium and sacrospinous ligament	Coccyx and last segment of sacrum	Supports pelvic floor against intra-abdominal pressure	Spinal nerves S4, S5
Deep transverse perinei	Inferior ramus of pubic bone.	Encircles urethra to insert upon pubis of other side	Constricts urethra	Pudendal nerve (S2 to S4)
Levator ani	Superior ramus of pubis and spine of ischium and arcus tendineus	Coccyx and raphe between anus and coccyx	Raises pelvic floor	Spinal nerves S3 to S5

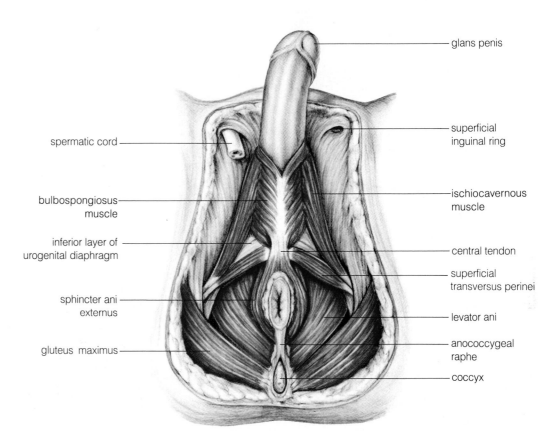

Figure 11-26   Muscles that close the pelvic outlet in the male. Inferior view.

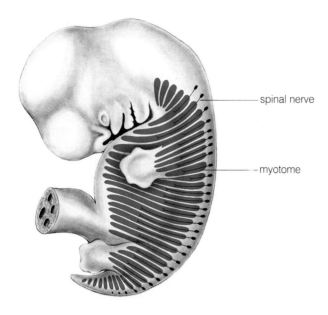

Figure 11-27 Lateral view of myotomes extending segmentally into the body wall.

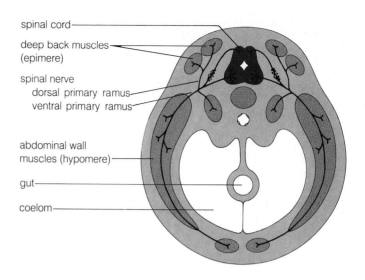

spinal cord

deep back muscles (epimere)

spinal nerve
dorsal primary ramus
ventral primary ramus

abdominal wall muscles (hypomere)

gut

coelom

Figure 11-28 Transverse section of an embryo, illustrating derivatives of the hypomere and epimere.

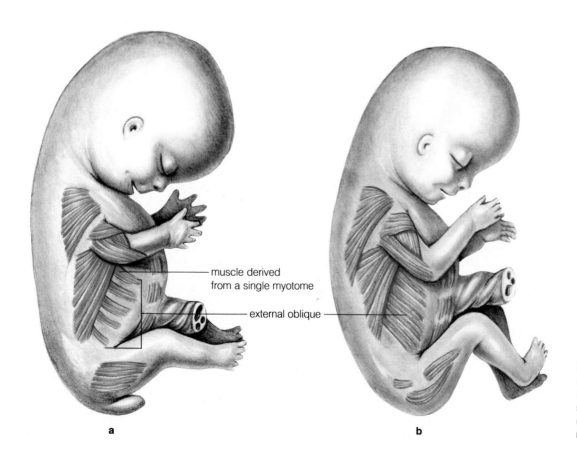

muscle derived from a single myotome

external oblique

a

b

Figure 11-29 Development of a simple muscle from several myotomes: (**a**) the segmented stage of differentiation, (**b**) muscle after the fusion of muscle slips.

Auricularis

Buccinator

Coccygeus

Corrugator supercilii

Deep transverse perinei

Depressor anguli oris

Depressor labii inferioris

Depressor septi

Diaphragm

External intercostals

External oblique

Infrahyoid muscles

Innermost intercostals

Internal intercostals

Internal oblique

Levator ani

Levator labii superioris

Levator veli palatini

Masseter

Mentalis

Nasalis

Superior oblique muscles

Inferior oblique muscles

Occipitofrontalis

Orbicularis oculi

Orbicularis oris

Palatoglossus

Palatopharyngeus

Prevertebral muscles

Procerus

Pterygoid muscles

Rectus abdominis

Rectus muscles

Rizorius

Sacrospinalis

Scalene muscles

Splenius muscles

Suboccipital muscles

Suprahyoid muscles

Temporalis

Tensor veli palatini

Transversospinalis

Transversus abdominis

Zygomaticus major and minor

SUMMARY OUTLINE

I. Intrinsic Muscles of the Head
    A. The muscles of facial expression are located in the skin and superficial fascia of the face and neck. They are primarily related to the mouth, nose, eyes, and ears, and function as dilators or sphincters of these openings.
    B. The muscles of mastication are four paired muscles that move the mandible during chewing. The lateral pterygoids assist in opening the mandible; the masseter, temporalis, and medial pterygoids close it.
    C. The muscles of the soft palate manipulate the muscular structure attached to the posterior border of the hard palate, especially during swallowing. The tensor veli palatini tightens the soft palate, preventing its prolapse into the nasal cavity. The levator veli palatini elevates the soft palate, and the palatoglossus depresses the palate or pulls the

tongue posteriorly. The palatopharyngeus depresses the palate or elevates the pharynx. The musculus uvulae thickens the soft palate.
    D. The suprahyoid muscles attach to the hyoid bone and to the skull, forming the major components of the floor of the mouth. The digastric, geniohyoid, mylohyoid, and stylohyoid either elevate the hyoid bone or depress the mandible. The genioglossus, hyoglossus, and styloglossus are extrinsic tongue muscles that manipulate the tongue and contribute to the structure of the floor of the mouth.
    E. The muscles of the orbit include the four rectus muscles (medial, lateral, superior, inferior), two oblique muscles (superior and inferior) that are attached to the eyeball, and the levator palpebrae superioris muscles of the eyelids.
II. The Trunk Muscles
    A. The layers of the abdominal wall include the trunk muscles, the skin and superficial fascia external to the muscles, and the transversalis fascia and peritoneum internal to the muscles.
    B. The flexors of the trunk are the muscles located anterior to the transverse processes of the vertebrae. They include the paired rectus abdominis muscles, which course longitudinally on each side of the anterior midline, and the external obliques, the internal obliques, and the transversus abdominis, arranged in layers (sheets) in the lateral abdominal wall.
    C. The muscles of the thorax include the external, internal, and innermost intercostals, which occupy the intercostal spaces and attach to adjacent ribs. The subcostals, transversus thoracis, and levatores costarum also attach to ribs but cross more than a single intercostal space.
    D. The extensors of the trunk, located posterior to the transverse processes of the vertebrae, consist of two muscle groups: a superficial group, the sacrospinalis, courses longitudinally, and a deep group, the transversospinalis, runs obliquely from lateral to medial as it ascends on the back.
    E. The muscles of the neck are the following: the superficial neck muscles, which include the sternomastoid anterolaterally and the trapezius posteriorly; the infrahyoid muscles, which are the paired "strap" muscles of the neck; the prevertebral muscles, which arise and insert on the vertebrae anterior to the transverse processes of the vertebrae; and the lateral vertebral muscles, the scalenes, which extend between the neck vertebrae and the first and second ribs.
    F. The muscles that close the pelvic outlet include the levator ani, a major component of the pelvic diaphragm, and the deep transverse perinei, the principal component of the urogenital diaphragm.
III. Development of the Axial Muscles
    A. The segmented myotomes segregate into a dorsal muscle mass (epimere) and a ventral muscle mass (hypomere).

B. A single muscle may develop from a single myotome or may develop from several myotomes that fuse into a single mass.

REVIEW QUESTIONS

1. What muscle functions as a sphincter that can close the mouth? the eyelids?

2. What are the attachments of each of the four muscles of mastication? Which of these muscles is essential for opening the mouth?

3. When swallowing fluid, an individual has a problem keeping the liquid from entering the nasal cavity. Which muscle of the soft palate is most likely to be paralyzed?

4. Which of the suprahyoid muscles does not have an attachment to the mandible?

5. What movement of the tongue would be most affected by bilateral paralysis of the genioglossus muscles?

6. Which of the muscles located superior to the eyeball tilts the front of the eyeball downward when it contracts?

7. Which layer of the lateral abdominal wall lies between the transversus abdominis muscle and the peritoneum?

8. Which of the muscles of the lateral abdominal wall is composed of fibers that course anteriorly and inferiorly from their origin to their insertion?

9. Which of the muscles of the thorax has fibers that course in the same direction as those of the external oblique of the abdominal wall?

10. What is the name of the superficial muscle mass of the deep back muscles?

11. What muscle is the major component of the pelvic diaphragm?

12. What is the superior attachment of most of the muscles that are commonly called the strap muscles of the neck?

13. What is the embryonic primordium of most of the skeletal muscle of the body?

SELECTED REFERENCES

Basmajian, J. V. *Surface Anatomy.* 2nd ed. Baltimore: Williams & Wilkins, 1983.

Gardner, E., *et al. Anatomy: A Regional Study of Human Structure.* 4th ed. Philadelphia: W. B. Saunders, 1975.

Hamilton, W. J., *et al. Surface and Radiological Anatomy.* 5th ed. Cambridge: Heffer, 1971.

Moore, K. L. *Clinically Oriented Anatomy.* Baltimore: Williams & Wilkins, 1980.

Rosse, C., and Clawson, D. K. *Introduction to the Musculoskeletal Systems.* 2nd ed. New York: Harper & Row, 1980.

Royce, J. *Surface Anatomy.* Philadelphia: Davis, 1965.

12

MUSCLES OF THE LIMBS

After completing this chapter you should be able to:

1. Name and give the location of each of the muscles of the pectoral girdle. Locate the joints of the pectoral girdle; describe the movements possible at each joint and list the muscles responsible for each movement.

2. Name and give the attachments of the shoulder muscles that move the arm and stabilize the shoulder joint. Describe the movements of the arm and demonstrate an understanding of the coordination of the muscles in producing these movements.

3. Name and give the attachments of the muscles of the elbow region. Describe the role of each of these muscles in producing the movements that occur at the elbow joint.

4. Describe the functional groups of extrinsic muscles of the hand, and say to which fascial compartment of the forearm each group belongs.

5. Describe the intrinsic muscles of the hand, relating the location of these muscles to the movements they produce when they contract.

6. List and give the relative positions of the functional groups of muscles that act on the hip joint. Describe the roles of these muscles in producing the various movements.

7. Name and give the attachments of the muscles that act on the knee joint. Describe the roles of these muscles in producing movements and in stabilizing the knee joint.

8. Describe the functional groups of extrinsic muscles of the foot, relating the location of these muscles to the fascial compartments of the leg.

9. Describe the intrinsic muscles of the foot, contrasting their actions with the actions of the intrinsic muscles of the hand.

Both the upper and lower limbs have many muscles that cross and can exert action on more than one joint. The upper limb is an instrument designed largely to hold the hand in position for manipulation; it is a system of joints and levers that moves the hand to any desired point in space and keeps it there as it performs its task. The muscles of the upper limb, therefore, are arranged with proximal origins that move very little and distal insertions where most of the movement occurs. Because the upper limbs are used for carrying loads, their attachments to the trunk must be strong enough to support considerable weight. In fact, the upper limbs may occasionally be called upon to support the entire weight of the body; it is necessary for the shoulder muscles to be strong enough for this task even though there are several joints that normally transfer the weight born by the limbs to the trunk.

Unlike the upper limb, the lower limb is designed for locomotion and support. The entire weight of the trunk is transferred to the limb at a single joint, the hip. Although the origins of the muscles are usually given as proximal and the insertions distal, the distal points may be nearly fixed during locomotion as the foot touches the ground, with the proximal points moving as the trunk swings forward.

THE UPPER LIMB

In the upper limb, a single joint does not usually move in isolation. The muscles of the limbs work together to move the limbs smoothly and keep them coordinated at all the joints. However, for descriptive purposes it is desirable to consider muscles as operating in functional groups at a single joint or at several joints that have coordinated movements. This is the approach that will be used in this chapter, beginning with the shoulder region.

The muscles of the shoulder region (Table 12-1) may be considered in three functional groups: (1) muscles that act upon the pectoral girdle, originating on the trunk and inserting on bones of the girdle; (2) muscles that move the arm in various directions, originating on the axial skeleton and inserting on the humerus; and (3) muscles that rotate and abduct the arm and provide for stability of the shoulder joint, originating on the pectoral girdle and inserting on the humerus.

Muscles That Act on the Pectoral Girdle

Muscles that act upon the pectoral girdle arise from the trunk and insert upon the scapula or clavicle. In addition to moving the scapula in relation to the trunk, these muscles keep the scapula close to the chest wall. On the back of the shoulder (Fig. 12-1) the most superficial muscle is the *trapezius,* a broad muscle that arises from the *skull,* the *nuchal ligament,* and the *spines* of all the *cervical* and *thoracic vertebrae,* and that inserts on the *spine* and *acromion process* of the *scapula* and the *lateral one-third* of the *clavicle.* Deep to the trapezius the *rhomboideus major* and *rhomboideus minor* are in the same plane as the

Table 12-1 Muscles of the Shoulder Region

Muscle	Origin	Insertion	Action	Innervation
Muscles Acting on the Pectoral Girdle				
Levator scapulae	Transverse processes of vertebrae C1 to C4.	Superior angle and vertebral border, scapula	Elevates scapula, extends head	Spinal nerves C3, C4
Pectoralis minor	Ribs 3, 4, 5	Coracoid process, scapula	Draws scapula inferiorly, elevates ribs	Medial pectoral C8, T1
Rhomboideus major	Vertebral spines, T2 to T5	Lower half, vertebral border, scapula	Adducts scapula	Dorsal scapular C5
Rhomboideus minor	Vertebral spines, C7 and T1	Root of spine of scapula	Adducts scapula	Dorsal scapular C5
Serratus anterior	First eight ribs	Vertebral border, scapula	Holds scapula against chest	Long thoracic C5 to C7
Subclavius	First rib	Inferior surface, clavicle	Stabilizes sternoclavicular joint	Spinal nerves C5, C6
Trapezius	Superior nuchal line, ligamentum nuchae, vertebral spines, C7 to T12	Lateral one-third of clavicle, acromion and spine of scapula	Rotates scapula	Spinal accessory
Muscles from Axial Skeleton to Arm				
Latissimus dorsi	Vertebral spines T6 to L5 and iliac crest	Humerus, intertubercular groove	Extends, rotates arm	Thoracodorsal (C6 to C8)
Pectoralis major	Medial half of clavicle, sternum, costal cartilages, aponeurosis of external oblique	Crest of greater tubercle of humerus	Flexes, adducts, and rotates arm	Medial and lateral pectoral (C5 to T1)
Muscles from the Pectoral Girdle to the Arm				
Coracobrachialis	Coracoid process	Medial surface, humerus	Adducts arm	Musculocutaneous (C6, C7)
Deltoideus	Lateral one-third of clavicle, acromion and spine of scapula	Deltoid tuberosity, humerus	Abducts arm	Axillary (C5, C6)
Infraspinatus	Infraspinatus fossa	Greater tubercle, humerus	Lateral rotator, arm	Suprascapular (C5, C6)
Subscapularis	Subscapular fossa	Lesser tubercle, humerus	Medial rotator, arm	Upper and lower subscapular (C5, C6)
Supraspinatus	Supraspinatus fossa	Greater tubercle, humerus	Initiates abduction, arm	Suprascapular (C5)
Teres major	Lateral border, scapula	Crest of lesser tubercle, humerus	Adducts and medially rotates arm	Lower subscapular (C5, C6)
Teres minor	Lateral border, scapula	Greater tubercle, humerus	Lateral rotator, arm	Axillary (C5)

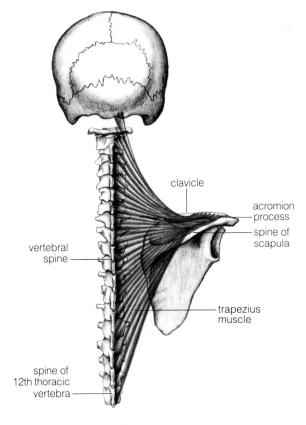

Figure 12-1 The trapezius muscle, which lies superficial to the levator scapulae and rhomboideus muscles.

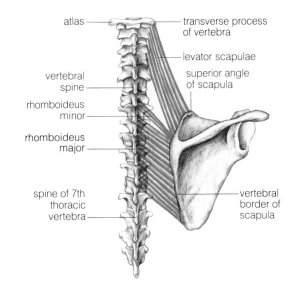

Figure 12-2 The levator scapulae and rhomboideus major and minor muscles

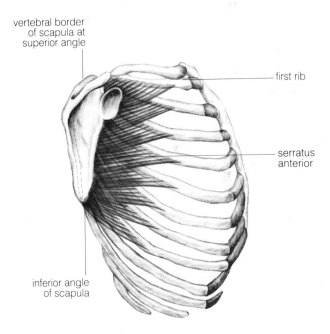

Figure 12-3 The serratus anterior muscle, which connects the scapula to the lateral wall of the thorax

levator scapulae (Fig. 12-2); all three muscles originate from *vertebral spines* and insert on the *vertebral border* of the *scapula*. The *serratus anterior* (Fig. 12-3) lies deep to the scapula, inserting on the *vertebral border* of the *scapula* and originating anteriorly on the adjacent *ribs*.

On the anterior surface of the trunk (Fig. 12-4) the *pectoralis minor* extends between the *rib cage* and the *coracoid process* of the *scapula*, pulling the shoulder joint forward. The small *subclavius* muscle (Fig. 12-4) lies deep to the clavicle, between the *clavicle* and the *first rib*, stabilizing the sternoclavicular joint and serving as padding to prevent injury to soft structures (nerves and blood vessels) beneath the clavicle in the event it is fractured.

Joints and Movements of the Pectoral Girdle

The joints of the pectoral girdle include the **acromioclavicular** (Fig. 12-5a) and **sternoclavicular** (Fig. 12-5b) **joints.** The sternoclavicular joint is a synovial joint with an articular disc, permitting rotation, elevation, depression, and anterior and posterior movements of the clavicle at the joint. The ligaments of the acromioclavicular joint allow little movement between the scapula and clavicle, so that the two bones move as a unit.

Movements of the pectoral girdle include *elevation* and *depression* in addition to *forward* and *backward* movement. Contraction of the upper fibers of the trapezius, the rhomboids, and the levator scapulae elevates the shoulders, as when one shrugs. Normally these muscles maintain the shoulder in level position by tonic contraction, but they may let the shoulders

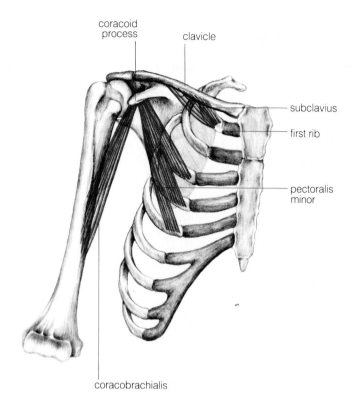

Figure 12-4 The pectoralis minor, subclavius, and coracobrachialis muscles, which lie deep to the pectoralis major muscle

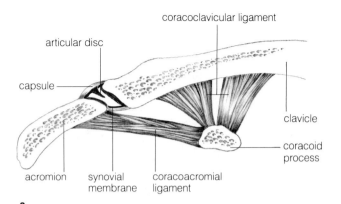

a

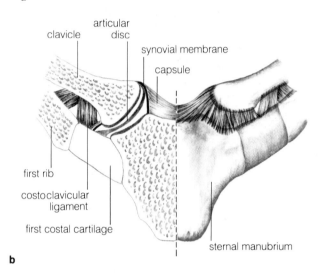

b

Figure 12-5 The joints of the pectoral girdle. (**a**) Acromioclavicular joint. (**b**) Sternoclavicular joint.

droop when fatigued by excessive use. They are important in lifting weights, supporting the muscles that pass from the girdles to the arm as the muscles contract. Elevation of the girdle is restricted by the *costoclavicular ligament* (Fig. 12-5B) which binds the clavicle to the rib cage. Depression of the shoulder girdle results from contraction of the lower fibers of the trapezius, the subclavius, and the pectoralis minor. The subclavius pushes the medial end of the clavicle against the sternum in any movement. These muscles are antagonists of the elevators of the pectoral girdle, and the position of the shoulders depends upon the balance between these two groups of muscles. The shoulders are pulled backward into the "attention" position by the central part of the trapezius and the rhomboids, which must relax when the shoulders are pulled forward by contraction of the serratus anterior and the pectoralis minor.

> The serratus anterior is important in holding the scapula against the posterior chest wall. If it is paralyzed the scapula protrudes posteriorly in a condition termed "winging" scapula, which causes difficulty with arm movements such as abduction. Without the serratus anterior to anchor it against the chest wall the scapula becomes freely mobile and does not provide a fixed base for arm movements.

Shoulder Muscles That Move the Arm

Shoulder muscles that act on the arm are of two types: those that originate on the axial skeleton, or trunk, and those that originate on the pectoral girdle. In both cases the insertion of the muscle is on the humerus.

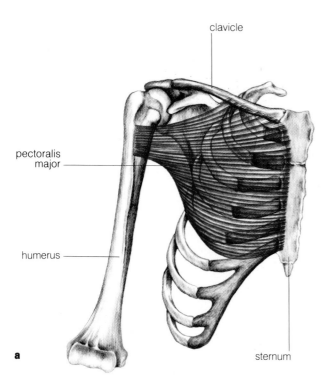

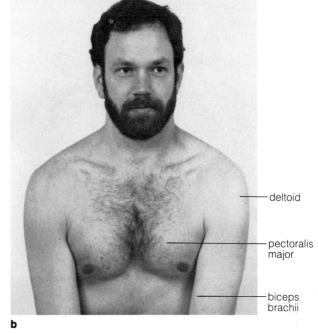

Figure 12-6 Pectoralis major muscle, the superficial muscle of the anterior chest wall. **(a)** Diagram. **(b)** Surface anatomy (photograph by R. Ross).

Muscles from the Trunk to the Arm Muscles originating on the trunk and inserting on the humerus provide a variety of arm movements. On the anterior chest wall the large *pectoralis major* (Fig. 12-6) originates from the *trunk* and clavicle and tapers to a narrow insertion on the crest of the *greater tubercle* of the humerus. On the back the *latissimus dorsi* (Fig. 12-7) arises from the *lower thorax,* the *lumbar regions* of the trunk, and the *iliac crest,* and ascends to insert on the floor of the *bicipital groove* of the *humerus.* As the tendon of the latissimus dorsi approaches its point of insertion, it twists so that the lower fibers of the muscle actually insert upon the humerus proximal to the insertion of the fibers of the superior part of the muscle.

Muscles from the Pectoral Girdle to the Arm Muscles that originate on the pectoral girdle and insert on the humerus serve to rotate the arm and stabilize the shoulder joint. These muscles include three superficial shoulder muscles—the deltoid (Fig. 12-8), teres major (Fig. 12-9), and coracobrachialis (Fig. 12-4)—and the deep shoulder muscles, or rotator cuff. The superficial muscles originate on the pectoral girdle and insert on the humerus at some distance from the joint. They abduct, flex, and extend the arm.

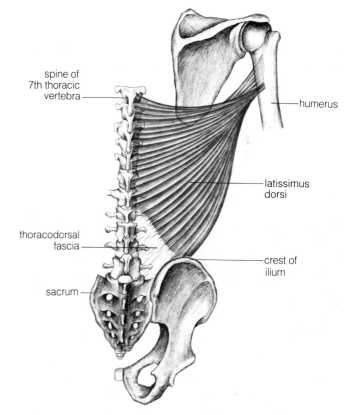

Figure 12-7 Latissimus dorsi muscle, the superficial muscle of the back

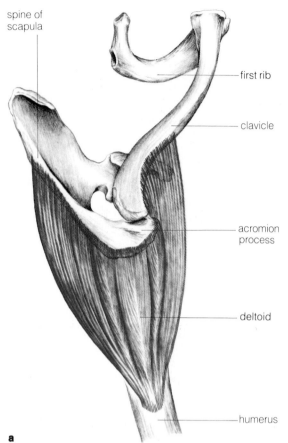

spine of
scapula

first rib

clavicle

acromion
process

deltoid

humerus

a

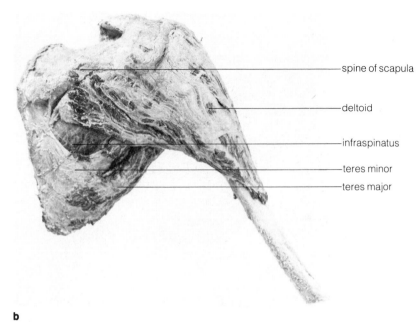

b

spine of scapula

deltoid

infraspinatus

teres minor

teres major

Figure 12-8 Deltoid muscle, the superficial muscle of the lateral surface of the shoulder. (**a**) Diagram of lateral view. (**b**) Posterior view of a cadaver preparation of deltoid and related muscles (photograph by R. Ross).

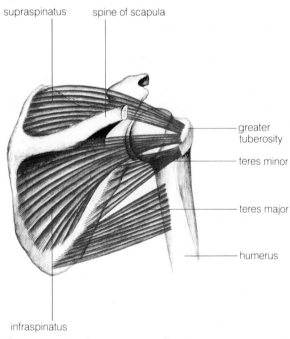

supraspinatus

spine of scapula

greater
tuberosity

teres minor

teres major

humerus

infraspinatus

a

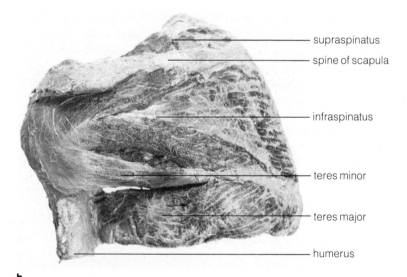

b

supraspinatus

spine of scapula

infraspinatus

teres minor

teres major

humerus

Figure 12-9 Supraspinatus, infraspinatus, and teres muscles of the dorsal aspect of the scapula. (**a**) Diagram. (**b**) Cadaver preparation (photograph by R. Ross).

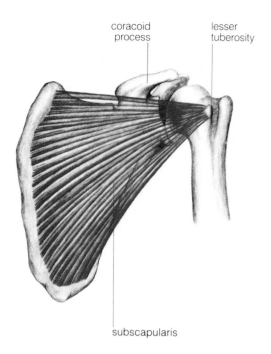

Figure 12-10 Subscapularis, muscle of the anterior aspect of the scapula

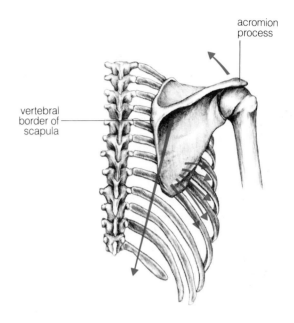

Figure 12-11 Diagram of the pattern of rotation of the scapula as the arm is elevated above the shoulder

The primary function of the *deep shoulder muscles*—subscapularis (Fig. 12-10), supraspinatus, infraspinatus, and teres minor (Fig. 12-9)—is to rotate the arm while maintaining the integrity of the shoulder joint. These muscles and their tendons form the **rotator cuff,** which surrounds the shoulder almost completely. Tendons of the rotator cuff muscles cross the shoulder joint to insert on the greater and lesser tubercles of the humerus just below the neck, providing support for the joint. The large *subscapularis* muscle (Fig. 12-10) takes origin from the anterior surface of the scapula and inserts on the lesser tubercle, while the three short muscles, the *supraspinatus, infraspinatus,* and *teres minor* (Fig. 12-9) originate on the posterior surface of the scapula and insert on the greater tubercle. The rotator cuff tendons function as collateral ligaments of the joint, and the muscles of the rotator cuff cooperate with other shoulder muscles in all movements by adjusting the tension on the joint appropriate for the movement.

The Shoulder Joint and Movements of the Arm

The **glenohumeral joint** (shoulder joint) is a ball-and-socket joint with no articular disc and with the socket (glenoid fossa) too shallow to enclose the head of the humerus (Fig. 9-13c) although it is deepened by the addition of a rim of hyaline cartilage around the edge of the articular surface. The fibrous joint capsule is loose without ligaments to restrict movement and thus allows freedom of movement in all planes.

Movements of the arm include flexion, extension, abduction, adduction, rotation, and circumduction. As the arm moves, the scapula is rotated to maintain the proper relationship of the glenoid fossa with the head of the humerus. This requires the coordination of the action of the shoulder muscles with the action of the muscles that move the arm; for example, abduction occurs by contraction of the deltoid, but it can proceed only as far as the horizontal without rotation of the scapula. For the arm to be elevated above the horizontal the scapula must rotate so that the glenoid fossa is tilted superiorly. This is accomplished largely through the action of the upper fibers of the trapezius, which pull the acromion process upward, and by the lower fibers of the trapezius and the lower fibers of the serratus anterior, which pull the vertebral border of the scapula downward and its inferior angle anteriorly (Fig. 12-11).

Muscles of the Elbow Region

Muscles that act principally at the elbow (Table 12-2) include arm muscles that cross the elbow joint and muscles of the forearm that act principally at the proximal radioulnar joint. Muscles that

Table 12-2 Muscles That Act Principally at the Elbow

Muscle	Origin	Insertion	Action	Innervation
Anconeus	Lateral epicondyle, humerus	Olecranon process and posterior surface of upper ulna	Abductor of ulna in pronation	Radial (C7, C8)
Biceps brachii	Long head, supraglenoid tubercle; short head, coracoid process	Radial tuberosity, bicipital aponeurosis into medial fascia, forearm	Flexes forearm, supinates hand	Musculocutaneous (C6, C7)
Brachialis	Anterior surface, distal two-thirds of humerus	Coronoid process and tuberosity, ulna	Flexes forearm	Musculocutaneous (C6, C7)
Pronator quadratus	Anterior surface, distal one-fourth of ulna	Anterior surface, distal one-fourth of radius	Pronates hand	Median (C8, T1)
Pronator teres	Medial epicondyle, humerus; coronoid process, ulna	Lateral surface, middle of radius	Pronates hand	Median (C6, C7)
Supinator	Lateral epicondyle, humerus; crest and fossa, ulna	Lateral surface, upper one-third of radius	Supinates hand	Radial (C6)
Triceps brachii	Long head, infraglenoid tuberosity; lateral head, lateral surface, humerus; medial head, posterior surface, distal one-half of humerus	Olecranon process of ulna	Extends forearm, aids in extension and adduction of arm	Radial (C7, C8)

cross the elbow (Fig. 12-12) are the biceps brachii and the brachialis within the anterior fascial compartment of the arm, and the triceps brachii and anconeus within the posterior fascial compartment. Forearm muscles that act primarily at the elbow are the pronator teres and pronator quadratus.

The biceps brachii overlies the brachialis and crosses the shoulder and elbow joints. It arises by two heads (Fig. 12-12), a short head from the tip of the *coracoid process* along with the *coracobrachialis* and a long head from the superior lip of the *glenoid fossa.* The long head arises by a slender tendon occupying the *bicipital groove,* where it is enclosed within a tendon sheath that communicates with the synovial cavity of the shoulder joint. This tendon provides some support for the shoulder joint, but the biceps muscle as a whole has little effect upon movement of the arm. Its primary effect is upon the forearm, where it inserts on the *radius* near its head by means of a slender tendon and on the fascia of the anterior forearm by an *aponeurosis* arising from the tendon and passing medially to blend with the fascia of the forearm. The tendon enables the biceps to *flex* and *supinate* the forearm while the aponeurosis acts as a check ligament on supination. The biceps is assisted in

supination by the *supinator*, a thin muscle that arises from the lateral epicondyle of the humerus and from the ulna and inserts on the radius. The *brachialis* (Fig. 12-12) has a broad origin from the anterior surface of the distal half of the shaft of the humerus and passes as a fleshy belly across the anterior aspect of the elbow joint to insert on the ulna distal to the coronoid process. The brachialis is a powerful flexor of the forearm.

The *triceps brachii* (Fig. 12-12) has three heads: a long head originating from a rough area at the inferior border of the glenoid fossa called the *infraglenoid tubercle,* a lateral head arising from a ridge extending down from the *lateral aspect* of the shaft of the *humerus* distal to the greater tubercle, and a medial head arising from the *posterior surface* of the lower half of the shaft. The three heads merge into a single muscle in the lower part of the arm and insert by means of a broad tendon into the *olecranon process* of the *ulna.* The triceps brachii is a powerful extensor of the forearm and stabilizes the elbow joint when the upper limb pushes against resistance. The small *anconeus,* which arises from the posterior aspect of the *lateral epicondyle* and inserts upon the *ulna,* assists the triceps in extending the forearm and abducts the ulna during pronation.

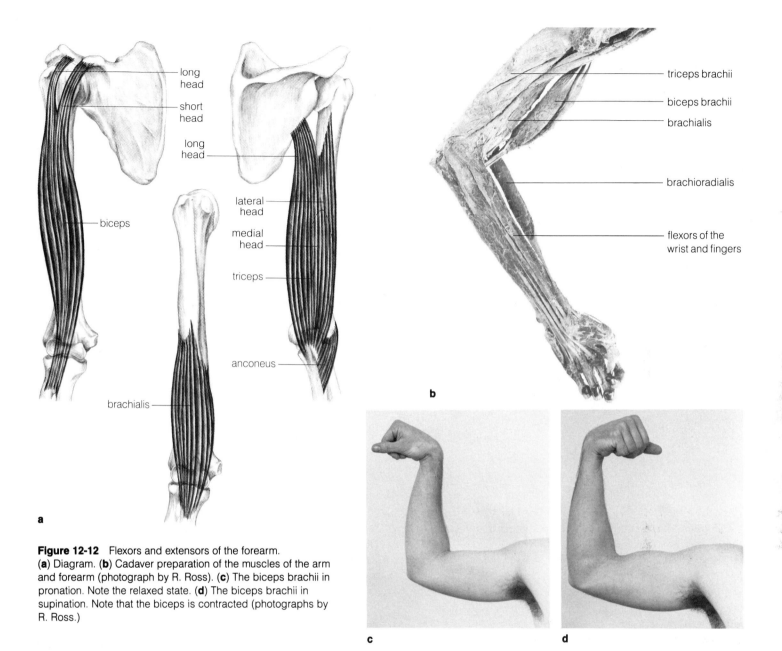

Figure 12-12 Flexors and extensors of the forearm.
(a) Diagram. (b) Cadaver preparation of the muscles of the arm
and forearm (photograph by R. Ross). (c) The biceps brachii in
pronation. Note the relaxed state. (d) The biceps brachii in
supination. Note that the biceps is contracted (photographs by
R. Ross.)

The Elbow Joint and Movements of the Forearm

The elbow, or *humeroulnar joint* (Fig. 9-14, 12-13), is a hinge joint in which the conformation of the articular surfaces of the bones virtually limits movement to flexion and extension. The joint capsule is loose anteriorly and posteriorly, but is reinforced laterally and medially by strong collateral ligaments. At the *proximal radioulnar joint* (Figs. 9-15, 12-14) the disclike head of the radius articulates with the radial notch of the ulna.

The *annular ligament* holds the head of the radius in place and permits rotation during pronation and supination of the forearm and hand.

Movements at the humeroulnar joint include *flexion* of the forearm, which results from contraction of the biceps and brachialis; when flexion is strongly resisted, these muscles are assisted by forearm muscles whose primary actions are upon more distal parts of the limbs. Flexion of the forearm brings the forearm into contact with the anterior surface of the arm. *Extension* of the forearm is accomplished by contraction of the triceps, assisted,

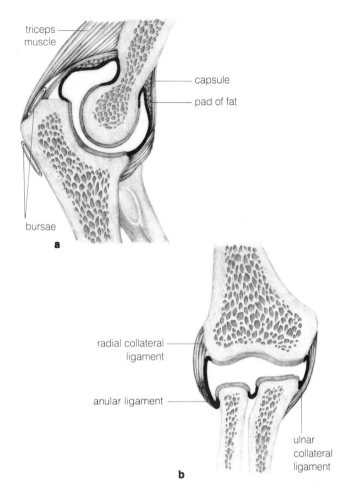

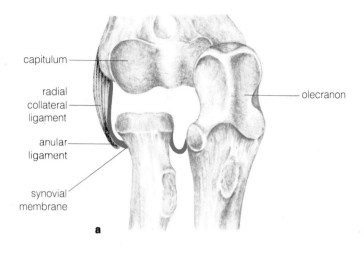

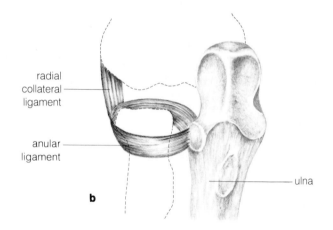

Figure 12-13 The elbow joint. (**a**) Sagittal section. (**b**) Coronal section.

Figure 12-14 The proximal radioulnar joint. (**a**) Radiohumeral and radioulnar joints. (**b**) Anular ligament.

when extension is strongly resisted, by the extensor groups of forearm muscles. Extension brings the forearm into approximately a straight line with the arm; the fit of the olecranon process into the olecranon fossa prevents any greater extension.

Movements at the proximal radioulnar joint are *supination,* bringing the radius and ulna into position parallel to one another, and *pronation,* bringing the radius across the ulna so that the palm is turned posteriorly. Supination is accomplished by the biceps brachii assisted by the supinator (Fig. 12-21b). The biceps is a powerful supinator with the forearm partially flexed and is the source of most of the force used in turning a screw threaded for use by right-handed individuals. The supinator muscles are more powerful than the two pronators.

Muscles That Act on the Hand

The hand, one of the most versatile parts of the body, can be compared to an unspecialized tool that serves as a hook, a ring, pliers, or a chuck (Fig. 12-15). These various functions are possible largely because the thumb can be brought into opposition with any of the other digits (Fig. 12-16a). Crude movements of the hand, such as the power grip (Fig. 12-16b) when an object is clamped against the palm, are accomplished mainly by extrinsic muscles of the hand. Precision movements of the fingers (Fig. 12-16c) are controlled mostly by the intrinsic muscles of the hand, which are used in such activities as typing and playing the piano.

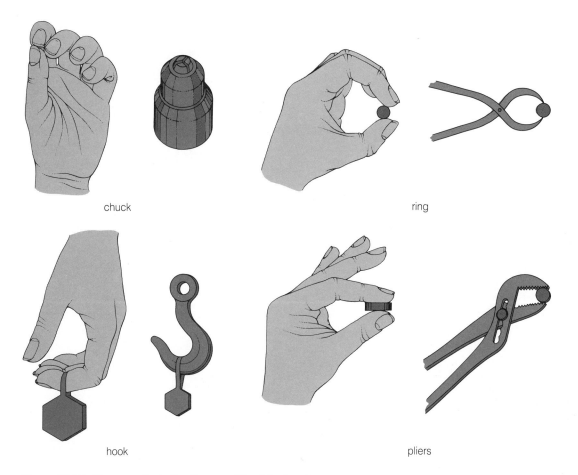

chuck

ring

hook

pliers

Figure 12-15 Diagrams illustrating the versatility of the hand

Extrinsic Muscles of the Hand   The **extrinsic muscles of the hand** (Figs. 12-17, 12-18) originate from skeletal elements of the arm and forearm, having muscle bellies principally in the forearm and inserting upon hand structures by means of long, slender tendons. These muscles are listed in Table 12-3.

Most of the *extrinsic muscles* of the *hand* (Fig. 12-19) originate proximally from the *medial* and *lateral epicondyles* of the humerus and cross the elbow joint as fleshy muscle before ending in slender tendons that insert in the hand. The natural position of rest with the hands down at the sides is with the thumbs forward, not turned outward as in the anatomical position. With the limbs at rest the lateral epicondyle of the humerus lies in line with and superior to the posterior or extensor surface of the forearm. Similarly the medial epicondyle has the same relation with the anterior or flexor surface of the forearm.

From the *lateral epicondyle* and the *supracondylar ridge* above it, a large group of superficial extensor muscles arises,

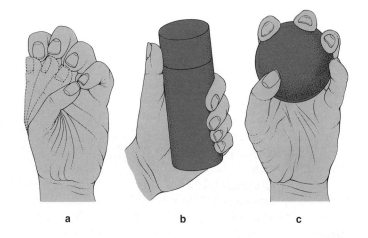

a   b   c

Figure 12-16 Basic hand movements. (**a**) Opposition of the thumb. (**b**) Power grip using fingers against the palm. (**c**) Precision grip using delicate control of all digits.

Figure 12-17 Forearm muscles.
(**a**) Anterior view. (**b**) Posterior
view. (**c**) Transverse section of a
cadaver preparation (photograph
by R. Ross).

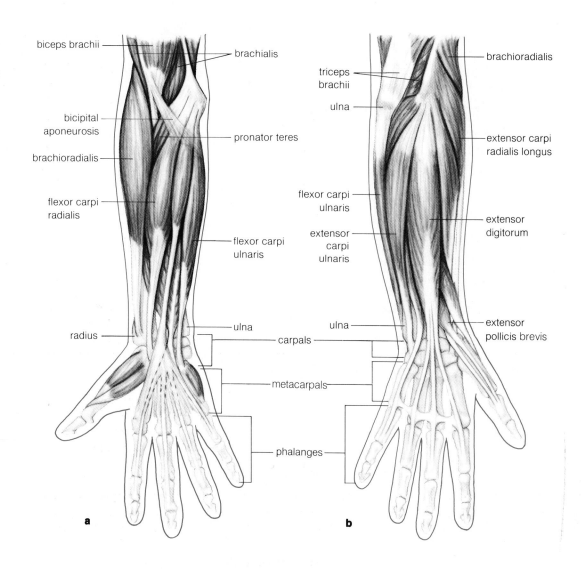

biceps brachii

brachialis

bicipital
aponeurosis

pronator teres

brachioradialis

flexor carpi
radialis

flexor carpi
ulnaris

radius

ulna

carpals

metacarpals

phalanges

a

triceps
brachii

brachioradialis

ulna

extensor carpi
radialis longus

flexor carpi
ulnaris

extensor
digitorum

extensor
carpi
ulnaris

ulna

extensor
pollicis brevis

b

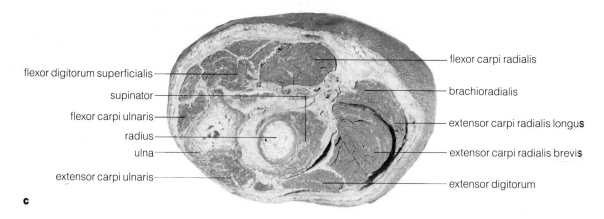

flexor digitorum superficialis

supinator

flexor carpi ulnaris

radius

ulna

extensor carpi ulnaris

flexor carpi radialis

brachioradialis

extensor carpi radialis longus

extensor carpi radialis brevis

extensor digitorum

c

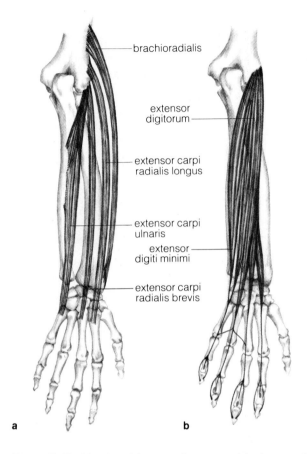

Figure 12-18 Muscles of the posterior aspect of the forearm. (**a**) Extensors of the wrist. (**b**) Extensors of the fingers.

including the *brachioradialis* that overlaps the *extensor carpi radialis longus* and *extensor carpi radialis brevis*. The origin of these muscles begins superiorly on the lateral supracondylar ridge and spreads distally to the posterior aspect of the lateral epicondyle, where several extensor muscles have a common origin. These include muscles that extend the wrist and have "carpi" in their names (Fig. 12-18a) as well as those that extend the fingers and have "digitorum" or "digiti" in their names (Fig. 12-18b).

The muscles with "carpi" in their names insert into the bases of the posterior aspect of the second, third, and fifth metacarpals. These muscles include the *extensor carpi radialis longus* and *brevis* and *extensor carpi ulnaris* (Fig. 12-18a), whose names not only describe their function but also indicate the location of the muscles in relation to the bones of the forearm.

The extensors of the fingers arise along with the extensors of the wrist; the *extensor digitorum communis* (Fig. 12-18b) splits into separate tendons that pass across the back of the wrist to insert, in part, into the base of the proximal phalanges and, in part, into an aponeurosis covering the dorsum of the phalanges. As the tendons cross the wrist, they are held firmly in place by a fascial band, the *extensor retinaculum* (Fig. 12-19b), oriented transversely across the back of the wrist. The tendons are enclosed in groups by *tendon sheaths* that prevent friction as the tendons slide over the underlying carpal bones. A separate extensor for the little finger, *extensor digiti minimi* (Fig. 12-18b), may be present.

> "Tennis elbow" is a painful condition common among tennis players and others, such as violinists, who use their elbow joints excessively. The condition is thought to be due to tearing the extensor muscles of the forearm from their origin at the lateral epicondyle of the humerus.

The flexors of the wrist and hand occupying the front of the forearm are located in fascial compartments and may be described in two groups, superficial and deep. All of the *superficial group* arise from the common flexor origin on the *medial epicondyle* of the *humerus* (Fig. 12-20). The most lateral of these muscles is the *flexor carpi radialis* (Fig. 12-20a) which inserts into the bases of the second and third metacarpals, and the most medial is the *flexor carpi ulnaris* (Fig. 12-20a) inserting at the base of the fifth metacarpal. The *flexor digitorum superficialis* (Fig. 12-20b) lies medially, deep to the wrist flexors, and splits into tendons to the four fingers. The tendons are bunched at the wrist, enclosed in a common tendon sheath, and then pass across the palm to reach the digits where they split around the tendons of the deep flexor (flexor digitorum profundis) before inserting onto the second phalanx. The tendons of the flexors of the digits pass deep to the **flexor retinaculum** (Fig. 12-19a) as they cross the wrist and are not easily palpated. However, the tendon of the wrist flexors and of the *palmaris longus* (Fig. 12-20d), a small muscle arising from the common flexor origin and absent in about 20% of individuals, pass across the wrist superficial to the flexor retinaculum and are palpable. The *palmaris longus* inserts into the **palmar aponeurosis,** a fibrous sheet beneath the skin of the palm.

The deep flexor of the fingers, *flexor digitorum profundus* (Fig. 12-21a), arises from the proximal ulna and splits into four tendons that cross the wrist and palm to enter the anterior

Table 12-3 Extrinsic Muscles of the Hand

Muscle	Origin	Insertion	Action	Innervation
Abductor pollicis longus	Posterior surfaces, ulna and radius	Base, first metacarpal	Abducts thumb and hand	Radial (C6, C7)
Brachioradialis	Lateral supracondylar ridge, humerus	Styloid process, radius	Flexes forearm	Radial (C5, C6)
Extensor carpi radialis brevis	Lateral epicondyle, humerus	Base, third metacarpal, posterior surface	Extends and abducts hand	Radial (C6, C7)
Extensor carpi radialis longus	Lateral supracondylar ridge, humerus	Base, second metacarpal, posterior surface	Extends and abducts hand	Radial (C6, C7)
Extensor carpi ulnaris	Lateral epicondyle, humerus; posterior surface, ulna	Base, fifth metacarpal	Extends and abducts hand	Radial (C6 to C8)
Extensor digiti minimi	From extensor digitorum communis	Proximal phalanx, fifth digit	Extends proximal phalanx of little finger	Radial (C6 to C8)
Extensor digitorum communis	Lateral epicondyle, humerus	Extensor expansion, fingers	Extends proximal phalanx of fingers	Radial (C6 to C8)
Extensor indicis	Posterior surface, ulna	Extensor expansion, second digit	Extends proximal phalanx of second digit	Radial (C6 to C8)
Extensor pollicis brevis	Posterior surface, radius	Base, proximal phalanx, thumb	Extends first phalanx and metacarpal	Radial (C6, C7)
Extensor pollicis longus	Posterior surface, ulna and interosseous membrane	Base, distal phalanx, thumb	Extends distal phalanx and metacarpal	Radial (C6, C7)
Flexor carpi radialis	Medial epicondyle, humerus	Base, second, and third metacarpals	Flexes and abducts hands	Median (C6, C7)
Flexor carpi ulnaris	Medial epicondyle, humerus; olecranon and posterior surface, ulna	Pisiform, hamate, fifth metacarpal	Flexes and adducts hand	Ulnar (C8, T1)
Flexor digitorum profundus	Ulna and interosseous membrane	Base, distal phalanges, fingers	Flexes fingers	Median (C8, T1)
Flexor digitorum superficialis	Medial epicondyle, humerus; coronoid process, ulna; radius	Palmar surface, middle phalanges, fingers	Flexes middle phalanges and hand	Median (C7 to T1)
Flexor pollicis longus	Interosseous membrane, radius	Base, distal phalanx, thumb	Flexes thumb	Median (C8, T1)
Palmaris longus	Medial epicondyle, humerus	Flexor retinaculum, palmar aponeurosis	Flexes hand	Median (C6, C7)

aspect of the fingers. They pass through pulleys created by the split tendons of the superficial flexors (Figs. 12-19c, d) and insert upon the distal phalanx of each finger. The deep flexor of the thumb, *flexor pollicis longus* (Fig. 12-21a), takes origin from the radius and remains separate from the remainder of the deep group, thus assuring independent action of the thumb.

The long pronator, *pronator teres* (Fig. 12-20), arises from the *medial epicondyle* and inserts by means of a flat tendon into the lateral surface of the *radius* a little below its midpoint. The short pronator, the *pronator quadratus* (Fig. 12-21a), arises from the anterior aspect of the distal quarter of the *ulna* and passes horizontally to insert into the anterior aspect of the *radius*.

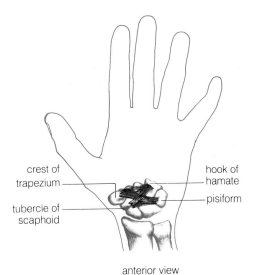

crest of
trapezium

tubercle of
scaphoid

hook of
hamate

pisiform

anterior view

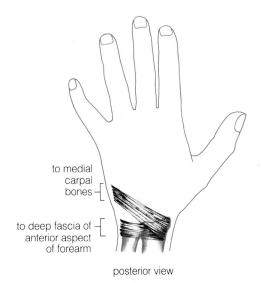

to medial
carpal
bones

to deep fascia of
anterior aspect
of forearm

posterior view

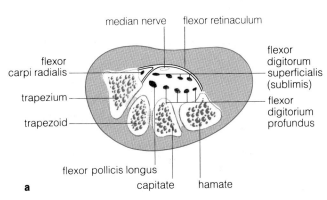

median nerve   flexor retinaculum

flexor
carpi radialis

trapezium

trapezoid

flexor pollicis longus

capitate   hamate

flexor
digitorum
superficialis
(sublimis)

flexor
digitorium
profundus

a

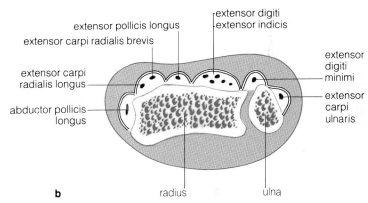

extensor pollicis longus

extensor carpi radialis brevis

extensor carpi
radialis longus

abductor pollicis
longus

extensor digiti
extensor indicis

extensor
digiti
minimi

extensor
carpi
ulnaris

radius   ulna

b

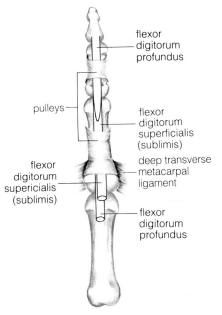

flexor
digitorum
profundus

pulleys

flexor
digitorum
superficialis
(sublimis)

flexor
digitorum
supericialis
(sublimis)

deep transverse
metacarpal
ligament

flexor
digitorum
profundus

c

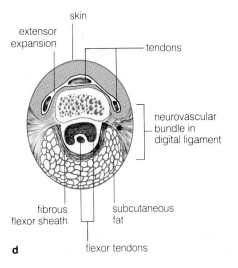

skin

extensor
expansion

tendons

neurovascular
bundle in
digital ligament

fibrous
flexor sheath

subcutaneous
fat

flexor tendons

d

Figure 12-19 Fasciae of the hand. (**a**) Flexor retinaculum. (**b**) Extensor retinaculum. (**c**) Palmar aspect of the finger showing the fibrous flexor sheath. (**d**) Transverse section of a finger showing fascial relations.

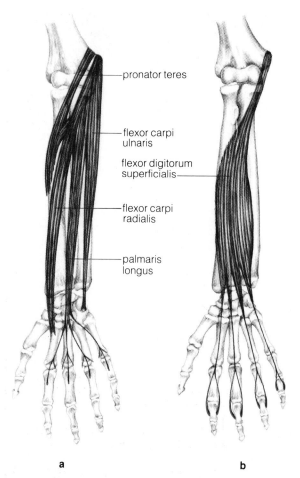

a   b

Figure 12-20   Forearm muscles. (**a**) Superficial muscles that terminate on the forearm, wrist, and palm. (**b**) Superficial flexor of the fingers.

pronator teres

flexor carpi ulnaris

flexor digitorum superficialis

flexor carpi radialis

palmaris longus

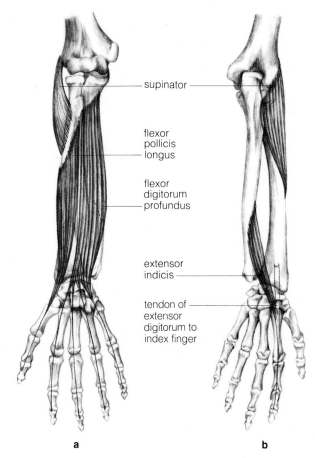

a   b

Figure 12-21   Deep muscles of the forearm. (**a**) Flexors. (**b**) Extensor indicis and supinator.

supinator

flexor pollicis longus

flexor digitorum profundus

extensor indicis

tendon of extensor digitorum to index finger

Intrinsic Muscles of the Hand   The **intrinsic muscles of the hand** (Fig. 12-22) are small muscles arising and inserting on skeletal and fascial components of the hand; they cooperate with the intrinsic muscles in producing the fine movements of the fingers. These include the *lumbricals, dorsal interossei* (sing, interosseus), and *palmar interossei* that terminate on the fingers, and the special muscles of the thumb and little finger that are responsible for the greater mobility of these digits. The four *lumbricals* (wormlike) (Fig. 12-22a) arise from the *tendons* of the *flexor digitorum profundus* and insert into the proximal phalanx at the metacarpophalangeal joint. Each muscle curves around the lateral side of the finger to reach its posterior aspect.

The four *dorsal interossei* (Fig. 12-22g) arise from the sides of adjacent *metacarpals* and pass distally to insert into the dorsal

aponeurosis of the *first phalanx* of the *second, third,* and *fourth fingers.* The first muscle inserts into the lateral surface of the second (index) finger, the second and third insert on each side of the middle (third) finger, and the fourth ends on the medial side of the fourth (ring) finger. These muscles separate the three fingers (abduction) by pulling the first and third fingers away from the middle finger. Due to the arrangement of the dorsal interossei on either side, the middle finger can be waggled from side to side.

The three *palmar interossei* (Fig. 12-22f) are arranged to adduct the fingers toward the middle finger. Each arises by a single head from the side of the metacarpal of the finger it adducts. The middle finger does not have a palmar interosseus associated with it, but the dorsal interossei move it in either direction.

Table 12-4   Intrinsic Muscles of the Hand

Muscle and Number	Origin	Insertion	Action	Innervation
Abductor digiti minimi	Pisiform and tendon, flexor carpi ulnaris	Medial side, base, proximal phalanx, fifth digit	Abducts little finger	Ulnar (C8, T1)
Abductor pollicis brevis	Scaphoid, trapezium, flexor retinaculum	Lateral side, base, proximal phalanx, thumb	Abducts thumb	Median (C8, T1)
Adductor pollicis	Capitate and second and third metacarpals; anterior surface, third metacarpal	Base, proximal phalanx, thumb	Adducts thumb	Ulnar (C8, T1)
Dorsal interossei, 4	By two heads from adjacent heads of metacarpals	Bases, proximal phalanges, fingers	Abducts digits 2, 3, 4; extends phalanges; flexes metacarpophalangeal joints	Ulnar (C8, T1)
Flexor digiti minimi brevis	Hamate and flexor retinaculum	Medial side, base, proximal phalanx, fifth digit	Flexes proximal phalanx, little finger	Ulnar (C8, T1)
Flexor pollicis brevis	Trapezium and flexor retinaculum	Base, proximal phalanx, thumb	Flexes thumb	Median, ulnar (C6 to T1)
Lumbricals, 4	Tendons, flexor digitorum profundus	Extensor expansion, phalanges	Flexes metacarpophalangeal joints; extends fingers	Median, ulnar (C6 to C8)
Opponens digiti minimi	Hamate and flexor retinaculum	Medial side, fifth metacarpal	Draws fifth metacarpal anterior and lateral	Ulnar (C8, T1)
Opponens pollicis	Trapezium and flexor retinaculum	Lateral border of first metacarpal	Draws first metacarpal toward palm	Median (C6, C7)
Palmaris brevis	Flexor retinaculum	Skin of palm	Causes creases in skin of palm	Ulnar (C8)
Palmar interossei, 3	Medial side, second, lateral sides, fourth and fifth metacarpals	Proximal phalanx	Adducts second, fourth, and fifth fingers	Ulnar (C8, T1)

The special muscles of the thumb create a thick pad on the lateral side of the palm at the base of the thumb termed the **thenar eminence.** Two superficial thenar muscles, the *abductor pollicis brevis* (Fig. 12-22d) and the *flexor pollicis brevis* (Fig. 12-22e), arise as a flat sheet from the carpal bones and from the lateral end of the flexor retinaculum to insert by a common tendon into the lateral side of the first phalanx of the thumb. A deep thenar muscle, the *opponens pollicis* (Fig. 12-22h), arises (from carpal bones and flexor retinaculum) and inserts into the lateral and anterior surface of the shaft of the first metacarpal. The *adductor pollicis* (Fig. 12-22i) is a deep muscle with a broad origin from the capitate, base of metacarpals II and III and shaft of metaacarpal III. As it approaches the thumb, it narrows to insert primarily into the base of the first phalanx on the medial side.

Special muscles of the little finger contribute to the **hypothenar eminence,** the fleshy pad on the medial side of the palm; these muscles include the superficial *abductor digiti minimi* and *flexor digiti minimi brevis* (Fig. 12-22b), which arise from the carpal bones and flexor retinaculum on the ulnar (medial) side of the wrist and insert on the medial side of the first phalanx of the little finger. The other hypothenar muscle is the *opponens digiti minimi* (Fig. 12-22c) arising deep to the muscles previously described and inserting into the anteromedial surface of the shaft of the fifth metacarpal.

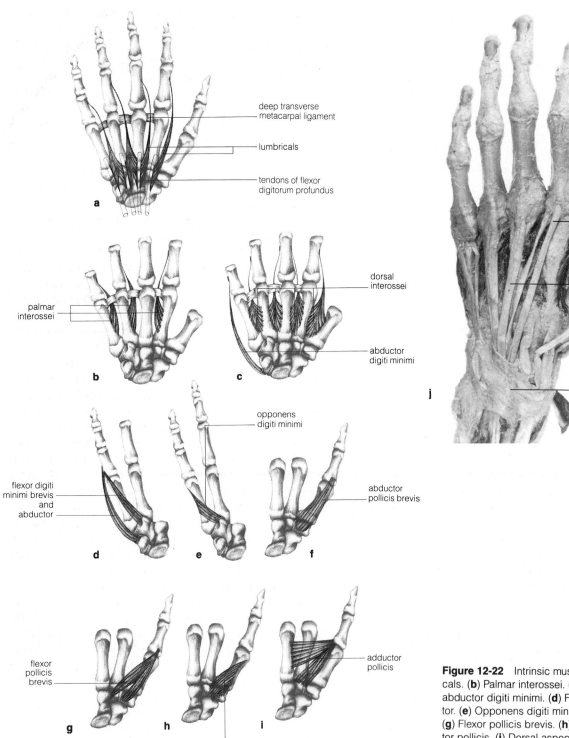

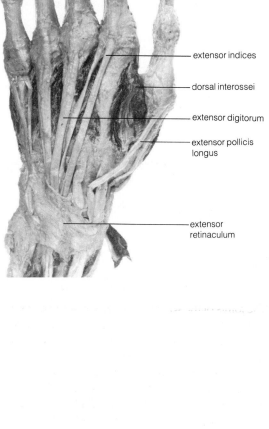

Figure 12-22 Intrinsic muscles of the hand. (**a**) Lumbricals. (**b**) Palmar interossei. (**c**) Dorsal interossei and abductor digiti minimi. (**d**) Flexor digiti minimi and abductor. (**e**) Opponens digiti minimi. (**f**) Abductor pollicis brevis. (**g**) Flexor pollicis brevis. (**h**) Opponens pollicis. (**i**) Adductor pollicis. (**j**) Dorsal aspect of the hand, cadaver preparation (photograph by R. Ross).

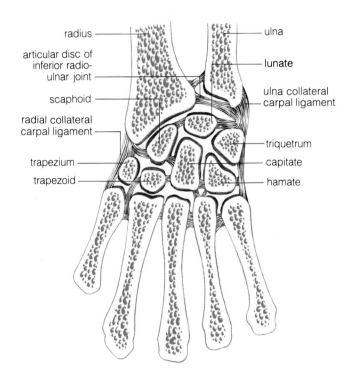

radius
articular disc of inferior radio-ulnar joint
scaphoid
radial collateral carpal ligament
trapezium
trapezoid

ulna
lunate
ulna collateral carpal ligament
triquetrum
capitate
hamate

Figure 12-23 Joints of the wrist, coronal section

Wrist Joints and Movements of the Hand

The joints of the wrist (Fig. 12-23) include the radioulnar, radiocarpal, middle carpal, and carpometacarpal joints. The distal head of the ulna is a disclike enlargement that contacts a depression in the side of the radius termed the *ulnar notch* of the radius. An articular disc of fibrocartilage attached to the styloid process of the ulna and to the margin of the ulnar notch of the radius prevents the ulna from articulating with the carpal bones. The joint capsule of the *radioulnar joint* is lax, permitting the rather extensive movement of pronation.

The *radiocarpal joint* is the articulation of the radius and the articular disc distal to the ulna with three carpals of the proximal row (scaphoid, lunate, and triquetrum). The capsule of the radiocarpal joint is strengthened by collateral ligaments (Fig. 12-23) attached to the styloid processes of the radius and ulna and passing distally to attach to the marginal carpal bones. The capsule is also reinforced anteriorly and posteriorly by ligaments that connect the radius with the carpals, forcing the wrist to move with the radius during pronation and supination. The carpal bones articulate with one another individually by means of hyaline cartilage articular surfaces and a single synovial cavity continuous throughout the wrist. Due to the arrangement of the transverse ligaments (**flexor** and **extensor retinacula** [Fig.

12-19]), little movement occurs between the individual bones of a row; instead, the proximal and distal rows move on one another as units. The term *midcarpal joint* is used for the S-shaped articulation between the *proximal* and *distal rows* of *carpals,* with the proximal row functioning more as an extension of the forearm and the distal row moving with the hand.

The *carpometacarpal joints* unite the distal row of carpals with the metacarpals by a series of synovial cavities. The joint at the metacarpal of the thumb differs from the others in having a saddle-shaped articular surface on which the thumb moves freely. The remaining four joints present relatively flat articular surfaces held together by a tight capsule and ligaments that allow little movement.

The *metacarpophalangeal joints* and the *interphalangeal joints* have essentially the same type of hinge joint construction with a pulley-shaped articular head fitting into the base of the adjacent phalanx, whose articular surface is oval and concave to match the sloped convex surface of the adjacent bone. The metacarpophalangeal joints permit abduction and adduction in relation to the middle finger; however, paired collateral ligaments on each side of the fingers restrict adduction and abduction in flexion of the fingers when one makes a fist.

Movements of the wrist include flexion, extension, adduction, abduction, pronation, and supination. *Flexion* is mainly due to the contraction of the flexor carpi radialis and flexor carpi ulnaris attached on the lateral and medial sides of the wrist. Any of the muscles crossing the anterior surface of the wrist may flex the wrist in addition to their other actions; for example, the palmaris longus tends to flex the wrist as it pulls on the palmar aponeurosis. *Extension* occurs mainly from the action of muscles antagonistic to the flexors: extensor carpi radialis longus and brevis and the extensor carpi ulnaris. Contraction of the muscles in unison extends the hand in line with the forearm. Unequal contraction of the flexors and extensors on either side leads to a deviation of the hand toward or away from the body (adduction or abduction). For example, simultaneous contraction of the flexor carpi ulnaris and extensor carpi ulnaris pulls the hand toward the thigh in the movement of adduction. *Pronation* and *supination* occur at both the proximal and distal radioulnar joints, with the hand turning as a unit with the radius. Pronation is accomplished by the pronator teres and pronator quadratus, and supination by the supinator and the biceps brachii.

THE LOWER LIMB

Because the lower limb supports the weight of the body, its joints and muscles are arranged to give maximum strength; but its movements are much more restricted than those of the upper limb. This is especially true of movement at the hip joint compared with that of the shoulder joint.

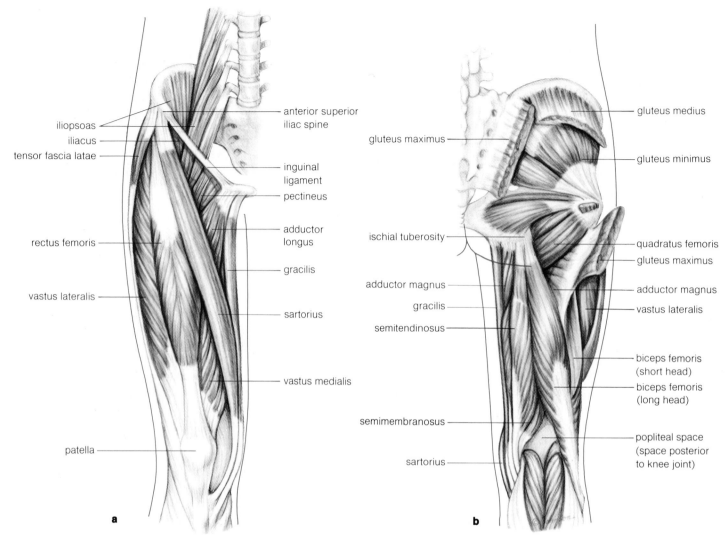

iliopsoas
iliacus
tensor fascia latae
rectus femoris
vastus lateralis
patella

anterior superior iliac spine
inguinal ligament
pectineus
adductor longus
gracilis
sartorius
vastus medialis

a

gluteus maximus
ischial tuberosity
adductor magnus
gracilis
semitendinosus
semimembranosus
sartorius

gluteus medius
gluteus minimus
quadratus femoris
gluteus maximus
adductor magnus
vastus lateralis
biceps femoris (short head)
biceps femoris (long head)
popliteal space (space posterior to knee joint)

b

anterior

vastus lateralis
vastus intermedialis
rectus femoris
vastus medialis
sartorius
adductor longus
adductor brevis
adductor magnus
biceps femoris
semitendinosus
semimembranosus
gracilis

medial

posterior

c

Figure 12-24 Muscles of the pelvic girdle and thigh. (**a**) Anterior view. (**b**) Posterior view. (**c**) Transverse section of the thigh, cadaver preparation (photograph by R. Ross).

Muscles of the Hip

In general terms, the muscles crossing the hip joint (Fig. 12-24) are arranged to provide movement of the lower limb while maintaining the stability of the joint as it is subjected to the stress of bearing the body's weight. Most of these muscles are large and are capable only of gross movements such as walking. In studying the arrangement and actions of the hip musculature it should be kept in mind that the body is kept erect as it moves through space. Consider for a moment the stress applied to the hip region as the trunk bends forward to the toes or returns to an upright position from deep knee-bends. Overcoming the

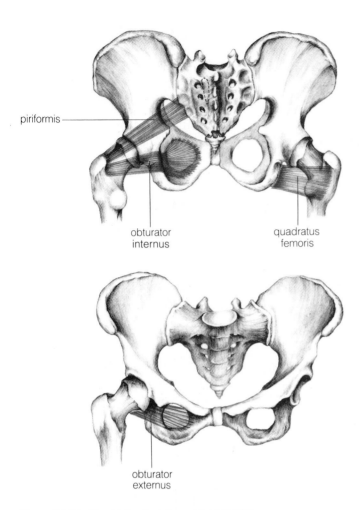

piriformis

obturator internus

quadratus femoris

obturator externus

Figure 12-25 Short lateral rotators of the hip joint

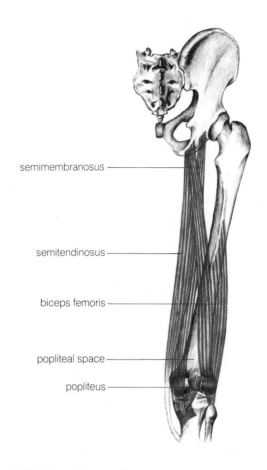

semimembranosus

semitendinosus

biceps femoris

popliteal space

popliteus

Figure 12-26 Hamstring muscles

force of gravity in such exercises requires bulky muscles attached close to the joint to maintain the integrity of the joint and to maintain balance. Muscles of the hip may be compared with those of the shoulder. Like the upper extremity which has some muscles that cross both the shoulder and the elbow joints, the lower extremity has muscles that act upon both the hip and the knee joints.

There are about twenty-five muscles in the hip region, which may be divided into five groups according to their actions: lateral rotators, extensors, flexors, adductors, and abductors (Table 12-5).

Lateral Rotators The lateral *rotators* (Fig. 12-25) are short muscles arising from the *sacrum* or coxal bone and inserting upon the *greater trochanter*. These are the *quadratus femoris,* the *obturator externus* and *internus,* the *gemellus superior* and

inferior, and the *piriformis;* all are located deep to the other hip muscles in close relation to the joint. They rotate the thigh so that the knee is turned outward.

Extensors The *extensors of the hip* (Fig. 12-26) also act on the knee joint. The short extensor, the *gluteus maximus* (Figure 12-27), is a powerful extensor of the thigh and is essential in movements involving simultaneous extension of the hip and knee joints, as in climbing stairs. It inserts not only on the femur but also into a heavy fascial sheet, the **iliotibial tract,** which extends down the lateral aspect of the thigh and terminates below the knee on the tibia. The long extensors, the *biceps femoris, semitendinosus,* and *semimembranosus,* collectively known as the **hamstring muscles,** flex the leg as they extend the thigh in walking. These muscles arise in part from the *ischial tuberosity* and insert on the bones of the *leg.*

Table 12-5 Muscles of the Hip

Muscle	Origin	Insertion	Action
Lateral Rotators of the Thigh	Innervation: Spinal nerves, L4 to S1		
Gemellus inferior	Ischial tuberosity	Greater trochanter, via tendon, obturator internus	Rotates thigh laterally
Gemellus superior	Spine of ischium	Greater trochanter, via tendon, obturator internus	Rotates thigh laterally
Obturator externus	Edge of obturator foramen and membrane	Trochanteric fossa, femur	Rotates thigh laterally
Obturator internus	Edge of obturator foramen and membrane	Greater trochanter	Rotates thigh laterally
Piriformis	Sacrum, internally sacrotuberous ligament	Greater trochanter	Rotates thigh laterally
Quadratus femoris	Ischial tuberosity	Intertrochanteric crest, shaft of femur	Rotates thigh laterally
Extensors of the Thigh	Innervation: Sciatic nerve (L5 to S3)		
Biceps femoris	Long head, ischial tuberosity; short head, linea aspera and lateral supracondylar ridge, femur	Head, fibula	Long head extends hip; flexes leg and knee, rotates leg laterally
Gluteus maximus	Ilium, sacrum, coccyx	Gluteal tuberosity, iliotibial tract	Extends and rotates thigh laterally
Semimembranosus	Ischial tuberosity	Medial tibial condyle	Extends thigh, flexes leg and knee, rotates leg medially
Semitendinosus	Ischial tuberosity	Medial surface, tibia	Extends thigh, flexes leg and knee, rotates leg medially
Flexors of the Thigh	Innervation: Femoral nerve (L2 to L4)		
Iliacus	Iliac fossa and sacrum	Lesser trochanter	Flexes thigh
Pectineus	Pubis, pectineal line	Lesser trochanter to linea aspera	Adducts and flexes thigh
Psoas major	Lumbar vertebrae	Lesser trochanter	Flexes thigh
**Rectus femoris	Anterior inferior iliac spine, rim of acetabulum	Tibial tuberosity	Flexes thigh, extends leg
Sartorius	Anterior superior iliac spine	Medial surface, tibia	Flexes thigh and leg

**Also extends the leg

Adductors of the Thigh	Innervation: Obturator nerve (L3 to L4)		
Adductor brevis	Body of pubis	Below lesser trochanter	Adducts, flexes, and rotates thigh medially
Adductor longus	Body of pubis	Linea aspera of femur	Adducts, flexes, and rotates thigh medially
Adductor magnus	Body of pubis	Linea aspera of femur and adductor tubercle	Adducts, flexes, and rotates thigh medially
*Gracilis	Body of pubis	Medial surface, tibia	Adducts thigh, flexes leg
Abductors of the Thigh	Innervation: Superior gluteal nerve (L4 to S1)		
Gluteus medius	Ilium	Greater trochanter	Abducts and rotates thigh medially
Gluteus minimus	Ilium	Greater trochanter, capsule of hip joint	Abducts and rotates thigh medially
Tensor fascia lata	Iliac crest	Iliotibial tract	Tenses fascia lata; assists in abduction, flexion, medial rotation of thigh

*Also flexes the leg

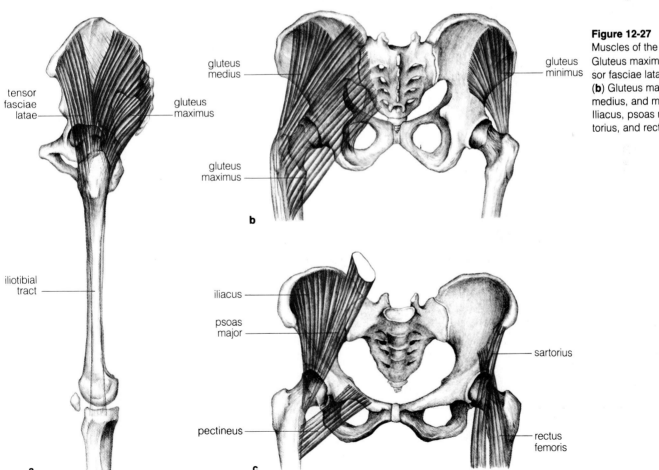

Figure 12-27
Muscles of the hip. (**a**) Gluteus maximus and tensor fasciae latae muscles. (**b**) Gluteus maximus, medius, and minimus. (**c**) Iliacus, psoas major, sartorius, and rectus femoris.

tensor fasciae latae

gluteus maximus

iliotibial tract

gluteus medius

gluteus maximus

gluteus minimus

a

b

iliacus

psoas major

pectineus

sartorius

rectus femoris

c

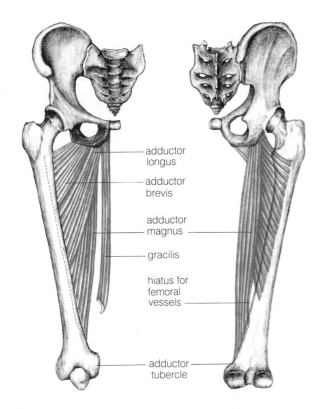

Figure 12-28 Adductor muscles of the hip

adductor
longus

adductor
brevis

adductor
magnus

gracilis

hiatus for
femoral
vessels

adductor
tubercle

The great thickness of the gluteus maximus makes it suitable for intramuscular injections of medicine. A needle can penetrate the coarse fasciculi of the muscle without causing appreciable damage. However, care must be taken not to injure the large sciatic nerve that passes deep to the muscle.

Flexors The short flexors of the hip (Fig. 12-27) originate either on the *vertebrae* or the *pelvic girdle* and insert on the *femur*. These include the *psoas major,* the principal flexor of the hip, and the *iliacus;* they unite as they cross anterior to the joint to form the ilipsoas, which inserts upon the *lesser trochanter*. The *pectineus* is a small muscle originating on the *pectineal line* of the *pubis* and inserting on the *femur* distal to the lesser trochanter. There are two long flexors, the *sartorius* and the *rectus femoris,* that cross both the hip and knee joints. The sartorius originates on the *ilium* and passes diagonally from the

lateral to the medial side of the thigh as it descends to attach to the *tibia* just below the knee. It flexes the leg and the thigh without much force because it is a strap muscle without enough fibers to exert much influence on the movement of the limb. The *rectus femoris* ends in a tendon that crosses the knee. It is part of the large muscle of the front of the thigh, the quadriceps femoris, which can be used to flex the thigh though it exerts its greatest effect in extending the knee.

Adductors The adductors of the thigh (Fig. 12-28) originate from the pelvic girdle and insert on the *linea aspera* of the *femur*. These muscles occupy the medial or adductor fascial compartment of the thigh and include the *gracilis* and the three muscles that have "adductor" as part of their names, the *adductor magnus, adductor longus,* and *adductor brevis*. The adductors function in locomotion as powerful muscles that draw the thigh medially and assist in maintaining the balance between flexion and extension during walking.

Abductors The abductors, the *gluteus medius* and *gluteus minimus* (Fig. 12-27b), take origin from the coxal bone and cross the lateral aspect of the hip joint to insert on the lateral side of the femur or overlying fascia. These muscles draw the thigh laterally as they contract and are antagonistic to the muscles of the adductor compartment.

The Hip Joint and Movements of the Thigh

The *hip joint* (Fig. 9-13a) is an excellent example of a ball-and-socket joint, with the round head of the femur fitting into the cup-shaped acetabulum. The depth of the acetabulum is increased by a rim of fibrocartilage that is incomplete inferiorly at the *acetabular notch,* which is closed by a *transverse ligament*. The joint capsule extending from the rim of the acetabulum to the neck of the femur is reinforced by thickenings of the capsule at its anterior portion; the capsule is weakest posteriorly.

Movement at ball-and-socket joints occurs freely in any direction, restricted only by ligaments of the joint and the length of the muscles that act at the joint. The main *flexors* of the hip joint are the *psoas major* and *iliacus;* assisted by the *rectus femoris, pectineus,* and *sartorius,* they are capable of flexing the thigh until it is stopped by contact with the anterior abdominal wall. Extension is more restricted: the thigh is almost fully extended in the standing position. The most powerful *extensor* muscle is the *gluteus maximus* which tightens the iliotibial tract, stabilizing the knee in extension as the thigh is extended. During walking the primary *extensors* are the *hamstring muscles* (biceps femoris, semimembranosus, and semitendinosus). The principal *abductors* are the *gluteus medius* and *minimus,* which

Table 12-6 Extensors of the Leg			Innervation: Femoral nerve (L2 to L4)
Muscle	Origin	Insertion	Action
Vastus intermedius	Proximal two-thirds of shaft of femur; distal one-half of inter-muscular septum	Common tendon into tibial tuberosity	Extends leg
Vastus lateralis	Intertrochanteric line, lateral intermuscular septum, greater trochanter, linea aspera	Common tendon into tibial tuberosity	Extends leg
Vastus medialis	Intertrochanteric line, spinal line, medial intermuscular septum, linea aspera	Common tendon into tibial tuberosity	Extends leg
Rectus femoris	Anterior inferior iliac spine	Common tendon into tibial tuberosity	Extends leg; flexes thigh

function chiefly as stabilizers of the pelvis during walking but which can rotate the femur laterally. The *adductors* on the medial side of the thigh (adductor longus, magnus, and brevis, and the gracilis) pull the thigh medially and function in both flexion and extension during walking. Adduction is not an important motion in walking, but it is important in counterbalancing the abductors to maintain an upright position.

In the upright position the hip joint is very difficult to dislocate due to the conformation of the bone. However, in the sitting position the head of the femur is almost out of the acetabulum, and a sudden jolt such as may occur in an auto accident may force the head of the femur through the relatively weak posterior portion of the joint capsule. The structure of the hip joint is a strong argument in favor of fastening automobile seat belts.

Muscles That Act at the Knee Joint

Most of the muscles acting at the knee joint also act at the hip joint. The extensor of the leg, the quadriceps femoris, has four heads of origin, of which only the rectus femoris head, mentioned as a flexor of the thigh, crosses the hip joint. The other three heads, the vasti, act only at the knee and are summarized in Table 12-6. (Muscles that act on both knee and thigh are included in Table 12-5).

The rectus femoris portion of the quadriceps femoris crosses anterior to the hip joint and the knee joint. It flexes the thigh at the hip joint but extends the leg at the knee. Posterior muscles of the thigh such as the hamstrings extend the thigh and flex the leg. This is in contrast to the situation in the upper extremity where the anteriorly placed biceps brachii crosses shoulder and elbow joints and flexes both arm and forearm; the triceps brachii of the posterior compartment extends the arm and the forearm.

The muscles of the thigh, some of which cross the knee joint to insert on bones of the leg, are ensheathed in deep fascia that sends extensions inward to fuse with the periosteum of the femur, dividing the thigh into three fascial compartments (Fig. 12-29). All of the muscles of the thigh except the tensor fasciae latae and the sartorius occupy one of the fascial compartments and belong to the functional group of that compartment.

Extensor Compartment The front of the thigh is the *extensor compartment,* or anterior compartment. It is occupied by the *quadriceps femoris,* a large muscle with four heads (Fig. 12-30): *rectus femoris, vastus lateralis, vastus intermedius,* and *vastus medialis.* The rectus femoris crosses the hip joint. The three vastus muscles arise separately from the shaft of the femur and fuse distally, along with the rectus femoris, into a common tendon (containing the patella) which inserts below the knee into the *tuberosity* of the *tibia.* The *sartorius* crosses the quadriceps femoris superficially as it descends diagonally down the front of the thigh to insert medially on the tibia.

Adductor Compartment The adductor muscles occupy the *adductor compartment,* or medial compartment of the thigh, described in the discussion of the hip region. The *gracilis,* a

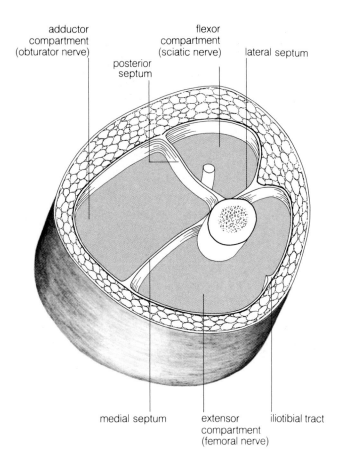

adductor compartment (obturator nerve)

posterior septum

flexor compartment (sciatic nerve)

lateral septum

medial septum

extensor compartment (femoral nerve)

iliotibial tract

Figure 12-29 Fascial compartments of the thigh

long, slender muscle arising from the lower margin of the pubis and inserting into the upper part of the medial surface of the tibia, is the only muscle of the adductor group that crosses the knee joint.

Posterior Compartment The *posterior compartment* of the thigh includes the hamstring muscles, all of which cross both the hip joint and the knee joint (Fig. 12-26). From their narrow origin at the *ischial tuberosity* they become broader as they descend together on the back of the thigh. As they approach the knee joint, the *biceps femoris* deviates laterally to insert on the proximal head of the *fibula*. Throughout its course the *semitendinosus* lies superficial to the *semimembranosus* as both of these muscles deviate medially to insert on the *medial surface* of the proximal end of the *tibial shaft* near its condyle. The tendons of the hamstring muscles may be felt as ridges behind

the knee, where they form the boundaries of the **popliteal space.** With the knee flexed this space is well defined, with the tendon of the biceps femoris forming the lateral boundary and the tendons of the semitendinosus and semimembranosus forming its medial boundary. The space (fossa) contains the major nerves and blood vessels supplying the leg. Within the floor of the popliteal fossa, the small *popliteus* muscle, arising from the *lateral femoral condyle* and inserting into the back of the *tibia* just below the knee, helps prevent dislocation of the knee joint. The other muscle crossing the back of the knee joint is primarily a leg muscle (gastrocnemius) that acts on the foot; it is described with the foot region.

Like all skeletal muscles the thigh muscles are subject to atrophy when they are not used. Muscular atrophy is a decrease in the size of muscle fibers due to a reduction in the number of myofibrils within the individual muscle fibers. For example, if the thigh is placed in a cast, its circumference is often reduced by more than one inch in a period of six weeks. When the cast is removed, the size of the thigh can be restored quickly by daily exercise of the muscles.

The Knee Joint and Movements of the Leg

The knee joint (Fig. 12-31) is a complicated hinge joint homologous to the elbow joint but with opposite orientation; the front of the knee is homologous with the back of the elbow. Unlike the elbow the knee joint is not stabilized by the conformation of bone but depends upon its ligaments and muscles to enable it to bear the weight of the body. Weight is borne by the flattened articular surfaces of the *tibial condyles,* separated by *intercondylar eminences,* which provide anchor points for ligaments. Each condylar articular surface is deepened by a half-moon-shaped *articular disc* of fibrocartilage attached at the periphery to the joint capsule. Between the two articular surfaces of the tibia, two heavy ligaments, *anterior* and *posterior cruciate,* arise from the intercondylar area and pass upward to attach to medial and lateral condyles of the femur. These two ligaments are important in holding the articular surface of the femur against the tibia, preventing dislocation. The *medial meniscus* (medial part of the articular disc) is more firmly attached to the joint capsule and collateral ligaments than is the *lateral meniscus* and is torn more frequently in knee injuries.

The synovial membrane lines the joint capsule, attaching to the margins of each meniscus. Anteriorly it extends between the extensor tendon and the femur above the condyles, forming the *suprapatellar bursa.* The membrane extends internally around

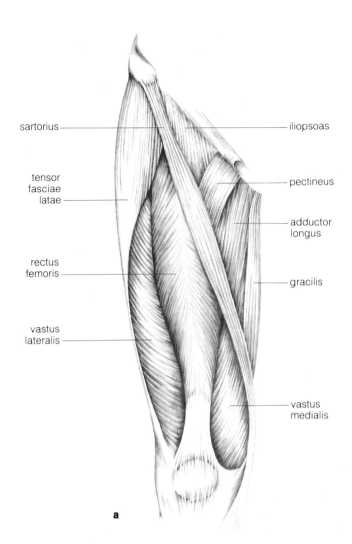

sartorius

iliopsoas

tensor
fasciae
latae

pectineus

adductor
longus

rectus
femoris

gracilis

vastus
lateralis

vastus
medialis

a

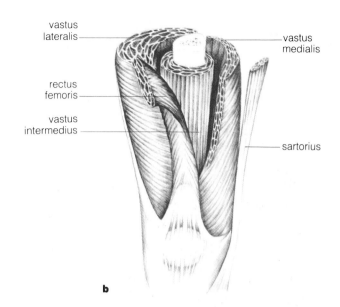

vastus
lateralis

vastus
medialis

rectus
femoris

vastus
intermedius

sartorius

b

Figure 12-30 Muscles of the thigh. (**a**) Anterior view of the thigh. (**b**) The rectus femoris muscle has been turned forward to reveal the vastus intermedius.

the cruciate ligaments, separating the knee joint into two synovial cavities on either side of the intercondylar eminence.

The fibrous capsule of the knee joint is reinforced anteriorly by the tendinous component of the quadriceps femoris, which below the patella is called the **patellar ligament** as it passes distally to insert on the tibial tuberosity. The capsule is strongly supported medially and laterally by the *tibial* and *fibular collateral ligaments,* preventing side-to-side movement of the knee. Posteriorly a strong ligamentous band, the *oblique popliteal ligament,* that crosses in back of the condyles and attaches just beyond the articular margins, provides support. The fibrous capsule, with its intra-articular ligaments (cruciate) and strong collateral ligaments reinforced by the tendons of muscles crossing the knee, stabilizes the joint and limits movement to flexion and extension, with slight rotation permitted to adjust the articular surfaces of the opposing bones on one another.

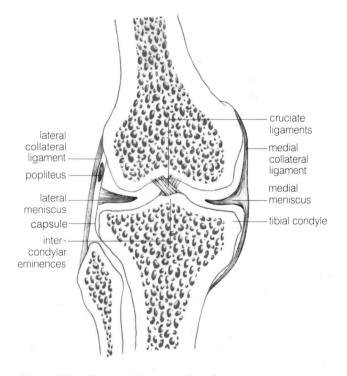

lateral
collateral
ligament

popliteus

lateral
meniscus

capsule

inter-
condylar
eminences

cruciate
ligaments

medial
collateral
ligament

medial
meniscus

tibial condyle

Figure 12-31 The knee joint, coronal section

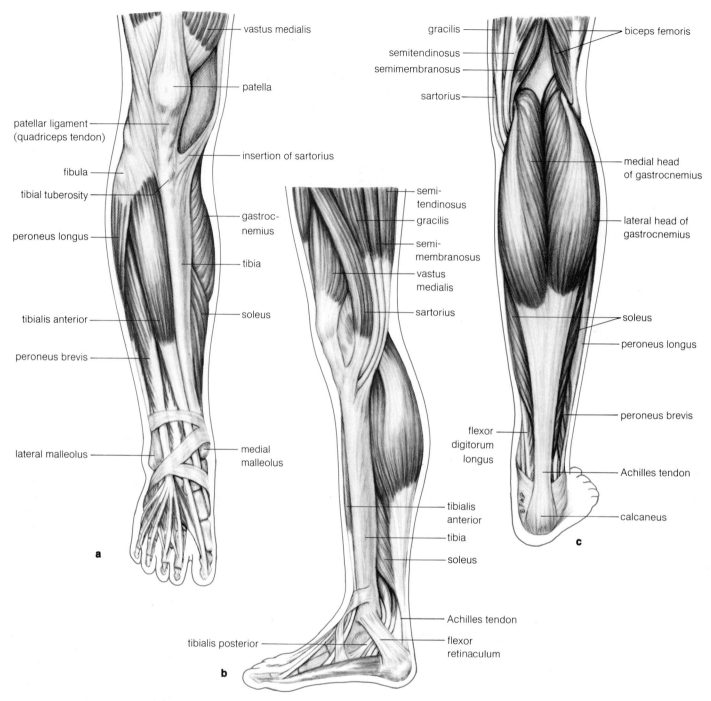

Figure 12-32 Muscles of the leg and foot. (**a**) Anterior view. (**b**) Medial view. (**c**) Posterior view. (**d**) Surface view, left leg. (**e**) Transverse section, cadaver preparation (photographs by R. Ross).

The *superior tibiofibular joint,* immediately distal to the knee, is a *plane synovial joint* in which the flat articular facets of the tibia and fibula slant to make close contact on the inferior surface of the tibial condyle. The tight joint capsule is reinforced by the anterior and posterior ligaments of the fibular head. In movements of the leg the tibia and fibula move as a unit at the knee joint.

At the knee the terms flexion and extension denote move-

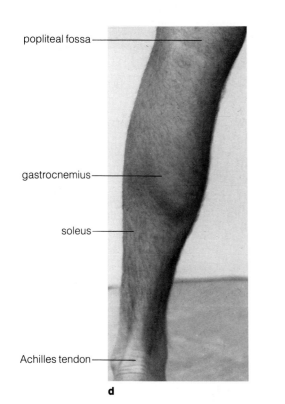

popliteal fossa

gastrocnemius

soleus

Achilles tendon

d

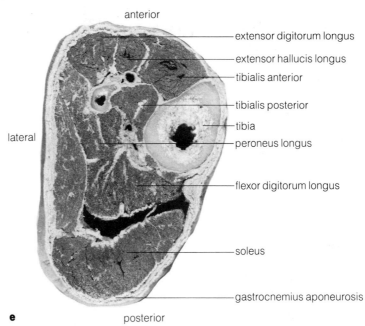

anterior

extensor digitorum longus
extensor hallucis longus
tibialis anterior
tibialis posterior
tibia
peroneus longus

flexor digitorum longus

soleus

gastrocnemius aponeurosis

lateral

e

posterior

ments in directions opposite to these same movements at the elbow, reflecting the opposite orientation of the upper and lower limbs during embryological development (see page 165). Flexion of the forearm is a forward motion, but flexion at the knee carries the leg backward; forward motion extends the leg. Extension brings the leg into approximately a straight line with the thigh and is accomplished largely by the quadriceps femoris muscle. Because the quadriceps tendon is an integral part of the knee joint capsule, contraction of the quadriceps tightens the capsule, tending to hold the knee in the fully extended position.

> Many football injuries occur when a player is blocked from the side unexpectedly when the thigh muscles are relatively relaxed. The blow stretches the medial side of the joint capsule, tearing the medial meniscus. Because cartilage repairs slowly, the injury is usually serious and may require surgical removal of the meniscus.

Flexion, resulting from contraction of the hamstring muscles, begins with the medial condyle of the femur sliding forward, bringing the rounded portion of the condyle into position prior to the hinge motion. Flexion of the leg is stopped by

contact with the back of the thigh. Extension at the hip, such as occurs when standing, tightens the fascia of the thigh and rotates the femur so that the flat anterior portion of its medial condyle presses against the flat surface of the tibial condyle. This causes the weight of the body to be supported by a flat table, with the knees in the locked position, minimizing the amount of energy required to maintain a fully upright position.

Muscles That Act on the Foot

The muscles that act on the foot include extrinsic muscles originating proximal to the foot, with tendons inserting on foot structures, and intrinsic muscles originating and inserting within the foot.

Extrinsic Muscles of the Foot The **extrinsic muscles of the foot** (Fig. 12-32) arise proximal to the foot from skeletal and fascial components of the thigh and leg. They are summarized in Table 12-7. These muscles are bound together into three functional groups in separate fascial compartments (Fig. 12-33). The anterior compartment, bounded by the interosseus membrane, the tibia, the fibula, and the anterior intermuscular septum, contains the *dorsiflexors* of the *foot* and *extensors* of the *toes* (Fig. 12-34). The *tibialis anterior* arises from the lateral tibial condyle and proximal shaft of the tibia and inserts near

Table 12-7 The Extrinsic Muscles of the Foot

Muscle	Origin	Insertion	Action
Muscles of the Anterior Compartment of the Leg		Innervation: Deep peroneal nerve (L4 to S1)	
Extensor digitorum longus	Lateral tibial condyle, proximal fibula, and interosseous membrane	By four tendons into extensor expansion of digits 2–5	Extends toes, dorsiflexes and everts foot
Extensor hallucis longus	Middle half, fibula; interosseous membrane	Base, distal phalanx, first digit	Extends big toe, dorsiflexes and inverts foot
Peroneus tertius	Distal one-fourth of fibula; interosseous membrane	Fifth metatarsal	Dorsiflexes and everts foot
Tibialis anterior	Upper two-thirds and lateral condyle of tibia; interosseous membrane	First cuneiform and first metatarsal	Dorsiflexes and inverts foot
Muscles of the Lateral Compartment of the Leg		Innervation: Superficial peroneal nerve (L4 to S1)	
Peroneus brevis	Lower two-thirds, fibula	Base, fifth metatarsal	Plantar flexes and everts foot
Peroneus longus	Lateral condyle, tibia; head and upper two-thirds, fibula	First cuneiform, first metatarsal	Plantar flexes and everts foot.
Muscles of the Posterior Compartment of the Leg		Innervation: Tibial nerve (L5, S1)	
Flexor digitorum longus	Middle one-half of tibia	By four tendons into distal phalanges, digits 2–5	Flexes four lateral toes, plantar flexes and inverts foot
Flexor hallucis longus	Lower two-thirds of fibula, intermuscular septa	Base, distal phalanx, first digit	Flexes big toe, plantar flexes and inverts foot
Gastrocnemius	Lateral and medial condyles, femur	Via calcaneal tendon onto calcaneus	Plantar flexes foot, flexes leg
Plantaris	Popliteal surface, femur	Medial side, calcaneal tendon	Plantar flexes foot
Popliteus	Lateral condyle, femur	Upper tibia	Rotates tibia medially
Soleus	Upper one-third of fibula, tibia	Via calcaneal tendon onto calcaneus	Plantar flexes foot
Tibialis posterior	Interosseous membrane and posterior fibula	Navicular, cuneiforms, cuboid, metatarsals II–IV	Plantar flexes and inverts foot

the base of the first metatarsal. The *extensor digitorum longus* muscle, arising from the lateral tibial condyle and upper fibula, splits into four tendons as it descends the leg to insert by means of slips into the middle and distal phalanges of the toes. The small *peroneus tertius* muscle is really the lower lateral part of the extensor digitorum longus, which inserts into the base of the fifth metatarsal bone. The *extensor hallucis longus* arises from the anterior surface of the fibula and inserts into the base of the distal phalanx of the big toe.

The *lateral compartment* (peroneal compartment) con-

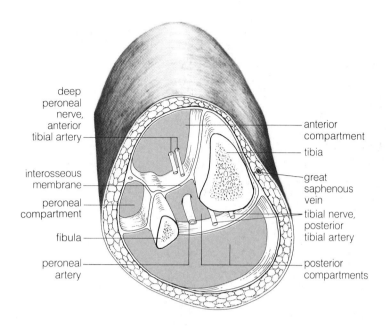

Figure 12-33 Fascial compartments of the leg

deep peroneal nerve, anterior tibial artery

interosseous membrane

peroneal compartment

fibula

peroneal artery

anterior compartment

tibia

great saphenous vein

tibial nerve, posterior tibial artery

posterior compartments

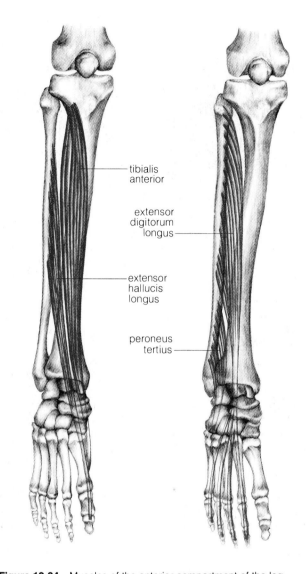

tibialis anterior

extensor digitorum longus

extensor hallucis longus

peroneus tertius

Figure 12-34 Muscles of the anterior compartment of the leg

tains the *peroneus longus* and *peroneus brevis* muscles (Fig. 12-35). The peroneus longus muscle arises from the lateral surface of the proximal two-thirds of the fibula; its tendon passes around the lateral border of the foot to enter the sole where it inserts into the base of the first metatarsal and the medial cuneiform bone. This muscle is important in maintaining the arches of the foot and is both a *plantar flexor* (plantar refers to the sole of the foot) and an *evertor* (eversion of the foot means turning the side outward). The peroneus brevis arises deep to the peroneus longus from the lower fibula and inserts into the fifth metatarsal bone, acting as an *evertor* of the foot.

The *posterior compartment* contains the most bulky muscles of the leg, forming the bulge on the back of the leg known as the calf. These muscles may be divided into two groups: a *superficial group,* consisting of the *gastrocnemius, soleus,* and *plantaris* separated by the *deep transverse intermuscular septum* from a *deep group* composed of the *popliteus, flexor digitorum longus, flexor hallucis longus,* and *tibialis posterior* muscles. The popliteus was described as a muscle of the knee joint because it is a short muscle acting at the knee and does not enter the foot.

The most superficial muscle of the calf is the *gastrocnemius* (Fig. 12-36a) which arises from the medial and lateral condyles of the femur and converges distally as a single muscle mass that

inserts along with the soleus into the *calcaneus tendon* (tendon of Achilles). The *soleus* (Fig. 12-36b) arises deep to the gastrocnemius from the proximal shafts of the tibia and fibula and from a fibrous band arching between the tibia and the fibula and inserts into the posterior surface of the calcaneus. These two muscles function together as powerful plantar flexors and are important in the maintenance of balance while standing. The remaining muscle of the superficial group, the *plantaris* (Fig. 12-36a), is a small muscle originating proximal to the lateral head of the gastrocnemius and inserting by means of a long,

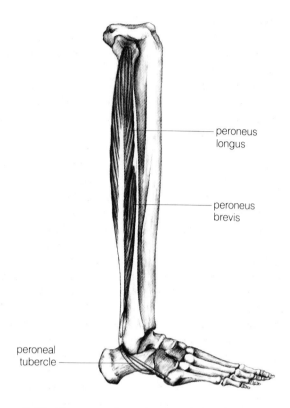

Figure 12-35 The peroneal muscles, lateral view of the leg

slender tendon onto the calcaneus. It is unimportant in humans though it may contribute somewhat to plantar flexion.

The *flexor digitorum longus* (Fig. 12-36c) is the most medial of the deep group of muscles in the posterior compartment, arising from the posterior surface of the tibia. Its tendon descends posterior to the medial malleolus (the medial eminence at the distal end of the tibia) and divides into four slender tendons that insert into the distal phalanx of all the toes except the first, flexing the toes. The most lateral of the deep muscles, the *flexor hallucis longus* (Fig. 12-36c), originates from the posterior surface of the distal part of the fibula and inserts at the base of the distal phalanx of the first toe, functioning as a flexor. The *tibialis posterior* muscle (Fig. 12-37) is intermediate in position, arising principally from the interosseous membrane between the other two muscles of the deep group to insert mainly into the tuberosity of the navicular, with tendinous slips passing to the cuneiforms and to the bases of the metatarsals. This muscle is the principal invertor of the foot, contributing also to plantar flexion.

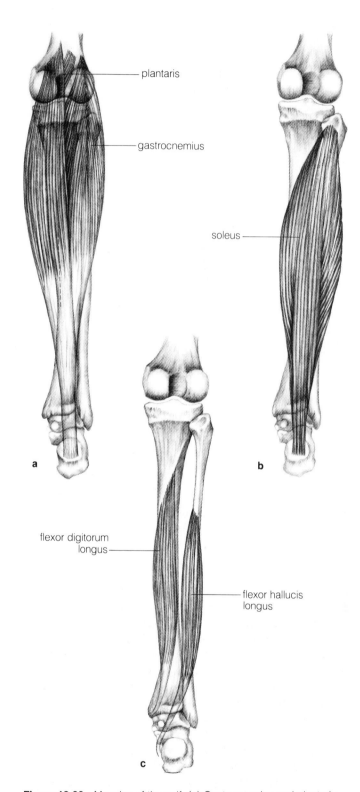

Figure 12-36 Muscles of the calf. (**a**) Gastrocnemius and plantaris. (**b**) Soleus. (**c**) Flexor digitorum longus and flexor hallucis longus.

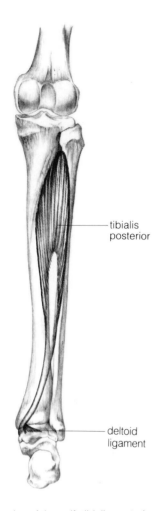

tibialis
posterior

deltoid
ligament

Figure 12-37 Muscles of the calf, tibialis posterior

Intrinsic Muscles of the Foot The **intrinsic muscles of the foot** (Table 12-8) are comparable to those of the hand, taking into consideration the fact that special muscles associated with the opposable thumb are absent. These short muscles of the foot do not actively contract during quiet standing but become important in providing the support needed while walking or standing on the toes. The *dorsum* (upper surface) of the foot has a single intrinsic muscle, the *extensor digitorum brevis*. The muscles of the plantar surface of the foot may be described in four layers, or planes. The superficial layer, or first plane (Fig. 12-38a), consists of the *abductor hallucis, flexor digitorum brevis*, and the *abductor digiti minimi*. The *abductor hallucis* arises from the calcaneus and adjacent fascia and inserts on the medial surface of the proximal phalanx of the big toe. The *flexor digitorum brevis* arises from the calcaneus and the central part of

the *plantar aponeurosis* and splits into four tendons that divide to insert on each side of the middle phalanx of the lateral four toes. The *abductor digiti minimi* muscle arises from the lateral surface of the calcaneus and plantar aponeurosis and inserts on the lateral side of the proximal phalanx of the little toe.

The second layer of intrinsic muscles (Fig. 12-38b) consists of the *quadratus plantae* (flexor accessorius) and the *four lumbricals*. The quadratus plantae originates by two heads from each side of the calcaneus, fusing into a single muscle prior to inserting into the tendon of the flexor digitorum longus, correcting the direction of pull of the latter. The four lumbricals arise from the tendons of the flexor digitorum longus and insert into the proximal phalanges of the lateral four toes and into the dorsal extensor expansion where the tendons of the extensors terminate. Consequently they have a double action: flexion of the proximal phalanx and extension of the distal phalanges.

The third layer of intrinsic muscles (Fig. 12-38c) includes the *flexor hallucis brevis, flexor digiti minimi brevis*, and the *adductor hallucis*. The *flexor hallucis brevis* arises principally from the cuboid and lateral cuneiform bone, passes forward beneath the first metatarsal, and splits into two tendons that insert on the medial and lateral surfaces of the first phalanx near its base. The *flexor digiti minimi brevis* arises from the plantar surface of the base of metatarsal V and inserts on the lateral surface of the base of the proximal phalanx of the little toe. The *adductor hallucis* arises by two heads: a transverse head from the ligaments of the metatarsophalangeal joints and an oblique head from ligaments of the midtarsal and adjacent tarsometatarsal joints. Both heads insert in common with the lateral head of the flexor hallucis brevis into the lateral surface of the proximal phalanx of the big toe.

The fourth layer of intrinsic muscles (Fig. 12-38d) includes the four *dorsal interossei* and the three *plantar interossei*. The interossei are arranged to hold the metatarsals firmly together and to produce side-to-side movement with the second toe as the axis. The *dorsal interossei* take origin from the adjacent sides of the metatarsals; they are located between and insert into the proximal phalanx on both sides of the second digit and the lateral side of the third and fourth digits, abducting the digits. The *plantar interossei* arise from the medial surface of the third, fourth, and fifth metatarsals and insert into the medial side of the proximal phalanx of the attached toes, adducting these digits. The actions of these muscles in the foot are similar to those of the hand.

Joints of the Ankle and Movements of the Foot

The joints of the ankle (Fig. 12-39) include the articulation of the distal ends of the tibia and fibula with each other and the

Table 12-8 Intrinsic Muscles of the Foot

Muscle	Origin	Insertion	Action
Muscles of the Dorsum of the Foot	Innervation: Deep peroneal nerve (L5, S1)		
Extensor digitorum brevis	Dorsal surface, calcaneus	By four tendons into tendons of extensor digitorum longus	Extends toes
Muscles of the First Plantar Plane	Innervation: Tibial nerve (L4, L5)		
Abductor digiti minimi	Medial and lateral tubercle, calcaneus	Lateral side, proximal phalanx, digit 5	Abducts little toe
Abductor hallucis	Medial tubercle, calcaneus, flexor retinaculum	Proximal phalanx, digit 1	Abducts and flexes big toe
Flexor digitorum brevis	Medial tubercle, calcaneus, plantar aponeurosis	By four tendons to middle phalanges, digits 2–5	Flexes lateral four toes
Muscles of the Second Plantar Plane	Innervation: Tibial nerve (L4 to S2)		
Lumbricals	Tendons of flexor digitorum longus	Proximal phalanx and extensor expansion, digits 2–5	Flexes metatarsophalangeal joints, extends distal two phalanges
Quadratus plantae	Medial and lateral sides, calcaneus; plantar fascia	Tendons of flexor digitorum longus	Assists flexor digitorum longus
Muscles of the Third Plane	Innervation: Tibial nerve (L4 to S2)		
Adductor hallucis	Plantar ligament, capsule of lateral four metatarsophalangeal joints	Proximal phalanx, digit 1	Adducts and flexes big toe
Flexor digiti minimi	Base, metatarsal V	Lateral side, base, proximal phalange, digit 4	Flexes little toe
Flexor hallucis brevis	Cuboid and third cuneiform	Via two tendons on each side, proximal phalanx digit 1	Flexes big toe
Muscles of the Fourth Plantar Plane	Innervation: Tibial nerve (S1, S2)		
Dorsal interossei	Adjacent metatarsals of intermetatarsal spaces	Both sides, proximal phalanx, digit 2; lateral side, digits 3 and 4	Abducts digits 2, 3, 4; flexes proximal phalangeal joints

talus. The *inferior tibiofibular* joint (distal tibiofibular joint) is a fibrous union (syndesmosis) of the distal ends of these bones by means of a strong interosseus tibiofibular ligament, reinforced by anterior and posterior tibiofibular ligaments that render the joint immobile. The interosseus ligament of the inferior tibiofibular joint is continuous superiorly with the interosseous membrane, uniting the shafts of the two bones and assuring a stable base for articulation of the tibia and fibula with the talus at the ankle joint. The distal end of the tibia together with the fibula forms an articular surface that is concave from front to back and fits the convex superior surface of the talus on which it rests, forming a pure hinge joint that permits only dorsiflexion

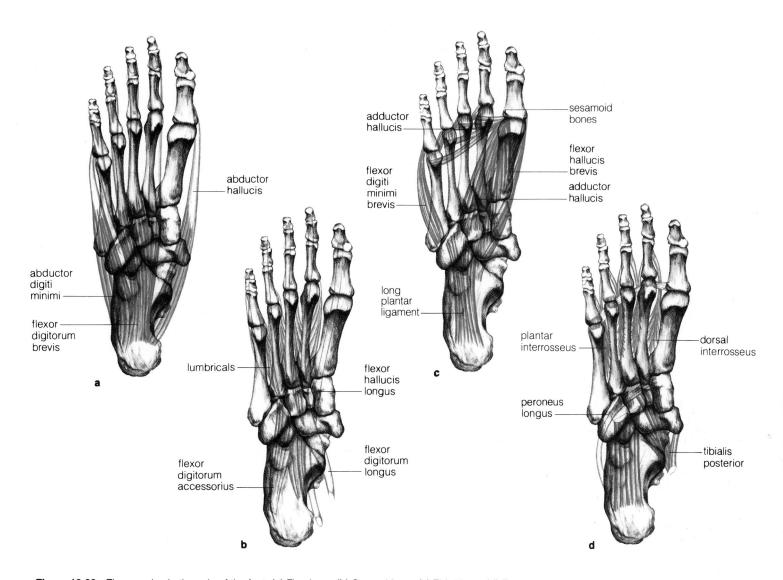

Figure 12-38 The muscles in the sole of the foot. (**a**) First layer. (**b**) Second layer. (**c**) Third layer. (**d**) Fourth layer.

and plantar flexion. Dorsiflexion means pulling the dorsum of the foot superiorly; *plantar flexion* means pulling the heel superiorly with the toes tilted downward, as occurs in standing on the toes.

As is usual with hinge joints the articular capsule of the ankle joint is thin anteriorly and posteriorly in the direction of movement and is reinforced on the sides by collateral ligaments that resist side-to-side movement. There are three lateral ligaments, *anterior talofibular, calcanofibular,* and *posterior talofibular,* attached superiorly on or near the lateral malleolus, and a single medial ligament, the *deltoid collateral,* or *medial*

collateral, attached superiorly to the margins and tip of the medial malleolus. These ligaments are assisted in preventing side-to-side motion by the projection of the lateral and medial malleoli along the sides of the talus. This joint is most secure when one is leaning forward (ankle dorsiflexed), engaging the broader anterior part of the articular surface of the talus between the two malleoli and causing them to be pushed slightly apart at the inferior tibiofibular joint. The joint is least secure when standing on the toes because the narrow posterior part of the articular surface lies between the malleoli, permitting slight side-to-side motion.

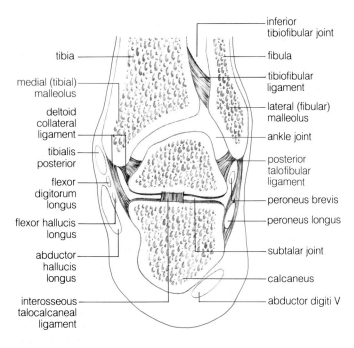

tibia

medial (tibial) malleolus

deltoid collateral ligament

tibialis posterior

flexor digitorum longus

flexor hallucis longus

abductor hallucis longus

interosseous talocalcaneal ligament

inferior tibiofibular joint

fibula

tibiofibular ligament

lateral (fibular) malleolus

ankle joint

posterior talofibular ligament

peroneus brevis

peroneus longus

subtalar joint

calcaneus

abductor digiti V

Figure 12-39 Joints of the ankle region, vertical section

subtaler joint, or talocalcaneal joint, between the inferior surface of the talus and superior surface of the calcaneus transfers the body's weight to the calcaneus. The subtaler is a synovial joint strongly supported by medial, lateral, and posterior talocalcaneal ligaments and by the interosseous talocalcaneal ligament. This joint permits the heel to take part in inversion and eversion of the foot. The *talocalcaneonavicular* joint, a synovial joint between three bones (talus, calcaneus, and navicular), forms part of the transverse tarsal joint. The head of the talus fits into a socket formed by the navicular anteriorly and the calcaneous posteriorly. The "spring" ligament, attaching the calcaneus to the navicular and helping maintain the arch of the foot, fills the space between the calcaneus and navicular.

Flat foot (*pes planus*) is a condition in which the longitudinal arch is depressed. The body weight forces the talus downward, causing the ligaments of the ankle joint to become permanently stretched. Flat feet are common in older persons, particularly if they gain weight rapidly. Usually, when the weight is removed the foot resumes its arched form.

Ankle sprains are usually caused by excessive inversion of the foot which may tear the ligaments on the lateral side of the ankle, causing pain and local swelling. The term "sprain" usually means ligaments are torn without fracture or dislocation of bones.

Dorsiflexion is accomplished principally by the *tibialis anterior,* assisted weakly by the extrinsic extensors of the toes, the extensor digitorum longus and the extensor hallucis longus. Plantar flexion with the knee extended is accomplished principally by the *gastrocnemius* and *soleus,* assisted by the deep extrinsic plantar flexors, flexor hallucis longus, flexor digitorum longus, and tibialis posterior. With the knee flexed the gastrocnemius is too short to be effective in plantar flexion, leaving the soleus as the principal muscle involved, assisted by the deep plantar flexors.

Intertarsal Joints and Movements of the Foot

The *intertarsal joints* of the foot allow slight adjustments of the bones on one another as the foot functions as a unit in the maintenance of balance, in weight-bearing, and in walking. The

Although the talocalcaneonavicular joint is shaped like a ball-and-socket joint, its short plantar ligaments permit only *sliding motion.* The synovial cavity of this joint is continuous with that of the transverse tarsal joint between the calcaneus and cuboid, which also permits only a slight sliding movement. Plane synovial joints with continuous synovial cavities occur between adjacent surfaces of all the other tarsal bones and at the tarso-metatarsal joints and intermetatarsal joints. Only *gliding motion* occurs at these articulations, with such movements contributing to the inversion and eversion of the foot as a unit. *Inversion* is accomplished by contraction of the tibialis anterior and posterior, whereas *eversion* is due to the action of the peronei (longus, brevis, and tertius).

Metatarsophalangeal Joints and Movements of the Toes

The *metatarsophalangeal* joints are condyloid joints between the heads of the metatarsals and the bases of the proximal phalanges, consisting of a synovial cavity enclosed within a loose capsule supported by a plantar and two collateral ligaments.

The plantar ligament, composed of fibrocartilage, supports the weight-bearing surface of the toes and restricts movement of the toes. The strong collateral ligaments on each side of the joint reinforce the capsule, limiting abduction and adduction. Flexion of the toes is less than that of the fingers, but extension of the toes is somewhat greater. *Flexion* of the toes is accomplished by the flexor digitorum longus and flexor hallucis longus (extrinsic muscles), assisted principally by the flexor digitorum brevis, flexor hallucis brevis, and flexor digiti minimi. The limited *adduction* of the toes is accomplished largely by the adductor hallucis and the plantar interossei, whereas *abduction* is the function of the abductor hallucis and the dorsal interossei.

WORDS IN REVIEW

Acromioclavicular joint	Iliotibial tract
Extensor retinaculum	Intrinsic muscles of the foot
Extrinsic muscles of the foot	Intrinsic muscles of the hand
Extrinsic muscles of the hand	Palmar aponeurosis
Flexor retinaculum	Patellar ligament
Glenohumeral joint	Popliteal space
Hamstring muscles	Rotator cuff
Hypothenar eminence	Sternoclavicular joint
	Thenar eminence

SUMMARY OUTLINE

I. The Upper Limb

   A. Muscles of the pectoral girdle are muscles that originate on the trunk and insert on the scapula or clavicle. These are the levator scapulae, pectoralis minor, rhomboideus major and minor, serratus anterior, and trapezius, which attach to the scapula, and the trapezius and subclavius, which insert on the clavicle. The acromioclavicular joint does not allow movement at that joint, so muscles attached to the scapula move the pectoral girdle in any direction at the sternoclavicular joint.

   B. Muscles that move the arm include muscles from the axial skeleton to the humerus (pectoralis major and latissimus dorsi) and muscles that originate on the pectoral girdle, such as the deltoid, and insert on the humerus. Special muscles of this group insert on the humerus near the glenohumeral joint. These muscles (supraspinatus, infraspinatus, subscapularis, and teres minor) rotate the humerus and stabilize the joint. Collectively they are called muscles of the rotator cuff.

   C. Muscles of the elbow region either flex or extend the forearm. The flexors are the biceps brachii and brachialis. The extensors are the triceps brachii and the small anconeus. In addition to flexion and extension, pronation and supination occur at the elbow (proximal radioulnar joint). Pronation is largely due to contraction of the pronator teres and pronator quadratus; supination is the result of contraction of the biceps brachii, assisted by the supinator.

   D. The extrinsic muscles of the hand arise chiefly from the medial and lateral epicondyles of the humerus. The extensors are in the posterior compartment; they arise from the lateral epicondyles and insert into bases of the second, third, and fifth metacarpals. The flexors of the wrist and fingers are arranged in a superficial and a deep group that occupy superficial and deep anterior compartments. The superficial group arises from the medial epicondyle, and the deep group arises from the proximal ulna. The deep flexors insert on the distal phalanx; the superficial flexors insert on the second phalanx.

   E. The intrinsic muscles of the hand assist in producing fine movements of the fingers. These include the lumbricals, dorsal and palmar interossei, which terminate on the fingers, and special muscles of the thumb and little finger that provide for the greater mobility of these two digits.

II. The Lower Limb.

   A. The muscles of the hip may be divided into five functional groups: (1) the lateral rotators (quadratus femoris, obturator externus and internus, gemellus superior and inferior, piriformis) arise from the sacrum or os coxa and insert on the greater trochanter; (2) the extensors (gluteus maximus and hamstrings) arise from the ilium (gluteus maximus) and insert on the femur, or arise from the ischial tuberosity and insert on bones of the leg; (3) the flexors (iliopsoas, pectineus, sartorius, and rectus femoris) arise from the vertebrae or os coxa and insert on the femur (short) or leg bones (long flexors); (4) the adductors (adductor magnus, longus, brevis, and gracilis) arise from the pelvic girdle and insert on the linea aspera of the femur; and (5) the abductors (gluteus medius and minimus) arise from the os coxa and insert on the lateral aspect of the femur or overlying fascia.

   B. The muscles that act at the knee joint occupy one of three compartments: the quadriceps femoris (rectus femoris, vastus lateralis, intermedius, and medialis) is in the extensor compartment of the thigh and inserts on the tuberosity of the tibia; the gracilis is a muscle of the adductor compartment that crosses the knee joint to insert on the medial surface of the tibia; the hamstring muscles (biceps femoris, semi-tendinosus, and semimembranosus) are in the posterior compartment and separate, with the biceps passing laterally to insert on the fibula and the other two inserting medially on the tibia; they flex the leg.

C. The muscles that act on the foot include extrinsic muscles that arise proximal to the foot and insert on foot structures and intrinsic muscles that originate and insert on foot structures.

   1. The extrinsic muscles lie together in three functional groups in separate fascial compartments: the anterior compartment of the leg contains the dorsiflexors of the foot and extensors of the toes; the lateral (peroneal) contains the peroneus longus and brevis muscles, which act as plantar flexors and evertors; the posterior compartment contains muscles that act as powerful plantar flexors and flex the toes.

   2. The intrinsic muscles of the foot are comparable to those of the hand and have similar actions.

REVIEW QUESTIONS

1. At what joint does contraction of the muscles that originate on the trunk and insert on the pectoral girdle produce movement?

2. List and describe the actions of the muscles of the rotator cuff.

3. What movements are possible at the articulation of the humerus with the ulna? What muscles are responsible for these movements? What movements are possible at the proximal radioulnar joint? What muscles are responsible for these movements?

4. The muscles of the posterior compartment of the forearm arise from what structure? What are the principal actions of these muscles?

5. Which group of flexors of the fingers inserts on the distal phalanx of each finger?

6. Name the three groups of intrinsic muscles of the hand that help manipulate all of the digits except the thumb.

7. List the five functional groups of muscles that act on the hip and name the muscles of each group.

8. List the muscles of three compartments of the thigh that act on the knee joint.

9. What are the principal actions of the muscles of the anterior compartment of the leg?

10. What are the principal functions of the muscles of the posterior compartment of the leg?

SELECTED REFERENCES

Basmajian, J. V. *Grant's Method of Anatomy*. 9th ed. Baltimore: Williams & Wilkins, 1975.

Crafts, R. C. *A Textbook of Human Anatomy*. 2nd ed. New York: John Wiley & Sons, 1979.

Romanes, G. J. *Cunningham's Textbook of Anatomy*. New York: Oxford University Press, 1981.

Snell, R. S. *Clinical Anatomy for Medical Students*. 2nd ed. Boston: Little, Brown & Co., 1981.

13

GENERAL ORGANIZATION OF THE NERVOUS SYSTEM

After completing this chapter you should be able to:

1. Describe the functional relations of the sensory, association, and motor components of the nervous system.

2. Compare and contrast the functional relations of the nervous system with the endocrine system in the regulation of effectors.

3. List the major components of the central nervous system and of the peripheral nervous system.

4. Describe the structural units of nervous tissue.

5. Classify neurons on the basis of the number of processes.

6. Name and describe the supporting cells of the central and peripheral nervous system.

7. Describe the anatomical relations of the components of the central nervous system to one another and to the peripheral nervous system.

8. Compare and contrast the components of cranial nerves with those of spinal nerves.

9. Diagram or describe the functional unit of the nervous system.

10. Locate the brain ventricles in relation to the parts of the central nervous system and to one another.

11. Name and describe the meninges; locate the cerebrospinal fluid and describe its circulation.

12. Name the five secondary brain vesicles of the embryo and list the derivatives of each vesicle.

The *nervous system,* together with the *endocrine system,* controls and integrates the activities of the body and provides the mechanisms for responding to changes in the external and internal environment. The activity of the nervous system enables the parts of the body to communicate by relaying information from one part of the body to another. A nerve cell's capacity to receive information (stimuli) is due to its high degree of irritability; its ability to transmit information over its length is due to its high degree of *conductivity.* The stimulus initiates an electrochemical impulse that may be transmitted the length of the cell, causing the release at the end of the cell of a minute amount of a chemical called a *neurotransmitter substance* that stimulates or inhibits another cell. The second cell stimulated by this process may be a nerve cell, muscle cell, or gland cell. A nerve cell conveys the impulse to its termination, a muscle cell contracts when stimulated, and a gland cell increases or decreases its secretion. These responses are elicited by the neurotransmitter substance produced and released by the nerve cell. Thus the nerve cell is basically an electrochemical conducting cell that influences the activities of other cells by means of its secretions. The neurotransmitter substances differ in various types of nerve cells, but the basic mechanism is the same for almost all of them.

The neurotransmitter substance was first detected in 1921 by Otto Loewi, who had observed that stimulation of the vagus nerve slowed the rate of the heartbeat in the frog. He perfused the heart of a frog with a saline solution while stimulating the vagus nerve. When he perfused a second frog heart with the saline solution from the first heart, there was a dramatic decrease in the rate of contraction of the second heart. Loewi called the active agent "vagus substance" and showed that acetylcholine has the same effect. We now know that *acetylcholine,* in fact, is the neurotransmitter substance released by the vagus nerve.

In lower animals with the simplest type of nervous systems, cells specialized as conductile cells have long, thin extensions or processes that extend from the surface of the animal to its deeper parts. A single cell transmits the surface stimuli to muscle or other cells internally. Thus a single cell serves both as a *receptor,* or *sensor cell,* and as an *effector,* or *motor cell.* In animals with a slightly more complex nervous system, the sensory and motor functions are performed by separate sets of nerve cells. Sensory cells receive stimuli due to changes in the external or internal environment of the animal and convey the messages to a central point where the messages are delivered to motor nerve cells. The motor cells, in turn, stimulate muscle or gland cells to contract or alter their rates of secretion. Humans and other vertebrates have a third set of nerve cells interposed between the sensory and motor cells that functions as *association,* or *integrative, neurons;* these have reached their highest degree of complexity in humans. Association neurons are capable of receiving tremendous amounts of information as stimuli, or impulses, and relaying the correct messages to motor neurons so that responses to environmental changes are right for the situation. Since the responses may involve many parts of the body, it is essential that all activity be properly integrated. In humans the nervous system not only has the capacity to integrate the activities of many parts of the body, but it also has developed to a high degree our memory, the ability to store information, and our reason, the ability to evaluate information and judge how it should be used.

The association neurons constitute the largest mass of human nervous tissue and are located entirely within the *brain* and *spinal cord,* which together form the *central nervous system* (CNS). The sensory and motor elements are located only partially within the CNS and have protoplasmic extensions that communicate with various parts of the body. These extensions, termed *nerve processes,* or *fibers,* are outside the CNS and constitute part of the *peripheral nervous system* (PNS). For the proper functioning of the body each of the three components (sensory, association, and motor) must be able to do its part; a breakdown of one of the components due to disease or trauma will prevent the others from working properly. For example, a loss of sensory information from the sensory receptors in the biceps brachii muscle and its tendon will become obvious if one contracts the muscle in order to lift something. The association component will not receive information about the weight of the object and will be unable to determine the force needed, so lifting will be difficult even if the motor supply of the muscle is intact. Similarly, destruction of the association component that integrates the activity of the biceps will cause problems even when the sensory and motor supply of the muscle is intact. For example, as the biceps is stimulated to contract to flex the forearm, the triceps brachii must be directed to relax. Without proper functioning of the CNS the activities of the two muscles will not be integrated, causing the movements to be jerky. Finally, if the motor nerve supply to the biceps is destroyed, the muscle will not contract, for skeletal muscle does not contract in the absence of nervous stimulation.

The activity of each organ system is influenced by the condition of the body as a whole and by the metabolic products of all the organ systems contributing to the internal environment of the cells. The condition of the body is maintained in the state intepreted as "healthy" as long as each organ system performs its functions at rates consistent with the maintenance of a constant internal environment. Each organ system has mechanisms for adapting to and compensating for internal and external environmental changes that usually involve changes in the rates of metabolism. Special **receptors** associated with either the nervous system or the endocrine system detect potentially harmful changes and initiate mechanisms eliciting the proper metabolic response. The cells whose metabolic rates are manipulated are **effectors** which may not in themselves be capable of detecting environmental changes. *Effectors* regulated by the nervous system are of two types: *muscles,* which respond to stimuli by contracting, and *glands,* which respond by altering their rate of secretion. Effectors regulated by the endocrine system include all the living cells of the body; they respond to hormonal influences by increasing or decreasing their rate of metabolism. Unlike the changes instigated by the nervous system, which are usually rapid and of brief duration, the changes produced by the endocrine system are usually slow to take effect and relatively long-

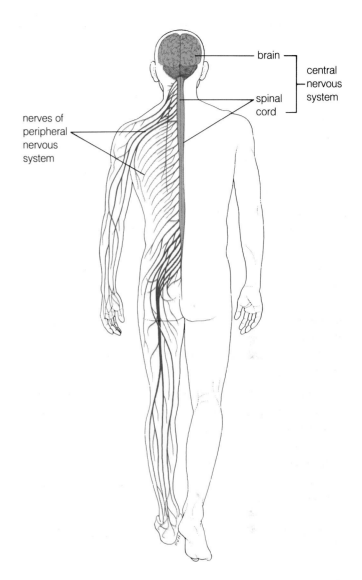

Figure 13-1 Diagrammatic representation of the nervous system, posterior view.

lasting. In some cases the nervous system and the endocrine system act together to integrate the activities of the body; because their effects cannot be separated, they may be considered to function as a *neuroendocrine system.* For example, the emotion of fear or anger mobilizes the autonomic nervous system (sympathetic division) for a "flight or fight" response and simultaneously increases the release from the suprarenal glands of the hormones norepinephrine and epinephrine, whose effects are similar to those of the sympathetic nervous system.

The nervous system (Fig. 13-1) is divided for descriptive purposes into a *central nervous system* (CNS) and a *peripheral nervous system* (PNS). The CNS consists of the *brain* (cerebrum,

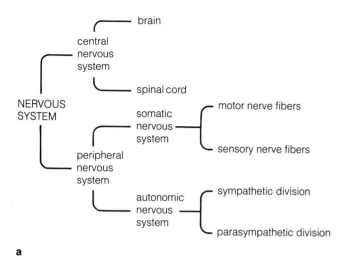

a

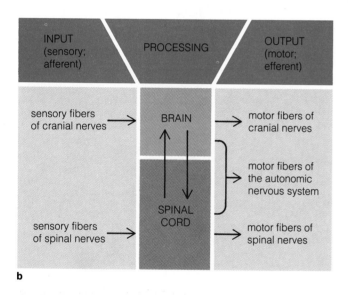

b

Figure 13-2 Organization of the nervous system. (**a**) Structural organization. (**b**) Functional organization.

Table 13-1	Nervous Tissue
Cell Type	Description
Neuron	Cell body containing nucleus; cytoplasmic extensions (axon and dendrites). May be unipolar, bipolar, or multipolar.
Neuroglia	Supporting cells within CNS. May be astrocytes with highly branched cytoplasmic extensions, oligodendrocytes which produce the myelin sheaths of nerve fibers, microglia which are small phagocytic cells, or ependyma that lines ventricles.
Neurilemma	Supporting cells outside the CNS. May be satellite cells in ganglia or sheath of Schwann cells which produce and maintain myelin around nerve fibers.

NERVOUS TISSUE

Nerve tissue is composed of three kinds of cells (Table 13-1). *Neurons,* the structural units of the nervous system, transmit nerve impulses. The supporting cells, the *neuroglia* and *neurilemma,* form a protective sheath for the neurons and deposit layers of protective membrane called myelin. Neuroglia cells are found in the CNS and neurilemma cells in the PNS.

Neurons

The structural units, the *neurons,* which transmit the impulses that regulate most of the activities of the body, all have the same essential components but differ in their morphology depending upon their location and functions. The essential components of all neurons (Fig. 13-3) are a *cell body* containing the *nucleus,* and *cytoplasmic extensions* (processes, fibers) named according to the direction of impulse flow. These extensions are either *dendrites,* which conduct the impulse toward the cell body, or an *axon,* which transmits the impulse away from the cell body.

Within the CNS, accumulations of nerve cell bodies, gray in color, are called *gray matter;* accumulations of myelinated nerve processes or fibers in the CNS are white and called *white matter.* A collection of functionally related nerve cell bodies within the CNS is a *nucleus;* a similar aggregation in the PNS constitutes a *ganglion.* A group of functionally related nerve fibers (axons or dendrites) within the CNS forms a *tract;* a bundle of nerve fibers in the PNS is a *nerve.*

cerebellum, diencephalon, and brainstem) and *spinal cord;* the PNS consists of *twelve pairs* of *cranial nerves* emerging from the *brainstem* and *thirty-one pairs of spinal nerves* arising from the *spinal cord.* The PNS includes the *somatic* system, which is under voluntary control, and the *autonomic* system, which is regulated by involuntary processes of the nervous and endocrine systems. The autonomic system in turn has two divisions, *parasympathetic* and *sympathetic.* See Fig. 13-2 for a summary of the organization of the nervous system.

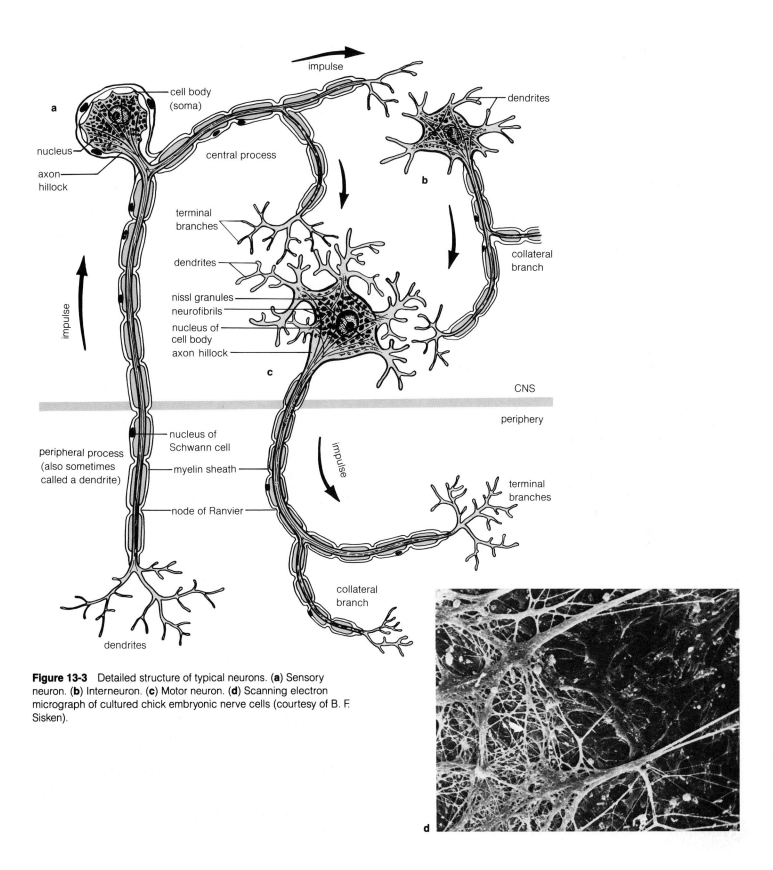

Figure 13-3 Detailed structure of typical neurons. (**a**) Sensory neuron. (**b**) Interneuron. (**c**) Motor neuron. (**d**) Scanning electron micrograph of cultured chick embryonic nerve cells (courtesy of B. F. Sisken).

The following labels appear in the figure:

- cell body (soma)
- a
- nucleus
- axon hillock
- central process
- impulse
- dendrites
- b
- collateral branch
- terminal branches
- dendrites
- nissl granules
- neurofibrils
- nucleus of cell body
- axon hillock
- c
- impulse
- CNS
- periphery
- nucleus of Schwann cell
- peripheral process (also sometimes called a dendrite)
- myelin sheath
- node of Ranvier
- impulse
- terminal branches
- collateral branch
- dendrites
- d

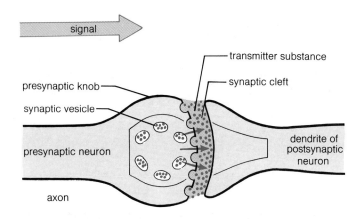

Figure 13-4 A synapse

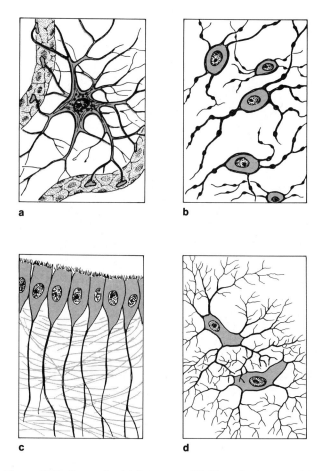

Figure 13-5 Neuroglia. (**a**) Astrocytes. (**b**) Oliogodendrocytes. (**c**) Ependymal cells. (**d**) Microglia.

The Synapse

The nerve impulse travels the length of a neuron and is transmitted to the next neuron in the chain across a junction called a synapse. The transfer of the impulse across the synapse depends upon the release of a chemical substance, called a neurotransmitter, by the neuron that brings the impulse to the synapse. This neuron, termed the presynaptic neuron, contains the neurotransmitter substance within synaptic vesicles of presynaptic knobs (Fig. 13-4); the transmitter is released into the cleft and initiates an impulse in the dendrite of the postsynaptic neuron that conveys the impulse to the next neuron in the neural pathway or to the effector (muscle or gland cell).

Classification of Neurons

Neurons are divided into three categories (Fig. 13-3) according to the number and arrangement of their processes (axons and dendrites).

Unipolar Neurons   A unipolar neuron (Fig. 13-3a) has a single process that bifurcates into a single dendrite and axon. All unipolar neurons are sensory neurons located in ganglia outside the CNS. The dendrite extends into the periphery, functioning as a receptor of stimuli that are capable of initiating an impulse. The axon extends into the CNS, conveying the impulse for relay by neurons of the association or projection systems.

Bipolar Neurons   A bipolar neuron has a single dendrite and axon extending from opposite sides of the cell body. This type of neuron is found only in the *olfactory nerves,* the *retina* of the eyeball, and the *vestibular* and *cochlear ganglia* of the *inner ear.* Bipolar neurons are sensory neurons that transmit impulses from these special sense organs.

Multipolar Neurons   The most numerous neurons are *multipolar* neurons (Fig. 13-3b, c), each with a single axon and several dendrites. Most of the neurons of the CNS are multipolar although they vary a great deal in size of the cell body and length of their processes. Multipolar neurons are also characteristic of the ganglia of the autonomic nervous system.

Supporting Cells

The supporting cells, the neuroglia of the CNS and neurilemma of the PNS, serve to protect (and support) the neurons.

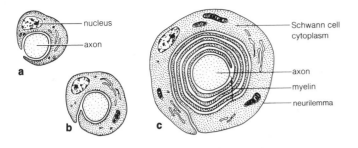

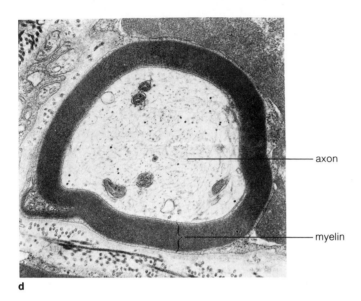

Figure 13-6 **(a–c)** A myelinated axon in cross section showing the myelin wrapping created by the Schwann cell as it develops. **(d)** Electron microscope photomicrograph of a cross section of a myelinated axon (courtesy of Art Nitz).

Neuroglia Supporting cells within the CNS are collectively called **neuroglia.** Neuroglia cells (Fig. 13-5) are of several morphological types classified as *astrocytes, oligodendrocytes, ependyma,* and *microglia.* The astrocytes have many highly branched cytoplasmic extensions that provide much of the supporting framework within the CNS. The oligodendrocytes are responsible for the production of the white fatty material called *myelin* that ensheaths the nerve fibers within the CNS. These cells maintain close association with the individual nerve fibers and adjacent blood capillaries, a close relationship that enables them to help regulate the permeability of the capillaries and prevent many potentially harmful substances from gaining access to the nerve cells. Thus the oligodendrocytes have an important role in the "blood–brain barrier" that shields CNS neurons from many toxic agents. The third type of neuroglia, the microglia,

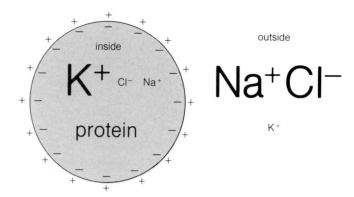

Figure 13-7 The nerve cell. The large number of negatively charged protein molecules ⁻P⁻ within the cell and the large number of Na₊ outside the cell are the major contributors to the potential difference between the inside and the outside of the membrane in a resting cell. The resting potential difference in most nerve cells is about −70 millivolts; that is, the inside is more negative than the outside by that amount.

are small phagocytic cells derived from embryonic connective tissue cells that migrate into the CNS during the early phases of its development. These cells function as scavengers, clearing away debris in a manner similar to that of all tissue macrophages. The ependyma is the lining of the fluid-filled cavity within the CNS.

Neurilemma Nerve cells and fibers outside the CNS have their own protective coverings, the **neurilemma** cells. Nerve cell bodies located in ganglia are surrounded by small neurilemma cells termed *satellite cells.* Many nerve fibers are enclosed by neurilemma cells, called *sheath-of-Schwann cells,* which produce myelin membranes on their inner surfaces. Each sheath-of-Schwan cell, along with its myelin, is wrapped around a portion of the nerve fiber (Fig. 13-6). In addition to these specialized coverings of myelin and neurilemma cells, peripheral nerve fibers have protective coverings formed by connective tissue elements.

THE NERVE IMPULSE

In a resting neuron that is not conducting an impulse there is a potential difference, or *resting potential,* due to the difference in concentration of ions inside the neuron and outside its membrane. The inside of the cell is negative with respect to the outside (Fig. 13-7). When the neuron is stimulated, its mem-

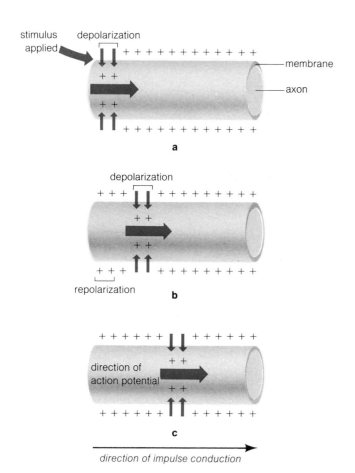

Figure 13-8 Initiation and conduction of an impulse

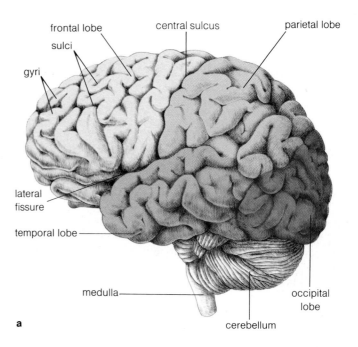

a

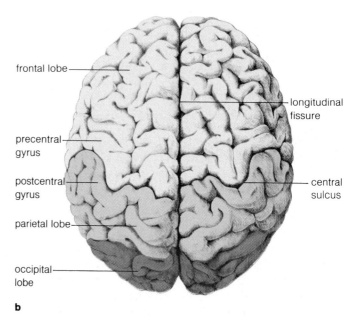

b

Figure 13-9 The external surface of the brain, showing the lobes of the cerebrum, gyri, sulci, and the relationships of the cerebrum to other parts of the brain. (**a**) Lateral view. (**b**) Superior view.

brane's permeability to sodium ions increases markedly at the point of stimulation, and sodium ions enter the cell for a brief period creating a positive potential difference at the point of stimulation (Fig. 13-8). This change in membrane potential is called *depolarization* because the potential difference is now less than it was in the resting state. Adjacent portions of the membrane become depolarized, and a wave of depolarization passes the length of the neuron. This is the *action potential;* its passage is the conduction of an impulse along the neuron. As the impulse is conducted along the neuron, the membrane is repolarized by the passage of potassium ions out of the cell, which restores the membrane potential to the resting state.

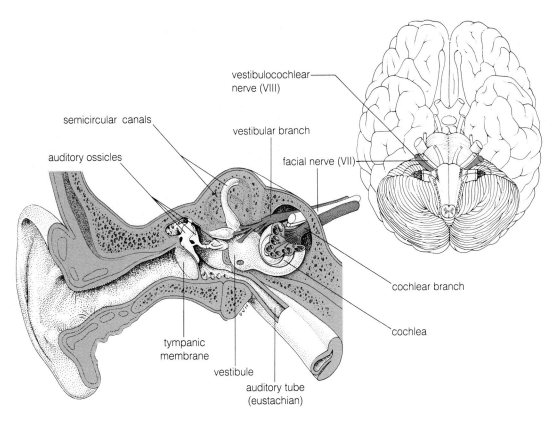

Figure 13-10 The auditory nerve and its relations to sensory receptor organs of the inner ear

BASIC PATTERNS OF THE CENTRAL AND PERIPHERAL NERVOUS SYSTEMS

The Central Nervous System

The central nervous system consists of the brain and spinal cord. The brain (Fig. 13-9) includes the **cerebrum,** the **diencephalon** (thalamus and hypothalamus), the **cerebellum,** and the **brainstem** (midbrain, pons, medulla). The most superior part of the brainstem, the midbrain, is continuous caudally with the pons, which in turn is continuous with the medulla. The cerebellum, located dorsal to the brainstem, has connections with all three parts of the brainstem but is considered a separate structure. All the components of the brain are housed within the *cranial cavity,* which is protected by bone and special connective tissue coverings called **meninges.** The cranial components of the CNS are thus directly continuous with one another

and join the spinal cord at the *foramen magnum* via the medulla. The spinal cord occupies the *vertebral canal,* protected by the vertebrae and the spinal meninges.

The Peripheral Nervous System

The PNS consists of cranial and spinal nerves that arise from the brainstem or spinal cord at specific points and pass into the periphery through foramina in the floor of the skull or between the vertebrae. The cranial nerves are of three types: sensory, motor, and mixed.

Cranial Sensory Nerves Some cranial nerves are *purely sensory,* conveying messages regarding stimuli from the periphery to the CNS. The cell bodies of these neurons are usually located outside the CNS in ganglia associated with the nerve. For example, the auditory nerve, cranial nerve VIII (Fig. 13-10), consists

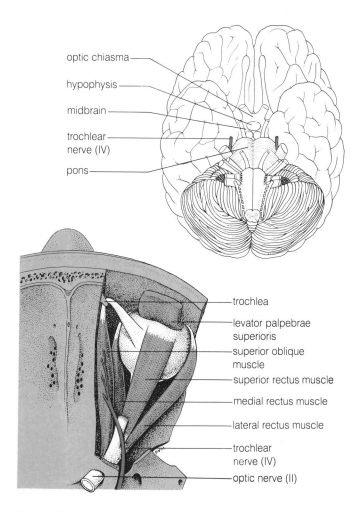

optic chiasma

hypophysis

midbrain

trochlear nerve (IV)

pons

trochlea

levator palpebrae superioris

superior oblique muscle

superior rectus muscle

medial rectus muscle

lateral rectus muscle

trochlear nerve (IV)

optic nerve (II)

Figure 13-11   The distribution of the trochlear nerve

of fibers that transmit impulses initiated in the organ of Corti, the hearing receptor. The cell bodies of these fibers are located in the spiral ganglion of the auditory nerve.

Cranial Motor Nerves   Other cranial nerves are almost *purely motor,* transmitting impulses from the CNS to the effectors (muscles or glands), which respond to the stimulus. These motor cranial nerves have their cell bodies within the CNS and send their axons into the periphery to terminate on the effector. For example, the trochlear nerve, cranial IV (Fig. 13-11), consists of axons that arise from cell bodies of the trochlear nucleus, located in the midbrain. These axons transmit impulses to a single eyeball muscle, the superior oblique, causing the muscle to contract, tilting the eyeball downward.

Cranial Mixed Nerves   The third type of cranial nerve consists of a strong *mixture* of *sensory* and *motor fibers.* The sensory fibers are dendrites conveying impulses toward their cell bodies, which are located in the ganglion of the nerve. From the ganglion, impulses are transmitted into the CNS by axons of the same neurons. The motor fibers arise from cell bodies comprising the nucleus of the cranial nerve and pass into the periphery in a single bundle along with the sensory fibers. For example, the trigeminal nerve, cranial V (Fig. 13-12), is a large mixed nerve responsible for sensations initiated in the skin of the face and for the motor supply of the muscles of mastication. Cell bodies of the sensory fibers are located in the trigeminal (semilunar) ganglion of the midbrain, from which impulses are conveyed into the pons. Cell bodies in the motor nucleus of the trigeminal nerve within the pons send their axons (motor fibers) from the pons to be incorporated in the trigeminal nerve for distribution to the muscles of mastication.

Spinal Nerves   Unlike the cranial nerves, which vary in their composition, all spinal nerves (with the exception, in some cases, of the first cervical nerve) are mixed nerves containing both sensory and motor fibers (Fig. 13-13). Each spinal nerve is attached to the spinal cord by two roots: a *dorsal root,* composed of sensory fibers with cell bodies located in a dorsal root ganglion, and a *ventral root,* composed of motor fibers that arise from cell bodies located in the *ventral* and *lateral horns of gray matter* of the spinal cord. The dorsal and ventral roots merge at the intervertebral foramen, forming the mixed spinal nerve for distribution to peripheral structures (see also Fig. 14-2).

THE FUNCTIONAL UNIT: THE REFLEX ARC

The functional unit of the nervous system is the *reflex arc,* diagrammed in a simplified form in Fig. 13-14. If your hand touches something hot, sensory fibers terminating in the skin of the hand transmit impulses that enter the dorsal side of the spinal cord via the dorsal root of the spinal nerve. The impulses are relayed by association neurons to motor neurons in the ventral gray matter of the spinal cord. These motor fibers convey impulses to skeletal muscles, stimulating their contraction, and the hand is withdrawn. The reflex arc just described is a simple response. The stimulus of burning the hand also leads to far more complex reactions. For example, within the spinal cord other branches of the same sensory neuron that forms part of the reflex arc, synapse with neurons that relay impulses to the cerebrum which results in the conscious awareness of pain. You may then decide to treat the burn, a decision brought about by neural activity within the cerebrum, which then controls the

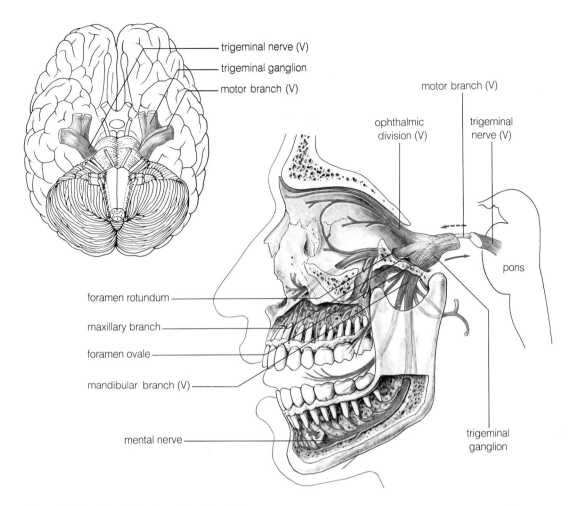

Figure 13-12   The distribution of the trigeminal nerve

additional procedures. However, the reflex action of withdrawing the hand did not have to wait for information to travel to the brain and back. The spinal cord serves as a reflex center for responses to changes in the environment and also provides pathways for conduction of impulses to and from the brain. But reflexes are not restricted to the spinal cord; they can occur at any level that receives afferent impulses from the periphery and sends efferent impulses to effector organs. Thus reflexes can occur at brainstem levels over cranial nerves; in this case the brainstem serves as the reflex center.

The essential components of the reflex mechanism, or arc, are a *receptor* to respond to environmental stimuli; an *afferent neuron* (sensory neuron) to conduct impulses from the periphery to the CNS (afferent impulses); a *central connection* (syn-

aptic) within the CNS; an *efferent neuron* (motor neuron) to conduct impulses from the CNS to the periphery (efferent impulses); and an *effector,* muscular or glandular, to react to impulses from the CNS. The receptor may be a single nerve ending that responds, for example, to pressure or a complex organ such as the eye. The central connection usually involves one or more association neurons but may consist only of a synapse between the afferent and efferent neurons.

The two most common basic reflexes are the "withdrawal," or "flexion," reflex (already described) and the "stretch" reflex, which most commonly involves extensors. The stretch (myotactic tendon jerk) reflex is the simpler, so we will use it for a more detailed description; all the components of the reflex mechanisms are illustrated in Fig. 13-14.

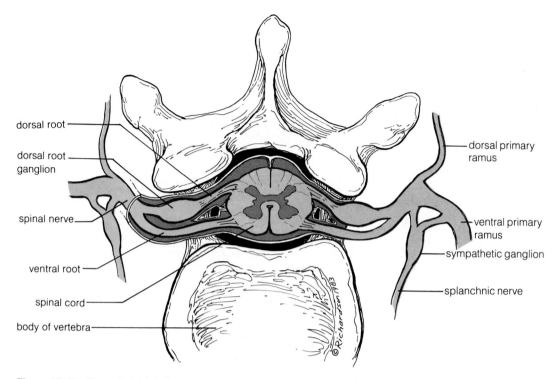

Figure 13-13 The spinal cord. Spinal nerves branch from the spinal cord, as shown

Labels for Figure 13-13:
- dorsal root
- dorsal root ganglion
- spinal nerve
- ventral root
- spinal cord
- body of vertebra
- dorsal primary ramus
- ventral primary ramus
- sympathetic ganglion
- splanchnic nerve

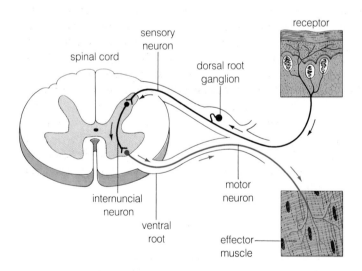

Figure 13-14 The flexor reflex.

Labels for Figure 13-14:
- spinal cord
- sensory neuron
- dorsal root ganglion
- receptor
- internuncial neuron
- motor neuron
- ventral root
- effector muscle

The skeletal muscles of the trunk and limbs are innervated by efferent neurons whose cell bodies are in the gray matter of the ventral horns of the spinal cord. When these cells, called "lower motor neurons," are stimulated, they cause the muscle fibers they supply to contract. If these cells or their axons are destroyed (by disease or trauma), the muscle fibers they innervate cannot contract, that is, they are paralyzed. When the muscle loses its innervation and its background nerve impulse input, it eventually loses its tone and becomes *flaccid*. In time there is atrophy, a reduction in contractile protein with a corresponding decrease in muscle fiber size.

Perhaps the best known example of the stretch reflex is the *knee jerk* (Fig. 13-15); a similar response can be elicited from several tendons including the achilles, biceps, triceps, and the masseter-temporalis complex. The stimulus for initiating the reflex arc is stretching of the *neuromuscular spindle,* an encapsulated sensory receptor located within small intrafusal muscle

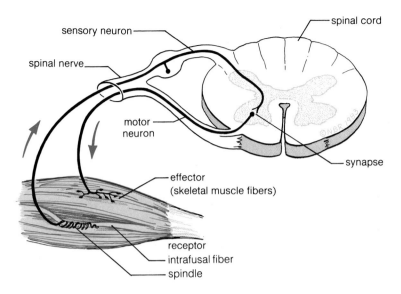

Figure 13-15 The stretch reflex (knee-jerk reflex). Note that there are no interneurons involved in this reflex.

fibers. In the case of the knee jerk, tapping the patellar tendon stretches the neuromuscular spindles in the quadriceps femoris muscle, causing impulses to travel from the muscle into the spinal cord where the sensory neuron synapses with the motor neurons supplying the quadriceps femoris muscle. These motor impulses cause the quadriceps to contract, resulting in a brisk kick. Note that the stretch reflex is a two-neuron reflex, in which the central connection is only a synapse. The flexion reflex is a three-neuron reflex, in which the central connection includes an association neuron and two synapses.

The nervous system works as a unit. The CNS receives information from sensory components of the PNS, uses this information to determine a proper response, and then obtains the proper response by having the motor fibers of the PNS stimulate the appropriate effectors. The determination of the response may be very simple as in the case of the reflex arc or highly complex as in the case of the conscious processes of the brain.

THE BRAIN VENTRICLES

The CNS begins in the embryo as a hollow tube that becomes filled with fluid, termed the **cerebrospinal fluid** (CSF). As the brain expands due to growth of cells adjacent to the cavity, the portion of the cavity within the brain forms compartments called **ventricles** that also expand, influencing the final form of the brain (Fig. 13-16). The cells lining the ventricles differentiate into a type of neuroglia called *ependyma,* which in some areas becomes thin and is invaginated into the ventricle by the growth of tufts of blood capillaries. This capillary network, covered by thin ependyma, constitutes the *choroid plexus,* which produces much of the CSF. Thus CSF is constantly elaborated into the ventricles and must exit at the same rate or pressure increases rapidly. The CSF passes from the ventricular system into the **subarachnoid space** through three openings in the roof of the *fourth ventricle:* the two *lateral apertures* (foramina of Luschka), located laterally, and a median aperture (foramen of Magendie). The characteristics of the ventricles are summarized in Table 13-2.

The outline of the ventricles and thus of adjacent brain tissue can be seen in radiographs taken after injecting contrast material into the ventricles. Air is a contrast material that is frequently used because it is harmless and has no density on x-ray films. A radiograph taken with this procedure is termed a *ventriculogram.*

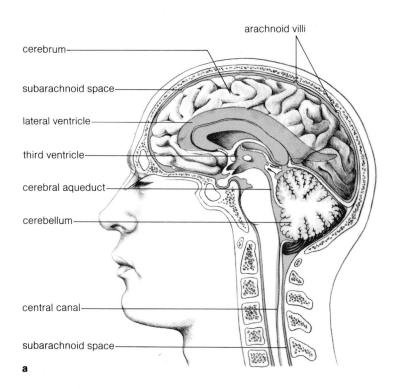

cerebrum

subarachnoid space

lateral ventricle

third ventricle

cerebral aqueduct

cerebellum

central canal

subarachnoid space

arachnoid villi

a

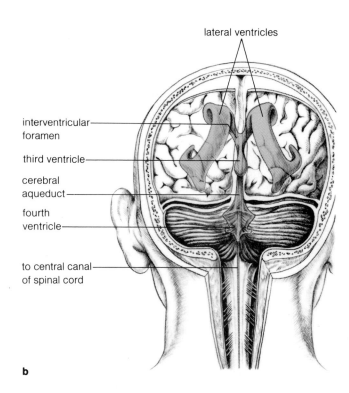

lateral ventricles

interventricular foramen

third ventricle

cerebral aqueduct

fourth ventricle

to central canal of spinal cord

b

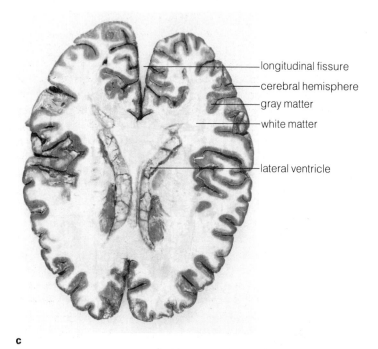

longitudinal fissure

cerebral hemisphere

gray matter

white matter

lateral ventricle

c

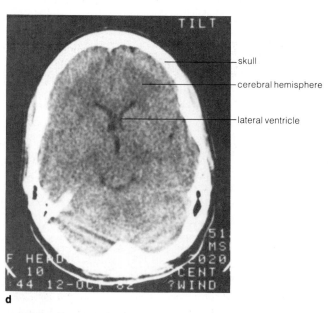

skull

cerebral hemisphere

lateral ventricle

d

Figure 13-16 The ventricles of the brain. (**a**) Lateral view. (**b**) Dorsal view, presented as if the brain were transparent. (**c**) Photograph of a thick slice of a human brain (horizontal section); gray matter appears dark; white matter is light (photograph by R. Ross). (**d**) CAT scan of the brain; horizontal view of the level of the lateral ventricles (courtesy of S. J. Goldstein).

Table 13-2 Brain Ventricles	
Lateral (I and II)	Located in the cerebral hemispheres. Body in the frontal and parietal lobes, posterior horn in the occipital lobe, inferior horn in temporal lobe.
Third	Located in the diencephalon. Connects with lateral ventricles and with cerebral aqueduct.
Fourth	Located in the pons and medulla. Connects with cerebral aqueduct and neurocoel of lower medulla. Opens into the subarachnoid space by means of two lateral apertures and a median aperture.

THE MENINGES

The CNS is protected not only by bone but also by three layers of connective tissue. The *dura mater,* or dura, is the tough, fibrous outermost layer. Within the vertebral canal (Fig. 3-17), dura mater is separated from the periosteum of the vertebrae by an *epidural space* filled with fat and a dense network of small veins; within the cranium it fuses with the periosteum of the bones of the skull, leaving no epidural space. In all regions the dura mater is in direct contact on its inner surface with a thin sheet of connective tissue termed the *arachnoid membrane* (arachnoid meaning spiderlike). The arachnoid has many thin projections (trabeculae) that connect with the *pia mater,* a thin transparent sheet of connective tissue that is applied directly to the surface of the brain and spinal cord and follows the irregularities of the brain's surface. The space between the pia mater and arachnoid is the *subarachnoid space,* which is filled with CSF.

Most of the CSF is formed in the brain ventricles and enters the subarachnoid space from the fourth ventricle. The fluid flows slowly through the subarachnoid space with the current created by the absorption of the CSF into a special venous channel called a **dural sinus** on the superior aspect of the cerebrum.

The dura mater is attached tightly to the bone around the openings in the cribriform plate through which olfactory nerves pass. A blow to the face that breaks the cribriform plate may tear the dura mater, allowing the cerebrospinal fluid to leak out of the subarachnoid space into the nasal cavity. This increases the danger of an infection in the meninges, a condition termed meningitis.

The subarachnoid space extends caudal to the lower end of the spinal cord, providing an area from which CSF may be withdrawn without appreciable risk. The spinal cord ends inferiorly at about the level of the *first lumbar vertebra.* The dura mater and the arachnoid end in the *sacrum;* thus the subarachnoid space can be entered by inserting a needle between vertebrae L4 and L5, for example, with no danger of injury to the spinal cord. CSF is often withdrawn for laboratory procedures useful in the diagnosis of disease.

EARLY DEVELOPMENT OF THE NERVOUS SYSTEM

The nervous system is an *ectodermal* derivative, originating in the early embryo as a thickening (called the neural plate) of the superficial ectoderm overlying the notochord (Figs. 13-18, 13-19a). The *neural plate* rolls into a tube, the **neural tube,** and loses its connection with the superficial ectoderm (Figs. 13-19b, c, d). Some of the cells of the neural plate remain outside the neural tube, forming a column of neural cells, the **neural crest,** in a position dorsal and lateral to the neural tube on each side (Fig. 13-19c). The neural crest cells give rise to the *dorsal root* (spinal) *ganglia* and *autonomic ganglia,* as well as to *neurilemma cells* and to *pigment cells* of the skin.

The cranial portion of the neural tube forms three dilations, the *primary brain vesicles* (Fig. 13-20), called the **prosencephalon, mesencephalon,** and **rhombencephalon.** With further growth of the embryo, the prosencephalon forms two secondary brain vesicles, the **telencephalon,** which undergoes tremendous growth and gives rise to the *cerebral hemispheres* (Table 13-3), and the **diencephalon,** whose principal derivatives are the *dorsal thalamus* and *hypothalamus.* The **mesencephalon** undergoes the least change, becoming the *midbrain.* The rhombencephalon forms two secondary vesicles, the **metencephalon,** from which the *pons* and *cerebellum* arise, and the **myelencephalon,** which becomes the *medulla.*

The final form of the human brain is achieved by three flexures (curvatures) that result in the orientation of the cerebral hemispheres almost at right angle to the brainstem. Curvatures at the level of the *mesencephalon* and the *upper cervical neural tube* bend forward. The *pontine flexure,* in contrast, is in the opposite direction, bending posteriorly and thus obliterating the cervical curvature. The *mesencephalic flexure* is the only curvature remaining in the adult.

During the growth of the CNS most of the cell division occurs adjacent to the neurocoel (the cavity of the nerve tube), with the cells being pushed outward as they begin differentiat-

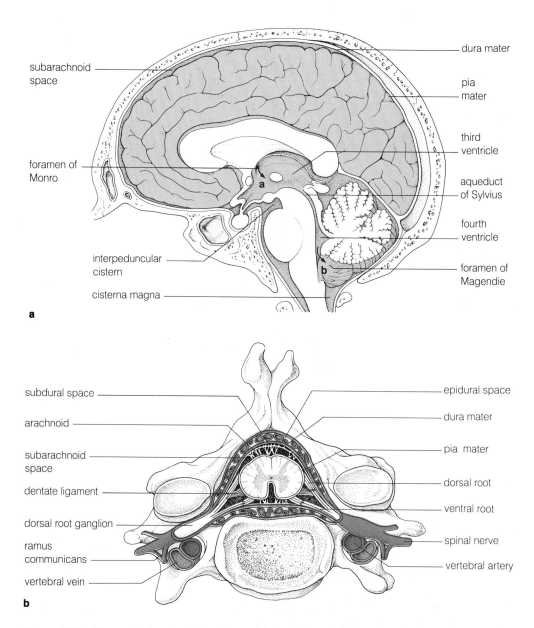

Figure 13-17 The cerebral and spinal meninges. (**a**) Sagittal section of cranium. (**b**) Transverse section of spine.

ing into neurons. In the development of the cerebral hemispheres many of the proliferating cells leave the deep area adjacent to the neurocoel and migrate to the surface. Here they continue multiplying, giving rise to the *cerebral cortex*. As the neurons differentiate, the processes lengthen rapidly as they extend from the cell bodies and pass through the gray matter, breaking it up into nuclei.

RECEPTORS

The regulation of the activities of the body by the nervous system depends on the central nervous system receiving input from stimuli initially received by receptors. The CNS then responds appropriately by sending impulses into the periphery to stimulate effectors.

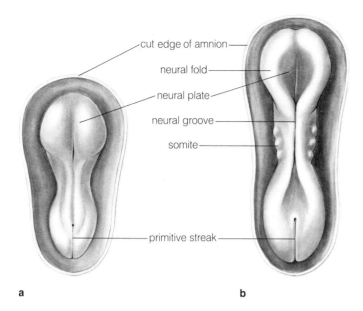

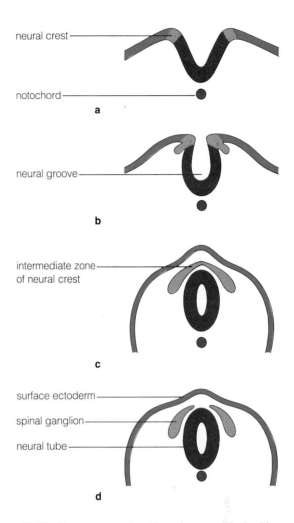

Figure 13-18 The origin of the central nervous system. (**a**) Dorsal view of an 18-day embryo illustrating the early neural plate. (**b**) Dorsal view of a 20-day embryo showing the neural groove and neural folds with associated somites.

Figure 13-19 Transverse sections through progressively older embryos, showing development of the neural tube. (**a**) Neural plate stage. (**b**) Neural groove stage. (**c**) Neural tube with unsegmented neural crest. (**d**) Neural tube with segmented neural crest (spinal ganglia).

Types of Receptors

Because receptors turn various forms of energy into nerve impulses, they are essentially *transducers* that respond readily to only one particular form of energy or stimulus. Receptors may be classified according to their location as *exteroceptors, interoceptors,* or *proprioceptors.*

Exteroceptors   Exteroceptors (Fig. 13-21) respond to an energy source external to the body such as light energy, sound energy, chemical energy (smell and taste), mechanical energy (touch and pressure), and thermal energy (heat and cold). There are two types: nonencapsulated receptors, which are free nerve endings (Fig. 13-18a), and encapsulated receptors. Examples of encapsulated receptors are *Meissner's corpuscles* (Fig. 18-21d), which are discriminatory (two-point or fine) tactile receptors located in dermal papillae projecting into the epidermis, and the *Pacinian corpuscles* (Fig. 13-21h), which are widely distributed in subcutaneous tissues, near joint capsules, and in the viscera and mesenteries. Pacinian corpuscles are thought to be mechanoreceptors sensitive to light pressures. Those in subcutaneous tissue are sensitive to pressure on the skin and may also serve as receptors for vibratory sensation. Those lying in the deep tissues (joint capsules, viscera, mesenteries) may require

a greater stimulus. These receptors consist of nerve fibers surrounded by a capsule of connective tissue. The morphology and functions of a group of receptors known as *end-bulbs* (Krause's, Ruffini's, Figs. 18-21c, b, g) are poorly understood, but they appear to be involved in the reception of several modalities of sensation (touch, pressure, and temperature). In general the end-bulbs consist of highly branched nerve endings surrounded by a thin connective tissue capsule.

Interoceptors   Interoceptors respond to forms of energy that act upon viscera. These receptors may be encapsulated nerve endings such as the Pacinian corpuscles or relatively undiffer-

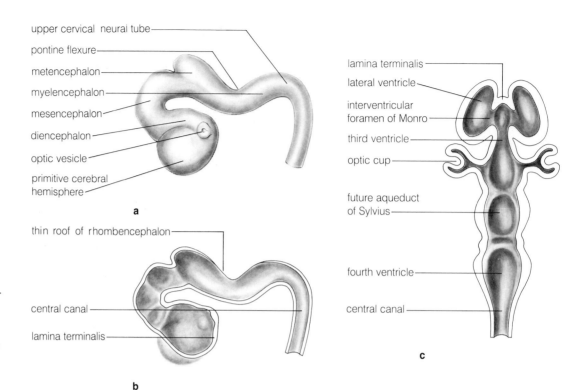

Figure 13-20 Development of the brain vesicles. (**a**) Brain vesicles of human embryo at beginning of sixth week (lateral view). (**b**) Brain vesicles and spinal cord of human embryo at beginning of sixth week (midline section). (**c**) Lumina of spinal cord and brain vesicles.

a

upper cervical neural tube
pontine flexure
metencephalon
myelencephalon
mesencephalon
diencephalon
optic vesicle
primitive cerebral hemisphere

b

thin roof of rhombencephalon
central canal
lamina terminalis

c

lamina terminalis
lateral ventricle
interventricular foramen of Monro
third ventricle
optic cup
future aqueduct of Sylvius
fourth ventricle
central canal

Table 13-3	Derivatives of the Embryonic Brain Vesicles
Brain Vesicle	Derivative
Prosencephalon	
Telencephalon	Cerebrum
Diencephalon	Thalamus, hypothalamus
Mesencephalon	Midbrain
Rhombencephalon	
Metencephalon	Cerebellum, pons
Myelencephalon	Medulla

entiated, highly branched, nonencapsulated nerve endings that resemble the free sensory terminations found in the epidermis. The nonencapsulated interoceptors are usually located in the adventitia connective tissue layer about blood vessels, and sensations from these receptors, interpreted by the brain as pain or pressure, are poorly localized.

Proprioceptors Proprioceptors respond to variations in mechanical energy in muscles, tendons, and joints. Pacinian corpuscles and other receptors located in and near joint capsules are responsible for the conscious sensation of position and movement. The receptors that mediate the motor reflex responses so essential to the maintenance of posture and muscular activity are encapsulated nerve endings called spindles in muscles or tendons (Fig. 13-22). The *neuromuscular spindles* vary in composition but essentially consist of from three to twenty intrafusal muscle fibers enclosed in a connective tissue capsule and supplied by two types of afferent fibers as well as by motor fibers. The neurotendinous spindles are simple, branched nerve endings enclosed in a thin connective tissue capsule of collagenous fibers.

The most obvious manifestation of neuromuscular spindle function is the stretch reflex. Slight stretching of the intrafusal muscle fibers of the spindle stimulates the *nerve endings* to transmit impulses directly to the cell bodies of *alpha motor neurons* in the spinal cord, which innervate the main mass of muscle in which the spindle is located; the impulse in the alpha motor neuron causes the muscle to contract. Thus by means of a two-neuron reflex arc, stretching causes muscle contraction.

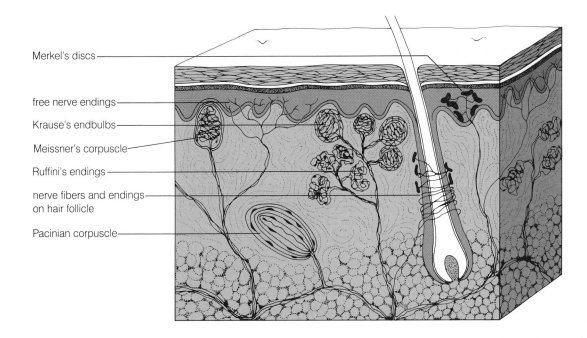

Merkel's discs

free nerve endings

Krause's endbulbs

Meissner's corpuscle

Ruffini's endings

nerve fibers and endings on hair follicle

Pacinian corpuscle

Figure 13-21 Composite diagram illustrating the receptors associated with human skin.

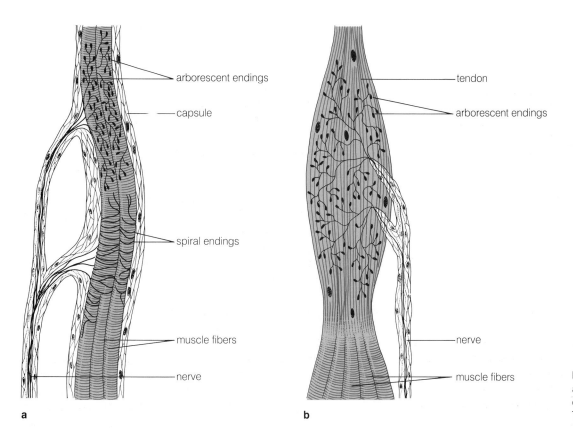

arborescent endings

capsule

spiral endings

muscle fibers

nerve

a

tendon

arborescent endings

nerve

muscle fibers

b

Figure 13-22 Sensory endings associated with muscles and tendons. (**a**) Muscle spindle. (**b**) Tendon spindle.

Clinical Applications: **Pain, CNS Lesions, and Loss of Motor Functions**

Motor and sensory functions of the nervous systems are closely interrelated, but some pathological states may predominantly affect one or the other. Complete destruction of a peripheral nerve results in loss of all sensation within the area supplied by the nerve. All modalities of sensation (pain, temperature, touch, and vibration and position sense) are completely eliminated.

Frequently, nerves are only partially damaged by injury or disease processes, resulting in distorted sensory perception (either *hypesthesia,* a decrease in sensation, or *hyperesthesia,* an increase in sensitivity). The most frustrating distortions to the patient are those that result in pain. A severe itch. for example, may lead to continued scratching until the onset of pain.

Types of Pain

The terms that are used to describe pain may be general, applying to pain from any cause, or they may pertain to pain with a neurological origin.

Local. This is somatic-type pain that occurs at the site of the problem if sensory receptors are present. The pain may be sharp, as in a pin-prick, or a dull ache, as occurs in the lower back.

Radicular. Radicular pain follows the distribution of a single spinal nerve. It is usually sharp and becomes worse with movement.

Radiating. Pain that arises in a local area and spreads to other areas is described as radiating. It may be sharp or dull.

Referred. Pain from an injury at one site may be referred to another site, usually from a visceral area to a somatic area. For example, pain that arises in the heart during a heart attack is frequently referred to the lower neck or arm, usually on the left side.

Paresthesia. This is the term used to describe abnormal sensations that occur spontaneously, such as tingling, burning, or crawling sensations that cannot be precisely located.

Dyesthesia. This is a general term referring to painful sensations produced by stimuli, such as clothes rubbing against the skin, that normally would not be painful.

Causalgia. Causalgia occurs when a peripheral nerve is damaged, causing a persistent and diffuse burning sensation. Even a slight stimulus will cause extreme pain, and the affected person becomes preoccupied with protecting the part of the body that is involved.

Phantom pain. Persons who have had a limb amputated may have sensations as if the limb was still present. Usually the sensations are of movement or itch; severe pain is rare.

Central Nervous System Lesions

A lesion of the spinal cord may affect only a portion of the cord and cause loss of sensation only in specific areas. In the *Brown-Sequard syndrome,* a lateral half of the cord is destroyed. The result is loss of pain and temperature sensation on the opposite side of the body below the level of the lesion, and loss of position sense and fine touch on the same side of the body below the level of the lesion. Some diseases, such as *tertiary syphilis,* may selectively destroy the posterior white columns of the spinal cord, resulting in the loss of position and vibratory sense.

A condition of the brainstem that results in severe pain is *tic douloureux,* a dysfunction of the sensory portion of the trigeminal nerve. Any stimulus applied to the face or mouth, such as swallowing food, may trigger pain.

Lesions of the parietal lobe may produce sensory changes. The person may have difficulty identifying objects by touch alone and may lose two-point discrimination. There also may be loss of the ability to localize any or all of the sensory modalities.

Motor Functions

Since skeletal muscles cannot contract unless stimulated by nerve impulses, any interruption of the motor innervation of a muscle or group of muscles will alter muscle activity. Proper innervation of skeletal muscles depends upon two neurons: a *lower motor* neuron whose cell body is within the spinal cord or brainstem and whose axon terminates upon skeletal muscle fibers, and an *upper motor* neuron whose cell body may be in the gray matter of the cerebral cortex and whose axons synapse on lower motor neurons. Destruction of these two neurons differs in the effects upon muscular activity.

Lower motor neuron paralysis. Lower motor neurons may be destroyed by cutting or injuring their axons in a peripheral nerve, or by directly affecting the cell bodies within the CNS. The *poliomyelitis* virus, for example, selectively attacks the ventral horn cells of the spinal cord, destroying the lower motor neurons. In both types of injury, the muscles are deprived of innervation and no longer contract. They become soft and flabby and begin to atrophy—the characteristics of a flaccid paralysis. The motor component of the reflex arc no longer exists, so the muscle can no longer respond reflexly to sensory stimuli; it is impossible for the muscle to contract.

Upper motor neuron paralysis. Damage to the cell bodies in the motor area of the cerebral cortex or their descending axons results in upper motor neuron paralysis. The most common site of injury is the cerebral hemisphere, before the fibers decussate. Injury usually results from a cerebrovascular accident (*stroke*) that deprives the neurons of oxygen supply so that the cells die. If the injury occurs rostral to the motor decussation, the muscles of the opposite side of the body will be affected. If the injury is caudal to the decussation (by a knife wound, perhaps) in one half of the spinal cord, then the muscles on the same side of the body will be affected. The muscles are paralyzed, but the paralysis is different from that of the lower motor neurons. Since the lower motor neurons are not affected, the reflex arc remains intact, and the muscles can respond to sensory stimuli. The inhibitory fibers originating in the premotor areas of the cerebral cortex are destroyed, and their braking effect on the lower motor neurons is eliminated. Without this braking effect, the lower motor neurons tend to fire excessively and spontaneously, causing the muscles to contract strongly, a condition known as *spasticity.* Thus, upper motor neuron paralysis is a spastic paralysis, in contrast to the flaccid lower motor neuron paralysis. A characteristic feature of upper motor neuron paralysis is the *Babinski reflex.* When the sole of the foot of a normal person is stroked from the heel toward the toes, the toes will curl. The same stroke in a person who has an upper motor neuron lesion will cause the great toe to reflex dorsally and the other toes to spread apart. The exact mechanism of this reflex is not understood.

Paralysis of varying degrees may occur as the result of a spinal cord injury: *monoplegia* is paralysis of a single upper or lower limb; *hemiplegia* is a paralysis of an upper and a lower limb on the same side; *paraplegia* is the condition on which both lower limbs are paralyzed; *quadriplegia* is paralysis of all four limbs.

The stretch reflex is responsible for the muscle tone or *tonus* that is important in maintaining posture. Muscle tone is a state of partial contraction of a muscle, a state of readiness. Neuromuscle spindles contribute to muscle function in several ways; the stretch reflex is their simplest role.

Neurotendinous spindles are also stimulated by tension although greater force is required for their excitation than is required by the neuromuscular spindles. Afferent impulses from the neurotendinous spindles inhibit the alpha motor neurons, causing the muscles to relax and thus protecting it from overloading.

Modality Specificity of Receptors

At physiological intensities of stimulation, receptors respond to only one form, or modality, of physical energy; therefore they are *modality specific.* Receptors respond to the application of the appropriate form of energy in one of two ways: a sensory cell (for example, a hair cell in the internal ear) may secrete a *chemical transmitter substance* onto its associated nerve terminal or the nerve ending itself may be the receptor, generating an impulse in response to the stimulus. The result in either case is the production of an impulse in the primary afferent nerve, which conveys it to the CNS.

General Properties of Receptors

Receptors have three general properties that affect their functions.

Threshold   Each type of receptor has a stimulation threshold, that is, it requires a certain minimum amount of energy to initiate an impulse in a primary afferent neuron. For example, if the skin temperature is lowered 1.0°C for three seconds, one experiences a sensation of cold due to stimulation of cold receptors.

Frequency Coding   Frequency coding is the variation in frequency of afferent nerve impulses that allows one to perceive variations in the intensity of stimuli. At physiological intensities there is a direct relationship between the intensity of the stimulus and the frequency of the impulses transmitted by the associated nerve. For example, the greater the pressure applied to a pressure receptor, the higher the frequency of the impulses. This correlation enables one to interpret the signal as a specific amount of pressure.

Adaptation   If an adequate stimulus is maintained, some receptors such as pain receptors continue to generate nerve

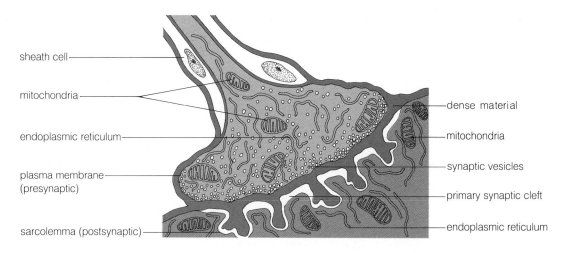

Figure 13-23 Diagram of the myoneural junction, as shown by an electron micrograph (magnification approximately × 15,000).

impulses more or less indefinitely. These receptors are said to be *slowly adapting,* or *tonic.* Other receptors generate impulses when the stimulus is initiated but stop quickly even if the stimulus is continued. These are said to be *rapidly adapting* or *phasic.* The secondary receptor of the muscle spindle is an example of a slowly adapting receptor; the Pacinian corpuscle is rapidly adapting.

EFFECTORS

The effectors that respond to nerve stimulation are *muscles* and *glands.* All three types of muscle (skeletal, smooth, and cardiac) are supplied with nerves.

Muscles

Skeletal The skeletal muscles of the body require neural stimulation for excitation to occur. Nerve fibers terminate on skeletal muscle fibers in highly branched arrangements termed *motor end plates* (Fig. 13-23). The terminations of the axon, lying in close contact with the sarcolemma of the muscle fiber, contain *acetylcholine,* a transmitter substance. The nerve impulse brings about the release of the acetylcholine, which crosses the gap between the motor end plate and the sarcolemma, causing an increase in the permeability of the sarcolemma to Na$^+$. An action potential thus originates in the same manner as in ner-

vous tissue and triggers contraction of the muscle fiber. A single alpha nerve fiber supplying large (extrafusal) muscle fibers may innervate from one to several hundred muscle fibers. One alpha nerve cell together with all the skeletal muscle fibers it supplies constitute a *motor unit,* which functions as a single entity, all the muscle fibers contracting more or less simultaneously. The muscle fibers contract maximally, or they do not contract at all. A muscle consists of many motor units, but not all of the units are brought into play each time the muscle goes to work; nor do all the units contract simultaneously. Muscular skill (fineness of control) is related to the size of the motor units. If a single nerve fiber controls a single muscle fiber, much finer control of movements is possible than if a single nerve controls one hundred muscle fibers.

Smooth Muscle Smooth muscle has an inherent ability to contract, and its nerves, which are autonomic, serve only to regulate the rate of contraction. Unlike skeletal muscle, which is completely paralyzed and atrophies after its nerve supply is severed, smooth muscle is still able to contract without innervation. Many smooth muscles have a double nerve supply, receiving one set of fibers from nerves that release *acetylcholine* (parasympathetic nerves) and another set from nerves that release *norepinephrine* (sympathetic nerves). The activity of these two divisions of the ANS oppose each other, one being *excitatory* and the other *inhibitory.* There is no general agreement about the way in which autonomic nerves terminate on smooth muscle cells. Some authorities believe that knotlike swellings in the muscles are synapses formed by individual nerve fibers. Others believe that the autonomic fibers form plexuses (networks of interlaced fibers) in the muscle, and that the fibers do not form synaptic junctions with the muscle cells but release the chemical transmitter substance into the extracellular space.

Cardiac Muscle Cardiac muscle also has the intrinsic capacity to contract without nerve stimulation. Its rate of contraction is regulated by autonomic fibers that terminate, in part, upon specialized conductile tissue of the heart known as the pacemaker. As with smooth muscle there are divergent opinions regarding the nature and distribution of nerve terminations in the heart. Because there are three sets of nerve fibers (sensory, parasympathetic, and sympathetic) innervating cardiac muscle, it is difficult to determine the relations of the terminations of each type of nerve fiber to the cardiac muscle fibers. It is generally considered that sensory fibers terminate principally in the *endocardium* (internal lining of the heart) and around the blood vessels supplying the heart. In addition to terminating upon the pacemaker, some *parasympathetic* and *sympathetic* fibers supply muscle fibers directly.

Glands

Most of the *exocrine glands* increase their rates of secretion as a result of *neural stimulation* by either *parasympathetic* or *sympathetic fibers*. The stimulation is by a neurotransmitter substance released by fibers whose endings form synaptic connections with the glandular cells. If glands receive a double innervation, the quality as well as the quantity of secretion is altered, according to whether the nervous impulses originate in the sympathetic or parasympathetic divisions of the autonomic nervous system. Some glandular cells (and also some smooth muscle cells) respond to *hormonal* as well as *neural stimulation*.

WORDS IN REVIEW

Axon	Metencephalon
Brainstem	Myelencephalon
Cerebellum	Neural crest
Cerebrospinal fluid	Neural tube
Cerebrum	Neurilemma
Choroid plexus	Neuroglia
Dendrite	Nucleus of CNS
Diencephalon	Prosencephalon
Dural sinus	Receptor
Effector	Rhombencephalon
Ganglion	Subarachnoid space
Meninges	Telencephalon
Mesencephalon	Ventricle

SUMMARY OUTLINE

I. General Functions of the Nervous System
   A. The nervous system receives stimuli from the environment (internal and external).
   B. The nervous system, together with the endocrine system, stimulates cells, tissues, and organs of the body to respond to changes in the environment.
   C. The basic plan of the nervous system involves a sensory (receptor) cell that receives stimuli from the periphery and transmits the information to association cells that then stimulate motor cells to effect a response.

II. Nervous Tissue
   A. The structural unit of nervous tissue is the neuron, a cell composed of a cell body and all of its processes.
   B. The neurons may be classified as unipolar with a single process, bipolar with one axon and one dendrite, and multipolar with one axon and many dendrites.
   C. The supporting cellular component of nervous tissue includes the neuroglia of the CNS and the neurilemma cells of the PNS. The neuroglia consists of astrocytes which form much of the supporting framework for nervous tissue, the oligodendrocytes which produce myelin, microglia which are phagocytic, and ependyma that lines the ventricles. The neurilemma cells of the PNS produce myelin and form a protective covering for peripheral nerve fibers.

III. Relations of the CNS and PNS.
   A. The CNS consists of the brain (cerebrum, thalamus, hypothalamus, midbrain, pons, medulla, and the cerebellum), and the spinal cord.
   B. The PNS is divided into the somatic and autonomic systems. The somatic system consists of twelve pairs of cranial nerves that arise from the brain and brainstem and thirty-one pairs of spinal nerves that arise from the spinal cord. The autonomic system consists of a parasympathetic division and a sympathetic division.

IV. The Functional Unit

    A. The reflex arc is the functional unit of the nervous system.

    B. The reflex arc consists of a sensory neuron that receives the stimulus in the periphery and usually transmits it to an association neuron within the spinal cord or brainstem, which relays the stimulus to a motor neuron that conveys the impulse back into the periphery to effect a response. Some reflexes (stretch) do not have an association neuron.

V. The Brain Ventricles

    A. The ventricles are cavities within the CNS filled with cerebrospinal fluid.

    B. The lateral ventricles of the cerebral hemispheres open into the third ventricle of the diencephalon. The third ventricle opens into the cerebral aqueduct (of Sylvius) of the midbrain, which leads into the fourth ventricle of the pons and medulla. CSF leaves the ventricular system via three openings in the roof of the fourth ventricle and enters the subarachnoid space.

VI. The Meninges

    A. The dura mater is the tough, fibrous outermost layer of the meninges.

    B. The arachnoid is a thin sheet in contact with the dura mater and with thin projections that connect with the pia mater.

    C. The pia mater is a thin, transparent sheet of connective tissue directly applied to the external surface of the CNS.

VII. Early Development of the Nervous System

    A. The CNS develops from an ectodermal plate overlying the notochord. The plate rolls into a longitudinally oriented tube.

    B. The cranial portion of the neural tube forms three dilations (prosencephalon, mesencephalon, and rhombencephalon) that later form five secondary brain vesicles (telencephalon, diencephalon, mesencephalon, metencephalon, and myelencephalon).

    C. The telencephalon becomes the cerebrum; the diencephalon gives rise to the thalamus and hypothalamus; the mesencephalon becomes the midbrain; the metencephalon gives rise to the pons and cerebellum; and the myelencephalon differentiates into the medulla.

REVIEW QUESTIONS

1. What are the parts of the central nervous system that are located within the cranial cavity?

2. What are the essential components of a neuron?

3. What is responsible for the gray color of gray matter and for the white color of white matter?

4. Where are bipolar neurons found in the human?

5. List and give the principal characteristics of the types of supporting cells in the CNS.

6. Name the supporting cell found in peripheral nerves and describe its relation to the nerve fiber.

7. What major component of the CNS is located dorsal to the pons?

8. What are the major structural and functional differences between the cranial and the spinal nerves?

9. What are the essential components of a reflex arc?

10. What is the stimulus for initiating the stretch reflex?

11. In what part of the CNS is the third ventricle located?

12. In what space is the cerebrospinal fluid found external to the CNS?

13. What structures are derived from the telencephalon of the embryo?

14. From what brain vesicle is the cerebellum derived in the embryo?

THE SPINAL CORD
AND SPINAL NERVES

OBJECTIVES

After completing this chapter you should be able to:

1. Describe the spinal cord and locate it within the vertebral canal.

2. Demonstrate an understanding of the way spinal nerves are named.

3. Relate the roots of a spinal nerve to the spinal cord and to the spinal nerve.

4. List the functional components of a spinal nerve and relate each component to the types of peripheral structures innervated.

5. Describe the distribution of a "typical" spinal nerve.

6. Define a nerve plexus. For each of the following plexuses describe the anatomical pattern of exchange of nerve fibers, name the major terminal branches, and give the general distribution of each branch: (1) cervical, (2) brachial, (3) lumbosacral.

7. Describe the development of nerve–muscle and nerve–dermatome relations.

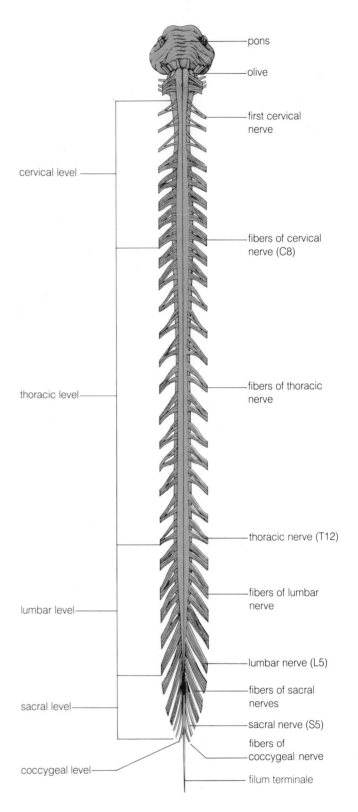

Figure 14-1 Ventral view of the spinal cord and spinal nerve rootlets

The peripheral nervous system consists of *twelve pairs* of **cranial nerves** and *thirty-one pairs* of **spinal nerves.** Each cranial nerve is unique in composition and in function, but the spinal nerves are serially repeated entities that are similar in composition and supply similar structures.

THE SPINAL CORD AND ROOTS OF THE SPINAL NERVES

The spinal cord (Fig. 14-1) is a more or less cylindrical structure located within the vertebral canal. Also within the vertebral canal are the roots of the spinal nerves, which consist of bundles of nerve fibers called rootlets. When the brain and spinal cord are exposed for study, we use the upper rootlet of the first cervical nerve to determine the point of their junction. With the brain inside the skull the lower border of the foramen magnum is considered the boundary between the brain (medulla) and the spinal cord. The spinal cord extends from the foramen magnum to about the level of the intervertebral disc between the first and second lumbar vertebrae.

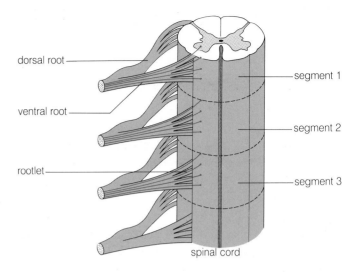

Figure 14-3 Diagram of a single spinal cord segment illustrating the composition of a spinal nerve

Labels in figure 14-3: dorsal root, ventral root, rootlet, spinal cord, segment 1, segment 2, segment 3

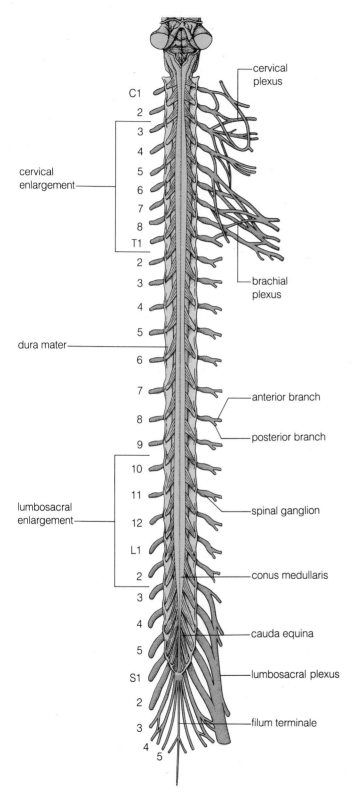

Figure 14-2 Dorsal view of the spinal cord showing the enlargements of the spinal cord

Labels in figure 14-2: cervical plexus, cervical enlargement, C1, 2, 3, 4, 5, 6, 7, 8, T1, 2, 3, 4, 5, 6, 7, 8, 9, 10, 11, 12, L1, 2, 3, 4, 5, S1, 2, 3, 4, 5, brachial plexus, dura mater, anterior branch, posterior branch, lumbosacral enlargement, spinal ganglion, conus medullaris, cauda equina, lumbosacral plexus, filum terminale

Two enlargements of the spinal cord (Fig. 14-2) are associated with the nerve supply to the upper and lower extremities. The *cervical enlargement* gives rise to nerves going to the upper extremity, and the *lumbrosacral enlargement* supplies the lower extremity. In addition to giving rise to the spinal nerves, the spinal cord serves as a *pathway* for relaying impulses to higher or lower levels of the CNS and as a central *coordinating* mechanism for spinal reflexes.

The thirty-one pairs of spinal nerves originate from individual segments of the spinal cord. A spinal cord segment (Fig. 14-3) is a portion of the cord that sends fibers into and receives fibers from a single pair of spinal nerves; each segment is associated with a single vertebra. Each spinal nerve arises from a dorsal root and a ventral root composed of rootlets emerging from the dorsolateral and vertrolateral surfaces of the cord, respectively (Fig. 14-4). The two roots unite at the intervertebral foramen where they leave the vertebral canal as a single spinal nerve enclosed in a connective tissue sheath, the **epineurium.** The spinal nerve thus appears as a cordlike structure as it passes into the body wall.

Because the spinal cord is considerably shorter than the vertebral column in the adult (Fig. 14-5), the **spinal nerve roots** arising from the lower spinal cord segments must travel a considerable distance vertically to reach the intervertebral foramina at which they exit. The upper cervical nerve roots extend almost straight laterally to reach their points of exit; the thoracic and lumbar nerves slant downward more and more;

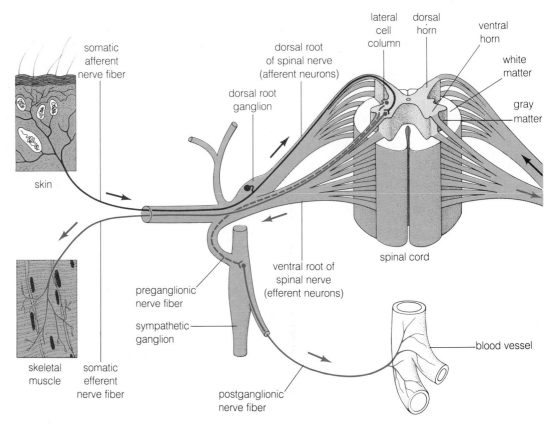

Figure 14-4 Diagram of the functional components of a spinal nerve

the sacral nerve roots must travel within the vertebral canal from the lower portion of the spinal cord (that is, from the upper lumbar level of the vertebral column) to the sacrum before the dorsal and ventral roots join to form the spinal nerve and make their exit. The spinal cord tapers into a cone-shaped structure called the **conus medullaris** at its lower tip. Below the conus medullaris the nerve rootlets within the vertebral canal form a brushlike bundle known as the *cauda equina* (horse's tail).

Within the spinal cord (Fig. 14-6) the white matter is on the outside, and the gray matter is centrally located. The white matter consists of fibers carrying impulses to or from neurons in the gray matter, providing a link between higher levels in the CNS and the spinal nerves that connect the cord and the periphery. The gray matter is H-shaped with dorsal and ventral horns housing columns of cells that extend the length of the spinal cord. The dorsal horns contain afferent neurons that relay sensory impulses brought into the spinal cord by way of the dorsal roots of the spinal nerves. The ventral horns contain efferent neurons whose axons leave the cord by way of the ventral roots to become the motor components of the spinal

nerves that innervate skeletal muscles. In addition, in certain segments (first thoracic to second lumbar nerves) there are lateral horns (Fig. 14-6) containing autonomic motor neurons whose axons form part of the spinal nerves of those segments.

> Up to the third month of intrauterine development the spinal cord extends the entire length of the vertebral canal. The vertebrae then grow faster than the spinal cord so that at birth the spinal cord only reaches the level of the third lumbar vertebra.

Figure 14-5 *(Facing page.)* Lateral view of the spinal cord within the vertebral canal

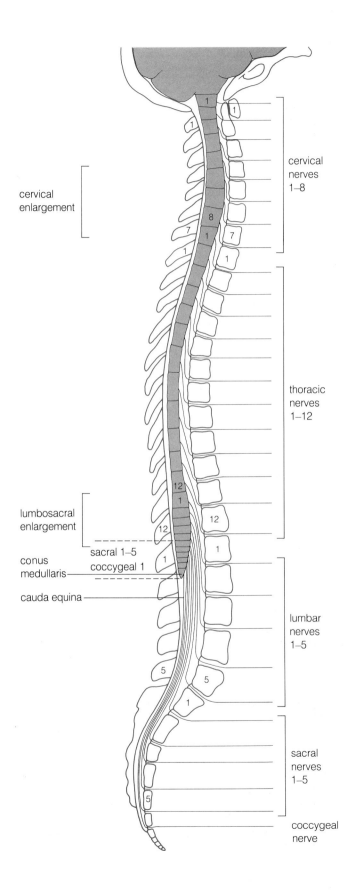

cervical nerves 1–8

cervical enlargement

thoracic nerves 1–12

lumbosacral enlargement

conus medullaris

cauda equina

sacral 1–5
coccygeal 1

lumbar nerves 1–5

sacral nerves 1–5

coccygeal nerve

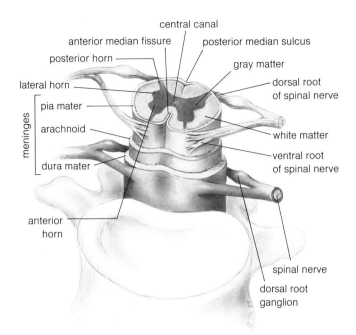

central canal
anterior median fissure
posterior median sulcus
posterior horn
gray matter
lateral horn
dorsal root of spinal nerve
meninges
pia mater
arachnoid
white matter
dura mater
ventral root of spinal nerve
anterior horn
spinal nerve
dorsal root ganglion

Figure 14-6 Cross section of the spinal cord, showing the internal anatomy of the cord with some surface of the cord visible to show meninges

A typical spinal nerve (Fig. 14-4), formed by the union of ventral and dorsal roots, is a mixed nerve containing both motor (ventral) and sensory (dorsal) fibers to visceral and somatic structures. (The exception is the first cervical nerve, which may be purely motor.) A small sensory ganglion, the **dorsal root ganglion,** is found just medial to the junction of the dorsal and ventral roots; this contains the cell bodies of the dorsal root fibers. The axons (proximal processes) of these neurons enter the spinal cord by way of the dorsal root and rootlets. The dendrites (distal processes) course in the spinal nerve; they begin in "free" nerve endings (unmyelinated and freely branching) or in specialized receptors. The cell bodies and dendrites of the ventral root fibers lie mainly in the ventral horn of the spinal cord; their axons course in the ventral rootlets, ventral root, and spinal nerve. The ventral rootlets within a spinal cord segment are somewhat closer together than the dorsal rootlets. The gaps between the rootlets of one segment and the next make the segmental character of the spinal cord much more evident ventrally than dorsally.

Each spinal nerve also contains fibers coming from viscera (visceral afferents). These run with the autonomic motor fibers in the spinal nerves, have cells of origin in the dorsal root ganglion, and terminate in the dorsal gray matter of the spinal cord.

Table 14-1	Composition and Distribution of Spinal Nerves
Component	**Function**
Dorsal root	Composed of sensory fibers with cell bodies in the dorsal root ganglion
Ventral root	Composed of motor fibers with cell bodies in the ventral horn of the spinal cord gray matter.
Spinal nerve	Formed by the junction of the dorsal and ventral roots. Contains both motor and sensory fibers (mixed nerve).
Dorsal primary ramus	Branch of the spinal nerve, containing both motor and sensory fibers. Distributed to the skin and muscles of the back.
Ventral primary ramus	Branch of the spinal nerve, containing both motor and sensory fibers. Distributed to the skin and muscles of the lateral and ventral body wall.

The visceral afferents and the autonomic motor fibers are described more fully in the chapter on the autonomic nervous system (ANS).

Naming the Spinal Nerves

The spinal nerves are named according to their regions and to the *vertebral levels* at which they leave the vertebral canal (Fig. 14-1). Cervical nerves (except the eighth) are designated by the vertebra *inferior* to the nerve. For example, spinal nerve C5, the fifth cervical nerve, leaves the vertebral canal *above* vertebra C-5. There are seven cervical vertebrae, but there are eight cervical nerves, with the first cervical nerve (C1) emerging above the first vertebra and the eighth emerging below the seventh vertebra. Spinal nerves from the first thoracic (T1) to the fifth sacral (S5) are designated by the vertebra superior to the nerve. Thus spinal nerve T5, the fifth thoracic nerve, leaves the vertebral canal *below* vertebra T5. In addition to the *eight cervical nerves* there are *twelve thoracic, five lumbar,* and *five sacral nerves.* The last spinal nerve is the *coccygeal* (Co1), making a total of *thirty-one pairs.* The number of spinal nerves may vary if the number of vertebrae is anomalous (variable).

Components and Distribution of a Spinal Nerve

The components and distribution patterns of the spinal nerves are summarized in Table 14-1. We have already described the nerve roots in connection with the spinal cord; here we shall describe the components and sources of the spinal nerves outside the vertebral canal.

Each spinal nerve branches just lateral to its point of exit from the vertebral canal into a dorsal and a ventral primary *ramus,* or branch. The **dorsal primary ramus** supplies motor fibers to the deep back muscles and sensory fibers to the skin of the back overlying the muscles (Fig. 14-7). The **ventral primary ramus** of the spinal nerve continues laterally, sending branches to the muscles and skin, and terminates near the anterior midline of the body wall. Each spinal nerve has branches containing both motor and sensory fibers that regulate the activity of the skeletal muscles supplied by the nerve. Branches of the spinal nerve penetrate the musculature of the body wall and enter the *dermis* (the superficial connective tissue layer of the skin) to supply strips overlying the muscles. Such a superficial branch, called a *cutaneous nerve,* contains no motor fibers to skeletal muscles. It does, however, contain autonomic (sympathetic) motor fibers to *smooth muscles* of the *blood vessels* and *hair follicles* and *sweat glands.*

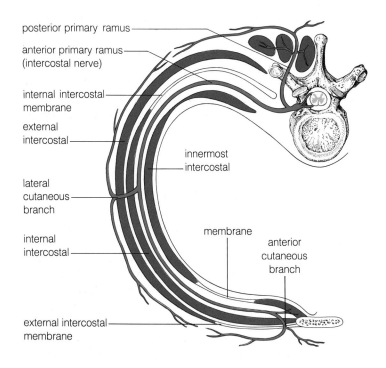

posterior primary ramus

anterior primary ramus (intercostal nerve)

internal intercostal membrane

external intercostal

innermost intercostal

lateral cutaneous branch

internal intercostal

membrane

anterior cutaneous branch

external intercostal membrane

Figure 14-7 Diagram of a thoracic spinal nerve illustrating the typical branching pattern of spinal nerves of the trunk wall

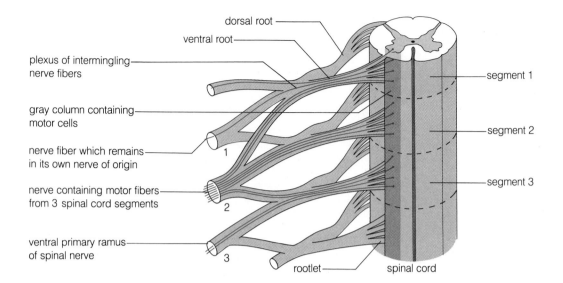

dorsal root

ventral root

plexus of intermingling
nerve fibers

gray column containing
motor cells

nerve fiber which remains
in its own nerve of origin

nerve containing motor fibers
from 3 spinal cord segments

ventral primary ramus
of spinal nerve

rootlet

spinal cord

segment 1

segment 2

segment 3

Figure 14-8 Diagram illustrating the exchange of nerve fibers between nerves of a plexus

Segmental Pattern of Spinal Nerves

There is a definite segmental pattern of muscle innervation that is established early and is carried with the individual muscles as they shift into their adult positions. In the thoracic region a single spinal nerve supplies the intercostal muscles of a single intercostal space. Other muscles also may be innervated by a single nerve arising from one cord segment, especially when the muscle is derived from a single **myotome.** Some muscles are supplied by more than one segment, particularly those muscles that serve more than one function.

PATTERNS OF DISTRIBUTION OF THE SPINAL NERVES

In the cervical, lumbar, and sacral regions the ventral primary rami of adjacent spinal nerves exchange nerve fibers in what is termed nerve plexuses. Nerves that arise as branches of these plexuses usually consist of nerve fibers from more than a single spinal cord segment (Fig. 14-8). Three major plexuses are formed that supply structures in the neck and the upper lower extremities. These plexuses each have a specific pattern of branching which will be described briefly.

Cervical Plexus

The cervical plexus (Fig. 14-9) is formed by the exchange of fibers between the ventral primary rami of the first four cervical

nerves (C1 to C4). Motor fibers of C1 and some from C2 join the hypoglossal nerve for a short distance prior to descending in the neck. Motor fibers from C2 and C3 merge into a single nerve that descends posteriorly in the neck to join the descending fibers of C1, forming a loop (ansa cervicalis) from which the motor supply of the infrahyoid muscles is derived. Other motor fibers from C3 and C4, with some contribution from C5, unite to form the phrenic nerve which supplies the diaphragm.

Sensory fibers form connecting loops between C2 and C3 and between C3 and C4, from which cutaneous nerves arise that supply sensory and autonomic fibers to the skin and superficial fascia of the posterior aspect of the head and anterior neck. They extend onto the anterior thoracic wall for a short distance below the clavicle.

Brachial Plexus

The anterior primary rami, called the *roots* of the *brachial plexus,* of the *last four cervical spinal nerves* (C5 to C8) and the *first thoracic* (T1) form the brachial plexus (Fig. 14-10) by exchanging fibers in a specific pattern. Through this interchange of fibers peripheral nerves are formed that innervate shoulder and upper extremity muscles. The five roots of the plexus unite to form three *trunks* that split into *anterior* and *posterior divisions,* which reunite as three *cords* (medial, lateral, and posterior) in relation to the axillary artery. The cords then split into five *terminal branches;* each branch contains nerve fibers from two or more of the spinal cord segments. In the process of forming divisions, cords, and peripheral nerves, the plexus gives rise to some small nerves that innervate the skin and some muscles of the region.

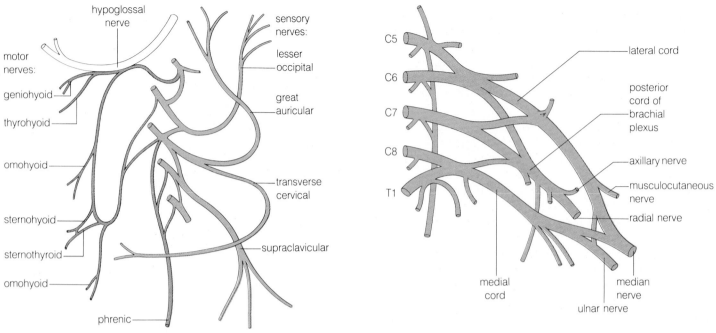

Figure 14-9 The cervical plexus

Figure 14-10 The brachial plexus

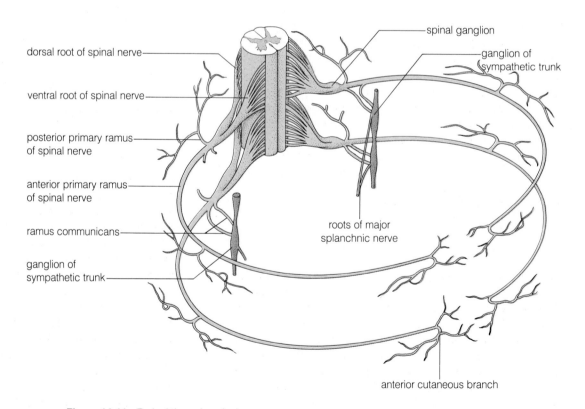

Figure 14-11 Typical thoracic spinal nerves

The major branches of the brachial plexus supply motor and sensory fibers to the muscles and skin of the arm, forearm, and hand. Peripheral nerves derived from the posterior cord include the *axillary nerve,* which supplies some shoulder muscles (deltoid and teres minor) and the overlying skin, and the *radial nerve,* which innervates the extensors of the forearm, wrist, and hand. The radial nerve also supplies the skin on the posterior aspect of the arm, forearm, and most of the hand. The muscles lying on the anterior aspect of the arm are supplied by the terminal portion of the lateral cord, known as the *musculocutaneous* nerve. This nerve supplies the coracobrachialis, biceps brachii, and brachialis in addition to the skin on the anterior surface of the arm and much of the forearm. The *median* nerve contains fibers from both the lateral and medial cords, and the *ulnar* nerve is the remainder of the medial cord. The ulnar and median nerves do not usually have branches in the arm but enter the forearm where they supply the flexors of the wrist and the muscles of the hand. The sensory innervation of the skin of the palm and the palmar surface of all the fingers is furnished by these two nerves, which overlap with branches of the radial nerve on the dorsum of the hand.

Thoracic Spinal Nerves

There are twelve pairs of spinal nerves that supply the muscles and skin of the intercostal spaces (Fig. 14-11). The upper six thoracic nerves continue to the anterior midline of the thoracic wall. The lower six nerves supply the intercostal spaces posteriorly and laterally, following the downward slope of the ribs until they reach the inferior margin of the rib cage. At this point, continuing the downward curving direction of the ribs, they enter the anterior abdominal wall and supply the muscles and skin of most of that region. It is convenient to remember that the tenth thoracic nerve lies at the level of the umbilicus anteriorly. The other thoracic nerves supplying the area are spaced evenly above and below the umbilicus.

Lumbosacral Plexus

The *five lumbar* and *first four sacral* spinal nerves interchange fibers in nerve plexuses (Fig. 14-12) that give rise to branches that supply the lower extremities and pelvic structures. The pattern of branching is more variable than that of the brachial plexus, but it may be described in two parts, a *lumbar plexus* and a *sacral plexus.* Branches of the upper part of the lumbar plexus (iliohypogastric, ilioinguinal, and genitofemoral) supply the lower anterior abdominal wall and genitals. The largest branches of the lumbar plexus, the nerves that supply the strong, massive muscles of the lower extremity, are the *obturator* to

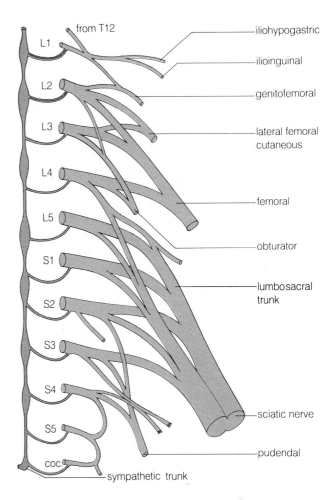

Figure 14-12 The lumbosacral plexus

the medial (adductor) side of the thigh and the *femoral* to the muscles and skin of the anterior thigh. The *sciatic* nerve, the largest nerve in the body, arises principally from the lower levels of the lumbar plexus (L4 and L5) and from segments S1 to S3 of the sacral plexus. The sciatic nerve passes down the back of the thigh, giving off branches to the muscles and skin of the posterior thigh, and terminates by dividing into the *common peroneal* and *tibial* nerves, which supply the entire leg and foot. The sacral plexus also sends branches to pelvic structures including the pelvic diaphragm and receives fibers from the lower lumbar nerves (L4 and L5) via the *lumbosacral trunk.*

The lumbar, sacral, and coccygeal nerves receive postganglionic sympathetic fibers from ganglia of the sympathetic trunk. In addition, the sacral spinal nerves (S2, S3, and S4) contain the sacral preganglionic parasympathetic outflow that enters the pelvic plexus of splanchnic nerves for distribution to pelvic viscera.

Clinical Applications: **Injuries to Peripheral Nerves**

Peripheral nerves in most areas of the trunk are well protected by fascia, muscles, and bones. In the limbs, nerves are frequently injured by laceration—particularly at the wrist, where the median and ulnar nerves are near the surface. Peripheral nerves may also be damaged by pressure following a fracture of the bone to which the nerve is closely related. For example, the radial nerve has a spiral course around the posterior aspect of the humerus as it passes distally. Fracture of the humerus may damage the nerve, or the nerve may be trapped in the bony callus as the broken bone heals. Stretching of a nerve may occur in industrial or traffic accidents; elbow injuries are especially likely to produce weakness or paralysis of muscles supplied by the ulna nerve.

Degeneration

Following injury of a peripheral nerve, such as severing of a nerve with a knife, the axis cylinder of the nerve fiber distal to the cut breaks up and the myelin sheath gradually disintegrates into oily droplets. This process, *known as Wallerian degeneration,* occurs for 2 to 3 cm around the site of the injury. The debris is cleared by phagocytic activity of tissue macrophages, and Schwann cells multiply to fill the endoneurial tubes within about 3 months.

Regeneration

If the severed ends of the nerves are in close apposition, the sprouting axons may find their way down the endoneurial tubes and reestablish connections with muscle fibers. Initially, several fibers may grow into each endoneurial tube, but soon all except one will degenerate. As the nerve fibers grow distally, the myelin sheaths begin to develop. Function is not restored unless the proper nerve ending forms and until the nerve fiber is myelinated.

Motor Effects of Peripheral Nerve Injuries

Interruption or injury of a motor nerve produces paralysis of the muscles it supplies. This results in muscle atrophy. The most commonly injured nerves are those of the extremities, and their injury may be readily diagnosed. The functional deficit caused by an injury to a nerve depends on the level of interruption of the activity of the nerve. To determine the level of the injury, knowledge of the levels at which particular muscular branches arise from the parent nerve is essential. For example, injury to the radial nerve at the elbow will not cause any loss of function of the ability to extend the forearm because the nerves to the tricep brachialis arise from the radial nerve high in the arm proximal to the injury. Examples of some of the common motor symptoms and tests for injury of the main nerves of the extremities are given below.

Radial nerve damage in the middle of the arm is usually made obvious by "wrist drop" because the flexors of the wrist are unopposed by the extensors that are innervated by the radial nerve. The extent of the nerve damage may be evaluated by trying to dorsiflex (extend) the wrist or to extend the metacarpophalangeal joints. Ulnar nerve damage at the wrist is checked by asking the patient to abduct and adduct the fingers; this tests the activity of the interossei muscles. Damage to the median nerve as it crosses the wrist may be tested by trying to abduct the thumb because the only muscle that can execute this movement, the abductor pollicis brevis, is innervated by the median nerve.

In the lower extremity, injury to the sciatic nerve in the gluteal region is likely to include the branches to the hamstring muscles. Thus the ability to extend the hip joint and to flex the leg at the knee is impaired. The muscles of the front and thigh, the sartorius and quadriceps femoris, are supplied by the femoral nerve, which also supplies the iliacus muscle that helps flex the hip. Paralysis of these muscles as a result of destruction of the femoral nerve will impair hip flexion and prevent extension at the knee joint. The nerve of the lower extremity that is most frequently injured is the common peroneal nerve that supplies the dorsiflexor and evertor muscles of the foot. Paralysis of these muscles produces a condition known as "foot drop." A person with this deficit raises his or her foot very high when walking so that the toes do not touch the ground. The foot is brought down suddenly and hits the ground with a slap that is characteristic of the condition.

DEVELOPMENT OF THE SPINAL NERVES

Development of Nerve-Muscle Relations

In the future spinal cord, cells within the ventral half of the neural tube (basal plate) begin to differentiate into multipolar neurons whose axons grow outward toward the periphery (Fig. 14-13a). These axons penetrate the external limiting membrane of the neural tube to become the ventral roots of the spinal nerves and extend into the surrounding mesenchyme until they reach the myotomes, the primordia of skeletal muscles (Fig. 14-13b). Within a myotome each axon becomes associated with a

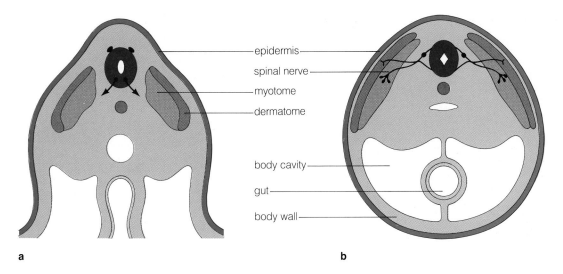

epidermis

spinal nerve

myotome

dermatome

body cavity

gut

body wall

a

b

Figure 14-13 The growth of nerve fibers into the somites. (**a**) Cross section of embryo as nerve fiber growth is initiated. (**b**) Cross section of an older embryo illustrating nerve fibers within the myotome and dermatome.

single myotome cell, and the association is retained for the life of the nerve fiber and of the myotome cell. In addition, whenever the myotome cell divides, the axon branches to provide each of the progeny with a nerve fiber. Later when the myotome cells fuse to form skeletal muscle fibers, each muscle fiber is supplied by the same nerve fiber that was associated with the cells from which it formed.

With rapid growth by cell proliferation the myotomes extend dorsally in between the spinal cord and the superficial ectoderm and also ventrally into the somatic mesoderm of the body wall. The nerve fibers lengthen to maintain their association with myotome cells. This means that nerve fibers may be dragged anteriorly or posteriorly, depending upon the direction of movement of the myotome cells with which they are destined to maintain contact. The myotomes segregate into a *dorsal muscle mass* and a *ventral muscle mass,* separated from one another by an *intermuscular fibrous septum* (Fig. 14-14). The *dorsal muscle mass* differentiates into the *deep back muscles* innervated by the *posterior primary rami* of the spinal nerves. The *ventral muscle mass* gives rise to the muscles of the *lateral* and *anterior body walls* and is supplied by the *anterior primary rami* of the spinal nerves (Fig. 14-14b).

The myotomes retain their segmented characteristics as they extend into the various regions of the body. The nerves also appear segmented because a single spinal cord segment sends fibers into each myotome. Therefore, when a single myotome gives rise to a single muscle or group of muscles, the muscles are innervated by fibers of a single spinal nerve. This

type of arrangement is seen in the pattern of innervation of the *intercostal muscles;* for example, intercostal nerve T9 innervates intercostal muscles in the ninth intercostal space.

Two patterns of development, as suggested by the adult nerve-muscle relations, may occur when a single muscle is derived from more than one myotome. In one pattern the segmental nature of the muscle may be obliterated, but the segmented pattern of innervation is retained. In other words, if a muscle is derived from nine myotomes, it is innervated by nine spinal nerves, each of which penetrates the muscle separately. For example, the *external oblique muscle* of the lateral abdominal wall does not appear segmented in the adult, but it is supplied by branches of the last seven intercostal nerves and branches of the first lumbar nerve.

In the other pattern of innervation a muscle derived from more than one myotome may be supplied by a single nerve containing fibers from more than one spinal cord segment. In most of these cases the embryonic muscle mass has migrated quite far from its point of origin, and the result has been a mixing of cells from more than one myotome. Because the nerve fibers follow their myotomal partners, nerve plexus formation results as the nerve fibers intermingle. A specific example is seen in the development of the *strap* (infrahyoid) *muscles* of the *neck,* which are derived from the first three cervical myotomes. As the neck elongates with the caudal shift of the heart and associated structures, the first three cervical myotomes migrate ventrally and caudally, differentiating into longitudinally oriented muscle masses. As the myotomes migrate, spinal

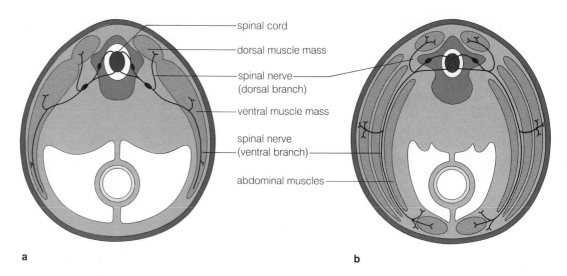

spinal cord

dorsal muscle mass

spinal nerve
(dorsal branch)

ventral muscle mass

spinal nerve
(ventral branch)

abdominal muscles

a

b

Figure 14-14 Transverse sections of embryos at two stages. (**a**) Earlier formation of dorsal and ventral primary rami. (**b**) Later stage in the development of the pattern of innervation of the trunk wall.

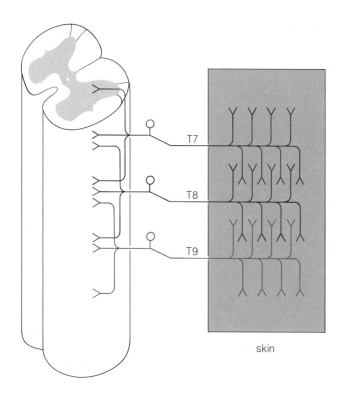

T7

T8

T9

skin

Figure 14-15 Diagram of the overlap in the innervation of dermatomes

nerves C2 and C3 combine into a single nerve (descendens cervicalis), and C1 joins the hypoglossal nerve for a short distance (Fig. 14-15). However, the hypoglossal and C1 soon part, because the hypoglossal nerve is associated with occipital myotomes destined for the tongue and C1 is associated with myotome cells that remain in the neck. Thus C1 passes inferiorly in the neck (descendens hypoglossi) to form a loop (ansa cervicalis) with the descendens cervicalis. Individual nerves arising from this loop to supply individual strap muscles may each contain fibers from all three spinal cord segments.

Development of Nerve-Dermatome Relations

While the efferent neurons are establishing a permanent relationship with myotome cells, the afferent neurons begin to differentiate from **neural crest cells** lateral to the neural tube. The proximal processes (axons) of these neurons extend into the dorsal aspect of the neural tube, and the distal processes (dendrites) grow laterally into the dermatome. As in the case of nerve-myotome relations the sensory fibers of a specific cord segment become associated with cells of a specific **dermatome** and retain their association with cells of the dermis derived from that dermatome. The dermatome grows rapidly, spreading dorsally and ventrally as the cells multiply. Dermatome cells intermingle with adjacent dermatomes as the cells migrate during development, so that three adjacent spinal nerves must be injured before loss of sensation can be detected clinically in the skin.

WORDS IN REVIEW

Conus medullaris   Myotome

Dermatome   Neural crest

Dorsal primary ramus   Spinal nerve

Dorsal root ganglion   Spinal nerve root

Epineurium   Ventral primary ramus

SUMMARY OUTLINE

I. The Spinal Cord

The spinal cord begins at its junction with the medulla at the level of the foramen magnum and extends to the level of the disc between the first and second lumbar vertebrae. It has a central core of gray matter surrounded by white matter. Thirty-one pairs of spinal nerves arise serially by dorsal and ventral rootlets.

II. The Spinal Nerves

The spinal nerves are mixed nerves composed of sensory fibers from the dorsal root and motor fibers from the ventral root.

A. The roots of the spinal nerves are both sensory, with cells of origin in dorsal root ganglia, and motor, with cells of origin in the ventral and lateral gray matter of the spinal cord.

B. The spinal nerves are named according to the vertebral region and numbered by vertebral level. There are eight cervical, twelve thoracic, five lumbar, five sacral, and one coccygeal nerve.

C. The components of a spinal nerve are motor (somatic) fibers to skeletal muscles, sensory (somatic) fibers to skin and other somatic structures, motor (visceral or autonomic) fibers to smooth muscle and glands, and sensory (visceral) fibers conveying afferent impulses from viscera.

D. In a typical pattern of distribution the spinal nerve leaves the vertebral canal and immediately gives off a dorsal primary ramus to the deep back muscles and overlying skin. The ventral primary ramus continues laterally and anteriorly, supplying muscles and skin of the lateral and anterior body wall.

III. Distribution Patterns of Spinal Nerves

There are regional differences in the patterns of distribution of spinal nerves.

A. In the neck the ventral primary rami of spinal nerves C1 to C4 form the cervical plexus, giving rise to motor nerves that supply infrahyoid muscles and the diaphragm and to sensory nerves to the skin of the neck.

B. The brachial plexus is formed by the ventral primary rami of spinal nerves C5 to T1. Branches of this plexus provide the motor and sensory innervation of the upper extremity.

C. The thoracic spinal nerves remain segmental, supplying structures of an intercostal space.

D. The lumbosacral plexus is formed by the ventral rami of the five lumbar nerves and the first four sacral nerves. Branches of this plexus supply pelvic structures and the lower extremity.

IV. Nerve-Muscle Relations

Nerve-muscle relations develop in the embryo by nerve fibers extending from cells of the basal plate into adjacent myotomes. A relationship is established in the embryo between nerve fibers and myotome cells which is maintained even as myotome cells fuse to form skeletal muscle fibers. Nerve fibers increase in length to remain associated with the muscle fibers and provide the motor innervation of the muscle fibers with which they became associated in the early embryo.

V. Nerve-Dermatome Relations

Nerve-dermatome relations are established in a similar fashion when neural crest cells differentiate into ganglion cells and send axons into the dorsal aspect of the spinal cord and dendrites into adjacent dermatome of the somites. Dermatome cells intermingle with adjacent dermatomes as the cells migrate during development so that three adjacent spinal nerves must be injured before loss of sensation can be detected clinically.

REVIEW QUESTIONS

1. What is the average vertebral level at which the spinal cord ends inferiorly?

2. What are the anatomical relations of the gray and white matter in the spinal cord?

3. Explain why there are eight cervical spinal nerves but only seven cervical vertebrae.

4. What is the name of the spinal nerve that exits between the twelfth thoracic and the first lumbar vertebra?

5. What is the location of the cell bodies of the neurons whose axons are in the dorsal roots of the spinal nerves?

6. Where are the cell bodies located for those neurons that innervate skeletal muscles of the trunk?

7. What is the name of the nerves that supply the deep back muscles?

8. Explain how the components of a motor unit become associated during development.

SELECTED REFERENCES

Bowsher, D. *Introduction to the Anatomy and Physiology of the Nervous System.* Philadelphia: F. A. Davis, 1974.

Chusid, J. G. *Correlative Neuroanatomy and Functional Neurology.* 17th ed. Los Altos, California: Lange Medical Publications, 1979.

DeCoursey, R. M. *The Human Organism.* 4th ed. McGraw-Hill, 1974.

Guyton, A. C. *Structure and Function of the Nervous System.* 2nd ed. Philadelphia: W. B. Saunders, 1976.

Hubel, D. H. *et al.* "Special Issue on Neurobiology." *Sci. Am.* 241: 45-232, September, 1979.

Stevens, L. A. *Neurons.* New York: Thomas Y. Crowell, 1974.

Watts, G. O. *Dynamic Neuroscience: Its Application to Brain Disorders.* Hagerstown: Harper & Row, 1975.

15

THE BRAIN
AND THE CRANIAL NERVES

OBJECTIVES

After completing this chapter you should be able to:

1. Name and give the boundaries of the lobes of the cerebral hemisphere.

2. Compare and contrast the arrangement (organization) of the gray and white matter in the cerebrum with that of the brainstem and spinal cord.

3. Demonstrate an understanding of functional areas of the cerebral cortex. Give the location of the principal motor, sensory, and association areas.

4. Name and locate the parts of the diencephalon and give the major functions of each part.

5. Name the parts of the brainstem and relate the parts to one another.

6. Describe the cerebellum and give its relations to the brainstem.

7. List and describe the course of the three major ascending pathways of the CNS.

8. Name and describe the two major descending pathways of the CNS. Explain the basis for the control of muscles of one side of the body by the cerebral hemisphere of the opposite side.

9. Describe the role of the cerebellum in the regulation of skeletal muscle contraction.

10. Demonstrate an understanding of the multiple influences upon the final common pathway.

11. Name the cranial nerves according to their number; give the opening through which each nerve exits from the cranial cavity; and describe the general distribution of each nerve.

The morphology of the central nervous system cannot be visualized adequately using the directional terminology with which the body as a whole is described. In gross anatomy the body is described in the anatomical position: the body is upright with the hands at the sides. Thus structures are said to be superior or inferior, dorsal or ventral, and medial or lateral. Because of the curvatures in the CNS these terms do not always adequately describe the relative positions of its parts. It is often more appropriate to describe structures relative to the *neuraxis,* an imaginary line extending from the tip of the frontal lobes of the cerebrum to the lower end of the spinal cord. With the neuraxis as the reference point the terms *rostral* (towards the rostrum, meaning the "beak," or tip of the forebrain) and *caudal* (toward

the lower tip of the spinal cord) are more descriptive. The terms medial and lateral are used as in gross anatomy.

The brain includes the cerebrum, the diencephalon, or thalamus, the cerebellum, and the brainstem which consists of the midbrain, pons, and medulla. The gross external features of these components will be considered first, followed by a brief account of the internal organization of each part. We will then describe the cranial nerves that arise from the brainstem. Finally we will discuss the conduction pathways of nerve impulses that travel between the brain and the cranial and spinal nerves.

THE CEREBRUM

The cerebrum is composed of the cerebral hemispheres, two bulky masses of nervous tissue that occupy most of the cranial cavity, separated from one another by a longitudinal fissure.

The brain is well protected by the skull and the fluid jacket, the cerebrospinal fluid that completely surrounds the brain. But repeated blows to the head such as those to which professional boxers are subjected may lead to permanent damage to the cerebral hemispheres. If the damage is severe, the person is described as "punch-drunk," a condition characterized by an unsteady gait, slow, halting speech, and slow muscular movements.

External Features of the Cerebrum

The surface of the cerebral hemispheres is marked by elevations called **gyri** (sing., gyrus), which are separated by shallow grooves called **sulci** (sing., sulcus) and deep grooves called *fissures* (Fig. 15-1). For descriptive purposes the cerebral hemispheres are divided into *lobes* (Fig. 15-2), the boundaries of which are formed (with one or more exceptions) by the grooves. Locations of the lobes are described in Table 15-1. The *central sulcus* (of Rolando) begins on the medial aspect of the hemisphere and passes over the vertex onto the lateral surface. Here it curves backward and then forward as it extends almost to the *lateral fissure*. The **frontal lobe** is rostral to the central sulcus and dorsal to the lateral fissure.

The lateral fissure (of Sylvius) is found on the lateral aspect of the cerebral hemisphere, beginning about a third of the way forward from the occipital (caudal) pole. It slants from behind forward, separating the *frontal lobe* from the **temporal lobe.** The other fissure can be seen effectively only on the medial

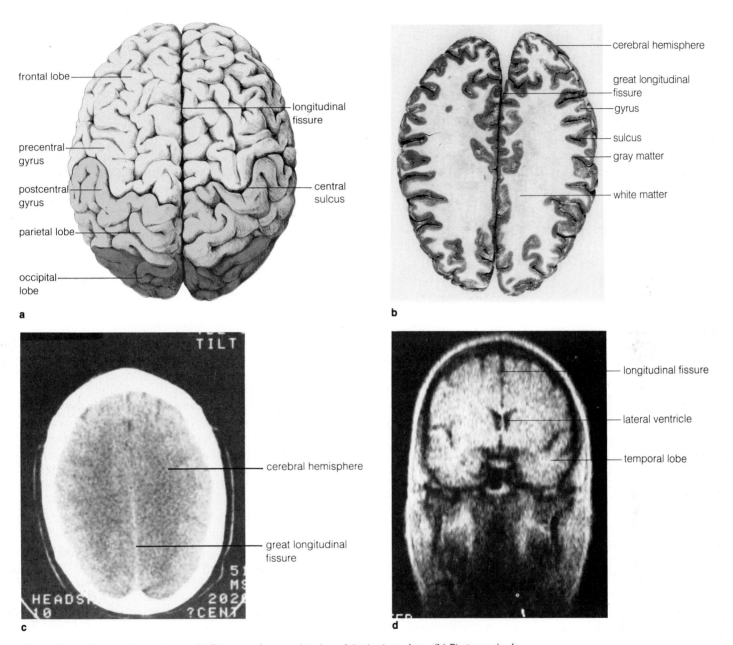

Figure 15-1 Cerebral hemispheres. (**a**) Diagram of a superior view of the brain surface. (**b**) Photograph of a thick horizontal slice of the upper part of the cerebral hemispheres. Note that the longitudinal tissue separates the cerebrum into two equal halves. (Photograph by R. Ross.) (**c**) CAT scan of the cerebrum, horizontal view. Note that the longitudinal fissure is in the midline and is straight. A tumor on the area would probably cause the fissure to deviate toward one side. (Courtesy of S. J. Goldstein.) (**d**) Nuclear magnetic resonance image of a frontal section of the head. Note the superior resolution of the soft tissues. (Courtesy of Paul Wang, Department of Diagnostic Radiology, University of Kentucky.)

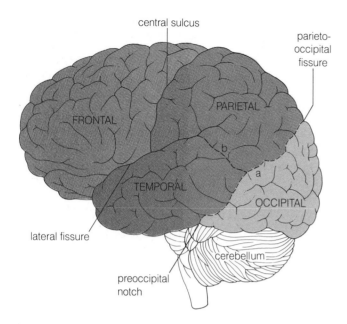

central sulcus

parieto-occipital fissure

PARIETAL

FRONTAL

b

a

TEMPORAL

OCCIPITAL

lateral fissure

cerebellum

preoccipital notch

Figure 15-2 Lobes of the cerebral hemisphere, lateral view. The line indicated as *a* extends from the parieto-occipital fissure to the pre-occipital notch; it represents the anterior boundary of the occipital lobe. The line labeled *b* is an extension of the lateral fissure to intersect *a*; it completes the separation of the parietal and temporal lobes.

Table 15-1	Lobes of the Cerebrum
Lobe	Location
Frontal	Anterior to the central sulcus
Temporal	Inferior to the lateral fissure
Parietal	Posterior to the central sulcus, superior to the lateral fissure, and anterior to a line from the pre-occipital notch to the parieto-occipital fissure
Occipital	Posterior to a line connecting the preoccipital notch and parieto-occipital fissure

surface (Fig. 15-3). One-third to one-fourth of the way forward from the occipital pole there is a very deep fissure, the **parieto-occipital fissure,** which tips over onto the lateral surface dorsally; ventrally a small notch, the *occipital notch,* can be seen on the ventrolateral surface. A line connecting these two structures forms the anterior (rostral) boundary for the **occipital lobe** on the medial surface. On the lateral surface a line drawn from the parieto-occipital fissure dorsally, forward and downward to the occipital notch ventrally, represents the anterior boundary of the occipital lobe. Finally a line on the lateral surface projected backward from the caudal extremity of the lateral fissure to a point that intersects the line delineating the occipital lobe completes the boundaries of the **parietal,** temporal, and occipital lobes. The cortical (surface) areas of the frontal, parietal, occipital, and temporal lobes correspond roughly to the areas covered by the frontal, parietal, occipital, and temporal bones of the skull.

The outstanding morphological feature of the medial aspect of the cerebral hemisphere is the presence of a large mass of fibers, called the **corpus callosum,** connecting the two cerebral hemispheres by crossing between them below the longitudinal fissure.

Internal Organization of the Cerebrum

The gross distribution of nerve cell bodies and of nerve fibers is unique in each part of the brain, brainstem, and spinal cord. This organization is visible grossly as *white matter* and *gray matter* (Fig. 15-4), named for the white appearance of the myelin covering of nerve fibers and the gray color of nerve cell bodies without a myelin covering.

In the cerebral hemisphere the gray matter, concentrated near the surface, is called the **cerebral cortex.** The total surface area, that is, the gray matter, is greatly increased by the development of many folds (gyri, or convolutions) with depressions (sulci and fissures) on both sides. The white matter of the cerebral hemispheres lies deep to the cortex and consists of nerve fibers conveying impulses to and from the cortex. During embryonic development most of the neurons migrate to the surface of the brain, but some cells remain near the cavities of the lateral ventricles to form the deep hemispheric (telencephalic) nuclei or **basal ganglia.**

The white matter consists of nerve fibers that connect different parts of the CNS and may be described as four types (Fig. 15-5): (1) *Short association fibers* connect adjacent gyri forming U-shaped bundles. (2) *Long association fibers* connect distant areas of the same cerebral hemisphere. (3) *Commissural fibers* cross between the two cerebral hemispheres, connecting areas that have similar functions; the largest mass of these fibers is the corpus callosum which forms a bridge between the right and left hemispheres. (4) *Projection fibers* connect the cerebral cortex with all other components of the CNS, conveying impulses either from the *cerebrum* to the *thalamus, basal ganglia, brainstem,* and *spinal cord* or from these structures to the cerebrum; these fibers pass to or from the cerebral cortex in a large mass of white matter, called the internal capsule, that is located within the cerebrum near the lateral ventricles.

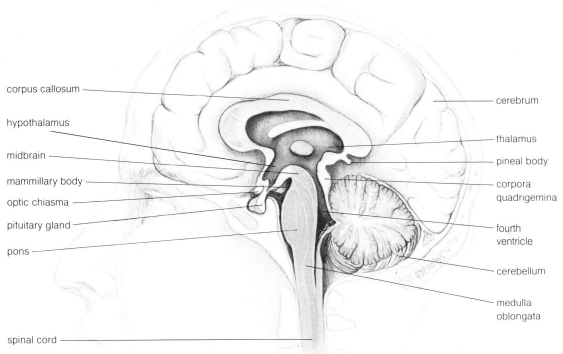

corpus callosum
hypothalamus
midbrain
mammillary body
optic chiasma
pituitary gland
pons
spinal cord

cerebrum
thalamus
pineal body
corpora quadrigemina
fourth ventricle
cerebellum
medulla oblongata

Figure 15-3 Midsagittal section of the brain. (**a**) Diagram of the medial aspect of the right half of the brain. (**b**) X ray of a thick slice of the head. Cadaver preparation, in which the arteries were injected with contrast material (from the University of Kentucky, Department of Anatomy, teaching collection). (**c**) Nuclear magnetic resonance image of a midsagittal section of the head of a normal male 26 years of age. Note that the detail seen is similar to that seen in a cadaver preparation. (Courtesy of Paul Wang, Department of Diagnostic Radiology, University of Kentucky.)

a

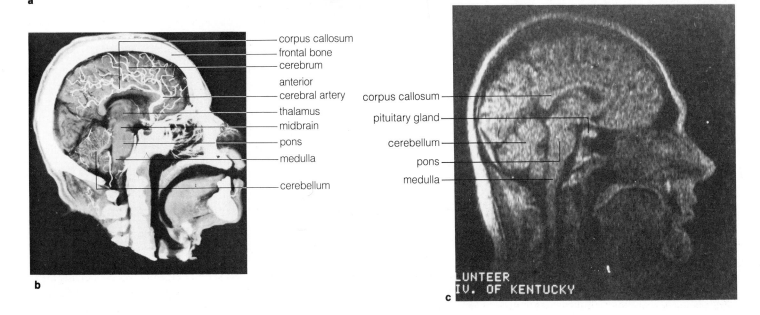

corpus callosum
frontal bone
cerebrum
anterior cerebral artery
thalamus
midbrain
pons
medulla
cerebellum

b

corpus callosum
pituitary gland
cerebellum
pons
medulla

LUNTEER
IV. OF KENTUCKY

c

Functional Areas of the Cerebrum

The localization of functions in specific areas of the cerebral cortex (Fig. 15-6) has been investigated extensively in the human and other mammals. There are *five main sensory areas: general sensation, visual, auditory, taste,* and *vestibular.* There are *two main motor areas,* the *primary motor area* (motor *cortex*) and the *premoter area* (*premotor cortex*), which control the activity of skeletal muscles. The remainder of the cerebral cortex falls into the category of *association cortex* and includes areas related to *behavior* and *intelligence.* The association cortex is responsible for the language functions that are unique to humans; these lie in *two special areas,* an area in the *temporal* and *par-*

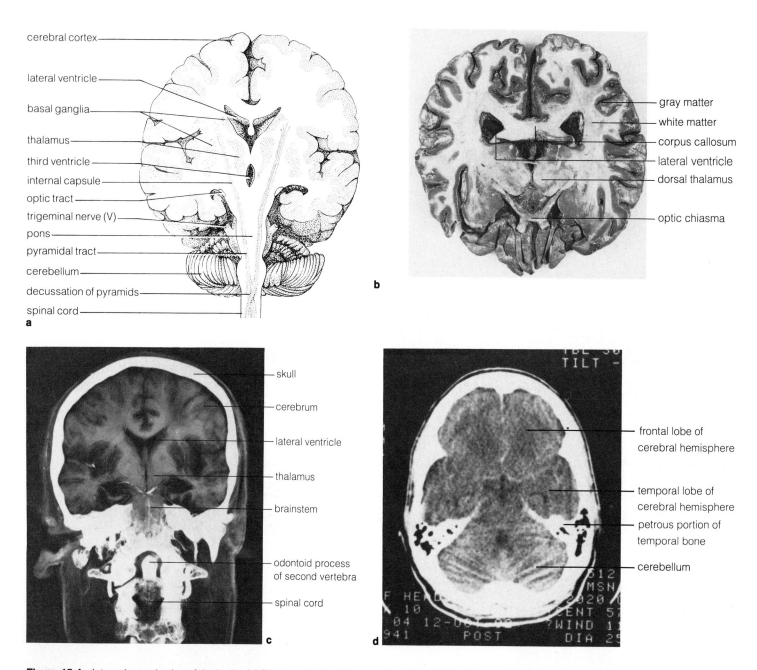

cerebral cortex
lateral ventricle
basal ganglia
thalamus
third ventricle
internal capsule
optic tract
trigeminal nerve (V)
pons
pyramidal tract
cerebellum
decussation of pyramids
spinal cord

a

gray matter
white matter
corpus callosum
lateral ventricle
dorsal thalamus
optic chiasma

b

skull
cerebrum
lateral ventricle
thalamus
brainstem
odontoid process of second vertebra
spinal cord

c

frontal lobe of cerebral hemisphere
temporal lobe of cerebral hemisphere
petrous portion of temporal bone
cerebellum

d

Figure 15-4 Internal organization of the brain. (**a**) Diagram of a frontal coronal section. (**b**) Photograph of a thick (frontal) slice of the human brain (photograph by R. Ross). (**c**) X ray of a thick coronal section of the head and upper neck, cadaver preparation (from the University of Kentucky, Department of Anatomy, teaching collection). (**d**) CAT scan of the brain, horizontal view (courtesy of S. J. Goldstein).

ietal lobes concerned with *sensory* aspects of *speech* and one in the *frontal lobes* involved in *motor speech.* The functional areas of the cerebrum are summarized in Table 15-2.

The general *sensory area* in the *postcentral gyrus* and the *primary motor area* in the *precentral gyrus* both control the

parts of the body in an inverted fashion. For example, in the primary motor area, often simply called the motor area (Fig. 15-7), the muscles of the leg and foot are controlled by neurons located on the medial aspect of the cerebral hemisphere, whereas those of the face are found in the lower part of the precentral

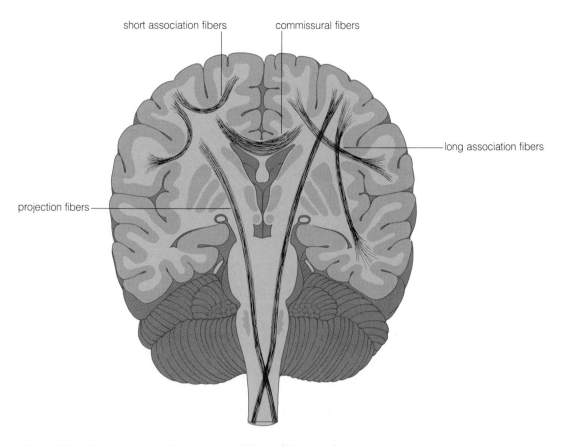

Figure 15-5 Diagram representing the types of fibers of the central nervous system.

Table 15-2	Functional Areas of the Cerebral Cortex
Function	Location
Primary motor area	Precentral gyrus
Premotor area	Frontal lobe, anterior to the precentral gyrus
General sensory area	Postcentral gyrus
Motor speech area	Frontal lobe, superior to lateral fissure
Sensory speech area	Temporal and parietal lobes
Auditory area	Temporal lobe
Visual area	Occipital lobe

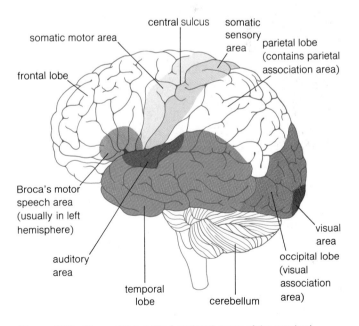

Figure 15-6 Some of the major functional areas of the cerebral cortex. Lateral view of the left cerebral hemisphere.

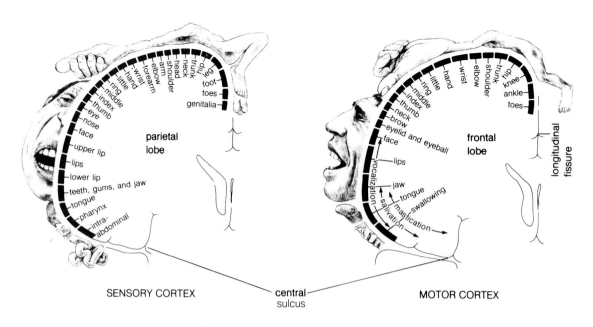

Figure 15-7 The location of sensory and motor areas of the cortex, with sketches of the body part that each portion serves. Each of the body parts is represented in proportion to the area of the cortex with which it is associated.

gyrus near the lateral fissure. Thus the body is represented in the motor area as upside down, suspended by the legs that hook over the top of the central hemisphere onto the medial side. The general sensory area, often simply called the sensory area, has a similar pattern of topographical representation of the body (Fig. 15-7). In both the motor and the sensory areas the proportion of cortex devoted to each part of the body is consistent with the neurological requirements of that part; for example, the head occupies a disproportionate part of both the motor and sensory areas because a very large number of neurons is required for regulating the muscles of the head and for relaying sensory information from the face and oral cavity. Information received in the sensory areas may be relayed via association neurons to the motor areas and used to modify motor activity, including contraction of skeletal muscles.

The brain is insensitive to pain, but the dura mater is well supplied with sensory nerve fibers, as are the blood vessels supplying the meninges and the brain. Many headaches appear to originate in the meninges or in blood vessels as a result of pressure changes within the cranial cavity.

Because voluntary movements are controlled by the brain, a lesion in the motor cortex results in paralysis of voluntary movements of a specific body region even though the muscles of that region are still capable of reflex responses.

THE DIENCEPHALON

The cerebrum and diencephalon together make up the *forebrain,* sometimes referred to simply as *the brain* (in contrast with the cerebellum and brainstem). The **diencephalon** (Fig. 15-3) consists principally of the right and left *dorsal thalami* and *hypothalami* separated in the midline by the *third ventricle.*

The dorsal thalami are slightly inferior and caudal to the cerebral hemispheres and can be seen by dividing the brain into two halves by a midsagittal cut. Each dorsal thalamus consists of a mass of gray matter that is actually an aggregation of several nuclei responsible for specific functions. The dorsal thalamus as a whole is the main *relay center* for all sensations except olfaction. Impulses from lower levels conveying a sensation, such as sense of position, reach a specific nucleus of the dorsal

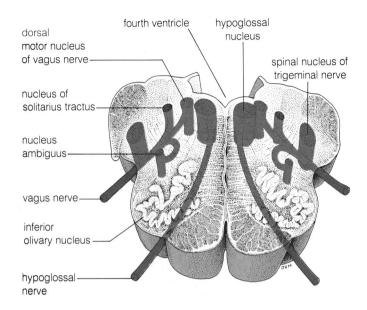

Figure 15-8 Transverse section of the medulla showing the internal organization of this level of the brainstem

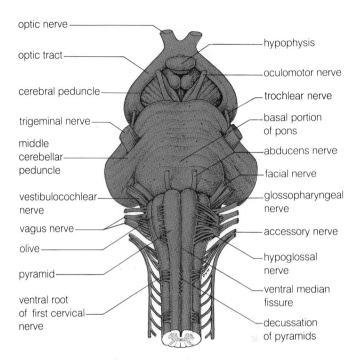

Figure 15-9 Ventral view of the brainstem showing the location of the cranial nerves

thalamus that contains neurons whose fibers project to the sensory area of the cerebral cortex. When the impulses reach the cortex, the position of the body is perceived. In addition to containing neurons whose fibers project to the cortex, the thalamus also contains neurons that relay impulses to the **basal ganglia** and the **hypothalamus.**

About midway down the lateral wall of the third ventricle is an indentation, the *hypothalamic sulcus,* that divides the dorsal thalamus above from the hypothalamus below. The hypothalamus is the center for the regulation of *autonomic activity;* it influences the endocrine system by controlling the *release* of *hormones* from the *anterior* lobe of the *pituitary gland,* which attaches by means of a stalk to the inferior aspect of the hypothalamus. *Hormones* produced by the hypothalamus travel down fibers of the stalk and are released in the *posterior lobe* of the *pituitary.* The regulatory activity of the hypothalamus will be discussed in Chapter 18.

THE BRAINSTEM

The arrangement of gray and white matter in the *brainstem* is distinctly different from that of the cerebrum. The gray matter of the brainstem is found, in large part, adjacent to the ventricular cavity of the nervous system or as scattered clumps of cells (nuclei) mixed with the white fibers (Fig. 15-8). The neurons making up these nuclei either relay impulses to other parts of the CNS or give rise to a motor component of a cranial nerve. The brainstem includes the midbrain, pons, and medulla (Figs. 15-9, 15-10), whose morphological features are summarized in Table 15-3.

The Midbrain

The **midbrain** lies caudal to the thalamus and bears on its ventral surface large paired *cerebral peduncles* (Fig. 5-9) composed of fibers passing to or from the cerebral cortex and lower levels. Dorsally the midbrain bears two pairs of elevations, the *superior* and *inferior colliculi* (Fig. 15-10), collectively called the *corpora quadrigemini,* which contain neurons involved in *visual* and *auditory reflexes.* In addition to providing connections between higher and lower parts of the CNS, the midbrain gives rise to the third and fourth cranial nerves, oculomotor and trochlear nerves, which supply the eyeball muscles. The midbrain contains a small cavity, the cerebral aqueduct (Fig. 15-3), that connects the third and fourth ventricles.

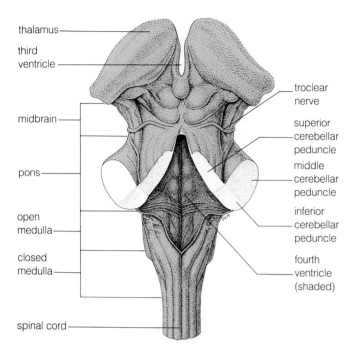

Figure 15-10 Dorsal view of the brainstem. The cerebellum has been removed and the fourth ventricle has been opened.

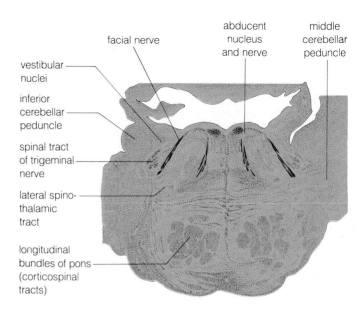

Figure 15-11 Transverse section of the pons

Table 15-3	Components of the Brainstem
Component	Morphological Features
Midbrain	Contains corpora quadrigemini dorsally and the cerebral peduncles ventrally. Gives rise to cranial nerves III and IV, the oculomotor and trochlear.
Pons	Contains connections between the cerebrum and cerebellum and between the higher and lower levels of the brain. Gives rise to cranial nerves V–VIII, the trigeminal, abducens, facial, and auditory.
Medulla	Contains fibers conveying impulses between higher and lower levels. Gives rise to cranial nerves IX, X, and XII, the glossopharyngeal, vagus, and hypoglossal.

The Pons

The **pons** (Figs. 15-9, 15-10) lies caudal to the midbrain and forms a ventral enlargement of the brainstem due to the presence of large numbers of nerve cells and fibers that serve as important connections between the cerebrum and cerebellum. (These fibers pass into the cerebellar peduncles, extensions of the cerebellum that adjoin the pons, Fig. 15-10.) Fibers also traverse the pons longitudinally to reach higher or lower levels (Fig. 15-11). Cranial nerves *V, VI, VII,* and *VIII* (trigeminal, abducens, facial, and auditory) arise from the pons.

The Medulla

The **medulla** oblongata, which serves as a pathway for longitudinal fibers that convey impulses to and from higher levels, is caudal to the pons and is continuous inferiorly with the spinal cord. Cranial nerves *IX, X,* and *XII* arise from nuclei of the medulla and exit from its lateral surface (Figs. 15-9, 15-12). Other nuclear masses, termed reticular nuclei, function as the respiratory center which regulates breathing.

The respiratory center is particularly vulnerable to compression. A lesion of the medulla, such as a tumor, may put pressure on the respiratory center with consequent respiratory failure.

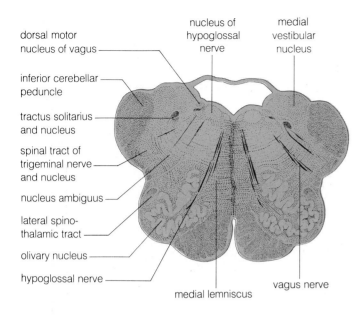

Figure 15-12 Transverse section of the medulla at the level of the fourth ventricle

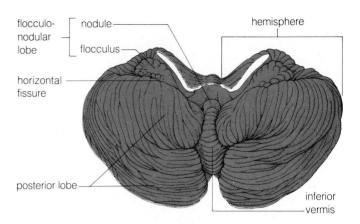

Figure 15-13 Inferior aspect of the cerebellum

THE CEREBELLUM

The **cerebellum** occupies the posterior cranial fossa below and posterior to the occipital lobe of the cerebrum and posterior to the pons and medulla. It consists of two **cerebellar hemispheres** (Fig. 15-13) continuous across the midline through a narrow medial portion, the **vermis.** Transverse ridges or **folia** (sing., folium) separated by deep fissures and shallow grooves mark the surface of each hemisphere. The cerebellum is connected with the brainstem by three paired bundles, or bridges, called **peduncles:** the *superior peduncles* run rostrally and join the *midbrain;* the *middle peduncles* run ventrally to connect the *cerebellum* and the *pons;* and the *inferior peduncles* extend caudally to the *medulla.* The superior peduncles consist principally of fibers that conduct impulses from the cerebellum

toward the brainstem; the middle and inferior peduncles contain mainly fibers providing input to the cerebellum.

The distribution of the neurons in the cerebellum resembles that in the cerebrum in that there is a concentration of cells at the surface (cerebellar cortex) and aggregations of cells in nuclei in its deeper portion. These nuclei relay impulses from the cerebellar cortex to the brainstem and thalamus. The cerebellum receives impulses from all parts of the nervous system and is important in the *regulation* (coordination) of *motor activity.*

THE CRANIAL NERVES

The cranial nerves are identified by Roman numerals and by names that suggest functional or anatomical features. Originating from the forebrain, brainstem, and upper cervical spinal cord, they emerge from the cranial cavity by passing through skull foramina to be distributed principally to head and neck structures. The distribution of each cranial nerve is summarized in Table 15-4.

> Cerebellar lesions may result in disturbances in the regulation of muscular coordination. Unilateral lesions give rise to symptoms and signs on the same side of the body as the lesion; one symptom is an unsteady gait.

Table 15-4 Cranial Nerves

Nerve	Function and Distribution
Olfactory (I)	Sense of smell. Neurons in nasal epithelium below the cribriform plate of the ethmoid; fibers extend to the olfactory bulbs.
Optic (II)	Vision. Fibers of ganglion cells of the retina extend to the thalamus.
Oculomotor (III)	From midbrain, motor to four eyeball muscles and the levator palpebrae superioris. Parasympathetics to ciliaris muscle and sphincter of the pupil.
Trochlear (IV)	From midbrain, motor to the superior oblique muscle.
Trigeminal (V)	From pons, sensory to the face, oral, and nasal cavities. Motor to muscles of mastication.
Abducens (VI)	From pons, motor to the lateral rectus muscle.
Facial (VII)	From pons, motor to muscles of facial expression; sensory for taste to anterior two-thirds of the tongue; parasympathetics to salivary and lacrimal glands.
Vestibulocochlear (VIII)	To medulla and pons, sensory from the organ of Corti and vestibular apparatus.
Glossopharyngeal (IX)	From medulla, sensory to posterior third of the tongue and to the pharynx; motor to stylopharyngeus; parasympathetics to the parotid gland.
Vagus (X)	From medulla, sensory to the lower respiratory tract and digestive tract from the esophagus to the end of transverse colon; motor to muscles of soft palate, pharynx, and larynx; parasympathetics to the respiratory system, heart, and digestive tract to the end of the transverse colon.
Spinal accessory (XI)	From the upper spinal cord, motor to the sternomastoid and trapezius.
Hypoglossal (XII)	From medulla, motor to muscles of the tongue.

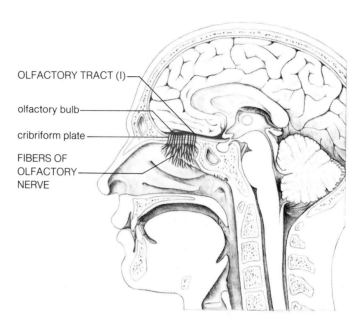

Figure 15-14 Midsagittal section of the head showing the location of the olfactory nerve

Cranial nerve I, the **olfactory** (Fig. 15-14), is seldom seen in gross specimens for the simple reason that there is no large compact nerve, and the olfactory filaments lying outside the cranial cavity (in the nasal cavity) are cut off when the brain is removed from the skull. Neurons whose cell bodies are situated in the *nasal epithelium* (olfactory epithelium) send their filamentous processes through perforations in the *cribriform plate* of the *ethmoid bone* to the *olfactory bulbs,* which lie on the ventral aspect of the frontal lobes of the cerebral hemispheres. The nerve cells function as receptors for the detection of chemical particles (odors) in the air passing over the olfactory epithelium. Impulses initiated by odors are relayed through the olfactory bulbs and olfactory tracts to the cerebral cortex for interpretation.

Cranial nerve II, the **optic** (Fig. 15-15), consists of fibers whose cell bodies (called ganglion cells) are actually located in the *retinal layer* of the eye. Impulses for visual images initiated in cells of the retina (rods and cones) are conducted over the fibers of the optic nerve to the *thalamus,* from which they are relayed over a long tract to the *visual cortex* in the occipital lobe of the brain. The optic nerves enter the cranial cavity through the *optic foramina* and then lie on the inferior aspect of the diencephalon. One half of each optic nerve crosses to the opposite side in the *optic chiasma* to form the *optic tract;* thus images from both eyes are received in each cerebral hemisphere.

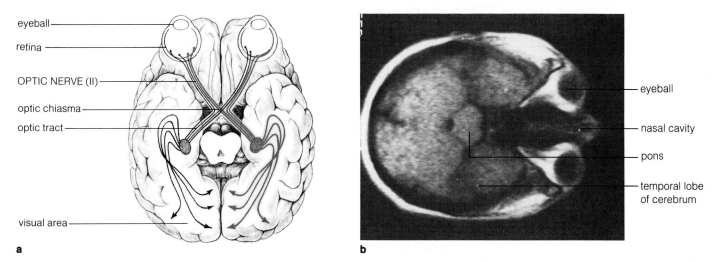

eyeball
retina
OPTIC NERVE (II)
optic chiasma
optic tract
visual area

a

eyeball
nasal cavity
pons
temporal lobe of cerebrum

b

Figure 15-15 (**a**) Ventral aspect of the brain showing the location of the optic nerve. (**b**) Nuclear magnetic resonance image of a horizontal section of the head showing the relations of the eyes to the brain. (Courtesy of Paul Wang, Department of Diagnostic Radiology, University of Kentucky).

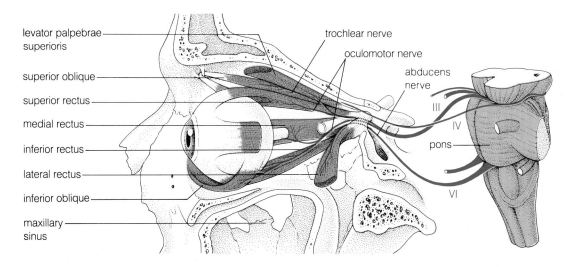

levator palpebrae superioris
superior oblique
superior rectus
medial rectus
inferior rectus
lateral rectus
inferior oblique
maxillary sinus

trochlear nerve
oculomotor nerve
abducens nerve
III
IV
pons
VI

Figure 15-16 Diagram of lateral view of the contents of the orbit showing distribution of the oculomotor, trochlear, and abducens nerves

The pituitary gland is located within the sella turcica immediately posterior to the optic chiasma. Tumors of the pituitary may extend superiorly and anteriorly causing visual symptoms such as double vision because of the pressure upon the optic chiasma.

Cranial nerve III, the **oculomotor** (Fig. 15-16), originates in the *midbrain* and passes into the orbit through the *superior orbital fissure*. It supplies motor fibers to a muscle in the upper eyelid (levator palpebrae superioris) and to four eye muscles (the superior, inferior, and medial rectus and the inferior oblique). It also provides *parasympathetic* (autonomic) *inner-*

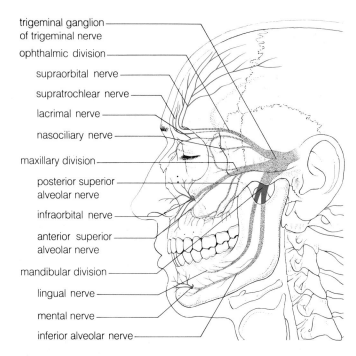

Figure 15-17 Diagram of a lateral view of the face showing distribution of the trigeminal nerve

trigeminal ganglion of trigeminal nerve
ophthalmic division
supraorbital nerve
supratrochlear nerve
lacrimal nerve
nasociliary nerve
maxillary division
posterior superior alveolar nerve
infraorbital nerve
anterior superior alveolar nerve
mandibular division
lingual nerve
mental nerve
inferior alveolar nerve

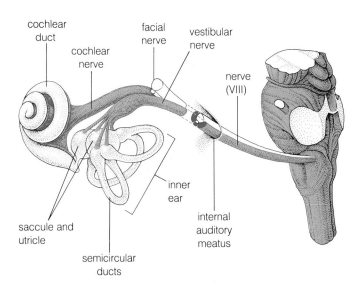

cochlear duct
cochlear nerve
facial nerve
vestibular nerve
nerve (VIII)
inner ear
saccule and utricle
semicircular ducts
internal auditory meatus

Figure 15-19 Diagram showing distribution of the vestibulocochlear nerve

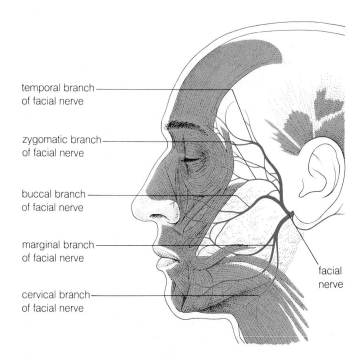

temporal branch of facial nerve
zygomatic branch of facial nerve
buccal branch of facial nerve
marginal branch of facial nerve
cervical branch of facial nerve
facial nerve

Figure 15-18 Lateral view of the head showing distribution of the facial nerve

vation via the ciliary ganglion to the *ciliaris muscle* and the *sphincter* of the *pupil*.

Cranial nerve IV, the **trochlear** (Fig. 15-16), is a small nerve arising from the dorsal aspect of the *midbrain* and entering the orbit through the *superior orbital fissure*. It supplies the *superior oblique muscle* of the eye.

Cranial nerve V, the **trigeminal** (Fig. 15-17), arises in the *pons* and branches prior to exiting from the cranial cavity. Each of its three branches, or divisions—ophthalmic, maxillary, and mandibular—exits through a separate opening to provide *sensory fibers* to the *face, oral,* and *nasal cavities,* and *motor fibers* to the *muscles* of *mastication*.

Cranial nerve VI, the **abducens** (Fig. 15-16), arises from the caudal portion of the *pons* and passes forward to enter the orbit through the *superior orbital fissure*. It provides motor innervation to the *lateral rectus muscle* of the eye.

Cranial nerve VII, the **facial** (Fig. 15-18), originates from the caudal end of the *pons* and passes from the skull through the *internal auditory meatus* and emerges exteriorly through the *stylomastoid foramen*. Its sensory component carries *taste* sensations from the *anterior two-thirds* of the *tongue*. Its largest component carries *motor fibers* to the *muscles* of *facial expression*. The *parasympathetic* component supplies the *salivary* and *lacrimal glands*.

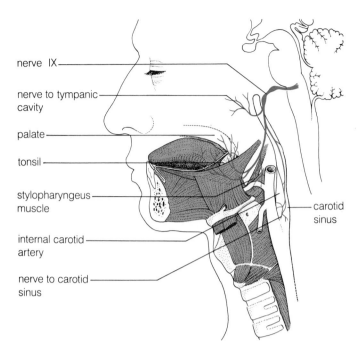

nerve IX

nerve to tympanic
cavity

palate

tonsil

stylopharyngeus
muscle

internal carotid
artery

nerve to carotid
sinus

carotid
sinus

Figure 15-20 Lateral view of the head showing distribution of the glossopharyngeal nerve

---

When the facial nerve is paralyzed, the orbicularis oculi muscle of the affected side is paralyzed. The eyelids then cannot be closed, because the levator palpebrae superioris, innervated by the oculomotor nerve, keeps the eyelid elevated. Thus, the individual must sleep with one eye open.

Cranial nerve VIII, the *vestibulocochlear* or *auditory* (Fig. 15-19), conveys different impulses for *hearing* and *equilibrium* from the auditory and vestibular receptors of the *inner ear* through the *internal auditory meatus* to the *medulla* and *pons*. Thus cranial nerve VIII is really two nerves conveying information on two very different sensations.

Cranial nerve IX, the **glossopharyngeal** (Fig. 15-20), arises in the *medulla* and exits from the cranial cavity through the *jugular foramen*. It is the only *sensory* nerve to the *posterior third* of the *tongue* and to the *pharynx* and sends a sensory nerve to the carotid sinus. It supplies a single skeletal muscle, the *stylopharyngeus,* and provides *parasympathetic fibers* to the *parotid gland*.

Cranial nerve X, the **vagus** (which means wandering), is a large nerve named for its wide distribution. Arising from the

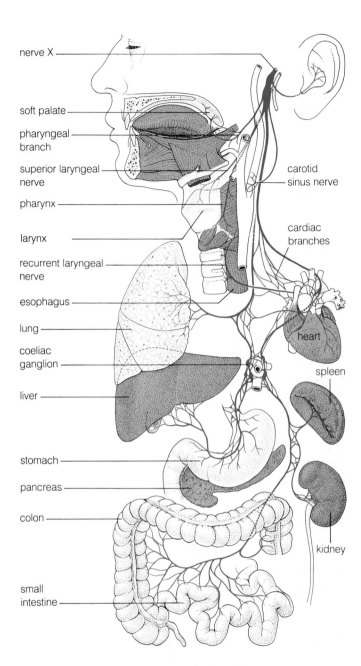

nerve X

soft palate

pharyngeal
branch

superior laryngeal
nerve

pharynx

larynx

recurrent laryngeal
nerve

esophagus

lung

coeliac
ganglion

liver

stomach

pancreas

colon

small
intestine

carotid
sinus nerve

cardiac
branches

heart

spleen

kidney

Figure 15-21 Diagram showing distribution of the vagus nerve

medulla and exiting through the *jugular foramen* (Fig. 15-21), it provides *motor* innervation to most of the muscles of the *soft palate, pharynx,* and *larynx,* and *sensory* fibers to the *mucosal lining* of the lower respiratory tract and *digestive tract,* from the *esophagus* to the beginning of the *descending colon*. It also

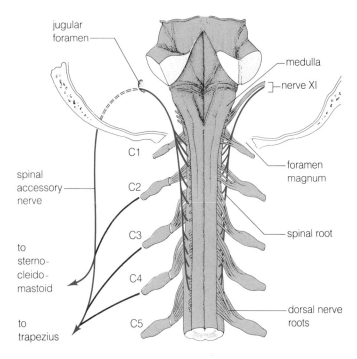

Figure 15-22 Diagram of the distribution of the spinal accessory nerve

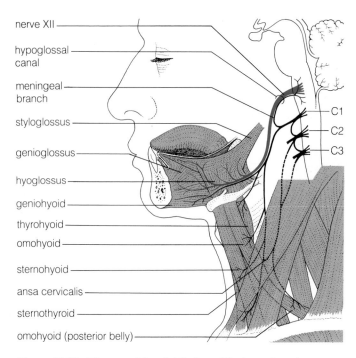

Figure 15-23 Diagram of the distribution of the hypoglossal nerve and the closely related motor nerves of the cervical plexus

provides *parasympathetic fibers* to the *glands* and *involuntary musculature* of most of the *digestive tract.*

Cranial nerve XI, the **spinal accessory** (Fig. 15-22), usually arises from the *first four cervical spinal cord segments* and ascends into the cranial cavity through the *jugular foramen,* where it provides the *motor* innervations to the *sternomastoid* and *trapezius* muscles, two muscles that also receive some direct fibers from the upper cervical spinal nerves.

Cranial nerve XII, the **hypoglossal** (Fig. 15-23), originates from the *medulla,* exits through the *hypoglossal canal,* and supplies *motor* fibers to the skeletal muscles of the *tongue.* It is joined by fibers from the first two cervical nerves that travel with the hypoglossal for distribution to neck muscles.

> When the hypoglossal nerve is paralyzed, the tongue deviates toward the affected side during protrusion of the tongue. A reliable test for the integrity of the hypoglossal is to ask a person to stick out the tongue. When he does this, the genioglossus of the unaffected side pulls the tongue toward the paralyzed side.

CONDUCTION PATHWAYS

The brain receives information from and transmits information to the spinal and cranial nerves. Impulses are carried in the spinal cord to or from higher levels by bundles of functionally related fibers and *tracts.* There are two systems of tracts: in *ascending systems* sensory impulses are carried from spinal nerves to the brain, and in *descending systems* motor impulses are conveyed from the brain to the spinal nerves. Most spinal cord tracts are named according to the origin and termination of the constituent fibers.

Ascending Systems

Ascending systems include the sensory, or afferent, pathways to the brain, which transmit impulses carried from peripheral receptors to the spinal cord by way of afferent neurons. Within the spinal cord the impulses are carried toward the brain over the appropriate *sensory tract.* The chain of neurons (usually three) required for the transmission of a sensory impulse from the receptor to its destination in the cerebral cortex constitutes a *sensory pathway,* or afferent pathway (Fig. 15-24). The *first-order neuron* in this chain is the *receptor neuron.* The cell body

Table 15-5   Ascending (Sensory) Neural Pathways

Sensation	Receptor	Location of Second-Order Neuron	Tract of Second-Order Neuron	Location of Third-Order Neuron	Cerebral Cortex Area
Proprioception, vibratory sense, 2-point discrimination	Muscle and tendon spindles and skin	Nucleus gracilis or cuneatus	Medial lemniscus	Dorsal thalamus	Postcentral gyrus
Pain and temperature	Free nerve endings	Dorsal gray horn	Lateral spinothalamic tract	Dorsal thalamus	Postcentral gyrus
Simple touch and pressure	Many types, including free endings of nerve plexuses and Pacinian corpuscles	Dorsal gray horn	Ventral spinothalamic tract	Dorsal thalamus	Postcentral gyrus

of the *second-order neuron* is located in a *sensory nucleus* in the spinal cord, and in most cases its axon extends to the *dorsal thalamus;* here it synapses with a *third-order neuron,* whose axon conveys the information to the *cerebral cortex.* The principal sensory pathways carrying general sensation from the periphery to the CNS and to the cerebral cortex are summarized in Table 15-5. Each pathway shares an ascending tract with other pathways; we will discuss the components of each tract together.

Proprioception, Vibratory Sense, and Two-Point Tactile Discrimination   (Fig. 15-25.) The receptors for conscious **proprioception** (sensations of position and movement) are endings in muscles and joints; those for **fine touch,** which includes the **vibratory sense** and the **two-point tactile discrimination,** are in the dermis of the skin. Two-point tactile discrimination is the ability to distinguish between being touched by one point or two, such as by a single pin or by two pins held apart. Impulses for these sensations travel via afferent neurons whose axons enter the dorsal region of the spinal cord. In the dorsal gray column, collaterals from these afferents synapse with *association* (internuncial) *neurons* for the *spinal reflexes* previously discussed. The main afferent fiber (still part of the first order neuron) enters the *posterior funiculus* (posterior or dorsal white matter) and continues on the same side until it reaches the *lower medulla.* There it synapses with second-order cells in either the *nucleus gracilis* (for fibers from the lower half of the body) or the *nucleus cuneatus* (upper half of the body).

Fibers of the second-order neurons *cross* to the opposite side of the spinal cord, where they make up a large bundle of fibers called the *medial lemniscus.* This tract passes through the pons and midbrain and terminates in the *dorsal thalamus.* Third-order neurons of the dorsal thalamus relay the impulses to the *postcentral gyrus* of the *cortex* (general sensory area) where the sensations come into consciousness.

Pain and Temperature   (Fig. 15-26.) The receptors for pain are *nonencapsulated* (so-called free sensory) *nerve endings* located in the epidermal layer of the skin and also in deeper structures such as the walls of blood vessels. The receptors for warmth and cold have not been established precisely, but these two modalities have the same pathway as pain within the CNS. Pain is one of the body's defense mechanisms, because the unpleasant stimulus that causes one to take evasive action is usually harmful. Pain receptors adapt slowly, preventing the individual from becoming accustomed to the stimulus.

The peripheral neurons conveying pain and temperature enter the spinal cord and the fibers split into an *ascending* and a *descending branch* that run up and down the spinal cord for one segment in either direction, giving off *collaterals* that synapse with internuncial neurons of a *reflex arc.* This reflex involves more than one spinal cord segment, because the muscles of the extremities are supplied by more than one spinal segment. This arrangement allows the withdrawal of an entire extremity (by flexing all joints) from the source of a painful stimulus. The pain (or temperature) impulse is also transmitted to *second-order*

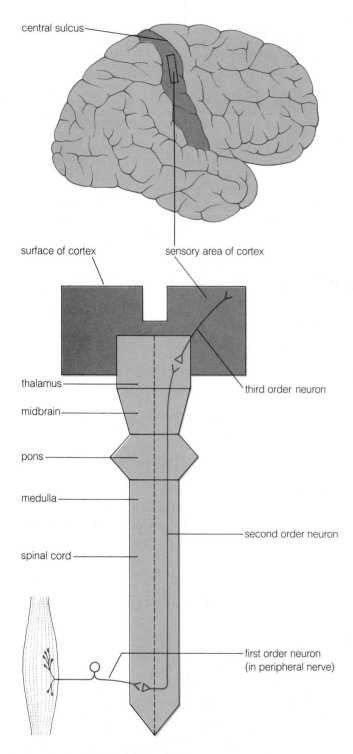

central sulcus

surface of cortex

sensory area of cortex

thalamus

third order neuron

midbrain

pons

medulla

spinal cord

second order neuron

first order neuron
(in peripheral nerve)

Figure 15-24 Diagram illustrating the components of the sensory pathways

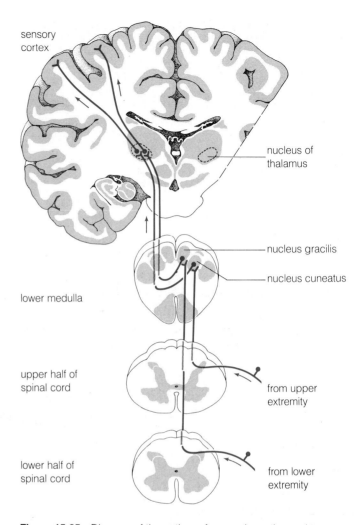

sensory cortex

nucleus of thalamus

nucleus gracilis

nucleus cuneatus

lower medulla

upper half of spinal cord

from upper extremity

lower half of spinal cord

from lower extremity

Figure 15-25 Diagram of the pathway for proprioception and two-point discrimination

neurons whose cell bodies are in the *dorsal horn gray* matter. Fibers of the second-order neurons *cross* to the opposite side

Intractable pain (pain that cannot be treated with drugs) can sometimes be relieved by transection of the spino-thalamic tract (cordotomy) on the side opposite the pain. The body will be insensitive to pain on the opposite side from and inferior to the level of the surgery. Because pain is an important body defense mechanism, this creates a potentially dangerous situation. For example, if one's foot is insensitive to pain, one could be stepping on broken glass without realizing it.

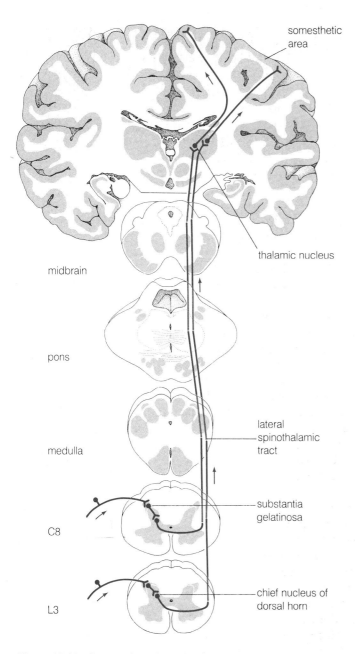

Figure 15-26 Pathway for pain and temperature

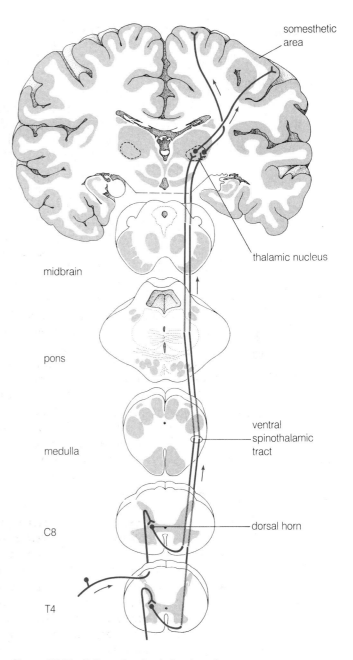

Figure 15-27 Pathway for simple touch and pressure

of the spinal cord, where they form the *lateral spinothalamic tract* in the white matter of the lateral aspect of the spinal cord, which ascends through the spinal cord and brainstem to terminate in the *dorsal thalamus*. *Third-order neurons* then relay the impulses to the *postcentral gyrus* of the *cerebral cortex,* where the sensation is perceived to its highest degree. Pain is

also felt by means of connections at the thalamic level, but it is not as discrete as that sensed by the cerebral cortex.

Simple (General) Touch and Pressure (Fig. 15-27.) The receptors for general tactile sensation and pressure are of many

Table 15-6 Descending (Motor) Neural Pathways

Location of First-Order Neuron	Pathway Crosses	Spinal Cord Tract	Location of Lower Motor Neuron	Effector Organ
Primary motor cortex	Medulla (80%)	Lateral corticospinal	Ventral gray horn	Skeletal muscle
Premotor cortex and subcortical nuclei	Chiefly midbrain	Chiefly rubrospinal	Ventral gray horn	Skeletal muscle

types including *nonencapsulated nerve endings,* such as the *nerve plexuses* around the bulbous portion of hair follicles, and some encapsulated receptors such as *Pacinian corpuscles* identified with pressure. Impulses initiated by stimulation of the receptors travel via fibers of first-order neurons that enter the spinal cord and *ascend* for *four* to *six segments* in the *spinal cord,* giving off numerous *collaterals* along the way that synapse with internuncial neurons for *reflexes* and with *second-order neurons* in the *dorsal horn gray* matter. The latter *cross* to the opposite side of the spinal cord and ascend to the dorsal thalamus in the *ventral spinothalamic tract. Third-order neurons* of the thalamus relay the impulses to the *postcentral gyrus* of the *cerebral cortex.* The result is a poorly localized conscious sensation of either touch or pressure. The ascension from the first-order neurons for a number of cord segments means that there is considerable overlapping of the spinal nerves within the cord, and loss of general tactile sensibility from a small lesion is rarely detectable.

These descriptions, with the exception of the pathway for visceral pain, complete the discussion of ascending pathways for specific modalities that reach the conscious level. These pathways are all crossed, and each requires three neurons to reach the cerebral cortex. The *visceral pain pathway* involves a large number of neurons with short axons that synapse at many levels to form a *multisynaptic chain* that eventually terminates in the *dorsal thalamus* and *hypothalamus.* The neuron chains of this pathway are both crossed and uncrossed.

Ascending pathways are also present in the spinal cord for the transmission of impulses to the cerebellum from receptors in muscles, tendons, and joints, as well as from touch receptors. These data enable the cerebellum to coordinate voluntary muscular activity by regulating the essential excitatory and inhibitory impulses for prime movers, antagonists, and synergists in various activities such as walking, standing, sitting, and skilled movements.

THE MOTOR SYSTEMS

The motor systems of the central nervous system translate sensations and thought into movement. The end product of the regulatory system is movement produced by the contraction of skeletal muscles. The muscle contraction may be controlled at the spinal cord level by a closed circuit in which sensory neurons carrying impulses from a given muscle are connected with motor neurons that transmit impulses back to the muscle. This type of simple reflex may function independently, but its activity is largely controlled by higher centers.

At the highest level of control is the *primary motor cortex,* located in the *precentral gyrus* of the cerebral hemisphere. From this region of the brain, signals are transmitted directly to the spinal cord that can consciously control the impulses that are transmitted to individual skeletal muscles. This is the pyramidal, or voluntary, motor pathway which presides over the entire motor system. All of the other tracts that descend into the spinal cord arise in nuclei of the brainstem. The nuclei of the brainstem, such as the red nucleus of the midbrain and the reticular nuclei scattered throughout the brainstem, receive signals from all higher centers and process them for transmission to the spinal cord. Interposed between the brainstem nuclei and the cerebral cortex are a group of nuclei in the cerebrum that are collectively called basal ganglia. These ganglia receive signals from the cerebral cortex and transmit them to the brainstem which sends impulses on into the spinal cord. This constitutes the *extrapyramidal* (semivoluntary) motor system, regulating activities we really do not have to think about, such as walking.

The cerebellum is interconnected with all levels of both pyramidal and extrapyramidal systems and functions as an overall coordinator of motor activities. It is a highly organized center which receives sensory input regarding all muscle activity. It then transmits signals to the motor pathways that smooth out movements.

Descending Systems Regulating Skeletal Muscular Activity

Descending systems include the motor, or efferent, pathways, which transmit impulses from the brain to effectors in the periphery. The descending systems that regulate skeletal muscle activity are diagrammed in Fig. 15-28 and summarized in Table 15-6. The figure shows only neuron chains that terminate in spinal nerves, but there are similar pathways ending in cranial nerves.

The pyramidal system is the principal connection between the cerebral cortex and the motor neurons of the cranial and spinal nerves, a connection that enables one to execute willed, or voluntary, movements. It consists of cells, called pyramidal cells, that originate in the *primary motor cortex* and descend into the brainstem, where those fibers that descend to the spinal cord and ultimately occupy an area of crossing nerve fibers in the ventromedial aspect of the medulla are referred to as the "pyramids." The pyramidal fibers are the *upper motor neurons,* which act on the *lower motor neurons* to produce skilled movement. The extrapyramidal system, so named because it is outside (*extra-*) the pyramidal tract, consists of motor fibers that also arise in the cortex and through multisynaptic connections act on the motor cells in the anterior horn of the spinal cord. These neurons regulate such semiautomatic movements as walking or chewing. The pyramidal system can take over and modify the activities of the extrapyramidal system at any time.

The Pyramidal System The **pyramidal system** is a *two-neuron pathway* that controls voluntary skeletal muscle activity. It includes the *corticospinal tracts,* which descend in the spinal cord and terminate on lower motor neurons of the spinal nerve, and the *corticobulbar* tracts, which terminate on lower motor neurons of the cranial nerves. "Willed" fine-skilled movements are controlled by neurons whose cell bodies are in the *precentral gyrus* of the *cerebral cortex,* the primary motor area. Although these neurons influence the contraction of individual muscles, control exerted at the conscious level is upon movements involving the cooperation of several muscles. Thus when you pick up a pencil, the entire grasping is willed, with no thought of contraction of the flexors of the finger. As part of this movement the extensors of the fingers are relaxed as the flexors of the fingers are contracted.

Cells of the corticospinal tracts (Fig. 15-28) have axons descending lateral to the thalamus through the brainstem to the lower medulla. Here 80% of them cross to the opposite side, forming the *lateral corticospinal tracts* which syanpse on *lower motor neurons* in the *ventral horn* of the *spinal cord.* The 20% of the upper motor neurons that do not decussate (cross) in the lower medulla descend in the spinal cord in the *ventral corticospinal tracts.* When they reach the spinal cord segment

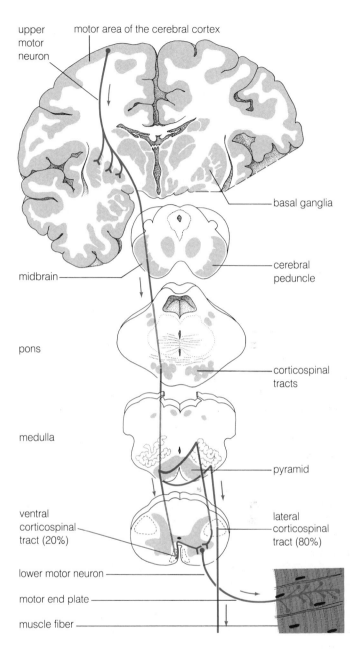

Figure 15-28 Diagram of the pyramidal system

in which they are to terminate, they also cross to the opposite side to synapse on lower motor neurons. Thus the upper motor neurons from one cerebral hemisphere control the skeletal muscles of the *opposite side* of the body.

The upper motor neurons not only stimulate lower motor neurons to cause muscles to contract, but at the same time they *inhibit* other motor neurons supplying *antagonistic muscles* (Fig. 15-29). For example, when the flexors of the forearm are stimulated to contract, the extensors of the forearm are relaxed, a process regulated by the upper motor neurons. The stimulatory effect is mediated by corticospinal fibers synapsing directly with the lower motor neurons that supply the flexors. The inhibitory effect is by collaterals from these same corticospinal fibers synapsing on small association neurons that synapse in turn on the lower motor neurons supplying the extensors. These association neurons are inhibitory; so the flow of impulses to the extensors is stopped, causing these muscles to relax. Because the same upper motor neurons control the two sets of antagonistic muscles, the movement is smooth with the proper balance between contraction of the flexors and relaxation of the extensors.

Voluntary control of the skeletal muscles innervated by cranial nerves is similar to that of muscles supplied by spinal nerves except that the upper motor neuron travels in the corticobulbar tract and terminates on a lower motor neuron in the brainstem. Some fibers of this tract are crossed, and some are not; it depends, for the most part, on whether the lower motor neurons are crossed or uncrossed. Because most fibers do cross over, it is still true that the *cerebral cortex* on one side *controls* the muscles on the *opposite side*. However, lesions destroying the upper motor neurons of one side may lead to muscle weakness (paresis) rather than to paralysis of the muscles involved.

An exception to the crossed and uncrossed pattern of the corticobulbar tracts is seen in those fibers controlling the neurons that innervate the muscles of facial expression of the lower half of the face. All the fibers supplying muscles below eye level are crossed, although those supplying the muscles in the upper half of the face (that is, those to the eyelids and the forehead) are both crossed and uncrossed. Thus a lesion destroying the corticobulbar tract on only one side causes paralysis of the muscles controlling the angle of the mouth on the opposite side of the face but leaves the orbicularis oculi capable of closing the eyelids.

The activity of the pyramidal system is influenced by the cerebellum, which modifies motor impulses at an unconscious level to ensure that the muscle action is appropriate to the situation (Fig. 15-30). Collaterals from the upper motor neurons synapse with neurons in *pontine nuclei* from which impulses are relayed to the cerebellum, conveying information from the pyramidal system. The cerebellum also receives, via the sensory neurons, proprioceptive and other data on the state of contraction of the skeletal muscles being activated by the pyramidal system. The cerebellum then sends impulses by way of the *thalamus* and *red nucleus* of the midbrain to the primary motor area of the cerebral cortex and to the spinal cord, modifying the original signals sent by the cortex. Thus cerebellar regula-

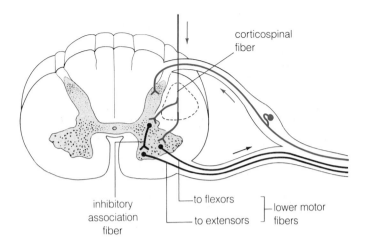

Figure 15-29 Mechanism of control of antagonistic muscles. Transverse section of the spinal cord and associated spinal nerve

tion of the pyramidal system is achieved at two levels, that of the cortex and that of the lower motor neuron. Adjustments in the position of the body while sitting and standing are usually made by reflexes that are modified as a result of sensory input into the cerebellum, without conscious involvement of the motor or sensory areas of the cerebral cortex. Movements such as those occurring during walking are usually accomplished with no conscious effort and are also influenced by the cerebellum acting through the motor neurons or motor cortex. However, such movements can be controlled by motor and premotor areas on the basis of direct sensory input whenever one decides to exert conscious control.

The Extrapyramidal System The **extrapyramidal system** (Fig. 15-31) includes the descending tracts that are not in the pyramids and the cells of origin of these tracts. This system, which coordinates the unconscious components of voluntary movement, involves the *cerebral cortex* and certain *subcortical nuclei,* with the cortex influencing the subcortical nuclei rather than acting directly on lower motor neurons. The fibers arising in the cortex originate in a large area anterior to the precentral sulcus, the so-called *premotor area.* The subcortical nuclei lie deep in the white matter of the cerebral hemisphere and are referred to collectively as the *basal ganglia.* They function as motor nuclei, relaying impulses to the thalamus for distribution to the motor and premotor cerebral cortex, which in turn projects to the red nucleus and scattered cells of the reticular formation in the midbrain, pons, and medulla. From the midbrain levels, with or without additional relays, the impulses reach the

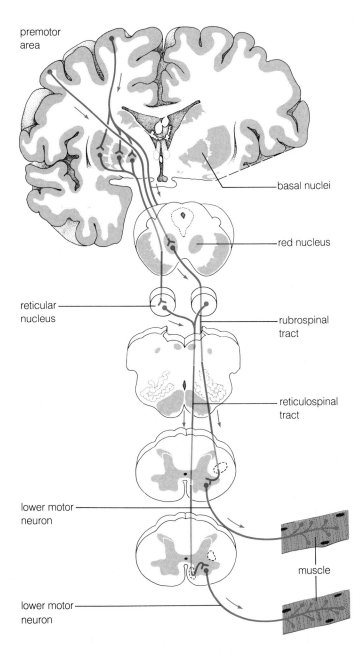

Figure 15-30 Diagram illustrating some of the connections by which the cerebellum influences motor activity of the cerebrum

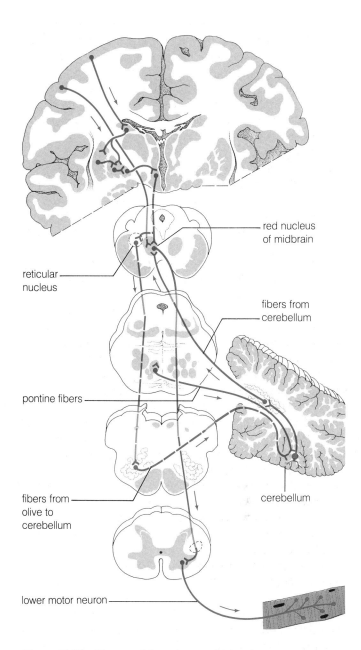

Figure 15-31 Diagram of the extrapyramidal system

spinal cord chiefly over the *rubrospinal tract*. Fibers from this tract terminate indirectly on anterior horn cell bodies of the *lower motor neurons,* the same neurons that receive impulses from the pyramidal system. Stimulation by fibers of either system results in the contraction of skeletal muscles supplied by the lower motor neurons.

The cerebral cortex is capable of modifying the activity of all the components of the extrapyramidal system. The cerebral cortex receives sensory input not only from the external environment but also from the nuclei of the extrapyramidal system, informing the cortex of the activity of the nuclei. The cerebral cortex can then change the activity of the nuclei by *inhibitory*

Clinical Applications: **Diseases of the Nervous System**

The CNS is well protected from traumatic injury by the bones of the skull and vertebral column and by the meninges containing the cerebrospinal fluid, which provides a liquid cushion around the brain and spinal cord. Further protection for the neurons is provided by the neuroglial cells, which surround the nerve cell bodies and fibers and line up along the capillaries within the CNS, helping to form a blood–brain barrier that prevents many potentially harmful substances from leaving the blood vessels and gaining access to the nerve cells. These protective features are counterbalanced by the lack of regeneration within the CNS. Nerve cells are sensitive to oxygen deficiency and begin to die in a few minutes when deprived of a blood supply. Neurons that are lost cannot be replaced because adult neurons do not undergo cell division. Thus the CNS is particularly susceptible to vascular lesions that interrupt the constant flow of blood.

Interruption of the flow of nerve impulses at any point along the pathway from receptor to effector results in alteration of motor response. This functional alteration may be interpreted as a sensory, motor, or sensory and motor disturbance. Many degenerative diseases of the central nervous system alter the flow of nerve impulses at various sites.

Disseminated (Multiple) Sclerosis

This is a degenerative disease of the central nervous system that is a major cause of chronic disability of young adults. It is considered a crippler, and not a killer. Victims of multiple sclerosis live about 85% of the life span of the general population.

The cause of multiple sclerosis is unknown, although the disease was first described by the French neurologist Jean Martin Charcot in 1868. It is characterized by scattered areas of demyelination along the long fiber tracts of the brain and spinal cord. The demyelinated nerve fibers degenerate, and the area becomes filled with a sclerotic patch of tissue. The scattering of the lesions in an unpredictable manner is responsible for the disease's name and for the variability of the symptoms. Both motor and sensory functions are usually affected, with motor loss varying from transient muscular weakness to paralysis of all four extremities. About 50% of those affected become incapacitated within 10 years. There may be periods of remission that may last from a few weeks to several years. The National Multiple Sclerosis Society, an organization founded to support research in the disease and provide services for patients, states that there is at present no treatment that will significantly alter the course of the disease.

Amyotrophic Lateral Sclerosis

A disease somewhat similar to multiple sclerosis that affects only the motor pathways is amyotrophic lateral sclerosis. The disease became well known to the public when Lou Gehrig, baseball star of the New York Yankees and now a member of the baseball Hall of Fame, became a victim of the disease and died in 1941. The symptoms vary depending on which motor neurons are involved. Frequently, there is atrophy of muscles of the forearms, and hands and leg muscles become progressively weaker until they become paralyzed. The condition is progressive and usually

---

or *stimulatory impulses,* adjusting the activity to the correct level. The activity of the extrapyramidal system is also influenced by the cerebellum, which influences all the components that have to do with voluntary muscles.

The *vestibular system,* a branch of the extrapyramidal system, regulates equilibrium for the body as a whole through the activity of the skeletal muscles. Impulses reach the *vestibular nuclei* at the medulla-pons level from the *semicircular canals* in the inner ear. The vestibular nuclei, which relay impulses to the ventral horn cells, constitute still another modifying influence over the lower motor neurons.

The lower motor neuron, with thousands of synaptic terminals impinging on it, is subject to the influence of *spinal nerve afferents, pyramidal tract fibers, extrapyramidal fibers* and discharges from the *cerebellum* and the *vestibular apparatus.* However, only the pyramidal and extrapyramidal systems can initiate a movement such as extending a knee. The extrapyramidal, cerebellar, and vestibular systems regulate and modify the response of the muscles by acting on and through the lower motor neuron. Because its axon is the only connection between the CNS and the skeletal muscle, this neuron is known as the **final common pathway.**

ORGANIZATION OF THE BRAINSTEM AND CRANIAL NERVES

Like the spinal cord, the **brainstem** provides for reflex activity involving its associated nerves and provides for the conduction

fatal within 3 to 20 years. As with multiple sclerosis, there is no effective treatment.

Syringomyelia

A more localized degenerative condition is syringomyelia, in which a cyst develops within the central core of the spinal cord or medulla and damages nerve fibers that cross within the cord or brainstem. It is progressive as the cyst or scar tissue expands. Within the spinal cord the crossing fibers that are destroyed are those conveying impulses for pain and temperature. Loss of the perception of pain and temperature makes these persons susceptible to burns and other injuries. For example, they can suffer a cut in the foot without realizing it. There is no specific treatment, although surgery or radiation therapy may be helpful.

Parkinson's Disease

One of the most crippling degenerative diseases of the nervous system is Parkinson's disease, named for the English physician James Parkinson, who described it as "shaking palsy" in 1817. This is a disease of the basal ganglia or the extrapyramidal motor system and is due to a serious deficiency of dopamine. This deficiency can be relieved by giving the patient L-dopa, a metabolic precursor of dopamine that crosses the blood-brain barrier and is converted into dopamine. A majority of patients with Parkinson's disease show marked improvement following this treatment.

Cerebrovascular Accident (Stroke)

Circulatory diseases of the central nervous system account for about 50% of all organic cerebral disease. More than 200,000 persons die each year in the United States as a result of cerebrovascular diseases; most of these die of a stroke. A stroke may be the result of cerebral hemorrhage, embolism, or thrombosis. If a cerebral vessel ruptures with massive cerebral hemorrhage, it is frequently fatal. *Embolism* is the occlusion of an artery by a mass, such as a blood clot or aggregation of bacteria, that cannot pass through the smaller vessels. *Thrombosis* is a narrowing of the lumen of an artery by disease that slowly occludes the lumen of the vessel. A stroke usually involves loss of consciousness followed by loss of function due to death of nerve cells in the affected area. The severity of the stroke depends upon the areas of the nervous system affected.

Viral and Bacterial Diseases

The nervous system is susceptible to various viral and bacterial organisms, which may be limited to the meninges or may invade the nervous tissue of the brain or spinal cord and affect the neurons themselves. *Meningitis* is the general term used when the infection is restricted to the meninges, and *encephalitis* includes diseases in which the infection involves the CNS. *Poliomyelitis* is a viral disease in which the anterior horn cells of the spinal cord are specifically destroyed as a result of a viral invasion, causing paralysis of the skeletal muscles supplied by the affected neurons.

of impulses to and from higher and lower levels. The form and structure of the brainstem are similar to that of the spinal cord in that there is a basic central core of gray matter continuous with the gray matter of the spinal cord and a peripheral coat of white matter which is also continuous with that of the spinal cord. The morphology of the brainstem is more complex than that of the spinal cord because its central core of gray matter is disrupted in some areas by large masses of white fibers crossing from one side to the other. Thus instead of having the gray matter in continuous columns as in the spinal cord, the gray matter of the brainstem is arranged in *discontinuous masses* of nerve cells (fragments of columns termed nuclei), many of which are associated with cranial nerves. The morphology of the brainstem is complicated still further by the addition of gray masses for which there are no corresponding structures in the spinal

cord. These nuclei may be considered as *relay stations* for impulses going to other centers such as the thalamus or to the cerebellum.

Another major difference between the brainstem and the spinal cord is related to the fact that the brainstem and its associated cranial nerves do not have the same orderly pattern of segmentation found in the cord. With the exception of the pharyngeal segments (represented by cranial nerves V, VII, IX, and X), there is no serial repetition of structures in the head comparable to that found in the trunk. Although each spinal nerve contains the same functional nerve components because each trunk segment contains comparable structures requiring innervation by these components, cranial nerves in the prevailing absence of segmentation vary in their functional components. There also are special receptors (organs of special senses) and

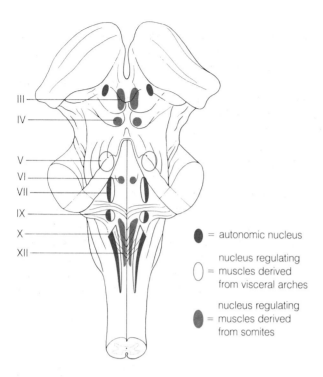

Figure 15-32 Diagram of a dorsal view of the brainstem depicting the relative position of the cranial nerve motor nuclei

= autonomic nucleus

= nucleus regulating muscles derived from visceral arches

= nucleus regulating muscles derived from somites

special effectors (skeletal muscles derived from visceral arch mesoderm rather than from somites) in the head that are not found in the trunk. This situation requires special functional components in addition to the general components that are found in spinal nerves.

The Efferents

The *efferent nuclei* (Fig. 15-32) situated in the more central regions of the brainstem are those whose components are the same as are found in spinal nerves innervating skeletal muscles of somite origin. These nuclei form a discontinuous column of cells located nearest the midline of the central gray matter of the midbrain, pons, and medulla. The extraocular muscles are supplied by the *oculomotor* and *trochlear* nerves originating from *nuclei* in the *midbrain* and by the *abducens* whose *nucleus* is in the *pons*. Muscles of the tongue are innervated by the *hypoglossal* nerve whose cells of origin lie in the *medulla*. The nuclei of these four cranial nerves are serially arranged in a straight line from the upper midbrain through the upper medulla, in an interrupted column of cells in line with the cell column of the anterior gray horn of the spinal cord.

The **motor nuclei** of cranial nerves V, VII, IX, and X, which send fibers to voluntary muscles derived from visceral arches,

form a discontinuous column of cells in the *pons* and in the *medulla,* ventral and lateral to the nuclei of cranial nerves that innervate skeletal muscles of somite origin (Fig. 15-32). The motor nucleus of the **trigeminal nerve** is in the midportion of the pons and contains the cell bodies of the neurons that innervate skeletal muscles derived from the *first visceral arch.* The motor nucleus of the **facial nerve,** located in the caudal portion of the pons in line with the nucleus of the trigeminal, supplies the muscles derived from the *second visceral arch. Nucleus ambiguus,* the motor nucleus common to both the *glossopharyngeal* and *vagus* nerves, forms a continuous column of cells through the upper and mid medulla. The **glossopharyngeal nerve** supplies only the single muscle, the stylopharyngeus, derived from the *third arch.* The **vagus** innervates all the muscles originating from the *fourth arch,* including all the voluntary muscles of the soft palate except the tensor veli palatini, all the pharyngeal muscles except the stylopharyngeus, and all the intrinsic muscles of the larynx. The caudal end of nucleus ambiguus (that is, the vagal end) is continuous inferiorly with the spinal accessory nucleus. Although listed as a cranial nerve the **spinal accessory nerve** arises from the spinal cord, travels upward through the vertebral canal, and passes through the foramen magnum into the cranial cavity. It then exits through the jugular foramen along with the vagus and glossopharyngeal nerves to enter the neck, where it supplies the sternomastoid and trapezius muscles.

The nuclei containing the cells of origin of the *cranial parasympathetic outflow* form a discontinuous column of cells located dorsolateral to the column that supplies skeletal muscles of somite origin (Fig. 15-32). The ciliaris muscle and sphincter of the pupil receive innervation from the oculomotor nerve, whose *preganglionic parasympathetic fibers* originate from the *Edinger–Westphal nucleus* in the *midbrain.* The lacrimal, submandibular, and sublingual glands are supplied by parasympathetic outflow carried by the *facial nerve,* with preganglionics originating in the *superior salivatory nucleus* in the *pons.* The parasympathetic outflow for the parotid gland travels in the *glossopharyngeal nerve,* arising from cell bodies in the *inferior salivatory nucleus* in the *upper medulla.* The *dorsal motor nucleus* of the *vagus* lies just caudal to the inferior salivatory nucleus. It contains parasympathetic cells serving as cells of origin for the preganglionic fibers distributed by the vagus to the small terminal ganglia that lie in the walls of the digestive tract and other viscera related to the digestive tract.

The Afferents

Afferents comparable to the somatic afferents of spinal nerves are carried mainly by the *trigeminal nerve.* Impulses for all modalities of general sensation from the skin of the face and the oral and nasal cavities of the anterior part of the head are conveyed by this nerve. The cell bodies for all the various sensations are in the *ganglion* (semilunar) of the trigeminal except for proprioceptive sense. Those neurons that conduct impulses for *proprioception* from the muscles of mastication are the only peripheral afferent neurons whose cell bodies are located within the CNS. These cell bodies are located mainly in the midbrain in a nucleus called the *mesencephalic nucleus* (of the trigeminal nerve).

The sensory fibers of the trigeminal nerve are *first-order neurons* and carry all of the various modalities: pain and temperature, general (light) tactile (not well localized) sensation, fine or discriminatory tactile sensation, and proprioception. These neurons synapse on *second-order neurons* of the sensory nuclei of the trigeminal, whose axons form the *trigeminothalamic tracts* that cross the midline and terminate in the opposite *dorsal thalamus*. The impulses are then relayed by *third-order neurons* located in the dorsal thalamus to the *postcentral gyrus* of the *cerebral cortex* for conscious appreciation of the sensation.

Cranial nerves VII, IX, and X convey impulses for *general sensation* from the *viscera* innervated by these nerves. It is perhaps coincidental that these same nerves contain fibers for *taste* and that taste and general visceral sensation have essentially the same pathway within the brainstem. First-order neurons for both general sensation and taste have their cells of origin in a ganglion on the cranial nerve in which the fibers are found (VII, IX, and X). Axons enter the brainstem at the pons (VII) or medulla (IX and X) and ascend or descend in the *tractus solitarius*. This tract is a compact bundle of fibers completely surrounded by cells of the nucleus solitarius, composed of second-order neurons of this pathway. Fibers of the second-order neurons are both crossed and uncrossed and join a large tract known as the *medial lemniscus* in which they ascend to the *hypothalamus* and *dorsal thalamus*. Third-order neurons from the dorsal thalamus relay the impulses to an area at the junction of the *precentral* and *postcentral gyri* of the cerebral cortex for conscious perception of *taste*.

WORDS IN REVIEW

Abducens nerve	Oculomotor nerve
Auditory nerve	Olfactory nerve
Basal ganglia	Optic nerve
Cerebellum	Parietal lobe
Cerebrum	Parieto-occipital fissure
Corpus callosum	Peduncle
Extrapyramidal system	Pons
Facial nerve	Proprioception
Folium	Pyramidal system
Frontal lobe	Rostral
Glossopharyngeal nerve	Spinal accessory nerve
Gyrus	Sulcus
Hypoglossal nerve	Temporal lobe
Hypothalamus	Thalamus
Medulla	Trigeminal nerve
Midbrain	Trochlear nerve
Neuraxis	Vagus nerve
Occipital lobe	Vermis

SUMMARY OUTLINE

I. The Cerebrum
   A. The cerebrum is composed of two cerebral hemispheres, two bulky masses of nerve tissue separated by a longitudinal fissure.
   B. The external features include elevations called gyri separated by grooves called sulci or fissures, which serve as landmarks for the boundaries of lobes.
      1. The frontal lobe is rostral to the central fissure and dorsal to the lateral fissure.
      2. The parietal lobe is posterior to the central fissure and rostral to the parieto-occipital fissure.
      3. The occipital lobe is posterior to the parietal lobe, and the temporal is inferior to the parietal and frontal lobes.
      4. The medial aspect of the cerebral hemisphere has the corpus callosum, a large mass of fibers crossing between the two hemispheres.
   C. Internally the cerebral hemispheres have a characteristic pattern of organization with the gray matter at the surface (cerebral cortex) and the white matter located deep to the cortex. Isolated patches of gray matter (basal ganglia) are located near the ventricles.
   D. Functional areas of cerebral cortex include five main sensory and two main motor areas. The remainder is the association cortex, which includes behavior, intelligence, and language functions.

II. The Diencephalon

   A. The diencephalon consists principally of right and left dorsal thalami and hypothalami.

     1. The dorsal thalami are relay centers for all sensations except olfaction.

     2. The hypothalami help regulate autonomic activity and influence the endocrine system.

III. The Brainstem

   A. The midbrain lies caudal to the thalamus, gives rise to the oculomotor and trochlear nerves, and provides a pathway for connections between the higher and lower parts of the CNS.

   B. The pons lies caudal to the midbrain, gives rise to the trigeminal, abducens, facial, and auditory nerves, and provides pathways for connections between the cerebrum and cerebellum and other parts of the CNS.

   C. The medulla lies between the pons and the spinal cord, gives rise to the glossopharyngeal, vagus, and hypoglossal nerves, and serves as a pathway for longitudinal fibers connecting higher and lower levels.

IV. The Cerebellum

   A. The cerebellar hemispheres are separated by the narrow median vermis.

   B. The superior, middle, and inferior cerebellar peduncles connect the cerebellum with the midbrain, pons, and medulla, respectively.

V. The Cranial Nerves

   A. Cranial nerve I, the olfactory nerve, has cell bodies located in the olfactory part of the nasal epithelium and sends processes to the olfactory bulbs through the cribriform plate of the ethmoid bone.

   B. Cranial nerve II, the optic, has cell bodies in the retina and sends processes through the optic foramen to the thalamus.

   C. Cranial nerve III, the oculomotor, passes into the orbit through the superior orbital fissure and supplies the levator palpebrae superioris and four eyeball muscles, the superior, inferior, and medial rectus, and the inferior oblique. It also provides parasympathetic fibers to the ciliaris muscle and the sphincter of the pupil.

   D. Cranial nerve IV, the trochlear, supplies the superior oblique eyeball muscle.

   E. Cranial nerve V, the trigeminal, has three major branches (ophthalmic, maxillary, and mandibular) that provide sensory fibers to the face and oral and nasal cavities, and motor fibers to the muscles of mastication.

   F. Cranial nerve VI, the abducens, supplies the lateral rectus eyeball muscle.

   G. Cranial nerve VII, the facial, supplies taste fibers to the anterior two-thirds of the tongue, motor fibers to the muscles of facial expression, and parasympathetic fibers to the salivary and lacrimal glands.

   H. Cranial nerve VIII, the vestibulocochlear, supplies sensory fibers to the auditory and vestibular receptors of the inner ear.

   I. Cranial nerve IX, the glossopharyngeal, provides taste and general sensory fibers to the posterior one-third of the tongue and general sensory fibers to the pharynx. It provides the motor innervation of the stylopharyngeus muscle.

   J. Cranial nerve X, the vagus, provides motor fibers to most of the muscles of the soft palate, pharynx, and larynx, and sensory fibers to the larynx and the digestive tract to the beginning of the descending colon. It provides parasympathetic fibers to the digestive tract for which it supplies sensory fibers.

   K. Cranial nerve XI, the spinal accessory, provides motor fibers to the sternomastoid and trapezius muscles.

   L. Cranial nerve XII, the hypoglossal, provides the motor innervation of the tongue muscles.

VI. Conduction Pathways

   A. Ascending systems carry sensory impulses from the periphery to the CNS for relay to the cerebral cortex.

     1. The pathway for proprioception, vibratory sense, and two-point tactile discrimination consists of first-order neurons that synapse in nucleus gracilis or cuneatus, second-order neurons that cross and synapse in the thalamus, and third-order neurons that terminate in the postcentral gyrus of the cerebral cortex.

     2. Pain and temperature are conveyed by a pathway that consists of first-order neurons that synapse in the dorsal horn of the spinal cord. Second-order neurons cross and ascend to the thalamus via the lateral spinothalamic tract. Third-order neurons of the thalamus terminate in the postcentral gryus of the cerebral cortex.

     3. The pathway for simple touch and pressure includes first-order neurons that synapse in the dorsal horn gray and second-order neurons that cross to the opposite side and ascend to the thalamus in the ventral spinothalamic tract. Third-order neurons of the thalamus relay the impulses to the postcentral gyrus of the cerebral cortex.

   B. Descending systems regulating skeletal muscles include the pyramidal and extrapyramidal systems.

     1. The pyramidal system is a two-neuron pathway. The upper motor neurons of the precentral gyrus have fibers that mostly cross to the opposite side in the medulla and descend to synapse on anterior gray horn cells of the spinal cord, which then send axons via the spinal nerve to the skeletal muscle supplied. Similar pathways regulate contraction of skeletal muscles innervated by cranial nerves.

     2. The extrapyramidal system consists of fibers that arise in the premotor area of the cerebral cortex or in subcortical nuclei and convey impulses that are relayed to the anterior horn cells, which in turn convey the impulses to skeletal muscles via spinal nerves.

REVIEW QUESTIONS

1. What is the anatomical landmark that serves as the boundary between the frontal and parietal lobes of the cerebral hemisphere?

2. What is the principal difference in the relative positions of the gray and white matter in the spinal cord and in the cerebral hemispheres?

3. Where in the cerebral hemisphere is the primary motor area of the cortex located?

4. What are the anatomical relations of the thalamus to the cerebral hemispheres, the third ventricle, and the hypothalamus?

5. What component of the brainstem contains the cerebral aqueduct?

6. What is the name of the median portion of the cerebellum? What structure connects the cerebellum with the pons?

7. Which of the cranial nerves provides parasympathetic fibers to the ciliaris muscle of the eyeball?

8. The trigeminal nerve provides motor fibers to what functional group of muscles?

9. Name the glands that receive parasympathetic fibers from the facial nerve.

10. What nerve supplies taste fibers to the anterior two-thirds of the tongue?

11. Name three structures that contain muscles that receive motor fibers from the vagus nerve.

12. The first-order neurons of the pathway for proprioception synapse in what nuclei?

13. What is the name of the tract in the spinal cord that conveys impulses for the sensation of pain to the thalamus?

14. Most of the fibers of the pyramidal tracts cross to the opposite side in what component of the CNS?

15. From what area of the cerebral cortex do fibers of the extrapyramidal system arise?

SELECTED REFERENCES

Bowsher, D. *Introduction to the Anatomy and Physiology of the Nervous System.* Philadelphia: F. A. Davis, 1974.

Chusid, J. G. *Correlative Neuroanatomy and Functional Neurology.* 17th ed. Los Altos, California: Lange Medical Publications, 1979.

DeCoursey, R. M. *The Human Organism.* 4th ed. New York: McGraw-Hill, 1974.

Guyton, A. C. *Structure and Function of the Nervous System.* 2nd ed. Philadelphia: W. B. Saunders, 1976.

Hubel, D. H. *et al.* "Special Issue on Neurobiology." *Sci. Am.* 241: 45-232, September 1979.

Stevens, L. A. *Neurons.* New York: Thomas Y. Crowell, 1974.

Watts, G. O. *Dynamic Neuroscience: Its Application to Brain Disorders.* Hagerstown: Harper & Row, 1975.

16

THE AUTONOMIC NERVOUS SYSTEM

After completing this chapter you should be able to:

1. Describe the outflow of the autonomic nervous system (ANS) as a two-neuron chain.

2. Contrast the sympathetic division with the parasympathetic division as to the site of origin of the preganglionic outflow and the neurotransmitter substance released by the postganglionic fibers.

3. Describe the innervation of the heart as an example of dual innervation of organs by the ANS.

4. Describe the relationship of the sympathetic trunk to the preganglionic outflow.

5. Demonstrate an understanding of the origin and distribution of the splanchnic nerves.

6. Name the cranial nerves that contain preganglionic parasympathetic fibers; give the nucleus of origin of the preganglionics of each nerve; name the ganglia containing the cell bodies of the postganglionic fibers; and give the general distribution of the postganglionic fibers of each of the nerves.

7. List the spinal cord segments from which parasympathetic preganglionic fibers arise and give the general distribution of these autonomic fibers.

8. Describe the prevertebral plexuses and demonstrate an understanding of their functional components.

The **autonomic nervous system** (ANS) is defined as a purely *motor system* that innervates all *smooth muscle, cardiac muscle,* and *exocrine glands* (glands that have excretory ducts); it works with the endocrine system to maintain the constancy of the internal environment of the body (**homeostasis**). The ANS is "automatic" in the sense that most of its functions are not carried out at the conscious level, but it is anatomically and functionally integrated with the rest of the nervous system. Although it is defined for descriptive purposes as a purely motor system, it receives constant input (by way of reflex arcs and higher levels of the CNS) from visceral afferents that course with the autonomic fibers in most of the nerves of the body.

The outflow of the ANS is a *two-neuron chain* (Fig. 16-1). A **preganglionic cell body** located within the CNS gives rise to a fiber (axon) that terminates upon a multipolar neuron in a peripheral *autonomic ganglion*. The axon of the ganglion neuron, called the **postganglionic fiber,** terminates upon smooth muscle, cardiac muscle, or the viscus supplied. Because

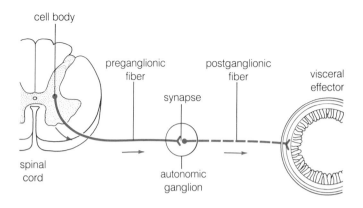

Figure 16-1 Preganglionic and postganglionic neurons of the autonomic nervous system

there are about thirty-two ganglion cells for each of the preganglionic neurons, a single preganglionic neuron may synapse on a number of ganglion cells. Thus a relatively few centrally located neurons may have a broad effect upon peripheral structures.

Anatomically and functionally the ANS forms two divisions (Fig. 16-2): the **sympathetic** nerves originate from *thoracic and upper lumbar spinal cord segments* (T1 to L2) and release **norepinephrine** at most of their *postganglionic terminals* on the effectors; the **parasympathetic** nerves arise from the *brainstem* and *sacral spinal cord segments S2 to S4,* and their *postganglionics* release **acetylcholine.** Norepinephrine and acetycholine are chemical mediators that carry the impulse from the nerve to the effector organ. The characteristics of the sympathetic and parasympathetic divisions are summarized in Table 16-1.

Many organs of the body are innervated by both divisions of the ANS, with one division stimulating or increasing the activity of the organ and the other having the opposite effect. For example, the sympathetic division sends fibers to the heart that terminate on the pacemaker and directly on cardiac muscle. Stimulation of the sympathetic outflow to the heart *increases* the rate of the heartbeat and the force of contraction of heart muscle, raising the cardiac output. The parasympathetic division also innervates the heart via cardiac branches of the vagus nerve. Stimulation of the vagus *reduces* the rate of the heartbeat and the force of contraction of heart muscle, lowering the cardiac output. Although the heart does not require innervation to beat, its rate of contraction is regulated by the ANS with some influence by hormones. The hormones **norepinephrine** and **epinephrine,** produced by the suprarenal glands, have effects similar to those of the sympathetics; these hormones increase the rate of heartbeat when they reach the heart via the blood stream.

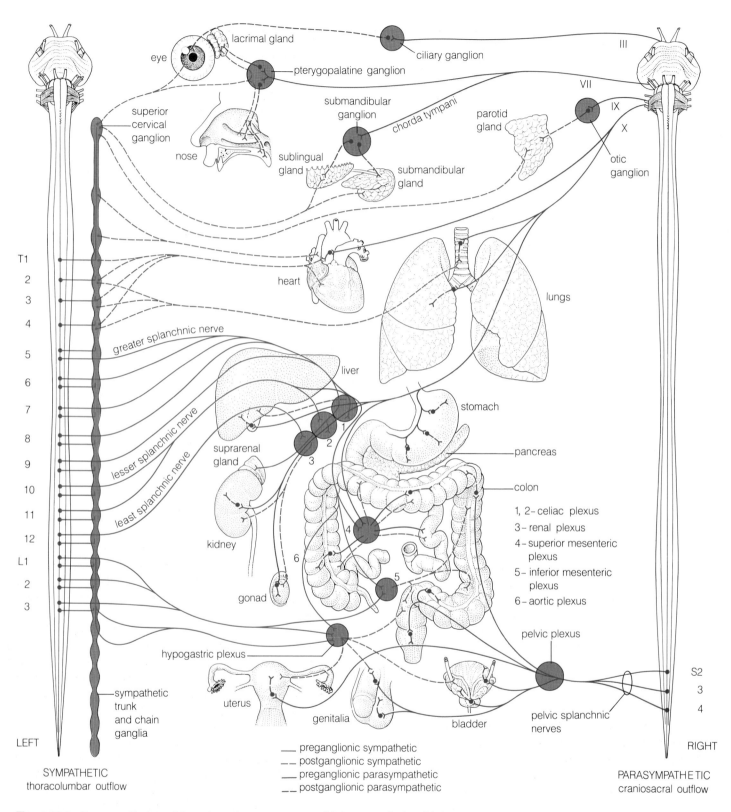

Figure 16-2 Summary diagram of the autonomic nervous system: (**a**) the sympathetics; (**b**) the parasympathetics.

Division	Preganglionic Outflow	Location of Cell Bodies of Postganglionics	Distribution
SYMPATHETICS	Spinal cord segments T1 to L2	Sympathetic trunk ganglia or collateral ganglia	Dilator of pupil, blood vessels, sweat glands, smooth muscle of hair follicles, heart
PARASYMPATHETICS			
Cranial outflow	Oculomotor (III)	Ciliary ganglion	Ciliaris muscle, sphincter of pupil
	Facial (VII)	Pterygopalatine ganglion	Lacrimal gland, mucosa of nasal cavity and palate
		Submandibular ganglion	Oral mucosa, submandibular and sublingual glands
	Glossopharyngeal (IX)	Otic ganglion	Parotid gland
	Vagus (X)	Organ supplied	Heart, lower respiratory tract, digestive tract to left colic flexure
Sacral outflow	Spinal cord segments S2 to S4	Organ supplied	Descending and sigmoid colon, rectum, urinary system, reproductive tract

Table 16-1   The Autonomic Nervous System

THE SYMPATHETIC DIVISION (THORACOLUMBAR OUTFLOW)

The **preganglionic sympathetic fibers,** or thoracolumbar outflow (Fig. 16-3), arise from cell bodies located in the inter-mediolateral gray columns of *spinal cord segments T1 to L2* and leave the spinal cord by way of the *ventral roots* of the spinal nerves. The preganglionic fibers of a spinal nerve leave the nerve just outside the vertebral column by way of a *white communicating ramus;* they then enter a ganglion of the *sympathetic trunk* at or near the level of their segment of origin. The sympathetic trunks with their chain (paravertebral) ganglia extend vertically on either side of the vertebral column adjacent to the vertebral bodies. Preganglionic fibers may synapse in the ganglia at the level of the origin of the fibers or in chain ganglia at higher or lower levels. Postganglionic fibers arising from the trunk ganglion enter the spinal nerves via a *gray communicating ramus* and are distributed to *blood vessels, smooth muscles* of *hair follicles,* and *sweat glands* in the skin of the trunk and limbs. Postganglionic fibers to the head (Fig. 16-4) arise in the *superior cervical ganglion* of the sympathetic trunk and are distributed chiefly through the nerve plexuses surrounding blood vessels.

Some preganglionic sympathetic fibers pass through the chain ganglia of the sympathetic trunk without synapsing and form part of the **splanchnic,** or visceral, **nerves** (Fig. 16-5). The splanchnic nerves penetrate the diaphragm to synapse on neurons in **collateral ganglia** (prevertebral, or outlying, ganglia) associated with the major branches of the aorta (celiac, superior and inferior mesenteric, and renal arteries). *Postganglionic fibers* from these ganglia are distributed to the *viscera* in *plexuses* surrounding *blood vessels* and innervate the smooth muscles of the blood vessels that supply the gut. Stimulation of the sympathetics causes contraction of the smooth muscle of the blood vessels. This reduces the size of the vessel lumen and reduces blood flow.

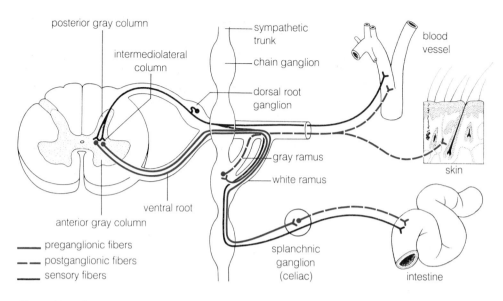

Figure 16-3 Two pathways for distribution of sympathetic fibers

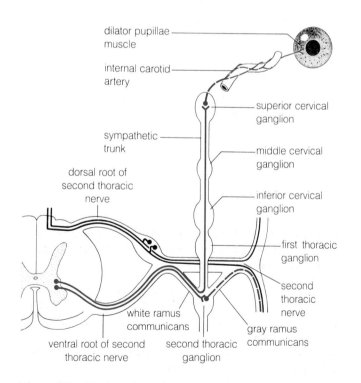

Figure 16-4 Diagram of the sympathetic pathway to the eyeball

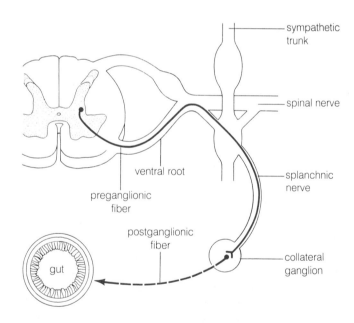

Figure 16-5 Diagram of the sympathetic pathway to the digestive tract

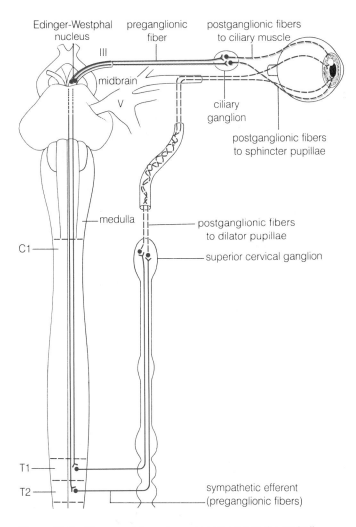

Figure 16-6 Diagram of the autonomic pathways to the eyeball

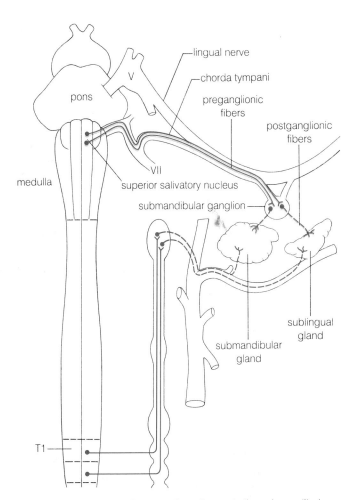

Figure 16-7 Diagram of autonomic pathways to the submandibular and sublingual salivary glands

THE PARASYMPATHETIC DIVISION (CRANIOSACRAL OUTFLOW)

Preganglionic parasympathetic fibers, frequently called the *craniosacral outflow,* arise from cells located in the brainstem and from spinal cord segments S2, S3, and S4. The preganglionic fibers synapse on neurons located in or near the viscus (visceral organ) supplied. Thus preganglionic fibers are *relatively long* and postganglionics are *relatively short* compared with those of the sympathetic division. The parasympathetics may be divided according to their place of origin into *cranial outflow* and *sacral outflow.*

Cranial Outflow

The parasympathetics arising from the brainstem, called the cranial outflow, are components of the oculomotor, facial, glossopharyngeal, and vagus nerves.

Oculomotor Nerve (III)   The preganglionic fibers arise from a nucleus of the oculomotor nerve (the Edinger–Westphal nucleus) in the *midbrain* and synapse with neurons in the **ciliary ganglion** in the *orbit* (Fig. 16-6). Postganglionic fibers supply the *sphincter muscle* of the *pupil* (responsible for reducing the amount of light entering the eye) and the *ciliaris muscle* (the muscle of accommodation of the lens of the eye).

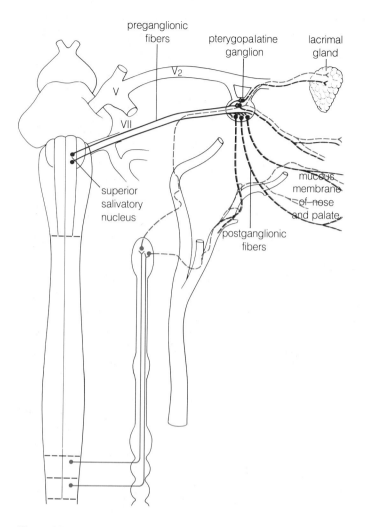

Figure 16-8 Diagram of autonomic pathways to the lacrimal gland

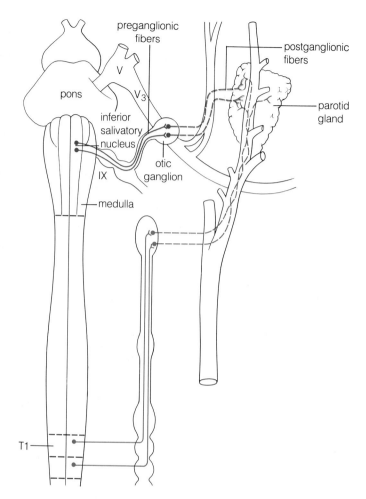

Figure 16–9 Diagram of autonomic pathways to the parotid salivary gland

Facial Nerve (VII) Preganglionic fibers arise from the **superior salivatory nucleus** of the facial nerve, situated in the *pons,* and synapse in the **submandibular ganglion** near the angle of the mandible (Fig. 16-7) and in the **pterygopalatine ganglion** in the *pterygopalatine fossa* (Fig. 16-8). Postganglionics from the submandibular ganglion supply the *submandibular* and *sublingual glands* (salivary glands), increasing their rates of secretion. Postganglionics from the *pterygopalatine ganglion* are distributed to the *lacrimal gland* as well as to secretory elements in the *mucous membrane* of the *nasal cavity.* The innervation of the lacrimal gland is extremely important because its failure to secrete tears results in a painful drying of the cornea that can cause its destruction.

Glossopharyngeal Nerve (IX) Preganglionic fibers from the **inferior salivatory nucleus** of the glossopharyngeal nerve in the *medulla* synapse in the **otic ganglion,** located anteromedial to the temporomandibular joint. Postganglionic fibers are distributed to the *parotid gland* (a salivary gland), increasing its rate of secretion (Fig. 16-9).

Vagus nerve (X) Preganglionic fibers arise from the *dorsal motor nucleus* of the *vagus* in the *medulla* and course with the main vagus nerve into the thoracic and abdominal cavities (Fig. 16-10). The ganglionic cells upon which the preganglionic cells terminate are located in tiny ganglia near or actually in the organs supplied. They are named according to the viscus sup-

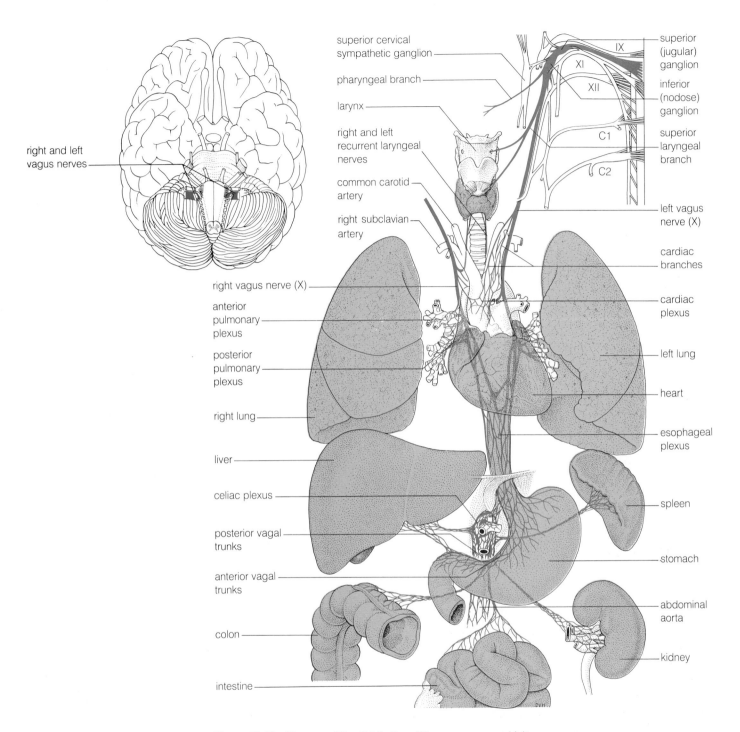

right and left
vagus nerves

superior cervical
sympathetic ganglion

pharyngeal branch

larynx

right and left
recurrent laryngeal
nerves

common carotid
artery

right subclavian
artery

IX

XI

XII

C1

C2

superior
(jugular)
ganglion

inferior
(nodose)
ganglion

superior
laryngeal
branch

left vagus
nerve (X)

cardiac
branches

cardiac
plexus

left lung

heart

esophageal
plexus

spleen

stomach

abdominal
aorta

kidney

right vagus nerve (X)

anterior
pulmonary
plexus

posterior
pulmonary
plexus

right lung

liver

celiac plexus

posterior vagal
trunks

anterior vagal
trunks

colon

intestine

Figure 16-10 Diagram of the distribution of the vagus nerve, which
carries preganglionic parasympathetic fibers. Postganglionic fibers
arise in the wall of the organ supplied. Inset is inferior view of the
brain showing origins of right and left vagi.

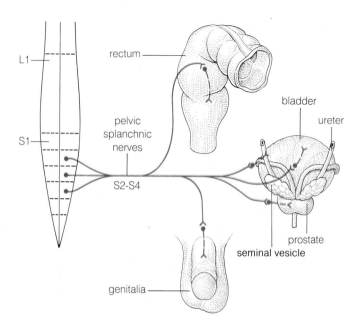

Figure 16–11 Diagram of the distribution of the sacral parasympathetics

rectum

pelvic
splanchnic
nerves

L1

S1

S2-S4

bladder

ureter

prostate

seminal vesicle

genitalia

plied; for example, there are ganglia called pulmonary, cardiac, and terminal (or enteric, meaning in the wall of the gut). The vagal parasympathetic fibers cause constriction of the bronchioles, slow the heart rate, increase production of digestive enzymes, and stimulate peristalsis of the digestive tract.

Sacral Outflow

Spinal cord sacral segments S2 through S4 give rise to preganglionic fibers that exit by way of the ventral root of the corresponding sacral spinal nerve. The preganglionics then leave the spinal nerves, forming part of the *nervi erigentes* (pelvic splanchnics or pelvic nerves) for distribution to the *pelvic viscera* (Fig. 16-11). The postganglionic cell bodies are found in very small ganglia in or near the organ innervated.

PREVERTEBRAL PLEXUSES

There are large networks of nerves that serve as pathways for the distribution of both sympathetic and parasympathetic fibers in the trunk region. These networks, called the **prevertebral plexuses,** lie in close proximity to the large blood vessels found

just anterior to the vertebral column. The plexuses are a mixture of sympathetic fibers from splanchnic nerves and parasympathetic fibers from branches of the vagus nerve. The ganglia present in the plexuses are sympathetic (collateral) ganglia. From these plexuses of the thorax, abdomen, and pelvis, autonomic fibers are distributed to all the viscera, mostly on the surface of blood vessels.

The plexuses are usually named according to the organs supplied or to the location; for example, the *cardiac* and *right* and *left pulmonary plexuses* in the thorax supply the heart and right and left lungs. In the abdomen the **celiac plexus** (solar plexus) is located near the origin of the celiac artery, a large artery that supplies the liver, spleen, pancreas, and the digestive tract of the upper abdomen. The autonomic fibers to these organs are distributed via branches of the celiac artery, for the nerve fibers course on the surface of these vessels.

The Hypothalamus

The **hypothalamus** is a portion of the diencephalon that regulates a great variety of autonomic activities. It not only receives data from other parts of the central nervous system; but it also contains cells that are sensitive to changes in the temperature of the blood and to substances in the blood such as glucose. The hypothalamus receives input from the *olfactory cortex* (smell and emotions), the *limbic lobe* (emotions), the *amygdala* (emotions), *prefrontal cortex* (moods), and the *dorsal thalamus* (visceral and somatic afferents). It exerts its influence over the *autonomic nervous system* and *endocrine system;* indeed, it is the principal integrator of these two regulatory systems. Influences of emotions upon such factors as the rate of heart beat and breathing are mediated by the centers for these activities in the pons and medulla.

Activity within the hypothalamus is transmitted through the dorsal thalamus to the cerebral cortex, where it has an influence upon emotions and moods. The hypothalamus thus has a role in determining whether an individual has a feeling of well-being or is "down in the dumps," emotional states determined chiefly by the prefrontal cortex and the components of the limbic system.

In addition to its neural connections affecting the brain the hypothalamus has a neurosecretory function that affects the endocrine system. Hormones released by the posterior lobe of the pituitary gland are synthesized in the hypothalamus; the release of hormones produced by the anterior lobe is controlled by chemical substances produced in the hypothalamus. On the other hand, the suprarenal medulla and the pineal gland (both endocrine glands) are regulated by direct nervous connections with the hypothalamus.

Clinical Applications: **ANS Disorders**

Autonomic innervation is not essential to the functioning of visceral tissues, but overaction or underaction of this innervation impairs normal function, at least until the intrinsic capacity of the organ to function without innervation takes over. There are many disorders, such as cardiovascular problems, that can be influenced by the autonomic nervous system, but in many cases the primary causes of the disorders are from a source other than the nervous system.

Underactivity of the sympathetic nerves to the smooth muscle of blood vessels produces a type of hypotension due to low vasomotor tone. If a person with low vasomotor tone is suddenly brought to an upright position and kept immobile, fainting is likely, due to insufficient blood flow through the brain because of the enforced immobility while erect.

Overactivity of vasomotor nerves can produce a neurogenic hypertension that may have harmful effects in specific parts of the body. In *Raynaud's disease* there is a spasmodic contraction of blood vessels supplying the digits, which consequently receive an inadequate supply of blood. The tissue may die and gangrene may set in. The condition may be corrected by cutting the preganglionic sympathetic fibers supplying the area.

Overactivity of the parasympathetic supply of the gastrointestinal tract can cause excessive production of hydrochloric acid, which when coupled with reduced production of mucin and changes in blood flow, may cause development of ulcers. Gastric ulcers are believed to result from reduction in secretion of mucin, while duodenal ulcers are probably the result of too much hydrochloric acid.

A few conditions are due specifically to disturbances of the autonomics. In a condition known as *Horner's syndrome* the patient has distinctive unilateral symptoms: a *drooping eyelid*, a *constricted pupil, dry skin,* and *flushed face*. These symptoms indicate failure of the sympathetics in the head. The drooping eyelid suggests the failure of the *superior tarsus* (smooth muscle supplied by sympathetics) to contract. The constricted pupil shows that contraction of the *sphincter* of the *pupil* (innervated by parasympathetics) is unopposed by contraction of the dilator of the pupil (supplied by sympathetics). The dry skin shows that the patient's *sweat glands,* innervated by sympathetic postganglionic fibers, are not functioning on the affected side. Finally, the flushed face indicates that the small *arterioles* in the skin of the face have *dilated,* increasing the blood flow in the skin capillaries. The blood in the superficial vessels imparts a reddish tinge to the skin, referred to as flushing or blushing.

All the symptoms can be explained on the basis of the interruption of sympathetic innervation. The lesion can be in the *thoracic spinal cord (spinal cord segments T1 and T2),* in the *cervical* **sympathetic trunk,** or in the *superior cervical sympathetic ganglion.* A lesion in any of these structures could block the flow of impulses of the sympathetics to the entire head region of one side and produce the symptoms described.

WORDS IN REVIEW

Acetylcholine	Preganglionic fiber
Celiac plexus	Prevertebral plexus
Ciliary ganglion	Pterygopalatine ganglion
Collateral ganglia	Splanchnic nerve
Epinephrine	Submandibular ganglion
Homeostasis	Superior cervical ganglion
Inferior salivary nucleus	Superior salivary nucleus
Norepinephrine	Sympathetic trunk
Postganglionic fiber	

SUMMARY OUTLINE

I. The Autonomic Nervous System (General Characteristics)
   A. The ANS is a purely motor system.
   B. The functional unit of the ANS is a two-neuron chain with the cell body of the first neuron (preganglionic) located in the CNS and that of the second neuron (postganglionic) located in a ganglion outside the CNS.

II. The Sympathetic Division
   A. Most sympathetic postganglionic fibers release norepinephrine as the neurotransmitter substance.
   B. The sympathetic preganglionic outflow arises from the intermediolateral cell column of spinal cord segments T1 to L2.

C. The sympathetic postganglionic fibers arise from cell bodies in the sympathetic trunk ganglia or in collateral ganglia associated with major branches of the aorta.

D. Most of the sympathetic postganglionic fibers reach their final destination in plexuses around blood vessels.

III. The Parasympathetic Division

A. Most parasympathetic postganglionic fibers release acetylcholine as the neurotransmitter substance.

B. The parasympathetic preganglionic fibers arise from the brainstem nuclei of four cranial nerves (cranial outflow) and from spinal cord segments S2 to S4 (sacral outflow).

C. Preganglionic fibers arising from the Edinger-Westphal nucleus of the oculomotor nerve synapse in the ciliary ganglion, from which postganglionics are distributed to the sphincter of the pupil and the ciliaris muscle (for accommodation of the lens).

D. Preganglionic fibers from the superior salivatory nucleus of the facial nerve are distributed to the submandibular and pterygopalatine ganglia, which send fibers to the submandibular and sublingual salivary glands and to the lacrimal glands in the orbits.

E. Preganglionic fibers from the inferior salivatory nucleus of the glossopharyngeal nerve are distributed to the otic ganglion, which sends postganglionic fibers to the parotid salivary gland.

F. Preganglionic fibers from the dorsal motor nucleus of the vagus are distributed to ganglia in or near the heart, lungs, and digestive tract, from which postganglionic fibers are distributed to these organs.

G. Preganglionic fibers from sacral spinal cord segments S2, S3, and S4 are distributed to ganglia found in close association with pelvic viscera. These ganglia provide postganglionic fibers to the digestive tract and to the urogenital organs of the pelvis.

IV. Prevertebral Plexuses

A. Prevertebral plexuses are networks that serve as pathways for the distribution of both sympathetic and parasympathetic fibers.

B. The ganglia present in plexuses are usually sympathetic (collateral) ganglia.

C. In the abdomen these plexuses are located near the point of origin of branches of the aorta, such as the celiac, and are distributed to abdominal viscera on the surface of branches of these arteries.

REVIEW QUESTIONS

1. What is the principal anatomical difference between the voluntary motor system, which stimulates skeletal muscles to contract, and the ANS, which regulates many involuntary activities?

2. Which of the divisions of the ANS has preganglionic fibers that are relatively much longer than their postganglionic fibers?

3. What is the neurotransmitter substance that is released by the postganglionic parasympathetic fibers?

4. What are the effects of the sympathetics and parasympathetics upon the rate of heartbeat?

5. What are the destinations of preganglionic sympathetic fibers that do not synapse in ganglia of the sympathetic trunk?

6. What are the principal functions of the motor fibers in the splanchnic nerves?

7. What structures are innervated by the autonomic fibers of the oculomotor nerve?

8. Name the ganglia associated with the parasympathetics of the facial nerve.

9. What is the major structure supplied by parasympathetics of the glossopharyngeal nerve?

10. What components of the digestive system receive parasympathetics from the vagus nerve?

11. From what spinal cord segments do the sacral preganglionic parasympathetics arise?

12. Where are the prevertebral plexuses located and what types of nerve fibers do they contain?

SELECTED REFERENCES

Bowsher, D. *Introduction to the Anatomy and Physiology of the Nervous System.* Philadelphia: F. A. Davis, 1974.

Chusid, J. G. *Correlative Neuroanatomy and Functional Neurology.* 17th ed. Los Altos, California: Lange Medical Publications, 1979.

DeCoursey, R. M. *The Human Organism.* 4th ed. New York: McGraw-Hill, 1974.

Guyton, A. C. *Structure and Function of the Nervous System.* 2nd ed. Philadelphia: W. B. Saunders, 1976.

Hubel, D. H., *et al.* "Special Issue on Neurobiology." *Sci. Am.* 241: 45-232, September 1979.

Stevens, L. A. *Neurons.* New York: Thomas Y. Crowell, 1974.

Watts, G. O. *Dynamic Neuroscience: Its Application to Brain Disorders.* Hagerstown: Harper & Row, 1975.

17

THE SPECIAL SENSE ORGANS

OBJECTIVES

After completing this chapter you should be able to:

1. List the components of the olfactory system and state the function of each component.

2. Describe the taste receptors and trace the flow of the impulses they initiate to the cortical area responsible for the perception of taste.

3. Describe the structure of the eyeball, relating structure to function.

4. Diagram or describe the principal visual pathways.

5. Diagram or describe the pathway responsible for accommodation in near vision.

6. Diagram or describe the light reflex.

7. Diagram or describe the visual reflexes involving skeletal muscles.

8. Describe the structure of the external and middle ears.

9. Describe the structure of the organ of hearing and trace the flow of impulse that results in the perception of sound.

10. Describe the vestibular apparatus and demonstrate an understanding of how it functions in maintaining equilibrium.

Changes in the external environment are detected by various receptors that are distributed throughout the body. Nerve impulses from receptors such as free or encapsulated nerve endings convey information to the brain that is interpreted as "general" sensations: pain, temperature, touch, and general muscle sense. Other sensations such as vision require highly specialized organs as receptors. These organs are restricted to the head and are called "special" senses.

The organs responsible for the five special senses (smell, taste, vision, hearing, and equilibrium) are protected by the skull. Impulses travel from the sensory receptors over certain cranial nerves and eventually reach the cerebral cortex where there are special functional areas that interpret the impulses. *Smell* and *taste* are related to the ingestion of food and are considered *visceral sensations; vision, hearing,* and *equilibrium* influence somatic structures such as skeletal muscles and are considered *somatic senses*.

The structure of each organ and some of the principal central connections of each organ will be discussed in this chapter. Table 17-1 summarizes some of the principal features of the special senses.

OLFACTION

The *olfactory system,* responsible for **olfaction,** the sense of smell, consists of the *olfactory epithelium, olfactory nerves, olfactory bulbs, olfactory tracts,* and the *olfactory areas* of the *cerebral cortex* (see Fig. 15-14). The olfactory epithelium is that portion of the nasal epithelium in the superior aspect of the nasal cavity beneath the cribriform plate of the ethmoid bone (Fig. 17-1a). The olfactory epithelium contains supporting columnar epithelial cells and scattered *bipolar neurons* with nuclei in the epithelial sheet. The dendrites of the neurons extend to the free surface of the mucous membrane from which long cilia,

Table 17-1	Special Senses			
Sensation	Receptor Structure	Location	Cranial Nerves	Cortical Area
Smell	Olfactory epithelium	Nasal cavity	Olfactory	Olfactory area, temporal lobe
Vision	Retina of eyeball	Eyeballs within orbits	Optic	Visual area of occipital lobe
Taste	Taste buds	Tongue, epiglottis	Facial, glossopharyngeal, vagus	Insular cortex
Hearing	Organ of Corti	Inner ear (cochlear duct)	Auditory part of cranial nerve VIII	Auditory area of temporal lobe
Balance	Vestibular apparatus (macula)	Inner ear (saccule, utricle, semicircular canals)	Vestibular portion of cranial nerve VIII	Lower part of postcentral gyrus of parietal lobe

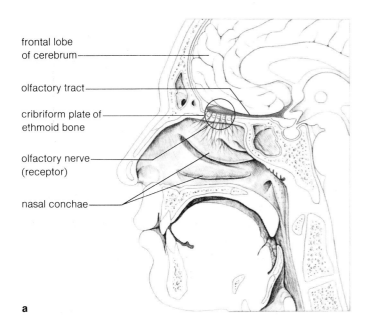

frontal lobe
of cerebrum

olfactory tract

cribriform plate of
ethmoid bone

olfactory nerve
(receptor)

nasal conchae

a

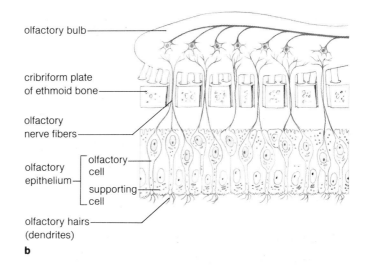

olfactory bulb

cribriform plate
of ethmoid bone

olfactory
nerve fibers

olfactory
epithelium — olfactory
cell

supporting
cell

olfactory hairs
(dendrites)

b

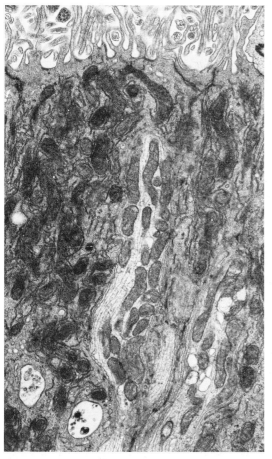

c

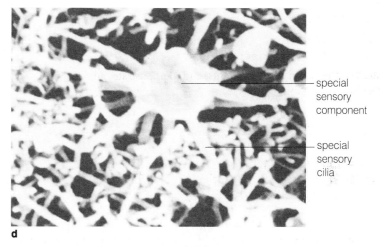

special
sensory
component

special
sensory
cilia

d

Figure 17-1   The olfactory receptors, located in the nasal cavity. (**a**)
Enlargement of a portion of the olfactory area. (**b**) The arrangement of
cells and nerve fibers. (**c**) Electron photomicrograph of the olfactory
epithelium of the mouse, × 13,000. (**d**) A scanning electron micrograph of the special sensory endings of olfactory neurons of the
mouse, × 7800. (Courtesy of D. H. Matulionis.)

called olfactory hairs, project into the nasal cavity (Fig. 17-1c). Volatile chemical substances in inhaled air are dissolved in the mucus of the olfactory epithelium, stimulating the ciliated dendrites of the bipolar neurons. The axons of these neurons convey impulses through the openings in the *cribriform plate* to the olfactory bulbs (Fig. 17-1b). The *olfactory bulbs* are central nervous system structures lying on the inferior aspect of the cerebral hemispheres, directly over the cribriform plate and on either side of the crista galli. From the olfactory bulbs, the *olfactory tracts* carry the impulses caudalward to *medial* and *lateral olfactory centers* (Fig. 17-2). Olfactory sensation comes into consciousness in the *lateral olfactory centers* in the *temporal cortex*. The *medial olfactory centers* located in the *frontal lobe* are for *reflexes* only. The sense of smell not only enables us to detect substances in the air that are pleasant or that might be harmful but also functions in triggering autonomic responses. For example, a pleasant odor stimulates the parasympathetic outflow to the salivary glands, increasing the production of saliva. If the odor is unpleasant, the response may be nausea. The sense of smell also influences the emotions, and the response to emotion is regulated by the hypothalamus.

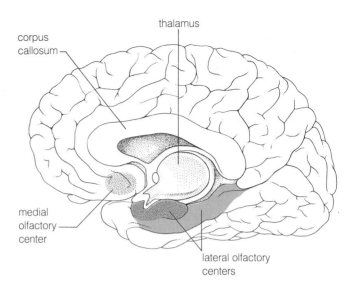

Figure 17-2 Midsagittal section of cerebrum illustrating olfactory areas of the cortex

The sense of smell is not very important in humans. However, it does enable us to detect some substances such as treated natural gas that are harmful. The sense of smell is closely related to the sense of taste, and the pleasant odor of many foods adds to our appreciation of their taste. The sense of smell gradually decreases with age.

Premature loss of the sense of smell results from damage to the receptor cells, the olfactory bulb, or the olfactory tract. The condition is known medically as *anosmia*. Many soldiers in World War I lost their sense of smell due to exposure to nitrogen mustard gas, which destroyed the olfactory epithelium.

Lesions of the temporal lobe in areas concerned with olfaction may produce olfactory hallucinations, epileptic seizures, or a combination of both. In experimental animals, lesions of parts of the brain that are known to be involved with olfaction often cause a greatly increased sex drive.

TASTE

The sense organs for *taste* are small clusters of cells called **taste buds,** usually located on the sides of the epithelial papillae on the dorsum (upper surface) of the tongue (Fig. 17-3a). Each taste bud consists of two types of cells, taste cells and supporting cells, that form a discrete oval mass within the tongue epithelium. It is exposed to the fluid in the oral cavity by means of a small opening, the *taste pore* (Fig. 17-3b, c). Although the specific functions of the taste bud cells are unknown, the cells of each type bear microvilli at the surface toward the pore, which may be involved in initiating taste impulses. Afferent fibers of *cranial nerves VII, IX,* or *X* terminate as fine filaments forming a *neural net* around the cells of the taste buds. Food in the mouth initiates impulses that are transmitted to the brainstem via the afferent fibers of one of the three cranial nerves. The impulses are then relayed by neurons of the **nucleus solitarius** in the medulla to the *dorsal thalamus*. Fibers from nucleus solitarius are both crossed and uncrossed and reach both sides of the thalamus. The impulses are then sent on to the taste area of the cerebral cortex, which may lie in the inferior part of the precentral and postcentral gyri or, more likely, in the adjoining portion of the insular cortex.

Loss of the sense of smell greatly reduces our appreciation of taste sensations, causing all food to seem bland. Some foods (for example, coffee) actually have no taste, and what we interpret as their taste is really odor. With a nose clip on and the eyes closed, raw onion and apple taste alike.

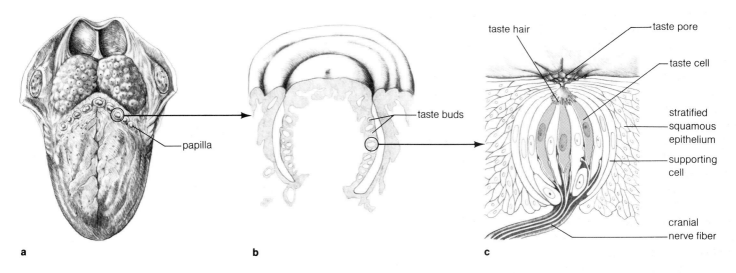

Figure 17-3 Most taste receptors are located in taste buds on the surface of the tongue (**a** and **b**). The structure of a taste bud is shown in (**c**).

VISION

The special sense organs for *vision* (sight) are located in the eyeballs, which occupy the bony orbits on either side of the nose. Each eyeball consists of a light-sensitive layer (retina) enclosed in protective coats. Anteriorly the protective coats are transparent to admit light, which is then focused on the *retina*. Developmentally the retina is part of the brain, and the protective coats of the eyeball are equivalent to the protective coverings of the brain, the meninges. The retina retains its connection with the brain by means of the *optic nerve*, which arises from the posterior aspects of the eyeball and passes into the cranial cavity through the *optic foramen*.

Structure of the Eyeball

The eyeball (Fig. 17-4a) is an oval structure with an outer covering of fibrous connective tissue, the **sclera,** enclosing approximately five-sixths of the posterior portion of the eyeball. At the back of the eyeball the sclera is continuous with the *dura mater* (external meninges), which forms the protective covering of the *optic nerve*. The muscles of the eyeball take origin from the fascia anchored to the bony wall of the orbit and insert into the sclera, upon which they exert tension during eye movements.

Anteriorly the sclera becomes transparent and is called the **cornea,** which is more convex than the sclera, causing the eyeball to protrude in front. The curvature of the cornea is part of the mechanism for focusing light rays upon the retina. The cornea is composed largely of collagenous fibers with epithelial layers covering its external and internal surfaces. It is avascular, receiving its nutrition principally from blood vessels near its scleral junction; however, it contains numerous afferent nerve endings, making it a very sensitive structure.

Corneal transplants for people with scarred or opaque corneas have a high success rate because the cornea has no blood supply and the antibodies responsible for rejection of foreign tissue do not reach it. It is common practice for people to arrange to have their eyes removed shortly after death so that their corneas may be used as transplants.

The cornea is prevented from becoming dehydrated by its contact with the fluid called **aqueous humor** internally and by constantly being moistened externally by tears produced by the **lacrimal gland.** The *lacrimal glands* (Fig. 17-4b) are located superiorly at the lateral corner of the orbit. Fluid (tears) constantly produced by this gland is discharged into the space (*conjunctival sac*) between the eyelid and cornea. As the eyelids blink, the tears are moved across the cornea toward the medial corner of the eye where they enter the *lacrimal sac* and drain into the nasal cavity through its *nasolacrimal duct*. The primary function of lacrimal fluid is to moisten the cornea, and the drainage system, assisted by evaporation, is capable of removing the normal production. However, this drainage system is inadequate in times of excessive production of tears, usually induced by emotion. When there is not enough fluid produced, the cornea becomes dry and painful. Impulses arise in the many sensory terminations in the cornea itself, apparently in response to both touch and pressure.

> The conjunctiva over the anterior surface of the sclera is colorless and appears white except when the blood vessels are dilated. Irritants such as dust or smoke may cause the vessels to dilate and give the conjunctiva a reddish appearance commonly called "bloodshot" eyes.

The fibrous sclera (Fig. 17-5) covering the external surface of most of the eyeball is in contact on its internal surface with a *choroid coat*, a vascular layer that represents the arachnoid and pia mater coverings of the brain. The choroid not only contains the blood vessels supplying much of the eyeball but also is pigmented, thus providing protection from excess light for the inner tissues. The choroid encloses the posterior part of the eyeball and is continuous anteriorly, near the corneal-scleral junction, with a thickened structure, the *ciliary body*, which encircles the junction (Fig. 17-6a). The ciliary body is attached to the corneal-scleral junction continuously around the circumference of the eyeball and also is attached to the lens by means of the *suspensory ligament* of the lens. The ciliary body contains smooth muscle, supplied by parasympathetic fibers, whose contractions slacken the suspensory ligament and reduce the tension on the lens. The ciliary body is continuous along its anterior surface with the iris, a thin sheet of tissue that forms a ledge in front of the lens. The **iris** contains pigment, which gives it color, and has an opening in its center called the **pupil.** The size of the pupil determines the amount of light that enters the eyeball and depends upon the relative contraction of antagonistic sets of smooth muscle within the iris. Circularly arranged smooth muscle cells of the iris, called the *constrictor* of the *pupil*, are supplied by *parasympathetic fibers* that react quickly to bright light to cause a rapid reduction in the diameter of the pupil (Fig. 17-7). Radially arranged smooth muscle cells, the *dilator* of the *pupil*, are innervated by the *sympathetics* and react much more slowly to darkness or dim light, causing the pupils to dilate gradually. Thus when the eyes are suddenly exposed to a bright

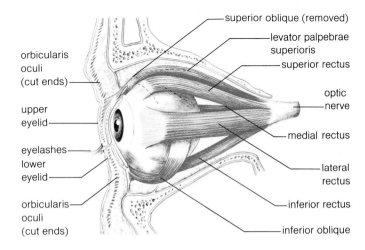

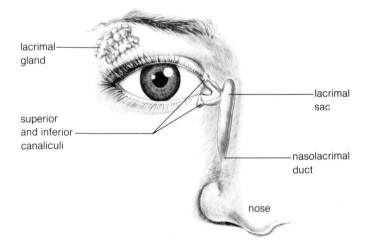

Figure 17-4   The external structure of the eyeball: (**a**) lateral view of the eyelids and extrinsic muscles of the eye; (**b**) the lacrimal apparatus of the right eye.

light, the adjustment is made quickly, preventing possible damage to the eyes; but if one leaves a brightly lit area to enter a dark one, it takes a relatively long time to become dark adapted.

Immediately behind the pupil is the *lens* (Fig. 17-6), a biconvex, oval, transparent elastic structure that is more convex anteriorly than posteriorly. It is held in place by the suspensory ligament that attaches the lens to the ciliary body. The lens is composed of elongated lens fibers derived from epithelial cells that lose most of their nuclei as they elongate and thus become transparent; it is covered by an epithelial capsule of cuboidal or columnar cells.

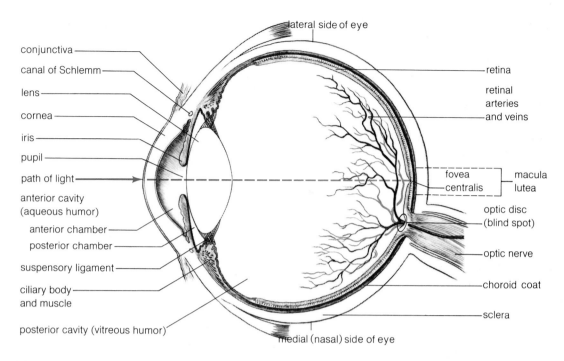

Figure 17-5 A horizontal section through the eyeball showing the layers of its wall and its internal structure

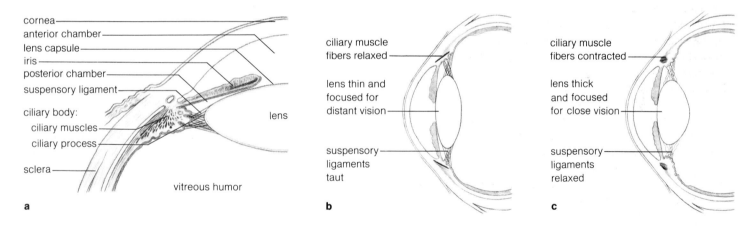

Figure 17-6 The components of the vascular layer: (**a**) The lens is attached to the ciliary body by the lens capsule and the suspensory ligament. (**b**) The relaxed state of the ciliaris muscle keeps the suspensory ligament taut and flattens the lens for distant vision. (**c**) The contraction of the ciliaris muscle relaxes the suspensory ligament, allowing the lens to round up for close vision.

Because of the constant tension on the lens due to its attachment to the ciliary body and because of pressure within the eyeball, the lens is normally flattened into a shape suitable for focusing on distant objects. But this tension can be relieved by contraction of the ciliaris muscle, which allows the lens to become more convex so that it can focus light from nearby objects upon the retina. This *mechanism* of *accommodation* (Fig. 17-6b, c) allows a rapid shift from distant to close objects while keeping the objects in sharp focus. Accommodation is under neural control, and the pathways comprise a visual reflex.

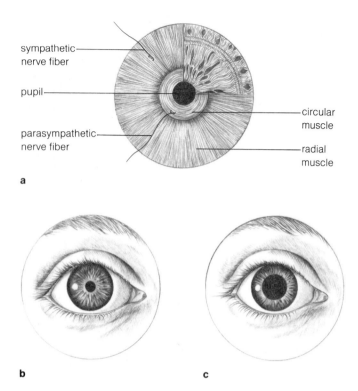

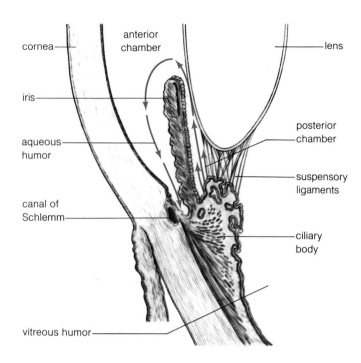

Figure 17-7 (a) The iris contains radial muscle fibers that are innervated by sympathetic nerves and circular muscle fibers that are innervated by parasympathetic nerves. (b) The circular muscles contract in bright light and constrict the pupil. (c) The radial muscles contract in dim light and dilate the pupil.

Figure 17-8 The aqueous humor secreted by the ciliary body travels from the posterior chamber through the pupil to the anterior chamber, where it drains into the canal of Schlemm

> The lens becomes less elastic with age and requires the assistance of glasses to help compensate for its inability to increase its convexity.

The lens, together with its attachment to the ciliary body, separates the eyeball into *anterior* and *posterior compartments*. The posterior compartment, called the **vitreous body,** is filled with an acellular transparent jelly, the *vitreous humor*, important in maintaining the shape of the eyeball. The anterior compartment contains the *aqueous humor,* an ultrafiltrate of blood elaborated principally by the ciliary body. The anterior compartment is divided for descriptive purposes into an *anterior chamber* between the iris and cornea and a *posterior chamber* between the iris and lens (Fig. 17-8). The aqueous humor is constantly produced and drained away at about the same rate, maintaining a relatively constant internal pressure within the eyeball.

> Fluid produced produced by the ciliary body is returned to the blood stream by a system of fine channels that open into a circular channel, the canal of Schlemm, located near the corneal-scleral junction. The canal of Schlemm opens into the venous system, returning to the aqueous humor to the blood stream. A blockage or any failure of this drainage leads to an increase in intraocular pressure, a condition known as *glaucoma,* and may, if not treated, cause degeneration of the retina and loss of vision.

The Retina

The **retina** is the light-sensitive layer and is situated inside the choroid coat of the posterior half of the eyeball (Fig. 17-5). It is continuous posteriorly with the optic nerve at a point called the **optic disc,** located medial to the midpoint of the back of the eyeball. The optic disc is not light-sensitive and is the "blind

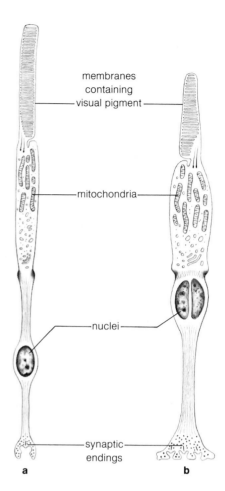

Figure 17–9 The receptor cells of the retina: (**a**) rod cell, (**b**) cone cell.

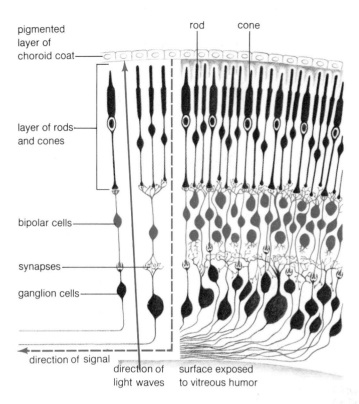

Figure 17-10 Light waves pass through the eye and through the transparent neurons of the retina to the back of the eye where they stimulate the rods and cones. Electrical signals are then conducted back through the retina by way of the bipolar cells to ganglion cells, whose axons are the fibers of the optic nerve.

spot" of the retina. Just lateral to the midpoint of the back of the retina is a thin, yellow spot, the **macula lutea,** with a depression in its center, the **fovea centralis,** that is the most sensitive part of the retina.

There are two distinct types of receptor cells in the retina, **rods** and **cones** (Fig. 17-9). In the human retina there are about 120 million rods and about 6 million cones. Only cones are found in the fovea centralis. The rods and cones are situated in a layer of the retina separated from the choroid only by a layer of epithelial cells containing pigment (Fig. 17-10). These cells send cytoplasmic extensions around the rods and cones to protect them from excessive exposure to light. The pigment cells also prevent the reflection of light back onto the rods and cones from the choroid coat and may have a

role in the biochemical events of photoreception.

The rod is a long, slender cell, oriented vertically within the retina with its long axis extending toward the center of the eyeball. The cone is a thicker cell with a similar orientation. The rods contain the visual pigment **rhodopsin,** which absorbs light readily. Rhodopsin reacts to light with a change in its molecular structure that produces an electric potential in the rods, setting up a nerve impulse. Because rhodopsin responds to small amounts of light, the rods make vision in dim light possible. The cones contain at least three different pigments, one being most sensitive to yellow, one to green, and one to blue light. Because these pigments do not react well to dim light, cones are used for vision in daylight or bright light and for color vision.

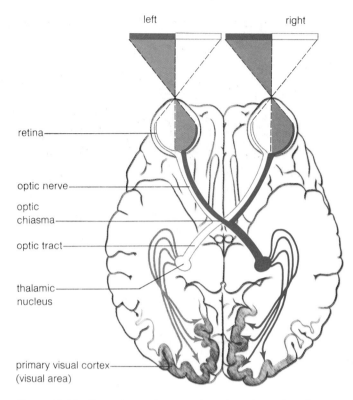

left   right

retina

optic nerve

optic
chiasma

optic tract

thalamic
nucleus

primary visual cortex
(visual area)

Figure 17-11 The pathway of neurons from the retina to the primary visual cortex

The Visual Pathway

Rods and cones develop in the embryo from undifferentiated nerve cells and may be considered *first-order neurons* in the *visual pathway* (optic pathway). When sufficient light strikes the rods and cones, it initiates a nerve impulse that is transmitted to *bipolar neurons* located entirely within the retina. The bipolar neurons are association neurons, receiving the impulse from the rods and cones and transmitting it along their axons to the next synapse in the pathway. A single bipolar neuron is activated by fewer cones than are required for its activation by rods; this means that cones give better resolution and thus sharper vision than rods, because the resolution of images depends upon the ability to discriminate between two closely adjacent points. Because the fovea centralis contains only cones, this part of the retina has the greatest resolving power; the eyeballs automatically move to focus the image on the midpoint of the back of the retina for greatest visual acuity.

The bipolar (second order) neurons transmit visual impulses to the *ganglion cells (third-order neurons)* located near the surface of the retina adjacent to the vitreous body. Fibers of the ganglion cells from all parts of the retina converge at the *optic disc* and leave the eyeball via the *optic nerve*. The physical

arrangement of the components of the retina are such that for light to reach the rods and cones, it must pass not only through the transparent parts of the eyeball—the cornea, lens, and aqueous and vitreous humors—but also through the layers of the retina containing ganglion cells and bipolar neurons. These latter cells are not light-sensitive and can be activated only by impulses initiated by the rods and cones.

The optic nerve (Fig. 17-11) is composed of all the *axons* of the *ganglion cells* from the retina of one eyeball. It passes back through the orbit and into the cranial cavity through the *optic foramen*. After a short course it converges with the nerve from the other eyeball to become the *optic chiasma*, in which some fibers of each optic nerve *cross* to the opposite side. The fibers from the *medial* (nasal) *half* of the *retina* of each eye cross, but those of the lateral (temporal) half remain on the same side. The fiber bundle extending from the optic chiasma to the brain, now called the *optic tract*, consists of fibers derived from the retina of the lateral half of the eyeball of the same side and the medial half of the eyeball on the opposite side. The fibers of each optic tract convey impulses from a *single visual field*, because the temporal half of the retina of the right eye receives the same image as that seen by the nasal half of the left eye. To demonstrate this, look straight ahead but think about an object located in the left half of your field of vision. Light from this object travels in a straight line to fall upon the retina of the right half of each eye. Maintaining your eyes in the same position, cover one eye at a time and note that your image of the object is essentially unchanged. The fibers conveying this image from both right and left eyes are traveling in the right optic tract.

Third-order neurons of the optic pathway convey visual impulses to their terminations in three locations: the *lateral geniculate body* (nucleus) of the *thalamus*, the *pretectal nucleus* of the *midbrain*, and the *superior colliculus* of the *midbrain*. The fibers terminating in the lateral geniculate body synapse on fourth-order neurons that transmit the impulses to the visual cortex of the *occipital lobe*, where the image is perceived at the conscious level. The visual cortex of each cerebral hemisphere receives impulses initiated in the retinae of both eyes. The pretectal nucleus and superior colliculus are central connections for various reflexes stimulated by light impulses.

The Light Reflex

The light reflex is a response to an increase in light intensity; it results in a rapid constriction of the pupils. Visual impulses reaching the pretectal nucleus of the midbrain by way of the optic tract are relayed by this nucleus to the *parasympathetic nucleus* (Edinger–Westphal) of the *oculomotor nerve* (Fig. 17-12). Preganglionic parasympathetic fibers from this nucleus transmit the impulses over the oculomotor nerve to the *ciliary ganglion* located within the orbit, which in turn sends postganglionic fibers to the *sphincter* of the *pupil*, reducing the diameter of the pupil. This is an *autonomic reflex pathway* for the response to light (pupillary light reflex). Because half of the optic fibers from each eye cross in the optic chiasma, the pupils of both eyes constrict when only one eye is exposed to light.

Visual Reflexes Involving Skeletal Muscles

Some of the third-order neurons of the visual pathway terminating in the *superior colliculus* of the *midbrain* convey impulses for *reflexes* involving skeletal muscles. Descending tracts arising in the superior colliculus, such as the *medial tectospinal tract*, terminate upon lower motor neurons in the cervical spinal cord that stimulate skeletal muscles of the neck and upper extremity. These reflexes are important in turning the head to follow a moving object with the eyes or in raising the arms to ward off a blow aimed at the head.

> If an optic nerve is damaged, both visual fields of that eye are affected and the person may have *anopsia* or blindness of that eye. An aneurysm of the right internal carotid artery as it lies lateral to the optic chiasma may put pressure on the nerve fibers from the temporal half of the retina of the right eye and cause *hemianopsia* (half-blindness) in the right eye. The pituitary gland lies between the two optic tracts near the optic chiasma; a pituitary tumor may expand anteriorly and press on the axons from the nasal half of the retina of both eyes, causing hemianopsia of the temporal visual fields of both eyes. A lesion of an optic tract, or of the optic radiation on one side, or of the visual cortex of cerebral hemisphere of one side can produce hemianopsia of the contralateral visual field (the visual field of the opposite side).

HEARING AND EQUILIBRIUM

The receptors for the special senses of *hearing* and *equilibrium* are located in the ear. The ear (Fig. 17-13) may be considered as consisting of three parts, two of which, the *external ear* and the *middle ear*, form the pathway for sound waves and for the conversion of sound energy into mechanical energy, the vibration of the eardrum. The third part, the *inner ear*, houses the *vestibular apparatus*, a major organ of equilibrium, and the *cochlear duct* which contains the *organ* of *Corti*, the organ of hearing.

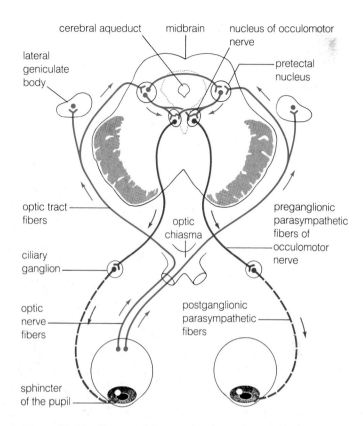

Figure 17–12 Diagram of the neural pathways involved in the pupillary light reflex

The External Ear

The external ear includes the appendage, the *auricle*, and the *external auditory canal*, or *meatus*, that leads to the *tympanic membrane*, or eardrum. The auricle is covered with skin and is made mostly of elastic cartilage which is continuous with the

> The external auditory canal curves slightly upward and anteriorly as it approaches the tympanic membrane. When the canal is straightened by pulling the auricle upward and backward, a special instrument (the otoscope) can be used to see the tympanic membrane.

elastic cartilage supporting the outer third of the external auditory canal. The skin covering the auricle is continuous with the skin lining the canal. The canal contains hairs with associated sebaceous glands as well as modified sweat glands called *ceruminous glands* that produce a waxy material. Secretions of the

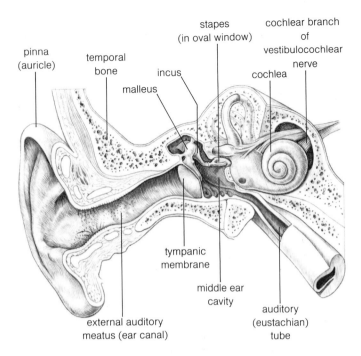

Figure 17-13   The structure of the ear

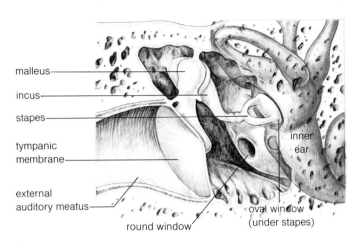

Figure 17-14   The bones of the middle ear. Not shown in the figure are the ligaments that hold the bones to each other and to the membranes.

sebaceous and ceruminous glands combine to form a sticky substance called *cerumen* (ear wax) which, in theory, prevents insects from entering the auditory canal. The inner portion of the canal has a rigid support of bone, in contrast to the flexible support of the cartilage in the peripheral region.

The Middle Ear

The external ear is separated from the middle ear by the *tympanum* (tympanic membrane), a thin connective tissue sheet covered by epithelium on its outer and inner surfaces. The middle ear (Fig. 17-14) is an air-filled cavity within bone, lined with epithelium. It contains three small bones, or *ossicles*, arranged in sequence from an attachment to the tympanic membrane in the lateral wall of the cavity to a point of contact on its medial wall. The middle ear cavity communicates superiorly and posteriorly with the *air cells* of the mastoid process of the temporal bone. It also communicates medially with the *nasopharynx* by means of the *auditory tube* (Eustachian tube), which opens during swallowing, allowing the air pressure within the middle ear cavity to equalize with the atmospheric pressure.

The vibrations created by sound waves reach the tympanic membrane and are transferred to the **malleus** (hammer), the club-shaped ossicle in contact with the membrane. Vibrations are transferred by the malleus to the **incus** (anvil) and by the

incus to the **stapes** (stirrup), the ossicle in contact with a membrane separating the middle ear cavity from the inner ear at the *oval window* (fenestra vestibuli). The vibrations are then transferred by the oval window to the fluid called *perilymph* of the bony labyrinth of the inner ear.

Earache is a common problem that has many causes, either in the ear or at some distant point. Infection of the external auditory meatus or within the middle ear cavity can cause pain locally. Infections or lesions in the mouth may also be interpreted as earache.

The Inner Ear

The inner ear (Fig. 17-15) consists of a *bony labyrinth*, a series of chambers in the petrous part of the temporal bone containing the *perilymph*. Inside the bony labyrinth is a matching *membranous labyrinth* containing another fluid called *endolymph*. The bony labyrinth houses the organ of hearing and the structures of equilibrium.

The Mechanism of Hearing   The portion of the bony labyrinth containing the auditory organ is the cochlea, a spiral structure in the shape of a snail shell. The portion of the membranous labyrinth enclosed by this structure is the *cochlear duct*, a triangular, ribbon-like tube that spirals within the bony labyrinth. Bony attachments at the base and tip of the cochlear duct divide the perilymph space within the bony labyrinth into

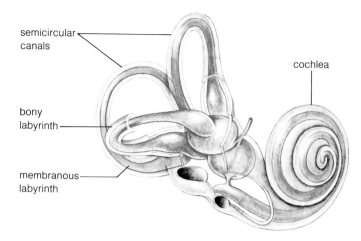

semicircular canals

cochlea

bony labyrinth

membranous labyrinth

Figure 17-15 The inner ear consists of a bony labyrinth within which is the membranous labyrinth

a **scala vestibuli** and a **scala tympani** (Fig. 17-16a). The perilymph of the scala vestibuli is in contact with the *oval window* and that of the scala tympani is in contact with the *round window*. The scala vestibuli and the scala tympani are continuous with one another only at the tip of the cochlear duct.

The endolymph in the cochlear duct is separated from the perilymph of the scala vestibuli by an epithelial sheet called the *vestibular* (Reissner's) *membrane* (Fig.17-16a). The *basilar membrane* on the scala tympani side of the cochlear duct supports a specialized receptor organ, the *organ of Corti* (Fig. 17-16b, c, d,), consisting of columnar epithelial cells that function as supporting cells with a few cells (hair cells) serving as the receptors for vibratory stimuli. Afferent nerve fibers of the cochlear portion of *cranial nerve VIII* (vestibulocochlear) that terminate around the hair cells convey nerve impulses for sound to the brain.

Sound waves in the air pass through the external auditory meatus and cause the tympanic membrane to vibrate. This vibration (Fig. 17-17), transmitted by the malleus, incus, and stapes to the oval window, sets up waves of vibration in the perilymph of the scala vestibuli. The vibrations caused by the movement of the fluid are transmitted by Reissner's membrane to the endolymph in the cochlear duct, which stimulates the hair cells of the organ of Corti to initiate impulses in the afferent nerve fibers of the auditory nerve. These impulses are conveyed by the *auditory nerve* to *cochlear nuclei* in the *medulla*, which relay the impulses to the *medial geniculate nucleus* in the *thalamus*. The pathway from the cochlear nuclei to the thalamus is multisynaptic and is both crossed and uncrossed. The *auditory area* in the *temporal lobes* of both cerebral hemispheres receives impulses from both ears. In addition to this pathway for the conscious perception of sound, connections are provided for reflex turning of the head and eyes toward the source of sound.

Since the auditory area in the temporal lobes of both cerebral hemispheres receives impulses from both ears, a unilateral lesion of the auditory pathway within the CNS will not cause a person to be deaf in one ear. If the auditory cortex in the right cerebral hemisphere is destroyed, the person will still be able to hear with both ears, using the intact auditory cortex of the left cerebral hemisphere. However, if the auditory nerve is severed or if the cochlear nuclei of one side are destroyed, then the person will be deaf in the ear of that side.

Deafness is usually divided into two types: (1) nerve deafness, due to impairment of the cochlea or auditory nerve, and (2) conduction deafness, due to impairment of middle ear mechanisms that transmit sound waves into the cochlea. If the cochlea and auditory nerve are intact, even if the ear ossicles are destroyed or have grown together (ankylosed), then sound waves may still reach the cochlea by bone conduction. Bone conduction may be tested by placing a tuning fork against the mastoid process of the temporal bone.

The Mechanism of Equilibrium The *vestibular apparatus* of the inner ear (Fig. 17-18) cooperates with visual stimuli and with proprioceptive information from the muscles and tendons in maintaining the orientation of the body in space (equilibrium or balance). The vestibular apparatus consists of *three semicircular canals*, a *utricle*, and a *saccule*, all part of the bony labyrinth. Inside there are three membranous semicircular ducts oriented at right angles to one another, with membranous dilations called *ampullae* where the ducts enter the utricle. These membranous structures contain endolymph and are surrounded by perilymph within the bony labyrinth of the petrous temporal bone. The vestibular apparatus communicates with the auditory apparatus by means of a slender duct connecting the saccule with the cochlear duct.

Impulses in response to movements of the head are initiated by the receptors in the semicircular ducts. The utricle functions in static equilibrium and is essential to the static postural, tonic neck, and righting reflexes. Its macula (receptor organ) is stimulated by changes in the position of the head and body in space, linear acceleration, gravity, and centrifugal force. The function of the saccule is uncertain but may be similiar to that of the utricle. The receptors in various parts of the vestibular apparatus differ in anatomical detail but are similar in principle

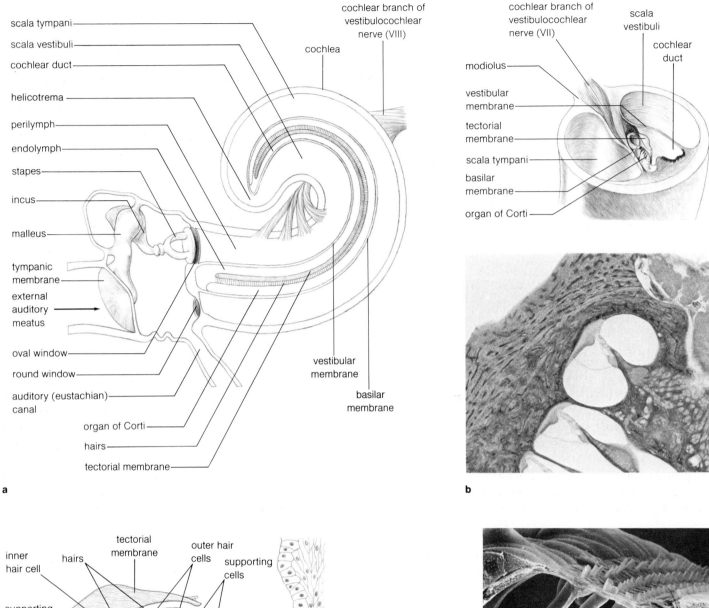

a

scala tympani

scala vestibuli

cochlear duct

helicotrema

perilymph

endolymph

stapes

incus

malleus

tympanic membrane

external auditory meatus

oval window

round window

auditory (eustachian) canal

organ of Corti

hairs

tectorial membrane

cochlear branch of vestibulocochlear nerve (VIII)

cochlea

vestibular membrane

basilar membrane

cochlear branch of vestibulocochlear nerve (VII)

scala vestibuli

cochlear duct

modiolus

vestibular membrane

tectorial membrane

scala tympani

basilar membrane

organ of Corti

b

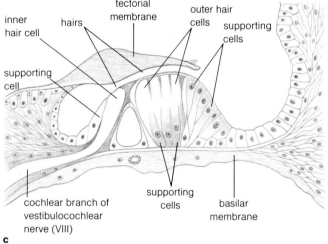

c

inner hair cell

supporting cell

hairs

tectorial membrane

outer hair cells

supporting cells

cochlear branch of vestibulocochlear nerve (VIII)

supporting cells

basilar membrane

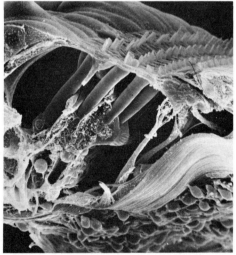

d

Figure 17-16 The structure of the cochlea: (**a**) unwound, and in longitudinal section; (**b**) cross-section diagram and photomicrograph, × 25; (**c**) the organ of Corti; (**d**) scanning electromicroscope photograph of organ of Corti. (Photograph b by Joan Creager; photograph d by Dr. Goran Bredbreg.)

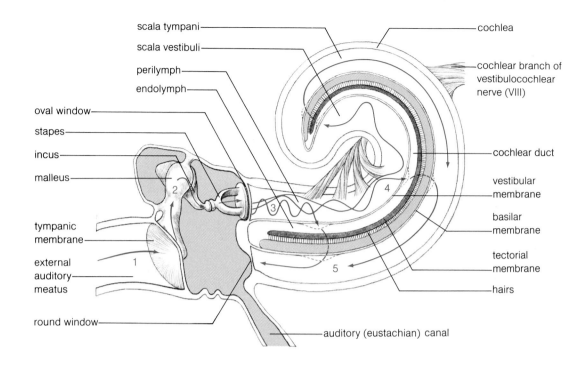

scala tympani

scala vestibuli

perilymph

endolymph

oval window

stapes

incus

malleus

tympanic membrane

external auditory meatus

round window

cochlea

cochlear branch of vestibulocochlear nerve (VIII)

cochlear duct

vestibular membrane

basilar membrane

tectorial membrane

hairs

auditory (eustachian) canal

Figure 17-17 The transmission of sound vibrations through the ear: (**1**) Sound vibrations in air pass through the external auditory meatus and set the tympanic membrane vibrating. (**2**) Vibrations are transferred to the bones of the middle ear and amplified as they reach the oval window. (**3**) Vibrations of the oval window set the fluid of the scala vestibuli in vibration. (**4**) Vibrations of this fluid cause certain portions of basilar membrane to bulge into the cochlear duct. The movement of the basilar membrane stimulates the hair cells of the organ of Corti and causes impulses to be conducted along the cochlear nerve. (**5**) Vibrations of the fluid of the scala vestibuli are transferred to the scala tympani and dissipated as the round window bulges out toward the middle ear.

(Fig. 17-19). In general the receptors are columnar epithelial cells bearing hairs that project into an overlying gelatinous mass called the **cupola**. The cupola may contain calcium deposits called **otoliths**. Any change in orientation of the head moves the cupola over the hair cells, stimulating the hair cells to initiate nerve impulses in the afferent fibers that terminate near the hair cells. These nerve fibers belong to the vestibular portion of cranial nerve VIII and convey the impulses to the *vestibular nuclei* of the *pons* and *medulla*.

Lesions of the vestibular system cause difficulty in maintaining one's balance while walking straight; since this system has connections with eyeball movements, lesions also cause abnormal movements of the eyes, called *nystagmus*. In nystagmus the eyes move constantly in a characteristic manner. First, they move slowly to one side as far as they can, and then they move very quickly back the other way and begin the process again.

A common symptom of vestibular damage is dizziness, although dizziness can be caused by many other conditions. *Meniere's disease* is a condition that some physicians consider to be brought on by elevated pressure within the perilymph and endolymph of the inner ear. Patients with this disease suffer severe attacks of dizziness, ringing in the ears, and nausea. There is no known cure and it can cause deafness.

Impulses received by the vestibular nuclei are relayed to the cerebral cortex by pathways that are poorly understood. However, most of the influences of the vestibular apparatus are mediated through connections of the vestibular nuclei with the cerebellum, where they modify motor activity as needed, and through direct connection with some cranial nerve nuclei and the lower motor neurons of the spinal cord. Through these latter connections the vestibular apparatus influences the proper state of muscular contraction in the maintenance of posture.

NEURAL REGULATION OF BEHAVIOR AT SUBCONSCIOUS LEVELS

The human's position as the most "advanced" organism on earth is due to the tremendous area of the human cerebral cortex whose capacity to interpret sensory data at the conscious level allows us to will behavior. Nevertheless much of our behavior is regulated by subconscious processes that we share with other animals. The ascending reticular system, the limbic system, and the hypothalamus are all important in the neural regulation of behavior at subconscious levels.

The Ascending Reticular System

The reticular formation originates in scattered cells of the various parts of the brainstem that occupy what might be consid-

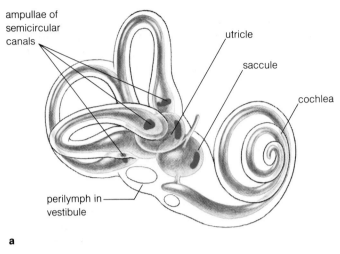

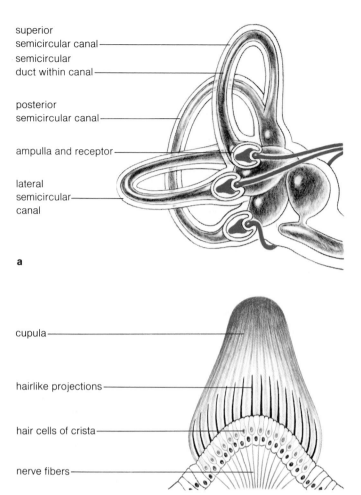

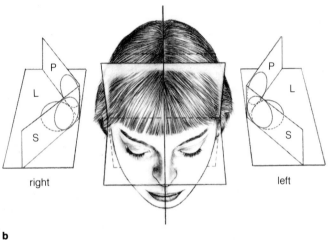

Figure 17-18 (**a**) The organs of static equilibrium are the utricle and possibly the saccule, located in the vestibule of the inner ear. The organs of dynamic equilibrium are the semicircular canals. (**b**) A diagram of the orientation of the semicircular canals.

Figure 17-19 (**a**) The organs of dynamic equilibrium are the semicircular canals. Each canal contains an ampulla (**b**) which contains hair cells on a crest (the crista). From the hair cells, projections extend into a gelatinous mass, the cupula.

ered the central core of the gray matter and give rise to ascending fibers that run rostrally in the central reticular area. The functions of the reticular formation are complex; among other things, it screens the multitude of signals received by sensory receptors so that those that are unimportant do not affect our behavior. This discussion will focus on the role of the reticular system in arousing or maintaining an alert state.

Although each of the ascending pathways for conscious sensation is restricted to impulses set up by a specific sensory modality, the ascending reticular system carries impulses elicited by most types of stimuli to the body wall or viscera. For example, stimuli interpreted at the conscious level as either pain or pressure can activate the ascending reticular system, with no differences in the resulting neural responses. Activation of the

ascending reticular system alone does not elicit any conscious sensation; however, when reticular fibers are stimulated by the same impulses that stimulate specific sensory pathways, such as the pathway for pain, reactions occur at both the conscious and subconscious levels.

In the specific pathway for pain the first-order neuron, upon entering the spinal cord, bifurcates into ascending and descending fibers that synapse with second-order cells of the dorsal gray column, which relay the pain impulses to the thalamus. Besides stimulating this specific pathway, the same first-order neurons have collaterals that transmit impulses to other neurons in the dorsal gray column that relay the impulses to the *reticular nuclei* of the brainstem. These latter ascending fibers, conveying impulses from the spinal cord to the brainstem retic-

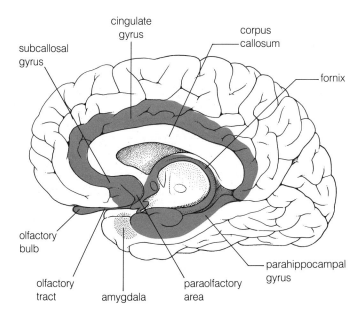

Figure 17-20 Midsagittal section of the cerebrum, illustrating the components of the limbic lobe

Labels in figure: cingulate gyrus, corpus callosum, subcallosal gyrus, fornix, olfactory bulb, olfactory tract, amygdala, paraolfactory area, parahippocampal gyrus

It should be noted that the arousal mechanism is not fully understood. The alerting mechanism may reside in the thalamus rather than the brainstem and may be influenced by conventional sensory pathways as well as by ascending fibers of the reticular formation.

The Limbic System

The *limbic system* (Fig. 17-20) appears to regulate *emotional states* and to influence a wide variety of activites related to the emotions, such as respiration, rate of heart beat, and sexual activities. It is composed of the *limbic lobes* of the cerebral hemispheres and certain closely linked structures of the diencephalon. The limbic lobe consists of *medial* and *lateral olfactory gyri*, *anterior* and *posterior paraolfactory areas* (septal area), *fornicate lobe* or *gyrus* (including the hippocampus), and the *olfactory stalk* and *bulb*. In addition to the limbic lobe the limbic system also includes portions of the *hypothalamus*, *thalamus*, and the *amygdala*, a gray mass in the temporal lobe that is closely associated with the olfactory area.

The limbic system is closely linked to the olfactory system and receives data from it. It receives afferents from all lobes of the cerebrum, especially from areas of the cortex adjacent to gyri of the limbic system. Somatic and visceral sensory modalities reach the limbic lobe by way of the thalamus and hypothalamus. The afferents conveying these sensory impulses ascend as a part of the ascending reticular system. All parts of the limbic system are initially interconnected. The connections bringing data to the limbic lobe are usually paralleled by efferent fibers through which influences of the limbic lobe are spread to *frontal* and *prefrontal association areas* of the cerebral cortex, areas involved in emotions. In addition to this "association" type of outflow there is a special outflow from the limbic lobe to the hypothalamus. This outflow is by a large bundle of fibers called the *fimbria fornix* and is the principal pathway for the influence of the limbic system upon the activity of the *autonomic nervous system*.

ular nuclei, constitute the *spinoreticular tracts*. Impulses in the spinoreticular tracts are transmitted to the dorsal thalamus by short chain relays through the central reticular formation of the brainstem. From the dorsal thalamus impulses are projected to wide areas of the cerebral cortex, including *association areas* and areas related to *emotion*. In this way the cortex is *alerted*, or *aroused* (you awaken if you have been asleep), and the brain becomes ready to make a conscious response. The level of excitation of the cerebral cortex is raised, resulting in a heightened state of awareness. Hence the ascending reticular system is sometimes called an *arousal mechanism*.

WORDS IN REVIEW

Aqueous humor	Cupola	Malleus	Pupil	Sclera
Auditory cortex	Fovea centralis	Nucleus	Retina	Stapes
Auricle	Incus	Olfaction	Rhodopsin	Superior colliculus
Cones	Iris	Olfactory bulb	Rods	Taste bud
Conjunctival sac	Lacrimal gland	Optic disc	Scala tympani	Vestibular apparatus
Cornea	Macula lutea	Otoliths	Scala vestibuli	Vitreous body

SUMMARY OUTLINE

I. Olfaction
   A. Neurons in the olfactory epithelium, located in the superior aspect of the nasal cavity, function as receptors.
   B. Axons of the olfactory neurons pass through the openings in the cribriform plates of the ethmoid to terminate in the olfactory bulbs.
   C. Neurons of the olfactory bulbs transmit impulses via the olfactory tracts to the lateral olfactory area in the temporal cortex where odor is perceived.

II. Taste
   A. Receptors are special cells in taste buds located in the epithelium covering the tongue.
   B. The receptor cells initiate impulses in nerve fibers that terminate in the taste buds.
   C. Nerve fibers conveying taste sensations are in cranial nerves VII, IX, and X which convey the impulses to the brainstem, where they are relayed by the nucleus solitarius to the thalamus.
   D. Impulses are relayed by the thalamus to the area of the cerebral cortex where the taste is perceived.

III. Vision
   A. The organ for vision is the eyeball, which has an outer connective tissue coat, the sclera, that encloses the vascular coat, the choroid. The sensory layer, the retina, is located internal to the choroid. Anteriorly the sclera is continuous with the transparent cornea; the choroid is continuous with the ciliary body and the iris. The elastic lens is held in place by the suspensory ligament attached to the ciliary body.
   B. The retina contains receptor cells, rods and cones, and bipolar neurons that relay impulses to ganglion cells in the inner layer of the retina. Fibers of the ganglion cells pass to the thalamus via the optic nerves and optic tracts.
   C. The visual pathway consists of rods and cones, bipolar neurons, ganglion cells, thalamus, and visual cortex.
   D. The light reflex pathway consists of rods and cones, bipolar neurons, ganglion cells, pretectal area, nucleus of Edinger–Westphal, ciliary ganglion, and the sphincter of the pupil.

IV. Hearing and Equilibrium
   A. The external ear is composed of the auricle and external auditory canal leading to the tympanic membrane.
   B. The middle ear contains bones, the malleus, incus, and stapes, that transmit vibrations to the fluid of the inner ear.
   C. The inner ear contains the cochlear duct bearing the organ of Corti (the organ of hearing); it also contains the vestibular apparatus, composed of the semicircular canals with connecting enlargements, the saccule and utricle, for response to movements of the head. The vestibular apparatus maintains equilibrium.

REVIEW QUESTIONS

1. What are the receptors for the sense of smell and where are they located?

2. In what lobe of the cerebral hemisphere are odors perceived?

3. Olfactory nerves enter the cranial cavity by passing through openings in what bone?

4. What cranial nerves transmit impulses for taste sensations to the brainstem?

5. Name the nucleus in the brainstem that relays taste impulses to the thalamus.

6. List the layers of the eyeball in their proper sequence from external to internal.

7. List the structures through which light must pass in order to reach the retina.

8. List in their proper sequence the components of the visual pathway.

9. List in their proper sequence the components of the light reflex.

10. Name the components of the middle ear that are involved in the transmission of sound waves to the inner ear.

11. The receptors for hearing are in what structure within the inner ear?

12. What are the essential components of the vestibular apparatus of the inner ear?

SELECTED REFERENCES

Bowsher, D. *Introduction to the Anatomy and Physiology of the Nervous System*. Philadelphia: F. A. Davis, 1974.

Chusid, J. G. *Correlative Neuroanatomy and Functional Neurology*. 17th ed. Los Altos, California: Lange Medical Publications, 1979.

DeCoursey, R.M. *The Human Organism*. 4th ed. New York: McGraw-Hill, 1974.

Guyton, A. C. *Structure and Function of the Nervous System*. 2nd ed. Philadelphia: W. B. Saunders, 1976.

Hubel, D.H., *et al.* "Special Issue on Neurobiology." *Sci. Am.* 241: 45–232, September 1979.

Stevens, L.A. *Neurons*. New York: Thomas Y. Crowell, 1974.

Watts, G.O. *Dynamic Neuroscience: Its Application to Brain Disorders*. Hagerstown: Harper & Row, 1975.

18

THE ENDOCRINE GLANDS

After completing this chapter you should be able to:

1. Describe the methods by which hormones may affect the cellular metabolism of target organs.

2. Locate the pituitary gland and list its parts based upon functions. Describe the position of the posterior lobe relative to the hypothalamus.

3. List the hormones released in the posterior pituitary, giving the source and function of each hormone.

4. List the hormones produced by the anterior pituitary and give the major functions of each hormone. Describe the control of the release of these hormones.

5. Locate the thyroid gland and describe its structure. Name the substance produced by the principal thyroid follicle cells; describe the storage and release of the hormone into the blood stream. Give the source and function of calcitonin.

6. Locate the parathyroid glands in relation to the thyroid. Describe the structure of the parathyroid and identify the hormone produced, giving its principal functions.

7. Locate the suprarenal glands in relation to the kidneys. Compare and contrast the structure of the suprarenal cortex with that of the medulla.

8. List the three major classes of hormones produced by the suprarenal cortex and give the chief functions of each.

9. Name the two hormones produced by the suprarenal medulla and state their general physiological effects.

10. Give the location and describe the structure of the islets of Langerhans. Name and give the general functions of the hormones produced.

11. Describe the ovaries and testes as endocrine glands. List the hormones produced and give the major functions of each.

12. Describe the development of the pituitary, thyroid, parathyroid, and suprarenal glands.

The **endocrine system** consists of glands that produce **hormones**—chemicals that in minute quantities influence the metabolism of cells other than those that produced them. Though the various endocrine glands produce hormones that differ in both chemical composition and function, they form part of an integrated system of chemical regulation whose activities are coordinated with those of the nervous system. Exocrine glands, by contrast, produce a variety of substances in relatively large quantities—mucus, sweat, and digestive enzymes, for example—and cannot be considered parts of a single organ system. Endocrine glands are further distinguished from exocrine glands (Fig. 18-1) by their lack of ducts; instead of funneling their products to specific places they discharge them into the blood stream, which carries the hormones to all parts of the body.

The simplest form of regulation of one tissue by another, and the first to have evolved, is by a diffusion of regulatory substances from one part of the body to another through the intercellular fluid. Control by this method is inefficient; it is slow and requires relatively large quantities of the regulatory substances, due to their rapid dilution as they diffuse. These problems were partially solved with the development of an efficient circulatory system, which eliminates the need for diffusion through many cell layers. During the course of evolution, however, a far more effective method of control developed through the specialization of cells (neurons) that could be stimulated by environmental changes and could conduct the stimuli from one part of the body to another, causing proper responses to the environment. But this neural regulation requires the termination of the nerve fiber upon the effector cell, limiting the number of cells that can be subjected to the influence of a single neuron. The endocrine system achieves much broader influence, using fewer cells than would be required by nervous control of all body functions, and thus has remained important in the regulation of processes that change comparatively slowly. The cells of endocrine glands must secrete larger amounts of regulatory substances than nerve cells; but since hormones are effective in very low concentrations, the quantities are still very small compared to those of most bodily products. A hormone may have specific effects on most of the cells of the body, as in the case of thyroid hormone (produced by the thyroid gland), or it may have a specific effect upon only a single structure called the **target organ**, as in the stimulation of thyroid secretion by the **thyroid-stimulating hormone** (produced by the anterior pituitary gland).

Some hormones regulate the secretion of others, as for example the thyroid-stimulating hormone. This regulation may involve stimulating either the production or the release of the second hormone. The hypothalamus, though not usually considered an endocrine gland, has modified nerve cells (neurosecretory cells) that produce hormones called **releasing hormones** whose function is to stimulate the release of other hormones from the anterior pituitary. You will encounter still other examples of the regulation of one hormone by another in the discussions of specific endocrine glands.

Hormones, along with the nervous system, play an important role in maintaining a constant internal environment for cells. This is achieved by interactions of the hormones with one another and with the nervous system to influence cellular metabolism. These interactions may reinforce one another; for

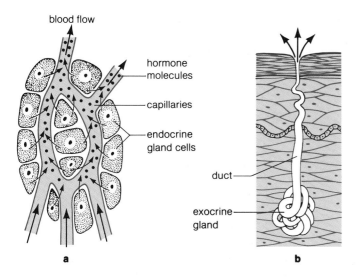

Figure 18-1 (**a**) An endocrine gland releases chemical substances called hormones into the capillaries that run through the gland. (**b**) An exocrine gland releases its secretions through a duct.

example, hormones produced by the suprarenal medulla have the same effects upon smooth muscle and some glands as does stimulation by the sympathetic nervous system, and the two forms of stimulation are brought about by the same environmental stimuli. Some metabolic reactions proceed under the dual influence of two hormones that have opposite effects; for example, the parathyroid hormone stimulates the resorption of calcium from bone whereas **calcitonin** enhances the deposition of calcium in bone.

Hormones may be divided into two broad categories according to their major effects upon cells: those whose primary function is the promotion of growth are called *trophic,* and those that modify the functioning of target tissues are called *tropic.*

The major endocrine glands include the **pituitary** (Fig. 18-2) located at the base of the brain; the *thyroid* in the neck on either side of the larynx; the four **parathyroids** in the posterior aspect of the capsule of the thyroid; the islets of Langerhans scattered throughout the pancreas; the **suprarenals** located on the superior poles of the kidneys; and the **ovaries** within the pelvis of the female or **testes** in the scrotum of the male.

TYPES OF HORMONES

Hormones are either steroids or polypeptides. Steroids such as testosterone are small molecules that are synthesized from the chemical precursor cholesterol obtained from food or synthesized in the body. All steroid hormones are lipid soluble, which

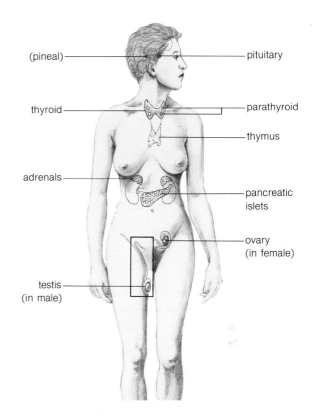

Figure 18-2 The glands of the endocrine system are located in various parts of the body.

enables them to pass readily through the cell membrane. Thus the first reaction to this type of hormone is probably the formation of a hormone receptor complex within the cell cytoplasm (Fig. 18-3). The polypeptide hormones include most of the hormones of the pituitary gland. They have difficulty passing across cell membranes and may be capable of altering the metabolism of target cells without entering the cell.

MECHANISMS OF HORMONE ACTION

Upon arrival in the tissues hormones may produce their effects in various ways. They may alter the permeability of the plasma membrane to a specific substance; this occurs in the action of insulin upon liver and muscle cells, resulting in the movement of glucose into them. In another mechanism, which appears to occur with protein hormones that act on specific target organs, the hormone may attach to the cell membrane and, without

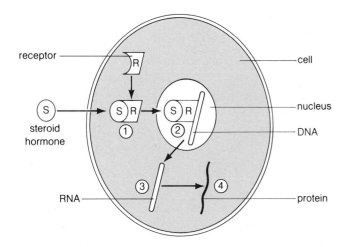

Figure 18-3 The possible action of steroid hormones is (1) the formation of a hormone-receptor complex inside the cell, (2) the movement of the complex to the nucleus where it activates a gene, (3) the production of RNA, and (4) the production of a protein.

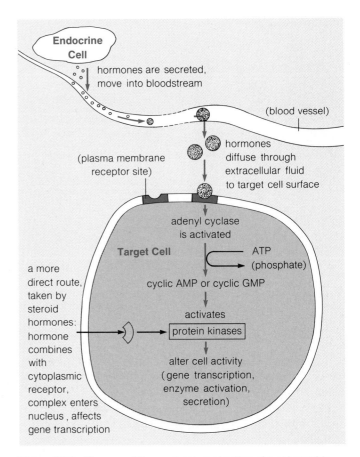

Figure 18-4 Diagram of the mechanism of action of a polypeptide hormone.

crossing it, alter the cell's metabolism (Fig. 18-4). An organ is a "target" for a hormone because the plasma membranes of the organ cells have *receptor sites* that will react specifically with that hormone, in much the same way that an enzyme reacts with its specific substrate. Upon attaching to the *receptor site* of the target cell membrane the protein hormone activates a membrane enzyme called *adenyl cyclase,* which stimulates the production of *cyclic adenosine monophosphate* (cAMP) from *adenosine triphosphate* (ATP) within the cell. Thus the hormone acts as the *primary messenger* at the cell membrane, dictating an increase in specific metabolic events within the target cell. Within the cell **cyclic AMP** functions as a **second messenger,** activating specific enzyme systems that bring about the required alteration of cellular metabolism. This general mechanism, mediated by cyclic AMP, is involved in the action of many hormones, such as **adrenocorticotropic hormone** (ACTH), which stimulates the suprarenals of the kidney to produce **glucocorticoid hormones,** and the thyroid-stimulating hormone (TSH), which stimulates the thyroid to produce thyroid hormones. The cellular response depends upon the specific enzymes available within the target cells; thus the differences in response to ACTH and TSH are due to differences within the cells of the suprarenal cortex and the thyroid gland rather than to the basic mechanism involved. However, each kind of cell is stimulated only by the appropriate hormone, because of the selectivity of the receptors on its membrane.

Hormones may also enter cells to affect cellular metabolism; this is the method of action of the steroids. Estrogen, for example, enters cells of the uterus and uterine tubes where it causes increased RNA and protein synthesis as well as cell proliferation. How it does this is not fully understood; but it appears that steroids increase the production of mRNA by binding to receptors on the DNA molecule, thus causing production of large amounts of certain proteins. The resulting growth of cells leads to their increased rate of cell division.

GENERAL STRUCTURE OF ENDOCRINE GLANDS

The characteristic morphological features of all endocrine glands include the absence of ducts and the presence of a rich supply of fenestrated blood vessels (those with pores in their lining)

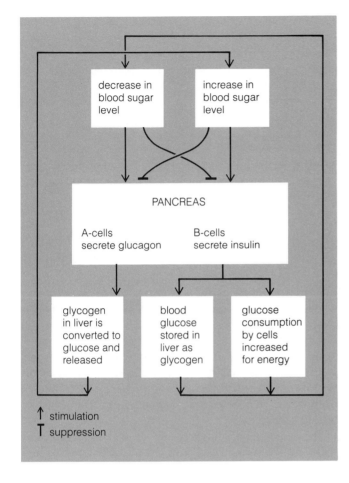

↑ stimulation
T suppression

Figure 18-5 Glucagon and insulin help to regulate the blood glucose concentration, and the blood glucose concentration in turn feeds back to regulate the secretion of glucagon and insulin.

in close relation to the secretory cells. Most of the endocrine and exocrine glands begin their development as outpocketings of tubular structures. For exocrine glands (Fig. 18-1) the original evagination branches repeatedly, with the terminal ends of the branches differentiating into secretory elements with a lumen (cavity); the original evagination becomes the duct that opens into the tubular structure from which the gland arose. Endocrine glands deviate from this pattern in that the secretory elements lose connection with their points of origin due to the degeneration of the duct. The secretory products are usually discharged toward blood vessels and enter the blood by diffusion.

The secretory cells of endocrine glands contain all of the cytoplasmic organelles associated with the secretory process, such as endoplasmic reticulum (ER) and a prominent Golgi complex. The ER is much less abundant in endocrine than in exocrine cells because the amount of hormone produced is

much less. In glands that produce a protein hormone, such as the insulin-producing beta cells of the islets of Langerhans, the hormone is synthesized on ribosomes of rough ER concentrated in the Golgi complex and is usually stored in secretory granules. The secretory granules are scattered throughout the cytoplasm until released at the cell surface by *exocytosis* (fusion of the secretory granule membrane with the plasma membrane and liberation of the granule contents outside the cell). All the tropic hormones of the anterior pituitary are secreted in this manner.

The *thyroid gland* produces protein hormones that are not stored within the secretory cells. Instead the secretions are enclosed in membranous vesicles that pass to the apex of the secretory cell, from which they are discharged into the lumen of the secretory follicle by exocytosis. Thus though the secretions do not accumulate within the cytoplasm of the thyroid cells, they do accumulate within the follicles bounded by the secretory cells.

Endocrine glands that produce steroid hormones (the suprarenal cortex, ovary, and testis) have secretory cells with *smooth* ER and a prominent Golgi complex that is not associated with secretory granules. The precursors of steroids are stored within the cytoplasm, but the hormones are released as quickly as they are produced by a mechanism that is not understood.

REGULATION OF HORMONE PRODUCTION AND RELEASE

The endocrine glands are subject to a variety of control mechanisms and in some instances are closely allied with the nervous system in the effects produced and in the detection of environmental changes. The most common method of control of hormone release is the **negative feedback** mechanism, in which hormone release causes a change in the environment of the cell that inhibits further hormone release. For example, the release of **insulin** by beta cells of the islets of Langerhans, stimulated by a rise in the glucose concentration of the blood, causes the uptake of glucose by liver and muscle cells and results in a drop in glucose blood level (Fig. 18-5). The low level of blood glucose inhibits further release of insulin, resulting in a decrease in the uptake of glucose by cells. The drop in blood glucose also stimulates the alpha cells to release *glucagon,* which promotes the breakdown of glycogen in the liver into glucose, raising the level of blood glucose to normal.

In some cases the release of a hormone is initiated by the nervous system. For example, in a lactating female the stimulus of sucking initiates a nerve impulse in the sensory nerve fibers supplying the nipple that is relayed to the hypothalamus (Fig. 18-6), causing the release of oxytocin by modified neurons. This

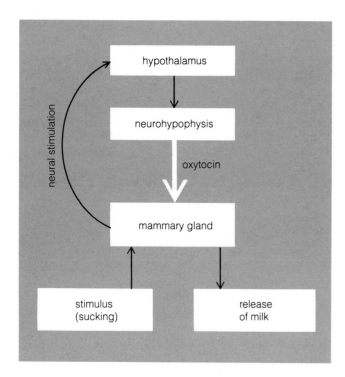

Figure 18-6 A neural-hormonal feedback loop. The sucking of an infant stimulates receptors in the nipple of the mammary gland. Impulses travel to the hypothalamus, where they stimulate the release of oxytocin from the neurohypophysis. Oxytocin travels in the blood to the mammary gland and stimulates the release of milk.

hormone reaches the mammary gland via the blood stream and stimulates the contraction of muscle cells around the secretory elements of the gland, facilitating the discharge of milk.

The nervous system also can respond to internal changes to produce endocrine effects. For example, loss of water from the blood is detected by neurons of the hypothalamus, causing the release of ADH (**antidiuretic hormone**) from the posterior pituitary. ADH stimulates the reabsorption of water from the glomerular filtrate in the kidney tubules, adding water to the blood. This is another example of negative feedback: a change (water loss) causes a reaction (hormone release) that counters the change.

TRANSPORT AND ACTION OF HORMONES

The small protein hormones are usually stored combined with larger protein molecules, called carriers, in secretory granules within the cytoplasm. When the granules are discharged by exo-

cytosis into the extracellular fluid, the hormones are split from the carrier protein, leaving the hormones free to enter the blood by diffusion. In the blood the hormones may again be combined with carrier proteins of the plasma or may remain free until they reach the target organ.

The thyroid hormone is a special case of a hormone stored outside the secretory cells. When thyroid hormone is discharged from the secretory cells into the lumen of the thyroid follicle, it combines with a colloidal substance called **thyroglobulin.** To enter the blood stream, thyroid hormone must be passed back across the follicle cells and discharged through the base of the cell.

THE PITUITARY GLAND

The **pituitary gland,** or the hypophysis (Fig. 18-7), is located in the *sella turcica,* below the **hypothalamus** to which it is attached by a stalk. It actually consists of two distinct organs: the **neurohypophysis,** or neural portion, and the **adenohypophysis,** or glandular portion. The neurohypophysis stores and secretes hormones produced by the neurosecretory cells of the hypothalamus; the adenohypophysis produces its own hormones, which it releases when stimulated by hypothalamic releasing hormones.

The neurohypophysis may be subdivided into the *median eminence of the* **tuber cinereum,** the *infundibular stem,* and the *posterior lobe,* also called the neural lobe or pars nervosa. (The term *posterior lobe* may be applied to the entire neurohypophysis.) The adenohypophysis also consists of three parts: the *pars tuberalis* which connects the adenohypophysis to the infundibulum, the *pars intermedia* which attaches to the posterior lobe, and the *anterior lobe,* or pars distalis. (The term *anterior lobe* is also applied to the entire adenohypophysis.)

> The pituitary gland is a good example of a "two-in-one" organ, of which nature seems to be very fond. Other examples are the cortex and medulla of the suprarenal gland and the exocrine and endocrine parts of the pancreas.

Blood is supplied to the pituitary gland (Fig. 18-7a) through branches of the *superior* and *inferior hypophyseal arteries* that arise from the internal carotids and supply the neurohypophysis, and through the capillary *plexus* of the *portal system* which receives releasing hormones from the hypothalamus and delivers them to the adenohypophysis via the *portal veins.* A

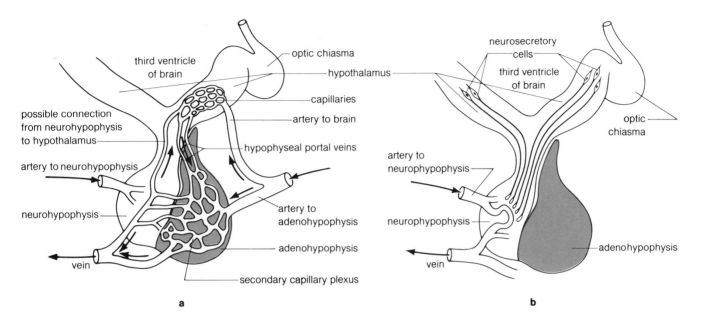

Figure 18-7 The pituitary gland really consists of two glands, the adenohypophysis and the neurohypophysis. It also has (**a**) circulatory and (**b**) neural connections with the hypothalamus. The circulatory connection constitutes the pituitary portal system.

portal system consists of two sets of capillaries connected by a vein. The only significant blood supply to the blood vessels of the anterior lobe is provided by the portal system.

The neural elements of the pituitary (Fig. 18-7b) consist of two specialized transportation systems supplying the neurohypophysis and adenohypophysis, respectively. The *neurosecretory pathway,* called the hypothalamic-neurohypophyseal system, extends from the *supraoptic* and *paraventricular nuclei* of the hypothalamus, where the hormones ADH and oxytocin are synthesized, to the *posterior lobe,* where they are stored until released. The *neurovascular pathway,* called the hypothalamic-adenohypophyseal system, extends from the *hypothalamus* via nerve fibers to the capillary plexus of the portal circulation and from there via the bloodstream to the *adenohypophysis.*

> The *optic chiasma* is located anterior to the pituitary gland and is an important neighbor, diagnostically speaking, because a pituitary tumor may indicate its presence by pressure upon the chiasma, causing visual symptoms. Tumors can also cause enlargement of the sella turcica that can be detected by x ray.

The Neurohypophysis

The neurohypophysis produces no hormones but is considered an endocrine gland, because hormones produced by the cells of the hypothalamus (Table 18-1) are released in the posterior lobe of the pituitary where they enter the blood stream. The cells of the posterior lobe, called *pituicytes,* are comparable in structure and origin to the neuroglia of the CNS. The secretory cells of the hypothalamus (neurosecretory cells) are modified neurons with axons extending into the posterior lobe of the pituitary. The neurosecretory cells produce **oxytocin** and **antidiuretic hormone** (ADH), which travel down their axons and are stored in the posterior lobe until they are released into the blood under the influence of nervous stimuli traveling down the same axons. Thus the hypothalamic cells have a dual function: the production and transport of hormones and the initiation and transmission of nerve impulses in response to a variety of stimuli.

The two hypothalamic-neurohypophyseal hormones are polypeptides, similar in structure and overlapping in function. The chief action of ADH is upon the *kidney* where it increases permeability to water of the cell membranes of the *distal convoluted tubules* and the *collecting tubules,* resulting in a reduction in the quantity of water excreted in the urine; hence its

Table 18-1   Hormones of the Neurohypophysis

Hormone	Abbreviation	Source	Target Organ	Major Effects
Oxytocin	—	Hypothalamus	Smooth muscle of uterus	Stimulates contraction of uterine muscle during labor
Antidiuretic hormone (vasopressin)	ADH	Hypothalamus	Kidney	Reduces volume of water excreted in urine

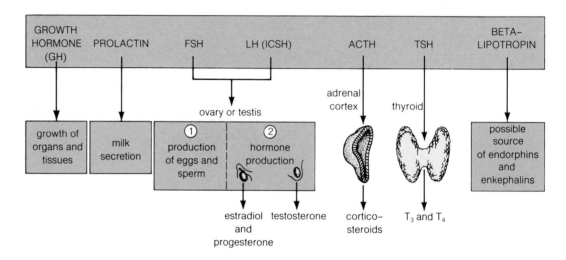

Figure 18-8   The hormones of the adenohypophysis and their effects.

name (*diuretic* means increasing the urine flow). The older name for ADH, *vasopressin* (from the Latin for *vessel* and *pressure*), refers to the hormone's effects on blood pressure, which are much weaker than its antidiuretic effects. The influence of ADH on blood pressure is due to the hormone's ability to stimulate contraction in smooth muscle, including that of blood vessels.

Oxytocin, a hormone released in large quantities by the posterior lobe of the pituitary late in pregnancy, specifically stimulates the smooth muscle of the uterus, called *myometrium,* playing a major role in the onset of labor as well as in the milk ejection reflex. The same hormone is released in smaller quan-

tities at certain times in the menstrual cycle and plays a role in the transport of the sperm and ovum preceding fertilization.

The Adenohypophysis

The adenohypophysis is the glandular part of the pituitary, with secretory cells arranged in anastomosing cords separated by sinusoids. At least six hormones (Table 18-2) are produced by the anterior pituitary (Fig. 18-8), and there is evidence that each is the product of a specific cell type. A seventh hormone, beta-lipotropin, may be produced by the adenohypophysis, although

Table 18-2 Hormones of the Adenohypophysis (Anterior Lobe of the Pituitary Gland)

Hormone	Abbreviation	Source (Cell Type)	Target Organ	Function
Growth hormone (somatropin)	GH	Acidophils	All cells	Maintains protein synthesis, releases fats and glucose
Thyroid-stimulating hormone	TSH	Basophils	Thyroid gland	Stimulates production and release of thyroid hormone
Adrenocorticotropic hormone	ACTH	Basophils	Suprarenal gland	Stimulates release of hormones from suprarenal cortex
Prolactin (luteo-tropic hormone)	LTH	Acidophils	Mammary gland	Stimulates growth of gland and milk production
Follicle-stimulating hormone	FSH	Basophils	Ovarian follicle	Stimulates growth of ovarian follicle and production of estrogen and production of sperm in the male
Luteinizing hormone (interstitial-cell-stimulating hormone)	LH (ICSH)	Basophils	Ovarian follicle (interstitial cells in male)	Stimulates development of corpus luteum and production of progesterone; (production of testosterone in male)

it has not been proven to have a specific effect in humans. This hormone may be a source of the sensory inhibitory substances, the endorphins and enkephalins. Morphologically the cells may be classified as *chromophils,* which are colored by histological stains, and as *chromophobes* (which may represent a phase of differentiation of the gland cells) whose cytoplasm fails to react with stains. The chromophils are readily classified as *acidophils* (stained by acid dyes) or as *basophils* (stained by basic dyes) on the basis of the staining qualities of their cytoplasmic granules. With the use of specialized staining techniques, at least two types of acidophils and four types of basophils have been identified. The acidophils are responsible for the production of *growth*

hormone (GH) and *prolactin,* or luteotropin (LTH); the basophils produce the *adrenocorticotropic hormone* (ACTH), *thyroid stimulating hormone* (TSH), and the gonadotropic hormones—*follicle stimulating hormone* (FSH) and *luteinizing hormone* (LH).

GH has a generalized effect upon metabolic processes including maintenance of the normal rate of protein synthesis, which increases the size of the body tissues. It inhibits fat synthesis and mobilizes fat stores, increasing free fatty acids in the plasma. Its effects upon carbohydrate metabolism include an anti-insulin action in the tissues that causes the blood glucose level to rise and may lead to the development of *diabetes mel-*

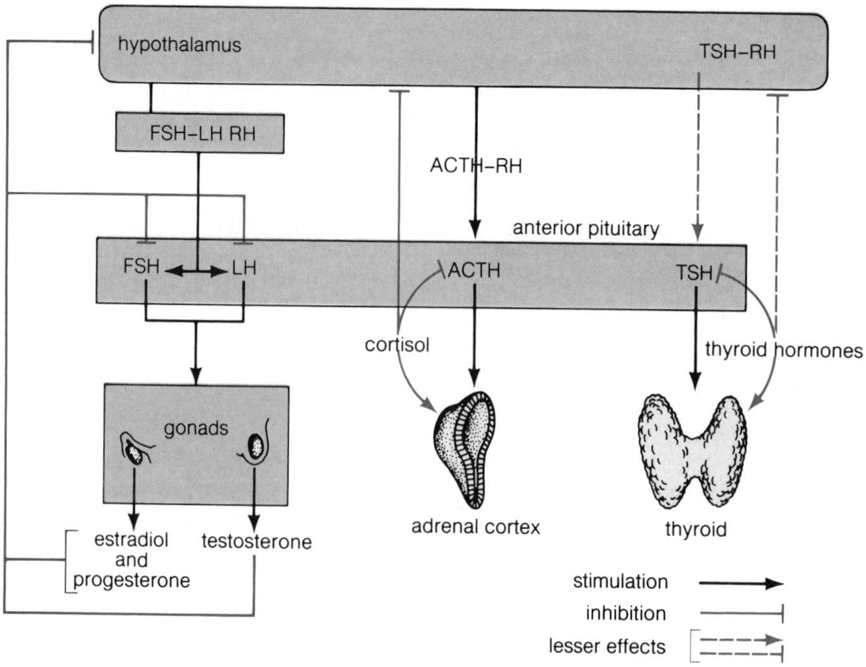

Figure 18-9 Examples of negative feedback in the regulation of the function of the anterior pituitary gland and glands it affects. Feedback into the hypothalamus is most important in the regulation of sex hormones, moderately important in the regulation of the adrenal cortex hormones, and only slightly important in the regulation of the thyroid hormones.

litus. Although GH affects the metabolism of all tissues, its most striking effect is seen in the *epiphyseal plates* of developing long bones; it stimulates the production of a substance in the liver called *somatomedin* that acts to stimulate the growth in length of these bones, thus determining the height of an individual.

ACTH is a small protein (polypeptide) that maintains the cells of the *suprarenal cortex* and stimulates them to synthesize *corticosteroid hormones.* Insufficient ACTH results in reduction in the production of corticosteroids, leading to low blood pressure, nausea, and hypoglycemia. The effects of corticosteroids will be discussed in the section on the suprarenal glands.

TSH is responsible for normal maintenance of the thyroid gland. The blood level of TSH is controlled by a negative feedback mechanism (Fig. 18-9) in which a rise in the blood level of *thyroxine* (thyroid hormone) causes a drop in TSH production and a drop in thyroxine stimulates the anterior pituitary to produce more TSH.

LTH, also called prolactin or lactogenic hormone, is essential for the development of the *mammary gland* during pregnancy and for the production of milk following parturition. The *gonadotropic hormones* include FSH, which stimulates the development of the follicle in the ovary and sperm production in the testes, and LH, which acts on the **corpus luteum** of the ovary. LH, so named in females, is the same hormone as ICSH (interstitial-cell-stimulating hormone) in the male, which stim-

ulates the interstitial cells of the testis to increase secretion of testosterone. The structure of the gonads (ovary and testis) and the role of the gonadotropins will be described further in the discussions of the ovary and testis as endocrine glands.

Releasing Hormones

The release of each of the hormones of the anterior lobe is controlled by a substance called a *releasing hormone* (RH), or releasing factor, produced by the *hypothalamus* (Table 18-3). For example, FSH is produced by acidophils of the anterior pituitary and is stored within these cells until they are stimulated by FSH-RH to discharge the hormone. The FSH-RH is produced by the hypothalamus and conveyed to the pituitary by axons of neurons that end close to the capillary plexus of the *hypophyseal portal system.* The capillary plexus not only supplies the anterior pituitary with nutrients but also receives the secretory products of the hypothalamus, including FSH-RH. Blood from the capillaries enters the *hypophyseal portal veins,* which empty into the sinusoids surrounding the secretory cells of the anterior lobe. The FSH-RH reaches the cells in which FSH has been synthesized and stimulates discharge of FSH into the blood stream for distribution to all parts of the body, including the ovary where it stimulates growth of the follicles. The hypophyseal portal sys-

Table 18-3   Releasing Hormones from the Hypothalamus

Hormone	Abbreviation	Function
Follicle-stimulating-hormone-releasing hormone	FSH-RH	Stimulates release of FSH
Luteinizing-hormone-releasing hormone	LH-RH (probably identical to FSH-RH)	Stimulates release of LH
Thyroid-stimulating-hormone-releasing hormone	TRH	Stimulates release of TSH
Growth-hormone-releasing hormone	GH-RH	Stimulates release of GH
Adrenocorticotropic-hormone-releasing hormone	CRH	Stimulates release of ACTH
Growth-hormone-inhibiting hormone (somatostatin)	GH-IH	Inhibits release of GH
Prolactin-releasing hormone	PRH	Stimulates release of prolactin
Prolactin-inhibiting hormone	PIH	Inhibits release of prolactin

tem provides the means by which substances produced in the hypothalamus reach the anterior lobe of the pituitary in high concentration without entering the general circulation.

THE THYROID GLAND

The **thyroid gland** (Fig. 18-10) is a bilobed structure located in the neck. Each lobe lies alongside the larynx with the upper poles overlapping the lower portion of the thyroid cartilage. The lobes are connected inferiorly by an isthmus of glandular tissue that crosses anterior to the second and third tracheal rings. The thyroid gland is almost completely covered anteriorly by the infrahyoid muscles.

Because the fascia enclosing the thyroid gland is much thicker anteriorly than posteriorly, any enlargement causes the gland to expand posteriorly. Due to the attachments of the fascial sheets superiorly, the gland may also expand inferiorly. Abnormal enlargement of the thyroid is called *goiter*.

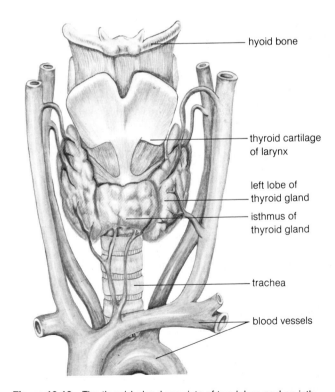

hyoid bone

thyroid cartilage of larynx

left lobe of thyroid gland

isthmus of thyroid gland

trachea

blood vessels

Figure 18-10   The thyroid gland consists of two lobes and an isthmus surrounding the trachea

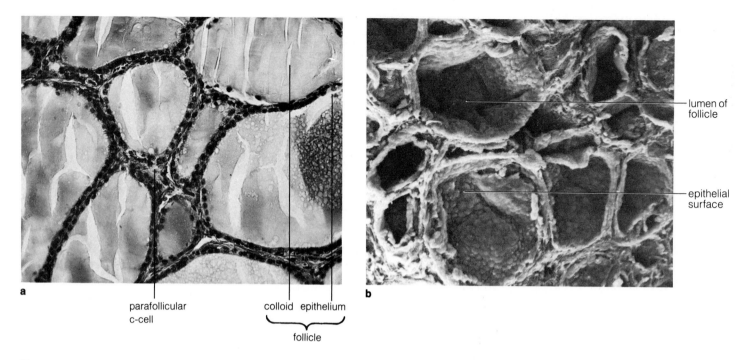

a

parafollicular
c-cell

colloid epithelium

follicle

b

lumen of
follicle

epithelial
surface

Figure 18–11   The thyroid gland. (**a**) The microscopic anatomy of the thyroid gland, × 235 (Armed Forces Institute of Pathology negative number 56-5264-738450). (**b**) A scanning electron micrograph of the thyroid gland × 340. (Photograph reproduced with permission from *Tissues and Organs: A Text-Atlas of Scanning Electron Microscopy,* by Richard G. Kessel and Randy H. Kardon. W. H. Freeman and Company. Copyright © 1979.)

The thyroid gland is reddish-brown in color due to the dense supply of blood capillaries investing its secretory elements. Its arterial supply is from the *superior thyroid* branch of the *external carotid artery* and the *inferior thyroid* branch of the *thyrocervical trunk* from the *subclavian artery*. The secretory elements are cystlike follicles, usually composed of simple cuboidal cells that enclose a lumen filled with colloid containing thyroglobin (Fig. 18-11).

The follicle cells in contact with the lumen, called *principal follicle cells,* produce *thyroid hormones;* those not in contact with the lumen, called *parafollicular cells,* produce *calcitonin,* a hormone that lowers blood calcium. The follicle cells vary in their morphology, depending upon the state of their activity, from low cuboidal cells with a centrally located nucleus in the nonsecreting state to a taller cell with the nucleus displaced toward the base of the cell in the actively secreting follicle. The thyroid hormones (Table 18-4) are *tetra-iodothyronine* (T_4) and *tri-iodothyronine* (T_3); they stimulate metabolism and oxygen consumption, thereby increasing the cells' heat production and affecting growth. These hormones, iodinated derivatives of the amino acid L-tyrosine, are secreted into the follicles where they are stored bound to *thyroglobulin,* a complex glycoprotein produced by the principal follicle cells and secreted into the lumen. When the thyroid hormones are to be secreted from the gland, thyroglobulin is taken into the follicle cells by pinocytosis, forming vesicles filled with thyroglobulin. Lysosomes fuse with these vesicles and modify the thyroglobulin, freeing the thyroid hormones, which then diffuse from the follicle cells into the blood vessels where they are combined with carrier protein molecules of the plasma. The hormone molecules are released from the carriers before entering tissue cells as free T_4 or T_3.

The major factor determining the rate of thyroid secretion is the production of thyroid stimulating hormone (TSH) by the anterior lobe of the pituitary. The amount of TSH released by the anterior pituitary is in turn regulated by the blood level of the thyroid hormones. A drop in thyroid hormone in the blood leads to an increase in the release of TSH, which causes secretion of thyroid hormone. As the level of thyroid hormone in the blood rises, there is a drop in the release of TSH, and the blood concentration of thyroid hormone stabilizes.

Table 18-4   Hormones of the Thyroid Gland

Hormone	Abbreviation	Source	Target Organ	Major Effects
Tetra-iodothyronine	T_4	Principal follicle cells	All cells	Stimulates metabolism and increases oxygen consumption
Tri-iodothyronine	T_3	Principal follicle cells	All cells	Stimulates metabolism and increases oxygen consumption
Calcitonin		Parafollicular cells	Blood	Decreases blood calcium

Calcitonin, produced by the parafollicular cells, works with the parathyroid hormone to maintain the calcium level of blood and is important in the maintenance of the skeleton and in calcium metabolism. When injected into an animal, calcitonin causes a rapid drop in blood calcium. The secretion of calcitonin normally is not influenced by pituitary secretions but depends upon the blood level of calcium.

THE PARATHYROID GLANDS

There are usually four **parathyroid glands,** located on the posterior surface of the lobes of the thyroid glands, with two of the parathyroids located superiorly above the middle of the lobes and two situated inferiorly near the lower edge of the lobes (Fig. 18-12). The glands are usually in the capsule of the thyroid gland but may be embedded in thyroid tissue. They are oval bodies about 5 mm in diameter enclosed within a connective tissue capsule that separates the parathyroid cells from thyroid tissue.

The secretory cells are arranged in dense oval masses or columns of cells supported by connective tissue, with numerous sinusoids between cell masses (Fig. 18-13). The secretory cells are of two types: *chief* (principal) *cells,* which secrete the parathyroid hormone, and the *oxyphil cells,* whose function is unknown.

The parathyroid glands are essential for the maintenance of blood calcium at a level that sustains life. Removal of the parathyroids is followed by a drop in blood calcium, resulting

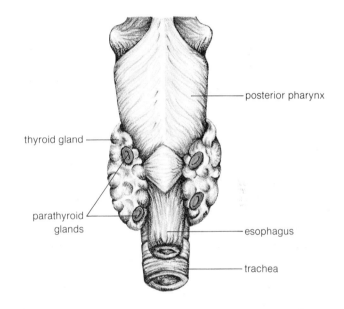

Figure 18–12   The parathyroid glands lie on the posterior surface of the thyroid gland.

in muscle spasms (tetany) and death. The parathyroid hormone acts in conjunction with calcitonin to maintain circulating calcium within narrow limits. Parathyroid hormone raises blood calcium by mobilizing calcium from bone and by stimulating tubular reabsorption of calcium in the kidney, whereas calcitonin lowers blood calcium by inhibiting the withdrawal of calcium from bone (Fig. 18-14).

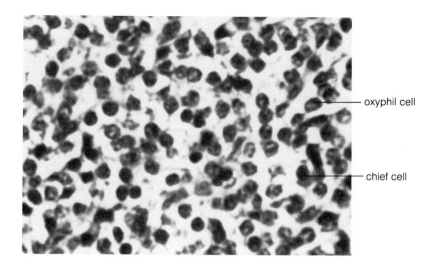

Figure 18-13 Microscopic structure of human parathyroid gland, ×800. (Courtesy of S. D. Smith.)

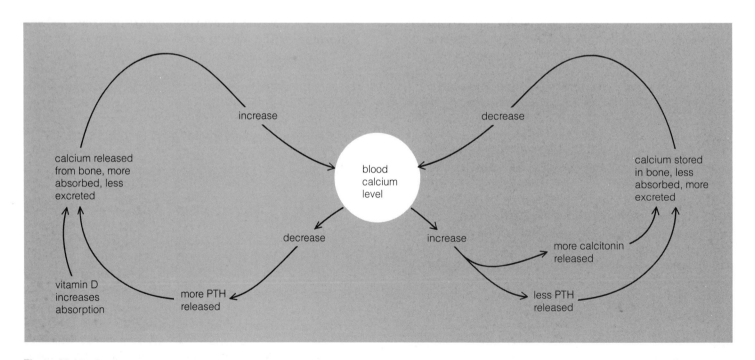

Figure 18-14 Blood calcium is regulated by the effects of parathormone (PTH), calcitonin, and vitamin D.

THE SUPRARENAL GLANDS

The **suprarenal glands** (adrenal glands) are paired organs located on the superior pole of each kidney where they appear to form a concave cap conforming to the convexity of the kidney surface (Fig. 18-15). Each gland is encased by a connective tissue capsule that separates it from the kidney. Like all endocrine glands the suprarenals have a good blood supply, each gland receiving blood through branches of three arteries: a branch of the *inferior phrenic* (superior suprarenal arteries), a direct branch of the *abdominal aorta* (middle suprarenal arteries), and a branch of the *renal artery* (inferior suprarenal artery). Although each suprarenal gland receives blood from as many as 6 arteries, it is drained by a single large vein; the left suprarenal vein empties into the left renal vein, and the right drains directly into the inferior vena cava.

A suprarenal gland measures approximately 5 × 3 × 1 cm and weighs about 7.5 gm. Each is really two separate endocrine organs with quite different structures and unrelated functions (Fig. 18-15). The *suprarenal cortex*, which forms the largest portion of the gland, surrounds the centrally located *suprarenal medulla*.

The Suprarenal Cortex

The secretory cells of the suprarenal cortex are arranged in clusters and columns of cells separated by sinusoids (Fig. 18-16b). The cellular arrangement and morphology are different near the surface and at each of two deeper levels, leading to the description of the gland in three concentric zones: a thin outer *zona glomerulosa* adjacent to the capsule; a thick middle layer, the *zona fasciculata;* and an inner *zona reticularis* adjacent to the medulla.

The zona glomerulosa consists of clusters of columnar secretory cells that contain small lipid droplets and abundant smooth endoplasmic reticulum with free polyribosomes in the cytoplasm. The clusters or oval masses of cells of the zona glomerulosa are continuous with the zona fasciculata, which consists of irregularly shaped secretory cells arranged in parallel columns or cords of cells oriented at right angles to the surface. The zona fasciculata is continuous with the zona reticularis in which the cords of secretory cells, usually only a single cell thick, form an anastomosing network separated by sinusoids. The secretory cells of the zona reticularis and zona fasciculata are also filled with numerous lipid droplets; they have more smooth ER (indicating steroid synthesis) than cells of the zona glomerulosa, and they have some rough ER (suggesting protein synthesis) as well.

The hormones produced by the suprarenal cortex (Table 18-5) are steroids and may be grouped as three general func-

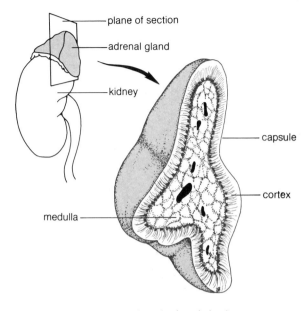

Figure 18-15 Each adrenal gland is located superior to the kidney and is composed of a cortex and a medulla.

tional types: **mineralocorticoids, glucocorticoids,** and **sex hormones.** The mineralocorticoids, produced by the zona glomerulosa, are concerned with fluid and electrolyte balance. The most important of these hormones is *aldosterone,* which causes retention of sodium in the body by stimulating tubular reabsorption of sodium in the kidneys and increasing the excretion of potassium. Thus the hormone is important in the maintenance of a constant internal environment for the cells, and the hormonal level must be carefully regulated in order to adjust quickly to changes.

A drop in the blood level of sodium, through a series of reactions in the kidney and blood, ultimately stimulates the suprarenal gland to secrete aldosterone. This results in increased reabsorption of sodium in the kidneys, conserving sodium until its normal concentration in the blood is restored (Fig. 18-17).

The glucocorticoids are produced by the zona fasciculata and zona reticularis, especially in response to stress; they stimulate the formation of glucose in the liver and the mobilization of fat. These hormones also cause a decrease in the rate of protein synthesis and an increase in protein degradation within the cells, increasing the level of amino acids in the blood; as the amino acids reach the liver, they are converted to glucose. The regulation of the secretion of the glucocorticoids is independent of that of the mineralocorticoids. In response to stress the hypothalamus discharges corticotropin-releasing hormone

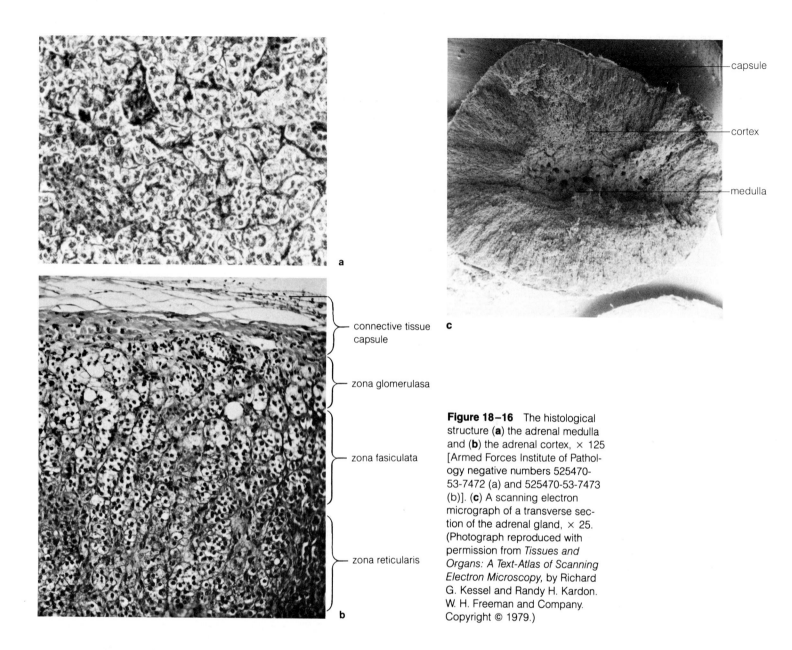

connective tissue
capsule

zona glomerulasa

zona fasiculata

zona reticularis

a

b

c

capsule

cortex

medulla

Figure 18–16 The histological structure (**a**) the adrenal medulla and (**b**) the adrenal cortex, × 125 [Armed Forces Institute of Pathology negative numbers 525470-53-7472 (a) and 525470-53-7473 (b)]. (**c**) A scanning electron micrograph of a transverse section of the adrenal gland, × 25. (Photograph reproduced with permission from *Tissues and Organs: A Text-Atlas of Scanning Electron Microscopy,* by Richard G. Kessel and Randy H. Kardon. W. H. Freeman and Company. Copyright © 1979.)

into the hypophyseal portal system, promoting the discharge of ACTH from the anterior lobe of the pituitary. The ACTH reaches the suprarenal cortex and causes the discharge of glucocorticoids, resulting in an increase of glucose and amino acids in the blood, which enables the body to respond to the stress. In addition to mobilizing nutrients in response to stress, the glucocorticoids also tend to depress the immune response in allergic reactions and inflammatory diseases.

The inner zones of the suprarenal cortex also produce small amounts of male sex hormones, or androgens. Although they are called "male" sex hormones because the much larger

quantities produced in the testes cause male sexual development, androgens are produced in the suprarenal glands of both males and females. Suprarenal androgen secretions are of interest primarily in diseases of the pituitary or suprarenal glands that result in hypersecretion or hyposecretion, causing altered sexual functions.

The Suprarenal Medulla

The secretory cells of the *suprarenal medulla* (Fig. 18-16a) are epithelioid cells filled with granules containing the hormones

Table 18-5 Hormones of the Suprarenal Glands			
Hormone	Source	Target Organ	Major Effects
Mineralocorticoid (aldosterone)	Zona glomerulosa	Kidneys	Causes reabsorption of sodium in kidneys, increases excretion of potassium
Glucocorticoid	Zona fasciculata, zona reticularis	Liver	Increases glucose and amino acid discharge into blood
Androgen	Suprarenal cortex	Male secondary sex characteristics	Alters sexual functions
Epinephrine	Suprarenal medulla	Mostly cardiovascular system	Responds to emergency situations
Norepinephrine	Suprarenal medulla	Mostly cardiovascular system	Responds to emergency situations

epinephrine (adrenalin) and **norepinephrine** (Table 18-5). They occur in clumps or short columns on which sympathetic preganglionic nerve fibers terminate. The secretory cells, derived from the embryonic neural crest, are modified sympathetic ganglion cells that do not form processes but become specialized to elaborate hormones. Interestingly, these hormones have the same effects as those elicited by postganglionic sympathetic neurons. Epinephrine and norepinephrine are both *catecholamines* and produce similar effects, most of which enable the body to respond to emergency situations; for example, they cause an increase in heart rate and a rise in blood pressure. Epinephrine also has effects upon carbohydrate metabolism.

THE ISLETS OF LANGERHANS

Islets of Langerhans are endocrine secretory cell masses scattered among the exocrine components of the *pancreas* (Fig. 18-18). The islets are richly vascularized and are separated from the exocrine elements by a thin reticular fiber network. There are estimated to be more than 200,000 islets in the pancreas (Fig. 18-19), with most located in the tail of the pancreas.

Three types of granule-containing cells can be identified with special stains: the *alpha cell* contains granules that are insoluble in alcohol; the *beta cell* contains granules that are

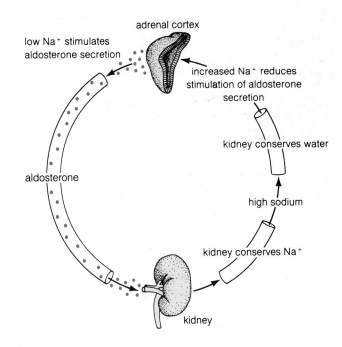

Figure 18-17 Aldosterone regulates the sodium and water in body fluids.

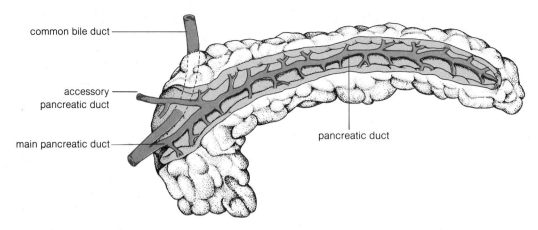

Figure 18-18 Gross anatomy of the pancreas.

common bile duct

accessory pancreatic duct

main pancreatic duct

pancreatic duct

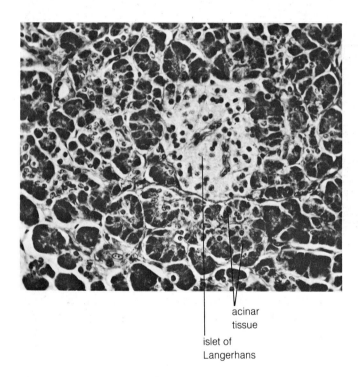

acinar tissue

islet of Langerhans

Figure 18-19 Microscopic anatomy of the pancreas, ×265 (Armed Forces Institute of Pathology negative number 71-9195).

soluble in alcohol; and the *delta cell* contains small basophilic granules. A nongranular cell has been designated the *C cell.* The hormone *insulin* (Table 18-6), secreted by the beta cells, is a polypeptide composed of 51 amino acids and was the first protein to be synthesized in the laboratory. It facilitates the removal of glucose from the blood by increasing the transport of glucose across cell membranes, especially the membranes of liver and muscle cells. Within these cells glucose is converted into glycogen for storage. Insulin therefore is important in preventing the blood glucose in the general circulation from exceeding the ability of the kidneys to reabsorb it. A deficiency of insulin results in *diabetes mellitus,* the disease in which glucose is detectable in the urine because the glomerular filtrate contains more glucose than can be reabsorbed by the uriniferous tubules. Insulin is the only hormone that lowers blood glucose and consequently may be considered the hypoglycemic hormone. Other hormones that affect blood glucose, such as growth hormone, epinephrine, glucagon, and thyroxine, raise the glucose level. Such mechanisms provide the central nervous system with the continuous supply of glucose necessary for survival.

Glucagon is secreted by the alpha cells and has an effect opposite to that of insulin (Fig. 18-5). It causes the breakdown of glycogen, releasing glucose into the bloodstream and elevating the blood glucose level. The delta cells secrete *somatostatin,* a powerful inhibitor of growth hormone secretion.

The functions of the C cells are not known, but the cells may represent different metabolic states of the alpha and beta cells.

THE OVARIES AS ENDOCRINE GLANDS

The structure of the ovary as a reproductive organ is described in Chapter 24. In addition to their primary function of producing germ cells for the perpetuation of the species, the ovaries pro-

Table 18-6 Hormones of the Pancreas			
Hormone	Source	Target Organ	Major Effects
Insulin	Beta cells	Liver, muscle	Converts glucose to glycogen
Glucagon	Alpha cells	Liver, muscle	Converts glycogen to glucose
Somatostatin	Delta cells	Anterior pituitary	Inhibits release of growth hormone

duce hormones that maintain the secondary sex characteristics; these hormones also induce cyclic changes in the ovaries in preparation for ovulation and in the uterus in preparation for pregnancy. The ovarian hormones are **estrogen,** secreted by the ovarian follicle cells (Fig. 18-20), and **progesterone,** secreted by the corpus luteum. The hormone cycle, called the menstrual cycle, takes an average of twenty-eight days.

The Ovarian Follicle

Following puberty the ovaries contain many ova in primary and secondary follicles in various stages of maturation, as well as ova within inactive primordial follicles. The primordial follicles consist of a single layer of flattened epithelial cells surrounding the ovum and show little evidence of secretory activity. Beginning in puberty, some of the follicles begin to mature each month, but usually only one completes its maturation. As the ovum of the maturing primordial follicle enlarges, the follicular cells also enlarge, becoming cuboidal secretory cells called *granulosa cells.* The granulosa cells proliferate, giving rise to stratified epithelium that forms the primary follicle. Secondary follicles (Fig. 18-21) are the result of secretion of a clear fluid, the *liquor folliculi,* into the follicle, causing rapid expansion of the follicle as it matures. In both primary and secondary follicles the granulosa cells are surrounded by a coat called the *theca interna,* composed of modified connective tissue containing secretory cells. The cells of the theca interna secrete the hormone *estrogen,* which is conveyed via the blood stream to the uterus where it stimulates the proliferation of cells of the endometrial (uterine lining) glands and associated connective tissue and blood vessels. The growth of the ovarian follicles and the consequent increase in estrogen production are stimulated by FSH released by the anterior lobe of the pituitary. Although estrogen is produced continuously by the ovarian follicles during the reproductive life of the female, it is secreted in larger quantities during the first half of each ovarian cycle, that is,

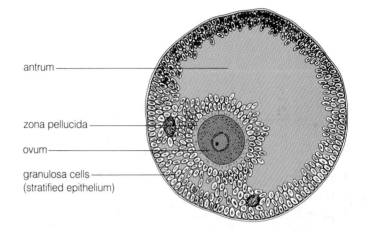

antrum

zona pellucida

ovum

granulosa cells
(stratified epithelium)

Figure 18-20 Diagram of a mature ovarian follicle.

during the time prior to ovulation, which occurs at about day 14 of a 28-day cycle. Ovulation is stimulated by a sudden surge in the release of LH by the anterior lobe of the pituitary, which causes the mature follicle (Fig. 18-21) to rupture, discharging the ovum.

The Corpus Luteum

As the follicle ruptures during ovulation, the liquor folliculi is expelled along with the ovum, causing the wall of the follicle to collapse. The granulosa cells of the follicle and the cells of the theca interna rapidly enlarge into pale-staining *lutein cells.* The lutein cells derived from the granulosa cells are larger and paler than those originating from the theca interna. The two layers of cells become highly convoluted, forming the secretory structure known as the *corpus luteum* (Fig. 18-22).

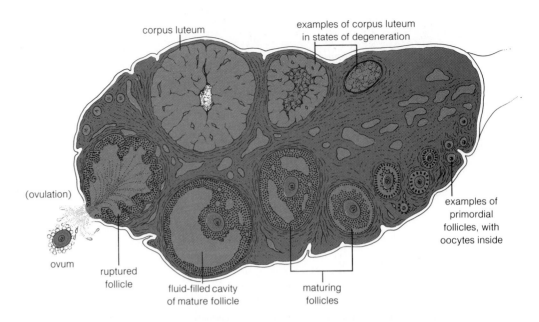

corpus luteum

examples of corpus luteum in states of degeneration

examples of primordial follicles, with oocytes inside

(ovulation)

ovum

ruptured follicle

fluid-filled cavity of mature follicle

maturing follicles

a

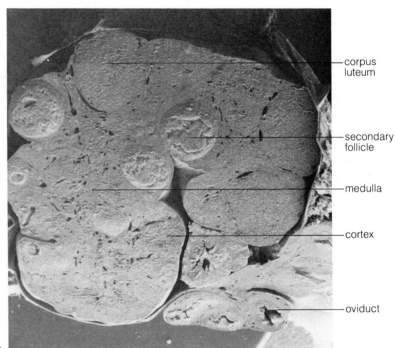

corpus luteum

secondary follicle

medulla

cortex

oviduct

b

Figure 18-21   The ovary (**a**) Diagram of a section of the ovary showing stages in the ovarian cycle. (**b**) A scanning electron micrograph of the ovary, × 30. (Photograph reproduced with permission from *Tissues and Organs: A Text-Atlas of Scanning Electron Microscopy,* by Richard G. Kessell and Randy H. Kardon. W. H. Freeman and Company. Copyright © 1979.)

Under the influence of LH the corpus luteum produces both estrogen and *progesterone,* which cause additional changes in the endometrium in preparation for pregnancy. As the endometrium thickens, there is accumulation of nutrients, particularly glycogen, with enlargement of the glands and an increase in interstitial fluid. If pregnancy occurs, LH release by the anterior pituitary continues; the corpus luteum is maintained during the first three months and continues to produce large amounts of estrogen and progesterone until the placenta takes over this function. If the ovum is not fertilized, the LH release by the anterior pituitary is halted, and the corpus luteum degenerates, with a subsequent drop in the secretion of ovarian hormones.

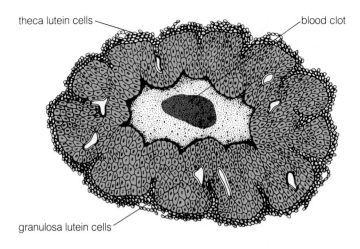

theca lutein cells

blood clot

granulosa lutein cells

Figure 18-22 Diagram of the corpus luteum.

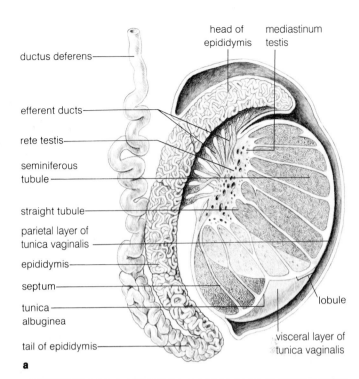

ductus deferens

head of epididymis

mediastinum testis

efferent ducts

rete testis

seminiferous tubule

straight tubule

parietal layer of tunica vaginalis

epididymis

septum

tunica albuginea

tail of epididymis

lobule

visceral layer of tunica vaginalis

a

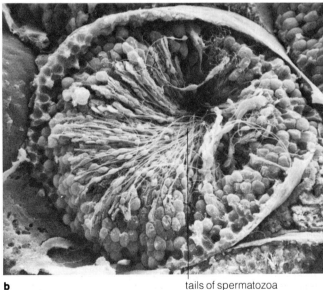

b

tails of spermatozoa

Figure 18-23 The testis. (**a**) A sagittal section through the testis. (**b**) scanning electron micrograph of the testis, × 330. (Photograph reproduced with permission from *Tissues and Organs: A Text-Atlas of Scanning Electron Microscopy*, by Richard G. Kessel and Randy H. Kardon. W. H. Freeman and Company. Copyright © 1979.)

The corpus luteum is gradually replaced with fibrous connective tissue, forming a white scar called the *corpus albicans*. The drop in the ovarian hormone level brings on menstruation, during which most of the endometrium sloughs off; it also affects the anterior pituitary and hypothalamus, causing a new cycle of growth of follicles. The first day of menstruation is considered the first day of the cycle.

The hypothalamus plays an important role in the menstrual cycle. The hypothalamic hormone LH-RH (luteinizing-hormone-releasing-hormone) stimulates the release of either FSH or LH from the anterior pituitary, depending on the conditions (such as the blood level of estrogen) that exist at a specific stage of the cycle. Some of these conditions also affect the anterior pituitary directly, so the regulatory system for ovarian hormones is quite complex. Many details of the system are still unknown; for instance, how specific follicles are selected to begin maturation each month, or why only one of them usually completes the process.

THE TESTES AS ENDOCRINE GLANDS

In addition to producing sperm the testes (Fig. 18-23) elaborate the androgen (male sex hormone) **testosterone,** which is important in the maintenance of the secondary sex characteristics that appear at puberty and for the production and maturation of sperm. Testosterone is produced by the *interstitial cells* (of Leydig) located between the seminiferous tubules in which the sperm are produced (Fig. 18-24). The interstitial cells occur in clusters, usually close to blood vessels. They are irregularly shaped cells with an oval nucleus, a well-developed Golgi complex, and extensive smooth ER. Unlike the ovary the testis is not cyclic in its hormone production. The interstitial cells are stimulated to secrete by the interstitial-cell-stimulating hormone (same as LH) of the anterior pituitary, which in the male is maintained at a relatively constant level, resulting in a constant level of testosterone secretion.

THE PINEAL GLAND

The **pineal gland** or **body** is a small structure, 5 to 8 mm in length, located above the diencephalon at the posterior end of the third ventricle (Fig. 18-25). The gland consists of cords of cells termed pinealocytes, or chief cells, with interstitial cells between the clusters of pinealocytes.

The precise function of the pineal gland in the human is still not firmly established. The hormone **melatonin** is produced by pineal glands and causes blanching of the pigment cells of amphibians; however, it has no influence on mammalian pigment cells (melanocytes). The hormone does appear to have an active role in the light-influenced reproductive cycles of rodents and birds. In humans it appears to inhibit the release of pituitary gonadotropins, preventing precocious puberty. Some substances produced by the pineal gland may be involved in regulating the rhythmic activities of endocrine glands. It remains to be shown which human physiological rhythms are influenced by the pineal.

THE THYMUS

The **thymus gland** may be considered as an endocrine gland; it produces blood-borne substances that influence lymphoid cells (a type of white blood cell). However, because its principal functions involve immunity, the thymus will be considered in Chapter 25 on the body defenses.

OTHER ENDOCRINE GLANDS

Many organs whose primary functions are not endocrine in nature produce hormones that are important in regulating activities of specific organs or systems. For example, the *gastrin* produced by the pancreas and intestinal mucosa stimulates glands

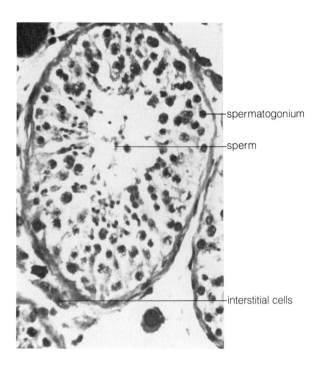

Figure 18-24 Microscopic structure of the human testis showing a seminiferous tubule and interstitial cells, × 200. (Courtesy of S. D. Smith.)

spermatogonium
sperm
interstitial cells

of the stomach to produce hydrochloric acid. The kidneys produce *renin,* an enzyme that raises blood pressure, and *erythropoietin,* a hormone that stimulates red blood cell production. Such hormones will be considered in the chapters on the systems in which these organs have major roles.

DEVELOPMENT OF THE ENDOCRINE GLANDS

Development of the Pituitary Gland

The pituitary gland arises from two distinct primordia, both hollow ectodermal evaginations (Fig. 18-26). The anterior pituitary arises from the *oral ectoderm,* but the posterior pituitary is a derivative of the floor of the *brain* (diencephalon). The first indication of the pituitary gland is an evagination called **Rathke's pouch,** projecting dorsally from the oral ectoderm near the degenerating oral plate. Soon after Rathke's pouch appears, a funnel-like evagination that will become the posterior lobe of the pituitary forms in the floor of the diencephalon and grows ventrally to approach Rathke's pouch. With further growth Rathke's pouch loses its connection with the oral ectoderm and projects

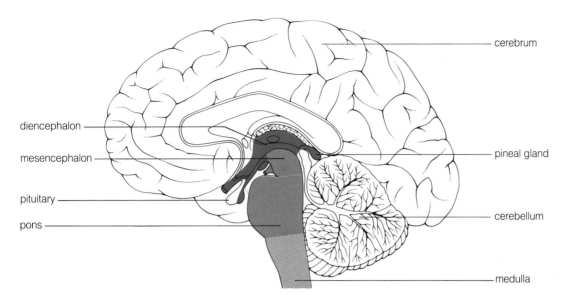

Figure 18-25 Midsagittal section of the brain showing the location of the pineal gland.

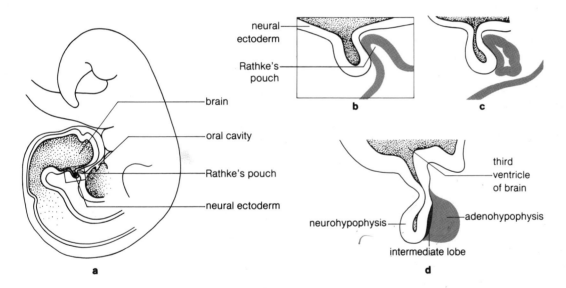

Figure 18-26 The pituitary gland is formed from Rathke's pouch, an ectodermal structure in the roof of the mouth, and neural ectoderm (**a**) and (**b**). Rathke's pouch separates from the mouth and comes to lie adjacent to the neural portion of the pituitary (**c**). The adenohypophysis and the intermediate lobe are derived from Rathke's pouch, and the neurohypophysis is derived from the neural ectoderm (**d**).

laterally and posteriorly to overlap the posterior lobe. As the secretory elements differentiate, Rathke's pouch forms the entire *anterior lobe,* including the pars distalis and pars tuberalis. It also gives rise to the pars intermedia, a part of the anterior lobe. The evagination from the diencephalon retains its connection with the brain, giving rise to the *posterior lobe* and to the infundibulum. The floor of the diencephalon differentiates into the lower part of the hypothalamus, to which the infundibulum remains attached. Neurons of the hypothalamus send axons through the infundibulum to terminate in the posterior lobe.

Clinical Applications: **Diseases of the Endocrine Glands**

Malfunctions of a single endocrine gland may have generalized effects upon the body as a whole and specific effects upon the target organs of the gland. In a disease or lesion affecting a gland that produces several hormones, the production of all of the hormones is likely to be altered, but the effect upon the production of one of the hormones may cause the predominant observable symptoms. Alterations in glandular function may be caused by a variety of lesions resulting from infection, tumor, or thrombosis (blood clot blocking the blood supply). Such lesions may lead to either an increase (hyperactivity) or a decrease (hypoactivity) in hormone production.

Diseases of the Pituitary Gland

Hypopituitarism of the pituitary gland may be caused by abnormal development and be present at birth, or it may be the result of destruction of tissue later in life. The specific symptoms depend on the time of onset of the condition and the degree to which different target organs are involved. Functions of the target organs are depressed by the reduction in the tropic hormones, resulting in multiple symptoms, such as failure of sexual development in congenital hypopituitarism, coupled with retarded growth and development in general. The clinical manifestations are usually those of a general inadequacy of pituitary hormones rather than of any one in particular.

In hyperpituitarism one or two effects usually overshadow the others. The predominant effect is due to overproduction of the growth hormone in most cases, with the adrenocorticotropic hormone frequently involved. Hyperproduction of the growth hormone prior to closure of the epiphyses results in *gigantism;* after epiphyseal closure it produces *acromegaly* (Fig. 18-27), enlargement of the extremities. In overproduction of ACTH the condition is called *Cushing's syndrome* (Fig. 18-28) and is characterized by hyperplasia (overgrowth) of the suprarenal cortex, resulting in obesity and round facial features (moon face), along with skeletal and sexual abnormalities. Hyperpituitarism is frequently caused by tumors of the anterior pituitary, which not only produce symptoms related to the overproduction of hormones but also lead to symptoms resulting from excessive enlargement of the gland. As a result of excessive growth the sella turcica may be eroded, and abnormal pressure may be exerted on the posterior lobe and the hypothalamus, producing symptoms related also to the posterior pituitary. The symptoms may include obesity, sexual underdevelopment, and mental deficiency.

Diseases of the Thyroid Gland

Under pathological conditions the thyroid gland may undergo either hyperplasia or involution. Hyperplasia may be due to an

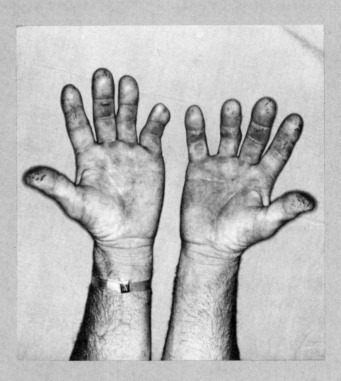

Figure 18-27 The hands of an individual suffering from acromegaly. Notice the unusually thick fingers. (Armed Forces Institute of Pathology negative number 72-14615.)

insufficiency of iodine, causing an increase in the size of the gland and in the amount of colloid it produces. Hyperplasia of the thyroid, or goiter (Fig. 18-29), was once common due to the lack of iodine in the diet, a situation that has been changed through the addition of iodine to common table salt. Hyperactivity of the thyroid gland produces a number of systemic effects, such as enlargement of the heart and increase in excitability of the nervous system.

Hypothyroidism may occur at any period of life; the results depend upon the age at onset. If it occurs during prenatal development, the result is *cretinism,* in which the individual is severely retarded, mentally and physically. If it occurs later in life, the condition is called *myxedema* and is characterized by a slowing of all physical and mental processes.

Diseases of the Islets of Langerhans

The most common pancreatic disease is *diabetes mellitus,* characterized by the increased use of proteins and fats and decreased

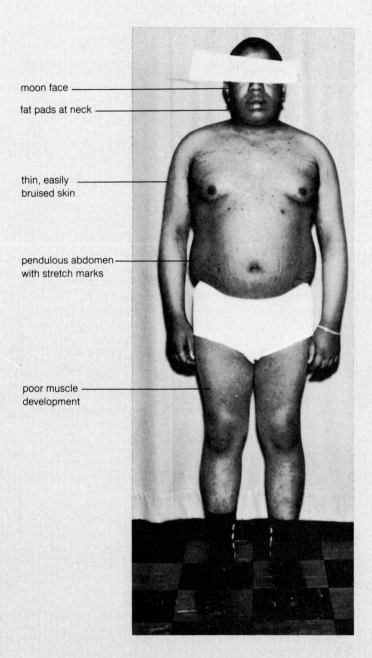

moon face ————————

fat pads at neck ————————

thin, easily ————————
bruised skin

pendulous abdomen ————————
with stretch marks

poor muscle ————————
development

Figure 18-28 Typical symptoms of Cushing's syndrome. (Armed Forces Institute of Pathology negative number 63-9617.)

metabolism of carbohydrates. The exact cause of diabetes is unknown, but it involves a production of insulin inadequate to meet the body's requirements. The blood level of glucose rises,

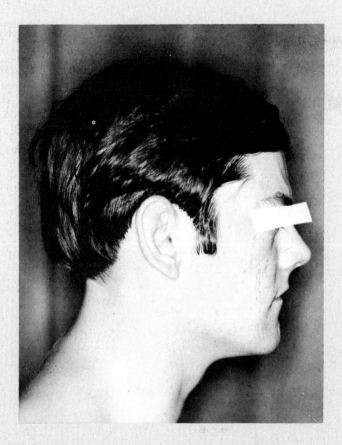

Figure 18-29 An individual with a goiter caused by hypothyroidism. (Armed Forces Institute of Pathology negative number 72-14994.)

exceeding reabsorption in the kidneys, with the result that glucose is excreted in the urine. The symptoms of diabetes can be largely controlled by insulin therapy.

Diseases of the Suprarenal Glands

The suprarenal cortex is essential for life, maintaining the electrolyte balance in addition to its important function in carbohydrate metabolism. Functions of the suprarenals may deviate from the normal either as a result of failure of the anterior pituitary to supply the proper quantity of ACTH or as a result of lesions that affect the suprarenal cortex directly. In either case the disease may cause overproduction or underproduction of the hormones. Hyperactivity may lead to *Cushing's disease,* with

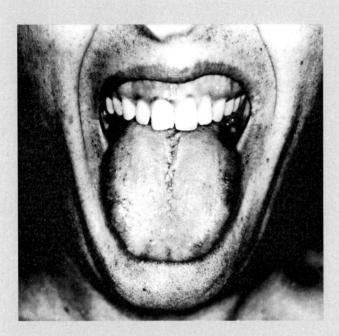

Figure 18-30 Spots of pigment on the skin, gums, and tongue are characteristic of an individual suffering from Addison's disease. (Armed Forces Institute of Pathology negative number 70-13320.)

suprarenal cortex produces masculinization of the female or premature sexual development in the male. Hypoactivity is known as *Addison's disease* (Fig. 18-30) and is characterized by loss of salt and water, with changes in general metabolism principally affecting carbohydrates. These metabolic effects cause hypoglycemia (low blood glucose level) and muscular weakness. Destruction or loss of function of the suprarenal cortex leads to increased blood levels of ACTH and MSH. Because these both stimulate the melanocytes in the skin, victims are usually tanned.

Diseases of the Parathyroids

The parathyroids do not appear to be related in function to other endocrine glands, and their diseases do not produce alterations in the functions of other glands. Because the parathyroid hormone regulates the metabolism of calcium, phosphorus, and bone, diseases of these glands may be expected to produce symptoms involving these functions. Clinical symptoms result from either hyperactivity or hypoactivity of these glands.

Hypoparathyroidism formerly occurred chiefly as a result of accidental surgical removal of the parathyroids during a thyroidectomy. Accidental removal of these glands is very rare at present, but hypoparathyroidism may result from hypoplasia, usually in young individuals. Underactivity of the glands produces hyperactivity of muscle due to the low level of blood calcium. A severe calcium deficiency may cause *tetany,* spasmodic muscle contraction, and convulsions that may prove fatal.

widespread symptoms. However, hyperactivity can also lead to more restricted symptoms, with the major effects upon the reproductive system: the overproduction of androgens by the

Development of the Thyroid Gland

The **thyroid gland** is the first endocrine gland to have evolved in vertebrates and is the first to develop embryologically in mammals. It begins as an evagination from the lining of the pharynx at the base of the tongue and grows rapidly, remaining connected with the pharyngeal floor by means of a slender tube, the *thyroglossal duct.* As the thyroid primordium grows caudally, the thyroglossal duct degenerates, and the lumen in the thyroid tissue is obliterated. The solid mass of cells differentiates into follicle cells, and colloid material soon begins to appear within these vesicles. Postnatal growth of the gland consists mainly in an increase in size of the follicles, with few new follicles formed after birth.

The point of origin of the thyroid is identifiable in the adult as the *foramen cecum,* a pit in the dorsum of the tongue near its base. The route of migration of the thyroid from the foramen cecum to its final location in the neck may occasionally be the site of formation of *thyroglossal cysts,* produced by persistent short segments of the thyroglossal duct.

Development of the Parathyroid Glands

The **parathyroid glands** develop from solid diverticula of the dorsal portions of the third and fourth visceral pouches. They begin as thickenings of the lining of the visceral pouches from which develop cords of cells. Fibrous tissue grows in between the fibrous cords, breaking them up into clumps of cells between which blood vessels grow.

As the parathyroids grow, their final location is influenced by the migration of the thyroid and the thymus gland. The thymus arises from the ventral portions of the third visceral pouches and migrates caudally into the thorax. Because of its origin from the same (third) pouch, parathyroid III is dragged caudal to parathyroid IV. As the thyroid gland migrates ventral to the gut (pharynx), the parathyroids become attached to the posterior surface of the thyroid gland, with parathyroid III located inferior to parathyroid IV.

During the course of development the parathyroids may adhere to the developing thymus. As the thymus descends, the parathyroids may be dragged inferiorly with it into the mediastinum (the area in the thorax between the lungs).

Development of the Suprarenal Glands

The **suprarenal gland** has a double origin: the cortex arises from the *coelomic epithelium* (mesoderm), and the medulla develops from *neural crest* (ectoderm). The cortex begins as a thickening of the splanchnic mesoderm lining the coelom near the mesonephros (an embryonic kidney). The cortical cells proliferate rapidly, forming cords of cells that lose continuity with the coelomic epithelium. At the same time neural crest cells migrate from their position alongside the neural tube and invade the cortical mass. The neural crest cells differentiate into secretory cells, forming the medulla located centrally within the cortical mass. The cells of the medulla represent neural crest cells that differentiate into gland cells instead of forming sympathetic ganglion cells.

The suprarenal cortex differentiates into cords of cells with vascular channels between the secretory cells. During fetal life two parts of the cortex may be distinguished: an inner fetal cortex that appears functional and an outer rim of relatively undifferentiated cells. During late fetal and early postnatal development the fetal cortex degenerates, and the outer, permanent cortex proliferates and differentiates. By the end of the third postnatal year the three zones characteristic of the adult suprarenal cortex are established.

WORDS IN REVIEW

Adenohypophysis	Norepinephrine
Adrenocorticotropic hormone	Parathyroid
Antidiuretic hormone	Pituitary
Calcitonin	Progesterone
Corpus luteum	Prolactin
Cyclic AMP	Ovary
Epinephrine	Oxytocin
Estrogen	Rathke's pouch
Glucocorticoid hormone	Releasing hormones
Growth hormone	Second messenger
Hormone	Suprarenal
Hypothalamus	Target organ
Insulin	Testis
Islets of Langerhans	Testosterone
Melatonin	Thyroid-stimulating hormone
Mineralocorticoids	Thyroglobulin
Negative feedback	Tuber cinereum
Neurohypophysis	Vasopressin

SUMMARY OUTLINE

I. Mechanisms of Action of Hormones
   A. Hormones may alter membrane permeability of target cells.
   B. Protein hormones may activate membrane enzymes that stimulate production of "second messenger" which activates enzyme systems in the cell. The cell has receptors on its membrane that recognize a specific hormone.
   C. Steroid hormones react with receptors on DNA, encouraging production of mRNA and proteins.

II. The Pituitary Gland
   A. The posterior lobe releases two hormones produced in the hypothalamus and delivered by nerve axons. Oxytocin stimulates smooth muscle contraction, and the antidiuretic hormone (ADH) decreases urine volume.
   B. The anterior lobe contains three types of cells, based upon their staining qualities, each producing different hormones.
      1. Acidophils produce growth hormone (GH) which has a generalized effect on metabolic processes and prolactin which stimulates development of mammary glands.
      2. Basophils produce thyroid-stimulating hormone (TSH) which stimulates thyroid gland, follicle-stimulating hormone (FSH) which stimulates development of ovarian follicles or sperm production, adrenocorticotropic hormone (ACTH) which influences the suprarenal cortex,

and luteinizing hormone (LH) which acts on corpus luteum of the ovary or interstitial cells of the testes. LH is also called ICSH (interstitial-cell-secreting hormone) in the male.

    3. Chromophobes may represent a phase of differentiation of the secretory cells.

  C. Releasing hormones are produced by the hypothalamus and travel via the bloodstream to the anterior lobe, causing release of specific anterior pituitary hormones.

III. The Thyroid Gland

The thyroid gland is a bilobed structure attached to the larynx and trachea; the thyroid hormones, thyroxine and tri-iodothyronine, are produced by principal follicular cells, bound to thyroglobulin, and stored in follicles until they enter the blood for distribution. These hormones increase the metabolic rate of cells. Calcitonin, produced by parafollicular cells, reduces the level of blood calcium.

IV. The Parathyroid Glands

The parathyroid glands are four small glands on the posterior surface of the lobes of thyroid gland. The parathyroid hormone raises blood calcium by stimulating resorption of bone.

V. The Suprarenal Glands

The suprarenal glands are located on the superior poles of the kidneys.

  A. The suprarenal cortex produces three functional types of steroid hormones.

    1. Mineralocorticoids are concerned with fluid and electrolyte balance.

    2. Glucocorticoids stimulate formation of glucose in the liver, glycogen deposition, and fat mobilization.

    3. Androgens are the same hormones as those produced in larger amounts by the testes.

  B. The suprarenal medulla produces epinephrine and norepinephrine, which function with the sympathetic nervous system to enable the body to respond to emergency situations.

VI. The Islets of Langerharns

The islets of Langerhans are scattered clumps of cells in the pancreas. They produce insulin, which maintains a constant level of blood glucose by promoting removal of glucose from blood, and glucagon, which promotes the release of glucose into the blood stream.

VII. The Ovaries

The ovaries are located in the pelvis. The production of ovarian hormones fluctuates over a 28-day cycle regulated by hormones of the hypothalamus and anterior pituitary.

  A. Ovarian follicles produce estrogen which promotes growth of glands of the endometrium.

  B. The corpus luteum produces estrogen and progesterone which prepare the uterus for pregnancy and maintain pregnancy.

VIII. The Testes

The testes are located in the scrotum. The interstitial cells produce the androgen testosterone which maintains secondary sex characteristics and stimulates the production of sperm.

IX. The Pineal Gland

The pineal gland is located above the diencephalon. It produces melatonin which may have a role in regulating the cyclic activities of endocrine glands.

X. Development of Pituitary Gland

  A. The posterior lobe arises from an evagination of the floor of the diencephalon.

  B. The anterior lobe arises from an evagination of the roof of the oral cavity (ectoderm).

XI. Development of Thyroid Gland

The thyroid gland develops from a midventral diverticulum from the floor of the pharynx.

XII. Development of Parathyroid Glands

The parathyroid glands develop from the third and fourth visceral pouches.

XIII. Development of Suprarenal Glands

The suprarenal glands have a double origin. The cortex arises from coelomic epithelium, mesoderm; the medulla arises from the neural crest, ectoderm.

REVIEW QUESTIONS

1. What are the major differences between the nervous system and the endocrine system and the methods by which they regulate specific organs?

2. What are the principal differences between exocrine and endocrine glands?

3. How does cyclic AMP function as a second messenger?

4. How does the hypothalamus affect the activity of the posterior lobe of the pituitary gland?

5. What is the principal mechanism by which the hypothalamus regulates the anterior lobe of the pituitary gland?

6. Name the hormones produced by the posterior lobe of the pituitary and give the principal functions of each.

7. Name the hormones of the anterior pituitary and describe the functions of each.

8. Name and give the functions of the hormones of the thyroid gland.

9. Describe the storage and release of the hormones of the thyroid gland.

10. Describe the relationship of the parathyroid glands to the thyroid.

11. What is the function of the parathyroid hormone?

12. How does the development of the parathyroids account for their location in the adult?

13. Describe the location and blood supply of the suprarenal glands.

14. Name the major classes of hormones produced by the suprarenal cortex and give the principal functions of each class.

15. What are the anatomical and functional relations of the suprarenal medulla to the sympathetic nervous system?

16. Name and describe the endocrine elements that produce insulin and glucagon. What are the functions of these hormones?

17. Describe the source of the hormones produced by the ovaries and testes and give their major functions.

SELECTED REFERENCES

Ezrin, C.; Godden, J. O.; and Volpe, R. *Systemic Endocrinology*. 2nd ed. Hagerstown, Md.: Harper and Row, 1979.

Hall, R.; Anderson, J.; Smart, G. A.; and Besser, M., *Fundamentals of Clinical Endocrinology*. 3rd ed. Kent, England: Pitman Medical Publishing Co., 1980.

Hershman, J. M., ed. *Practical Endocrinology*. New York: John Wiley & Sons, 1981.

Ingbar, S. H., ed. *Contemporary Endocrinology*. Vol. 1. New York: Plenum Medical Book Co., 1979.

Weiss, L., ed. *Histology. Cell and Tissue Biology*. 5th ed. New York: Elsevier Publishing Co., 1983.

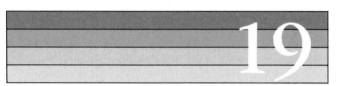

THE CIRCULATORY SYSTEM

19

OBJECTIVES

After completing this chapter you should be able to:

1. State the general functions of the circulatory system.

2. List and describe the factors that dictate that the blood flows in only one direction.

3. Define pericardial cavity, fibrous pericardium, and visceral pericardium. Locate these structures relative to the heart, lungs, and diaphragm.

4. Locate the heart relative to ribs and sternum.

5. Describe the structure of the heart. Describe the structures preventing the backflow of blood within the heart and in the arteries conveying blood from the ventricles.

6. Define or describe the cardiac skeleton.

7. Describe the mechanism of the intrinsic regulation of the heartbeat.

8. Describe the cardiac cycle, using the terms systole, diastole, rapid ejection phase, and slow injection phase.

9. State Starling's law of the heart.

10. Give the points at which the stethoscope is placed on the chest wall for listening for the sound produced by closure of each of the valves of the heart and great vessels.

11. Describe the partitioning of the heart during embryonic development.

The circulatory system consists of a pump, called the *heart,* and two systems of tubes, or vessels, that convey fluids from one part of the body to another. The *blood vessels* carry a complex fluid, *blood,* containing cells and proteins as well as dissolved respiratory gases and nutrients. The *lymph vessels* carry *lymph,* which is essentially the same as intercellular tissue fluid, from the tissues to the blood. The lymph vessels, often considered a separate system, are called the *lymph system;* the heart and blood vessels are called the *cardiovascular system.*

For descriptive purposes the cardiovascular system may be divided into the heart and two circuits (Fig. 19-1): the **pulmonary circuit,** which conveys blood from the heart to the lungs and back, and the **systemic circuit,** which distributes blood to all the other tissues of the body. Each circuit includes *arteries,* the vessels that carry blood away from the heart, and *veins,* the vessels through which blood returns to the heart. Thus the cardiovascular system is a *closed system* consisting of the heart, arteries, capillaries, and veins.

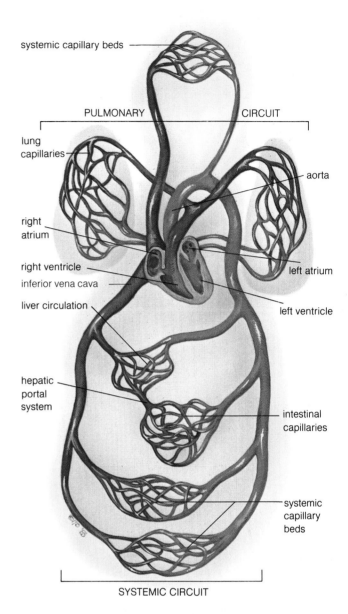

Figure 19-1 The general plan of the circulatory system.

The chief function of blood is to distribute substances throughout the body, helping to maintain a constant environment compatible with the life of the cells. Fluid and small dissolved molecules such as oxygen, carbon dioxide, and nutrients pass through the capillary walls into the tissues, but the blood cells and large proteins normally remain within the closed system. Excess fluid in the tissues is returned to the blood via the lymph vessels. Blood also has an important role in the regulation of body temperature and in the body's defense mechanisms. This chapter emphasizes the anatomy of the circulatory system as it affects the distribution function of blood.

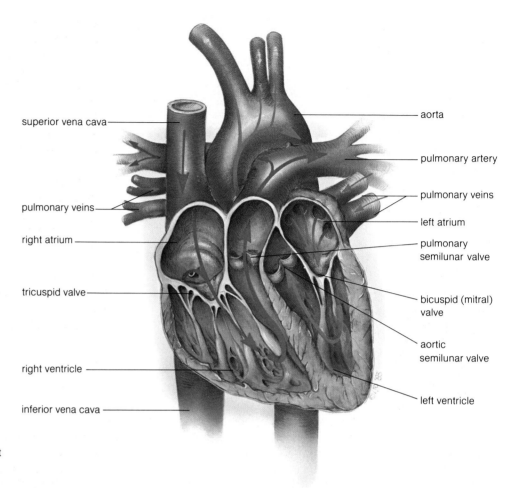

superior vena cava

pulmonary veins

right atrium

tricuspid valve

right ventricle

inferior vena cava

aorta

pulmonary artery

pulmonary veins

left atrium

pulmonary semilunar valve

bicuspid (mitral) valve

aortic semilunar valve

left ventricle

Figure 19-2   The anterior heart wall removed to show the pathway of blood through the heart.

GENERAL PLAN OF THE CIRCULATORY SYSTEM

The heart (Fig. 19-2) is a double pump. It is divided into right and left sides, with no blood flow between them; the right half pumps blood into the pulmonary circuit, and the left half into the systemic circuit.

Each half of the heart is composed of two chambers: the **atrium,** which receives blood into the heart, and the **ventricle,** which pumps blood away from the heart. The right and left ventricles contract simultaneously, sending blood into the arteries.

In the pulmonary circuit, contraction of the right ventricle sends deoxygenated blood (indicated in blue in Fig. 19-1) through the **pulmonary arteries** to the lungs, where it picks up oxygen, which combines with hemoglobin in the red blood cells. The blood changes from the dark red color of the deoxygenated state to the bright red color of oxygenated blood (shown as red

in Fig. 19-1). The oxygenated blood is returned to the left atrium by way of the **pulmonary veins,** completing the pulmonary circuit.

The left ventricle receives the oxygenated blood from the left atrium and pumps it through the *aorta* (the main systemic artery) to the tissues, including the wall of the heart, where it gives up oxygen to the cells for use in metabolic processes. The deoxygenated blood returns to the right atrium through the *superior and inferior venae cavae* (the main systemic veins), completing the systemic circuit. Thus the left side of the heart is the pump for the systemic circuit, the right for the pulmonary circuit.

Blood is carried away from the heart by arteries that branch (bifurcate) repeatedly, becoming smaller in diameter, and finally terminate as small vessels called **arterioles** that distribute blood to the tissues. The arterioles open into anastomosing networks

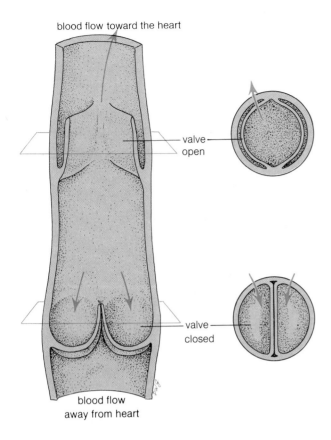

blood flow toward the heart

valve open

valve closed

blood flow away from heart

Figure 19-3   The location and structure of the valves of a vein.

of **capillaries,** called *capillary beds,* located in the organs and tissues of the body in close proximity to cells. The thin walls of the capillaries, consisting of simple squamous epithelium termed **endothelium** (the lining of all blood vessels), permit the passage of substances in and out of the blood with little energy expenditure. Nutrients and oxygen, more concentrated in the blood than in the tissues, diffuse out of the blood stream into the tissue fluid that surrounds the tissue cells and are taken into the cells as needed. Metabolic wastes such as urea and carbon dioxide are discharged from the cells and accumulate in the tissue fluid; because these wastes are at a higher concentration in the fluid than in the blood, they diffuse through the capillary walls into the blood and are carried to the kidneys and lungs for elimination from the body. From the capillaries the blood is carried back to the heart by the **veins.** The small veins that communicate with the capillaries, called **venules,** become larger

veins as they anastomose and merge into main channels as they approach the heart.

When substances in the blood pass through the capillary walls into the intercellular tissue fluid, they always are accompanied by some fluid from the blood stream. This noncellular portion of blood, or *plasma,* contains water, small molecules, and proteins called plasma proteins; the blood fluid that diffuses into the tissues is plasma minus the proteins, which are too large to pass through capillary walls. This fluid is similar in composition to intercellular fluid and is colorless (the color of blood is derived from red blood cells). Some intercellular fluid also passes back into the capillaries along with the wastes and other materials produced in the tissues; however, more fluid leaves the blood than is returned by this route. The excess fluid in the tissues enters vessels of the lymph system and is then called **lymph.** The lymph system, which feeds into the blood stream, does not form a circuit but provides a one-way passage from the tissues to the blood. The anatomy of the lymph system and its role in distribution will be considered later in this chapter. Its role in immunity will be considered in Chapter 25.

REGULATION OF BLOOD FLOW

Because blood is a liquid, it flows within the closed system of vessels along the path of least resistance as dictated by pressure differences. These differences are determined largely by the *autonomic nervous system,* which affects the rate and force of the heartbeat and the diameters of the blood vessels. The rate and force of the heart's contraction determine the amount of blood that flows through the system (cardiac output), and the diameter of the vessel determines the amount of blood entering any particular blood vessel. The diameter of the arteries is readily enlarged or reduced as their circular smooth muscle relaxes or contracts, primarily as a result of neural stimulation; hormonal influences are generally less significant.

Valves located at specific points within the circuits direct blood flow in a particular direction. The valves of veins are usually half-moon shaped (semilunar), thin, fibrous sheets that are covered with endothelium. They usually occur in pairs and project into the lumen of veins in an arrangement that permits blood to flow only toward the heart (Fig. 19-3). The veins of the extremities contain valves located at frequent intervals, but many of the veins of the head and neck and the large veins of the trunk have no valves. The arteries have valves only in the aorta and pulmonary trunk as they exit the heart (Fig. 19-2). These valves consist of three semilunar cusps that are similar to the venous valve cusps. The only other valves in the cardiovascular system are found between the atria and ventricles within the heart (Figs. 19-2, 19-4). Here also the valve cusps are fibrous sheets covered with endothelium; however, these cusps are held in place by fibrous cords called **chordae tendineae** that attach to special muscle bundles of the heart wall.

Within the heart, blood passes from the right atrium into the right ventricle when pressure within the atrium is above that of the ventricle. The **tricuspid valve,** or right atrioventricular valve (Figs. 19-2, 19-4), located at the orifice between these two chambers, is constructed so that as blood passes from the right atrium into the right ventricle, the valve leaflets are shoved aside, allowing blood to flow freely through the opening. Upon contraction of the right ventricle, pressure rises and forces blood against the valve leaflets, which come together to block the *right atrioventricular orifice.* With the valve closed blood can only flow into the *pulmonary trunk,* the large artery conveying blood toward the lungs. As the right ventricle completes its contraction, pressure within the ventricle begins to drop, and pressure within the pulmonary trunk quickly becomes greater than in the ventricle. Blood cannot run back into the right ventricle due to the closure of the **pulmonary semilunar valve** (Figs. 19-3, 19-4) near the origin of the pulmonary trunk; thus the blood can flow in one direction only: toward the lungs. The force created by the contraction of the right ventricle propels the blood through the capillaries of the lungs into the veins draining the lungs. Within the veins blood flow toward the heart is dictated by the pressure initiated by contractions of the ventricle and by the presence at frequent intervals of valves that prevent backflow.

If the edges of the valves become thickened, they may be unable to close adequately to prevent the regurgitation of blood from the ventricles into the atria. The regurgitation sets up vibrations that may be detected with a stethoscope as a *cardiac murmur.*

On the left side of the heart a similar arrangement of valves and pressure changes also ensures the flow of blood in a single direction. At the *left atrioventricular orifice* the **mitral valve,** also called the bicuspid or left atrioventricular valve (Fig. 19-2), allows the passage of blood from the left atrium into the left ventricle when the ventricles are relaxed, but it closes upon contraction of the ventricles. With the mitral valve closed, blood can only flow into the *aorta,* the large artery conveying blood away from the left ventricle in the systemic circuit. Backflow of blood from the aorta into the left ventricle is prevented by the **aortic semilunar valve** (Figs. 19-2, 19-4), which is similar to the pulmonary semilunar valve. Blood flows away from the heart

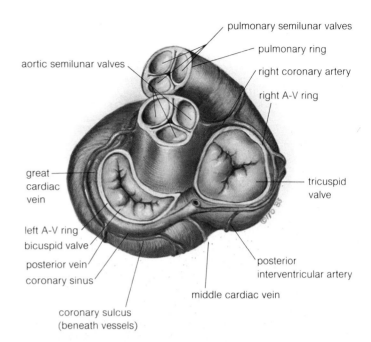

Figure 19-4 The valves of the heart and their supporting fibrous rings as seen in a cutaway superior view after the removal of the atria.

because the contraction of the left ventricle temporarily elevates the pressure in the aorta. This temporary pressure elevation passing along the arterial system is known as the *pulse;* the effect is gradually lost as the vessels branch and become smaller. Blood flow through the capillaries and veins is a steady flow and under less pressure than in the arteries.

You can demonstrate the presence of valves in a superficial vein of the forearm in the following manner. Press two adjacent fingers against a vein that is visible on the forearm. Move the proximal finger along the vein, forcing the blood from approximately two inches of the vein. Lift the distal finger and observe that the vein fills immediately with blood. Now repeat the procedure and lift the proximal finger, but keep the blood flow blocked with the distal finger. This time the segment of vein will remain empty, demonstrating the position of a valve that prevents backflow. Such valves are especially important in the return of blood to the heart from the extremities.

THE HEART

The heart is a muscular structure that contracts rhythmically, serving as a simple pump to force blood through the blood vessels. A person's heart is normally about the size of his or her

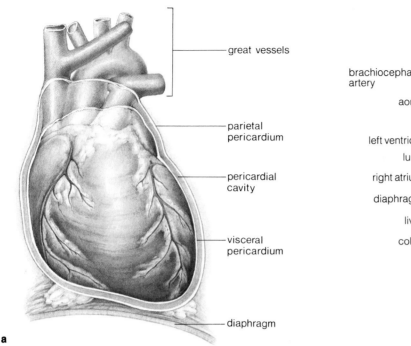

great vessels

parietal
pericardium

pericardial
cavity

visceral
pericardium

diaphragm

a

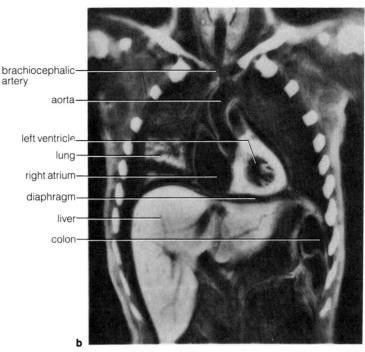

brachiocephalic
artery

aorta

left ventricle

lung

right atrium

diaphragm

liver

colon

b

Figure 19-5 Relationship of the heart to the pericardial cavity. (**a**) Diagram of an anterior view. (**b**) X-ray of a thick frontal section of the thorax and abdomen. Note the relationship of the heart to the lungs and diaphragm. (From the University of Kentucky, Department of Anatomy teaching collection).

closed fist, averaging about 12 cm long, 9 cm wide, and 6 cm in anteroposterior diameter in the adult. However, the heart frequently undergoes *hypertrophy* (enlargement) in those individuals who exercise vigorously at regular intervals or who have vascular problems that increase the work load of the heart.

> A physician can estimate the size of a person's heart using the technique of percussion. Tapping on the chest wall, the physician listens for the change in sound as the density changes from the heart to the lungs.

Protective Coverings of the Heart

The heart is protected from external environmental forces by the bony *rib cage* and from its internal environment by a covering of *serous membranes* that form the **pericardium,** or *pericardial sac* (Fig. 19-5). The inner membrane, the *visceral pericardium,* lies directly on the heart muscle; in fact it is regarded

as the outer layer of the heart wall and in that connection is usually called the *epicardium* (Fig. 19-6). In the region of the great vessels the visceral pericardium is continuous with the outer membrane, the *parietal pericardium,* but around most of the heart there is a potential space between the two membranes called the *pericardial cavity,* which contains a small amount of serous fluid. The epithelial layers of the parietal and visceral layers of serous membrane, in contact with one another and moistened by serous fluid, provide smooth surfaces that eliminate friction as the heart beats. The serous membrane forms a closed sac that lines the heavy fibrous sheet called the fibrous pericardium.

> *Pericarditis* is a general term used to indicate an infection (inflammation) of the serous membranes covering the heart. It usually causes pain and may result in the accumulation of fluid within the pericardial cavity. If a considerable amount of fluid is present, pressure on the pulmonary veins may retard venous return from the lung.

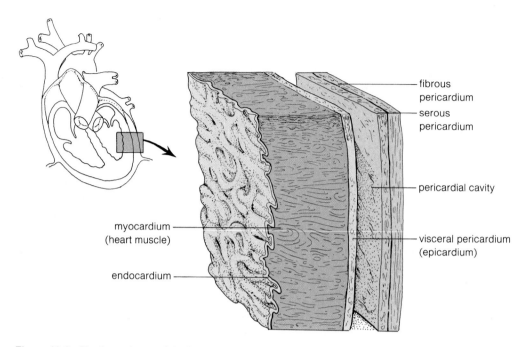

fibrous pericardium

serous pericardium

pericardial cavity

myocardium (heart muscle)

visceral pericardium (epicardium)

endocardium

Figure 19-6   The tissue layers of the heart and pericardium.

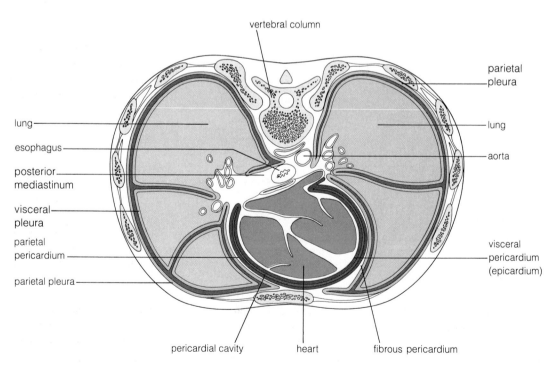

vertebral column

parietal pleura

lung

lung

esophagus

aorta

posterior mediastinum

visceral pleura

parietal pericardium

visceral pericardium (epicardium)

parietal pleura

pericardial cavity

heart

fibrous pericardium

Figure 19-7   Relationship of the heart and pericardial cavity to the lungs; transverse section.

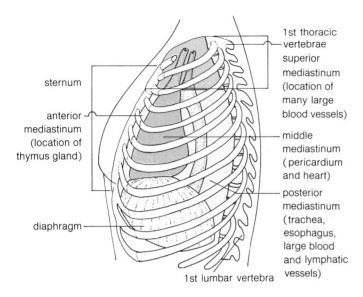

Figure 19-8 The location of the heart within the mediastinum, as seen in side view. The lungs have been removed in this figure.

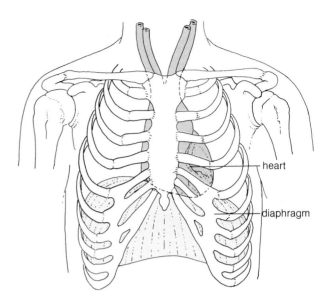

Figure 19-9 The location of the heart in relation to the ribs and sternum; anterior view.

Location of the Heart

The heart, enclosed by the serous membranes, is located in the *middle mediastinum* of the thorax (Fig. 19-7). The **mediastinum** is the region between the external coverings of the lungs (parietal pleurae), extending from the lower neck to the diaphragm and lying between the vertebral column and the sternum. This region is further divided into areas called anterior, superior, posterior, and middle mediastinum (Fig. 19-8). The middle mediastinum contains the pericardial sac enclosing the heart and the origins of the great vessels. The *anterior mediastinum* is the space anterior to the heart; it contains no structures except the internal thoracic arteries and veins, lymph nodes, and a portion of the thymus gland in young individuals.

Although the position of the heart is not static, its approximate position can be described in reference to superficial landmarks (Fig. 19-9). The pericardium extends in the median plane from about the level of the *sternal angle* to the level of the *xiphisternal junction.* Two-thirds of the pericardial sac lies to the left of the midline, and one-third lies to the right, overlapped anteriorly and to a variable extent on each side by the parietal pleura (outer membrane) of the lung (Fig. 19-10). Though the pericardial sac extends from about the level of the diaphragm to the second rib, the heart extends superiorly only to the level of the third rib, where it is fixed relatively securely by the attachment of its base (broad upper portion) to the great vessels. The apex (lower tip) of the heart hangs free in the pericardial sac. The fibrous pericardium is attached to most of the structures with which it has a direct relationship, including the diaphragm. When the diaphragm moves downward during inspiration, the heart assumes a rather elongated appearance, with the apex moving closer to the midline as the pericardium pulls it downward. During expiration the diaphragm is forced upward, largely by the body wall muscles, causing the heart to shift to the left.

External Anatomy of the Heart

The heart is conical in shape and has a *base,* an *apex* (Figs. 19-10a, 19-11), and three surfaces: *sternocostal, diaphragmatic, and pulmonary* (left). The base is the most posterior part of the heart; it projects superiorly and is formed chiefly by the left atrium. The sternocostal surface is formed chiefly by the right ventricle, since the right ventricle is the most anterior part of the heart. Both ventricles contribute to the diaphragmatic surface at the inferior border as they rest on the central tendon of the diaphragm. The pulmonary, or left, surface at the left border is formed mainly by the left ventricle, which causes the *cardiac notch* on the lower lobe of the left lung. The right border is formed by the right atrium.

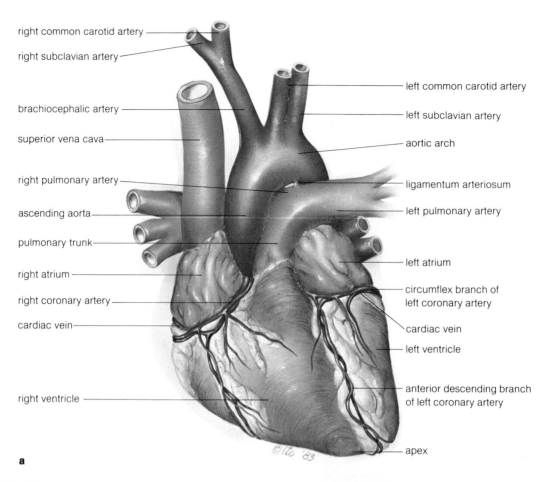

right common carotid artery

right subclavian artery

brachiocephalic artery

superior vena cava

right pulmonary artery

ascending aorta

pulmonary trunk

right atrium

right coronary artery

cardiac vein

right ventricle

left common carotid artery

left subclavian artery

aortic arch

ligamentum arteriosum

left pulmonary artery

left atrium

circumflex branch of
left coronary artery

cardiac vein

left ventricle

anterior descending branch
of left coronary artery

apex

a

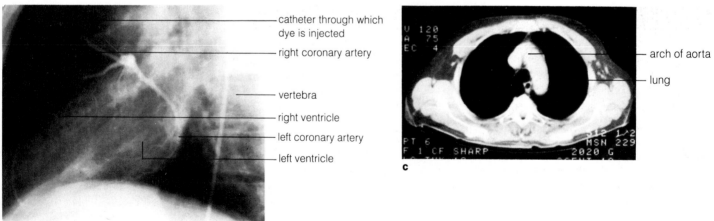

catheter through which
dye is injected

right coronary artery

vertebra

right ventricle

left coronary artery

left ventricle

arch of aorta

lung

b

c

Figure 19-10   The heart and great vessels. (**a**) An anterior external view of the heart. (**b**) Angiogram showing a lateral view of the right and left coronary arteries. (From the University of Kentucky, Department of Anatomy teaching collection.) (**c**) CAT scan of the thorax. Transverse view at the level of the arch of the aorta. (Courtesy of S. J. Goldstein.)

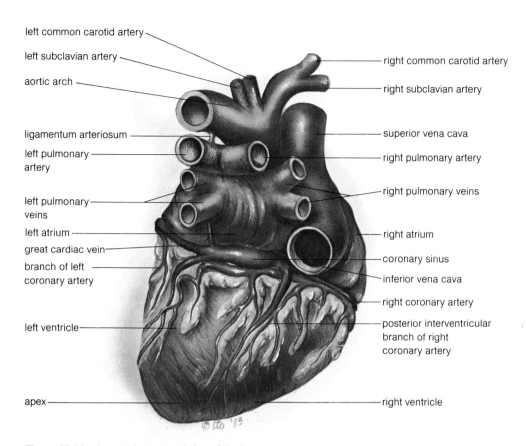

left common carotid artery

left subclavian artery

aortic arch

ligamentum arteriosum

left pulmonary artery

left pulmonary veins

left atrium

great cardiac vein

branch of left coronary artery

left ventricle

apex

right common carotid artery

right subclavian artery

superior vena cava

right pulmonary artery

right pulmonary veins

right atrium

coronary sinus

inferior vena cava

right coronary artery

posterior interventricular branch of right coronary artery

right ventricle

Figure 19-11   A posterior external view of the heart.

Each atrium has a thin-walled extension, called the **auricle,** located on its anterior surface on either side of the aorta and pulmonary trunk. Clinicians frequently use the term auricle to mean the entire chamber, calling the auricle the *auricular appendage.* The separation of the atria and ventricles is marked at the surface by a groove, the *coronary* (atrioventricular) *sulcus,* or *coronary groove,* in which blood vessels that supply the heart are located. The groove is most obvious on the posterior heart surface; it is obscured anteriorly by the pulmonary trunk and aorta. The ventricles are separated by shallow *anterior* and *posterior interventricular grooves* containing branches of the *left* and *right coronary arteries,* which descend from the coronary sulcus toward the apex and supply the heart muscle (Fig. 19-10a and b).

Internal Structure of the Heart

The Heart Wall   The wall of the heart consists of three layers: an inner lining called **endocardium,** composed of simple squamous epithelium (endothelium) and associated connective

tissue; the middle thick portion, the **myocardium,** made up of cardiac muscle; and the outer covering of serous membrane, the **epicardium** *(visceral pericardium),* which is the inner membrane of the pericardial sac (Fig. 19-6). The layers of the wall are similar in all chambers of the heart. The greatest difference between chambers is in the myocardium, where a difference in thickness is correlated with the force required for the work of each chamber.

The Atria   The wall of the right atrium is the thinnest part of the heart, and its posterior and septal internal surfaces are smooth (Fig. 19-12). The *superior* and *inferior venae cavae* open into the upper and lower portions of the right atrium. The opening of the superior vena cava has a semilunar fold of endocardium that sometimes partly covers the opening. A ridge of muscle, the **crista terminalis,** runs vertically between the orifices of the two veins and is a landmark for locating the specialized conductile tissue, the **sinoatrial node,** which serves as the pacemaker of the heart and is located in the upper end of the crista terminalis. The **coronary sinus,** a venous channel that

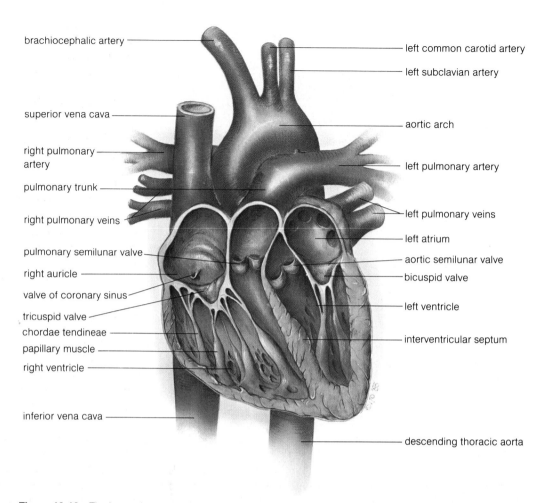

Figure 19-12 The internal anatomy of the heart in cross section, viewed anteriorly.

returns blood from the heart wall to the right atrium (Fig. 19-11), opens into the chamber between the orifice of the inferior vena cava and the right atrioventricular orifice. The medial wall of the chamber, the **interatrial septum,** is smooth, with an oval depression, the **fossa ovalis,** marking the point of communication of the right and left atria in fetal life.

A slit may persist in the fossa ovalis, resulting from failure of the two parts of the interatrial septum to overlap adequately. This slit, called the **foramen ovale,** allows the deoxygenated blood of the right atrium to mix with the oxygenated blood of the left atrium, which in the newborn causes the baby to be "blue."

The left atrium has extensions or pouches on each side that contain the openings of the four *pulmonary veins.* Its internal wall is mainly smooth, with ridges of muscle, the *musculi pectinati,* prominent only in the auricle.

The Ventricles The atrioventricular orifices are oval openings supported in a rim of dense connective tissue in the form of atrioventricular rings that comprise the **cardiac skeleton,** to which the bases of the cusps of the valves attach (Fig. 19-4). The valve cusps are thin folds of endocardium strengthened internally by connective tissue, with fibrous strands, the *chordae tendineae,* attached to the free border of each cusp (Fig. 19-12). The chordae tendineae are fibrous strands that connect the valve cusps to *papillary muscles.* These strands allow the valves to close the orifices as the ventricles contract but prevent prolapse of the cusps into the atria. The tricuspid valve on the right side

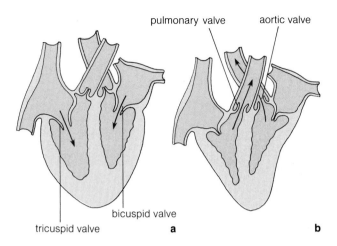

Figure 19-13 (**a**) The tricuspid and bicuspid valves open when the atria contract. (**b**) The tricuspid and bicuspid valves close, and the semilunar valves open, when the ventricles contract.

of the heart has three cusps, while the mitral, or bicuspid, valve on the left has two cusps.

The inner surface of the walls of both ventricles is irregular due to the presence of ridges of muscles called **trabeculae carnae,** a general name for any muscle ridge projecting from the inner surface (including the papillary muscles). The wall of the left ventricle is usually more than double the thickness of the right ventricle, because the left ventricle does more work than the right, and arterial pressure is much higher in the aorta and systemic arteries, in general, than it is in the pulmonary arteries. This extra work not only causes the wall of the left ventricle to thicken but also pushes the interventricular septum to the right, so that the lumen of the left ventricle is round in transverse section while that of the right ventricle is triangular.

The **interventricular septum** extends from the apical region of the heart and has a *muscular* and a *membranous* part. The muscular part of the septum extends superiorly from the apex. It is composed of cardiac muscle and contributes to the pumping action of the ventricles. The membranous part is thin and smooth-surfaced and consists largely of fibrous connective tissue that is part of the **cardiac skeleton.**

In addition to communicating with an atrium each ventricle contains the orifice for the large artery into which the blood passes during ventricular *systole* (contraction of the ventricular heart muscle). The right ventricle pumps into the *pulmonary trunk,* whose orifice can be closed by the three thin cusps of the pulmonary semilunar valve (Fig. 19-13), preventing backflow of blood into the ventricle during *diastole* (relaxation of the ventricular heart muscle). The pulmonary trunk arises ante-

rior to the origin of the aorta, since the right ventricle lies anterior to the left ventricle. The opening of the aorta, leading from the left ventricle, is guarded by the aortic semilunar valve, identical in structure to the valve of the pulmonary trunk.

The Cardiac Skeleton As with skeletal muscle, cardiac muscle fibers must have attachments in order to be effective. The fibers attach to a condensation of dense irregular connective tissue around the base of the valves at the atrioventricular orifices and around the semilunar valves of the aorta and pulmonary trunk at their exit from the ventricles. These fibrous connective tissue rings constitute the *cardiac skeleton* (Fig. 19-6). From their attachments to the skeleton the cardiac muscle fibers take a spiral course around the ventricles or the atria and then return to the insert on the skeleton near their points of origin. As they contract, the muscle fibers exert a "milking" action on cardiac contents, squeezing out the blood.

Regulation of the Activity of the Heart

The heart has the capacity to contract in the absence of any outside nervous or hormonal stimulation. This intrinsic capacity is due to the presence in the heart of a system of modified cardiac muscle fibers, specialized for conduction of electrical impulses. A portion of this muscle tissue, called the sinoatrial node, or pacemaker, generates impulses that spread through the heart by way of the conductile tissue pathways, stimulating the heart muscle to contract. Regulated by this internal conduction system, first the atria contract and then the ventricles; a heart that has had its nerve connections severed will continue to beat. However, the **intrinsic** (internal) **system** alone cannot respond effectively to changes in the internal environment as the body's activities alter; **extrinsic** (external) **regulation** by the nervous system is required for adjusting the rate of the heartbeat. We shall consider first the heart's intrinsic conduction system, and then the system of neural regulation.

Intrinsic Conduction System of the Heart Near the point of entrance of the superior vena cava is a specialized oval mass of modified cardiac muscle tissue called the **sinoatrial node,** or pacemaker (Fig. 19-14). From the sinoatrial node, which fires rhythmically, impulses pass through the ordinary cardiac muscle of the atria and cause the atria to contract, forcing the blood into the ventricles. Conveyed from cell to cell by the atrial muscle fibers, the impulse reaches the **atrioventricular node,** a mass of specialized conductile tissue near the right atrioventricular orifice. From the atrioventricular node, conductile fibers known as the **bundle of His,** which consists of **Purkinje fibers,** convey the impulses to the cardiac muscle of the ventricles. The large bundles of Purkinje fibers that run down through the inter-

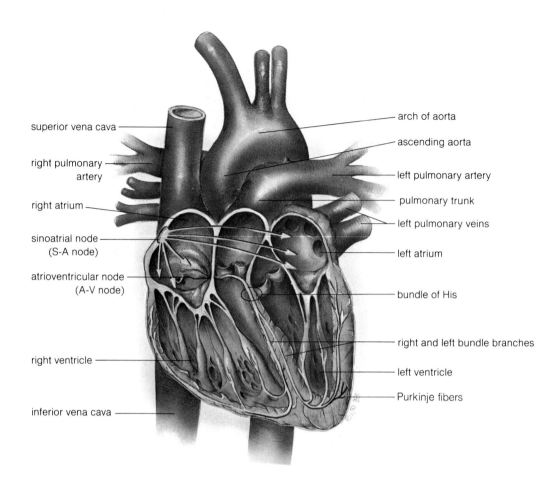

superior vena cava

right pulmonary artery

right atrium

sinoatrial node (S-A node)

atrioventricular node (A-V node)

right ventricle

inferior vena cava

arch of aorta

ascending aorta

left pulmonary artery

pulmonary trunk

left pulmonary veins

left atrium

bundle of His

right and left bundle branches

left ventricle

Purkinje fibers

Figure 19-14 The conduction system of the heart: Impulses are initiated by the S-A node and pass through the muscle cells of the atria, causing the atria to contract. Some of these impulses reach the A-V node and stimulate it to send impulses along the bundle of His and Purkinje fibers. Impulses pass from the Purkinje fibers through the muscle cells, causing a wave of contraction to pass over the ventricles.

ventricular septum are sometimes called the branches of the bundle of His; they run just beneath the endocardium and send branches out into the ventricular heart muscle. The impulses from the atrioventricular node pass down the Purkinje fibers toward the apex of the ventricles, causing the muscle at the apex to contract first. The impulses then spread back toward the base of the ventricles, passing from Purkinje fibers into ordinary cardiac muscle, so that a wave of contraction spreads upward from the apex, forcing the blood from the ventricles into the arteries.

An electrocardiograph can be used for recording the potential of the electrical currents that pass through the heart from the sinoatrial node. This gives valuable information regarding the normal functioning of the heart and can also reveal abnormal functioning of the impulse-conducting system, giving clues as to the exact causes of irregularities in heartbeat.

Battery-powered cardiac pacemakers can be implanted in the heart; these artificial pacemakers give off electrical impulses that produce ventricular contractions at a specific rate. The terminal of the electrode is inserted into the heart via a vein that opens into the right atrium. The electrode is passed through the atrium and into the right ventricle, where it is attached to the endocardium covering the trabeculae carnae.

Extrinsic Neural Regulation of the Heart Rate The function of the cardiovascular system is to deliver blood to all parts of the body at the proper rate. This is accomplished largely by control of the diameter of the blood vessels but is facilitated by changes in the heart rate as regulated by the nervous system. Though the pacemaker fires without nervous stimulation, its rate of firing is regulated in part by its innervation. The *cardiovascular center* in the *reticular formation* of the *medulla*

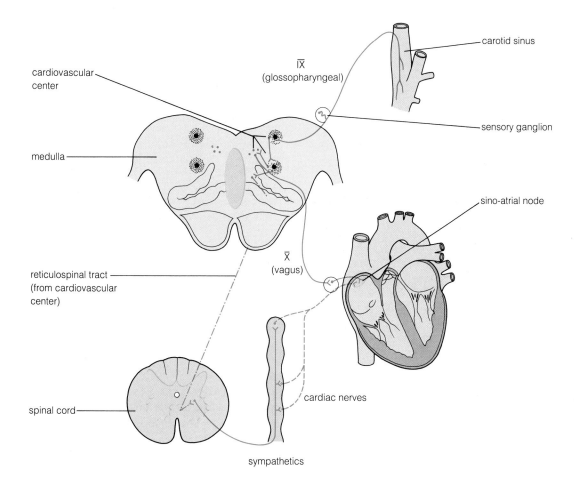

cardiovascular center

IX (glossopharyngeal)

carotid sinus

medulla

sensory ganglion

sino-atrial node

reticulospinal tract (from cardiovascular center)

X (vagus)

spinal cord

cardiac nerves

sympathetics

Figure 19-15 Diagram of the autonomic innervation of the heart. See text for description.

responds to input from the periphery by stimulating the *parasympathetics* to reduce the rate of heartbeat or the *sympathetics* to increase it.

The most important afferent input to the cardiovascular center is from the *carotid sinus,* a dilation near the bifurcation of the common carotid artery in the neck. An increase in blood pressure in the carotid artery initiates afferent impulses that are conveyed to the cardiovascular center, triggering impulses that stimulate the *dorsal motor nucleus* of the *vagus nerve* (Fig. 19-15). The preganglionic parasympathetic fibers of the vagus transmit impulses to postganglionic fibers of the *cardiac plexus* near the heart. The postganglionic fibers terminate on the *sinoatrial node;* impulses brought by these fibers cause a *reduction* in the *heart rate* with a resulting drop in blood pressure. As the blood pressure drops, afferents ending in the carotid sinus activate the cardiovascular center to stimulate the sympathetic outflow to the heart, resulting in an increase in heart rate and rise in blood pressure. The *sympathetic outflow* is by *preganglionics* arising from the first *four thoracic spinal cord*

segments and by postganglionics whose cell bodies are mainly in the *cervical sympathetic ganglia.* The postganglionic sympathetics stimulate the sinoatrial node by releasing the neurotransmitter substance, *norepinephrine,* at the ends of the fibers. An increase in heart rate can also be produced by the hormones *epinephrine* and *norepinephrine* produced by the *suprarenal medulla.* Release of these hormones is under the control of the sympathetic division of the autonomic nervous system.

THE CARDIAC CYCLE

The *cardiac cycle* begins with the initiation of impulses in the sinoatrial node, followed by contraction of the atria. Filling of the ventricles is completed by the atrial contraction, and the period of ventricular contraction, known as *systole,* begins with the pressure in the atria and ventricles approximately equal. As the ventricles contract, blood forces the atrioventricular valves to close, and pressure within the ventricles rises. During the *rapid ejection phase,* pressure within the ventricles forces blood into the aorta and pulmonary trunk more rapidly than it runs into the periphery. This causes an increase in pressure within these arteries until near the end of the ventricular contraction, the *slow ejection phase,* when blood is passing through the arteries more rapidly than it enters. The pressure within the aorta and pulmonary trunk then drops as ventricular contraction ceases. As the ventricles begin the period of relaxation, *diastole,* the passive semilunar valves are closed by reversal of flow in the arteries, preventing the blood from re-entering the ventricles, where the pressure is significantly lower than it is in the arteries. During ventricular contraction blood flows from the great caval veins (superior and inferior venae cavae) into the atria as the atria relax. Then during diastole, when the pressure in the ventricles drops below that of the atria, the atrioventricular valves open, allowing the blood to flow from the atria into the ventricles; the next cardiac cycle begins with the initiation of atrial contraction.

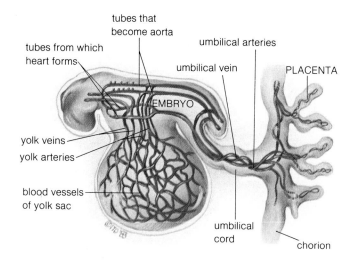

Figure 19-16 The structure of the circulatory system of the human embryo at about three weeks.

Ventricular Volume

The volume of blood discharged from the left ventricle during systole, known as the *stroke volume,* is from 60 to 100 ml. The *cardiac output* per minute is the stroke volume multiplied by the heart rate per minute, usually about 5 liters/minute. At the end of the systole, from 20 to 50 ml of blood usually remains in the ventricle; this is the *residual volume.* At rest the residual volume is greater than during exercise.

The volume of blood in the ventricles influences the force of contraction. According to *Starling's law* of the heart the force of contraction is directly proportional to the volume of blood in the ventricle. When cardiac muscle fibers are stretched, their contraction is more forceful. The fibers can be elongated or stretched by increased venous return during diastole, increasing the quantity of blood in the ventricles; a larger stroke volume results.

Heart Sounds

Closure of the valve causes vibrations, producing *heart sounds* that are transmitted through the chest wall and can be heard with a stethoscope. There are two distinct heart sounds produced by valve closure: the first caused by closure of the mitral and tricuspid valves and the second by closure of the aortic and pulmonary semilunar valves. Closure of individual valves can be heard by placing the stethoscope over the chest at a spot in the path of the vibrations from a particular valve. For example, closure of the mitral valve may be heard best over the apex of

the heart at the fifth intercostal space to the left of the midline, and the tricuspid valve may be heard at the right edge of the sternum just above the xiphoid process. The aortic valve is heard near the sternal angle on the right, and the pulmonary valve is checked on the left over the second intercostal space.

Abnormal sounds, called *murmurs,* are due to vibrations caused by turbulence in the blood flow and may be produced by diseased valves that fail to close completely.

DEVELOPMENT OF THE CARDIOVASCULAR SYSTEM

The first evidence of a cardiovascular system in the embryo is the appearance of oval masses of cells, called *blood islands,* in the mesoderm on the cranial surface of the yolk sac. The blood islands develop into blood-containing tubes (blood vessels) through the differentiation of peripheral cells into endothelium and of centrally located cells into blood cells. Initially these vessels grow in length, principally through the coalescence of adjacent blood islands; later expansion occurs largely by means of local proliferation of the endothelium at a "growing tip." Blood vessel development spreads rapidly from the yolk sac mesoderm throughout the embryonic mesoderm, becoming continuous with the heart tube now developing in the cranial region of the embryo (Fig. 19-16). With changes in body form the heart position is gradually shifted to its more caudal final location.

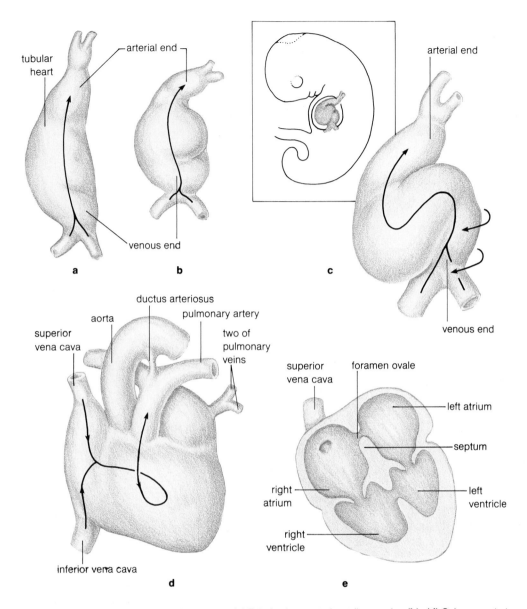

Figure 19-17 The development of the heart. (**a**) Tubular heart at about five weeks. (**b**)–(**d**) Subsequent stages in the development of the heart over the next several weeks. (**e**) A section through the heart showing the development of the septum that divides the heart into four chambers.

Development of the Heart

A cavity in the mesoderm, the future *pericardial cavity,* appears rostral to the head prior to the development of the heart primordium. The heart is initially a straight tube consisting of two dilated chambers, with the single atrium lying inferior (caudal) to the single ventricle (Fig. 19-18a). However, the tube soon begins to fold into an S-shape (Fig. 19-18b, c, d), so that the atrium eventually comes to lie dorsal and superior to the ventricle. At first the atrium and ventricle are separated by a narrow **atrioventricular canal.** Then local proliferation of the *endo-cardium* produces *endocardial cushions* that project from the dorsal and ventral borders of the canal (Fig. 19-17e). These cushions meet and fuse in the middle of the canal, forming a

Clinical Applications: **Diseases of the Cardiovascular System**

Diseases of the heart are the most common cause of death in developed countries and are responsible for the eventual death of one out of three people who live to the age of ten. There are numerous diseases involving structural lesions; a few of the most common will be described briefly.

Valvular lesions, such as *mitral stenosis* (narrowing) or *mitral insufficiency* (weakening) generally cause blood and fluid to collect in the lungs, because the left ventricle has trouble discharging all the blood it receives. Both the right and left ventricles are overworked and undergo hypertrophy; in severe cases this leads to *congestive heart failure* in which more blood enters the heart than leaves it, which may cause death.

Myocarditis is a general term denoting the presence of inflammatory lesions in the heart muscle. Such lesions may be caused by various bacteria or toxic substances and may cause death in acute attacks. *Pericarditis* is inflammation of the pericardium.

Nearly one-third of all male deaths from all causes are the result of coronary artery disease. When cardiovascular deaths are considered separately from other causes, coronary artery disease is the cause in two-thirds of the male victims. The female fatality rate amounts to about 35 percent of all cardiovascular deaths.

Atherosclerosis

Atherosclerosis is the most common coronary artery disease. It is characterized by the deposition of fatty substances, including cholesterol, in the walls of arteries varying in size from the aorta to the coronary arteries. These deposits usually take the form of plaques that reduce the lumen of the artery or may completely occlude it. They may erode the arterial lining, creating a roughened area that promotes the formation of clots. When a coronary artery or one of its branches is narrowed, the cardiac muscle supplied by the artery is inadequately supplied with blood, a condition termed *ischemia.* If the narrowing occurs gradually over a long period of time, other arteries in the vicinity may enlarge and bypass the obstruction, establishing collateral circulation to the area. If the main artery becomes partially or totally occluded before the collateral circulation develops, the cardiac muscle it supplies begins to die; the necrotic region is termed a *myocardial infarct.*

Myocardial Infarction

When cardiac muscle is deprived of oxygen, there frequently is anginal pain that may be localized in the chest or referred to the lower neck and shoulder, where it may radiate down the arm. Usually the pain is referred to the left side of the body. In a myocardial infarction, commonly called a heart attack, the patient usually feels a tightness in the chest and is cold and sweating, pale and anxious. The heart rate may be too slow (*bradycardia*) or too fast (*tachycardia*) at the onset of the infarction. Blood pressure is usually high due to overactivity of the sympathetics. The diagnosis of a heart attack is confirmed by changes in the electrocardiogram and by elevation in the blood of certain enzymes that are released from the necrotic heart muscle.

Coronary-Arterial Surgery

In patients with coronary artery disease who have severe angina brought on by physical effort, it may be desirable to bypass the narrowed regions of the coronary arteries. The positions of the arterial lesions are located precisely by cineangiography, in which radiographic contrast material is injected through catheters into each coronary artery and the passage of the material recorded as a motion picture. The patient's great saphenous vein from the thigh is usually used as the graft. One end of the vein is anastomosed to the aorta and the other end to the coronary artery distal at the point at which that artery is narrowed or occluded. This allows blood to flow from the aorta into the portion of the coronary artery that previously had been inadequately supplied. The relief from angina in such cases has been dramatic.

Heart Block

Although heart block may be caused by factors other than a heart attack, one of the dangers of a heart attack is that the areas deprived of oxygen may include parts of the conduction system of the heart. Damage to the sinoatrial node may cause its impulses to be too weak to activate the atrioventricular node. This results in failure of impulses to be transmitted to the ventricles. If the failure occurs only occasionally in a partial heart block, the heart will beat in a rhythm that includes missing a beat at regular intervals. If no impulses reach the atrioventricular node from the sinoatrial node, the ventricles contract slower than the atria and are not coordinated. This is complete heart block.

Heart block, complete or partial, is an indication for use of a cardiac pacemaker. Electrode wires are normally passed into the right side of the heart through a vein and lodged in the right ventricle. The pacemaker may be implanted beneath the skin on the anterior chest wall. Most modern pacemakers are powered with lithium batteries that will function at a normal heart rate for 5 to 10 years.

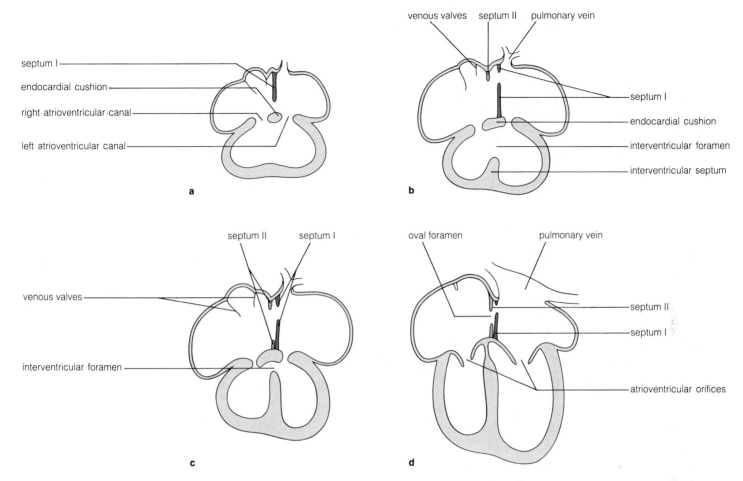

Figure 19-18 Diagram of frontal sections of the heart illustrating partitioning of the heart. (**a**) Early phase of formation of Septum I. (**b**) Fusion of Septum I with the endocardial cushion. (**c**) Growth of Septum II. (**d**) Completion of interventricular septum.

partition between the future atrioventricular orifices. At the same time, septa growing inferiorly and superiorly begin to partition the atrium and ventricles into two chambers each.

The atrium is partitioned by a thin sickle-shaped membrane called the primary interatrial septum, or *Septum I,* which grows down from the middorsal wall of the atrium (Fig. 19-18a), forming a temporary complete separation of the right and left atria. Shortly after Septum I is completed it begins to break down, and the secondary septum, or *Septum II,* begins to develop to the right of Septum I (Fig. 19-18b). The opening thus created is called the *foramen ovale.* Because most of the blood enters on the right side of the heart, Septum I is pushed toward the left, keeping the foramen ovale unobstructed (Fig. 19-18c) and

allowing blood to pass directly from the right atrium into the left atrium, bypassing the fetal lungs. (Fetal circulation will be described later.) The opening persists until after birth when increased pulmonary circulation equalizes the pressure on the two sides (Fig. 19-18d) and brings Septum I into contact with Septum II, closing the *foramen ovale.* If the two septa are too short and do not overlap, the baby has a patent (open) foramen ovale, which allows oxygenated blood to mix with deoxygenated blood, a congenital malformation that can be corrected surgically.

Partitioning of the ventricle occurs in part by means of an interventricular septum (*medial septum*) projecting inward from the apex of the ventricle. For a time there is continuity between right and left ventricles through the *interventricular foramen* (Fig. 19-18b, c). Closure of this opening is completed as a result of the splitting of the single *ventral aorta* (truncus arteriosus) into the aorta and pulmonary trunks, which is accomplished by the formation of a *spiral septum* from thickenings of the endocardium. In the adult the spiral septum is responsible for the fact that the pulmonary trunk leaves the right ventricle anterior to the aorta, becoming posterior to the aorta at its bifurcation into the *right* and *left pulmonary arteries*.

Differential growth of the trunk of the embryo contributes to the shift in the relative position of the heart from the cranial region to the cervical region ventral to the gut and then to its final position in the thorax, where it is protected by serous membrane and the rib cage.

WORDS IN REVIEW

Arteriole	Mitral valve
Atrioventricular canal	Papillary muscle
Atrium	Pericardium
Auricle	Pulmonary circuit
Capillary	Purkinje fibers
Cardiac cycle	Semilunar valves
Cardiac skeleton	Sinoatrial node
Chordae tendineae	Starling's law
Crista terminalis	Stroke volume
Diastole	Systole
Endothelium	Trabeculae carnae
Foramen ovale	Vein
Fossa ovalis	Ventricle
Mediastinum	Venule

SUMMARY OUTLINE

I.   The Circulatory System
     The circulatory system is a double pump and a system of tubes that form two circuits.
     A. The direction of blood flow is controlled by pressure due to heart contraction and by valves that prevent backflow.
     B. The circulatory system functions in distribution, temperature regulation, and body defenses.

II. The Heart
     A. Protective coverings include the visceral and fibrous pericardium and the ribcage.
     B. The heart is located in the middle mediastinum.
     C. The principal external anatomical features are a conical structure with a base and apex, and with sternocostal, diaphragmatic, and pulmonary surfaces.
     D. Internal features include a wall of three layers: endocardium, myocardium, epicardium.

1. The atria are thin-walled chambers. The right atrium receives the superior and inferior venae cavae and the coronary sinus; the left atrium receives four pulmonary veins.
2. The ventricles are thick-walled chambers. The right ventricle has tricuspid valve cusps attached to papillary muscles by means of chordae tendineae; the right ventricle sends blood into the pulmonary artery. The left ventricle has a mitral valve and gives rise to the aorta.
3. As the blood flows through the heart, the right atrium receives deoxygenated blood that passes into the right ventricle. This blood passes through the pulmonary circuit to the lungs where it is oxygenated. Blood is returned to the left atrium. It passes to the left ventricle and through the aorta to the organs and tissues of the body where it gives up oxygen and is returned to the right atrium.
4. The cardiac skeleton consists of fibrous connective tissue around the base of the valves; it serves for the attachment of cardiac muscle fibers.
5. The conduction system consists of modified muscle fibers. The sinoatrial node is the pacemaker; the atrioventricular node sends Purkinje fibers to the muscle of the ventricles.

III. Development of the Cardiovascular System
     A. Development of the heart begins as a single tube. The atria are partitioned by two septa that overlap; ventricles are partitioned by a septum from the apex together with the spiral septum of the truncus arteriosus.

REVIEW QUESTIONS

1. What are the principal functions of the circulatory system?

2. What are the chief forces that dictate the direction of blood flow?

3. What is the potential space between the visceral and parietal pericardium? What layer of pericardium is fused with the parietal pleura?

4. What proportion of the pericardial sac is normally located to the left of the midline in the thorax?

5. The tricuspid valve is located between what two heart chambers?

6. The dense irregular connective tissue surrounding the base of the cardiac valves forms part of what structure?

7. What structure within the heart is responsible for the initiation of contraction?

8. The final opening in the interventricular septum is closed by the septum that splits what embryonic blood vessel?

SELECTED REFERENCES

Anderson, J. E. *Grant's Atlas of Anatomy.* 8th ed. Baltimore: Williams & Wilkins, 1983.

Goss, C. M. *Gray's Anatomy.* 29th ed. Philadelphia: Lea & Febinger, 1973.

Lachman, E., and Faulkner, K. K. *Case Studies in Anatomy.* New York: Oxford University Press, 1981.

Moore, K. L. *Clinically Oriented Anatomy.* Baltimore: Williams & Wilkins, 1980.

Romanes, G. C., ed. *Cunningham's Textbook of Anatomy.* 12th ed. New York: Oxford University Press, 1981.

Woodburne, R. T. *Essentials of Human Anatomy.* 7th ed. New York: Oxford University Press, 1983.

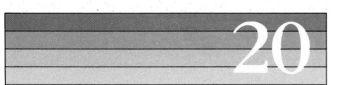

20

THE PERIPHERAL CIRCULATION

OBJECTIVES

After completing this chapter you should be able to:

1. Correlate the structure of arteries, capillaries, and veins with their role in the functions of the circulatory system.

2. Give examples that illustrate methods by which blood vessels are named.

3. Define by means of an example the concept of "end" arteries.

4. Define by means of an example the concept of collateral circulation.

5. Define and state the function of arterio-venous shunts.

6. Define a portal system.

7. Trace the flow of blood through the pulmonary circuit.

8. Trace the flow of blood through the systemic circuit.

9. Describe the coronary circulation, locating the major vessels with respect to the heart chambers.

10. Describe the distribution of the carotid arteries.

11. Describe the basic pattern of arterial supply of the superior extremity.

12. Describe the blood supply of the brain, naming the major arteries.

13. Name the blood vessels that enter and leave the heart and give the major branches of these vessels within the thorax.

14. Describe the arteries of the pelvis.

15. Describe the basic pattern of arterial supply of the inferior extremity.

16. Describe the mechanisms for the regulation of heartbeat and for the control of blood flow in peripheral blood vessels.

17. Name and locate the paired branches of the abdominal aorta.

18. Describe the venous drainage of the posterior body wall.

19. Describe the venous drainage of the abdominal portion of the digestive system.

20. Describe the venous drainage of the upper extremity.

21. Trace the pathway of lymph flow from the tissues of any region of the body into the blood stream.

22. Describe the development of the aortic arches.

23. Describe the changes in circulation that occur at birth.

FUNCTIONAL ANATOMY OF THE BLOOD VESSELS

The blood vessels have a dual role, as containers for *storage* of blood and as a system of tubes for its *distribution* throughout the body. The large arteries, particularly the aorta, have elastic walls that expand during systole, storing blood, and recoil during diastole, propelling blood forward, which results in a pulsating blood flow.

The openings of the first branches of the aorta, the *right and left coronary arteries,* lie immediately above two of the valve cusps. During systole the blood rushing from the left ventricle pushes the valve cusps against the wall of the aorta, partially covering the entrances of the coronary arteries. During diastole the elastic recoil of the aorta fills the cusps with blood, and they pull away from the arterial wall, contacting one another in the lumen. This prevents blood from returning to the ventricle but allows blood to enter the coronary arteries. Thus the coronary arteries are subjected only to the force of the elastic recoil of the aorta and not to the force of ventricular contraction. Ventricular contraction supplies the energy for expansion of the arterial wall during systole; the elastic recoil during diastole does not require the expenditure of energy.

Veins also have dual storage and conduit functions. Their diameters are, on the average, larger than those of the arteries, allowing a slower passage of a larger volume of blood. Because the total capacity of the veins is greater than that of the arteries, at any given time more blood volume is in the veins, through which it flows continuously toward the heart under less pressure than in the arteries.

Vessel Walls

The walls of the arteries and veins consist of three coats, or tunics (Fig. 20-1): the *tunica intima* is the endothelial lining and associated connective tissues; the *tunica media* is the thick middle portion, consisting of one or more layers of smooth muscle and elastic fibers; the *tunica adventitia* is the outer layer of fibrous connective tissue. The large arteries have an abundance of elastic fibers that in the aorta and its major branches are arranged in the form of elastic membranes between the layers of smooth muscle. These membranes are stretched during systole and recoil to their unstretched state during diastole.

As the arteries branch they diminish in size, and the amount of elastic tissue diminishes relative to the amount of smooth muscle. The tunica media of the arterioles contains circular smooth muscle under sympathetic nervous control, capable of reducing the size of the lumen of the vessel (**vasoconstriction**) and thus reducing blood flow through the arteriole. This is the

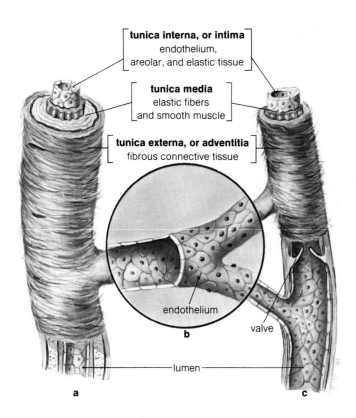

tunica interna, or intima
endothelium,
areolar, and elastic tissue

tunica media
elastic fibers
and smooth muscle

tunica externa, or adventitia
fibrous connective tissue

endothelium

valve

lumen

a

b

c

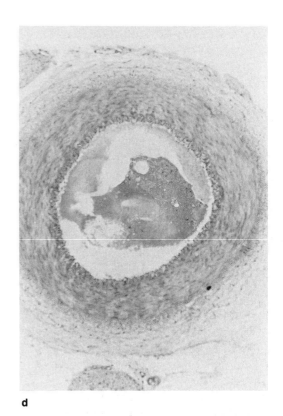

d

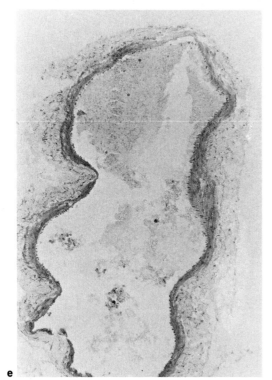

e

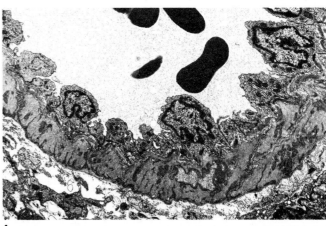

f

Figure 20-1 The general structure of (**a**) an artery, (**b**) a capillary, (**c**) a vein; the capillary is shown larger than is proportional to the size of the artery and vein. (**d**) Cross section of an artery, × 100. (**e**) Cross section of a vein, × 100. (Photographs by Joan Creager.) (**f**) An electron photomicrograph of a transverse section of an arteriole. Note the irregular appearance of the endothelium, × 4200. (Courtesy of D. H. Matulionis.)

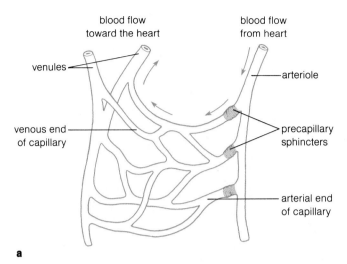
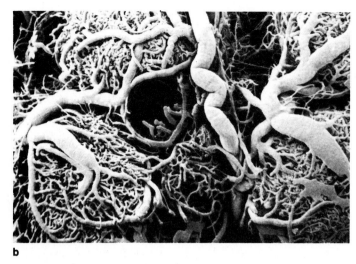

Figure 20-2 Capillaries. (**a**) A capillary network, showing the relationship of arterioles, precapillary sphincters, capillaries, and venules. (**b**) A scanning photomicrograph of casts of the capillary networks in the corpora lutea of the ovary. (Photomicrograph reproduced with permission from *Tissues and Organs: A Text-Atlas of Scanning Microscopy,* by R. G. Kessel and R. H. Kardon. W. H. Freeman and Company © 1979.)

principal mechanism regulating the blood flow to the tissues of the body. Precapillary sphincters at the junctions of the arterioles with the capillaries control precisely the amount of blood allowed to enter the capillaries (Fig. 20-2).

Neural Regulation of Vessel Diameter

The smooth muscle in the walls of peripheral blood vessels is innervated by *sympathetic fibers* that respond to stimuli initiated by local conditions such as oxygen deficiency. In most regions of the body, stimulation of the sympathetic fibers causes vasoconstriction, reducing the blood flow in the vascular bed. In these same blood vessels increased blood flow may result from inhibition of the sympathetics. This is the pattern of innervation that regulates the diameter of the arterioles and precapillary sphincters, shunting blood flow to areas of the body having a great need for oxygen or nutrients at a particular moment.

The Role of Pressure in Circulation

The pressure caused by the pumping of the heart ensures the flow of blood through the arteries to the capillary beds and into the veins, which is why arteries and capillaries need no valves to prevent backflow (except where the arteries leave the heart). Pressure within the veins is lower than in the arteries or capillaries because of their greater distance from the pump; further-

more the blood must flow against the pull of gravity in the veins of the lower body. Nevertheless, with the help of valves to prevent backflow, of voluntary muscle action such as walking, and of pressure on veins from adjacent arteries, venous pressure remains high enough to cause the blood to flow toward the heart and fill the atria.

Because veins do not shunt blood into and away from capillary beds, the tunica media of the veins contains relatively little smooth muscle; so the walls of veins are more flexible than those of arteries, and a vein is easily occluded by external pressure. The brachial vein, for example, is completely occluded by a blood pressure cuff inflated to an external pressure of 40 mm mercury, whereas blood flow continues in the brachial artery against a pressure of 120 mm Hg or more. Though the small amount of smooth muscle in the walls of veins allows them to be easily collapsed, it is adequate to regulate their pressure and capacity by influencing the diameter of the lumen.

GENERAL FEATURES OF THE PERIPHERAL CIRCULATION

There is considerable variation among individuals in the position and distribution of their blood vessels. These differences become greater as the blood vessels continue to branch. Some organs, such as the kidney, have a specific point for the entry

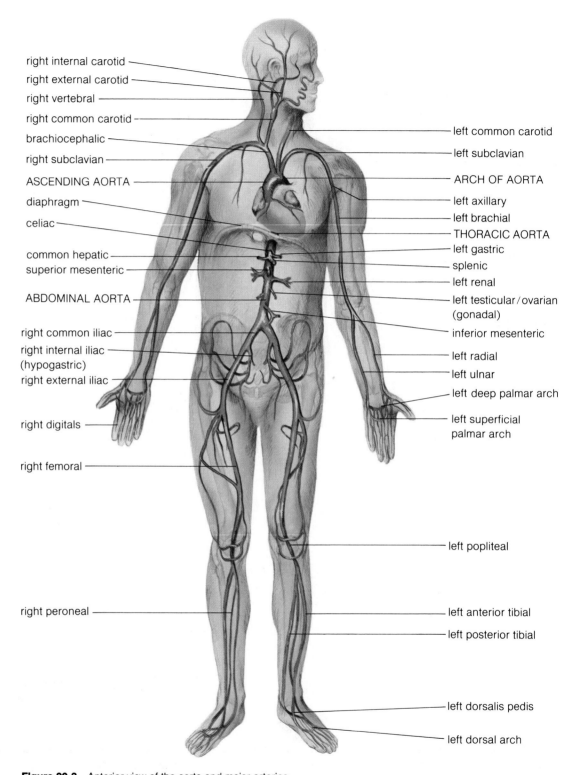

right internal carotid

right external carotid

right vertebral

right common carotid

brachiocephalic

right subclavian

ASCENDING AORTA

diaphragm

celiac

common hepatic

superior mesenteric

ABDOMINAL AORTA

right common iliac

right internal iliac
(hypogastric)

right external iliac

right digitals

right femoral

right peroneal

left common carotid

left subclavian

ARCH OF AORTA

left axillary

left brachial

THORACIC AORTA

left gastric

splenic

left renal

left testicular/ovarian
(gonadal)

inferior mesenteric

left radial

left ulnar

left deep palmar arch

left superficial
palmar arch

left popliteal

left anterior tibial

left posterior tibial

left dorsalis pedis

left dorsal arch

Figure 20-3 Anterior view of the aorta and major arteries.

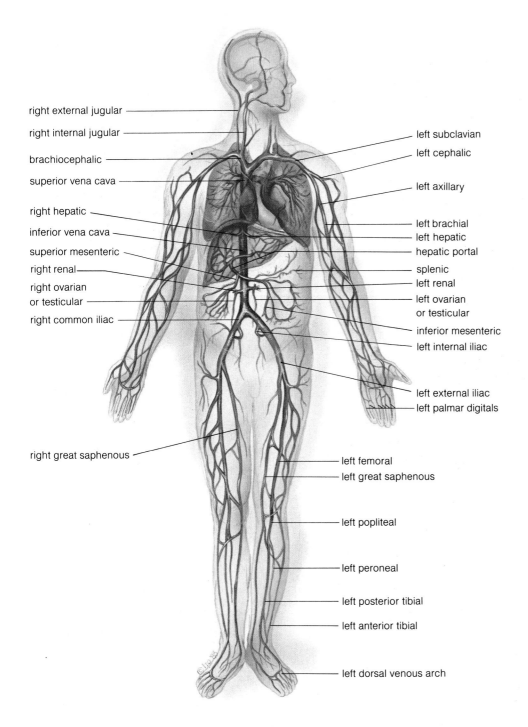

right external jugular

right internal jugular

brachiocephalic

superior vena cava

right hepatic

inferior vena cava

superior mesenteric

right renal

right ovarian
or testicular

right common iliac

right great saphenous

left subclavian

left cephalic

left axillary

left brachial

left hepatic

hepatic portal

splenic

left renal

left ovarian
or testicular

inferior mesenteric

left internal iliac

left external iliac

left palmar digitals

left femoral

left great saphenous

left popliteal

left peroneal

left posterior tibial

left anterior tibial

left dorsal venous arch

Figure 20-4 The major veins of
the human body in anterior view.

of blood vessels and a definite pattern of distribution of the vessels within the organ, but many structures such as skeletal muscles may be penetrated by blood vessels at almost any point by branches from the closest vessel, making it difficult to establish a definite pattern for "normal" blood supply. Consequently few arterial branches to muscles are named, being called simply "muscular branches" to the named muscle. However, those arteries and veins that do have a dependable pattern are named regardless of their size. Figs. 20-3 and 20-4 show the major named vessels of the body.

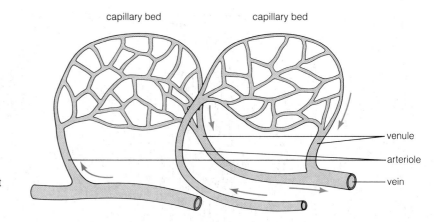

Figure 20-5 Diagram of the relationship of end arteries to capillaries and venules. Note that a single arteriole supplies a single capillary bed.

capillary bed   capillary bed

venule

arteriole

vein

The Naming of Blood Vessels

Most major blood vessels are given specific names related either to their location or to the organ supplied. For example, the *profunda* (deep) *brachial* (arm) *artery* lies in a deeper plane than the *brachial artery* as these two vessels descend toward the elbow. A main artery may give rise to numerous branches without its name being changed, but some branching situations produce a name change. For example, the *internal thoracic artery* descends alongside the sternum on the anterior chest wall, giving off anterior intercostal arteries to the intercostal spaces. At the diaphragm it branches into the *musculophrenic,* supplying the diaphragm, with the apparent continuation of the main channel passing through the diaphragm to descend along the inner surface of the anterior abdominal wall. The abdominal vessel is named *superior epigastric artery,* and the internal thoracic is considered to terminate by branching into the musculophrenic and superior epigastric arteries. There are numerous examples of this type of name change in the body.

In some cases the name of a blood vessel is changed simply because it passes into a different region of the body. For example, the *subclavian artery* lies beneath the clavicle in the root of the neck. When it crosses the first rib to enter the axilla (arm pit), its name is changed to the *axillary artery.* This same vessel becomes the *brachial artery* when it enters the arm at the lower border of the teres major muscle. It is much like the names of streets in some cities: at a certain intersection North Main Street suddenly becomes South Main Street.

In general, veins accompany arteries and are given the same name, but there are several important exceptions to this rule. For example, the inferior vena cava parallels the aorta in much of its course but is not called the aortic vein.

Variations in Vascular Patterns

The standard pattern of peripheral blood flow consists of the blood flowing into a capillary network from small arteries (arterioles) and being carried away from the capillaries by small veins called venules. There are, however, variations on this typical vascular pattern.

"End" Arteries   If adjacent capillary beds are supplied by separate arteries, with no flow of blood between the capillary beds, the tissue involved is said to be supplied by end arteries (Fig. 20-5). Because blood can enter the capillary bed only from the single artery by which it is supplied, occlusion of the artery results in the death from lack of oxygen of the tissue it supplies. This pattern of circulation occurs in the kidneys.

Collateral Circulation   If capillary beds interconnect or overlap to some extent, there is a possibility of the mixing of blood between beds, depending upon the pressure relations at any given moment (Fig. 20-6). When two adjacent arterioles supply interconnecting beds, the blood takes the path of least resistance, so that each arteriole supplies blood only to the capillaries with which it is most closely related. However, if one of the arteries is blocked, the artery remaining functional supplies blood to all the tissues through the connections between the capillary beds. Furthermore, because arteries have no valves to prevent backflow, blood can enter the occluded artery and flow in a reverse direction to the point of occlusion, sending blood into any branches that arise from the artery up to that point. This type of arrangement is called **collateral circulation,** and is important in many areas of the body, such as around the joints

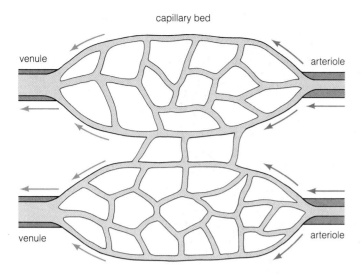

Figure 20-6 Diagram of connections between capillary beds that permit collateral circulation.

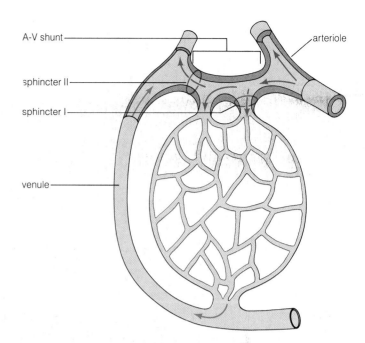

Figure 20-7 Diagram of an arterio-venous shunt. Sphincter I represents a precapillary sphincter that prevents blood from entering the capillary bed. Sphincter II is a postcapillary sphincter that relaxes when sphincter I contracts, allowing blood to pass directly from the arteriole to the vein without entering capillaries.

of the extremities, in maintaining an uninterrupted blood supply to the tissues in spite of the occlusion of blood vessels by trauma or vascular "accidents."

Arterio-Venous Shunts The body does not have enough blood to fill all the capillaries at one time. So blood is shuttled to the places where the need is immediate and prevented from entering tissues that temporarily have a reduced demand for oxygen and nutrients. One of the anatomical features of the peripheral circulation (particularly in skeletal muscles) that makes the process possible is that of the **arterio-venous (A-V) shunt** (Fig. 20-7). In many places in the body there are direct channels that connect arteries and veins, allowing the blood to pass from the artery to the vein without going through the capillaries that normally receive blood from the artery. Sphincters located at the junction of the arteriole and the capillaries and at the junction of the arteriole and vein may close at either point, determining the direction of blood flow. Contraction of the sphincter at the entrance to the capillary bed shunts the blood directly into the vein, sending it back to the heart without passing through capillaries. Contraction of the sphincter at the vein shunts the blood into the capillaries. Partial constriction of the sphincters regulates precisely the volume of blood entering the capillaries at that point. Thus the A-V shunt is part of the mechanism for controlling the volume of blood flowing through the capillary beds of the body and for bypassing any given capillary bed that is not in need of oxygenated blood

at that moment. The shunt is regulated by sympathetic nerves responding to stimuli created by local conditions.

Portal Systems In contrast with the vascular pattern in most parts of the body, in which a single capillary bed is interposed in the arterio-venous circuit, the vascular pattern of a **portal system** includes two capillary beds (or two groups of capillary beds) connected by a vein (Fig. 20-8). Thus blood from the heart through a portal system passes through the following sequence of vessels: *arteries—capillaries—veins—capillaries—veins.* This arrangement allows substances to enter the blood at one set of capillaries for direct transport to a second set of capillaries where the substances may be removed without having entered the general circulation. The human has two important portal systems: the *hepatic portal system,* which takes the blood with nutrients absorbed in the various capillary beds of the digestive tract to the capillary beds of the liver where the nutrients are removed for storage, and the *hypophyseal portal system* by which blood conveys releasing hormones from the capillaries of the hypothalamus directly to those of the pituitary gland, where they regulate the gland's activity.

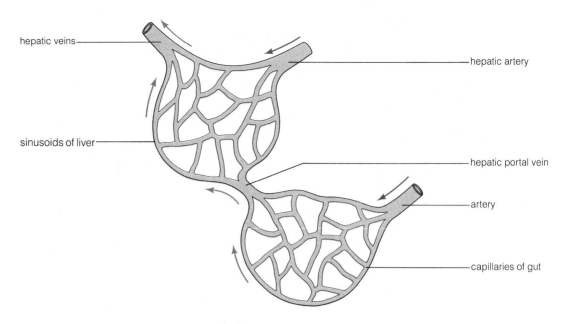

hepatic veins

sinusoids of liver

hepatic artery

hepatic portal vein

artery

capillaries of gut

Figure 20-8 Diagram of the hepatic portal system.

THE PULMONARY CIRCUIT

The *pulmonary circuit* (Fig. 20-9), beginning in the *right ventricle* and ending in the *left atrium,* includes the *pulmonary arteries,* the *capillaries* of the *lungs,* and the *pulmonary veins.* The pulmonary trunk, the most anterior of the great vessels, arises from the *conus arteriosus,* the tapering upper portion of the right ventricle, from which it is separated by the pulmonary semilunar valve. It ascends on a spiral course in relation to the aorta, passing to the left and posteriorly before bifurcating into right and left pulmonary arteries inferior to the arch of the aorta. The right pulmonary artery passes posterior to the ascending portion of the aorta to enter the root of the right lung; the left pulmonary artery passes anterior to the descending portion of the aorta to reach the root of the left lung. The **ligamentum arteriosum** (Fig. 19-10) connects the point of bifurcation of the pulmonary trunk to the arch of the aorta. This ligament is a fibrous strand that represents the obliterated *ductus arteriosus,* a fetal vessel connecting the pulmonary and systemic circuits, allowing the fetal blood to bypass the lungs prior to birth. Fetal circulation will be discussed later in the chapter.

Upon entering the lungs, the arteries branch, paralleling the branches of the bronchial tree (air passageways of the lungs) until they terminate as arterioles that open into the capillary beds of the *alveoli* (sing., alveolus), the sites of gas exchange in the lungs (Fig. 20-10). Blood is oxygenated as it passes through the lung capillaries and enters the venules to begin its return to the heart. The veins become larger as they anastomose, finally joining near the root of the lung into two pulmonary veins from each lung. Usually each of the four pulmonary veins opens separately into the left atrium.

THE SYSTEMIC CIRCUIT

The *systemic circuit,* which conveys oxygenated blood to all the organs and tissues of the body, consists of the aorta and all of its branches, the capillary beds into which the arterial blood flows, and the veins that return the deoxygenated blood to the heart. The systemic circuit begins in the *left ventricle,* from which the aorta arises, and ends in the *right atrium,* into which the coronary sinus and the superior and inferior venae cavae open. We shall discuss first the coronary circuit, the branch of the systemic circuit that supplies the heart wall, then the rest of the systemic circuit, considering first the arteries and then the veins.

The Coronary Circuit

The *coronary circuit,* which supplies the heart muscle with oxygenated blood from the left ventricle, consists of the two

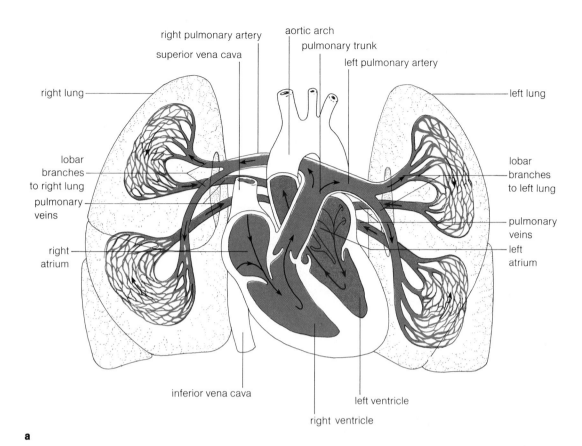

right pulmonary artery

aortic arch

pulmonary trunk

superior vena cava

left pulmonary artery

right lung

left lung

lobar
branches
to right lung

lobar
branches
to left lung

pulmonary
veins

pulmonary
veins

right
atrium

left
atrium

inferior vena cava

left ventricle

right ventricle

a

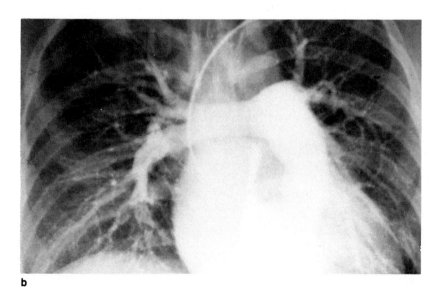

b

Figure 20-9 The pulmonary circulation. (**a**) Diagram of arteries carrying deoxygenated blood are shown in blue. Veins carrying oxygenated blood are shown in red. (**b**) X ray of pulmonary trunk and pulmonary arteries (from the University of Kentucky Department of Anatomy teaching collection).

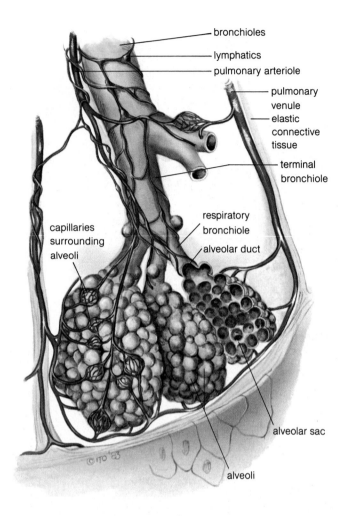

bronchioles
lymphatics
pulmonary arteriole
pulmonary venule
elastic connective tissue
terminal bronchiole
respiratory bronchiole
capillaries surrounding alveoli
alveolar duct
alveolar sac
alveoli

Figure 20-10 Relationship of the capillary networks to the lung alveoli at the end of the air passages.

coronary arteries and their branches, the *capillaries* of the *heart,* and the *coronary veins* which empty deoxygenated blood into the right atrium via the *coronary sinus*. The coronary arteries, right and left, arise from the ascending aorta near its point of exit from the heart. The openings of these vessels are partially covered at systole by the semilunar valves, and the arteries are filled during diastole by the force of the elastic recoil of the aorta. The right coronary artery arises behind the right cusp of the aortic semilunar valve and passes to the right in the coronary sulcus, sending branches to the right ventricle as it descends to the diaphragmatic surface of the heart, where it ends as the *posterior interventricular artery* (posterior descending). The left coronary artery arises behind the left valve cusp, enters the left coronary sulcus and divides into the *circumflex* branch,

which continues in the sulcus, and the *anterior interventricular* branch (anterior descending), which descends on the anterior surface of the heart (Fig. 19-10). From the capillary beds in the cardiac muscle, blood passes into veins paralleling the arteries. These veins drain into a channel, the coronary sinus, which opens into the right atrium (Fig. 19-11).

The branches of the right and left coronary arteries anastomose with one another on the surface of the heart and within the heart muscle, providing collateral circulation. Unfortunately the collateral circulation is usually inadequate for the survival of cardiac muscle if a large vessel is blocked suddenly, usually by a clot (embolism). When a vessel is occluded, depriving an area of cardiac muscle of oxygen (myocardial infarct), the muscle begins to undergo necrosis (cell death) (Fig. 20-11). At the same time blood vessels in the area begin to expand, and the growth of new vessels into the area soon restores the blood supply. If the person survives long enough for this process to be completed, the blood supply of the heart wall is usually restored; but the degenerated cardiac muscle is usually replaced by fascia (scar tissue), making the heart less effective as a pump.

Arteries of the Head, Neck, and Superior Extremities

The head, neck, and superior extremities receive all their blood from paired *common carotid* and *subclavian arteries* (Fig. 20-12). The first major branch of the aorta after the coronary arteries is the *brachiocephalic* trunk arising from the superior surface of the **aortic arch.** This vessel divides quickly into the *right subclavian* and *right common carotid*. The second major branch of the aortic arch is the *left common carotid,* followed by the *left subclavian*. The two common carotid arteries send blood to the neck and head. The two subclavian arteries have branches supplying the upper extremities, axillary and shoulder regions, and parts of the neck. In addition each subclavian has a branch (the vertebral) that supplies the head.

The Subclavian Artery and Its Branches The subclavian artery passes laterally across the root of the neck toward the shoulder. The first part of the artery is medial to the anterior scalene muscle (Fig. 20-13) and has three branches: the *vertebral,* the *internal thoracic,* and the *thyrocervical trunk.* The vertebral artery, usually the first branch of the subclavian artery, ascends in the neck to pass into the *transverse foramen* at the sixth cervical vertebra; it then runs superiorly through the foramina of the transverse processes of the first six cervical vertebrae and enters the skull through the *foramen magnum*. Within the cranial cavity the right and left vertebrals unite to form the *basilar artery* (Fig. 20-16). The basilar gives rise to vessels supplying the cerebellum and terminates by bifurcating into the

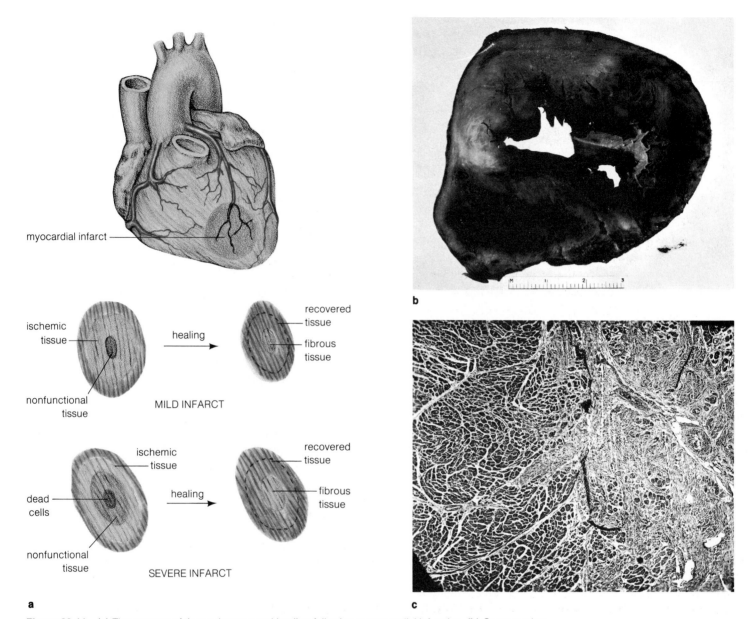

Figure 20-11 (a) The process of tissue damage and healing following a myocardial infarction. (b) Cross section of the heart following the death of the individual three days after a myocardial infarction. Notice the white areas in the wall of the heart where blood supply was impaired. (Armed Forces Institute of Pathology negative number 60-6123.) (c) Photomicrograph of a healed myocardial infarction, × 42. Notice the fibrous tissue on the left side of the figure where the infarction has healed. (Armed Forces Institute of Pathology negative number 61-3009.)

two posterior cerebral arteries that are part of the blood supply of the brain.

The internal thoracic, arising from the inferior aspect of the subclavian artery, passes deep to the first rib to enter the thorax where it runs along the sternum (Fig. 20-18), giving off branches to each of the intercostal spaces, and goes on into the abdominal region as the *superior epigastric artery,* supplying the anterior abdominal wall. We shall return to the branches of the internal thoracic when we discuss the arteries of the trunk and lower extremities. The thyrocervical trunk divides into three

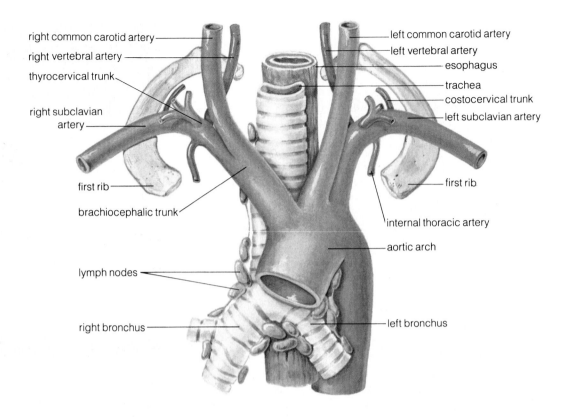

Figure 20-12 The branches of the arch of the aorta, anterior view.

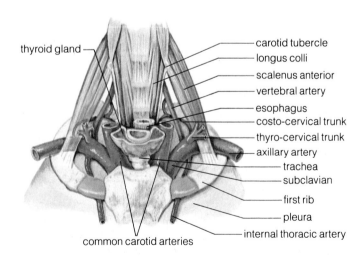

Figure 20-13 The relationship of the subclavian arteries and their branches in the root of the neck.

branches: the *suprascapular* and the *transverse cervical* to the shoulder and lateral neck regions, and the *inferior thyroid* to the thyroid gland and anterior neck region.

The second part of the subclavian artery, passing posterior to the anterior scalene (Fig. 20-13), gives off the *costocervical trunk,* which has two branches: a *deep cervical* to the back of the neck and the *supreme intercostal,* which sends branches to the posterior portion of the first two intercostal spaces. The third part of the subclavian artery lies lateral to the anterior scalene muscle, and as it passes across the first rib to enter the axilla, it becomes the *axillary artery.* At the lower border of the teres major muscle the axillary artery becomes the *brachial artery* of the upper limb (Fig. 20-14). On its way to the arm the axillary artery gives rise to six branches that supply the chest wall (including the breast), axilla, shoulder, scapula, and upper humerus. These branches become an elaborate network of anastomosing arteries that provides alternate routes for blood flow in cases of occlusion.

The *brachial artery* (Fig. 20-14) may be palpated the entire length of the arm in the *medial bicipital furrow.* As it descends in the arm, it gives off the *profunda* (deep) *brachial* artery that

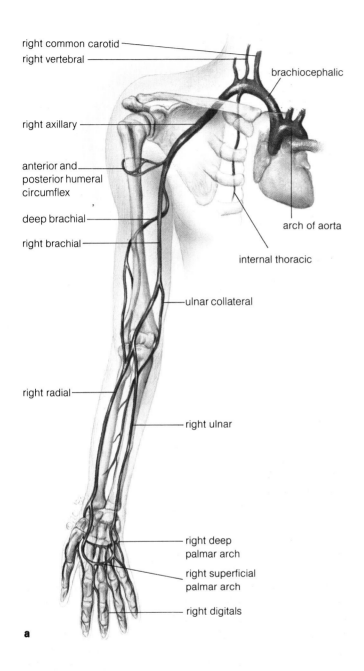

right common carotid

right vertebral

brachiocephalic

right axillary

anterior and posterior humeral circumflex

deep brachial

right brachial

arch of aorta

internal thoracic

ulnar collateral

right radial

right ulnar

right deep palmar arch

right superficial palmar arch

right digitals

a

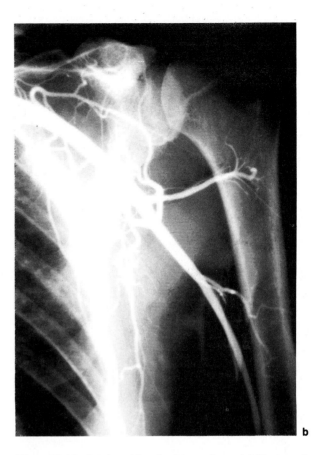

b

Figure 20-14 Arteries of the shoulder and arm. (**a**) Diagram of an anterior view. (**b**) Angiogram of the arteries of the shoulder (from the University of Kentucky Department of Anatomy teaching collection).

takes a spiral course around the humerus as it passes down the posterior aspect of the humerus to reach the elbow. In addition to *ulnar collateral* branches to the elbow region the brachial artery has *muscular* branches and supplies a *nutrient artery* to the humerus. The *ulnar* and *radial arteries* arise in the **cubital fossa** by bifurcation of the brachial artery. As the radial and

There is considerable individual variation in the contribution of the radial and ulnar arteries to the blood flow into the palmar loops or arches. You can check the pattern of filling the loops in your palm by the following procedure. With your left hand compress both the ulna and radial arteries as they cross your right wrist. Close your right fist, forcing the blood from your palm, and then open your fist as you release the pressure over one of the arteries. Observe the color change as the blood flows into the loop. Repeat the process, releasing the pressure over the other artery. Unequal filling indicates that one of the arteries is larger than the other.

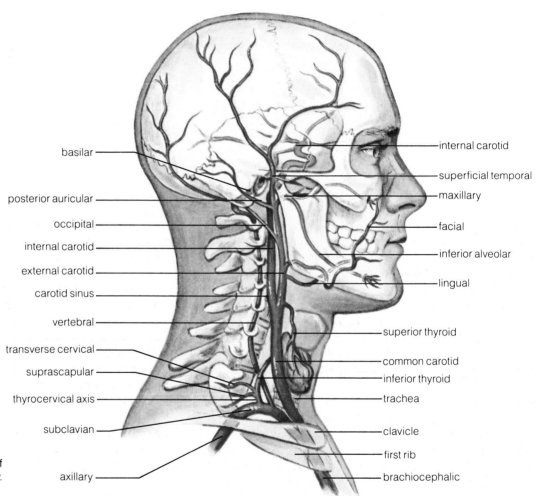

basilar

posterior auricular

occipital

internal carotid

external carotid

carotid sinus

vertebral

transverse cervical

suprascapular

thyrocervical axis

subclavian

axillary

internal carotid

superficial temporal

maxillary

facial

inferior alveolar

lingual

superior thyroid

common carotid

inferior thyroid

trachea

clavicle

first rib

brachiocephalic

Figure 20-15 Major arteries of the head and neck, lateral view.

ulnar arteries descend in the forearm, they give off *recurrent arteries* that pass back to the elbow region to anastomose with the collateral branches of the brachial.

At the wrist the radial artery (on the thumb side) is the vessel from which the pulse is usually taken. Near the wrist both the radial and the ulnar have *superficial* and *deep branches* that

join to form *superficial* and *deep loops,* or palmar arches, in the palm (Fig. 20-14). These loops give rise to *digital arteries* that supply the fingers.

Arteries of the Head and Neck　Most of the blood supply of the head and upper neck is derived from branches of the

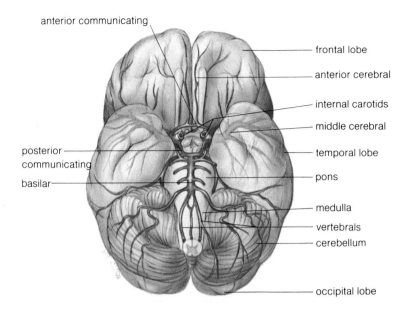

anterior communicating

frontal lobe

anterior cerebral

internal carotids

middle cerebral

temporal lobe

posterior communicating

pons

basilar

medulla

vertebrals

cerebellum

occipital lobe

Figure 20-16 Major arteries at the base of the brain, ventral view.

common carotid arteries (Fig. 20-15). The common carotids arising from the *brachiocephalic* artery on the right and from the arch of the aorta on the left have no branches in the neck other than their terminal branches, the *internal* and *external carotids*. The internal carotid, as a rule, has no branches in the neck. In the head it gives rise to the *ophthalmic branch,* which supplies the orbit and the eye, and to branches that go to the brain.

The external carotid usually has five branches in the neck and three in the head. The exact sequence by which the five neck branches arise may vary somewhat, but usually the first branch, the *ascending pharyngeal* which supplies the lateral wall of the pharynx, arises from the posterior aspect of the external carotid and ascends between the external and internal carotids. The *superior thyroid* comes off the anterior aspect of the external carotid and sends branches into the larynx and thyroid gland, where it anastomoses with the *inferior thyroid* branch of the *thyrocervical trunk* from the *subclavian.* The *lingual* artery passes deep to enter the root of the tongue, and the *facial* crosses the mandible to supply the superficial aspect of the face. The *occipital* arises from the posterior aspect of the external carotid and passes deep to the sternomastoid to reach the back of the head. The *posterior auricular* arises at the angle of the mandible and supplies the external ear. The *maxillary* and the *superficial temporal* are the terminal branches of the external carotid. The maxillary supplies the deep structures of the face, including the teeth of both jaws. The superficial temporal supplies the scalp covering the side of the head.

Blood Supply of the Brain The *internal carotids* enter the cranial cavity through the *carotid canal.* They give off the *ophthalmic* arteries to the orbit and the *anterior* and *middle cerebral* arteries (Fig. 20-16), which supply the medial and lateral aspects of the cerebral hemispheres, respectively. These cerebral arteries are a part of an arrangement of arteries, termed the **circle of Willis,** that encircles the pituitary gland and optic chiasma at the base of the brain. The circle is completed by the internal carotids, the *anterior communicating* artery which connects the two *anterior cerebrals,* the *posterior cerebrals,* and the *posterior communicating* arteries which connect the *internal carotids* and the *posterior cerebrals.* The posterior cerebrals are branches of the *basilar* artery (Fig. 20-16) formed by the union of the right and left vertebrals. In general, blood from the vertebral arteries supplies the brainstem, cerebellum, and posterior part of the cerebrum. The internal carotid supplies the rest of the cerebrum and the eyes.

The circle of Willis would seem an ideal arrangement for collateral circulation. For example, if the vertebral or basilar were blocked, eliminating the normal blood supply of the posterior cerebrals, blood could reach the posterior cerebrals through the posterior communicating artery and, by all appearances, should be adequate to supply the entire brain. However, there is a great deal of individual variation in the size of the communicating vessels; in most people, if one of the major vessels is occluded abruptly, the collateral circulation is insufficient to maintain the brain in that area. The posterior communicating artery usually cannot supply the posterior cerebral

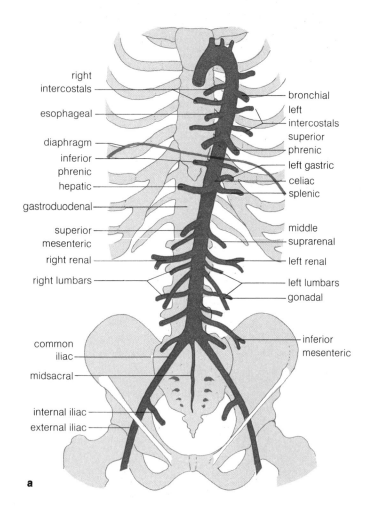

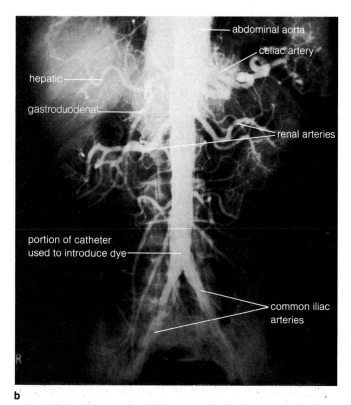

Figure 20-17 The major arteries branching from the descending aorta. (**a**) Diagram of the positions of the arteries. (**b**) Arteriogram showing some of the major branches of the abdominal aorta. An arteriogram is made by injecting a dye that is opaque to x rays into an artery and x-ray photographing the artery and its branches as the dye passes through them. (Arteriogram courtesy of Alexandria Hospital, Alexandria, Virginia, Joyce R. Isbel, R. T.)

artery with enough blood for the brain stem if the vertebrals are blocked. Likewise the anterior communicating may not be large enough to let one anterior cerebral artery take care of both sides of the cerebrum. Collateral circulation may be adequate if a vessel is occluded slowly, allowing time for the cross links to enlarge. Within the substance of the brain the smaller arteries are functionally more like end arteries than like the anastomosing arteries characteristic of collateral circulation, so a sudden occlusion of a vessel (stroke) results in cell death and loss of brain function.

Arteries of the Trunk and Lower Extremities

The descending aorta branches to supply the trunk and lower extremities. The posterior body wall receives blood from arter-

ies that arise directly from the aorta, called *parietal* branches, and the anterior body wall is supplied by arteries that originate from the large arteries supplying the superior and inferior extremities. The unpaired viscera (digestive tract, liver, and pancreas) are supplied by *visceral* arteries arising from the anterior surface of the aorta, and paired viscera of the thorax and abdomen (lungs, suprarenal glands, kidneys, gonads) are supplied by right and left branches of the aorta, which may not arise in a bilaterally symmetrical fashion. The pelvic viscera (rectum, urinary bladder, and uterus) are supplied, at least partially, by branches of the paired internal iliac arteries.

Arteries of the Trunk Wall　The posterior body wall is supplied by the posterior intercostals, subcostals, lumbars, and phrenics, all parietal branches of the aorta. Because the aorta ascends in the thorax only to about the level of the fourth tho-

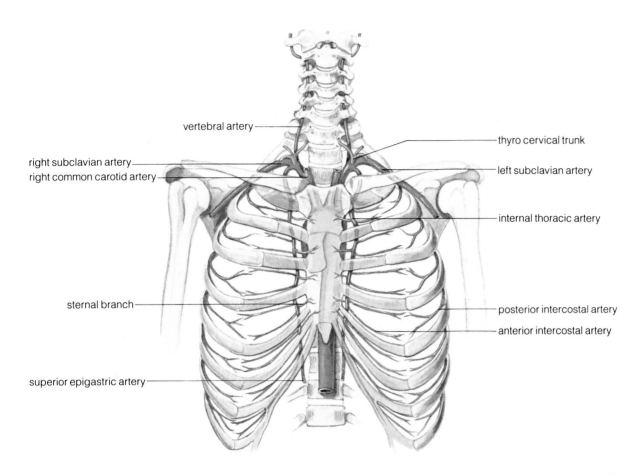

vertebral artery

thyro cervical trunk

right subclavian artery

left subclavian artery

right common carotid artery

internal thoracic artery

sternal branch

posterior intercostal artery

anterior intercostal artery

superior epigastric artery

Figure 20-18   Arteries of the thoracic wall, anterior view.

racic vertebra, the first two intercostal spaces are supplied posteriorly by branches of the subclavian and axillary arteries. The remaining intercostal spaces, III to XI, receive parietal branches of the thoracic aorta that arise in sequence from each side of the aorta (Fig. 20-17). These vessels, the *posterior intercostal* arteries, send branches posteriorly to supply the spinal cord and deep back muscles before passing laterally within the intercostal space. Below the twelfth rib this paired series of arteries supplying the body wall is continued by the *subcostals,* arising from the thoracic aorta just superior to the diaphragm, and three or four *lumbar* arteries, originating in sequence from the abdominal aorta inferior to the diaphragm. In the vicinity of the diaphragm *superior* and *inferior phrenics* arise and supply the superior and inferior surfaces of the diaphragm. The *subcostals* descend on the body wall below the diaphragm to assist the *lumbars* in supplying the posterior abdominal wall. The pattern

of distribution is similar to that of the posterior intercostal arteries: posterior branches supply the deep back muscles and spinal cord, and anterior branches pass into the lateral abdominal wall to end anteriorly near the lateral edge of the rectus abdominis muscle, where they anastomose with blood vessels of the anterior abdominal wall.

The anterior trunk wall receives blood from branches of the arteries to the superior extremities (subclavian arteries) and inferior extremities (external iliac arteries). Recall that the *internal thoracics* (Fig. 20-18) arise from the right and left subclavians and descend on the anterior thoracic wall on either side of the sternum, giving off branches to the sternum medially and to each intercostal space laterally. The lateral branches are the *anterior intercostal* arteries, which supply the structures of the intercostal space; they anastomose with the posterior intercostal arteries, providing excellent opportunity for collateral circula-

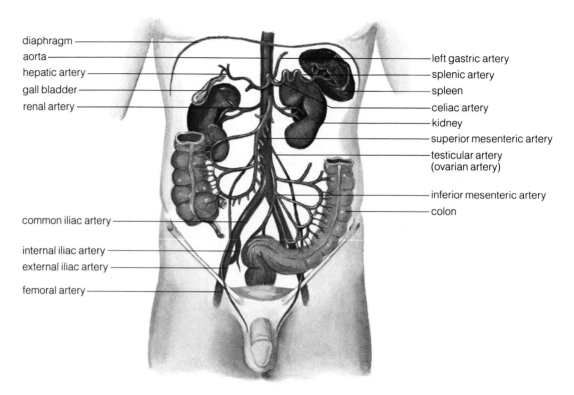

diaphragm
aorta
hepatic artery
gall bladder
renal artery

left gastric artery
splenic artery
spleen
celiac artery
kidney
superior mesenteric artery
testicular artery
(ovarian artery)
inferior mesenteric artery
colon

common iliac artery

internal iliac artery
external iliac artery

femoral artery

Figure 20-19 The abdominal aorta and its branches, anterior view.

tion between the thoracic aorta and the subclavian artery. For example, if the subclavian artery is occluded proximal to the origin of the internal thoracic artery, the posterior intercostal arteries become a source of blood for the superior extremity, with reversal of flow in the anterior intercostals and internal thoracic, sending blood into the subclavian artery for distribution through its branches in the usual fashion.

At the diaphragm the internal thoracic bifurcates into two terminal branches: the *musculophrenic* to the diaphragm and the *superior epigastric,* which continues below the diaphragm on the anterior abdominal wall. Inferiorly it is directly continuous with the *inferior epigastric,* which originates from the external iliac and ascends on the deep surface of the anterior abdominal wall to meet the superior epigastric. These two arteries provide for collateral circulation between the upper and lower extremities. By anastomosing with the subcostals and lumbars the epigastrics also provide for collateral circulation between the aorta and the blood vessels supplying both the upper and lower extremities.

Arteries of the Thoracic Viscera The visceral branches of the thoracic aorta are the *bronchial* and *esophageal* arteries (Fig. 20-17). There are usually two left bronchial arteries arising from

the aorta to supply nutrient vessels to the bronchial tree of the left lung. The single right bronchial artery usually arises from the third right posterior intercostal artery to be distributed to the tissue of the right lung. These vessels anastomose with the pulmonary vessels and provide for collateral circulation between the pulmonary and systemic circuits.

The esophageal arteries are several small unpaired arteries arising from the anterior surface of the thoracic aorta and passing forward to enter the esophagus.

Arteries of the Abdominal Viscera. Below the diaphragm the abdominal aorta gives off three *paired visceral branches,* serving the suprarenal glands, kidneys, and gonads, and three *unpaired visceral branches* serving the digestive organs (Fig. 20-19). The paired branches are the *middle suprarenal* arteries (one of three arteries to the suprarenal glands), the *renal* arteries to the kidneys, and the *testicular* (spermatic) or *ovarian* arteries to the testes or ovaries. The suprarenal glands also receive branches from the inferior phrenic and renal arteries. The middle suprarenal arteries arise from each side of the aorta just above the origins of the renal arteries. The right renal artery is longer than the left; either or both may branch before entering the kidney. In the male the testicular artery descends into the

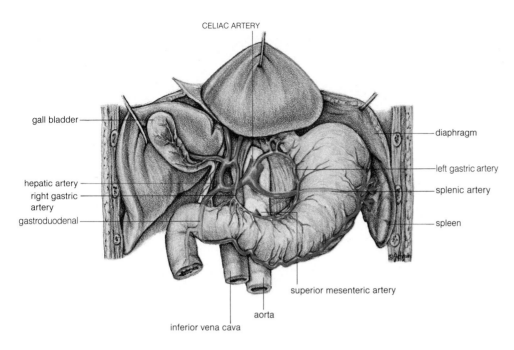

CELIAC ARTERY

gall bladder

hepatic artery

right gastric
artery

gastroduodenal

inferior vena cava

aorta

superior mesenteric artery

diaphragm

left gastric artery

splenic artery

spleen

Figure 20-20   The celiac artery and its branches, anterior view.

pelvis and continues inferiorly to pass through the *inguinal canal,* becoming a component of the *spermatic cord* (a fascial cord containing the vas deferens, nerves, and blood vessels) as it descends within the scrotum to end in the testis.

The three unpaired visceral branches of the abdominal aorta are the celiac, superior mesenteric, and inferior mesenteric arteries (Fig. 20-19), which arise in sequence from the anterior surface of the aorta for distribution to the digestive tract, liver, and pancreas. The *celiac* artery (Fig. 20-20) arises just below the diaphragm as a short trunk that branches almost immediately into three arteries, the left gastric, splenic, and hepatic. The *left gastric* courses to the left toward the esophageal–stomach junction, supplying branches to the left portion of the lesser (inside) curvature of the stomach. The *splenic* passes to the left along the superior margin of the pancreas, giving off branches to the pancreas, spleen, and stomach. A large branch of the splenic, the *left gastroepiploic,* supplies the greater (outer) curvature of the stomach and *pancreatic* branches to the body and tail of the pancreas. The pattern of branching of these vessels is highly variable, and the names of the individual arteries indicate the organ supplied. The *right gastric* branch of the hepatic supplies the lesser curvature of the stomach and anastomoses with the left gastric. The *gastroduodenal* is the common stem of origin of the *right gastroepiploic* to the greater curvature of the stomach, where it anastomoses with the left gastroepiploic branch of the splenic artery and the *superior pancreaticoduodenal* artery that descends between the head of the pancreas and the duodenum, supplying these structures and anastomosing with the *inferior pancreaticoduodenal* branch of the superior mesenteric artery.

The second unpaired visceral branch of the abdominal aorta, the *superior mesenteric* artery, arises at the inferior border of the pancreas from the anteror surface of the aorta, sending branches to the lower part of the duodenum and pancreas and to all of the gut proximal to the left colic flexure, where the transverse colon becomes continuous with the descending colon (Fig. 20-21). The gut branches unite to form a *marginal* artery that courses parallel to the colon and gives rise to anastomosing loops (arcades), from which arteries enter the gut wall. The usual branches of the superior mesenteric are the *inferior pancreaticoduodenal, jejunal, ileal, ileocecal, right colic,* and *middle colic.*

The third unpaired visceral branch, the *inferior mesenteric* artery, supplies the descending colon, the sigmoid colon, and the superior part of the rectum, where it anastomoses with the *rectal* branches of the *internal iliac.* Anastomoses also occur

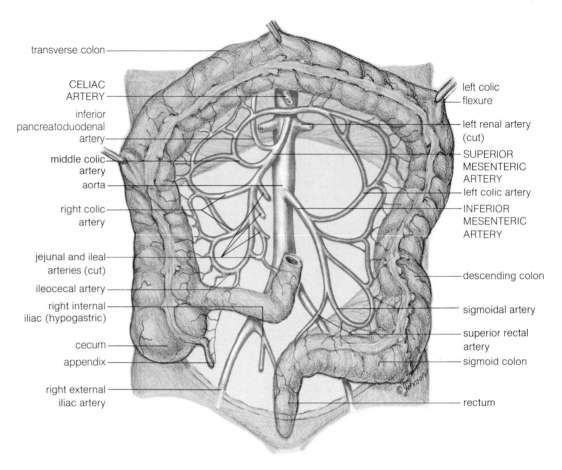

transverse colon

CELIAC ARTERY

inferior pancreatoduodenal artery

middle colic artery

aorta

right colic artery

jejunal and ileal arteries (cut)

ileocecal artery

right internal iliac (hypogastric)

cecum

appendix

right external iliac artery

left colic flexure

left renal artery (cut)

SUPERIOR MESENTERIC ARTERY

left colic artery

INFERIOR MESENTERIC ARTERY

descending colon

sigmoidal artery

superior rectal artery

sigmoid colon

rectum

Figure 20-21 The branches of the superior and inferior mesenteric arteries.

between the inferior mesenteric and superior mesenteric arteries in the vicinity of the left colic flexure (Fig. 20-21). The superior mesenteric and celiac have anastomoses at the pancreas and duodenum, providing the abdominal digestive tract with collateral circulation through its entire extent.

Arteries of the Pelvis The abdominal aorta ends inferiorly at the level of the fourth lumbar vertebra by bifurcating into the *right* and *left common iliac* arteries (Fig. 20-22). The common iliacs are short, branching at the upper border of the sacrum into the *internal* and *external iliac* arteries.

The internal iliacs course medially, giving off *superior* and *inferior gluteal* branches to the gluteal region and the *obturator* artery to the medial aspect of the thigh. Within the pelvis the internal iliac gives off a *uterine* artery and the *superior vesical* artery to the urinary bladder. It has a *middle rectal* branch to the rectum and an *internal pudendal artery* to the pelvic floor.

The *external iliac* artery gives off the *inferior epigastric* artery that ascends on the inner surface of the rectus abdominis

muscle, where it anastomoses with the superior epigastric, and the *deep circumflex iliac,* which anastomoses with the *iliolumbar artery* in the lateral abdominal wall. It then passes behind the *inguinal ligament* to enter the thigh where it is called the *femoral artery.*

Arteries of the Lower Extremity The femoral artery (Fig. 20-23) gives off the *profunda* (deep) *femoral,* as well as branching to the hip region and descending in the thigh. It passes medially and posteriorly to the femur and gives off muscular branches and collateral branches to the knee. It enters the **popliteal fossa** posterior to the knee, where it is called the *popliteal* artery. Distal to the knee the popliteal artery terminates by bifurcating into the *anterior* and *posterior tibial* arteries. The anterior tibial artery gives off *recurrent* branches to the knee and descends on the front of the leg, giving off *muscular* and *cutaneous* branches. It passes onto the dorsum of the foot, becoming the *dorsalis pedis* artery, which supplies the vascular network of the dorsum and communicates with the *plantar arch* of the sole.

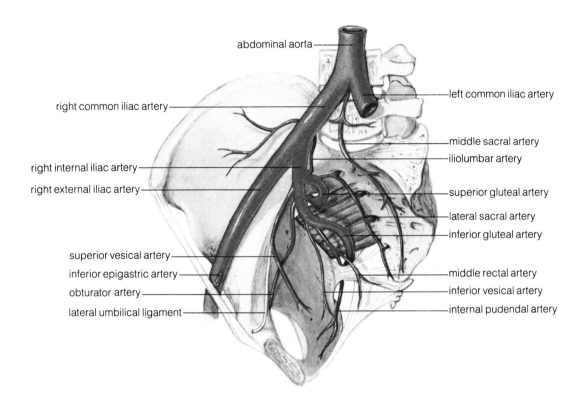

Figure 20-22 Arteries of the pelvis.

The posterior tibial artery gives off the *peroneal* artery which sends branches to the lateral compartment of the leg; the posterior tibial then courses down the back of the leg, supplying muscles and skin of the calf. It terminates by bifurcating into *medial* and *lateral plantar* arteries, which anastomose with one another by way of the plantar arch.

Systemic Veins

The veins will not be described in as much detail as the arteries in this book, since they are usually parallel to the arteries and have the same names. We will discuss the major differences between the pattern of arterial distribution and that of venous drainage in specific areas of the body, naming only the larger veins. First, however, we shall consider some general principles of venous drainage.

Principles of Venous Blood Flow Veins begin in the capillary beds and carry deoxygenated blood toward the heart. Blood flows in one direction, toward the heart, and is prevented from flowing in the opposite direction by the presence of valves within most of the veins, particularly those of the extremities. The valves are folds in the internal lining (tunica intima) of the veins and usually are semilunar in shape (Fig. 19-2), occurring in pairs arranged in series throughout the length of the vein. The large veins and some of the veins of the head do not have valves. The valves are an important factor in the return of blood to the heart: the contractions of adjacent skeletal muscles apply pressure on the veins, moving blood forward, and the valves prevent backflow as relaxation of the muscles reduces the pressure. Venous return is facilitated by the decrease in intrathoracic pressure during expiration. As pressure within the thorax drops, decreasing the resistance, the resulting suction draws blood toward the heart.

Blood flow through veins is much slower and under less pressure than in the arteries; the blood moves constantly, without the pulsating spurts typical of arterial flow. The diameter of the vein is usually larger than that of the artery it accompanies unless there are two veins alongside the artery. The walls of the

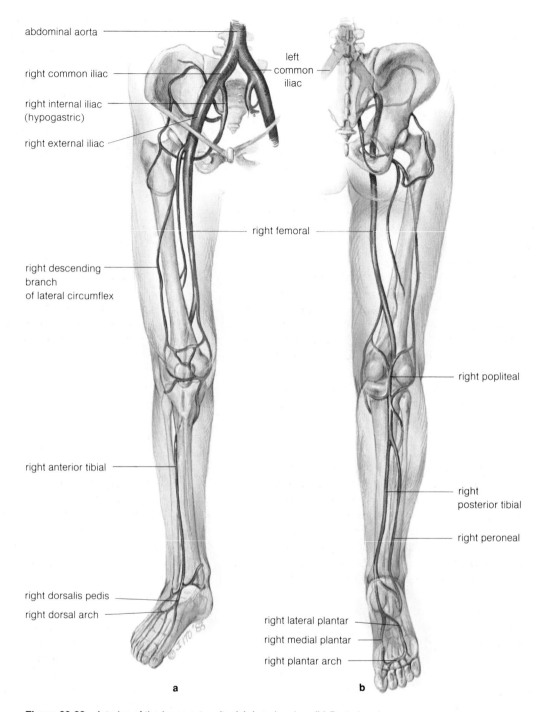

abdominal aorta

right common iliac

right internal iliac
(hypogastric)

right external iliac

left
common
iliac

right femoral

right descending
branch
of lateral circumflex

right popliteal

right anterior tibial

right
posterior tibial

right peroneal

right dorsalis pedis

right dorsal arch

right lateral plantar

right medial plantar

right plantar arch

a

b

Figure 20-23 Arteries of the lower extremity. (**a**) Anterior view. (**b**) Posterior view.

veins are thinner than the arterial walls and tend to collapse when empty. The thickness of the wall correlates very well with the pressure to which it is subjected.

The veins usually anastomose at Y-shaped junctions, becoming larger as they unite. Small branches of larger arteries, in contrast, usually come off at right angles, so that the smaller artery does not receive the full force of the pressure in the larger artery. Pressure is not a factor in the pattern of anastomosis of veins, because the venous pressure declines as the vein approaches the heart; so veins unite in patterns that do not retard blood flow.

The venous pattern is more variable, as a rule, than that of the arteries. This is particularly true of the superficial veins draining the skin, with each individual having a unique pattern of veins that may differ on the two sides of the body.

In some areas of the body there are two sets of veins draining the blood from the capillaries supplied by a single set of arteries. In the head, neck, and extremities, *deep veins* parallel the arteries supplying the area; *superficial veins* draining the same region are not usually accompanied by arteries. The deep and superficial veins communicate with one another freely, so that occlusion of a single vein has little effect upon venous return.

Veins may form an anastomosing network, termed a **venous plexus,** resembling a network of extremely large thick-walled capillaries. An example of this arrangement, called the *vertebral venous plexus,* is seen around the external covering of the spinal cord. Veins of a plexus have no valves, so blood can flow through them in either direction. Venous channels, called **venous sinuses,** may also be present; these are vessels with atypical walls containing connective tissue and no muscle. For example, the veins draining the brain empty into channels called *dural venous sinuses* between the two layers of the *dura mater,* the outer protective connective tissue covering of the brain. The dural sinuses converge at the base of the skull, emptying the blood into the large *internal jugular* veins, the major deep veins of the neck.

Veins of the Head and Neck The veins of the head and neck are illustrated in Fig. 20-24. The variously named *dural sinuses* are a major part of the deep venous drainage of the head, emptying into the internal jugular vein that parallels the common carotid artery in the neck. Some of the superficial veins of the head parallel branches of the external carotid artery, but they do not parallel arteries in the neck. As they descend into the neck, the superficial veins communicate with the internal jugular and anastomose superficially, forming the *anterior jugular* and the *external jugular* which join together to empty into the *subclavian vein* just superior to the clavicle. The subclavian vein from the superior extremity joins the internal jugular to form the *brachiocephalic vein.* The junction of the right and left brachiocephalics forms the *superior vena cava,* which conveys the blood to the *right atrium.*

Venous Drainage of the Extremities The major veins of both the upper and lower extremities occur as *deep* veins that parallel the arteries (Figs. 20-24 and 20-25). Typically, two deep veins accompany each major artery in close contact with it on either side and wrapped tightly by deep fascia. This close anatomical relationship is responsible for part of the force propelling the blood forward in the veins; the pulsation of the artery as blood courses through in spurts exerts external pressure on the veins, tending to collapse their walls. Because blood can move in only one direction due to the presence of valves, compression of the vein walls moves the blood toward the heart.

The *superficial* veins of the extremities (Fig. 20-25) form a highly variable network, readily visible in most people on the dorsum of the hand and anterior surface of the forearm. They are located in the superficial fascia just beneath the skin. There are frequent communications between the superficial and deep veins by means of *perforating* veins, which have valves arranged to prevent blood flow from deep to superficial veins. The veins of the upper extremity all flow into the subclavians, and those of the lower extremity empty into the external iliacs (Fig. 20-26).

The superficial veins that cross the *cubital fossa* of the upper limb are quite large and close to the surface, making this a suitable site for withdrawing blood by venipuncture or for injecting medication. Many medicines that are harmless to cells when diluted by the blood are toxic at the concentration used for injection. When such a substance is injected into a venous channel, it is rapidly diluted as it passes toward the heart and is reduced to a harmless level before it reaches the capillary beds. However, the same material injected into an artery may pass directly into the capillaries before being diluted sufficiently to render it nontoxic, with the result that the endothelium lining the capillaries may die, eliminating the microcirculation of the tissues. Deprived of a blood supply, the tissues soon become necrotic and are sloughed. This can happen when a substance such as the anesthetic nembutal is injected into the cubital fossa, because the brachial artery is in close proximity to the superficial vein usually selected as the injection site. The *cephalic* vein on the lateral side of the cubital fossa communicates with the *basilic* vein by means of a straight or Y-shaped vein that crosses the fossa anteriorly; this *median cubital vein* is the one usually used for injections. The exact pattern of the median cubital vein is highly variable; but as it passes medially, it usually crosses superficial to the aponeurosis of the biceps brachii muscle. The brachial artery passes just deep to the aponeurosis, so it is possible to insert the injection needle through the vein and

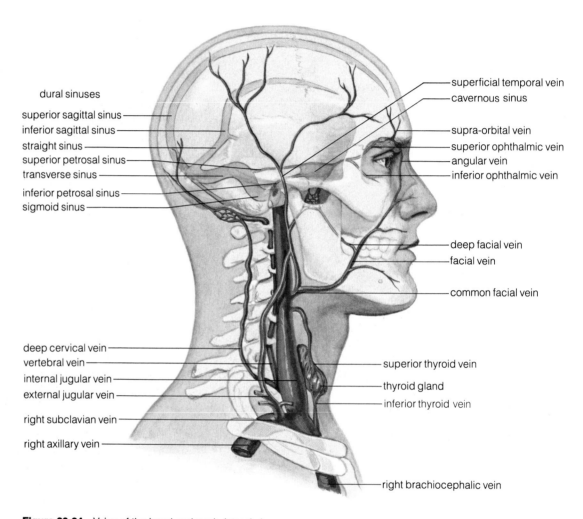

dural sinuses

superior sagittal sinus

inferior sagittal sinus

straight sinus

superior petrosal sinus

transverse sinus

inferior petrosal sinus

sigmoid sinus

deep cervical vein

vertebral vein

internal jugular vein

external jugular vein

right subclavian vein

right axillary vein

superficial temporal vein

cavernous sinus

supra-orbital vein

superior ophthalmic vein

angular vein

inferior ophthalmic vein

deep facial vein

facial vein

common facial vein

superior thyroid vein

thyroid gland

inferior thyroid vein

right brachiocephalic vein

Figure 20-24   Veins of the head and neck, lateral view.

very thin aponeurosis into the artery. Injection of nembutal into the brachial artery has resulted in loss of the entire hand.

Veins of the Trunk   The veins of the trunk are illustrated in Fig. 20-27. The veins draining the trunk wall above the dia-phragm empty into the *superior vena cava,* and those below the diaphragm drain into the *inferior vena cava.* Veins that parallel the arteries of the same names drain the thoracic wall. The veins paralleling the anterior intercostal arteries, which drain the anterior wall, join the *internal thoracic* vein, which

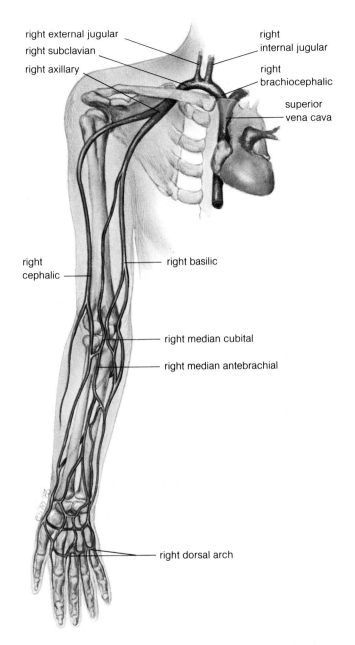

right external jugular
right subclavian
right axillary
right internal jugular
right brachiocephalic
superior vena cava
right cephalic
right basilic
right median cubital
right median antebrachial
right dorsal arch

Figure 20-25 Veins of the upper extremity. Most of the veins shown are superficial veins.

empties into the *brachiocephalic* vein. Veins that parallel the posterior intercostal arteries drain the lower intercostal regions of the posterior wall. Those of the right side empty into a vertically oriented vein, the *azygos,* usually lying to the right of the bodies of the thoracic vertebrae. On the left the posterior inter-

costal veins join another vertical vein, the *hemiazygos,* which ascends to the left of the vertebral column to about the level of the sixth thoracic vertebra. The upper intercostal spaces are drained by posterior intercostal veins that empty into the *accessory hemiazygos,* which descends to the middle of the thorax, where it joins the hemiazygos to cross the vertebral column and either opens into the *azygos* vein or remains separate as the two veins cross the vertebral column. The azygos courses superiorly to empty into the *superior vena cava* just before it enters the *right atrium.*

Veins of the pelvis anastomose to form the *right* and *left internal iliac* veins, which in turn unite with the *external iliacs* from the lower extremities to form the *common iliacs* which join to form the *inferior vena cava* that lies on the posterior abdominal wall to the right of the aorta. The inferior vena cava receives the *lumbar* veins from the body wall and the *paired visceral* veins from the gonads, kidneys, and suprarenal glands. (It should be noted that the *left testicular* or *ovarian* vein empties into the *left renal* vein and not directly into the inferior vena cava.) The inferior vena cava ascends on the posterior body wall and passes through the posterior aspect of the liver to penetrate the diaphragm.

When the inferior vena cava is narrowed by the development of a fold below the renal veins, the lower limbs may be drained through connections between the lumbar veins and the azygos system of veins in the thorax. Blood from the lower extremities passes through ascending lumbar veins that lie anterior to the transverse processes of the lumbar vertebrae and pass through the diaphragm to join the azygos or hemiazygos vein.

The Hepatic Portal System   The hepatic portal system (Fig. 20-27) consists of the capillary beds of the digestive tract, pancreas, and spleen, the veins connecting these beds to those of the liver, and the capillary beds of the liver. The veins draining the digestive tract, pancreas, and spleen (Fig. 20-28) drain into either the *superior mesenteric* vein or the *splenic* vein before these two veins join to form the *hepatic portal,* or *portal,* vein. Note that the names of these smaller veins do not correspond exactly to those of the arteries supplying the same organs. All of the blood from the unpaired abdominal viscera is taken by the hepatic portal vein to the liver, where it passes through the *hepatic sinusoids* and is collected into the two large *hepatic* veins that open into the *inferior vena cava* near the diaphragm. The *hepatic portal system* is essential in conveying nutrients from the digestive tract to the liver before they enter the general circulation.

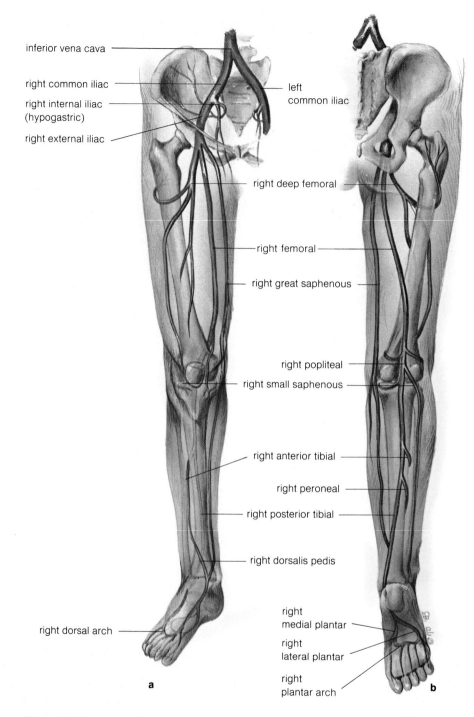

inferior vena cava

right common iliac

right internal iliac
(hypogastric)

right external iliac

left
common iliac

right deep femoral

right femoral

right great saphenous

right popliteal

right small saphenous

right anterior tibial

right peroneal

right posterior tibial

right dorsalis pedis

right dorsal arch

right
medial plantar

right
lateral plantar

right
plantar arch

a

b

Figure 20-26 Veins of the lower extremity. (**a**) Anterior view. (**b**) Posterior view.

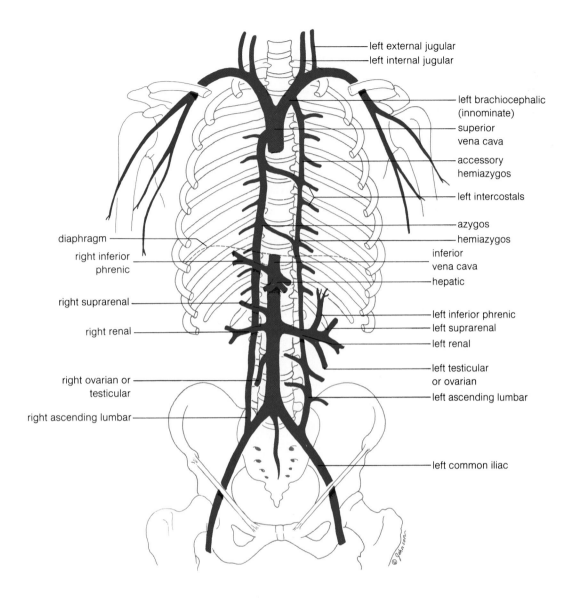

left external jugular
left internal jugular
left brachiocephalic (innominate)
superior vena cava
accessory hemiazygos
left intercostals
azygos
hemiazygos
inferior vena cava
hepatic
left inferior phrenic
left suprarenal
left renal
left testicular or ovarian
left ascending lumbar
left common iliac

diaphragm
right inferior phrenic
right suprarenal
right renal
right ovarian or testicular
right ascending lumbar

Figure 20-27   Veins of the thorax and abdomen.

THE LYMPHATICS

Lymph flow is *one-way* drainage (not a circulation) from the tissue into the blood vessels. Lymph capillaries begin blindly in the tissues (Fig. 20-29) and join an anastomosing network of lymph vessels that ultimately empty into the great veins carrying the blood (Fig. 20-30). Because fluid constantly diffuses from the blood stream in the capillary beds, the blood vessels would quickly become depleted of blood and the tissues swollen (edema) with excess fluid without a mechanism for returning the fluid to the circulation. The lymphatics provide a mechanism not only for returning fluid to the blood stream but also for supplying the blood tissue-manufactured substances and cells that do not penetrate the endothelium of the blood capillaries, or else penetrate only slowly.

Water and other small molecules such as urea and salts leave and enter the blood capillaries by *simple diffusion* according to the *concentration gradient* across the capillary lining. However, the rate at which both water and small molecules leave the capillaries is greatly increased by *hydrostatic pressure*

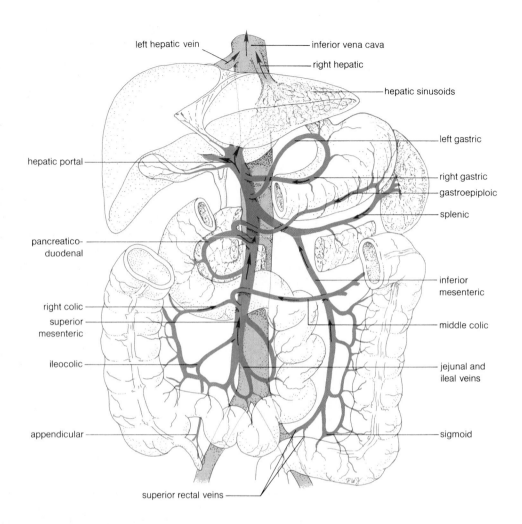

left hepatic vein

inferior vena cava

right hepatic

hepatic sinusoids

left gastric

hepatic portal

right gastric

gastroepiploic

splenic

pancreatico-
duodenal

inferior
mesenteric

right colic

middle colic

superior
mesenteric

ileocolic

jejunal and
ileal veins

appendicular

sigmoid

superior rectal veins

Figure 20-28 Veins of the hepatic portal system. Arrows show the direction of the flow of blood.

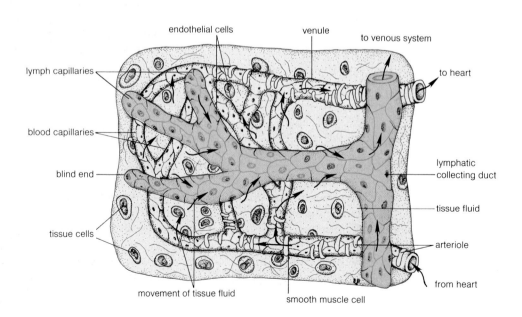

endothelial cells

venule

to venous system

lymph capillaries

to heart

blood capillaries

lymphatic
collecting duct

blind end

tissue fluid

tissue cells

arteriole

from heart

movement of tissue fluid

smooth muscle cell

Figure 20-29 The location and structure of lymph capillaries, collecting ducts, and lymphatics.

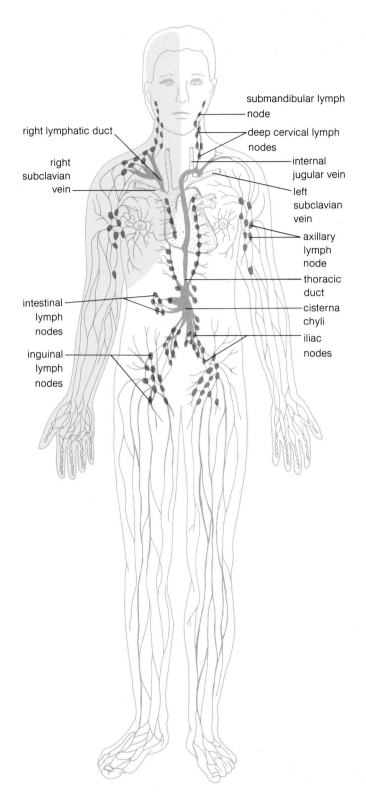

Figure 20-30 The location of the major lymphatic vessels and lymph nodes. The shaded area shows the portion of the body drained by the right lymphatic duct.

(compression pressure) caused by the pumping action of the heart. Hydrostatic pressure is high on the arterial side of the capillary bed, causing fluid to leave the blood in this region. The fluid loss means that blood on the venous side of the capillary bed has a higher concentration of large molecules such as plasma proteins than the blood on the arterial side, so some fluid from the tissues diffuses back into the capillaries of the venous side. However, more fluid leaves blood capillaries than is returned to them, and the excess fluid in the tissues diffuses into the lymph capillaries. In addition large molecules, such as some proteins elaborated by tissue cells, do not readily penetrate the endothelium of the blood capillary wall but pass into the lymph capillaries instead. The lymph capillaries converge, forming larger lymph vessels that carry lymph through specialized structures called lymph nodes before joining the blood stream. Lymph nodes, discussed in Chapter 25, occur in aggregates or chains along the lymph channels in specific locations such as the inguinal (groin) region.

The lymph vessels of the lower extremities follow the blood vessels into the abdomen where they join the lymph vessels draining the digestive tract. All these vessels converge in a dilated vessel on the posterior wall of the abdominal cavity called the *cisterna chyli.* The cisterna chyli marks the beginning of the *thoracic duct,* a single channel coursing up the midline of the posterior wall of the thorax. In the superior region of the thorax the thoracic duct is joined by a *left lymphatic vessel* from the left side of the head and neck and a vessel from the left upper extremity. After all three join, the thoracic duct opens into the blood stream near the junction of the left internal jugular and subclavian veins. On the right side, vessels draining the right half of the thorax, the right upper extremity, and the right side of the head and neck join together to form a single vessel called the *right lymphatic duct.* It opens into the junction of the right internal jugular and right subclavian veins. Thus the entire body inferior to the diaphragm and the left half of the body superior to the diaphragm drain into the thoracic duct and join the blood stream in the *left venous channel;* the right side of the head and neck, the right upper extremity, and the right thorax drain into the *right venous channel.*

DEVELOPMENT OF THE GREAT VESSELS

Prior to the partitioning of the heart chambers into right and left halves the blood leaves the heart tube through the *ventral aorta,* which is ventral to the gut and later gives rise to the ascending aorta. The blood passes into the *dorsal aorta* (the future *descending aorta),* dorsal to the gut, through a series of

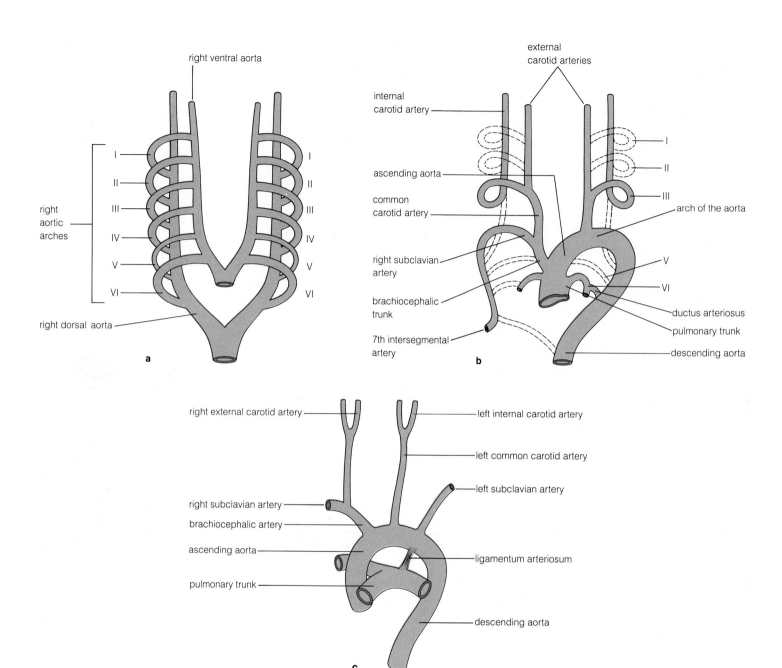

Figure 20-31 The fate of the aortic arches. (**a**) Diagram of the aortic arches before transformation into the definitive vascular pattern. (**b**) Diagram of the aortic arches after transformation; the obliterated components are indicated by broken lines. (**c**) The great arteries in the adult.

Clinical Applications: **Peripheral Vascular Disease**

Peripheral vascular disease is usually closely associated with coronary artery disease and heart function.

An essential function of blood vessels is to perfuse the tissues with blood, conveying oxygen and nutrients to the tissue cells and removing carbon dioxide and other wastes. The presence of a disease such as *arteriosclerosis* (a general term for thickening and loss of elasticity of the arteries) or of *atherosclerosis* (the deposition of fatty material in the walls of arteries so that the lumen is reduced) will interfere with blood flow. Any obstruction such as a clot will also interfere with blood flow and reduce the oxygen provided to the tissue cells. Any obstruction of a vein will retard venous return of blood to the heart, causing the accumulation of fluid in the tissues (*edema*). This will decidedly interfere with exchange in the tissues.

Shock

An extreme example of the failure of adequate perfusion of tissues occurs in the condition known as shock. A person in shock can deteriorate rapidly due to death of tissue cells. Shock is a clinical condition in which the blood flow in organs and vital tissues is inadequate to meet minimal requirements. The low blood flow leads to rupture of lysosome membranes and the subsequent release of enzymes that cause cell death. Shock is essentially inadequate capillary perfusion and is associated with low cardiac output and peripheral vasoconstriction. A characteristic symptom is cold skin. Many capillaries become occluded by clots, adding to the problem. Acids accumulate within the tissues and inactivate enzymes, increasing cell death. Patients in shock usually respond to blood transfusion.

Aneurysm

An aneurysm is a ballooning out of a small area of the wall of a blood vessel. An aneurysm may occur anywhere in the body but most commonly occurs in the aorta. A frequent site is just inferior to the origin of the renal arteries. An aneurysm of the arch of the aorta may occur slowly and has been known to dissect its way into the neck without causing severe pain. The danger of an aneurysm lies in the fact that as the wall of the artery pushes outward, it becomes progressively thinner and may eventually rupture. An aneurysm can be repaired surgically.

Hypertension

Elevation of the blood pressure above the normal is referred to as *hypertension*. This may be directly related to malfunction of the kidney (renal hypertension) or may be due to multiple extrarenal problems (essential hypertension).

Blood Diseases

The blood itself is subject to a wide variety of conditions that alter its capacity for carrying on normal functions. Some of the blood diseases are the *anemias,* in which there is a deficiency or abnormality of hemoglobin; the *purpuras,* characterized by bleeding into the skin and mucous membranes of the body; and *hemophilia,* in which there is a failure of the blood clotting mechanisms. In *leukemia,* a cancer of the blood-forming tissues, there is an abnormal increase in white blood cells in the bone marrow, in lymphoid tissue, and in the circulation. An unusual disease that resembles leukemia is *infectious mononucleosis,* which is caused by a virus and is characterized by an increase in numbers of one type of leukocyte in the blood. Unlike leukemia, infectious mononucleosis is self-limiting, that is, it is cured without treatment.

---

paired arteries, called the *aortic arches,* that pass on either side of the pharynx in the mesenchyme of the visceral arches (Fig. 20-31). There are a total of six aortic arches present during development, but only portions of three of them (3, 4, and 6) persist as major arteries of the adult. The ventral aorta develops as the ascending aorta, the brachiocephalic trunk, and the common and external carotid arteries; the third aortic arches become the internal carotids. The fourth arch on the left persists as the arch of the aorta and the sixth arch on the left becomes the pulmonary trunk. The fourth arch on the right becomes the right subclavian artery. In the embryo the *ductus arteriosus* connects the sixth arch (pulmonary artery) and the fourth arch (aortic arch); like the foramen ovale of the heart the ductus arteriosis shunts blood away from the lungs. At birth the ductus arteriosus collapses as the lungs expand, and it eventually becomes the *ligamentum arteriosum.*

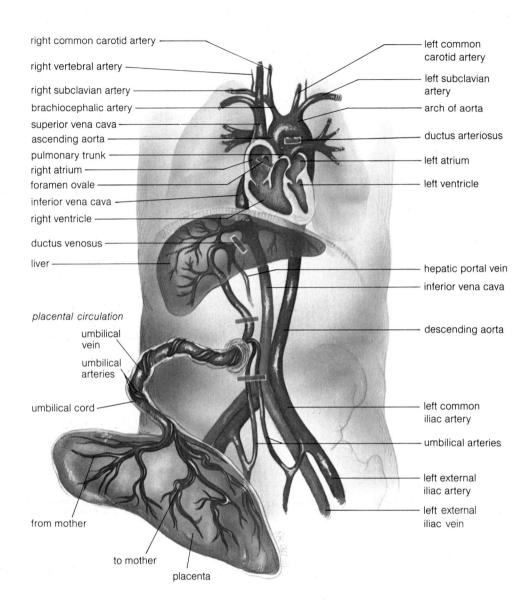

right common carotid artery

right vertebral artery

right subclavian artery

brachiocephalic artery

superior vena cava

ascending aorta

pulmonary trunk

right atrium

foramen ovale

inferior vena cava

right ventricle

ductus venosus

liver

placental circulation

umbilical vein

umbilical arteries

umbilical cord

from mother

to mother

placenta

left common carotid artery

left subclavian artery

arch of aorta

ductus arteriosus

left atrium

left ventricle

hepatic portal vein

inferior vena cava

descending aorta

left common iliac artery

umbilical arteries

left external iliac artery

left external iliac vein

Figure 20-32 The circulation of blood through a fetus. The bars across blood vessels indicate sites through which blood no longer flows after birth.

Fetal Circulation and Changes at Birth

In fetal life the internal iliac arteries send *paired umbilical* arteries up the inside of the anterior abdominal wall into the *umbilical cord* for distribution to the *placenta* where the blood is oxygenated (Fig. 20-32). Blood returns via a *single umbilical* vein into the *inferior vena cava*, which opens into the *right atrium*. Most of the oxygenated blood from the placenta passes through into the *left ventricle* to be pumped through the aorta for distribution to the tissues. At birth the patent foramen ovale is closed as the pressure of the two sides of the heart is equalized. Blood flow to the placenta ceases; the ductus arteriosus closes, increasing blood flow to the lungs; and the lungs take over the job of the exchange of gases.

WORDS IN REVIEW

Anastomose	Hepatic portal system
Aortic arch	Hypophyseal portal system
Arterio-venous shunt	Lymph
Bifurcate	Portal system
Circle of Willis	Thoracic duct
Cisterna chyli	Tunica adventitia
Collateral circulation	Tunica intima
Coronary artery	Tunica media
Coronary sinus	Vasoconstriction
Cubital fossa	Venous plexus
Ductus arteriosis	Venous sinus
Edema	

SUMMARY OUTLINE

I. The Blood Vessels
The walls of blood vessels are described as tunics: intima, media, and adventitia.
  A. The large arteries have much elastic tissue in the tunica media; the small arteries have muscle that controls diameter.
  B. The veins have a larger lumen and less muscle than arteries.
  C. The capillaries have endothelium only, which allows exchange by diffusion.

II. General Principles of Peripheral Circulation
  A. Blood vessels are named either according to location or according to the organ supplied.
  B. Variations in vascular pattern:
    1. "End" arteries have a single artery supplying a single capillary bed.
    2. In collateral circulation capillary beds are supplied by more than one artery.
    3. In an arterio-venous shunt there is a connection between the artery and the vein that bypasses the capillary bed.
    4. In a portal system a capillary bed connects two veins.

III. The Pulmonary Circuit
Blood in the pulmonary circuit passes in sequence through the right ventricle, pulmonary arteries, lungs, pulmonary veins, and left atrium.

IV. Arteries of the Systemic Circuit.
  A. Blood in the coronary circuit passes in sequence through the aorta, right and left coronary arteries, capillaries in cardiac muscle, coronary veins, coronary sinus, and right atrium.
  B. Arteries of the superior extremity include the subclavian with branches to the neck and thorax, which becomes the axillary with branches to the shoulder, and becomes the brachial with branches to the arm; it bifurcates into the radial and ulnar supplying the forearm and forms deep and superficial loops in the hand.
  C. The arteries of the head and neck include the common carotid, which bifurcates in the upper neck into external and internal carotids. The external carotid supplies the neck, face, and oral cavity; the internal carotid enters the cranial cavity.
  D. Blood is supplied to the brain by vertebrals that arise from subclavians that anastomose at the pons, forming the basilar which supplies the brainstem and cerebellum. The basilar terminates as the posterior cerebrals; the internal carotids contribute to the circle of Willis, which consists of the anterior cerebrals, anterior communicating, middle cerebrals, and posterior communicating arteries which join the posterior cerebrals.
  E. The arteries of the thorax include the thoracic aorta, which has paired posterior intercostals to intercostal spaces; the anterior chest wall receives blood from internal thoracic branches of subclavians; the thoracic viscera receive unpaired esophageal and bronchial arteries.
  F. Arteries of the abdominal viscera include paired branches of the abdominal aorta to kidneys and gonads and unpaired branches to the digestive tract.
  G. The arteries of the pelvis include the internal iliacs from the common iliac branches of the aorta.
  H. The arteries of the lower extremity include the external iliac, which sends branches to the trunk wall and becomes the femoral which supplies the hip and thigh; the femoral becomes the popliteal behind the knee and bifurcates into the anterior and posterior tibial to the leg. The posterior tibial gives off the peroneal artery to the lateral compartment and forms the plantar arch of the foot.

V. Veins of the Systemic Circuit
  A. Veins begin in capillary beds and carry deoxygenated blood toward the heart; blood flow is slower and under less pressure than in arteries. Usually two sets of veins drain capillaries supplied by a single artery.
  B. The veins of the head and neck include dural sinuses which drain blood from the brain into the internal jugular; superficial veins of the head and neck empty into the subclavian veins.
  C. The veins of the extremities include a deep set that parallels arteries and a superficial set in the superficial fasciae.
  D. The veins of the trunk above the diaphragm empty into the superior vena cava; below the diaphragm they empty into the inferior vena cava.

VI. Regulation of the Cardiovascular System

    A. Intrinsic regulation of the heartbeat is by the conduction system; extrinsic regulation is by parasympathetics that reduce the rate of contraction and sympathetics that increase it.

    B. Blood flow in peripheral vessels is regulated by vasoconstriction, which is controlled by sympathetic stimulation of the smooth muscle of arteries.

VII. The Lymphatics

    A. Lymph vessels begin as blind-ending capillaries in tissues.

    B. All lymph vessels below the diaphragm and from the left side above the diaphragm drain into the thoracic duct, which empties into the point of junction of the left internal jugular and the left subclavian veins.

    C. Lymph from the right side of the body above the diaphragm drains into the junction point of the right subclavian and right internal jugular veins.

VIII. Development of the Great Vessels

The great vessels develop from aortic arches connecting the ventral and dorsal aortae; the first and second arches degenerate, and the ventral aorta becomes the common carotid and external carotid; the third arch becomes the internal carotid; the fourth becomes the aorta; the sixth is the pulmonary artery.

IX. Changes in the circulation at birth.

At birth the foramen ovale closes, blood flow to the placenta ceases, and the ductus arteriosus closes, increasing blood flow to the lungs.

REVIEW QUESTIONS

1. What morphological characteristic of capillaries facilitates their performance of the function of filtration?

2. What is the morphological relationship of an "end" artery to its capillary beds? How does this affect the functioning of the beds?

3. Give an example of an area that has the potential for collateral circulation.

4. What is the principal role of arterio-venous shunts?

5. What is a portal system?

6. The pulmonary veins empty into what chamber of the heart?

7. The systemic circuit begins in what chamber of the heart?

8. The circumflex artery is a branch of which major artery that supplies blood to the heart?

9. What major branch of the common carotid contributes to the blood supply of the larynx and thyroid gland?

10. What are the principal morphological features of the pattern of distribution of arteries to the superior extremity?

11. What artery supplies blood to the lateral surface of the parietal lobe of the cerebral hemisphere?

12. The junction of what two veins forms the superior vena cava?

13. Name the visceral branches of the internal iliac artery.

14. Compare the pattern of distribution of the arteries of the inferior extremity with that of the superior extremity.

15. What is the role of the carotid sinus in the regulation of blood pressure?

16. What are the paired visceral branches of the abdominal aorta?

17. What lymph vessel conveys lymph from the lower part of the body to the blood stream?

SELECTED REFERENCES

Anderson, J. E. *Grant's Atlas of Anatomy.* 8th ed. Baltimore: Williams & Wilkins, 1983.

Goss, C. M. *Gray's Anatomy.* 29th ed. Philadelphia: Lea & Febinger, 1973.

Lachman, E., and Faulkner, K. K. *Case Studies in Anatomy.* New York: Oxford University Press, 1981.

Moore, K. L. *Clinically Oriented Anatomy.* Baltimore: Williams & Wilkins, 1980.

Romanes, G. C., ed. *Cunningham's Textbook of Anatomy.* 12th ed. New York: Oxford University Press, 1981.

Woodburne, R. T. *Essentials of Human Anatomy.* 7th ed. New York: Oxford University Press, 1983.

21

THE RESPIRATORY SYSTEM

OBJECTIVES

After completing this chapter you should be able to:

1. List in proper sequence the organs through which air passes to reach the lung alveoli.

2. Give anatomical boundaries of the nasal cavity and describe its location relative to the oral cavity and the pharynx.

3. Correlate the morphology of the mucous membrane lining the nasal cavity with its functions of warming and cleaning the air.

4. Describe the anatomical boundaries of the three parts of the pharynx and the positions of the parts relative to the nasal cavity, the oral cavity, and the larynx.

5. Describe the structure of the larynx that enables it to maintain an open airway, to voluntarily close the airway, and to produce sound.

6. Locate the trachea in relation to the larynx, esophagus, and primary bronchi. Describe the structure of the wall of the trachea and correlate its structure with the functions of maintaining a patent airway and cleaning the air as it passes.

7. Describe the arrangement of the hyaline cartilage in the bronchi located within the substance of the lung.

8. Locate the lungs relative to the chest wall, the heart, and the diaphragm. Describe the location of each lung. Define a bronchopulmonary segment.

9. List the structures found in the root of the lung.

10. Discuss the significance of elastic fibers in the walls of the alveolar ducts.

11. Describe the anatomical (cellular) components of the alveoli, through which gases must pass for exchange between the air and the blood.

12. Discuss the role of the macrophages in the retention of particulate matter within the substance of the lungs.

13. Describe the layers of the pleura in relation to the lung, chest wall, diaphragm, and mediastinum. Define a pleural recess and discuss its role in breathing.

14. Describe the embryonic development of the lower respiratory tract and the pleural cavities.

Respiration is the exchange of gases (oxygen and carbon dioxide) between an organism and its environment. The term includes

ventilation, or *breathing,* meaning the inhalation and exhalation of air; *external respiration,* the exchange of gases between the air and the blood in the lungs; *internal respiration,* the exchange of gases between the blood and the tissue cells; and *cellular respiration,* the utilization of oxygen by the cell to produce energy (and carbon dioxide). The respiratory system includes the organs of ventilation and external respiration. The structures for internal respiration were discussed in Chapter 19, and cellular respiration was presented in Chapter 2.

The organs of ventilation and external respiration form a system of conduits, or air passages (Fig. 21-1), consisting of the nose, pharynx, larynx, trachea, and bronchi, through which air passes into the lungs, the organs in which gas is exchanged between the atmosphere and the blood. Each respiratory organ is modified structurally to meet one or more of the following requirements of respiration: (1) The air passages must be kept open most of the time. (2) The air should be cleansed, warmed, and moistened by the time it reaches the delicate lung tissues. (3) A mechanism must be provided for the bellows-like action of the thorax in alternately drawing air into the lungs and expelling it. (4) Close contact must be maintained between the air spaces of the lungs and the blood vessels; the lining of both components must be structurally suited to the passage of gases but must prevent the passage of particulate matter that might be harmful if allowed to enter the blood stream.

Though the primary functions of the respiratory system are to supply the body with oxygen and rid it of carbon dioxide, it has accessory functions as well. Temporarily closing the air passage (holding one's breath) increases intrathoracic pressure, providing a stable platform against which to brace the upper limbs (as in lifting) and the abdominal muscles (as in defecation or childbirth). Increased intrathoracic pressure also provides the force for the sudden expulsion of air that occurs in such defensive reactions as coughing and sneezing. The respiratory organs provide the anatomical basis for vocalization, an important function, though not essential for life.

THE UPPER AIR PASSAGES

The upper air passages include the nose, pharynx, larynx, and trachea. Air enters the respiratory tract at the nose and crosses the common air–food passage (the pharynx) to reach the larynx, located anterior to the digestive tract. The larynx and trachea lie anterior to the pharynx and esophagus, with the trachea ending in the thorax at about the level of the sternal angle by bifurcating into right and left primary bronchi, the tubes that enter the lungs.

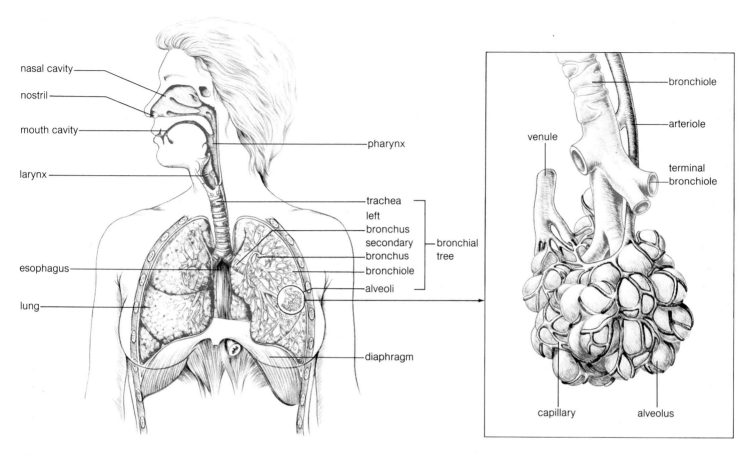

Figure 21-1 The respiratory system, anterior view. The inset is an enlarged diagram of alveoli and associated blood capillaries.

The Nose

The nose (Fig. 21-2) includes the visible *external nose* and the *nasal cavity,* considerably more extensive posteriorly. The external nose has two openings inferiorly, the **nostrils** (external nares), separated by the *nasal septum* and bounded laterally by the *alae* (wings, sing. *ala*) of the nose. It is supported by a skeletal framework composed of cartilages anteriorly and of bones in its posterior portion. The cartilages of the anterior nose consist of a *septal cartilage* medially and a variable number of *alar cartilages* laterally. Although there are voluntary muscles attached to the ala (Chapter 11), the cartilages prevent the nostrils from being closed by the nasal muscles. Thus the air passage cannot be closed by the action of voluntary muscles at this point.

The nostrils open into the vestibules that lead into the *nasal cavity,* which is divided into right and left halves by a median partition, the *nasal septum.* The nasal cavity opens posteriorly into the **nasopharynx** through the **choanae,** or internal nares, at the posterior end of the nasal septum. The nasal septum consists anteriorly of the *septal cartilage,* centrally and superiorly of the *perpendicular plate* of the *ethmoid,* and posteriorly of the *vomer.*

The very narrow roof of the nasal cavity is cartilaginous anteriorly; posterior to the cartilage the following bones are arranged in sequence from anterior to posterior: *nasal, frontal, cribriform plate* of the *ethmoid,* and the *body* of the *sphenoid.* The floor of the nasal cavity is the *hard palate,* formed anteriorly by the *palatine processes* of the *maxilla* and posteriorly by the *horizontal plate* of the *palatine bones.* The lateral walls are formed by parts of the *nasals, maxillae, lacrimals, ethmoid, palatines, sphenoid,* and *inferior nasal conchae.* There are three *conchae* (sing., concha), curved ledges projecting into the cavity from the lateral wall, each with a space beneath it termed a **meatus.** The superior and middle conchae are part of the ethmoid bone. Some of the bones contributing to the lateral nasal wall (maxilla, ethmoid, and sphenoid) contain large air-filled spaces communicating with the nasal cavities through openings

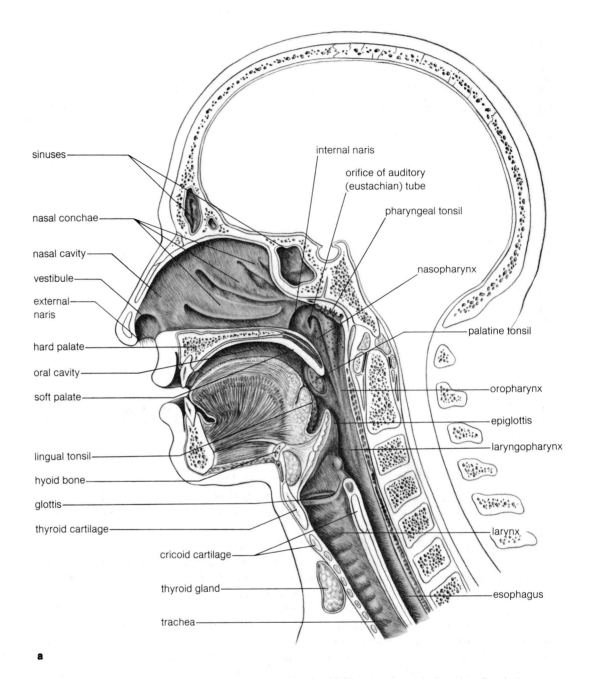

sinuses

internal naris

orifice of auditory
(eustachian) tube

pharyngeal tonsil

nasal conchae

nasopharynx

nasal cavity

vestibule

external
naris

palatine tonsil

hard palate

oral cavity

oropharynx

soft palate

epiglottis

laryngopharynx

lingual tonsil

hyoid bone

glottis

thyroid cartilage

larynx

cricoid cartilage

thyroid gland

esophagus

trachea

a

Figure 21-2 The nasal cavity, oral cavity, pharynx, and larynx. (**a**) Diagram of a sagittal section. *Top, facing page:* (**b**) X ray of thick sagittal section, to the right of the midline. Cadaver preparation. (From the University of Kentucky Department of Anatomy teaching collection.)

in the lateral wall. These are the *paranasal sinuses* (Fig. 21-3), which include the *frontal sinus* (in the frontal bone superior to the eye orbit), the *anterior ethmoid air cells* (ethmoid sinuses), and the *maxillary sinus,* all opening into the *middle meatus;* the *posterior ethmoid air cells,* which open into the *superior meatus;* and the *sphenoid sinus,* which opens into the *spheno-ethmoid recess,* the small space superior to the superior concha. The sinuses reduce slightly the weight of the facial bones and serve to add resonance to the voice but do not contribute significantly to respiration.

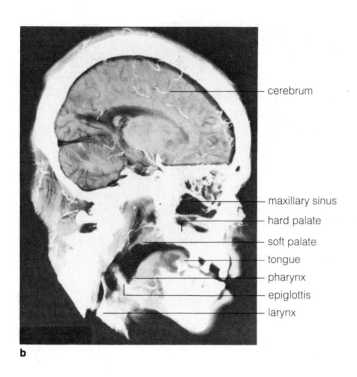

cerebrum

maxillary sinus

hard palate

soft palate

tongue

pharynx

epiglottis

larynx

b

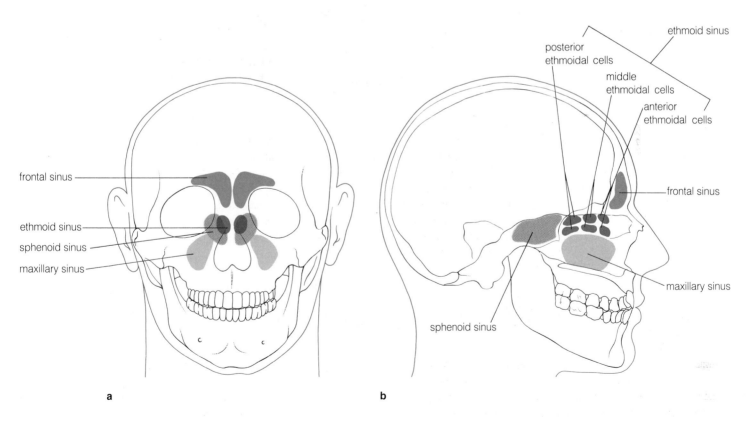

ethmoid sinus

posterior
ethmoidal cells

middle
ethmoidal cells

anterior
ethmoidal cells

frontal sinus

frontal sinus

ethmoid sinus

sphenoid sinus

maxillary sinus

maxillary sinus

sphenoid sinus

a

b

Figure 21-3 (below)   Diagram of the paranasal sinuses projected to the surface of the skull. (**a**) Anterior view. (**b**) Lateral view.

The skeletal framework of the nose provides a rigid support for the air passage, keeping it open and providing a strong base for the mucous membrane that lines the nasal cavity. The irregularities of the lateral wall, the conchae, not only serve to increase the surface area of the mucous membrane but also cause the air to swirl as it passes through the cavity, increasing the contact between the air and the mucous membrane. This contact is essential for removing particulate matter and for warming and moistening the air, functions for which the mucous membrane is adapted.

The epithelium of the mucous membrane begins inside the nostrils as an abrupt transition from the stratified squamous epithelium of the skin to the pseudostratified ciliated columnar epithelium that lines much of the respiratory tract (Fig. 21-4). At the nostrils the skin epithelium bears long hairs that may contribute to cleaning the air by blocking large particles. Within the nasal cavity the swirling air brings particulate matter into contact with the sticky mucus on the surface of the lining. The mucus is produced by numerous goblet cells scattered throughout the epithelium and by mucous glands in the lamina propria of the membrane. The particles trapped in the mucus are carried to the oral pharynx by a current created by the beating of cilia along the membrane surface. When the mucus reaches the oral pharynx it is swallowed, and foreign particles are either digested in the intestine or eliminated from the body in the feces.

Infections, such as the common cold, cause the mucous membranes to swell, which may occlude the sinus opening. Failure of the sinuses to drain properly leads to sometimes painful accumulation of fluid within the sinuses. The openings of the maxillary and sphenoid sinuses are in a position better suited for drainage when the body is on all fours, rather than upright; that is, these openings are on the superior aspect of the anterior wall of the sinuses with the body upright but are on the inferior aspect of the wall with the body on all fours.

The warming of the air that passes through the nasal cavity is due principally to the unique vascular pattern of the lamina propria, the thin connective tissue sheet of the mucous membrane that adheres tightly to the underlying periosteum. Within the lamina propria, blood flows through dilated thin-walled vessels, called venous sinuses, bringing a large quantity of blood close to the membrane surface. The blood is a constant source of heat, warming the air and enabling it to pick up moisture as its temperature increases (since warm air holds more water than

cold air). Thus the air is cleansed, warmed, and moistened as it passes through the nasal cavity, an air passage that is always kept open by its skeletal framework.

The mucous membrane is modified in the upper nasal cavity to form a small patch of *olfactory epithelium* immediately below the cribriform plate of the ethmoid (Fig. 17-1). The olfactory epithelium contains nerve cells capable of responding to chemicals in the air, relaying impulses that are interpreted by the brain as *odors*. Though the sense of *smell* serves other functions as well, it may be considered part of the body's defenses, warning when the air being breathed contains potentially harmful substances. The nasal mucous membrane is continuous with the mucous membranes lining the paranasal sinuses at their openings into the nasal cavity. The mucous produced by the membranes of the sinuses is added to that of the nasal cavity by ciliary action within the sinuses as long as the sinus opening remains patent (open).

The Pharynx as an Air Passage

As the air passes through the choanae at the posterior end of the nasal cavity, it enters the *nasopharynx,* a wide, muscular tube (part of the pharynx) superior and posterior to the soft palate (Fig. 21-2). The lateral walls of the nasopharynx contain the openings of the *auditory tubes* by which the middle ear cavity communicates with the pharynx. Opening the auditory tube during the process of swallowing makes the air pressure within the middle ear cavity the same as that of the external environment. The mucous membrane (Fig. 21-4) of the nasopharynx is of the respiratory type (pseudostratified ciliated columnar), with a lamina propria containing an abundance of elastic fibers that condense to form an elastic membrane in the relative position of the muscularis mucosa. In the midline of the posterior nasopharyngeal wall, just below the sphenoid sinus, the lamina propria contains an accumulation of lymphatic tissue, the **pharyngeal tonsil.** When enlarged, it is given the name *adenoids* and may partially obstruct the upper air passage, leading to mouth breathing which may alter the growth patterns of the face in children. External to the lamina propria the skeletal muscles provide support for the pharynx and keep it open. These muscles are active during the passage of food but do not contract during the passage of air. The pharynx serves only as a tube through which the air passes, entering the larynx in much the same condition as it left the nasal cavity.

The Larynx

The larynx (Fig. 21-5) is the respiratory organ between the pharynx and the trachea. It serves as a valve guarding the lower air passages against foreign particles, maintains an open airway,

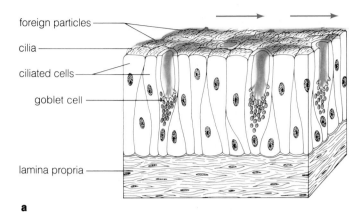

foreign particles
cilia
ciliated cells
goblet cell
lamina propria

a

Figure 21-4 Respiratory epithelium. **(a)** Diagram of pseudostratified ciliated columnar epithelium that is typical of the respiratory tract. **(b)** Scanning electron micrograph of the surfaces of ciliated tracheal epithelium of the monkey, × 15,000 (courtesy of M. Greenwood and P. Holland). **(c)** Photomicrograph of the ciliated pseudostratified columnar epithelium of the rabbit trachea (courtesy of W. K. Elwood).

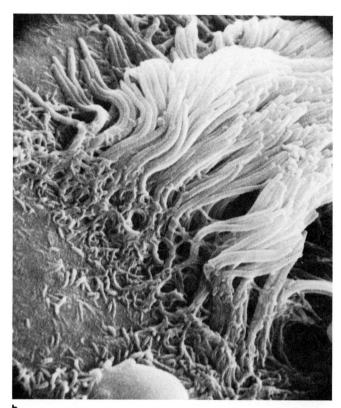

b

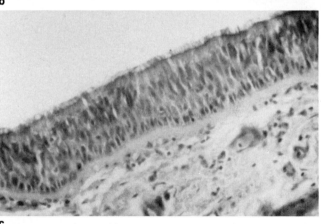

c

provides a mechanism for temporarily closing it, and provides for sound production. The superior opening of the larynx, termed the **aditus,** cannot be closed voluntarily, but swallowing pushes the larynx upward, so that the epiglottis (epiglottic cartilage) closes the opening to keep food from entering the larynx, as described in Chapter 22. You can demonstrate elevation of the larynx by placing your fingers on your laryngeal prominence (Adam's apple) and swallowing.

The cartilages of the larynx (Fig. 21-5) provide skeletal support, assuring an open airway. There are three unpaired cartilages—the epiglottic, thyroid, and cricoid—and three paired cartilages—the arytenoid, corniculate, and cuneiform. The thyroid, cricoid, and arytenoids are of *hyaline cartilage* and may undergo calcification, or ossification, with advancing age, but flexibility of these cartilages is not essential for life. The others are of *elastic cartilage,* which remains flexible throughout life. Flexibility of the *epiglottis* helps prevent food from entering the larynx during swallowing. The *thyroid* (Gr., *thyreos* = shield, *eidos* = shape) *cartilage* is composed of two plates, or *laminae,* fused anteriorly but opened posteriorly. The laminae fail to fuse superiorly in front, leaving a palpable gap, the *superior thyroid notch.* The posterior border of each lamina extends upward as the *superior cornu* (horn) and downward as the inferior cornu. The thyroid cartilage is attached superiorly to the hyoid bone by means of a fascial sheet, the **thyrohyoid membrane,** and inferiorly to the cricoid cartilage by the **cricothyroid membrane.** The cricoid is a complete ring and articulates with the

small, paired *arytenoid cartilages* resting on its superior surface. The *vocal folds* (vocal cords) attach to the arytenoids posteriorly and extend anteriorly to attach to the thyroid cartilage (Fig. 21-6). The cricoid and thyroid cartilage maintain an open airway at all times unless the passage is closed by bringing the vocal folds together.

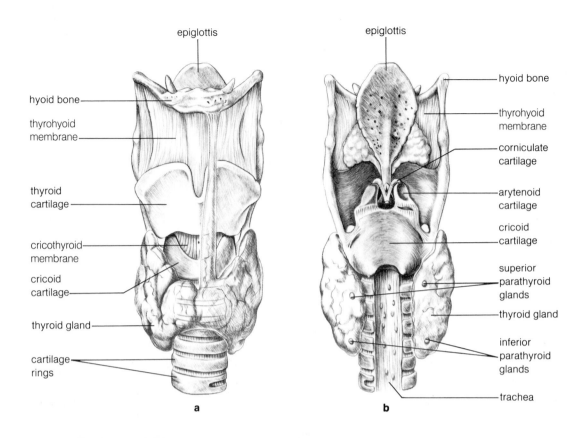

epiglottis

hyoid bone

thyrohyoid
membrane

thyroid
cartilage

cricothyroid
membrane

cricoid
cartilage

thyroid gland

cartilage
rings

a

epiglottis

hyoid bone

thyrohyoid
membrane

corniculate
cartilage

arytenoid
cartilage

cricoid
cartilage

superior
parathyroid
glands

thyroid gland

inferior
parathyroid
glands

trachea

b

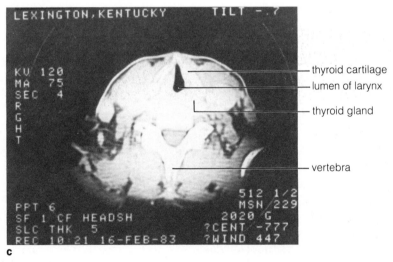

LEXINGTON,KENTUCKY   TILT -.?

KV 120
MA 75
SEC 4
R
G
H
T

PPT 6
SF 1 CF HEADSH
SLC THK 5
REC 10:21 16-FEB-83

512 1/2
MSN 229
2020 G
?CENT -777
?WIND 447

thyroid cartilage

lumen of larynx

thyroid gland

vertebra

c

Figure 21-5 The larynx. (**a**) From the anterior surface. (**b**) From the posterior surface. (**c**) CAT scan of a transverse view of the neck at the level of the larynx (courtesy of S. Goldstein)

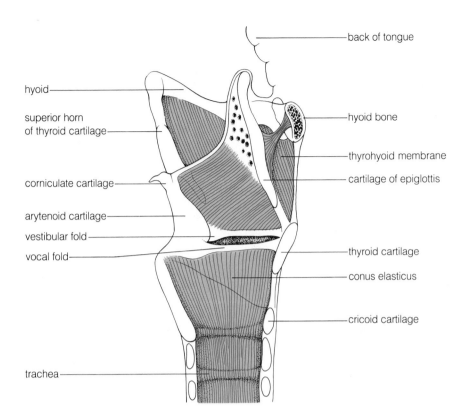

Figure 21-6 Internal view of the left half of the larynx.

Labels for Figure 21-6:
- back of tongue
- hyoid
- superior horn of thyroid cartilage
- corniculate cartilage
- arytenoid cartilage
- vestibular fold
- vocal fold
- trachea
- hyoid bone
- thyrohyoid membrane
- cartilage of epiglottis
- thyroid cartilage
- conus elasticus
- cricoid cartilage

The thyroid is the largest of the laryngeal cartilages and is responsible for the laryngeal prominence, commonly called the Adam's apple, which is larger in the male than in the female. This sex difference is not apparent until puberty, at which time rapid growth of the thyroid cartilage in the male lengthens the vocal folds, causing the initially poorly controlled voice change to a deeper pitch.

The *vocal folds* are structures whose anatomy does not really suggest their unusual capabilities. Viewed from inside the larynx (Fig. 21-7) they appear as rather thick folds in the mucous membrane, with a slit, the *glottis,* between them. Most of the larynx is lined by pseudostratified columnar ciliated epithelium typical of the respiratory tract, but the vocal folds are covered at the surface by stratified squamous epithelium. The vocal folds form the inferior edge of lateral extensions of the lumen of the larynx, called the ventricles, and do not appear to differ greatly in morphology from the **vestibular folds** (false vocal folds) located at the superior edge of the *ventricles* and separating them from the *vestibule*. However, the vocal folds contain an

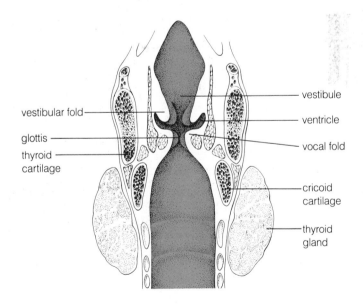

Figure 21-7 Diagram illustrating the shape of the lumen of the larynx.

Labels for Figure 21-7:
- vestibular fold
- glottis
- thyroid cartilage
- vestibule
- ventricle
- vocal fold
- cricoid cartilage
- thyroid gland

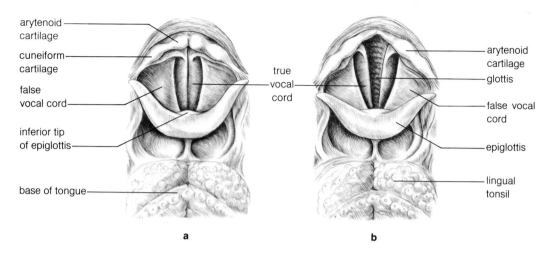

Figure 21-8 A superior view of the larynx, showing the vocal cords. The glottis is closed in (**a**) and open in (**b**)

elastic ligament, the *vocal ligament,* which may be regarded as the superior border of a sheet of elastic tissue, the **conus elasticus,** extending superiorly from the cricoid cartilage in the wall of the larynx. The elastic vocal ligament may be tightened or slackened, in short segments or throughout its entire length, by the short fibers of the vocalis muscle within the vocal folds and by the action of the larger intrinsic muscles of the larynx as they manipulate the laryngeal cartilages. The alterations in the length and tension of the vocal folds influence the pitch of the sound produced as the vocal folds are vibrated by the passage of air. Other structures, such as the tongue, are involved in the final sound produced.

The production of sound is not essential for life, but the vocal cords do have a vital role in activities essential in daily living. They may be brought together (adducted) to close the airway or separated (abducted) to open it (Fig. 21-8). The importance of closing the airway temporarily may perhaps be best emphasized by the fact that five muscles are arranged to close the openings whereas only one pair of muscles opens it. The larynx is the only place the air passage can be closed by the action of voluntary muscles, and it is sometimes essential to close it to prevent the entrance of water or of noxious substances in the air. The passage must also be closed to increase the intrathoracic pressure enough to provide a stable platform for the upper limbs. To understand the significance of this, try to pick up a heavy object while continuing to breathe (it is difficult). Maximum force can be achieved by inhaling, closing the vocal folds to prevent air from leaving the lungs, and contracting the abdominal and thoracic muscles to increase the pressure within the thorax. Then the weight can be lifted without collapse of the thorax, to which the force is transmitted from the limbs.

Other vital activities, such as coughing, sneezing, and defecation, are facilitated by closing the vocal folds to allow an increase in the intrathoracic or intra-abdominal pressure. The ability to close the vocal folds is also the last line of defense against foreign objects entering the lungs, for the airway cannot be closed below the larynx. Since the normal air flow into and out of the lungs is passive (that is, requires no effort), a single pair of muscles is adequate to open the folds with no pressure being applied. However, this creates a dangerous situation in the event of muscle spasm resulting from a drug reaction or local irritation. The muscles that close the vocal folds are much stronger than the single pair of muscles whose contraction abducts them. Thus laryngeal muscle spasm may block the airway, necessitating the insertion of a tube in the trachea (tracheotomy) or in the inferior portion of the larynx.

The Trachea

The *trachea* (Fig. 21-9) is a straight tube beginning at the lower border of the *cricoid cartilage* of the larynx and extending inferiorly, anterior to the esophagus, into the thorax, where it terminates at the level of the *sternal angle* by bifurcating into the right and left primary bronchi. It is a thin-walled, noncollapsible yet flexible tube (Fig. 21-10) lined with mucous membrane and supported by sixteen to twenty horseshoe-shaped hyaline cartilage rings that are open posteriorly. The mucous membrane is pseudostratified columnar ciliated epithelium, containing numerous goblet cells, and a lamina propria with an abundance of elastic fibers and many small mucous glands. The elastic fibers bind the tracheal rings together, allowing more flexibility than

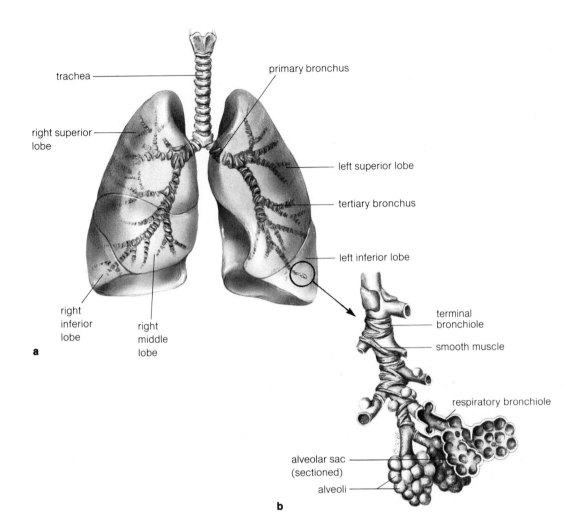

trachea

primary bronchus

right superior lobe

left superior lobe

tertiary bronchus

left inferior lobe

right inferior lobe

right middle lobe

a

terminal bronchiole

smooth muscle

respiratory bronchiole

alveolar sac (sectioned)

alveoli

b

Figure 21-9 The bronchial tree. (**a**) The major branches of the bronchial tree. (**b**) An enlarged portion of (a) with smaller branches of the tree.

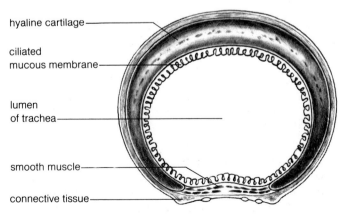

hyaline cartilage

ciliated mucous membrane

lumen of trachea

smooth muscle

connective tissue

a

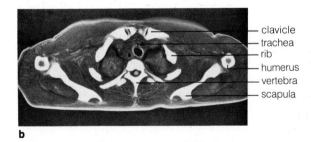

clavicle

trachea

rib

humerus

vertebra

scapula

b

Figure 21-10 The trachea. (**a**) A cross section through the trachea, showing the location of cartilage and other tissues. (**b**) X ray of a thick section of the thorax superior to the bifurcation of the trachea (from the University of Kentucky Department of Anatomy teaching collection).

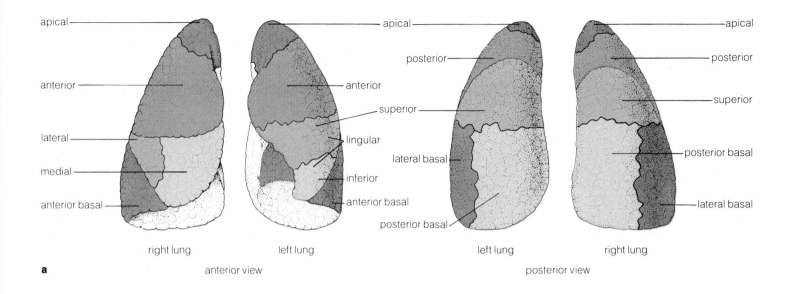

right lung   left lung

anterior view

left lung   right lung

posterior view

a

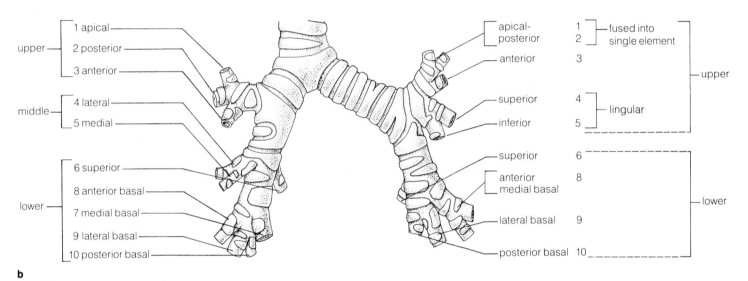

b

Figure 21-11 The bronchopulmonary segments, surface views with the segments shown on the lung surfaces and at the segmental bronchi.

would be possible with collagenous fibers. The mucus provides the sticky fluid for trapping particles to be moved towards the mouth by ciliary action. Posteriorly, in the area of contact between the trachea and esophagus, the gap in the tracheal ring is filled in by smooth muscle, providing the flexibility needed to permit the passage of a bolus of food down the esophagus, indenting the tracheal wall as it passes but without shoving the trachea forward. The smooth muscle fibers are mostly oriented trans- versely (trachealis muscle) but cannot close the lumen; the tra- chea always remains open unless blocked by trauma (injury).

> The trachea is easily seen on x-ray films, for it contains air that is more radiotranslucent than adjacent structures. In old people the tracheal rings may calcify, making it more difficult to interpret the x ray.

THE LUNGS

The *lungs* are paired organs filling most of the thorax on either side of the heart, with each lung receiving a primary bronchus entering the substance of the lung at the *hilus,* or **root of the lung** (Fig. 21-9).

Until it enters the substance of the lung, each bronchus is identical in structure to the trachea, but after it enters the lung, the incomplete cartilage ring becomes broken up into irregular plates of hyaline cartilage.

> The right bronchus is wider and descends more vertically than the left. This accounts for the greater tendency for foreign bodies to fall into the right bronchus.

The *primary bronchi* divide into *secondary bronchi,* each of which supplies a *lobe* of the lung (Fig. 21-9). The secondary bronchi branch into *tertiary bronchi,* which supply units of lung tissue called **bronchopulmonary segments** (Fig. 21-11). The right lung is composed of ten segments, whereas the left is usually considered to be divided into nine segments.

> Bronchopulmonary segments are important clinically, because a tumor or an abscess may be localized in a single segment. A bronchopulmonary segment may be removed surgically without great harm to the surrounding lung tissue.

The bronchi continue branching in the lung; those with a diameter of about 1 mm or less are called *bronchioles.* The bronchioles, like the bronchi, are lined with ciliated epithelium, but they have no cartilage in their walls. The bronchioles continue to branch, finally ending as *respiratory bronchioles,* from which a few thin-walled air sacs called **alveoli** project (Fig. 21-12). The respiratory bronchioles are the last parts of the bronchial tree that contain cilia; they terminate by branching into two or more highly distensible **alveolar ducts,** into which many alveoli open. Most of the alveoli open directly into the ducts, but some open indirectly by means of spaces called *atria,* each of which bears several alveoli. The wall of the alveolar ducts contains elastic fibers coursing longitudinally in a spiral fashion, so that the fibers are stretched during inspiration and recoil to their unstretched state upon relaxation of the respiratory muscles, providing much of the force that expels air from the lungs.

Blood capillaries surround the alveoli, and gas exchange occurs through the epithelia of the alveoli and the adjacent capillaries (Fig. 21-1). The alveolar epithelium contains two cell types that form a complete lining of the air space: *Type I cells* are thin, simple squamous cells through which the gases diffuse readily; *Type II cells* are cuboidal cells that secrete a substance called **surfactant,** which decreases surface tension at the air–cell interface, preventing the collapse of the alveoli. Gases must pass through the squamous cells lining the alveolus, the intercellular fluid of the thin connective tissue of the alveolar wall, and the squamous epithelium (endothelium) of the capillary in order to enter the blood. Oxygen diffuses from the air into the blood, and carbon dioxide diffuses from the blood into the alveolar air to be discharged into the external environment.

The connective tissue of the alveoli in which the capillaries are embedded contains *macrophages* capable of ingesting particles that penetrate the alveolar lining. The accumulation of carbon particles in lung macrophages, chiefly as a result of smoking, causes the lungs to change from the normal pink color due to the blood circulating in the capillaries to gray or black as the accumulation of carbon masks the color. When carbon particles have entered the lung tissue, there is really no effective means of eliminating them, because as the macrophages die, they release their particulate matter into the intercellular fluid of the alveolus, where other macrophages ingest it. Thus the particulates remain in the substance of the lungs. Smoking also inhibits ciliary action and may even destroy cilia, thereby allowing potentially harmful substances to reach the lungs. The inability of the lungs to rid themselves of foreign particles makes it essential that the air we breathe be kept as free as possible from pollutants that are not biodegradable.

THE PLEURA

The lungs are enclosed within a double-layered sac of serous membrane called **pleura** (Fig. 21-13). The outer layer, the *parietal pleura,* lines and is fused with the thorax wall, the diaphragm, and the fibrous pericardium of the heart. The inner layer of serous membrane, the *visceral pleura,* is directly applied to and fused with the lung surface, following the curvature of the lungs and dipping into the fissures between the lobes. The parietal and visceral pleurae are continous with one another at the root of the lung, where the bronchi and pulmonary blood vessels enter the lung substance. The potential space between these two layers of pleurae constitutes the *pleural cavity* and contains only a thin film of serous fluid moistening the membranes, enabling them to slide on one another without friction during inspiration and expiration. The potential space may

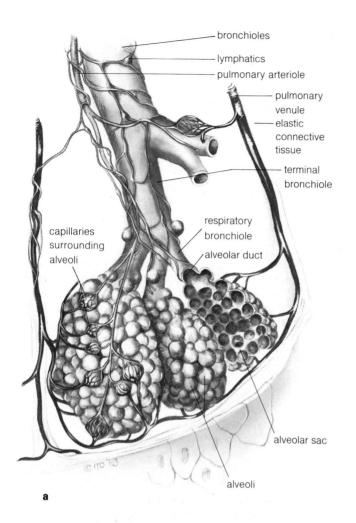

bronchioles

lymphatics

pulmonary arteriole

pulmonary venule

elastic connective tissue

terminal bronchiole

respiratory bronchiole

alveolar duct

capillaries surrounding alveoli

alveolar sac

alveoli

a

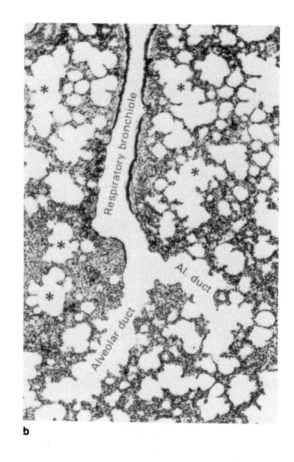

Respiratory bronchiole

Alveolar duct

Al. duct

b

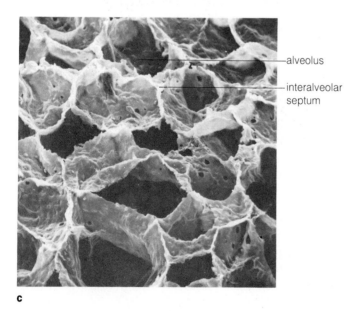

alveolus

interalveolar septum

c

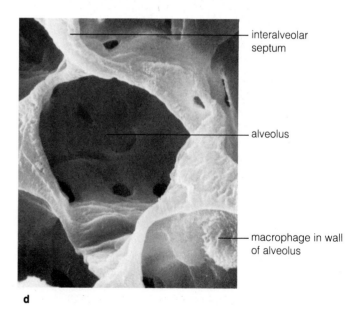

interalveolar septum

alveolus

macrophage in wall of alveolus

d

(Facing Page) **Figure 21-12** (**a**) The branching of terminal bronchioles to form the respiratory membranes and the arrangement of blood and lymph vessels supplying these tissues. (**b**) A very low-power photomicrograph of the lung of a young child. The respiratory bronchiole is cut longitudinally and may be seen opening into two alveolar ducts. The asterisks indicate alveolar sacs, which open into round sacs called alveoli. (Photomicrograph reproduced with permission from A. W. Ham, *Histology*, J. P. Lippincott, 1979.) (**c**) Scanning electron photomicrograph of a section of monkey lung ×300. (**d**) Scanning electron photomicrograph of a higher magnification of a single alveolus of a section of monkey lung ×2200 (courtesy of M. Greenwood and P. Holland).

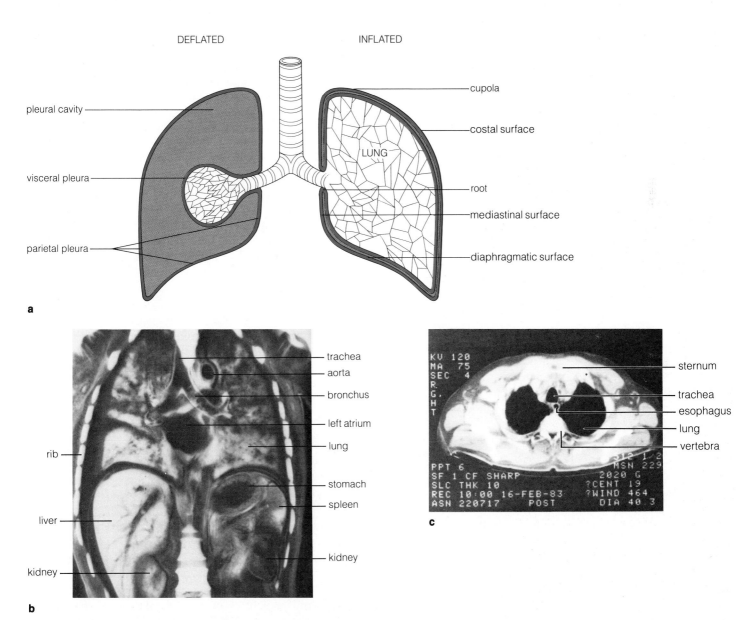

(Above) **Figure 21-13** The lungs within the thorax (**a**) Diagram illustrating the relations of the pleura. The right lung is shown collapsed to illustrate the pleural cavity. (**b**) X ray of a thick frontal section of the trunk, cadaver preparation (from the University of Kentucky Department of Anatomy teaching collection). (**c**) CAT scan of the thorax, transverse view (courtesy of S. Goldstein).

become a real space if filled with air (pneumothorax), blood (hemothorax), or pus (empyema).

> Fluid is usually drained from the pleural cavity by inserting a wide-bore needle into the 7th intercostal space posteriorly. Below the 7th intercostal space there is danger of running the needle through the diaphragm. The needle should be inserted along the superior border of the lower rib to insure that the intercostal nerves and blood vessels are not injured.

The pleurae are named according to the structures to which the parietal pleurae are fused: the pleura lining the diaphragm is known as the *diaphragmatic pleura,* with the *mediastinal pleura* next to the mediastinum and the *costal pleura* next to the rib cage. At the junction of the costal, mediastinal, and diaphragmatic pleurae (that is, at the junction of the thorax wall with the diaphragm), the parietal pleura is folded back on itself to form a **pleural recess,** not filled with lung, within the confines of the pleural cavity. It is into the pleural recesses that the lungs expand during inspiration; in all other areas the visceral pleura is in contact with the parietal pleura, with contact maintained principally by tension between the two moist membranes and by the negative pressure within the pleural cavity.

> Since the parietal pleura is segmentally innervated by somatic intercostal nerves, inflammation (pleurisy) results in pain that is usually sharply localized. It may, however, be felt in the thoracic wall or, in the case of the lower nerves, in the anterior abdominal wall.

THE MECHANICS OF BREATHING

Inspiration is the result of expansion of the thorax, which increases the volume of the air spaces within the lungs, causing air to enter through the open air passages. Because the airway opens externally, the pressure within the filled lungs is equal to that of the external atmospheric environment, except when the vocal folds are closed. The thorax is expanded in normal quiet breathing chiefly by contraction of the external intercostal muscles, which elevate the ribcage, increasing both the transverse and the anteroposterior diameters of the thorax, and by

contraction of the diaphragm, increasing the superior-inferior volume (Fig. 21-14). Inspiration is an *active* process requiring the expenditure of energy. However, *expiration* is a *passive* process accomplished by the elastic recoil of the tissue stretched during inspiration and does not require energy in quiet breathing. In *forced expiration,* muscles such as the internal intercostals and abdominal wall muscles may be used to further reduce the thorax, expelling air. This is an active process requiring energy.

The volume of air moved in and out of the lungs with each normal breath (about 500 ml) is called the *tidal volume.* The volume drawn into the lungs with the deepest possible inspiration is known as *inspiratory capacity* and averages about 3500 ml. The volume of air expelled in a forced expiration following a maximal inspiration is known as the *vital capacity* and is about 4700 ml. The *residual volume* is the amount of air left following a maximal expiration and is usually about 1200 ml. The *total lung capacity* is about 6000 ml in healthy young men and is approximately the sum of the vital capacity and the residual volume.

REGULATION OF THE RESPIRATORY SYSTEM

Respiration involves two movements: expansion of the thorax as a result of muscle contraction during inspiration and reduction in volume of the thorax as a result of the elastic recoil of the thoracic wall and the lungs as the muscles relax during expiration. Contraction of the muscles during inspiration is due to *neural stimulation;* relaxation results from sudden *inhibition* of the *neural* stimulation. The rate at which the nerves are alternately stimulated and inhibited is regulated by chemical changes in the blood, such as the oxygen and carbon dioxide concentrations and blood pH.

Quiet Breathing

The respiratory muscles such as the intercostals and the diaphragm are skeletal muscles that can be controlled voluntarily. You can hold your breath until you lose consciousness if you are determined to do so. However, we ordinarily do not think about breathing, and the process operates as a reflex, regulated by the respiratory centers in the *reticular formation* of the *medulla* and *pons.* The "excitatory" respiratory center initiates inspiration by sending impulses into the spinal cord, stimulating motor neurons of the *phrenic nerves* which supply the diaphragm and the neurons of the *intercostal nerves* which innervate the intercostal muscles. Contraction of these muscles expands the thorax wall, lowering the intrathoracic pressure and allowing air to enter the lungs until the pressure within the lungs is

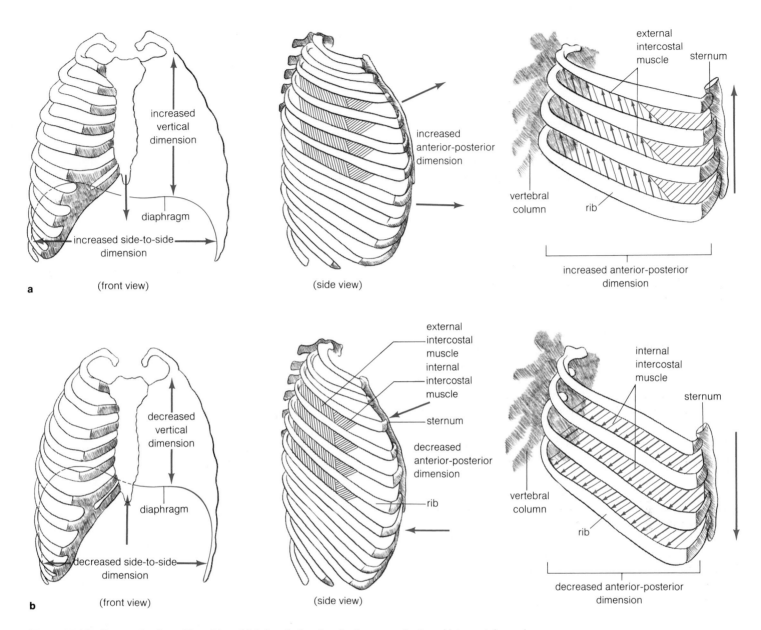

Figure 21-14 The mechanism of breathing. (**a**) In inspiration the diaphragm and external intercostal muscles contract. (**b**) In expiration those muscles relax, and in forceful expiration the internal intercostal muscles contract.

equal to the atmospheric pressure. As the lungs expand, the *small bronchioles* are stretched, initiating nerve impulses in *sensory fibers* of the *vagus nerve* that terminate on the bronchioles. These sensory impulses are conveyed by the vagus nerve to an "inhibitory" respiratory center in the medulla, *inhibiting* the discharge of impulses from the "excitatory" center. The muscles relax, and the thorax returns passively to the unex-

panded state, expelling air from the lungs. Reduction in the size of the lungs initiates sensory impulses in the vagus nerve that *activate* the *excitatory center* to begin the process of inspiration. This mechanism is known as the **Hering–Breuer reflex,** named for the two men who first described it, and is largely responsible for the regular pattern of quiet breathing. *Stretching the lungs stops inspiration; the unstretched lung starts inspiration.*

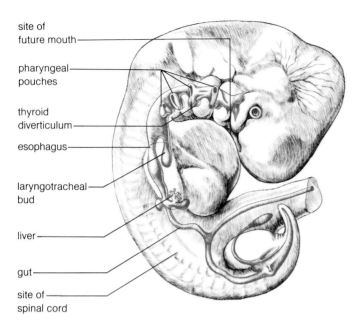

site of
future mouth

pharyngeal
pouches

thyroid
diverticulum

esophagus

laryngotracheal
bud

liver

gut

site of
spinal cord

Figure 21-15 Side view of an embryo early in the second month of development, showing the pharynx and laryngotracheal buds that contribute to the development of the respiratory system.

Factors Modifying the Rate of Respiration

Many reflexes initiated by stimuli in the mucous membrane lining the respiratory tract may disrupt the normal pattern of breathing. For example, irritants that cause a person to sneeze or cough initiate the response by halting breathing, allowing a buildup in intrathoracic pressure by muscle contraction prior to the sudden expulsion of air under pressure. These are complicated reflexes involving many muscles and nerves.

Other reflexes that influence breathing are initiated by *chemical changes* in arterial blood. A slight change in the *carbon dioxide* concentration in the blood is detected by special receptors in the **carotid body,** located in the wall of the internal carotid artery near its origin. A decrease in carbon dioxide increases sensory impulses from the carotid body to the respiratory center, resulting in a decrease in the rate of breathing; a rise in carbon dioxide concentrations results in more rapid breathing. Oxygen concentrations have similar, although opposite, effects on the same neural pathway, but the oxygen level in the blood must change much more than the carbon dioxide level in order to be detected by the carotid body. These reflexes are important in maintaining respiration at the proper level to remove carbon dioxide at the same rate it is produced in the tissues and to provide adequate oxygen for the activities of the body at any given time. Thus during periods of violent exercise

muscle contraction uses more oxygen and gives off more carbon dioxide than in periods of rest. As the carbon dioxide accumulates in the blood, the rate of breathing increases, providing increased airflow through the lungs, facilitating elimination of the excess carbon dioxide, and promoting the passage of more oxygen into the blood for distribution to the tissues.

In addition to having an effect upon the carotid body carbon dioxide in the blood has a direct effect upon receptor cells in the medulla near the respiratory center, so that a rise in carbon dioxide concentration in the blood reaching the medulla stimulates the excitatory center to fire more rapidly, thus increasing the rate of respiration.

DEVELOPMENT OF THE RESPIRATORY APPARATUS

The Lower Respiratory Tract

The larynx, trachea, bronchi, and lungs are ventral derivatives of the foregut. An outgrowth called the respiratory diverticulum (laryngotracheal bud) arises from the floor of the pharynx (Fig. 21-15) and is pinched off from the gut by a mesenchymal septum (the esophagotracheal septum), except in the area of the aditus where the larynx will open into the pharynx. Failure of this septum to develop properly may result in communication between the trachea and esophagus (**tracheoesophageal fistula**), a congenital malformation that is usually fatal unless surgically corrected shortly after birth.

The laryngotracheal tube elongates inferiorly and bifurcates to form two lung buds, which give rise to successive divisions of the bronchial tree (Fig. 21-16a). About seventeen orders of branches of the bronchial tree (Fig. 21-16b) develop during approximately the first six months of gestation, as blood vessels grow into the lung tissue in association with the bronchial branches. The air passages end blindly at the alveolar ducts; alveoli do not form until later (Fig. 21-16c). Until about six months of intrauterine age, the alveolar ducts are usually lined with cubiodal epithelium and do not permit sufficient gas exchange to support life if the baby is born at this stage. By six months the alveolar ducts begin to give rise to alveoli lined with simple squamous epithelium (Fig. 21-16d) and usually will sustain life if the baby is born two or three weeks after the start of the process. Type II cells also differentiate by six months and usually produce enough surfactant to prevent the collapse of the alveoli; the capillary networks associated with the air sacs are also usually adequate by this time. In the normal nine-month gestation period about twenty-four orders of branches are formed prior to birth, with alveolar–capillary membranes sufficiently thin to allow exchanges of gases immediately. At birth the lungs are filled with fluid that is cleared by expelling some of the fluid through the nose and mouth, and the remainder is absorbed into the blood stream and lymph vessels. The fluid is quickly replaced with air following rapid expansion of the lungs as the infant begins to breathe. Respiratory movements begin before birth at about five months of fetal age, drawing amniotic fluid in and out of the lungs, so the newborn baby is well prepared for the mechanical work of breathing. If death occurs from respiratory failure in premature infants, it is usually the result of inadequate differentiation of the alveoli, with the consequence of insufficient gas exchange to sustain life.

The Pleural Cavities and the Diaphragm

At the lung bud stage of development (four weeks) the body cavity is continuous lateral to the gut from the neck region, where the heart begins development (Fig. 21-17a), into the abdominal region, where the peritoneal cavity forms. A sheet of mesenchyme called the **septum transversum,** lying transversely at the caudal end of the heart on the cranial face of the yolk sac, partially separates the future thorax from the future abdomen. The septum transversum ends dorsally at the ventral edge of the foregut, leaving the embryonic coelom open on either side of the gut. These spaces connecting the thoracic (pericardial) cavity with the abdominal (peritoneal) cavity are the *pericardioperitoneal canals.*

The lung buds project ventrally into the visceral (splanchnic) *mesoderm* surrounding the esophagus and are located dorsal to the developing heart (Fig. 21-17b). As the lung buds expand rapidly, pushing into the *pericardial cavity,* they reach the lateral thoracic wall, causing it to erode in advance of the expanding lung. Cavitation of the thoracic wall creates folds of mesoderm, the pleuropericardial folds, which project medial and ventral to the lung, separating the lung from the heart. These folds soon fuse with the mesoderm ventral to the future roots of the lungs, completing the pleuropericardial membrane which separates the pericardial and pleural cavities (Fig. 21-17c). As the lungs continue to grow laterally and ventrally around the heart, the *pleuropericardial membrane* differentiates into the parietal and fibrous pericardium next to the heart and into the parietal (mediastinal) pleura adjacent to the lungs.

At the caudal end of the pleural cavities the pleuropericardial membrane extends ventrally into the pleuroperitoneal canals, completing the separation of the pleural cavities from the peritoneal cavities by fusing with the dorsal aspect of the septum transversum. The septum transversum becomes the *central tendon* of the diaphragm; muscle is contributed by the pleuroperitoneal membranes and the dorsal mesentery of the esophagus in the dorsal part of the diaphragm and by the body wall in the lateral and ventral portions of the diaphragm.

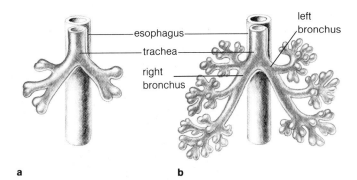

a

b

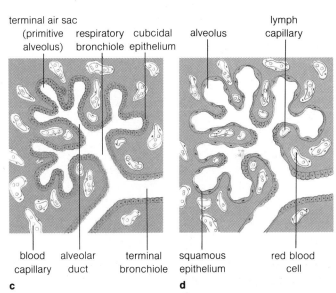

c

d

Figure 21-16 Development of bronchioles and alveoli. (**a**) Laryngotracheal buds at five weeks. (**b**) Further development of buds at eight weeks. (**c**) Cuboidal epithelium in cross section of alveoli at twenty-four weeks. (**d**) Squamous epithelium in cross section of alveoli at full-term development.

TOPOGRAPHICAL ANATOMY OF THE LUNGS

A knowledge of the topographical anatomy of the lungs is useful in locating *auscultation points* during a physical examination (auscultation is listening to the sounds of internal viscera) and in determining the location of tumors or sites of infection by means of x-ray films.

For descriptive purposes each lung may be considered to have an apex and three surfaces, costal, diaphragmatic, and mediastinal, named according to their proximity to other structures. The lines of junction of these three surfaces are described as the three borders of the lungs: the *anterior, posterior,* and *inferior borders.*

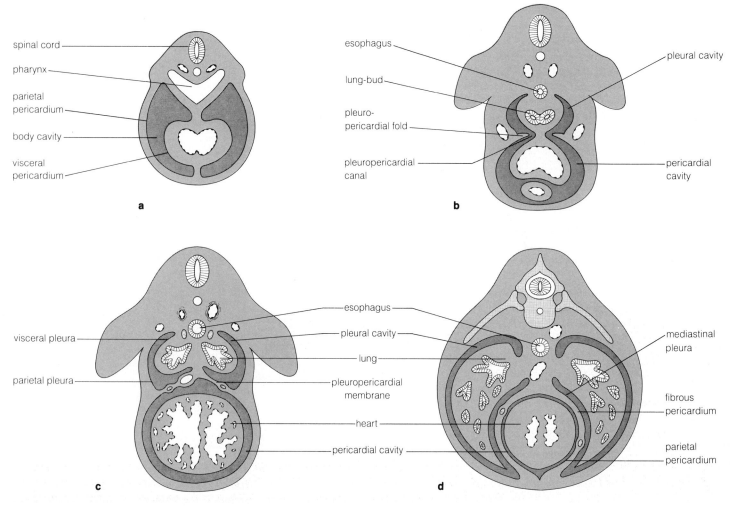

Figure 21-17 Development of the lungs and pleural cavity, transverse sections. (**a**) Relations of pharynx and heart prior to formation of lung buds. (**b**) Early development of the lung buds. (**c**) Separation of the pleural and pericardial cavities by the pleuropericardial membrane. (**d**) Differentiation of mediastinal pleura, fibrous pericardium, and parietal pericardium from the pleuropericardial membrane.

The apex extends into the neck about one inch above the medial third of the clavicle during quiet respiration (Fig. 21-18). In deep inspiration elevation of the ribcage raises the sternum, causing the sternal end of the clavicle to move superiorly. In quiet respiration the lung apex is protected anteriorly by the sternomastoid muscle which attaches to the sternum and clavicle. Posteriorly the apex is protected by the ribs and the muscles attached to the vertebrae as its summit reaches the level of the spine of the first thoracic vertebra.

The surfaces of the lung vary in their extent, depending upon the position of the body (upright or lying down) and upon the state of respiration (expiration or inspiration). The posterior border of the lung, at the posterior junction of the costal and mediastinal surfaces, is located about three-fourths of an inch from the midline and in quiet breathing extends from the level of the second thoracic vertebra to about the level of the eleventh thoracic vertebra. From that point the inferior border passes horizontally to join the anterior border at the sixth costal cartilage. Laterally the inferior border is well above the costal margin of the ribcage, crossing the eighth rib in the mid-axillary line. In deep inspiration the inferior border may descend at least two inches as the lung expands into the diaphragmatic pleural recess.

The anterior borders of the two lungs begin superiorly behind the sternoclavicular joints and approach one another as they lie behind the manubrium, remaining close together until

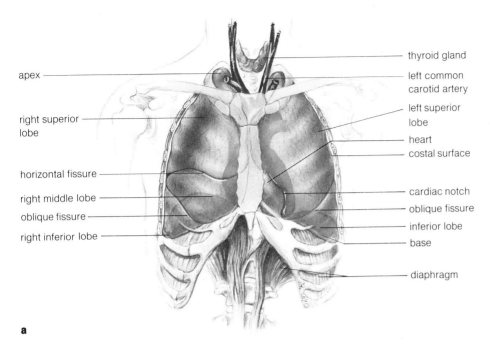

apex

right superior lobe

horizontal fissure

right middle lobe

oblique fissure

right inferior lobe

thyroid gland

left common carotid artery

left superior lobe

heart

costal surface

cardiac notch

oblique fissure

inferior lobe

base

diaphragm

a

Figure 21-18 **(a)** The structure and location of the lungs. **(b)** A chest x ray (x ray courtesy of Alexandria Hospital, Alexandria, Virginia, Joyce R. Isbel, R. T.).

b

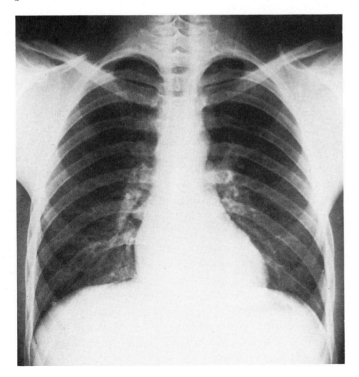

the level of the fourth costal cartilage is reached. Here the anterior border of the left lung swings laterally, forming the *cardiac notch* before reaching the inferior border of the lung at the sixth costal cartilage.

Each lung is divided into *lobes* by *fissures* that vary in depth among different individuals. The right lung usually has three lobes and the left has two, but other fissures may be present in either lung, creating an additional lobe. Each lung has an *oblique fissure,* beginning posteriorly at the level of the third thoracic vertebral spine and passing obliquely anteriorly along the line of the fifth intercostal space to end at the inferior border of the lung at the sixth costal cartilage. This fissure separates the upper and lower lobes of the left lung and the upper and middle lobes from the lower lobe of the right lung. The upper and middle lobes of the right lung are separated by a *horizontal fissure,* corresponding to a line drawn transversely at the level of the fourth costal cartilage.

Each lobe of the lung is supplied by secondary branches of the primary bronchi. The tertiary branches (segmental bronchi) supply a portion of lung tissue known as the bronchopulmonary segment (Fig. 21-11), which is not separated from other segments by fissures but can be removed surgically as a unit.

Clinical Applications: **Diseases of the Lungs**

The lungs are the only visceral organs of the body with direct exposure to the air of the external environment. The thin epithelial lining of the alveoli affords little protection against infectious agents or harmful substances present in the air. The greatest protection comes from the sticky mucus that traps foreign particles before they reach the alveoli. Once trapped in the mucus, particles such as bacteria are removed by ciliary action, which creates a current in the mucus flow toward the mouth, where the material is swallowed. Bacteria or viruses that do reach the alveoli frequently find the environment compatible with growth and produce clinical symptoms characteristic of the organism.

Pneumonia

Pneumonia is a general term for acute inflammation of the lungs, whether caused by infection or by irritation. The pneumonias are characterized by congestion (accumulation of fluid) of the lungs, leading to difficulty in breathing and in the exchange of gases. The most common bacterial diseases of the lung are the *bacterial pneumonias,* usually caused by pneumococcal infections but sometimes by streptococcus or staphylococcus. The pneumococcus group of bacteria includes about 75 recognized types that produce similar symptoms, and infection with one does not produce immunity to all of the others.

Pneumonia may also result from viral infections, such as influenza and measles. Viral infections of the upper respiratory tract (including the many types known as "colds") are the most common illnesses of man, and these may spread to the lower respiratory tract, leading to the symptoms of pneumonia. The most typical form of viral pneumonia is characterized by extreme tracheal and bronchial congestion accompanying the accumulation of fluid in the alveoli.

Tuberculosis

Another infectious disease of the lungs is tuberculosis, caused by the tubercle bacillus. Unlike most disease-causing organisms, this bacterium does not produce toxic substances but damages lung tissues by stimulating the immune system to react locally to the presence of the bacilli, a reaction akin to allergy. This leads to an accumulation of fluid and reduction in function of the infected parts of the lungs.

Asthma and Emphysema

Asthma is a condition in which there is an inability to exhale as much air as is inhaled. This is due to congestion or spasm of the bronchioles as a result of allergies or infection.

Emphysema is a general term for any chronic disease of the lungs in which the alveoli are stretched and distended. Its most characteristic feature is hyperinflation after inspiration (Fig. 21-19), which increases the peripheral resistance to blood flow in the pulmonary vessels, resulting in an increase in the carbon dioxide content of the blood. The affected person responds by devoting greater effort to breathing.

Circulatory Disturbances

The most common conditions due to disturbances in the pulmonary circulation of the lungs are *congestion,* usually resulting from hypertension in the pulmonary vessels, and *pulmonary embolism,* resulting from a clot in the pulmonary artery. Congestion occurs when the blood flow to the lungs exceeds the venous return to the heart, causing the lung capillaries to become engorged with blood and the lung tissues to become swollen from the accumulation of fluid (edema). When a large embolus (blood clot) blocks one of the main branches of the pulmonary artery, the remaining blood vessels that are not blocked are taxed beyond their limit, and death results because of an extreme rise in pulmonary arterial pressure with right-sided heart failure.

Pleurisy

Inflammation of the pleura, or *pleurisy,* is due to bacterial infection that may originate in the lung, the pleura, or the chest wall. The inflammation may cause *adhesions* of stringy connective tissue strands between the visceral and parietal layers of pleura, which may make it difficult and painful to breathe deeply. There may be an increase in the serous exudate collecting in the pleural recesses, reducing the areas into which the lungs expand and reducing the individual's vital capacity.

The *segmental bronchus* is accompanied by a branch of the *pulmonary artery* that supplies the segment without crossing into adjacent segments. Drainage is by veins that course between bronchopulmonary segments, receiving blood from adjacent segments. The left lung is usually considered to be composed of nine bronchopulmonary segments (four segments in the upper lobe and five in the lower, and the right lung consists of ten segments (three in the upper, two in the middle, and five in the lower lobe). In a chest film some of the segments are superimposed on one another.

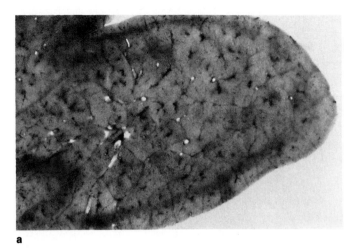
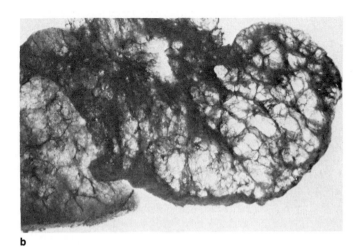

Figure 21-19 The lung of (**a**) a normal person and (**b**) a person with emphysema (Webb-Waring Institute for Medical Research, courtesy Environmental Protection Agency).

WORDS IN REVIEW

Aditus

Alveolar duct

Alveoli

Bronchopulmonary segment

Carotid body

Choanae

Conus elasticus

Cricothyroid membrane

Hering–Breuer reflex

Meatus

Nasopharynx

Nostrils

Pharyngeal tonsil

Pleura

Pleural recess

Root of lung

Septum transversum

Surfactant

Thyrohyoid membrane

Tracheoesophageal fistula

Ventilation

Vestibular folds

SUMMARY OUTLINE

I. Definition of Respiration
   A. Ventilation includes inhalation and exhalation.
   B. External respiration is the exchange of gases between the lungs and the blood.
   C. Internal respiration is the exchange of gases between the blood and tissue cells.
   D. Cellular respiration is the utilization of oxygen by cells to produce energy.

II. The Air Passages
   A. The nose (nasal cavity) is supported by bone and cartilage. The lateral walls have three curved ledges called conchae. The cavity is lined with pseudostratified ciliated columnar epithelium of the mucous membrane that cleanses and warms the air. The paranasal sinuses are lined by mucous membrane that is continuous with the lining of the nasal cavity and open into the nasal cavity via separate openings. Air enters the nasal cavity through the nostrils and passes through the posterior choanae into the nasopharynx.
   B. The pharynx is divided into three parts, defined by their relation to the nose, oral cavity, and larynx.
      1. The nasopharynx is a muscular tube superior and posterior to the soft palate. It contains openings to the auditory tubes in its lateral walls and the pharyngeal tonsil in its posterior wall. The mucous membrane is lined with ciliated respiratory epithelium.
      2. The oral pharynx extends from the level of the soft palate to the level of the aditus, the opening of the larynx. It is the common passage for both air and food and is lined with stratified squamous epithelium.
      3. The laryngopharynx is inferior to the glottis and is posterior to the larynx. It is only a food passage.
   C. The larynx lies between the pharynx and the trachea. It is supported by three unpaired cartilages (epiglottic, thyroid, and cricoid) and three paired cartilages (arytenoid, corniculate, and cuneiform) and contains the vocal folds, which are responsible for sound production. Adduction of the vocal folds closes the air passage.

D. The trachea extends inferiorly from the lower border of the larynx (cricoid cartilage) to the point of bifurcation into primary bronchi at the level of the sternal angle. It is supported internally by a series of horseshoe-shaped cartilages (tracheal rings) that are open posteriorly where the trachea lies against the esophagus. The gaps in the tracheal rings are filled by smooth muscle. The epithelium of the mucous membrane is of the respiratory type.

E. The lungs are paired organs; each receives a primary bronchus that enters at the hilus (as part of the root of the lung) and branches into secondary bronchi supplying the lung lobes. These bronchi branch into tertiary bronchi that supply the bronchopulmonary segments. The bronchi continue to branch and become smaller until bronchioles of 1 mm or less in diameter are produced. Bronchioles contain no cartilage and end as respiratory bronchioles that open into alveolar ducts. The alveolar ducts open into air spaces called atria from which thin-walled air sacs called alveoli, surrounded by capillaries, project. Gas exchange between the air and blood takes places through the alveolar and capillary epithelium. The epithelium of the alveoli is of two cell types: squamous (Type I), through which exchange takes place, and cuboidal (Type II), which produces surfactant, a substance that lowers surface tension and prevents the collapse of the alveoli.

III. The Pleura

The pleura consists of a double-layered sac of serous membrane; the parietal layer is fused to the thoracic wall, diaphragm, and fibrous pericardium, and the visceral layer is fused to the external surface of the lung. The space between the two layers is the pleural cavity.

IV. Regulation of the Respiratory System

A. Quiet breathing is regulated principally by the Hering–Breur reflex; stretching the lungs stops inspiration, whereas unstretched lungs start inspiration.

B. Increased CO_2 is detected by the carotid body and increases the rate of breathing; decreased CO_2 decreases the rate of breathing. Changes in O_2 concentration have the opposite effect from changes in CO_2 concentration. Other factors that modify the rate of respiration include irritants that may cause sneezing or coughing.

V. The Lower Respiratory Tract

The lower respiratory tract (larynx, trachea, bronchi, and lungs) develops as a midventral tube from the floor of the pharynx; the tube elongates and bifurcates, forming lung buds that grow in length and continue branching. Alveoli differentiate to a functional state by about six months in utero.

REVIEW QUESTIONS

1. What is the name of the air passage that lies between the larynx and the primary bronchi?

2. What is the posterior boundary of the nasal cavity?

3. What morphological features of the mucous membrane of the nasal cavity facilitate removal of particulate matter from the air being breathed?

4. What anatomical landmarks are used as the dividing line between the nasopharynx and the oral phaynx?

5. What structures are essential in closing the larynx voluntarily?

6. What structures maintain the trachea as an open passageway at all times?

7. What is the arrangement of cartilage in the bronchi within the substance of the lung?

8. What is a bronchopulmonary segment?

9. What are the contents of the root of the lung?

10. Why does expiration in normal, quiet breathing require very little expenditure of energy.

SELECTED REFERENCES

Anderson, J. E. *Grant's Atlas of Anatomy.* 8th ed. Baltimore: Williams & Wilkins, 1983.

Goss, C. M. *Gray's Anatomy.* 29th ed. Philadelphia: Lea & Febiger, 1973.

Lachman, E., and Faulkner, K. K. *Case Studies in Anatomy.* New York: Oxford University Press, 1981.

Moore, K. L. *Clinically Oriented Anatomy.* Baltimore: Williams & Wilkins, 1980.

22

THE DIGESTIVE SYSTEM

After completing this chapter you should be able to:

1. List in sequence and give the general locations of the organs of the digestive system.

2. Describe the structure of the oral cavity, giving the anatomical boundaries. Correlate the morphology of each of the major components of the oral cavity with its role in the mechanical breakdown of food.

3. Name the organs of the digestive system that transport food rapidly from the oral cavity to the stomach, describing the morphological features that facilitate this function.

4. Describe the modifications of the mucosa of the stomach that provide for increased numbers of secretory cells.

5. Describe the modifications of the mucosa of the small intestine that provide for increased numbers of secretory cells and increased surface area for the absorption of nutrients.

6. Describe the muscle layers of the stomach and small intestine involved in mixing the food with enzymes and in moving the food down the digestive tube.

7. Describe the anatomical organization of the liver into lobules and relate this organization to the liver's dual role as a biochemical storage facility and an exocrine digestive gland.

8. Describe the pancreas as an exocrine gland; describe the connection of its duct to the duct of the liver and to the digestive tract.

9. List in sequence the parts of the large intestine and give the approximate position of each part. Correlate the anatomy of the mucosa with its role in dehydration of the chyme. Correlate the anatomy of the muscularis mucosa with its roles in compaction and movement of the fecal masses through the colon.

10. Describe the morphological features of the rectum and anal canal that are involved in the retention of feces and in the process of defecation.

11. Describe the regulation of muscle contraction and enzyme production in all parts of the digestive system.

The *digestive system* (Fig. 20-1) consists of the *alimentary canal,* a tube extending from the mouth to the anus, and the *digestive glands,* which empty their secretions into the tube. The functions of the alimentary canal, also called the digestive tube or digestive tract, include *ingestion* of food; *digestion,* the mechan-ical and chemical conversion of food into absorbable material; and *absorption* of the digested material into the blood stream for distribution to the cells of the body. The digestive tract is modified in structure along its course, forming specialized parts called *digestive organs,* which have specific roles in the functions of the system.

The digestive tube is continuous with the external environment at the *mouth* and *anus,* and its lining becomes continuous at the *lips* and at the edge of the *anus* with the *epidermis* covering the surface of the body. Food is taken into the tract (ingested) at the mouth and passes through the tube where digestion and absorption occur; undigested material, called feces, is eliminated through the anus. Each of the digestive organs performs its special task in sequence and at a rate that allows digestion and absorption to proceed to completion. The *mouth,* or oral cavity, is specialized for *ingestion* and for the *mechanical breakdown* of food by chewing. The muscular *pharynx* assists the oral structures in swallowing and transfers the food to the *esophagus* for rapid transport to the stomach. This process occurs too quickly to permit enzymatic breakdown of food to occur during its passage through the esophagus. The process of *enzymatic hydrolysis* (breakdown by enzymes) begins in the *stomach,* a large organ capable of retaining the food for a period of time, whose muscular wall facilitates mixing the food with *enzymes* and *acid.* The partially digested food that leaves the stomach is called *chyme.* Digestion continues as the chyme passes through the *small intestine.* The lining of the small intestine is specifically modified to facilitate the absorption of nutrients, and the highly coiled, slender tube is long enough to complete the process by the time the material reaches the *large intestine.* The term *gut* refers to both the small and large intestines. The large intestine has structural features specialized for the *removal* of *excess water* from the undigested material without allowing the feces to become excessively hard. The lower end of the digestive tract, the *sigmoid colon,* the *rectum,* and *anal canal,* retain the feces until they are eliminated by the process of *defecation.*

GENERAL ORGANIZATION OF THE DIGESTIVE TUBE AND ASSOCIATED GLANDS

The digestive glands that produce the enzymes involved in the chemical breakdown of food include the *accessory digestive glands,* located external to the digestive tract, and the *digestive glands proper* whose secretory components are located in the wall of the organ receiving the secretions. The accessory digestive glands are the *salivary glands,* which empty into the mouth (oral cavity), and the *liver* and *pancreas,* which supply their products to the small intestine. Most of the digestive glands proper consist of secretory cells scattered in clusters within the highly folded epithelial lining of the alimentary canal. Their secretions pass directly from the secretory cells into the lumen (cavity) of the digestive tract, in most cases without a complicated duct system. Nevertheless the digestive glands are considered *exocrine glands,* because they produce enzymes rather than hormones.

The digestive tract is a tube within an outer tube (the body wall), extending the length of the body. In the early embryo the gut tube is essentially a straight tube, approximately the same length as the body. As the digestive tube develops, it grows faster than the trunk, ending as a highly coiled tube about twenty-five feet long, housed principally within the abdomen, which is usually no more than one foot in vertical length. Throughout most of its length the wall of the digestive tube is composed of tissues that form distinct layers arranged in a definite sequence (Fig. 22-2). The lumen of the digestive tube is lined with epithelium that is structurally modified in each digestive organ to perform its specialized function. The epithelium rests on connective tissue termed the **lamina propria** (layer proper), which in most organs of the digestive tract is separated from an additional layer of connective tissue by a thin layer of smooth muscle called the **muscularis mucosa.** These three layers, the epithelial lining, lamina propria, and muscularis mucosa, form the **mucous membrane,** or **mucosa,** of the digestive tract; the mucosa is kept moist by mucus, a thick, viscous fluid containing the protein mucin. The connective tissue immediately external to the muscularis mucosa, called **submucosa,** serves as the pathway for the larger nerves and blood vessels supplying the wall of the tube. External to the submucosa the **muscularis externa** usually consists of two layers of smooth muscle, an inner circular layer and an outer longitudinal layer, whose contractions move food down the tube. The outer surface of the tube is covered in many areas by a **serous membrane,** or **serosa,** which is kept moist by a thin, watery fluid called *serous fluid* that prevents the gut from sticking to adjacent structures. This layer, called the *visceral peritoneum,* is connected to the *parietal peritoneum,* the membrane lining the body wall, by serous membranes called **mesenteries** that help support the visceral organs. Each layer of the gut wall has an important role in the basic function of the digestive system; that is, each layer has a role in providing the body with nutrients derived from external sources.

The specializations of the digestive organs make the human digestive tract far more efficient than the digestive tracts of invertebrate animals. For example, a human can readily survive by spending a few minutes each day ingesting food that weighs only a small fraction of his body weight; but an earthworm, with a straight, unspecialized gut tube, must spend all of its time each day passing soil weighing many times its body

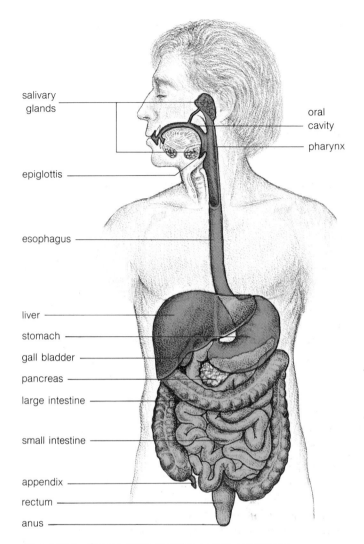

Figure 22-1 Organs of the digestive system, anterior view.

weight through its digestive tract to obtain enough nutrients for survival. The specializations adding to the efficiency of the human digestive system are principally modifications of the mucous membrane and muscularis externa.

MECHANICAL BREAKDOWN OF FOOD

The breakdown of large chunks of food into small particles, to facilitate swallowing and to increase surface area for more rapid enzymatic action, occurs as a result of chewing, the process of

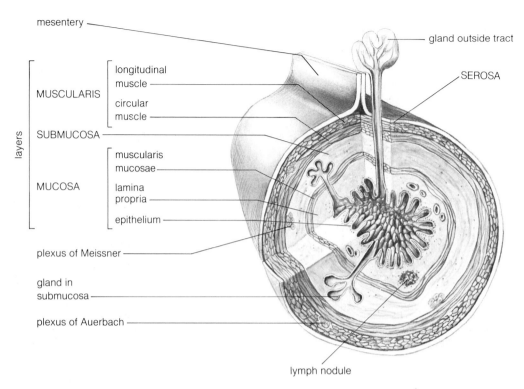

mesentery

gland outside tract

MUSCULARIS
longitudinal muscle
circular muscle

SEROSA

layers

SUBMUCOSA

MUCOSA
muscularis mucosae
lamina propria
epithelium

plexus of Meissner

gland in submucosa

plexus of Auerbach

lymph nodule

Figure 22-2 Generalized diagram of the structure of the digestive tube, transverse section.

grinding food between the teeth. The mouth (Fig. 22-3) is the first part of the digestive tract, and it is the only part provided with skeletal support within its boundaries. It is bounded anteriorly by the muscular *lips,* laterally by the muscular *cheeks,* inferiorly by the muscular *floor* which contains the *tongue,* and superiorly by the bony *hard palate* and muscular *soft palate;* it is continuous posteriorly with the *pharynx* at the *anterior pillars* of the *fauces,* a fold in the mucous membrane that extends from the soft palate to the tongue. The mouth may be divided into two parts: the **oral vestibule** is the space between the lips and cheeks externally and the teeth and gingivae (gums) of the upper and lower jaws internally, and the *oral cavity proper* is the space inside the teeth that, with the mouth closed, is virtually filled with the tongue, upon which most of the *taste buds* are located. The mouth is lined partly by a mucous membrane composed of *keratinized stratified squamous epithelium* capable of resisting the friction of dry food. The epithelium is thick and is constantly moistened by *saliva.* Because no muscularis mucosa is present in the mouth, the lamina propria is continuous with the submucosa. The muscularis externa, the layer described earlier as moving food along the digestive tract, is not present; however, it is represented functionally by the skeletal muscles

of such structures as the lips, cheeks, and floor of the mouth, and by the bone of the jaws and hard palate. These muscles are not the primary muscles of mastication (see Chapter 10) but assist the muscles that move the jaws in the process of chewing.

The Lips

The upper and lower lips bound the opening of the mouth, meeting laterally at the angle of the opening. The lips are covered externally by skin and internally by mucous membrane, with a red transitional zone between the skin and mucous membrane that is moistened by licking the lips with the tongue. Both the upper and lower lips have a small midline fold, termed the **frenulum,** which connects the lips to the *gums.* The mucous membrane inside the lips is moistened by saliva produced by the *major salivary glands* (parotid, submaxillary, and sublingual) and by *minor salivary glands,* some of which are found beneath the epithelial lining of the lips. These minor glands may be felt by running the tip of the tongue along the inner surface of the lip. Embedded within the connective tissue (fascia) between the skin and mucous membrane are the *orbicu-*

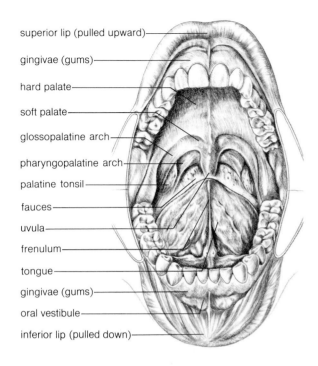

superior lip (pulled upward)
gingivae (gums)
hard palate
soft palate
glossopalatine arch
pharyngopalatine arch
palatine tonsil
fauces
uvula
frenulum
tongue
gingivae (gums)
oral vestibule
inferior lip (pulled down)

Figure 22-3 Anatomy of the oral cavity, anterior view of the open mouth.

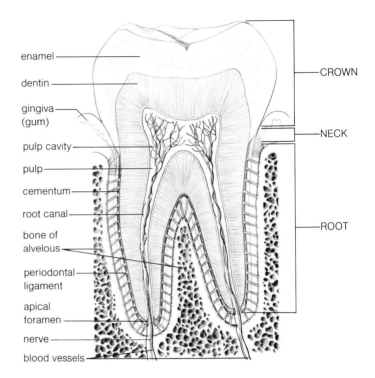

enamel
dentin
gingiva (gum)
pulp cavity
pulp
cementum
root canal
bone of alvelous
periodontal ligament
apical foramen
nerve
blood vessels

CROWN
NECK
ROOT

Figure 22-4 A longitudinal section through a tooth.

laris oris muscle and the terminations of other *muscles* of *facial expression* inserting into the lips (Figs. 11-1, 11-2). These muscles, together with the *buccinator* muscles of the cheeks, provide the human with a sucking mouth useful in the intake of fluid and capable of compressing the lips against the anterior teeth, preventing the accumulation of food in the vestibule.

The Cheeks

The structure of the cheek is similar to that of the lips. Its principal muscle is the *buccinator,* which plays an important role in chewing by helping to keep food between the molar teeth. Its significance in the normal process of chewing is best revealed in individuals whose buccinator muscle is paralyzed; the accumulation of food in the vestibule causes their cheeks to bulge outward, like the cheek pouches of a hamster. The cheek contains the duct of the *parotid* salivary *gland* that courses anteriorly to open through the mucous membrane into the vestibule opposite the upper second molar tooth. Minor salivary glands, the **buccal glands,** located in the connective tissue beneath the epithelial lining of the cheek, may be felt with the tip of the

tongue as small irregularities on an otherwise smooth surface near the angle of the mouth.

The Teeth

The teeth are arranged in upper and lower *dental arches,* securely set into bony sockets in the *maxilla* and *mandible.* Each tooth (Fig. 22-4) consists of a *crown* projecting above the *gingiva* (gum), a *root* embedded within the *bony socket,* and a *neck* at the point of junction between the crown and root. The surface of the crown is covered with *enamel,* the hardest substance in the body; enamel is composed principally of crystals of inorganic salts (apatite crystals) arranged as minute rods and packed tightly together. The deep surface of the enamel attaches to the **dentin,** a hard, calcified substance that encloses the dental pulp cavity and extends into the root, where it is covered externally by the **cementum,** or cement. The *pulp cavity* or *chamber* consists of connective tissue containing nerves and blood vessels that supply the tooth. The area of contact between the enamel and cementum constitutes the neck of the tooth and lies at the level of the gingiva. Within the bony socket the cementum encir-

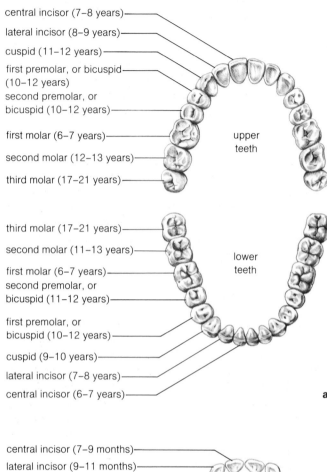

central incisor (7–8 years)

lateral incisor (8–9 years)

cuspid (11–12 years)

first premolar, or bicuspid (10–12 years)

second premolar, or bicuspid (10–12 years)

first molar (6–7 years)

second molar (12–13 years)

third molar (17–21 years)

upper teeth

third molar (17–21 years)

second molar (11–13 years)

first molar (6–7 years)

second premolar, or bicuspid (11–12 years)

first premolar, or bicuspid (10–12 years)

cuspid (9–10 years)

lateral incisor (7–8 years)

central incisor (6–7 years)

lower teeth

a

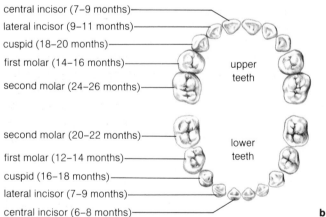

central incisor (7–9 months)

lateral incisor (9–11 months)

cuspid (18–20 months)

first molar (14–16 months)

second molar (24–26 months)

upper teeth

second molar (20–22 months)

first molar (12–14 months)

cuspid (16–18 months)

lateral incisor (7–9 months)

central incisor (6–8 months)

lower teeth

b

Figure 22-5 (**a**) Permanent teeth. (**b**) Deciduous teeth.

cling the root attaches to the surrounding bone by means of a fibrous sheet of connective tissue, the **periodontal membrane** or *periodontal ligament.* Cementum is a modified type of bone containing cells embedded in a calcified matrix. Collagenous fibers of the periodontal membrane become entrapped within the matrix and extend across the gap between the tooth and the bone of the socket as long as the membrane is intact. In the normal, healthy situation, the stratified squamous epithelium of the gingiva attaches to the enamel near its junction with the cementum, protecting the periodontal membrane from exposure to the environment of the oral cavity.

With proper dental hygiene, the attachment of gingiva to the enamel should remain intact for the life of the individual; however, the accumulation of *dental calculus,* a hard substance, in the area of the epithelial attachment may break the attachment, allowing food particles and accompanying bacteria to collect. This frequently leads to destruction of part of the periodontal membrane, creating a "periodontal pocket" between the tooth and bone. The resulting infection and erosion of bone may lead to loss of the tooth. This can be avoided in most cases by good home care and regular dental treatment.

The teeth vary in shape and are named in accordance with their functions and location. There are typically thirty-two permanent teeth (Fig. 22-5a); each quadrant (half arch) includes two *incisors,* one *canine,* two *premolars,* and three *molars.* On their occlusal (contacting) surfaces the crowns of the premolars and molars have rounded elevations, termed *cusps,* that are useful in grinding food. Each tooth has a characteristic number and arrangement of the cusps, from two on the premolars to as many as five on the lower second molars. The *deciduous teeth* (baby teeth) begin development during prenatal life but usually do not erupt until a baby is about six months old. The *permanent teeth* (second teeth) also form partially during prenatal development but do not begin erupting until an individual is about six years old. There are twenty deciduous teeth (two incisors, one canine, and two molars in each quadrant) that erupt at variable ages by rapid growth of the roots of the teeth. Eruption of the upper teeth commonly occurs in the following sequence at about the ages indicated: central incisors, 7 to 9 months; lateral incisors, 9 to 11 months; canine, 18 to 20 months; first molars, 14 to 16 months; and second molars, 24 to 26 months (Fig. 22-5b). The crowns of the permanent teeth form alongside the roots of the deciduous teeth, with no root development until shortly before their eruption. As their roots begin to grow, the crowns of the permanent teeth push against the deciduous teeth, facilitating resorption of their roots, which eventually causes the deciduous teeth to fall out. The times of eruption of the permanent teeth also vary greatly but generally proceed in the following sequence and at the following ages for the lower teeth: central incisors, 6 to 7 years; lateral incisors, 7 to 8 years; canines, 9 to 10 years; premolars, 10 to 12 years; first molars, 6 to 7 years; second molars, 11 to 13 years; and third molars (wisdom teeth) 17 to 21 years (Fig. 22-5a).

The Palate

The palate may be divided into an anterior *hard palate* and a posterior *soft palate* (Fig. 22-3). The hard palate separates the mouth from the nasal cavity; it consists of bone (palatine processes of the maxilla and horizontal plates of the palatine bones) covered on its inferior surface by the mucous membrane lining the oral cavity. The thin connective tissue component of the mucous membrane fuses tightly with the periosteum of the palatal bones. The soft palate is a flexible muscular sheet separating the oral cavity from the nasopharynx. It is continuous with the posterior border of the hard palate and consists of several muscles covered by mucous membrane. Elevation of the soft palate during swallowing prevents food from entering the nasal cavity (Fig. 22-6).

The Tongue

The *tongue* (Figs. 22-3 and 22-7) is a motile muscular organ, important in speech and ingestion. It extends upward from the floor of the mouth and works with the cheeks to keep food between the teeth during chewing. The tongue consists of a mass of skeletal muscles (four pairs of muscles originating outside the tongue and three sets of muscles entirely within the tongue) covered by mucous membrane bearing specialized sensory receptors, the *taste buds,* responsible for the sense of taste.

The mucous membrane covering the upper surface (dorsum) of the tongue is marked by elevations called *papillae,* which bear the taste buds (Fig. 17-3) and which vary in size and shape from pointed *filiform* papillae and small rounded *fungiform* papillae to large oval **circumvallate papillae.** The circumvallate papillae are arranged in a row, forming a V-shape just anterior to a V-shaped groove called the **sulcus terminalis.** The sulcus terminalis marks the junction of the freely motile anterior two-thirds of the tongue with the more stable posterior third, called the *base* or *root,* through which the extrinsic muscles enter the substance of the tongue. At the apex of the V of

the sulcus terminalis a pit, the *foramen cecum,* marks the point of embryonic origin of the *thyroid gland.* The surface of the base of the tongue bears a series of round elevations caused by aggregations of lymphoid tissue, called the *lingual tonsils,* in the submucosa.

The tongue not only has a significant role in chewing but also is essential in the initial phases of swallowing (deglutition). It prepares the food for swallowing by forming it into a *bolus* (rounded mass) and passing it posteriorly and superiorly against the soft palate as the elevation of the pharynx encloses it.

Beneath the tongue a fold of the mucous membrane, the *lingual frenulum,* connects the tongue to the floor of the mouth. The frenulum may be too short in some individuals, restricting tongue movement in speech, a condition referred to as "tongue-tied." This condition can usually be corrected surgically by clipping the frenulum.

The Salivary Glands

In addition to the numerous minor salivary glands in the mucous membrane, which secrete constantly to keep the membrane moist, there are three pairs of large compound salivary glands (Fig. 22-8), principally of the tubuloalveolar type, which are activated to secrete by the presence or even the anticipation of food. The *parotid gland* is located anterior to the ear, filling the space between the styloid process of the temporal bone and the ramus of the mandible, which it overlaps superficially. The parotid gland is a *compound alveolar gland* that produces a watery fluid. It empties its secretions into the oral cavity by means of the *parotid duct* (Stensen's duct), which passes lateral to the ramus of the mandible and the overlying masseter muscle to pierce the mucous membrane lining the cheek opposite the upper second molar tooth. The duct contains smooth muscle that contracts when stimulated by nerve impulses. Oral pain, as from dental work, may initiate reflexes that stimulate contraction of the duct musculature, causing saliva to spurt from the duct.

The parotid gland can be infected by the virus that causes mumps. Viral infections tend to produce swelling of the infected organ, which is a problem for the parotid, because the gland is encased by a tight connective tissue capsule that resists swelling. As the accumulation of fluid (swelling, or edema) continues within the gland, pressure on the nerve endings can cause severe pain.

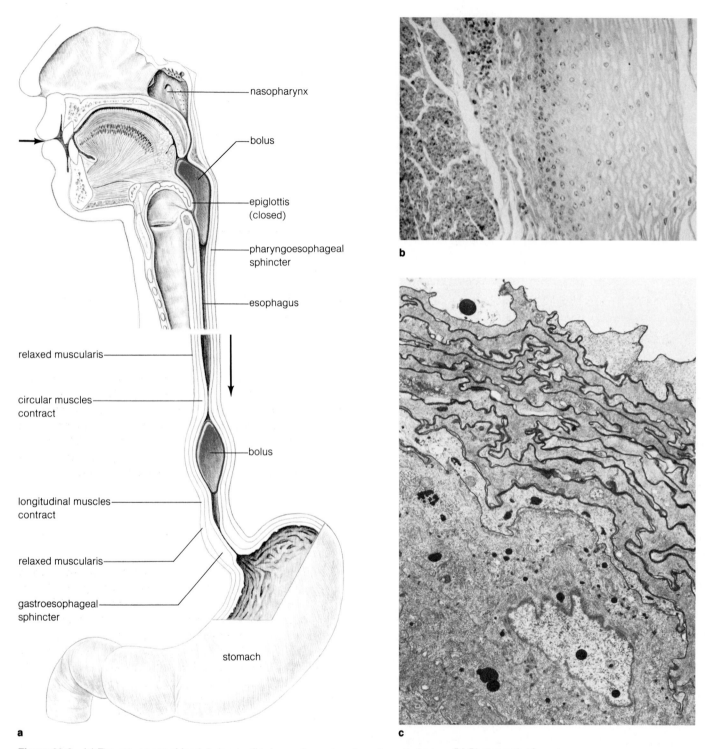

nasopharynx

bolus

epiglottis
(closed)

pharyngoesophageal
sphincter

esophagus

relaxed muscularis

circular muscles
contract

bolus

longitudinal muscles
contract

relaxed muscularis

gastroesophageal
sphincter

stomach

a

b

c

Figure 22-6 (**a**) The movement of food during swallowing and passage down the esophagus. (**b**) Photograph of the microscopic anatomy of the esophagus, ×140 (courtesy of W. K. Elwood). (**c**) Electron photomicrograph of the surface of mouse esophageal epithelium, ×5900. Note the tightly packed layers of cell membranes. (Courtesy of D. H. Matulionis.)

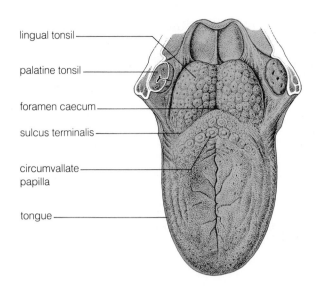

Figure 22-7 Dorsal view of the tongue and associated structures.

lingual tonsil

palatine tonsil

foramen caecum

sulcus terminalis

circumvallate papilla

tongue

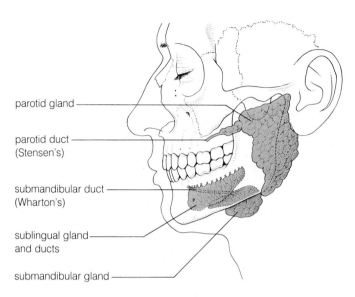

Figure 22-8 The location of the salivary glands, lateral view.

parotid gland

parotid duct (Stensen's)

submandibular duct (Wharton's)

sublingual gland and ducts

submandibular gland

The *submandibular gland* (formerly called the submaxillary gland) is a tubuloalveolar salivary gland that produces a mixture of serous and mucous secretions. It is located inferior to the body of the mandible, and can usually be palpated just anterior to the angle of the mandible. The submandibular is a large gland that wraps around the posterior border of the mylohyoid muscle of the floor of the mouth, so that part of the gland lies superficial and part lies deep to the muscle. The *submandibular duct (Wharton's duct)* arises from the deep portion of the gland and passes anteriorly beneath the mucous membrane of the floor of the mouth to open lateral to the lingual frenulum.

The *sublingual glands* lie at either side of the floor of the mouth, creating a fold or ridge, the *sublingual fold,* in the mucous membrane that can be felt with the tip of the tongue. These glands are of the mixed type and have several small ducts opening into the oral cavity along the summit of the sublingual fold.

> Hard deposits of calcium salts may accumulate as "stones" in the ducts of the salivary glands and cause severe pain, particularly during eating, by blocking the flow of saliva.

The fluid produced by the salivary glands is termed saliva and contains proteins such as *mucin* and the enzyme ptyalin (salivary amylase), in addition to water. The fluid is important in moistening the food during chewing and in preparing the bolus for swallowing. Ptyalin is an enzyme that can convert starch into sugar, but it does not appear to be of major importance in the digestive process of humans. The odor or sight of food stimulates the secretory cells of the salivary glands; even thinking of food can cause an increase in the flow of saliva.

RAPID TRANSPORT OF FOOD

When food is swallowed, it passes rapidly down the *pharynx* and *esophagus,* reaching the stomach without substantial physical or chemical changes. Thus the significant structural features of the pharynx and esophagus are for moving the bolus of food rapidly, not for breaking down the food. Because this movement is caused by muscle contraction, it can be anticipated that the principal features to be examined are the muscles in the walls of the organs involved. Smooth muscle contracts slowly, skeletal muscle contracts rapidly; thus it can also be anticipated that the muscles involved are chiefly skeletal muscles.

The Pharynx

The pharynx (Fig. 22-9) serves as both an *air passage* (respiratory) and a *food passage* (digestive); it extends from the base

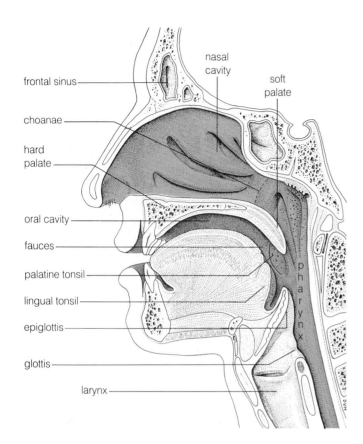

frontal sinus

choanae

hard palate

oral cavity

fauces

palatine tonsil

lingual tonsil

epiglottis

glottis

larynx

nasal cavity

soft palate

pharynx

Figure 22-9 Midsagittal section of the oral cavity, nasal cavity, and pharynx.

caused by the passage of inadequately moistened food, are lined with *nonkeratinized stratified squamous epithelium* that is moistened by mucus secreted by glands located in the underlying connective tissue, the submucosa. The connective tissue component of the mucous membrane is relatively thin but includes a dense sheet of fascia externally that may be considered the functional equivalent of the muscularis mucosa of the gut wall. This fascial sheet, called pharyngobasilar fascia, contains numerous elastic fibers that allow flexibility and increase the strength of the mucous membrane.

> If a lump of food enters the larynx, the laryngeal muscles contract spasmodically, closing the larynx so that air cannot reach the lungs. The person is in danger of choking to death unless the passage is cleared quickly.

The wall of the pharynx assists in the complicated process of swallowing (deglutition) and propels the bolus of food into the esophagus. The wall contains two sets of skeletal muscles located external to the pharyngobasilar fascia in the position equivalent to that of the muscularis externa of the gut tube. These muscles are *vertically oriented* paired muscles (the stylopharyngeus, salpingopharyngeus, and palatopharyngeus), which elevate the pharynx during swallowing (Fig. 11-9), and *circularly oriented* paired muscles (the superior, middle, and inferior pharyngeal constrictors), which move the bolus into the esophagus. Swallowing begins as a voluntary act with the elevation of the root of the tongue and of the hyoid bone to which it is attached, which passes the bolus of food posteriorly into the oral pharynx. At the same time, elevation of the soft palate by contraction of the levator veli palatini and tensor veli palatini (Fig. 11-9) prevents food from entering the nasal cavity. Elevation of the hyoid bone, accomplished principally by contraction of the suprahyoid muscles (mylohyoid and geniohyoid) and the simultaneous contraction of an infrahyoid muscle (the thyrohyoid) elevates the larynx. This pushes the epiglottis against the base of the tongue, effectively closing the glottis and preventing food from entering the larynx. The position of the long tongue prevents food from returning to the mouth. All these actions, except the first, are controlled automatically by reflexes, and any attempt to consciously alter the procedure once it is initiated may end unpleasantly with food in the larynx. With all the openings of the pharynx closed except the opening into the esophagus, the bolus of food can only pass inferiorly as the pharyngeal constrictors contract in a peristaltic wave from above downward, propelling the bolus into the relaxed esophagus (Fig. 20-6).

of the skull to the level of the cricoid cartilage of the larynx, at which point it becomes continuous inferiorly with the esophagus. It is continuous anteriorly with the nasal cavity via the choanae, the oral cavity via the fauces, and the larynx via the glottis. Based upon its association with these structures, the pharynx may be divided into three parts: the *nasopharynx,* superior to the level of the soft palate; the *oral pharynx,* extending from the soft palate inferiorly to the level of the glottis; and the *laryngopharynx,* directly posterior to the larynx and ending inferiorly by joining the esophagus. From this point on, the digestive and respiratory tracts are separate; food continues through the esophagus and air through the larynx.

The nasopharynx serves only for the passage of air, which does not cause friction, and is lined with the *respiratory* type of *epithelium* (pseudostratified columnar ciliated). In contrast both the oral pharynx and laryngopharynx, subjected to the friction

The Esophagus

The esophagus is a straight, slender tube modified to deliver a bolus of food to the stomach quickly. The bolus passes into the lumen of the stomach gently, so that the sensitive lining of the stomach is not injured. The mucous membrane of the esophagus is similar to that of the pharynx, except that the esophagus has a muscularis mucosa of longitudinally oriented smooth muscle in the position equivalent to that of the pharyngobasilar fascia of the pharynx. The muscularis externa, external to the submucosa, consists of outer longitudinal and inner circular layers of muscle continuous with the inferior pharyngeal constrictors. In the *upper one-third* of the esophagus this muscle is of the *skeletal* type, whose contractions continue the rapid propulsion of the bolus on its way down the tube. The muscularis externa in the *middle one-third* of the esophagus is a *mixture* of the *skeletal* and *smooth* muscle; it becomes entirely *smooth* muscle in the *lower one-third*. The slower contractions of the smooth muscle in the lower esophagus pass the bolus into the stomach without the force that would be generated by a muscularis externa of skeletal muscle. The esophagus is not covered with serous membrane except near its junction with the stomach below the diaphragm. The rest of its surface is covered with connective tissue continuous with the visceral fascia of the neck and thorax.

Movement of the food down the pharynx and esophagus (Fig. 20-6) takes only a few seconds, and the bolus of food changes little in the process. It is kept moist by the secretions of the mucous glands of the pharynx and esophagus, so that it enters the stomach as a soft mass of relatively small particles in condition to be acted upon by enzymes.

ENZYMATIC HYDROLYSIS AND ABSORPTION

The *chemical breakdown* of food, splitting large molecules into small molecules that can be absorbed, is catalyzed by *enzymes.* This process begins in the stomach (except for the minor action of ptyalin that begins in the mouth) and is completed in the small intestines. The enzymes and other substances such as *hydrochloric acid* and *bile,* which facilitates specific biochemical processes, are produced by secretory cells in the lining of the stomach and small intestines or in glands that empty their secretions into the small intestine by means of ducts. The main problems of enzymatic hydrolysis relate to exposing the mass of food to the proper concentration of enzymes for the correct period of time: digestion must be completed in time for the nutrients to be absorbed before reaching the end of the small intestine. This requires a large number of secretory cells and a mechanism to ensure the passage of food at the optimal rate. The

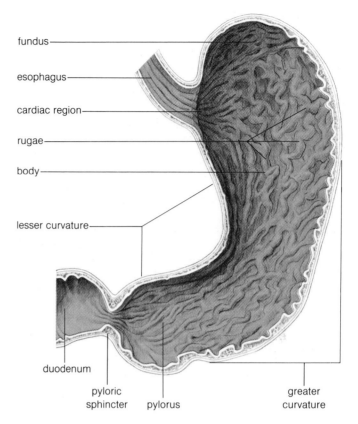

Figure 22-10 Macroscopic anatomy of the stomach, longitudinal view of the internal lining.

stomach functions only in enzymatic hydrolysis, with no appreciable absorption of nutrients; the small intestine is the major organ both of enzymatic hydrolysis and of absorption.

The Stomach

The *stomach* (Fig. 22-1), an organ concerned with the storage and digestion of food, projects to the left of the midline in the superior part of the abdomen. It varies in size and shape depending upon its contents: its average capacity is about one liter, but it can expand to hold more. When full, it is J-shaped, projecting to the left from its junction with the esophagus just inferior to the diaphragm. As seen in an anterior view, the stomach possesses a *greater curvature* toward the left and a *lesser curvature* to the right (Fig. 22-10). The area of the stomach around the esophageal orifice is the *cardiac portion;* the *body* of the stomach is below the esophageal orifice; the *fundus*

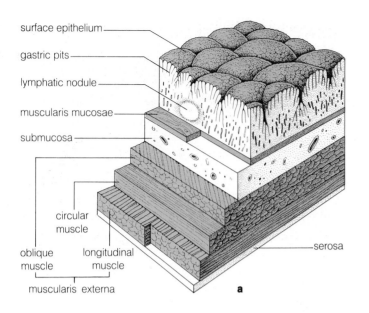

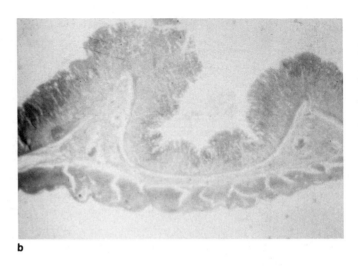

surface epithelium

gastric pits

lymphatic nodule

muscularis mucosae

submucosa

circular
muscle

oblique
muscle

longitudinal
muscle

muscularis externa

serosa

a

b

Figure 22-11 Microscopic anatomy of the stomach. (**a**) Diagram showing layers of the stomach wall. (**b**) Photomicrograph of a section of the stomach wall (courtesy of W. K. Elwood).

projects superior to the esophageal orifice; the **pyloric** *region,* or *pylorus,* is the narrow portion joining the small intestine at the *pyloric sphincter.*

The circular muscle of the pyloric sphincter may undergo hypertrophy during embryonic development, resulting in a severe narrowing (stenosis) of the pyloric canal. This congenital defect is more common in males than in females and appears to have a genetic basis, since it usually occurs in both members of a pair of identical twins. Pyloric stenosis can be corrected surgically.

Except at each end where it is fixed at the junctions with the esophagus and small intestine, the stomach is freely movable, suspended from the posterior body wall by a double layer of serous membrane called a *mesentery.* The outer surface of the stomach is covered by a serous membrane, termed *visceral peritoneum,* which prevents it from sticking to adjacent structures.

The wall of the stomach consists of layers typical of the digestive tract: mucous membrane, submucosa, muscularis externa, and serosa (Fig. 22-11). The mucous membrane is specialized for the production of enzymes and other substances

involved in digestion. This specialization includes not only the differentiation of cells capable of synthesizing the various substances secreted but also the modifications of membrane structure that provide for the increase in cell numbers required for the production of the large quantity of secretions involved. After a meal the stomach may produce up to a liter of fluid, which requires many more cells than could be accommodated in a smooth lining. Therefore the surface area of the stomach lining is increased by the presence of folds in the mucous membrane, called **rugae,** which become less pronounced as the stomach fills. The development of *gastric glands* (Fig. 22-12) as tubular extensions from the surface epithelium into the lamina propria provides for an even greater increase in cell numbers. Beginning abruptly at the esophageal-cardiac junction the stomach is lined with *simple columnar epithelium* that at the surface appears to be of a single cell type, producing mucus of a special type. The surface epithelium dips into the stomach wall, forming numerous *gastric pits* (estimated at 3.5 million) into which the tubular gastric glands open. The glands differ morphologically and functionally in the three parts of the stomach. In the cardiac and pyloric regions the secretory cells of the glands are chiefly mucus-secreting cells. Glands of the body and fundus of the stomach are the most important in the production of the gastric juice, which contains the substances (enzymes and hydrochloric acid) involved in the process of digestion. These glands are relatively straight, branched tubules extending into the lamina propria to the muscularis mucosa.

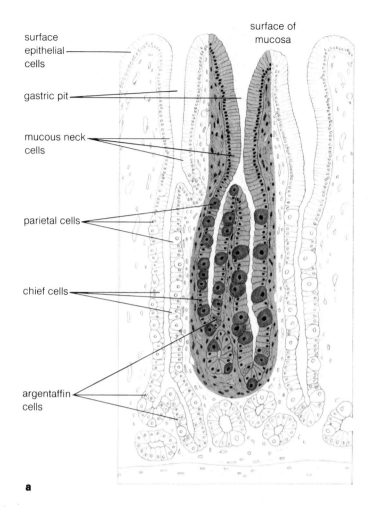

surface
epithelial
cells

gastric pit

mucous neck
cells

parietal cells

chief cells

argentaffin
cells

surface of
mucosa

surface of
mucosa

a

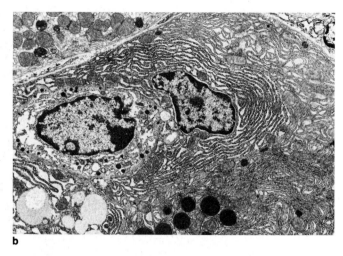

b

Figure 22-12 (**a**) Microscopic anatomy of a gastric gland. (**b**) Electron photomicrograph of a portion of a gland cell of the mouse stomach, × 4800. Note the endoplasmic reticulum. (Courtesy of D. H. Matulionis.)

The secretory cells of gastric glands are of four types: the *neck mucous cells* located near the gastric pits, which in addition to elaborating mucus appear to function as undifferentiated cells capable of replacing lost cells of either the surface or the glands; the *zymogenic cells,* or *chief cells,* which secrete pepsin, an enzyme capable of breaking down, or hydrolyzing, protein (proteolytic enzyme); the **parietal cells**, responsible for the production of hydrochloric acid; and the *argentaffin cells* near the base of the gland, which produce serotonin, a hormone that stimulates the activity of parietal cells, increasing the acidity of the stomach contents.

The secretion of stomach acid is largely under the control of the vagus nerve. It is possible to reduce the acid in individuals with peptic ulcers by cutting the vagus nerve as they pass through the diaphragm, a procedure called vagotomy.

Beneath the mucous membrane the *submucosa* provides the pathway for the larger blood vessels and nerves within the stomach wall. The submucosa separates the muscularis mucosa from the *muscularis externa,* which consists of three layers of smooth muscle (inner oblique, middle circular, outer longitudinal). At the junction of the stomach with the small intestine, the middle circular layer is thickened to form the *pyloric sphincter,* which helps control the emptying of the stomach. Within the wall of the stomach the three layers are arranged to mix the stomach contents and to move the contents out of the stomach by peristaltic waves. Food that enters the stomach may remain for a time in the body of the stomach as a relatively compact mass. The mass loosens as it mixes with gastric juice, and the material gradually moves into the pylorus. By the time it is discharged from the stomach, it is a thick fluid called *chyme.* Peristaltic waves originate in the fundus and spread downward, discharging from 5 to 15 ml of chyme into the small intestine with each wave of contraction.

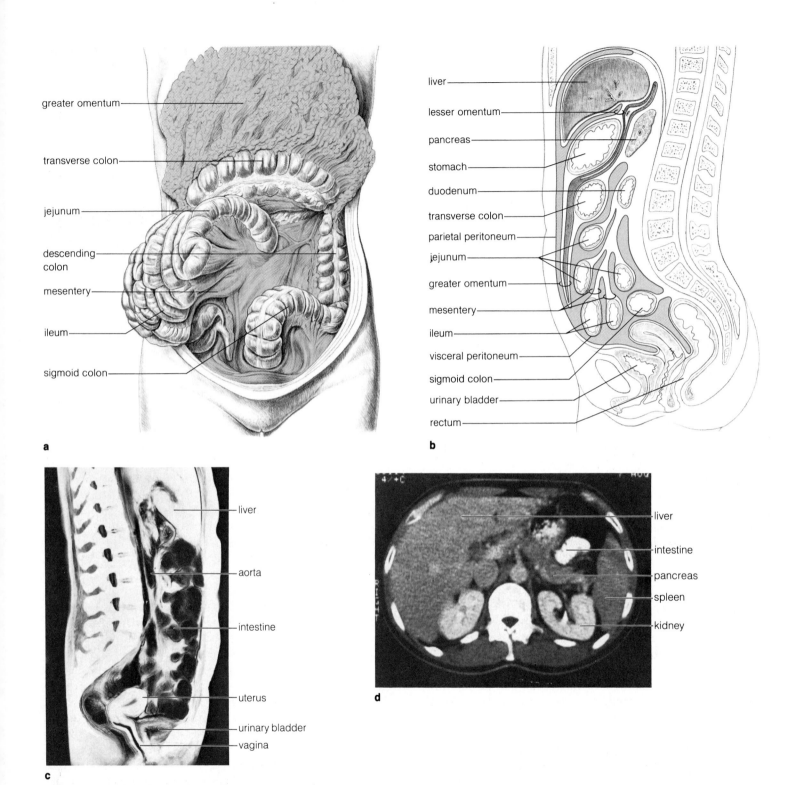

Figure 22-13 Membranes of the abdominal cavity. (**a**) Anterior view with greater omentum reflected (lifted out of the way of other structures). (**b**) Lateral view of midsagittal section. (**c**) X ray of a thick sagittal section of the trunk, cadaver preparation (from the University of Kentucky, Department of Anatomy teaching collection). (**d**) CAT scan of the abdomen, transverse view (courtesy of S. Goldstein).

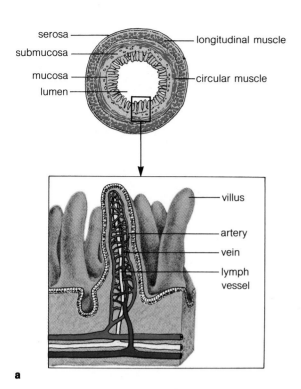

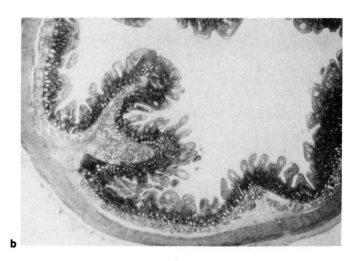

b

Figure 22-14 (**a**) Diagram of the microscopic structure of a section of the duodenum. (**b**) Photomicrograph of a histological section of the duodenum (courtesy of W. K. Elwood).

The Small Intestine

The *small intestine* is the part of the digestive tract between the stomach and large intestine. This highly coiled structure begins at the pyloric sphincter to the right of the midline in the upper abdomen, occupying the central portion of the abdominal cavity and ending at the *cecum,* its junction with the large intestine in the lower right quadrant of the abdomen.

The small intestine may be divided into three parts: *duodenum, jejunum,* and *ileum* (Fig. 22-1). The duodenum is about 10 inches long and is fixed in a **retroperitoneal** position, lying embedded in the fascia of the posterior abdominal wall and covered anteriorly by *peritoneum.* The jejunum is the proximaal two-fifths of the highly coiled, freely motile part of the small intestine and occupies the upper left portion of the abdominal cavity. It is **intraperitoneal** (within the peritoneum) in position, and its motility is due to its suspension from the posterior abdominal wall by a mesentery (Fig. 22-13). The small intestine varies in length depending upon the state of contraction of the smooth muscle in its wall, but it is about twenty feet long in the cadaver and is packed into the abdominal cavity as a series of coils crossed anteriorly only by the transverse portion of the large intestine.

All three portions of the small intestine have essentially the same structure, but the duodenum contains special *submucosal glands,* or Brunner's glands (Fig. 22-14), that produce mucus, whereas the ileum has prominent aggregations of lymphoid tissue called *Peyer's patches.* All parts of the small intestine are modified for the chemical breakdown of food and for its absorption through the intestinal mucosa into the blood stream. The requirements for increased surface area of the epithelium are met by permanent folds in the lining that are far more extensive in the small intestine than in the stomach.

The largest of the folds in the small intestine are the **plicae circularis** (valves of Kerckring), grossly visible folds involving both the mucous membrane and the submucosa (Fig. 22-15). These are highest near the junction of the duodenum and jejunum and gradually become less prominent until they are no longer present by the middle of the ileum. Numerous fingerlike projections of mucosa, called **villi,** cover the entire surface of the small intestine, giving the mucous membrane a velvety appearance and increasing its surface area greatly. The villi (Fig. 22-16) have a core of lamina propria containing the *blood capillaries* and *lacteals* (lymph capillaries of the gut), both of which transport nutrients being absorbed from the intestinal lumen through the surface epithelium. Usually there is only one lacteal per villus. The surface area is further increased by the invagination of the epithelium between the bases of the villi to form tubular intestinal glands, the *crypts of Lieberkuhn,* that extend

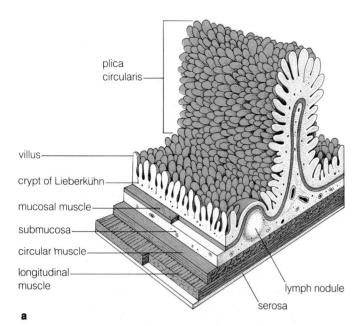

plica
circularis

villus

crypt of Lieberkühn

mucosal muscle

submucosa

circular muscle

longitudinal
muscle

lymph nodule

serosa

a

b

c

Figure 22-15 (**a**) Diagram of the microscopic structure of a section of the ileum. (**b**) Photomicrograph of a histological section of the jejunum × 100. (**c**) Photomicrograph of a histological section of the ileum × 115. Note that the plicae circulares are longer in the jejunum than in the ileum. (Courtesy of W. K. Elwood.)

into the lamina propria, almost reaching the muscularis mucosa; these glands produce enzymes that split protein, fat, and carbohydrates.

The free surfaces of the cells covering the villi (absorptive cells) have *microvilli,* fingerlike extensions of the plasma membrane that increase the surface area exposed to the lumen of the gut about thirty-fold. The plasma membrane covering the microvilli functions in the transport of substances across the membrane and serves as the framework to which some enzymes attach; thus digestion occurs partially on the cell surface, rather than entirely in the lumen of the gut as was formerly assumed. Some of the enzymes produced in the crypts of Lieberkuhn are not discharged into the lumen of the gut but become incorporated into the plasma membrane of the microvilli. Here the enzymes complete the hydrolysis of partially digested nutrients, producing small molecules such as glucose and amino acids

that can be moved by active transport across the epithelial cells for passage into the circulatory vessels.

The duodenum receives *bile* from the liver and *pancreatic juice* from the pancreas (Fig. 22-17). Bile is not an enzyme; it consists principally of the salts of bile acids produced by liver cells from cholesterol and of the bile pigment bilirubin (the waste product of dead red blood cells). Acting with the mechanical mixing action of peristalsis, bile emulsifies the fat in the chyme, facilitating its breakdown by enzymes to free fatty acids that can be absorbed into the lacteals of the villi. Pancreatic juice contains enzymes that act on all classes of food: proteins, carbohydrates, and fats. As peristaltic waves produced by the contractions of the double-layered muscularis externa (Fig. 22-18) move the chyme down the gut, the products of digestion (amino acids, glucose, and free fatty acids) come into contact with the absorptive cells covering the villi and are trans-

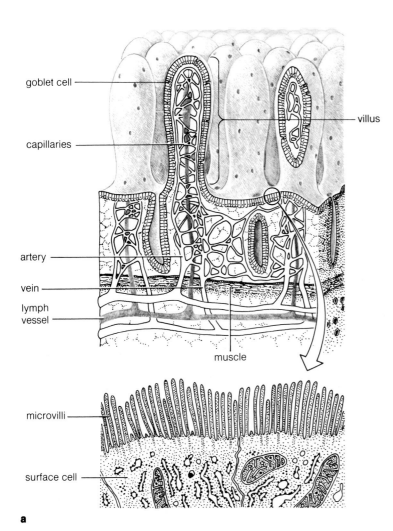

goblet cell

villus

capillaries

artery

vein

lymph
vessel

muscle

microvilli

surface cell

a

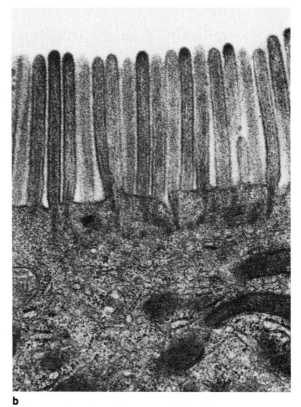

b

Figure 22-16 (**a**) Diagram of the microscopic anatomy of the villi. (**b**) Electron photomicrograph of villi of mouse small intestine, × 31,000. (Courtesy of D. H. Matulionis.)

ported across the epithelium into the intercellular fluid of the lamina propria. Amino acids and glucose pass into blood capillaries adjacent to the epithelium covering the villus; fatty acids pass into the lacteal, located centrally in the connective tissue core of the villus.

Protein molecules, including antibodies, may be absorbed through the gut mucosa for about the first three months after birth; so during this time the infant's defenses against infection may be supported by antibodies obtained from its mother's milk. After about three months the mucosa becomes an effective barrier, excluding large molecules and particulate matter.

By the time the chyme reaches the distal end of the ileum at its junction with the cecum of the large intestine, absorption of all substances is complete, except for water remaining to be absorbed through the mucosa of the large intestine. Undigested nutrients that enter the colon are not absorbed but are passed from the body with the feces.

The Liver

The liver is the largest gland in the body, occupying the upper right quadrant of the abdominal cavity, where it is in contact with the diaphragm and protected by the ribcage (Fig. 22-19). It is a soft organ, and although it has a sharp border about an inch below the lower costal margin, it is not normally palpable. A palpable liver usually indicates a diseased state such as cir-

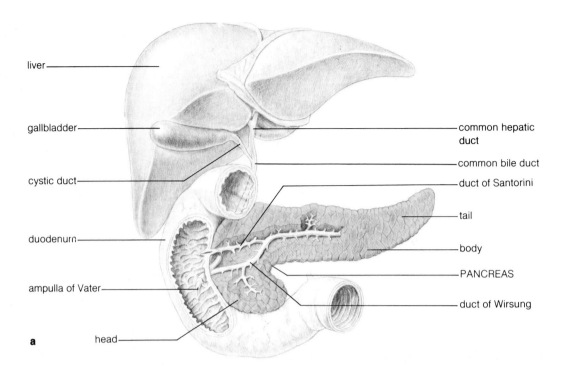

liver

gallbladder

cystic duct

duodenum

ampulla of Vater

head

common hepatic duct

common bile duct

duct of Santorini

tail

body

PANCREAS

duct of Wirsung

a

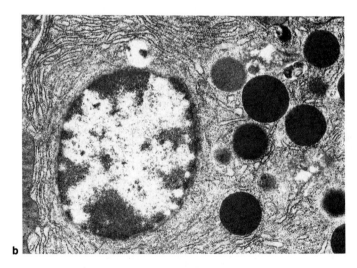

b

Figure 22-17 (**a**) Macroscopic anatomy of the pancreas and its ducts. (**b**) Electron photomicrograph of an exocrine gland cell of the mouse pancreas, × 11,600. Note the extensive endoplasmic reticulum and secretion granules. (Courtesy of D. H. Matulionis.)

rhosis, which causes the liver to be abnormally hard and distended.

The liver consists of a large *right lobe,* extending to the right of the gall bladder; a smaller *left lobe,* to the left of the fissure for the **ligamentum teres hepatis** (fibrous remains of the umbilical vein of the fetus); the *quadrate lobe,* lying between the gall bladder and the ligamentum teres; and the *caudate lobe,* posterior to the quadrate lobe (Fig. 22-20). From one point of

view the internal structure of the liver should be considered to be organized around the blood vessels draining the liver. The *hepatic artery* conveys oxygenated blood from the aorta to the liver, and the *hepatic portal vein* carries blood containing recently absorbed nutrients from the digestive track to the liver cells where nutrients such as glucose are removed from the blood, converted to a storage form such as glycogen, and held in the liver cells until needed by the body. Upon entering the liver at

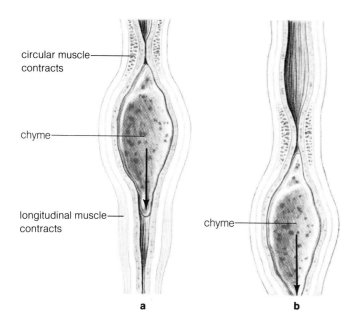

Figure 22-18 Peristalsis. (**a**) Prior to contraction of longitudinal muscle. (**b**) Position of chyme following contraction of longitudinal muscle.

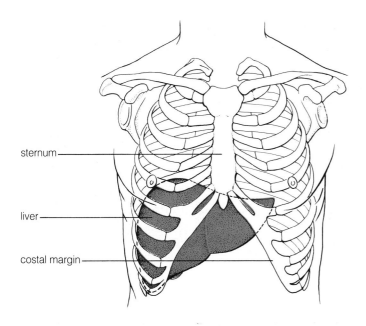

Figure 22-19 The location of the liver in relation to the ribcage.

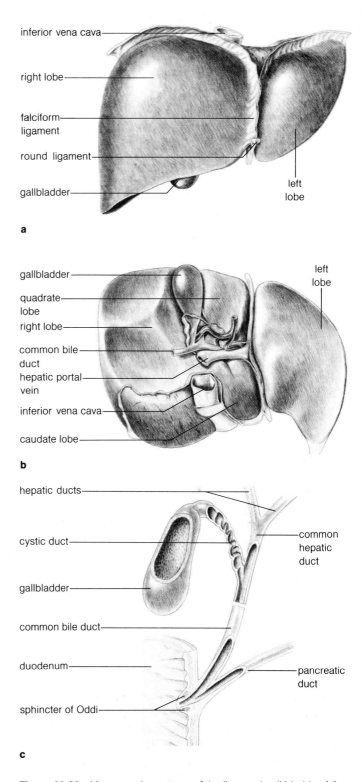

Figure 22-20 Macroscopic anatomy of the liver and gall bladder. (**a**) The anterior surface of the liver. (**b**) The interior surfaces of the liver and gall bladder. (**c**) The bile ducts.

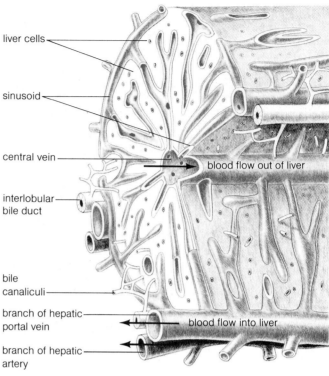

liver cells

sinusoid

central vein

blood flow out of liver

interlobular bile duct

bile canaliculi

branch of hepatic portal vein

blood flow into liver

branch of hepatic artery

a

Figure 22-21 Microscopic anatomy of the liver. (**a**) Diagram. (**b**) Photomicrograph, ×20 (photograph by Joan Creager). (**c**) Photomicrograph, ×85. (Photograph reproduced with permission from *Tissues and Organs: A Text-Atlas of Scanning Electron Microscopy*, by Richard G. Kessel and Randy H. Kardon. W. H. Freeman and Company. Copyright © 1979.)

the *porta* (a transverse fissure on the inferior surface of the liver), the artery and vein parallel one another as they branch repeatedly within the connective tissue extensions of the capsule. The terminal branches of both the hepatic artery and hepatic portal vein open into special vessels called *sinusoids* that bring the blood into close proximity with the gland cells. The sinusoids are lined in part by cells, called Kupffer cells, that are phagocytic and function in removing particulate matter, including bacteria, from the blood stream. The gland cells associated with all the sinusoids that empty into a single *central vein* constitute a *liver lobule* (Fig. 22-21), a hexagonal column of cells supplied peripherally with blood from several branches of the hepatic artery and hepatic portal vein. The gland cells of a lobule are arranged as plates, a single cell in thickness, between adjacent sinusoids; they are oriented in rows, radiating like the spokes of a wheel from the central vein. All the blood is carried away from the lobule by the central vein, which anastosmoses with other central veins, finally joining the hepatic veins.

The vascular pattern of the liver suggests greater exchange of substances between the blood and gland cells than is required simply for the maintenance of an exocrine gland, and in fact the gland cells have essential functions other than bile production. The liver is a constant source of glucose for the rest of the

body; glucose from the intestine is converted by liver cells into glycogen, which is broken down into the glucose again as needed, maintaining a fairly constant blood level of glucose. Glycogen is a less soluble carbohydrate than glucose and is thus more suitable for storage in the cells. Because glucose is our most important energy source, the importance of taking glucose directly from the gut to the liver for storage rather than dumping it into the general circulation is easy to understand. For example, suppose that blood loaded with glucose, following a big meal, passed through skeletal muscles before going to the liver. This might lead to capacity for great physical exertion for a limited period of time until the glucose was exhausted, followed by a period of weakness until another meal was eaten. The maintenance of a constant level of blood glucose makes physical work possible even while fasting, unless the fasting is carried to the extreme of exhausting all available stored sources of energy. Thus the liver functions as a biochemical factory and storage facility from which products of the metabolism of the gland cells are conveyed away from the blood stream. The blood sugar level is regulated by the pancreatic hormones insulin and glucagon, which promote storage of glycogen when the blood level is high (insulin) and stimulate release of glucose when the level of glucose (glucagon) drops.

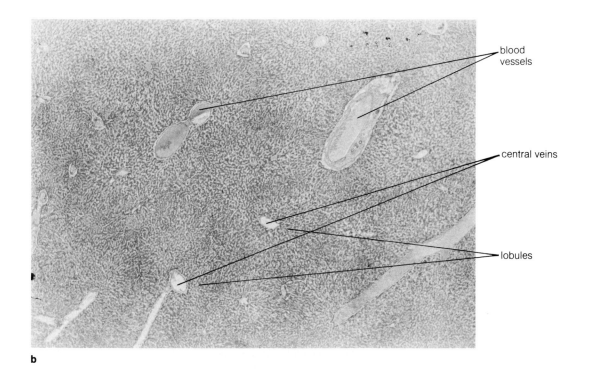

blood
vessels

central veins

lobules

b

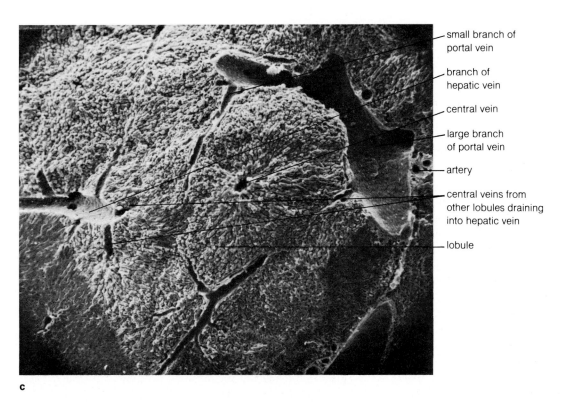

small branch of
portal vein

branch of
hepatic vein

central vein

large branch
of portal vein

artery

central veins from
other lobules draining
into hepatic vein

lobule

c

The liver functions as an exocrine gland by producing *bile,* which is delivered to the duodenum by means of the *common bile duct.* Bile is secreted by liver cells into a network of minute canals, called *canaliculi,* which are expanded intercellular spaces between the plasma membranes of adjacent liver cells. These canaliculi communicate with *small bile ducts* accompanying the hepatic portal vein at the periphery of the lobule. The small bile ducts anastomose, becoming larger as they approach the hepatic porta until they finally emerge at the porta as a single duct, the *hepatic duct,* which sends a branch, the *cystic duct,* to the *gall bladder* where bile is stored and concentrated (Fig. 22-20). The presence of fat within the duodenum causes the release of a hormone, *cholecystokinin* (pancreozymin) from the intestinal mucosa, which stimulates smooth muscle of the gall bladder to contract, forcing bile down the cystic duct toward the duodenum. From the junction of the cystic duct with the hepatic duct, bile is conveyed to the duodenum by the *common bile duct* which joins the pancreatic duct just prior to opening into the duodenum. Bile therefore enters the small intestine mixed with the digestive enzymes of the pancreatic juice elaborated by the pancreas.

The gall bladder is not an essential structure and can be removed surgically in a procedure called cholecystectomy if problems develop making this desirable. In such cases the hepatic ducts and the common bile duct may dilate sufficiently to store bile.

The Pancreas

Next to the liver the *pancreas* (Fig. 22-17) is the largest digestive gland, lying on the posterior abdominal wall at about the level of the second and third lumbar vertebrae and extending transversely from the duodenum on the right to the spleen on the left. The pancreas consists of a *head* with a hooklike extension, the *uncinate process,* from its inferior surface and the *body,* continuous with the *tail* that ends near the spleen. The pancreas has both endocrine and exocrine functions, which are performed by two separate sets of cells. The endocrine cells are in scattered clusters called the *islets of Langerhans,* described in Chapter 18. The exocrine cells collectively constitute a *compound acinar gland,* which elaborates more than a liter per day of pancreatic juice, a fluid containing enzymes that act on all three classes of foods (proteins, fats, and carbohydrates). The exocrine secretory cells (acinar cells) are arranged in clusters called *acini;* each acinus consists of a single layer of cells around a lumen at the end of a small tubule, into which the cells discharge their secretions. These tubules are the terminal branches

of a duct system communicating with the *main pancreatic duct* like the branches on a tree. The main pancreatic duct (of Wirsung) joins the common bile duct as previously described. There is sometimes a second (accessory) pancreatic duct (of Santorini) opening separately into the duodenum superior to the opening of the main pancreatic duct.

The discharge of pancreatic juice is stimulated principally by the hormones *secretin* and *cholecystokinin;* the latter is often called *pancreozymin* in discussions of its effects on the pancreas, as opposed to its effects on the gall bladder. Secretin, produced by the stomach and duodenal mucosae, is released by the presence of food in the pyloric region of the stomach and stimulates the pancreas to elaborate an alkaline fluid that increases the volume of pancreatic juice and can neutralize the acid in the chyme entering the duodenum. Cholecystokinin, produced by the intestinal mucosa, is released largely due to the presence of chyme in the duodenum, and it stimulates the secretion of pancreatic enzymes. In addition, cholecystokinin stimulates contraction of the gall bladder, as described earlier. The enzymatic action of the pancreatic juice takes place as the chyme is moved slowly by peristaltic waves through the twenty feet of small intestine, with absorption of the breakdown products taking place at the same time. Undigested wastes and water are the chief substances remaining in the material entering the large intestine at the ileocecal junction.

DEHYDRATION AND COMPACTION

The material entering the large intestine, or colon, is converted into feces by the processes of dehydration and compaction. In addition to water and undigested particulate matter of food, the feces also contain bacteria, intact blood cells (leukocytes and epithelial cells), cellular debris, glandular secretions, and salts of such minerals as calcium, magnesium, iron, and phosphate. The *large intestine* is so constructed that it removes much of the water, leaving a solid or semisolid mass to be eliminated from the body by way of the rectum and **anal canal.**

The large intestine, or colon (Fig. 22-22), is about five feet long and consists of the *cecum* with its attached *appendix* (or vermiform appendix), the *ascending colon, transverse colon, descending colon,* and *sigmoid colon.* The cecum is a blind-ending pouch extending inferior to the ileocecal junction, usually located in the lower right iliac region of the abdominal cavity (Fig. 22-31). The opening of the ileum is a transverse slit guarded by two mucosal folds, the *ileocecal valve,* that prevent regurgitation of cecal contents into the ileum. The cecum also opens into the appendix, a fingerlike, blind-ending projection of the cecum that varies in length from two to about six inches. The appendix is a vestigial structure with no digestive function.

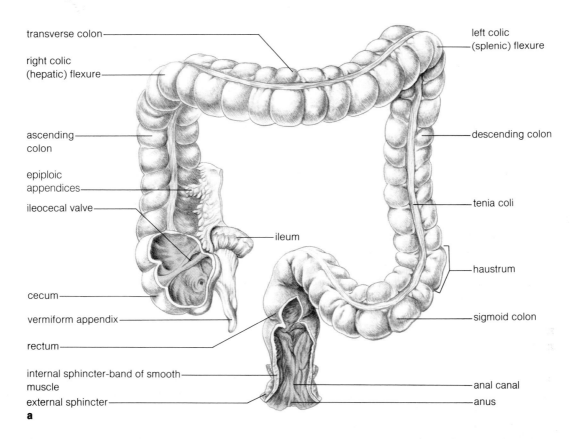

transverse colon
right colic (hepatic) flexure
ascending colon
epiploic appendices
ileocecal valve
cecum
vermiform appendix
rectum
internal sphincter-band of smooth muscle
external sphincter
a

left colic (splenic) flexure
descending colon
tenia coli
haustrum
sigmoid colon
anal canal
anus
ileum

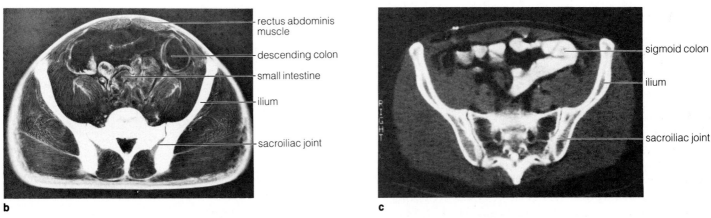

rectus abdominis muscle
descending colon
small intestine
ilium
sacroiliac joint
b

sigmoid colon
ilium
sacroiliac joint
c

Figure 22-22 The colon. (**a**) Macroscopic anatomy of the large intestine. (**b**) X ray of a thick transverse section of the pelvis, cadaver preparation (from the University of Kentucky, Department of Anatomy teaching collection). (**c**) CAT scan of the pelvis, transverse view. The lower digestive tract contains contrast medium (courtesy of S. Goldstein).

Superiorly the cecum is continuous with the colon as it ascends the right side of the posterior body wall in a retroperitoneal position (between the body wall and the parietal peritoneum lining the body cavity). The ascending colon is the shortest part of the colon, ending in a right-angled turn, the *right colic flex-*

ure, where it becomes continuous with the transverse colon at the inferior surface of the liver. The ascending colon has no mesentery and is fixed in position due to its location against the posterior body wall, but the transverse colon is mobile, hanging from its mesentery as it crosses the abdomen below

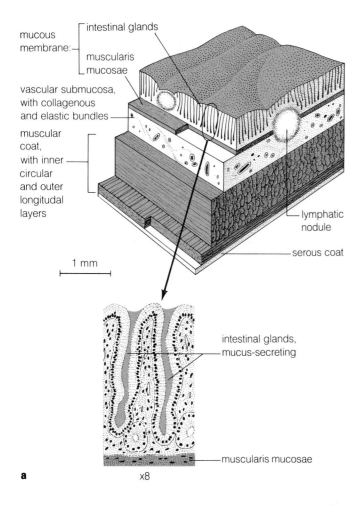

mucous membrane:
— intestinal glands
— muscularis mucosae

vascular submucosa, with collagenous and elastic bundles

muscular coat, with inner circular and outer longitudal layers

— lymphatic nodule
— serous coat

1 mm

intestinal glands, mucus-secreting

muscularis mucosae

a   ×8

b

Figure 22-23 The colon. (**a**) Diagram of the microscopic structure of a block of the wall of the colon and a section of the epithelium of the colon. (**b**) A histological section of the colon ×365 (courtesy of W. K. Elwood).

the stomach and anterior to the loops of the small intestine. The transverse colon is the longest part of the colon, ending at the *left colic flexure* (splenic flexure) near the spleen in the left hypochondriac region (Fig. 22-31). Beginning at the left colic flexure the descending colon passes inferiorly in a retroperitoneal position along the left side of the posterior abdominal wall to reach the pelvic brim (superior boundary of the true pelvis), where it becomes the sigmoid colon. The S-shaped sigmoid colon is suspended by a mesentery as it passes medially and inferiorly to become continuous with the rectum.

In some diseases, such as cancer of the colon, it may become necessary to remove much of the colon. As a result it may be impossible to connect the remaining colon with the anus. In such cases an opening (colostomy) is established in the anterior abdominal wall, creating an artificial anus though which the feces can be discharged.

The mucosa of the colon (Fig. 22-23) can readily be distinguished from that of the small intestine by the absence of villi and the presence of numerous mucous-secreting cells (goblet cells) in the glands. The tremendous surface area required for the absorption of nutrients in the small intestine is not essential in the large intestine, where the only significant absorption is of a relatively small quantity of water. The large quantity of mucus produced by the *goblet cells* facilitates the formation of fecal masses and prevents the masses from becoming dehydrated. The fecal masses are moved along the colon not only by peristaltic waves but also by special processes accomplished by the unique arrangement of the smooth muscle (the muscularis externa) in the wall of the colon. The inner circular muscle is capable of local constriction, forming, in effect, a valve blocking the lumen of the tube. The outer longitudinal muscle is in the form of three strips of muscle, the **taeniae coli,** which are somewhat shorter than the wall as a whole, causing pouches in the wall in the form of sacculations called **haustra** that are characteristic of the colon. Accumulations of fat, called **epiploic appendices**, are also characteristic of the colon. Contraction of the longitudinal muscle shortens the colon, a process that proves useful in moving feces. Assume, for example, that a fecal mass is located at the left colic flexure. The circular muscle contracts, forming a valve behind the mass, blocking the mass at that point. Contraction of the taeniae coli of the descending colon pulls the valve and fecal mass in front of it a short distance down the colon, after which the taeniae coli relax, allowing the valve to return to the region of the left colic flexure but leaving the fecal mass in its new location, ready for a new valve to form. Repetition of this process gradually moves the mass into the sigmoid colon. The circular muscle not only plays an important role in moving the fecal mass but also facilitates compaction and mixing of the feces with mucus by forming valves within the mass, breaking it up (segmentation). Normally most of the feces accumulate in the sigmoid colon prior to elimination.

DEFECATION

Defecation is the process of elimination of the feces from the *rectum* through the *anal canal*. The rectum (Fig. 22-22) is about six inches long; it lies in the midline anterior to the sacrum and follows the curvature of the concave anterior surface of the sacrum. The mucosa of the superior part of the rectum is similar to that of the colon; in the lower portion the rectal mucosa is thrown into longitudinal folds, the *rectal columns* (Fig. 22-24), containing veins that may become abnormally distended, causing the condition known as *hemorrhoids*. At the anal canal there is an abrupt transition from the simple columnar epithelium of the mucosa to stratified squamous epithelium of the skin.

The muscularis externa of the rectum has continuous circular and longitudinal layers of smooth muscle that end at the anal canal with thickening of the circular layer to form the *internal anal sphincter*. The anal canal is lined by epithelium resembling the skin which, like the mucosa of the lower rectum, is thrown into longitudinal folds (anal columns) containing veins that may become hemorrhoidal. The muscle surrounding the anal canal is skeletal muscle (part of the levator ani) and forms the voluntary external anal sphincter.

Defecation occurs by relaxation of the internal and external anal sphincters, coincident with massive peristaltic waves beginning in the sigmoid colon. The force of the peristaltic wave, augmented by closing the vocal folds (holding the breath) and contracting the abdominal wall muscles, increases the intra-abdominal pressure, forcing the fecal mass downward.

REGULATION OF THE DIGESTIVE SYSTEM

The digestive system contains two types of effectors that respond to neural stimuli: *muscles* and *glands*. The muscle in the wall of the pharynx and upper esophagus is skeletal, or voluntary, muscle and that of the remainder of the digestive tract is smooth, or involuntary, muscle. The contraction of both types of muscle is regulated by the *nervous system*, and though the mechanisms

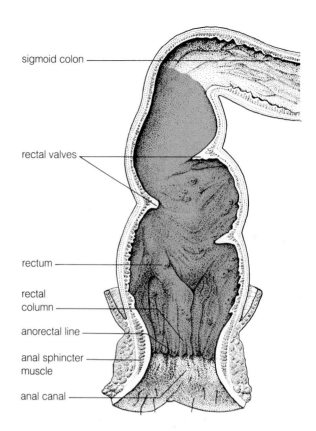

sigmoid colon
rectal valves
rectum
rectal column
anorectal line
anal sphincter muscle
anal canal

Figure 22-24 Diagram of the rectum opened to show the lining.

of regulation differ somewhat, the principal difference is in the speed with which the muscle fibers contract: skeletal muscle contracts much more rapidly and with more force than does smooth muscle.

The digestive glands vary in size and location—from unicellular glands, such as mucous-secreting (goblet) cells scattered among the epithelial lining cells of the gut, to large compound exocrine glands, such as the pancreas and salivary glands located outside the wall of the digestive tube. These glands are subject to influences or regulation by both the *nervous system* and *hormones*.

Regulation of Muscle Contraction

The functions of muscle contractions in the wall of the digestive tube are principally to transport the food down the tube and to mix the food thoroughly with enzymes, a feat accomplished by a churning action in the abdominal portion of the digestive tract. In the oral cavity the food is broken into small pieces and moistened with saliva by the process of chewing, with the teeth

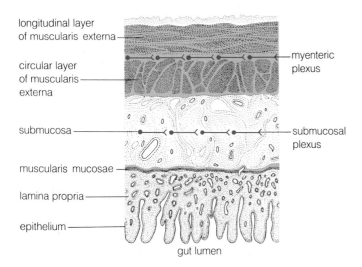

longitudinal layer
of muscularis externa

circular layer
of muscularis
externa

submucosa

muscularis mucosae

lamina propria

epithelium

myenteric
plexus

submucosal
plexus

gut lumen

Figure 22-25 Diagram of a section of the gut wall illustrating the intrinsic innervation of the gut. See text for description.

serving as grinders. The muscles involved in chewing are skeletal muscles under *voluntary control*. The most important muscles of mastication are supplied by the *trigeminal* (cranial V) *nerve,* whose motor nucleus is located in the pons. It contains the cells of origin (cell bodies) of the peripheral fibers that terminate on the muscle fibers. The activity of the trigeminal nerve, and thus of the muscles, may be regulated at the conscious level by corticobulbar fibers from the primary motor area of the cerebral cortex. In other words, you may chew or not as you like. Most of the time chewing is initiated consciously and is then continued automatically by the extrapyramidal system, with the aid of proprioceptive stimuli from receptors in the periodontal ligaments, adjusting the force of the bite to the requirements.

The process of *swallowing* is initiated voluntarily as the bolus of food is pushed upwards and backwards by the tongue (Fig. 22-6). Then reflexes take over, elevating the soft palate and constricting the superior portion of the pharynx (the nasopharynx). The afferent limb of the reflex arc consists of sensory fibers in *cranial nerve IX* (glossopharyngeal). The impulses are relayed within the medulla to a motor nucleus (nucleus ambiguus) of *cranial nerve X* (vagus), whose fibers supply the constrictors of the pharynx and most of the muscles elevating the soft palate and larynx. A similar series of neural reflexes propels the bolus of food down the esophagus. The presence of the bolus in the esophagus stimulates sensory nerve endings of afferent fibers of the vagus nerve, which relay the impulses within the medulla to the motor nuclei of the vagus, which stimulates the muscles of the esophagus to contract. In the upper third of the esoph-

agus, which contains skeletal muscle fibers, the motor nucleus involved is the nucleus ambiguus; it sends out fibers that terminate directly on the skeletal fibers. Thus the contraction of the upper esophagus is potentially a voluntary motor mechanism, but actually it always operates on a *reflex* basis, never reaching the conscious level. In the lower two-thirds of the esophagus which contain smooth muscle, the afferent limb of the reflex is the same. The efferent limb, however, is from the dorsal motor nucleus (parasympathetic) of the vagus, and the peripheral fibers (preganglionic parasympathetic) terminate on the cell bodies of postganglionic parasympathetic neurons scattered between the layers of smooth muscle in the wall of the esophagus. The postganglionic fibers transmit the impulses to the smooth muscle cells, causing them to contract. This *extrinsic regulation* by the vagus nerve provides the means of control of peristalsis for the lower esophagus, stomach, small intestine, and colon as far as the junction of the transverse colon at the left (splenic) colic flexure. The visceral, or autonomic, peristaltic reflex produces waves of constriction of circular muscle that churn the food and propel it down the gut. Stimulation of the vagus nerve by higher centers, especially the hypothalamus, may cause hyperactivity of the gut musculature. Influences arising from a higher center may be initiated by emotional states, such as anxiety or worry, resulting in digestive symptoms that can be severe.

The extrinsic control of the gut by the vagus nerve can be eliminated surgically without abolishing peristalsis, clearly illustrating that there is also a mechanism for intrinsic control of the gut musculature. The *intrinsic regulation* of the contraction of the smooth muscle of the gut wall is the function of the postganglionic parasympathetic neurons, which form a plexus of neurons called the *myenteric plexus,* located between the outer longitudinal and inner circular layers of muscle (Fig. 22-25). These are the postganglionic neurons on which the preganglionic fibers of the *vagus* and *sacral nerves* synapse, increasing their activity. When the influence of the preganglionic fibers is removed, the muscle still possesses an inherent slow rhythmic contractility. The postganglionic neurons are also still capable of being stimulated by the presence of food in the gut. The postganglionic neurons not only stimulate smooth muscle cells but also stimulate one another within the myenteric plexus, which extends the length of the gut. Thus a bolus of food in the duodenum, for example, may initiate impulses that spread through the intrinsic neural plexus, resulting in a wave of muscle contraction that may travel the entire length of the gut.

In addition to the extrinsic and intrinsic neural control of contraction, the smooth muscle has the unique ability to contract upon stimulation by factors other than nerve impulses. Unlike skeletal muscle, which cannot contract unless stimulated by the nervous system, smooth muscle contracts in response to *direct stimuli* such as pressure or stretch. A bolus of food entering the gut exerts pressure against the gut wall, stretching the muscle cells. The smooth muscle responds by contracting, forcing the bolus down the tube. This mechanism does not involve the nervous system and is considered *myogenic regulation* of peristalsis.

Regulation of Enzyme Production

The glands of the digestive system include the salivary glands, pancreas, liver, and glands that are totally within the wall of the digestive tract, such as the gastric glands of the stomach and the intestinal glands of the small intestine. All of these glands except the liver are regulated chiefly by nerves, with hormonal influences on the activity of some glands. The liver is an unusual gland in that the product important in the process of digestion (bile) is a by-product of the liver's metabolic activity. That is, the production of bile is regulated only by the metabolism of the liver itself (subject to many nutritional and biochemical variables), though the discharge of bile into the digestive tract is under hormonal regulation. The influence of the nervous system on all the other glands is simply a matter of increasing the secretory activity of the cells.

Salivary Glands   The quantity of secretion produced by the salivary glands is regulated by the parasympathetics, whereas the sympathetics influence the quality of the material elaborated. Secretory activity of the submandibular and sublingual glands, as well as the minor salivary glands of the oral mucosa of the floor of the mouth, is increased by parasympathetic stimulation, whose preganglionic fibers arise in the lower pons in the facial nerve (cranial VII). The parotid gland receives a similar parasympathetic supply from the glossopharyngeal nerve (cranial IX). Stimulation of the sympathetic fibers innervating the smooth muscle of the arteries that supply the salivary glands reduces blood flow to the glands, causing the secretions to be more viscous. As a result of the reduced blood flow the volume of the secretions is drastically reduced. Thus the sympathetics have an indirect effect upon the production of saliva but do not innervate the secretory cells.

The influence of higher centers on the activity of the autonomic nervous system is clearly demonstrated by observable psychic effects upon salivary secretion. For example, the thought of food results in stimulation of the parasympathetics, increasing the quantity of saliva produced, whereas fear inhibits the parasympathetics and stimulates the sympathetics, resulting in a dry mouth.

Mucosal Glands of the Stomach and Small Intestine

These glands are regulated by the parasympathetics of the *vagus nerve,* which conveys preganglionic fibers to ganglionic neurons in the submucosa (submucosal plexus of Meissner) of the stomach and small intestine. The postganglionic fibers of these neurons are distributed to the secretory cells, and stimulation of the fibers increases the cells' activity. Hyperactivity, especially of the parietal cells of the gastric glands, which produce hydrochloric acid, is frequently a contributing cause of gastric or duodenal ulcers.

The Pancreas   Secretory activity of the pancreas is regulated by the *nervous system* (vagus nerve) and by *hormones.* The relative importance of these two factors is still a matter of debate. Stimulation of the *vagus nerve* causes an increase in the production of enzymes by the pancreatic cells, without much increase in fluid volume. Hormonal influences are initiated by the presence of food in the stomach and of chyme in the duodenum, which stimulate the release of *secretin* in the stomach mucosa and cholecystokinin (*pancreozymin*) from the duodenum. These two hormones stimulate the pancreas to increase its secretory activity. Secretin stimulates the production of a large volume of fluid containing bicarbonate that neutralizes the acidity of the duodenal contents, whereas cholecystokinin, like neural stimulation, increases production of enzymes with little increase in fluid volume. In addition to these two hormones, *gastrin* elaborated by the stomach mucosa has a slight effect similar to that of cholecystokinin.

The Liver   The liver receives sensory and autonomic nerve fibers from the *vagus nerve,* but the influence of these fibers upon the metabolic activity of liver cells has not been determined. The sensory fibers are, however, important in initiating reflexes that stimulate the smooth muscle in the biliary ducts, facilitating the transport of bile from the liver to the gall bladder. Bile is produced at a relatively constant rate and is stored in the gall bladder. Discharge of the bile from the gall bladder into the duodenum is controlled by the hormone *cholecystokinin.* The presence of fat in the duodenum stimulates the duodenal mucosa to release cholecystokinin, which stimulates the contraction of the smooth muscle in the wall of the gall bladder. The bile is discharged through the common bile duct into the duodenum, where it facilitates the digestive process by emulsifying fat.

DEVELOPMENT OF THE DIGESTIVE SYSTEM

The digestive system begins as a simple *endoderm-lined tube* formed by the closing of the body wall (Fig. 22-26). Closure of

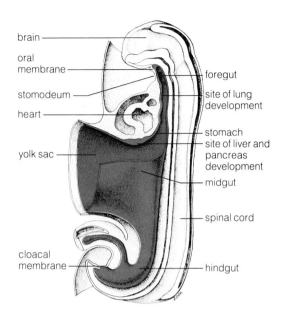

brain

oral membrane

stomodeum

heart

yolk sac

cloacal membrane

foregut

site of lung development

stomach

site of liver and pancreas development

midgut

spinal cord

hindgut

Figure 22-26 The digestive tract at the beginning of the fourth week of development.

the wall leaves the *yolk sac* suspended from the middle portion of the tube (see Chapter 3). Using the yolk sac attachment as a landmark, we can divide the embryonic gut for descriptive purposes into *foregut, midgut,* and *hindgut.* The *embryonic pharynx* is relatively more extensive than the pharynx of the adult and can be equated roughly with the foregut. Derivatives of the foregut include the *oral cavity, pharynx, esophagus,* and *stomach,* and the *duodenum* to the point of entrance of the common bile duct and main pancreatic duct. The *liver* and *pancreas* arise as evaginations of the foregut. Derivatives of the midgut include the remainder of the *small intestine* and the *colon* as far as the *splenic flexure.* The hindgut gives rise to the *descending* and *sigmoid colon,* the *rectum,* and parts of the *urinary* and *reproductive systems.*

The Oral Cavity

The tongue develops in the floor of the oral cavity from swellings associated with the visceral arches. The anterior part of the tongue develops mostly from *lateral lingual swellings* of the first arch, and the rest of the tongue develops from the *hypobranchial eminence* of the third arch. The line of fusion between the lingual swellings and the hypobranchial eminence is indicated in the adult by the sulcus terminalis. The skeletal muscle of the tongue is derived from *occipital myotomes.*

Salivary glands all begin their development as solid epithelial buds from the oral epithelium that branch repeatedly as they lengthen. The solid cords canalize (become hollow), forming ducts, and the cells at the terminal ends of solid cords canalize, forming ducts, and the cells at the terminal ends of the ducts specialize as secretory cells. The ducts open into the oral cavity at the points from which they originated from the oral epithelium.

The Pharynx

The embryonic pharynx (Fig. 22-27) lies ventral to the brain and thus is in the head region. However, most of the pharyngeal derivatives grow or migrate to a more caudal (inferior) definitive position.

The endodermal lining of the embryonic pharynx projects laterally as a series of visceral pouches between the visceral arches. The dorsal recesses of the *first* and *second visceral pouches* elongate, and the terminal portion dilates to become the *middle ear cavity,* which remains continuous with the pharynx via the auditory tube derived from the constricted portion of the first visceral pouch. The ventral recess of *visceral pouch II* remains shallow, forming the epithelium of the *tonsillar fossa* and the covering of the palatine tonsil. The *third* and *fourth visceral pouches* give rise to the *inferior* and *superior parathyroid glands.* Most of the *thymus* is derived from the *ventral portion* of the *third pouches,* with perhaps some contribution from the fourth pouches.

The wall of the pharynx is thickened by skeletal muscles, such as the pharyngeal constrictors, that differentiate from the mesoderm of the visceral arches. With the growth of these and other muscles of the neck, the visceral clefts initially present between the visceral arches gradually disappear.

The Esophagus

In early embryos there is no esophagus. As the neck grows, the esophagus forms as a tube between the pharynx and stomach, growing in length to keep pace with adjacent structures. The epithelial lining becomes stratified and is ciliated for a time. Muscle layers of the wall differentiate from the unsegmented mesoderm of the lateral plate. The esophagus, developing dorsal to the heart in the posterior mediastinum and embedded in mesenchyme, is never covered by serous membrane and does not have a mesentery.

The Stomach

The stomach begins as a slight dilation of the gut tube. From its earliest position in the neck it is shifted to its permanent posi-

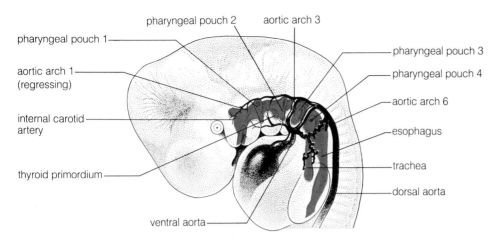

Figure 22-27 Diagram of a lateral view of the embryonic pharynx and related aortic arches.

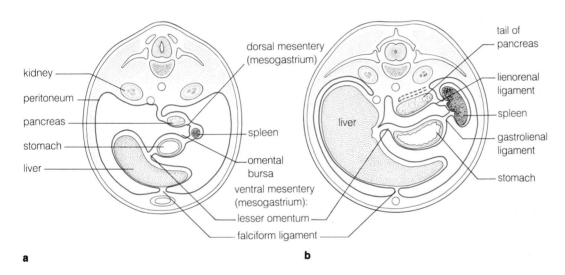

a   **b**

Figure 22-28 (**a**) and (**b**) Diagrams of transverse sections of two stages in the development of the stomach and related structures.

tion in the abdominal region as a result of differential growth of the neck and thorax. The stomach tube is flattened laterally and is suspended from the body wall by *dorsal* and *ventral mesenteries* (Fig. 22-28a), in which the liver, pancreas, and spleen develop. The dorsal aspect of the stomach expands at a more rapid rate than the ventral surface, creating the greater and lesser curvatures of the stomach. As the greater curvature expands, the stomach rotates so that the original dorsal surface (greater curvature) shifts toward the left, pulling the dorsal mesentery and tail of the pancreas toward the left. The *spleen* begins to develop in the dorsal mesentery in its new position, remaining on the

left in the adult. As the dorsal mesentery expands, its attachment to the posterior abdominal wall shifts from the midline to the left side (Fig. 22-28b), and the pancreas becomes *retroperitoneal* (dorsal to the peritoneum of the posterior wall). That portion of the ventral mesentery between the lesser curvature of the stomach and the liver becomes the *lesser omentum;* that portion between the liver and the ventral body wall becomes the *falciform ligament.*

With the enlargement of the stomach in the formation of the greater curvature, there is some increase in surface area of the epithelial lining. However, this increase is very slight in

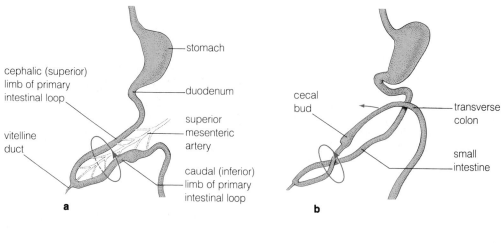

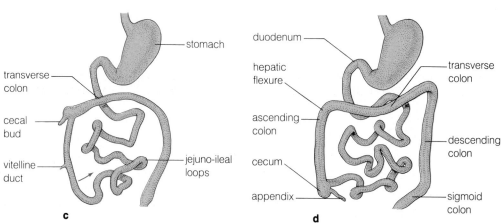

Figure 22-29 Development of the intestines illustrating rotation of the gut. (**a**) Protrusion of the gut into the umbilical cord. (**b**) Rotation of the caudal limb of the intestinal loop toward the right. (**c**) Withdrawal of the cephalic limb into the body cavity. (**d**) Withdrawal of the caudal limb into the body cavity so that the transverse colon lies anterior to the small intestine.

comparison with the increase in area created by the mucous membrane folds (rugae) and by the development of the gastric glands from localized proliferation of the epithelium.

The Intestines

The midgut elongates more than either the foregut or the hindgut and gives rise to the *duodenum, jejunum, ileum, appendix, cecum,* and the *colon* to the *left (splenic) flexure.*

Initially the *yolk sac* is very large and is open wide at its junction with the gut. This changes rapidly as the yolk sac regresses until its connection narrows to a slender stalk, the vitelline duct (Fig. 22-29a). The gut tube elongates more rapidly than the abdominal cavity, and with the enlargement of the liver there is not enough room within the trunk to house everything, so a portion of the gut including the yolk sac and vitelline stalk herniates (pushes out) into the umbilical cord (Fig. 22-29a). While still herniated into the umbilical cord, the gut rotates so that

the inferior limb comes to lie on the right and the superior limb on the left (Fig. 22-29b). Return of the gut to the abdomen follows a decrease in relative size of the liver and kidneys. The originally superior limb returns first (Fig. 22-29c) and begins to develop into the highly coiled small intestine; it pushes against the future descending colon (derived from the hindgut) on the left, causing it to lose its mesentery and become retroperitoneal. When the inferior limb of gut is pulled back into the abdominal cavity and becomes the ascending colon, it descends into the iliac (lower abdominal) region by a slow process of growth, with the cecum ending up in the lower right quadrant (Fig. 22-29d). The ascending colon loses its mesentery, but the transverse colon crosses anterior to the small intestine and retains its mesentery.

The wall of the small intestine does not expand as does that of the stomach. However, the epithelial lining does proliferate rapidly; it thickens until the lumen of the gut is obliterated by endodermal cells. As development proceeds, the gut begins

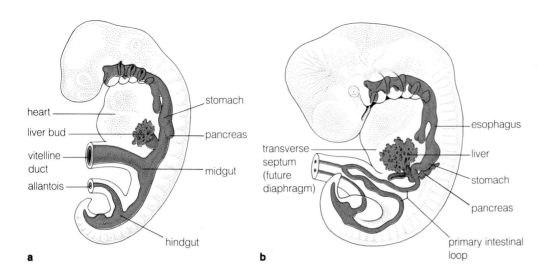

Figure 22-30 Early development of the liver and pancreas. (**a**) Diagram of a 25-day embryo. (**b**) Diagram of a 32-day embryo.

The Liver and Pancreas

The liver and pancreas are derived during embryonic development as outgrowths from the digestive tube and retain their connection with the gut by means of ducts that usually join one another just before opening into the medial surface of the duodenum.

The tremendous capacity of the endoderm to proliferate is best seen in the development of the liver which begins as a sacklike *evagination* of the duodenal lining into the *ventral mesentery* (Fig. 22-30a). The original evagination remains hollow and becomes the *common bile duct*. The blind end of the sack grows rapidly, forming a solid mass of cells. As blood vessels penetrate the gland, the mass breaks up into plates, associated with blood vessels, a process that continues until the adult morphology is attained. At first the liver is completely enclosed in the ventral mesentery, but it expands rapidly, becoming the largest abdominal organ, and comes into contact with the developing diaphragm (septum transversum) superiorly. The liver fuses with the diaphragm, giving rise to the "bare area" of the liver.

The pancreas begins as two evaginations from the duodenum, a *dorsal* and a *ventral* outgrowth into the dorsal and ventral mesenteries (Fig. 22-30b). The ventral pancreas arises in close association with the embryonic common bile duct, which is also the liver primordium. With further growth there is a shift in the relative position, so that the ventral pancreas grows into the dorsal mesentery, joining the dorsal pancreas. The mesen-tery is resorbed as the duodenum becomes retroperitoneal, leaving the pancreas embedded in the dorsal body wall. The duct of the ventral pancreas becomes the *main pancreatic duct,* which opens into the duodenum along with the common bile duct, and the ducts of the dorsal pancreas become connected to it. The original duct of the dorsal pancreas may remain patent as an *accessory pancreatic duct*. The ventral pancreas gives rise to the *uncinate process* and part of the *head,* and the dorsal pancreas becomes the *body* and *tail* and contributes to the *head.*

TOPOGRAPHICAL ANATOMY OF THE ABDOMINAL VISCERA

The parts of the digestive tract have no absolutely fixed position or shape. Although some parts are more fixed than others, the location and shape of any part of the abdominal digestive tract depends upon the position of the body (standing or lying down) and the contents (full or empty). The surface of the abdominal region has visible and palpable landmarks that can be used as reference points in locating structures approximately. The *linea alba,* extending vertically in the midline, is usually visible, and in slender individuals the outlines of the *rectus abdominis* muscles are usually visible when the muscles are contracted. It is always possible to palpate the *costal margin* and the *iliac crest.* The *anterior superior iliac spines* and the *pubic tubercles,* to which the inguinal ligaments are attached, provide additional palpable landmarks that can be used in drawing lines, marking off the abdomen into regions that may prove useful in locating the abdominal viscera. The simplest regional subdivisions are obtained by drawing a *vertical line* down the linea alba and a

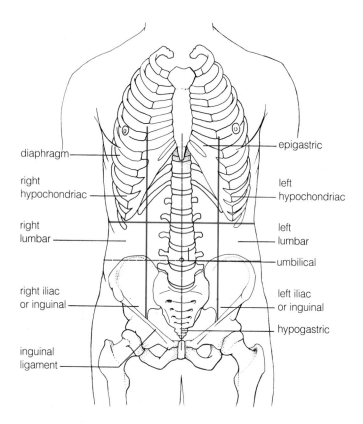

Figure 22-31 The nine topographical regions of the abdomen.

diaphragm

right hypochondriac

right lumbar

right iliac or inguinal

inguinal ligament

epigastric

left hypochondriac

left lumbar

umbilical

left iliac or inguinal

hypogastric

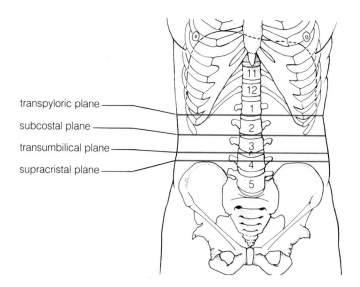

Figure 22-32 Transverse planes of the abdomen based on vertebral levels.

transpyloric plane

subcostal plane

transumbilical plane

supracristal plane

transverse line through the *umbilicus,* dividing the abdomen into the right and left upper and lower *quadrants.* More precise locations may be designated by dividing the abdomen into nine regions according to the following description.

Abdominal Regions Based on Surface Landmarks

The abdomen may be divided into nine regions by drawing two vertical and two transverse lines (Fig. 22-31). The vertical lines (right and left lateral lines) extend parallel to the midsagittal plane from the *midclavicular point* to the *midlinguinal point.* The latter point may be obtained by measuring the distance between the anterior superior iliac spine and the pubic tubercle. The upper transverse line is drawn through the *transpyloric plane,* located at half the distance between the jugular notch (upper border) of the sternal manubrium and the pubic symphysis. The lower transverse line passes through the *iliac tubercles* (palpable prominences near the highest point of the iliac crest) of either side. This divides the abdomen into three central subdivisions named *epigastric, umbilical,* and *hypogastric* and six lateral subdivisions designated right and left *hypochondriac, lumbar,* and *iliac* (inguinal).

These subdivisions of the abdomen may be used to indicate the approximate location of the abdominal viscera. For example, the esophageal-stomach junction is relatively fixed in position in the epigastric region slightly to the left of the midline, whereas the fundus extends into the left hypochondriac region following the dome of the diaphragm. The pylorus and first part of the duodenum are usually at the transpyloric plane to the right of the midline. The junction of the duodenum with the jejunum is usually inferior to the transpyloric plane in the left part of the umbilical region. The remaining small intestine has no fixed position but occupies much of the umbilical and hypogastric regions, ending at the ileocecal junction in the right iliac region. The ascending colon ascends lateral to the right lateral line, usually ending in the right lumbar, whereas the transverse colon is more variable in position. It may descend into the hypogastric region as it crosses the abdominal cavity. The descending colon passes from the left hypochondriac and lumbar into the iliac region where it becomes the sigmoid colon as it turns medially into the hypogastric regions before terminating at its junction with the rectum.

Transverse Planes Based on Vertebral Levels

Abdominal viscera may also be located in reference to transverse planes (Fig. 22-32). Though the planes are defined by specific vertebral levels, they can be located anteriorly by palp-

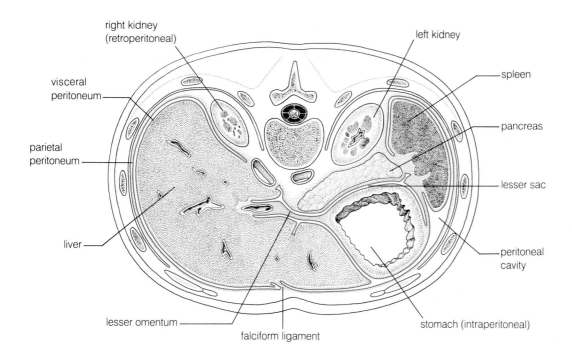

right kidney (retroperitoneal)

visceral peritoneum

parietal peritoneum

liver

lesser omentum

falciform ligament

left kidney

spleen

pancreas

lesser sac

peritoneal cavity

stomach (intraperitoneal)

Figure 22-33 Diagram of retro-peritoneal structures, transverse section.

able or visible landmarks. The planes are variously designated, but usually the *transpyloric plane* lies at the level of the first lumbar vertebra (L-1). The *subcostal plane* connecting the lowest points of the costal margins lies at the level of L-2; the plane through the *umbilicus* usually lies at the intervertebral disc between L-3 and L-4. The *supracristal plane* connecting the highest points of the iliac crest crosses the level of the spine of L-4.

Surgeons take advantage of these and other landmarks in planning their incisions for abdominal operations. For example, *McBurney's point,* one-third of the way from the right anterior superior iliac spine to the umbilicus, usually overlies the appendix and is the point at which the incision is made for an appendectomy.

THE PERITONEAL RELATIONS OF THE ABDOMINAL VISCERA

The abdominal viscera, contained within the abdominal cavity, are prevented from adhering abnormally to one another and to the body wall by a *serous membrane,* called the *peritoneum.* A serous membrane is a smooth, glistening sheet of a simple squamous epithelium, termed *mesothelium,* and an associated thin layer of loose connective tissue containing numerous blood

capillaries. The membrane is kept moistened by the elaboration of *serous fluid,* which enables the membranes to slide on one another.

General Arrangement of the Peritoneum

The abdominal cavity, enclosed by the body wall, extends from the diaphragm to the pelvic floor but is usually divided for descriptive purposes into the *abdominal cavity proper* and the *pelvic cavity* (inferior to the pelvic brim). The serous membrane lining the inner surface of the body wall and the diaphragm, termed *parietal peritoneum,* forms the outer boundary of the peritoneal cavity or the coelom. The membrane covering the organs within the peritoneal cavity is termed *visceral perito-neum* and may be considered the inner lining of the *peritoneal cavity,* the potential space between the visceral and parietal layers of the peritoneum (Fig. 22-13). Because the abdominal viscera are closely packed within the available space, the peritoneal cavity is a potential space only, containing nothing but a small quantity of serous fluid between the two membranes, which are pressed closely together.

Some of the abdominal viscera, for example, the pancreas, lie on the posterior abdominal wall, covered by peritoneum only on their anterior surfaces. Organs in this position are said to be *retroperitoneal* in location and are more or less fixed in place (Fig. 22-33). Other organs, such as most of the small intes-

Clinical Applications: **Diseases of the Digestive System**

The stomach and that part of the duodenum proximal to the entrance of the common bile and pancreatic ducts are derived in the embryo from the foregut, and their exposure to the acid contents of the stomach leads to certain similarities in responses to ulcers. The remainder of the small intestine and most of the colon, which are derived from the embryonic midgut, are exposed in the adult to alkaline digestive juices and respond in a similar manner in diseased states.

Diseases of the Stomach and Duodenum

The stomach and first part of the duodenum are susceptible to the onset of *peptic ulcers* (ulcers of the digestive tract), caused by the exposure of sensitive mucosa to the combination of pepsin and hydrochloric acid, which can erode the mucous membrane and denude the blood vessels and nerve endings of the underlying connective tissue. Bleeding may result, and the exposure of the nerves of the stomach may cause severe pain. Acute peptic ulcers usually heal quickly, but chronic ulcers usually persist, refusing to heal over long periods of time. *Inflammatory diseases* (gastritis) may be caused by surface irritants such as aspirin or pepper, or by bacterial and viral infections, producing cell death and leading to lesions involving edema and cellular infiltrate typical of inflammatory reactions.

Almost all of the *tumors* of the stomach (98%) are carcinomas (epithelial in origin). A tumor may occur as a large mass projecting into the lumen of the stomach or may be of a diffuse, infiltrating form, spreading rapidly throughout the stomach, to the lymph nodes, and to other organs.

Diseases of the Intestines

Lesions of the mucous membrane of the intestine may be produced by a variety of causes, including bacterial infections and chemical poisons. *Chronic ulcerative colitis* is a distressing inflammatory lesion that may be caused by an overproduction of lysozyme, an enzyme that can destroy the protective coat of mucus within the colon. The lesions usually involve only the mucous membrane. *Dysentery* may be caused by bacteria or by *Entamoeba histolytica,* a parasitic ameba (a protozoan). In the bacterial type of dysentery, necrotic lesions (regions of dead tissue) are common in the mucous membrane, whereas in amebic dysentery the lesions are usually submucosal. The two kinds of dysentery can be differentiated by microscopic examination of the feces.

The intestines may fail in the primary responsibility of absorbing digested nutrients, especially fatty acids, in a general condition referred to as malabsorption. In some cases the defect appears to have a morphological basis with atrophy of the intestinal villi; in other cases, such as the disease known as *sprue,* the defect may be biochemical, involving the vitamin B-12.

---

tine, are almost completely invested by visceral peritoneum and are suspended from the posterior abdominal wall by a *mesentery* composed of two layers of serous membrane. Such organs are considered to be *intraperitoneal* in location. The mesenteries are continuous with the parietal peritoneum at the body wall and with the visceral peritoneum on the surface of the organ. Blood vessels and nerves coursing from the body wall to the viscus (organ) pass within the mesenteries connecting the two layers of serous membrane.

The coelom may be considered as a *peritoneal sac* that is subdivided into a *greater sac* and a *lesser sac* (omental bursa) by the stomach and two special mesenteries, the **greater** and **lesser omenta.** The greater omentum is a double layer of serous membrane attached to the greater curvature of the stomach, extending inferiorly like an apron and lying between the anterior abdominal wall and the intestines. It folds back on itself to pass superiorly, becoming attached to the surface of the transverse colon, which it crosses before reaching the posterior body wall behind the stomach. Usually the double folds of the greater omentum fuse inferior to the transverse colon, so that no space exists within the "apron" lying in front of the intestine. From the lesser curvature of the stomach the lesser omentum extends as a double sheet of serous membrane, connecting the stomach and the liver. Near the stomach–duodenal junction the lesser omentum ends abruptly with a free border containing the common bile duct and the blood vessels supplying the liver. The space posterior to the lesser omentum and stomach is the lesser sac, communicating with the space anterior to the lesser omentum and stomach by an opening, the *epiploic foramen,* between the free border of the lesser omentum and the posterior body wall. The lesser sac is bounded inferiorly by the greater omentum and its attachment to the posterior body wall. The potential space of the lesser sac is important in providing room for expansion of the stomach when a large volume of food is consumed.

The greater sac consists of all the coelom except the lesser sac, extending anterior to the stomach and small intestine from the diaphragm superiorly to the pelvic floor inferiorly. The lesser sac contains no viscera within its potential space; the greater sac is considered to contain within its boundaries all of the intraperitoneal organs of the abdominal cavity.

WORDS IN REVIEW

Anal canal	Mucosa
Buccal glands	Mucous membrane
Cementum	Muscularis externa
Circumvallate papillae	Muscularis mucosa
Deglutition	Oral vestibule
Dentin	Parietal cells
Falciform ligament	Periodontal membrane
Foramen cecum	Plicae circularis
Frenulum	Pyloric
Haustra	Retroperitoneal
Hypobranchial eminence	Rugae
Intraperitoneal	Serosa
Lacteal	Serous membrane
Lamina propria	Submucosa
Lesser omentum	Sulcus terminalis
Ligamentum teres hepatis	Villi
Mesentery	

SUMMARY OUTLINE

I. The Digestive System
The digestive system consists of the alimentary canal and the digestive glands; digestion is the mechanical and chemical (enzymatic) conversion of food into absorbable material.

II. The Alimentary Canal
The alimentary canal (digestive tract) is the tube from the mouth to the anus; it includes organs specialized for ingestion and mechanical breakdown, rapid transport, enzymatic breakdown and absorption, compaction and dehydration, and elimination.

III. Digestive glands
Digestive glands produce enzymes and other substances involved in digestion; they include the liver and pancreas (accessory glands) and smaller glands located within the alimentary canal.

IV. Layers of the Gut Wall
A. Mucosa (mucous membrane) includes the epithelium, lamina propria, and the muscularis mucosa.
B. Submucosa is the connective tissue layer between the mucosa and the muscle layer.
C. The muscularis externa is smooth muscle in the abdominal portion of the digestive tract and consists of an inner circular and an outer longitudinal layer.
D. The serosa is the visceral peritoneum in contact with the outer surface of the wall of the gut.

V. Mechanical Breakdown of Food
Mechanical breakdown of food (chewing) takes place in the oral cavity (mouth).
A. The oral cavity is bounded by the lips, cheeks, anterior pillars of the fauces, floor of the mouth, and palate.
B. The lips contain skeletal muscles that close the mouth.
C. The cheeks contain skeletal muscles (the buccinators) that help keep the food between the teeth.
D. The teeth have a crown of enamel over dentin and a root of cementum over dentin, with a neck where the enamel meets the cementum. They bite and grind the food.
E. The palate is the roof of the mouth; it separates the oral and nasal cavities.
   1. The hard palate is bone.
   2. The soft palate consists of skeletal muscles covered by mucous membrane.
F. The tongue contains three pairs of extrinsic and three sets of intrinsic muscles that function in speech, help keep food between the teeth, and are essential in swallowing.
G. The salivary glands include the parotid, submandibular, sublingual and minor salivary glands of the mucous membrane; they moisten food and prepare it for swallowing.

VI. Rapid Transport of Food

The region of rapid transport of food includes the pharynx and esophagus.

   A. The epithelium that lines the pharynx and esophagus is stratified squamous.

   B. The muscle consists of skeletal muscles in the pharynx and upper esophagus, with smooth muscle in the lower esophagus.

   C. The pharynx is an organ of swallowing, which is initiated voluntarily and continued as a reflex action.

VII. Enzymatic Hydrolysis and Absorption

Enzymatic hydrolysis and absorption take place in the stomach and small intestine.

   A. The stomach is specialized for enzymatic hydrolysis (chemical breakdown of enzymes). The surface area of the epithelium is increased by mucosal folds called rugae and gastric glands that produce enzymes and hydrochloric acid. The muscularis externa consists of three layers of smooth muscles arranged to facilitate the mixing of food, which is termed chyme when it is liquefied.

   B. The small intestine consists of the duodenum, jejunum, and ileum, which are specialized for both enzymatic hydrolysis and absorption.

   1. Specializations for enzymatic hydrolysis include the intestinal glands (crypts of Lieberkuhn) throughout the small intestine and the submucosal glands (of Brunner) in the duodenum, and the accessory glands (liver and pancreas). The liver produces bile which acts on fat, and the pancreas produces enzymes that act on all classes of foods.

   2. Specializations for absorption include increased surface area, such as the valves of Kerckring (folds including submucosa), villi (fingerlike extensions of mucosa), and microvilli (fingerlike extensions of plasma membranes).

VIII. Dehydration and Compaction

Dehydration and compaction of unabsorbed material takes place in the large intestine (colon).

   A. The large intestine includes the cecum, the appendix, and the ascending, transverse, descending, and sigmoid colon.

   B. Specializations include mucus-secreting cells (goblet cells), taenia coli (three bands of outer longitudinal smooth muscle), and inner circular smooth muscle capable of constricting locally to form valves.

IX. Defecation

Defecation, the elimination of feces (compacted wastes), occurs by way of the rectum and anal canal, which have an internal anal sphincter of smooth muscle and an external anal sphincter of skeletal muscle (levator ani).

X. Regulation of the Digestive System

   A. Regulation of the contraction of the muscles of mastication, which are voluntary, is by the trigeminal; the pharynx, which functions on a reflex basis, is supplied principally by the vagus; the gut tube from the esophagus to the splenic flex-ure of the colon is supplied by parasympathetics of the vagus; and the remainder of the gut is supplied by sacral parasympathetics. Wavelike movements of the smooth muscle of the esophagus, stomach, and intestine, called peristalsis, are regulated by parasympathetic (extrinsic regulation) or by postganglionic fibers (intrinsic regulation).

   B. Regulation of enzyme production is principally by parasympathetics.

   1. Salivary glands include the parotid, which is supplied by parasympathetics of the glossopharyngeal, and the submandibular and sublingual, supplied by parasympathetics of the facial nerves.

   2. The mucosal glands of the stomach and small intestine are supplied by parasympathetics of the vagus nerve.

   3. The pancreas is supplied by parasympathetics of the vagus which cause an increase in fluid volume. The hormones, secretin from the stomach and cholecystokinin (pancreozymin) from the duodenum, also increase the secretory activity of the pancreas.

   4. The liver elaborates bile; bile production is not under hormonal control, but cholecystokinin stimulates the smooth muscle of the gall bladder, which facilitates the discharge of bile.

XI. The Development of the Digestive Tract

The digestive tract arises as a simple tube lined with endoderm.

   A. The oral cavity contains the tongue which develops from the lateral lingual swellings of the first arch and the hypobranchial eminence of the third arch.

   B. Salivary glands develop as outgrowths of the oral epithelium.

   C. The embryonic pharynx has a series of four paired visceral pouches from which develop the middle ear cavity (first pouch), the tonsillar fossa (second pouch), the parathyroid glands (third and fourth pouches), and the thymus gland (third and fourth pouches).

   D. The stomach develops as a dilatation of the gut tube.

   E. The small intestine initially is solid but canalizes as villi and intestinal glands differentiate.

   F. The liver and pancreas develop as outgrowths of the gut tube.

   G. Derivatives of the midgut include the small intestine and the ascending and transverse colon. The hindgut gives rise to the descending and sigmoid colon and the rectum.

REVIEW QUESTIONS

1. What component of the digestive tract is located between the pharynx and the stomach?

2. What is the posterior boundary of the oral cavity?

3. Name the parts of the digestive tract that are responsible for the rapid transport of food.

4. What morphological features of the mucosa of the stomach provide for increased numbers of cells in the epithelial lining of the stomach?

5. What morphological features of the small intestine facilitate absorption of nutrients?

6. What is the arrangement of smooth muscle in the muscularis externa in the stomach?

7. What constitutes a liver lobule?

8. What is the relationship of the common bile duct to the main pancreatic duct?

9. What prevents undigested food material in the colon from being completely dried or dehydrated?

10. Give the distribution of the nerves responsible for regulating muscle contraction and enzyme production of the digestive system.

SELECTED REFERENCES

Anderson, J. E. *Grant's Atlas of Anatomy*. 8th ed. Baltimore: Williams & Wilkins, 1983.

Goss, C. M. *Gray's Anatomy*. 29th ed. Philadelphia: Lea & Febiger, 1973.

Lachman, E., and Faulkner, K. K. *Case Studies in Anatomy*. New York: Oxford University Press, 1981.

Moore, K. L. *Clinically Oriented Anatomy*. Baltimore: Williams & Wilkins, 1980.

Woodburne, R. T. *Essentials of Human Anatomy*. 7th ed. New York: Oxford University Press, 1983.

23

THE URINARY SYSTEM

OBJECTIVES

After completing this chapter you should be able to:

1. Diagram and describe the pathway of urine flow from the nephron to the urethra.

2. Locate the kidneys in relation to the ribcage, suprarenal glands, and vertebral column.

3. Describe the relations of the renal fascia to the peritoneum and to the transversalis fascia.

4. Define a lobe of the kidney. Relate the lobation of the kidney to its blood supply and collecting duct system.

5. Name the parts of a nephron and describe its blood supply. Relate the morphological features of each part to its role in filtration of the blood and concentration of urine.

6. Locate the urinary bladder in relation to the peritoneum, urogenital diaphragm, and anterior abdominal wall.

7. Name the parts of the male urethra and relate these parts to structures surrounding the urethra.

Excretion is the process of eliminating metabolic wastes from the body. It includes the elimination of carbon dioxide, accomplished chiefly by the respiratory system, and the elimination of nitrogenous wastes (urea) and other metabolic products, performed chiefly by the kidneys. This chapter will consider the anatomy of the organs concerned with the latter functions: the *urinary system* (Fig. 23-1) consists of the **kidneys**, **ureters**, **urinary bladder**, and **urethra**. This system not only clears the blood of substances such as urea that are toxic at high concentration, but also helps keep the concentrations of many constituents of body fluids within the narrow range compatible with the normal functioning of cells. For example, inorganic salts and water are excreted or conserved as needed to maintain the proper salt concentrations and acid-base balance of the blood. The kidneys remove wastes from the blood, and the ureters convey the wastes to the urinary bladder for storage until they are discharged to the outside through the urethra. In addition to their excretory function the kidneys produce the hormone **erythropoietin**, which stimulates the production of red blood cells, and the enzyme **renin**, which influences blood pressure.

THE KIDNEYS

The paired kidneys (Fig. 23-1) are bean-shaped organs approximately 11 cm long, 6 cm wide, and 4 cm thick and are indented on their medial aspects. They are *retroperitoneal* in position, usually located on either side of the vertebral column between the *twelfth thoracic* and *third lumbar vertebrae.* The left kidney is usually somewhat higher than the right and may be found as high as the eleventh rib, from which it is separated by the intervening diaphragm. The superior pole of each kidney is capped by the *suprarenal gland,* an endocrine gland with no connection to the urinary system.

The concavity on the medial surface of the kidney is termed the **hilus,** and is the area from which the ureter emerges along with the blood vessels. The kidney is enclosed within a connective tissue coat, the **renal capsule,** directly applied to the surface of the kidney (Fig. 23-2). It is further protected by being embedded in fat, the *perirenal fat,* and enclosed within a heavy fascial covering, the **renal fascia** and *pararenal fat* outside the renal fascia. This fascia is a special thickening of the internal fascia of the abdominal wall, the *transversalis fascia,* which is quite thin except in the area of the kidney where it splits to enclose the kidney and perirenal fat within a closed sac, isolating the kidney from adjacent structures. The renal fascia not only affords protection to the kidney but also tends to prevent kidney infections from spreading to other structures.

At the hilus the kidney tissue surrounds a large cavity, the **renal sinus,** extending inward from the hilus and containing loose connective tissue in which are embedded the renal blood vessels and the dilated end of the ureter, the **renal pelvis** (Fig. 23-3). The funnel-shaped renal pelvis sends two or three branches into the substance of the kidney, forming the **major calyces,** which branch again, giving rise to a number of **minor calyces,** usually from eight to eighteen.

Internal Organization

The internal substance of a slice of kidney, as viewed with the naked eye, is divided into a dark **renal cortex** and a lighter **renal medulla** (Fig. 23-3). The medulla is composed of subdivisions called **renal pyramids** that are conical in shape with the base of the cone at the cortex and the apex of the cone projecting into the minor calyx. The pyramids, separated from one another by extensions of the cortical tissue called **renal columns** (of Bertin), are composed principally of *kidney tubules* (collecting tubules) that drain the associated renal cortex and empty the urine into the lumen of the *minor calyx.* Each minor calyx receives the urine from a *lobe* of the kidney, which consists of a *pyramid* and its associated *cortical substance* (renal

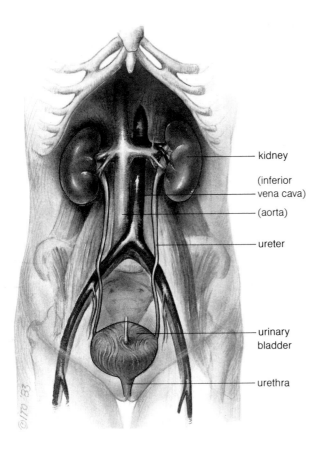

kidney

(inferior
vena cava)

(aorta)

ureter

urinary
bladder

urethra

a

Figure 23-1 The urinary system. (**a**) Diagram of an anterior view. (**b**) X ray of a thick frontal section of the trunk. Note that the left kidney lies somewhat superior to the right kidney. (From the University of Kentucky, Department of Anatomy teaching collection.) (**c**) CAT scan of the abdomen, transverse view. Note the location of the kidneys adjacent to the vertebra. (Courtesy of S. J. Goldstein.)

lung

descending
aorta

spleen

liver

kidney

vertebral
column

b

liver

small intestine

spleen

kidney

rib

c

columns and overlying cortex). The collecting tubules extend from the medulla into the cortex, where they are connected to the **nephrons** (uriniferous tubules) in which the urine is formed.

The nephron is a specialized kidney tubule, regarded as the *functional unit* of the kidney. The formation of urine in the nephron involves the removal of most of the liquid and dissolved solutes from the blood (filtration), followed by the return of most of the useful substances, including water, to the blood stream (resorption), leaving potentially harmful wastes concentrated in the urine. Urine production not only requires specialized kidney tubules but also requires a special arrangement

of the blood supply in relation to the nephrons to permit the exchange of substances between the blood and urine in different parts of the nephron.

Vascular Pattern

Each kidney receives blood from a *renal artery,* branching from the abdominal aorta and entering the kidney at the hilus, where it gives off *interlobar arteries* that pass between the pyramids toward the cortex (Fig. 23-4). At the junction of the cortex and medulla the interlobar arteries give rise to *arcuate arteries* that

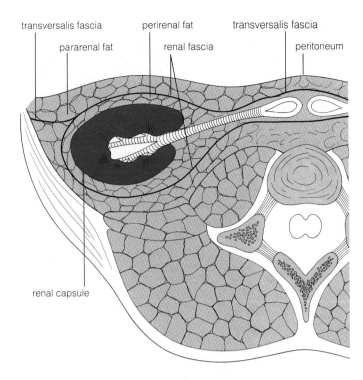

Figure 23-2 Diagram of the renal fascia in a transverse section through the posterior abdominal wall.

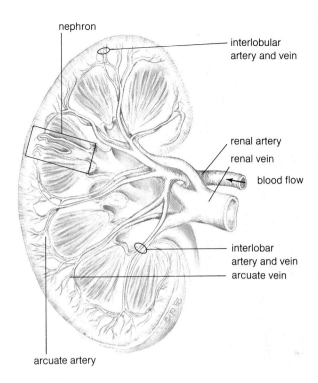

Figure 23-3 A longitudinal section through the kidney, showing its gross anatomy.

run parallel to the kidney surface at the base of the pyramid. The arcuate arteries are "end" arteries that do not anastomose but give off *interlobular arteries* that ascend into the cortex. The interlobular arteries give off branches called *afferent arterioles* that end in a tuft of *capillaries,* the **glomerulus,** located in a cup-shaped terminal expansion of the nephron called **Bowman's capsule** (Fig. 23-5). An *efferent arteriole* carries the filtered blood away from the glomerulus to another network of capillaries surrounding the remaining parts of the nephron (Fig. 23-6), where the blood enters *venules* to begin its passage out of the kidney. The efferent arterioles are smaller in diameter than the afferent arterioles, causing high blood pressure in the glomerulus, an important factor in filtration. The capillary network surrounding the hairpin loop (loop of Henle) formed by the nephron in the medulla is called the *vas recta.*

Structure and Functioning of the Nephron

Nephrons vary somewhat in structure according to their location within the renal cortex (Fig. 23-6). A "typical" nephron (Fig. 23-6), located near the junction of the renal cortex and the

medulla, is composed of a Bowman's capsule, proximal convoluted tubule, loop of Henle, and distal convoluted tubule; the loop of Henle extends deep into the medulla where it is surrounded by the vas recta. More superficial nephrons have a short loop of Henle and generally do not extend into the medulla. In both kinds of nephrons, filtration occurs at the glomerulus in Bowman's capsule, and concentration of the urine occurs in the rest of the nephron and the collecting tubule.

Bowman's capsule (Fig. 23-7) is a cup-shaped structure consisting of a *visceral layer* of *epithelium* collapsed on the surface of the capillaries and a *parietal layer* that forms the outer wall of the cup, becoming continuous with the epithelium of the proximal convoluted tubule. The visceral layer of Bowman's capsule and the endothelial lining of the capillaries contain pores that allow the passage of fluid and small molecules from the blood into the lumen of Bowman's capsule.

Because of the high blood pressure in the glomerulus, fluid is continually forced out of the capillaries. About 1200 ml of blood flows through the glomeruli of the two kidneys each minute, with about 125 ml of fluid passing out of the blood into

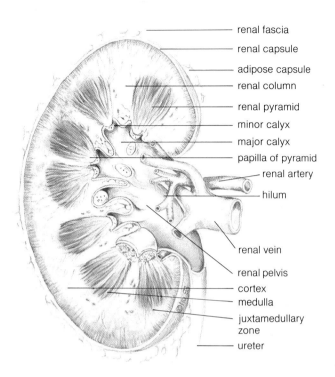

renal fascia
renal capsule
adipose capsule
renal column
renal pyramid
minor calyx
major calyx
papilla of pyramid
renal artery
hilum
renal vein
renal pelvis
cortex
medulla
juxtamedullary zone
ureter

Figure 23-4 The vascular pattern of the kidney, longitudinal section.

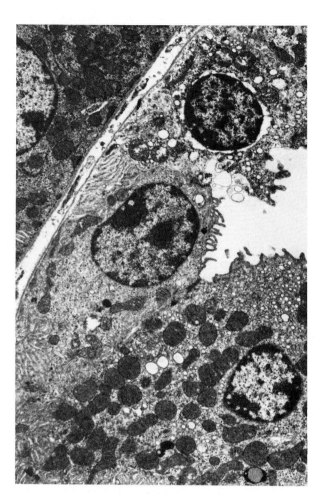

Figure 23-5 Electron photomicrograph of a distal convoluted tubule of the mouse renal cortex, ×6142. (Courtesy of D. H. Matulionis.)

Bowman's capsules. This fluid, termed the **glomerular filtrate,** contains all the components of plasma except the plasma proteins. In other words the filtration process is not very selective; it filters out blood cells and proteins but allows other useful substances to pass into the nephron along with the wastes. However, of the 125 ml of filtrate produced each minute, all except 1 ml is eventually returned to the blood as the filtrate passes through the nephrons, a process called **resorption** that begins in the proximal convoluted tubules. Resorption serves both to concentrate the urine and to conserve useful substances, including water, for the body.

The **proximal convoluted tubules** are composed of a single layer of cuboid epithelium (Fig. 23-8) whose cells have a *brush border* (microvilli) on their luminal surfaces and contain

numerous mitochondria. The *microvilli* provide the increased surface area needed for active transport across the cell. As the glomerular filtrate passes through the proximal convoluted tubule, all the useful nutrients such as glucose and amino acids cross the epithelial cells by active transport, entering first the intercellular fluid and then the capillaries surrounding the tubule; by the time the filtrate reaches the loop of Henle all the nutrient molecules have been returned to the blood. Sodium and other ions essential to the body's function are actively transported out of the nephron all the way along its course. As a result of this resorption of nutrients and other solutes, the filtrate in the proximal convoluted tubules becomes more dilute than the surrounding fluid, so that water diffuses passively out of the tubule and is returned to the blood stream.

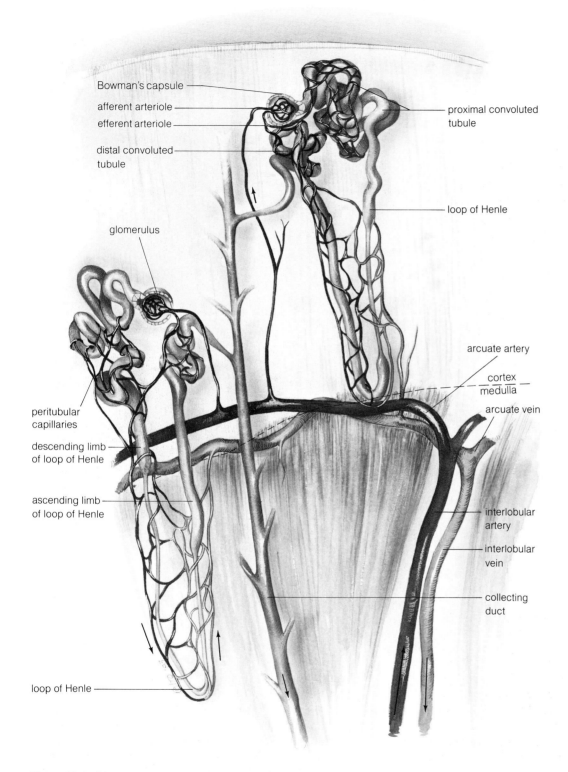

Figure 23-6 The structure and blood supply of nephrons.

Bowman's capsule

afferent arteriole

efferent arteriole

distal convoluted tubule

glomerulus

peritubular capillaries

descending limb of loop of Henle

ascending limb of loop of Henle

loop of Henle

proximal convoluted tubule

loop of Henle

arcuate artery

cortex
medulla

arcuate vein

interlobular artery

interlobular vein

collecting duct

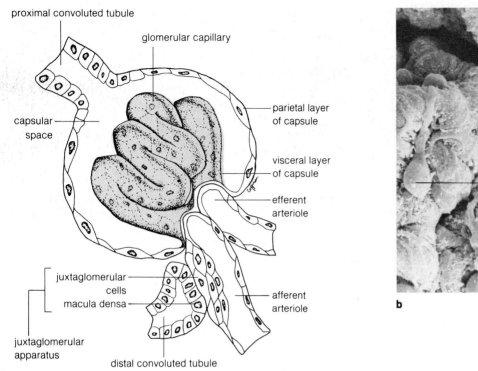

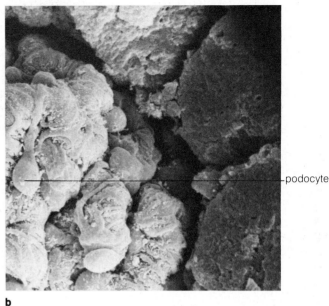

b

Figure 23-7 Bowman's capsule. (**a**) Diagram of the relationship of the glomerulus and Bowman's capsule to the juxtaglomerular apparatus. (**b**) Scanning electron photomicrograph of the visceral layer of Bowman's capsule ×1250. (Photograph reproduced with permission from *Tissues and Organs: A Text-Atlas of Scanning Electron Microscopy,* by Richard G. Kessel and Randy H. Kardon. W. H. Freeman and Company. Copyright © 1979.)

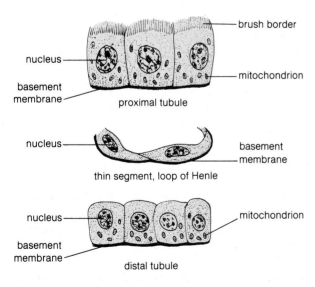

Figure 23-8 The characteristics of tubular epithelium in different regions of the renal tubule.

The **loop of Henle** is composed of descending and ascending limbs of cuboid epithelium connected by a thin limb of simple squamous epithelium (Fig. 23-8). The descending limb is permeable to both water and ions; but the ascending limb is impermeable to water, so the filtrate becomes more dilute as it ascends, because it is retaining water while losing sodium and other ions. In the **distal convoluted tubule**, composed of cuboid epithelium (Fig. 23-8), and in the collecting tubule, water resorption is regulated by the **antidiuretic hormone** (ADH) of the posterior pituitary, which increases the permeability of the tubule wall. Normally, nearly all of the water originally present in the filtrate is resorbed by the time the filtrate reaches the end of the collecting tubules in the renal pelvis, so that a highly concentrated urine is produced. However, when the body needs to rid itself of excess water, a feedback mechanism inhibits the production of ADH, water is retained in the tubules, and a dilute urine is produced.

Although resorption of water is the main factor in the concentration of urine, it is not the only one. As water and solutes are lost from the filtrate, other materials, including hydrogen and potassium ions, are actively transported into it from the surrounding fluid, adding to the concentration of solutes in the tubules and acidifying the filtrate. This process of secretion, which occurs chiefly in the proximal and distal tubules, helps maintain the proper acid–base balance of the body.

Urine passes from the distal convoluted tubules into **collecting tubules** of the cortex, where resorption of water is completed. The collecting tubules of the cortex convey the urine toward the renal pelvis. The tubules coalesce as they pass through the pyramid and converge at the apex of the pyramid, where the collecting tubules open into the **minor calyx** with which the pyramid is associated. The twelve to fifteen calyces coalesce into the two to three **major calyces** that in turn empty into the **renal pelvis**. The renal pelvis joins the **ureter** at the hilus of the kidney.

Regulation of Kidney Function

The production of urine, which is vastly more complicated than the preceding account suggests, is regulated by a complex interaction of hormones that both affects the kidney directly and affects it indirectly through changes in its blood supply. The kidney receives *sympathetic fibers* that can cause an elevation of blood pressure by constricting the arterioles, but the main control of kidney function depends upon three hormone systems: **antidiuretic hormone, aldosterone,** and **parathyroid hormone** with **calcitonin** acting as an antagonist.

The Antidiuretic Hormone The hypothalamus detects an increased concentration of inorganic ions, especially sodium ions, in plasma and signals the posterior lobe of the pituitary to release the antidiuretic hormone (ADH). ADH enters the blood, and upon reaching the kidneys causes the resorption of water from the glomerular filtrate by increasing the permeability to water of the plasma membranes of the distal convoluted tubules; the loss of water from the tubules results in a more concentrated urine. The addition of the water to the blood lowers its concentration of inorganic ions, resulting in a decrease in ADH release from the posterior pituitary. This feedback mechanism maintains the concentration of plasma ions at a fairly constant level (Fig. 23-9).

The Aldosterone System *Aldosterone,* provided by the *cortex* of the *suprarenal gland,* stimulates the resorption of sodium by the renal tubules. The amount of aldosterone is regulated by the kidney and depends upon its internal environment (Fig. 23-10). A reduction in blood flow to the kidney, caused by low blood pressure, stimulates the release of the hormone *renin* by specialized secretory cells associated with the nephron near

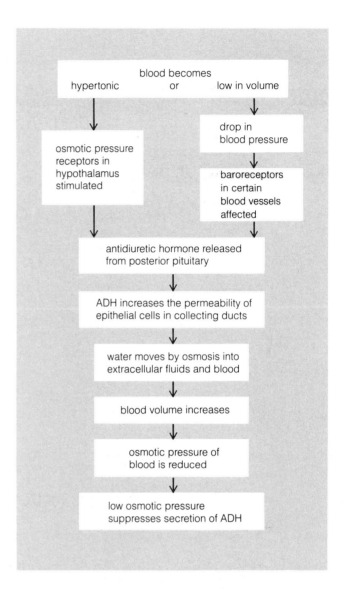

Figure 23-9 The role of ADH in regulating blood volume and, to some extent, blood pressure.

the glomerulus. These cells, termed the **juxtaglomerular apparatus** (Fig. 23-7), release *renin* into the blood where it facilitates the conversion of a plasma protein, called *angiotensin I,* into *angiotensin II,* a substance that constricts arterioles, elevating the blood pressure. Angiotensin II also stimulates the secretion of aldosterone by the suprarenal cortex, which in turn increases the resorption of sodium in the nephrons, raising the sodium level of the blood and promoting the passage of water

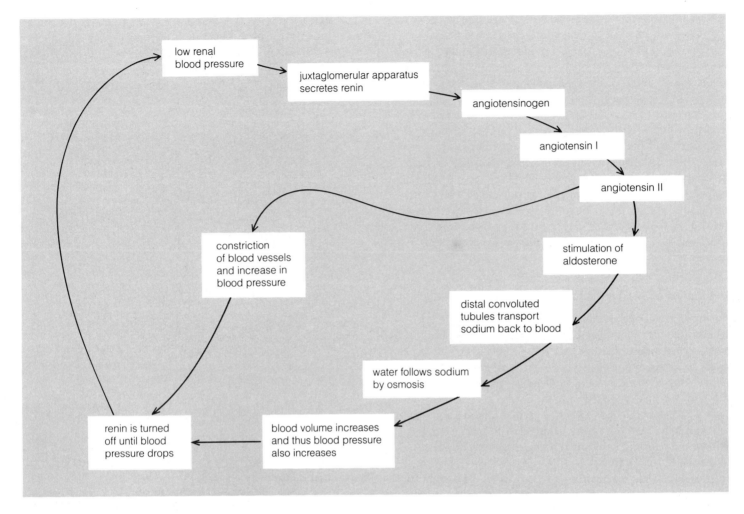

Figure 23-10 The aldosterone system and its relationship to renal blood pressure.

from the tissue fluid into blood. As a result the blood volume rises along with elevation of the blood pressure. The resulting increased blood flow to the kidneys has a negative feedback effect upon the secretion of renin.

The Parathyroid Hormone and Calcitonin The principal function of the *parathyroid hormone* is to *elevate* the calcium level of the blood. As the concentration of the hormone in the blood rises, calcium is withdrawn from the bone, raising the blood calcium level. At the same time the parathyroid hormone promotes the retention of calcium in the kidneys. As the blood calcium rises, production of *calcitonin* is stimulated, facilitating the deposition of calcium in bone and the excretion of calcium by the kidneys.

THE URINARY PASSAGES

The urinary passages, which lead from the renal pelvis to the outside of the body, are the *ureters, urinary bladder,* and *urethra.*

The Ureters

The *renal pelvis* narrows rapidly to form the *ureter,* a slender tube that descends retroperitoneally on the posterior abdominal wall to enter the pelvis, where it terminates by opening into the *urinary bladder.* The ureter is lined with transitional epithelium, which is relatively impermeable to water. The wall of the ureter has a lamina propria of loose connective tissue that blends with the smooth muscle of the outer wall (Fig. 23-11).

The smooth muscle consists of circularly and longitudinally oriented muscle cells, forming two layers that are distinctly separated as in the gut; contractions (peristalsis) of the smooth muscle propel the urine through the ureters and into the urinary bladder. The ureters open bilaterally into the inferior part of the urinary bladder by passing obliquely through its posterior wall (Fig. 23-12). This oblique course functions as a valve, preventing the backflow of urine into the ureters. As the urinary bladder fills, the lips of the opening of the ureters are pushed together, effectively sealing the openings except during the final phase of a peristaltic wave, during which urine passing down the ureter is forced into the urinary bladder.

> The living ureter often undergoes wormlike, writhing movements when it is observed during an operation, particularly if it is stroked gently. In its course on the posterior abdominal wall and in the upper pelvis, it adheres closely to the overlying peritoneum, through which it can usually be seen. These characteristics help the surgeon identify the ureters with little difficulty.

The Urinary Bladder

The urinary bladder (Fig. 23-12) is a midline structure located posterior to and in contact with the pubic symphysis. Its size and the thickness of its wall depend upon its contents, because it is subject to considerable expansion as it fills. It is lined with transitional epithelium that becomes thinner as the bladder expands (Fig. 23-14). The connective tissue contains an abundance of elastic fibers, causing the lining to form numerous folds in the contracted state. The smooth muscle forms three indistinct layers, inner and outer longitudinal layers and a middle circular layer that thickens around the opening of the urinary bladder into the urethra, forming the *internal sphincter* (involuntary sphincter). The bladder narrows inferior to the pubis. The opening of the urethra and the bilateral openings of the ureters form the points of a triangular area of the mucosa, termed the *trigone,* that is smoother than the remainder of the lining of the bladder. The retention of urine within the bladder is partially a function of the internal sphincter, with final control exercised by the *external sphincter* (voluntary sphincter) of skeletal muscle (deep transverse perinei) encircling the urethra. Urination, or **micturition,** occurs as a result of reflexes initiated as the bladder becomes full. When the bladder contains about 300 ml of urine there is a sensation of fullness, initiating a reflex that causes the relaxation of the internal sphincter. Upon voluntary relaxation of the external (voluntary) sphincter, the smooth muscle of the urinary bladder contracts and, assisted

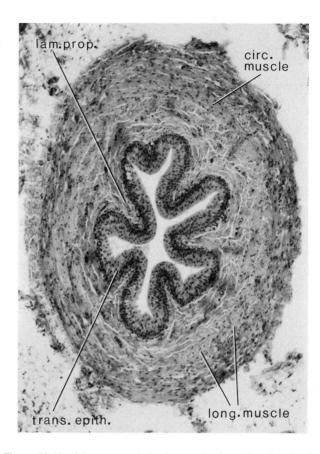

Figure 23-11 A low-power photomicrograph of a ureter, showing the layers of its wall. Note from the lumen outward: the mucosal layer, the longitudinal and circular smooth muscle layers, and the fibrous outer layer. (The lamina propria is a thin layer of connective tissue between the epithelium and muscle layer.) (Reproduced with permission from A. W. Ham, *Histology,* J. P. Lippincott, 1979.)

by contraction of the abdominal wall muscles, expels the urine.

> The interior of the urinary bladder can be examined by inserting an instrument called a cystoscope through the urethra. It also can be examined by inserting a catheter and filling the bladder with a radio-opaque material that will enable the outline of the inside of the bladder to be seen on a radiograph.

Urination cannot be initiated unless there is a sensation that the bladder is full. The internal sphincter cannot be forced

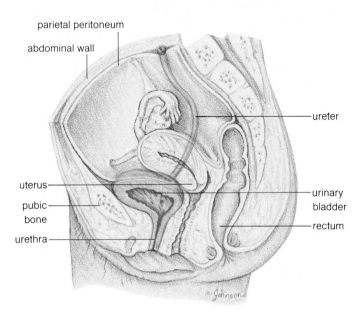

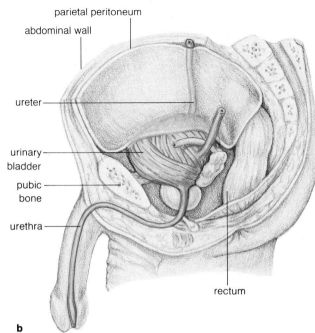

Figure 23-12 The pelvic viscera, sagittal section. (**a**) Female. (**b**) Male.

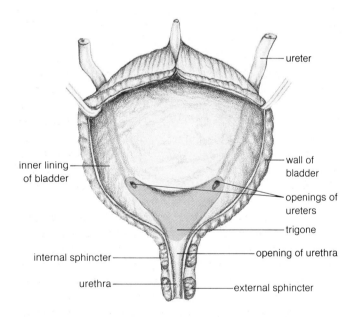

Figure 23-13 The wall and internal structure of the urinary bladder. The bladder has been opened to expose the internal aspect of the posterior wall.

open by contraction of the abdominal muscles, but these muscles can be used to increase intra-abdominal pressure and to apply pressure on the urinary bladder, causing the same sensation as the presence of urine in the bladder and eliciting the same neural response.

The Urethra

The urinary bladder opens inferiorly into the urethra, a tubular structure differing in structure and functions in males and females. The male urethra is longer than that of the female and conveys reproductive products in addition to urine, whereas the female urethra has only an excretory function.

The *male urethra* (Fig. 23-12b) is approximately 20 cm long and may be divided into three parts on the basis of location and morphology. The proximal part, the prostatic urethra, passes through the *prostate gland,* receiving *prostatic secretions* via numerous openings of the gland ducts. It also receives the *ejaculatory ducts* through bilateral openings on either side of an elevation in the posterior urethral wall, the *colliculus seminalis.* The *prostatic urethra* is lined with transitional epithelium, with a lamina propria containing many elastic fibers.

The *membranous urethra* is the part of the male urethra that passes through the *urogenital diaphragm* (deep transverse

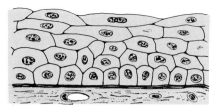

wall of bladder in distended condition

a

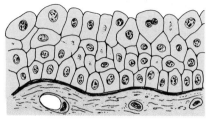

wall of bladder in collapsed condition

b

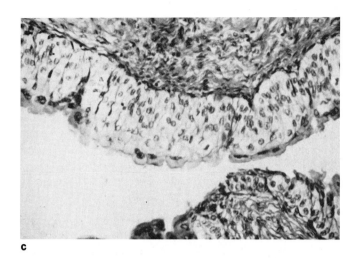

c

Figure 23-14 The epithelial lining of the urinary bladder has (**a**) flattened, almost squamous cells when it is distended and (**b**) rounded, almost cuboidal cells when it is empty. (**c**) Photomicrograph of transitional epithelium, contracted ×300 (courtesy of W. K. Elwood).

perinei muscle and associated fasciae). This part is lined with stratified or pseudostratified columnar epithelium with recesses receiving tubular mucous-secreting glands (of Littre).

The third part of the male urethra, the *penile urethra* (cavernous urethra), passes longitudinally through erectile tissue of the penis, termed the **corpus spongiosum,** to open externally at a dilated part of the urethra called the **fossa navicularis.** The penile urethra also has many mucous glands and is lined with stratified or pseudostratified columnar epithelium, changing to stratified squamous at the fossa navicularis. The lamina propria is loose connective tissue with numerous elastic fibers and scattered bundles of smooth cells.

Enlargement of the prostate such as occurs in old men may obstruct the urethra in such a way that the bladder does not empty completely. This necessitates frequent urination, a common characteristic of old age in men.

The *female urethra* (Fig. 23-12a) is from 2.5 to 3 cm long and passes through the *urogenital diaphragm* to open into the vestibule. It is lined principally with stratified squamous epithelium from which mucous glands project into the lamina propria. The connective tissue of the lamina propria contains elastic fibers and is bounded externally by two layers of smooth muscle, an inner longitudinal and an outer circular layer.

Micturition

The process of the discharge of urine from the bladder, called micturition, or urination, is controlled entirely by the autonomic nervous system in infants and small children. In adults the emptying of the bladder is controlled voluntarily up to a point. The smooth muscle of the urinary bladder is supplied by *parasympathetics* from the *sacral cord segments* S2, S3, and S4; afferents from the urinary bladder travel with the parasympathetics. The *pudendal nerve,* a branch of the lumbrosacral plexus containing somatic efferents and afferents, innervates the skeletal muscle of the urogenital diaphragm, which functions as the external sphincter of the bladder. *Sympathetics* supply smooth muscle in the *urethra* and the *trigone,* but these fibers are not important in normal emptying of the bladder. Afferents that travel with the sympathetics to the urethra near the neck of the bladder are important in initiating reflexes in normal micturition.

When the volume of fluid in the bladder exceeds about 300 ml, impulses are initiated in *visceral afferents,* and one has the desire to micturate. At this point, centers in the cerebral cortex may inhibit the process of micturition by maintaining contraction of the external sphincter and inhibiting the parasympathetics supplying the bladder wall. If the decision is made to urinate, higher centers allow the parasympathetics to *stimulate* the **detrusor muscle** of the urinary bladder to contract and the *internal sphincter* to relax. When urine enters the urethra, afferent fibers to the sacral spinal cord transmit impulses that *inhibit* the somatic efferent fibers to the *external sphincter* of the bladder, causing it to relax. Normally the bladder is emptied

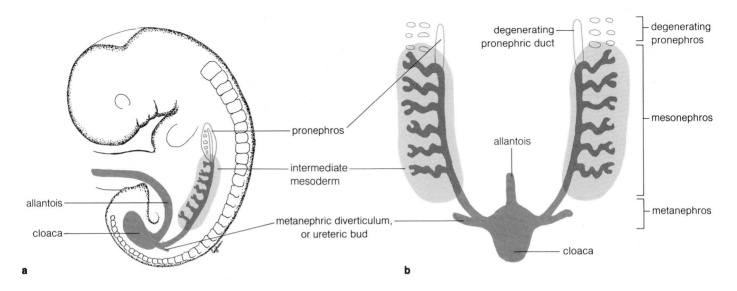

Figure 23-15 The development of the kidney. (**a**) Lateral view. (**b**) Enlarged anterior view.

completely without interruption, but the process can be stopped by higher centers in the cortex that control the skeletal muscle of the external sphincter.

DEVELOPMENT OF THE URINARY SYSTEM

Two embryonic kidneys develop from paired longitudinal cords of mesenchyme called *intermediate mesoderm;* they disappear during the course of development but are important in contributing ducts to the urogenital system. In the neck region tubules develop in the intermediate mesoderm constituting the **pronephros,** which never functions and begins to degenerate almost as soon as the tubules are formed (Fig. 23-15). Their significance lies in the fact that in conjunction with the tubules a duct forms and begins to grow caudally in the intermediate mesoderm. Cranially, in the area of the pronephros, this duct is called the *pronephric duct.* Caudally, it is associated with the developing **mesonephros** and is known as the *mesonephric duct.*

As the pronephric duct grows caudally into the trunk region, becoming the mesonephric duct, a more complex series of tubules develops and opens into it. The *mesonephric tubules* are functional in the embryo, producing urine. The two mesonephric ducts extend caudally and open bilaterally into the hind gut (Fig. 23-15b), which is thus a **cloaca,** a common receptacle for fecal, excretory, and reproductive products, such as is found in birds and reptiles.

By means of a caudal growth of a mass of mesenchyme termed the *urorectal septum,* a wedge of tissue forms between the *allantois* and *hindgut* that divides the cloaca into a *dorsal* **rectum** and a *ventral* **urogenital sinus,** into which the mesonephric ducts open (Fig. 23-16a, b) and from which the urinary bladder and urethra develop.

During development the mesonephric duct may give off two metanephric buds, resulting in drainage of one kidney by two ureters. These ureters may fuse prior to opening into the urinary bladder or may open separately into the bladder.

The *adult kidney,* or **metanephros,** develops in part from an outgrowth from the mesonephric duct and in part from the intermediate mesoderm (Fig. 23-16c). An evagination of the mesonephric duct called the *ureteric bud* (Fig. 23-15a, b) extends into the condensed mesenchyme of the intermediate mesoderm and begins to bifurcate repeatedly. The first branches form the major calyces of the metanephros (Fig. 23-16b), the second branches form the minor calyces (Fig. 23-16c), and further generations of branches become the collecting ducts of the renal medulla (Fig. 23-16d). The *nephrons* (uriniferous tubules) differentiate in the dense mesenchyme surrounding the ureteric bud. They begin as solid cords of cells that canalize and

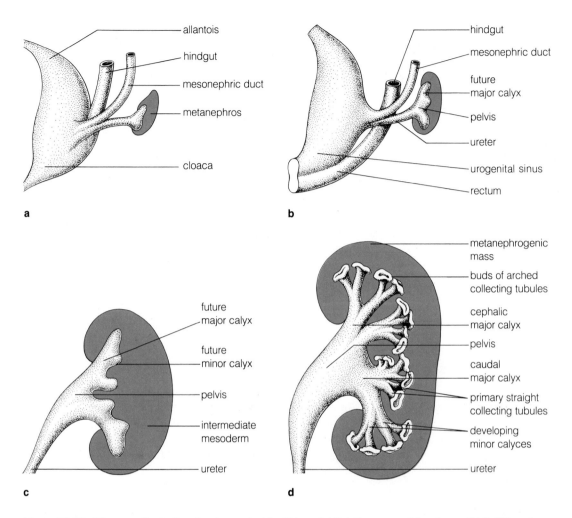

Figure 23-16 Diagrams illustrating development of the kidney. (**a**) Relations around the cloaca. (**b**) Splitting of the cloaca into rectum and urogenital sinus. (**c**) Origin of the minor calyces. (**d**) Origin of the collecting tubules.

elongate, becoming convoluted in the process, and connect secondarily with the collecting tubules to complete the pathway of urine flow.

FUNCTIONAL ANATOMY OF THE NEPHRON

The cortex of each kidney contains about one million nephrons whose function is to eliminate wastes and conserve useful substances. In blood filtration about one-fifth of the liquid leaves the blood stream as it passes through the glomeruli. Liquid is forced from the capillaries due to the difference in pressure in the blood vessels and in the capsular space. This pressure difference, termed *filtration pressure,* is about 18 mm Hg and is

adequate for the production of about 125 ml of glomerular filtrate each minute (by both kidneys). The *glomerular filtrate* is produced by a *filtration apparatus* composed of the endothelial lining of the capillaries, the basal lamina, and the visceral layer of Bowman's capsule. The unique structure of this apparatus permits the passage of small molecules but not of cells and larger molecules.

Structure of the Filtration Apparatus

The *endothelial lining* of the glomerular capillaries is fenestrated by *oval pores* about 700 A° in diameter (Fig. 23-17). These pores are too large to prevent the passage of any molecules other than large proteins, but they do exclude the passage of blood cells. External to the capillary lining the *basal lamina* is

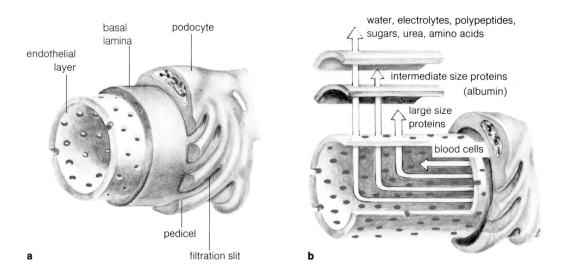

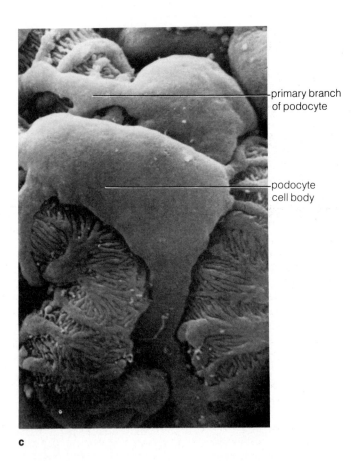

Figure 23-17 (**a**) The structure of a glomerular capillary and associated podocyte (**b**) The nature of substances that pass through the various layers. (**c**) Scanning electron photomicrograph of the fenestrated endothelium of the glomerous ×4600. (Photograph reproduced with permission from *Tissues and Organs: A Text-Atlas of Scanning Electron Microscopy,* by Richard G. Kessel and Randy H. Kardon. W. H. Freeman and Company. Copyright © 1979.)

Clinical Applications: **Acute Renal Failure**

Many conditions that interfere with renal function can cause acute renal failure. Usually the failure is due either to alteration or to destruction of the glomeruli or the tubules. In *acute glomerular nephritis* many glomeruli become markedly inflamed and may be blocked due to deposition of antigen–antibody complexes within the glomerular membrane. These complexes usually result from an infection in another part of the body. In combating the infection, the body produces antibodies that react with the bacterial antigens, forming complexes that precipitate in the glomerular membranes. The severity of the condition depends on the number of glomeruli involved and whether or not they are destroyed permanently. The glomeruli that are not blocked become much more permeable, allowing large amounts of protein to enter the glomerular filtrate. In severe cases the glomerular membrane may rupture allowing red blood cells to enter the filtrate.

Another common cause of acute renal failure is *tubular necrosis* which is usually the result of poisons, such as carbon tetrachloride and heavy metals, that specifically destroy tubular epithelial cells. As the epithelial cells die, they slough away from the basal lamina and may block the tubules. If the basal lamina remains intact, the tubules may be restored in a few days by multiplication of cells that were not killed by the poison. Similar effects upon the renal tubules may be produced by inadequate blood flow within the kidney (ischemia). If the heart fails to pump enough blood, tubular epithelial cells will die from lack of adequate nutrition, producing blockage similar to that produced by poisons.

In acute renal failure kidney functions may be inadequate to maintain homeostasis. There may be an abnormal accumulation of fluid in the tissues (edema) with abnormal retention of urea and other metabolic products that are normally excreted in the urine. This condition is called *uremia* and may cause death within a short time unless the products are eliminated. One method by which the accumulated excretory products may be discharged from the body is by hemodialysis with the artificial kidney (Fig. 23-18). Artificial kidneys have been used for about 30 years to treat patients with severe renal failure and are now used in some cases to maintain individuals indefinitely without kidneys. Basically the artificial kidney is a machine containing very minute channels bounded by thin membranes. As blood is passed through the channels, waste products in the blood pass by diffusion into a dialyzing fluid outside the channels. The membrane does not permit the passage of cells or large plasma protein molecules. Small molecules, such as urea, that are at a higher concentration in the plasma than in the

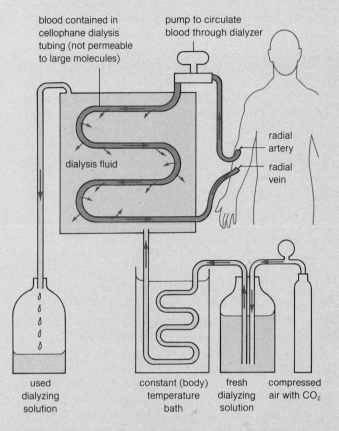

Figure 23-18 A schematic diagram to illustrate the operation of an artificial kidney (hemodialysis unit). In dialyzer large arrows show direction of blood flow; small arrows show removal of wastes from blood.

dialyzing fluid diffuse from the blood into the fluid, thus eliminating the wastes.

The artificial kidney is usually used only for a short period of time while the individual recovers from acute renal failure or waits for a kidney to become available for a transplant. Long-term use of the artificial kidney not only is very expensive but also results in a risk of infections and problems with blood clots. Kidney transplants have become a more practical long-term solution for individuals with kidney failure; the success rate for transplants has increased as procedures for suppressing transplant rejection have improved.

the only continuous layer in the filtration apparatus (Fig. 23-17a). It is composed of a network of fine filaments embedded in a glycoprotein matrix, appearing as a fine line in electron micrographs. The basal lamina prevents the passage of large protein molecules but is readily penetrated by small proteins and other small molecules, such as glucose and amino acids (Fig. 23-17b).

The final barrier to be penetrated by the glomerular filtrate is the *visceral layer* of *Bowman's capsule* (Fig. 23-17c). This layer is composed of specialized cells that lose their epithelial characteristics, becoming highly branching cells called *podocytes* (Gr., *podos* = foot). The cytoplasmic processes of the podocytes extend from the cell bodies to the outer surface of the basal lamina where they form an interlacing net, leaving slits between the footlike processes (pedicels). These slit pores are bridged by a very thin membrane (60 A°), which with the slit pores (about 250 A° wide) serves to block the passage of large molecules. Thus the glomerular filtrate is a dilute fluid containing only small molecules.

Tubular Transport

Certain components of the glomerular filtrate are resorbed by active transport mechanisms capable of transporting a fixed amount in a unit of time. One such component is *glucose,* which is always present in the glomerular filtrate but is completely resorbed, so that no glucose is present in the urine until the plasma level of glucose exceeds about 180 mg per 100 ml. Above this critical level of concentration, termed "threshold," the nephrons cannot resorb all the glucose, and it appears in the urine at a concentration directly related to the plasma level. Other sugars, such as fructose, that use the same transport mechanism as glucose will interfere with glucose transport, causing glucose to appear in the urine even with a plasma concentration below the critical level.

Phosphates have such a low threshold that some phosphate always appears in the urine. The amount of phosphate in the glomerular filtrate varies with diet and other factors, such as the activity of the parathyroid glands which increases the excretion of phosphates by reducing its resorption.

Amino acids are normally excreted only in trace amounts but are present in large amounts in glomerular filtrate. Each amino acid probably has its own critical concentration, which when exceeded will result in failure of the amino acid to be resorbed completely. *Proteins* in the glomerular filtrate, such as *albumin,* are normally resorbed in the proximal convoluted tubule, but resorption is incomplete, leaving an amount in the urine below the concentration detectable in routine clinical tests.

The fluid in the kidney tubules may be isotonic, hypotonic, or hypertonic relative to the intercellular fluid outside; that is, it may have an equal, lesser, or greater concentration of solutes. Though the glomerular filtrate entering the proximal convoluted tubule is isotonic, urine is usually a hypertonic solution of salts by the time it leaves the kidney. Much of the adjustment of the salt concentration, particularly of sodium, occurs in the loop of Henle as it descends into the renal medulla before ascending to join the distal convoluted tubule in the cortex. As the filtrate enters the descending limb of the loop of Henle, it is hypotonic; but it becomes hypertonic by losing water more rapidly than sodium as it passes toward the thin limb of the loop. By the time it reaches the bottom of the loop, it is hypertonic; but it becomes more dilute as it passes through the ascending limb because the ascending limb is impermeable to water but actively transports sodium into the intercellular fluid. In the distal convoluted tubule and the collecting tubules urine is further concentrated by loss of water, a process that is under hormonal regulation. The antidiuretic hormone of the posterior pituitary gland, when present, increases the permeability of the tubules to water, causing resorption of the water, which produces a hypertonic urine. Absence of the hormone reduces the permeability of the tubules to water, so the urine becomes hypotonic.

WORDS IN REVIEW

Aldosterone	Micturition
Antidiuretic hormone	Minor calyx
Bowman's capsule	Nephron
Cloaca	Pronephros
Corpus spongiosum	Renal capsule
Detrusor muscle	Renal column
Erythropoietin	Renal cortex
Fossa navicularis	Renal fascia
Glomerular filtrate	Renal medulla
Glomerulus	Renal pelvis
Hilus	Renal pyramid
Juxtaglomerular apparatus	Renal sinus
Major calyx	Renin
Mesonephros	Ureter
Metanephros	Urethra
	Urogenital sinus

SUMMARY OUTLINE

I. Excretion
   Excretion is the elimination of metabolic wastes and includes the elimination of CO_2 by the respiratory system.

II. The Urinary System
   The urinary system consists of the kidneys and the urinary passages (ureters, urinary bladder, and urethra); it eliminates toxic wastes other than CO_2 and regulates the concentrations of essential substances in the body.

III. The Kidneys
   A. Kidneys are paired bean-shaped organs on the posterior body wall of the upper abdomen, encased by a capsule, perirenal fat, and fascia, and capped by the suprarenal glands.
   B. Internally the kidneys have a cortex composed principally of urine-producing tubules called nephrons and a medulla containing collecting tubules and the lower portions of nephrons (loops of Henle). The medulla is divided into sections called pyramids by extensions of the cortex called renal columns; a pyramid and the related cortical tissue constitute a lobe of the cortex.
   C. The collecting tubules of a pyramid drain into a minor calyx that drains a single lobe; minor calyces empty into the major calyces which empty into the renal pelvis (the dilated end of the ureter).

D. Renal arteries branch into interlobar arteries between pyramids; these give rise to arcuate arteries, end arteries that curve between the cortex and medulla. From the arcuates, interlobular arteries of the cortex give rise to afferent arterioles that supply glomeruli (capillaries within Bowman's capsule of the nephron); capillaries of the glomeruli drain into afferent arterioles that supply capillaries around tubules of nephrons, which in turn drain into veins conveying the blood away from the kidneys.
   E. The nephron consists of Bowman's capsule where filtrate is collected from blood, the proximal convoluted tubule where nutrients are returned to blood, and the loop of Henle and distal convoluted tubule, where water is removed and salt content is adjusted, concentrating the urine.

IV. The Urinary Passages
   The urinary passages consist of the paired ureters, the urinary bladder, and the urethra.
   A. The ureters begin at the renal pelvis and transport urine to the urinary bladder; they are lined with transitional epithelium and have smooth muscle in their walls that propels urine to the bladder.
   B. The urinary bladder is located posterior to the pubic symphysis; it is an expandable sac lined with transitional epithelium. Its smooth muscular wall tapers to the neck, where it opens into the urethra.
   C. The urethra conveys urine outside the body. The male urethra is in the penis; the much shorter female urethra opens into the vestibule.

V. Development of the Urinary System
   A. The urinary system originates from intermediate mesoderm. There are three pairs of kidneys during development: the pronephros never functions; the mesonephros functions in the embryo and empties urine into the bladder via the mesonephric duct; the metanephros is the adult kidney.
   B. The ureteric bud evaginates from the mesonephric duct and penetrates the mesenchyme, giving rise to the collecting tubules; metanephric nephrons differentiate from mesoderm around the ureteric bud.

VI. Regulation of the Urinary System
   A. Antidiuretic hormone (ADH) concentrates urine by promoting the resorption of water by the distal convoluted tubules.
   B. Aldosterone stimulates the resorption of sodium by kidney tubules. Renin, an enzyme produced by the juxtaglomerular apparatus of the nephron in response to low blood pressure, converts the plasma protein angiotensin I into angiotensin II, which raises the blood pressure and stimulates the secretion of aldosterone by the suprarenal cortex. Aldosterone increases tubular resorption of sodium, which results in decreased secretion of renin.

C. The parathyroid hormone and calcitonin are antagonistic to one another. Parathyroid hormone promotes the retention of calcium by the kidneys; calcitonin promotes the excretion of calcium.

D. The process of micturition begins with the sensation of the bladder being full, which initiates a reflex via afferents to parasympathetics that stimulate smooth muscle of the bladder to contract; at the same time neural outflow to the external sphincter of the bladder is inhibited, allowing it to relax.

REVIEW QUESTIONS

1. List in sequence the structures through which urine passes from the nephron to outside the body.

2. What structure is located in contact with the superior pole of the kidney?

3. What layer of fascia of the abdominal wall thickens to form the renal fascia?

4. Explain why the renal circulation is considered to be an example of "end" arteries.

5. What constitutes a renal lobe?

6. List the parts of a nephron in sequence from the region of filtration to the point of exit from the nephron.

7. In the male the neck of the urinary bladder is separated from the urogenital diaphragm by what structure?

8. Name the part of the male urethra that passes through the urogenital diaphragm.

9. The penile urethra is located within what component of the penis?

SELECTED REFERENCES

Anderson, J. E. *Grant's Atlas of Anatomy.* 8th ed. Baltimore: Williams & Wilkins, 1983.

Langman, J. *Medical Embryology.* 4th ed. Baltimore: Williams & Wilkins, 1981.

Moore, K. L. *Clinically Oriented Anatomy.* Baltimore: Williams & Wilkins, 1980.

Romanes, G. J. *Cunningham's Textbook of Anatomy.* 12th ed. New York: Oxford University Press, 1981.

THE REPRODUCTIVE SYSTEM

OBJECTIVES

After completing this chapter you should be able to:

1. Name and describe the components of the testis. Correlate the structure of these components with their role in the production and transport of mature sperm.

2. List in their proper sequence the male reproductive ducts. Correlate the structure of each duct with its role in the storage and transport of sperm.

3. Locate and describe the spermatic cord, listing its major components.

4. Describe the course of the vas deferens from its origin at the epididymis to its termination at its junction with the duct of the seminal vesicles.

5. List and describe the accessory male reproductive glands, giving the principal contributions of each gland to semen.

6. Locate the ovaries and describe their principal attachments. Describe the structural entities of the ovary that are essential in the production of mature ova.

7. Locate the uterine tubes in relation to the ovaries and to the uterus. Describe their role in fertilization and in the transport of ova.

8. Locate the uterus and give its principal attachments. Describe the structure of the endometrium and myometrium.

9. Describe the structure of the vagina and give its relations to the uterus, urethra, and rectum.

10. List and give the approximate time sequence of the major events of the ovarian cycle.

11. List and give the approximate time sequence of the major changes of the endometrium during the uterine cycle. Correlate these events with those of the ovarian cycle.

12. List the major events leading to fertilization of the ovum and its implantation, giving the approximate times involved in each phase.

13. List the most common contraceptive measures, giving the basic procedure for each method.

The *reproductive system* is the only major system of the body that is not essential to the life of the individual. It is, however, essential to the perpetuation of the species, as each generation produces offspring continuing the genetic lines. Offspring are the result of the fusion of two germ cells, or **gametes,** produced in adult males and females by the process of *meiosis* (see Chap-

ter 3), in which the number of chromosomes is reduced to one-half that of the other cells of the body. The male gamete, or *spermatozoon,* unites with the female gamete, or *ovum,* in *fertilization* to form a **zygote,** in which the full complement of chromosomes is restored. The zygote is capable of developing into a new individual.

The reproductive system is also the only organ system that differs in its gross structure in males and females. The functions of the male reproductive system are to produce large numbers of gametes continuously and to deliver them to the site of fertiization; those of the female reproductive system are to produce gametes singly on a cyclical basis and to nurture the embryo after fertilization. Accordingly the two systems are quite different in their anatomy and physiology.

THE MALE REPRODUCTIVE SYSTEM

The reproductive system of the male (Fig. 24-1) consists of the *testes* in which spermatozoa (*sperm*) and *testosterone* are produced, a system of *ducts* conveying the sperm to the outside, and *glands* that produce secretions that are added to the sperm to form a fluid called **semen.** The testes are located outside the body wall in the *scrotum,* a pouch of skin and fascia located posterior to the penis and inferior to the pubic symphysis; the ducts course from the testes through the inferior anterior abdominal wall to open into the *prostatic urethra.* Specific names are applied to the components of the duct system, depending upon their histological structure and relative position. The urethra is incorporated into the penis as it continues toward its external opening. The glands, whose secretions are important in providing a proper environment for the sperm in the semen, are the **seminal vesicles,** the **prostate,** and the **bulbourethral** (Cowper's) **glands.**

The Testis

The scrotum is divided into two compartments by a fascial septum. Each testis (Fig. 24-2) is located within a compartment and is incompletely enclosed by a double layer of serous membrane called the *tunica vaginalis,* representing peritoneum that has lost its connection with the peritoneal cavity during development. The descent of the testes from the abdominal region into the scrotum during fetal life is preceded by an outpocketing of the peritoneum lining the abdominal wall, which pushes through the body wall to create the *inguinal canal* (Fig. 24-19).

As the testis descends, it slides through the inguinal canal behind the fingerlike extension of the peritoneal membrane, which becomes a cup-shaped, double-layered sac almost completely enclosing the testis within it. Normally the connection

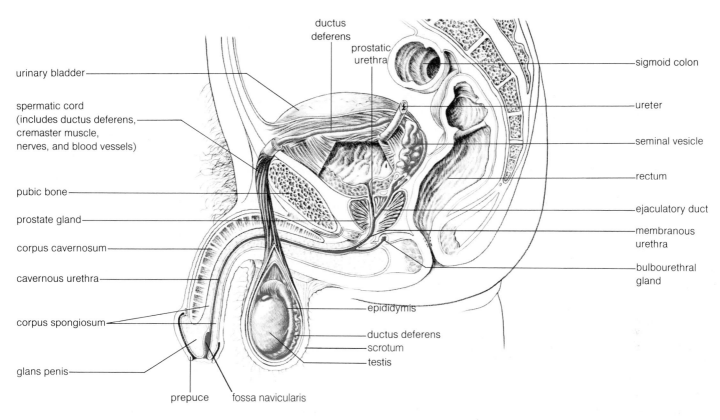

Figure 24-1 The male reproductive system, medial aspect of the right half of the male pelvis.

of this membrane with the peritoneal cavity is obliterated, leaving the testis partially enclosed in its own sac of serous membrane, the tunica vaginalis.

In addition to producing sperm the testis functions as an endocrine gland. It is a compound tubular gland covered by a heavy fibrous sheet, the **tunica albuginea** (Fig. 24-2), which forms an unusually tough, glistening capsule. On the inner aspect of the tunica albuginea a looser layer of connective tissue provided with blood vessels constitutes the **tunica vasculosa.** From this layer connective tissue septa push in, separating the testis into lobules, each containing two to four *seminiferous tubules,* with the total number of tubules in each testis probably exceeding 800. Between the tubules, secretory cells called *interstitial cells* (of Leydig) are present in small clumps; they function as an endocrine gland, producing the male sex hormones, termed *androgens,* of which *testosterone* is the most important.

The seminiferous tubules are convoluted tubules containing the germ cells (Fig. 24-3), which multiply rapidly, producing numerous sperm. The immature germ cells (spermatogonia) are located at the periphery of the tubules, where they retain their capacity to undergo *mitosis,* providing a constant source of the cells that undergo meiosis and differentiate into mature

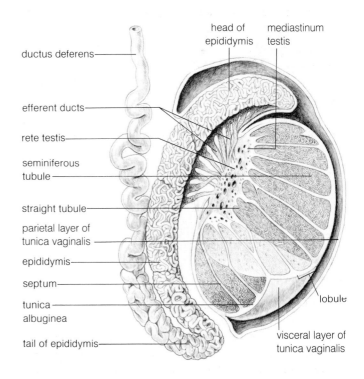

Figure 24-2 The testis and associated structures, sagittal section.

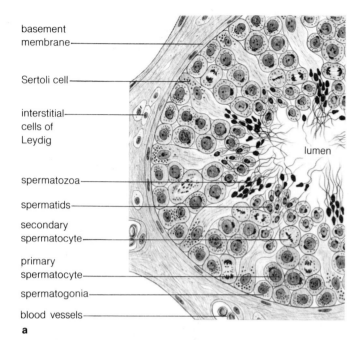

basement membrane

Sertoli cell

interstitial cells of Leydig

lumen

spermatozoa

spermatids

secondary spermatocyte

primary spermatocyte

spermatogonia

blood vessels

a

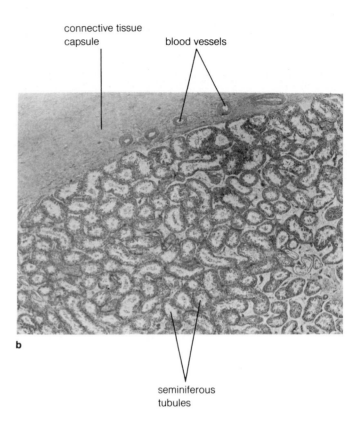

connective tissue capsule

blood vessels

seminiferous tubules

b

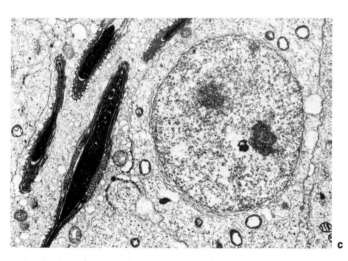

c

Figure 24-3 (a) Diagram of the cell types found within a seminiferous tubule; also shows the interstitial cells of Leydig adjacent to the tubule. (**b**) Low-power photomicrograph of the human testis, showing seminiferous tubules. (Reproduced with permission from A. W. Ham, *Histology,* J. B. Lippincott, 1979.) (**c**) Electron photomicrograph of mouse testis, showing developing sperm, × 6,600. (Courtesy of D. H. Matulionis.)

sperm. At intervals between the *spermatogonia* are *sustentacular cells* (of Sertoli) which function in support of the germ cells and are perhaps important in providing nutrients to these rapidly dividing cells. While the cells are maturing, they are moved toward the lumen of the seminiferous tubules. Mature sperm (Fig. 24-4) are gradually moved through these tubules toward the **mediastinum testis** (Fig. 24-2), a mass of fibrous tissue near the posterior margin of the testis that is continuous with the tunica albuginea. The seminiferous tubules straighten to become the *tubuli recti* (straight tubules) which open into

the **rete testis,** an elaborate network of fine tubules traversing the mediastinum testis. The tubuli recti and rete testis are lined with a single layer of cuboidal epithelium. Sperm entering the rete testis are delivered into the *ductuli efferentes* (efferent ducts), a series of fifteen to twenty minute tubules that exit from the testis at the mediastinum testis and enter the head of the **epididymis,** which is partially within the tunica vaginalis. The *ductuli efferentes* are lined with a single layer of columnar cells, with some of the cells bearing cilia that beat toward the epididymis, moving the sperm through the tubules (sperm do not

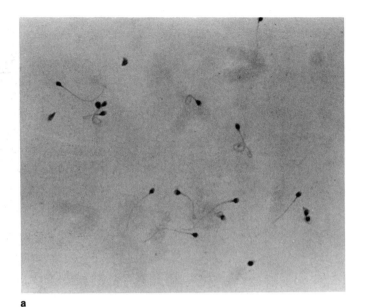

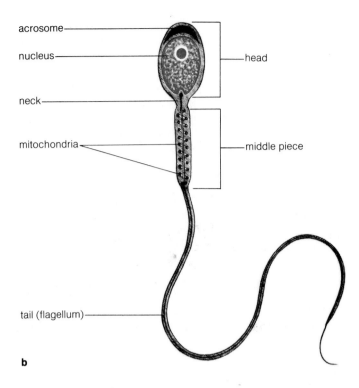

acrosome

nucleus

neck

mitochondria

head

middle piece

tail (flagellum)

b

Figure 24-4 (**a**) Photomicrograph of sperm, ×340. (Photograph by Joan Creager.) (**b**) The structure of a mature sperm.

become motile until they enter the vagina). The ductuli efferentes have a thin layer of circular smooth muscle outside the basal lamina of the epithelium to facilitate movement within the tubules.

The inguinal canal represents a weak area in the inferior region of the anterior abdominal wall where layers from the transversus abdominis, internal oblique, and external oblique muscles are pushed into the future scrotum by the projecting peritoneum. The resulting gaps in the three muscle layers or their aponeuroses become filled with loose connective tissue, but it remains a potentially weak area through which the intestine may herniate. The most superficial part of the inguinal canal passes through the aponeurosis of the external oblique, which forms a palpable rim of the canal, termed the *superficial inguinal ring,* which is investigated as part of a physical examination. The physician inserts a finger in the superficial inguinal ring and asks you to cough. A loose inguinal ring suggests the possibility of the development of a hernia.

The Epididymis

The *epididymis* (Fig. 24-2) is a single tubule about six meters in length, but it is so highly coiled that it forms a compact mass only about 2.5 cm long within the scrotum on the posterior aspect of the testis. The epididymis may be regarded as having a *head, body,* and *tail* as it leads away from the testis. It is lined with pseudostratified columnar epithelium bearing nonmotile *stereocilia,* which are huge *microvilli* associated with absorption. Outside the epithelium there is a layer of circular smooth muscle that becomes progressively thicker distally, reaching its maximum thickness in the tail. Rhythmic contractions of this muscle move the sperm slowly through the epididymis to the tail, where they are stored until ejaculation.

The Ductus Deferens

The **ductus deferens** (Figs. 24-1 and 24-2) is a continuation of the epididymis, with a much larger lumen and a thicker wall due to the increase in the smooth muscle present. As it ascends from the testis toward the body wall, the ductus deferens (vas deferens, deferent duct) is a component of the *spermatic cord,* a structure consisting (in addition to the ductus) of the nerves and arteries supplying the ductus and testis, a plexus of veins, and lymph vessels draining the testis. All the components of the spermatic cord are bound together in a fascial sheath, the **external spermatic fascia,** which is continuous with the fascia of the external oblique muscle. The middle covering of the *spermatic cord* contains slips of skeletal muscle, the *cremaster,* derived from the internal oblique muscle; the inner covering is **internal spermatic fascia** derived from the fascia of the transversus abdominis muscle. Thus each layer covering the spermatic cord structures represents a layer of the abdominal body wall. These layers were acquired as the testis descended through the body wall to reach the scrotum. The spermatic cord as an anatomical unit terminates at the body wall, but the ductus deferens and nerves continue inside the abdomen, passing inferiorly and medially through the inguinal canal towards their terminations. The ductus deferens passes between the ureter and the bladder, where it widens into an *ampulla* just before it is joined by the *seminal vesicles* (Fig. 24-2) posterior to the prostate gland. After it is joined by the seminal vesicles, it is called the *ejaculatory duct,* and it pierces the substance of the prostate gland to open into the urethra. The ejaculatory ducts extend only from the seminal vesicles to the urethra.

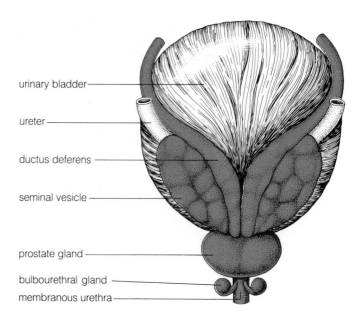

Figure 24-5 The accessory male reproductive glands shown in relation to the urinary bladder, posterior view.

The inguinal canal provides the pathway for the ductus deferens and may also be the pathway along which a portion of the gut works its way through the body wall into the scrotal sac. This is considered an *indirect inguinal hernia,* because of its oblique course through the body wall; in a *direct inguinal hernia* the gut pushes through the body wall directly deep to the superficial inguinal ring without following the course of the inguinal canal. A congenital inguinal hernia may be present at birth if the connection between the tunica vaginalis and the peritoneal cavity fails to degenerate. In such cases the persistent **processus vaginalis** is an open invitation for the gut to enter, and the condition must be corrected surgically. The danger of a hernia lies in the possibility that the gut may be strangulated (compressed so that the blood supply is cut off) and become sphacelous (gangrenous) as a result of the loss of its blood supply. The danger of a hernia is much less in a female, because the ovaries normally do not descend and the inguinal canal is relatively smaller.

The ductus deferens is a thick-walled tube lined with pseudostratified columnar epithelium and containing smooth muscle in its wall that becomes thicker in the abdominal portion. The muscular coat consists of inner and outer longitudinal layers and a well-developed intermediate layer of circular muscle, which is a major factor in ejaculating sperm. The thick muscular coat imparts a firmness to the ductus deferens that makes it palpable through the skin and fascia covering the spermatic cord. As it ascends within the spermatic cord, it is accessible for the simple procedure of vasectomy, a birth control measure in which the ductus deferens is transected and tied off to prevent the passage of sperm.

Accessory Glands of the Male Reproductive Tract

The accessory glands include the *seminal vesicles, prostate,* and *bulbourethral glands.* Their secretions are added to the sperm during ejaculation, making semen, and are important in providing an environment in which the sperm are motile.

Seminal vesicles   The seminal vesicles (Figs. 24-1 and 24-5) are unusual glands in that there is no distinct differentiation into a system of ducts draining a system of secretory units. Each vesicle is an elongated saccular organ, with the central lumen and irregular outpocketings all lined with pseudostratified columnar epithelium containing secretory cells. The walls of the gland contain smooth muscle cells that contract during ejaculation, forcing its secretions into the *ejaculatory duct* (Fig.

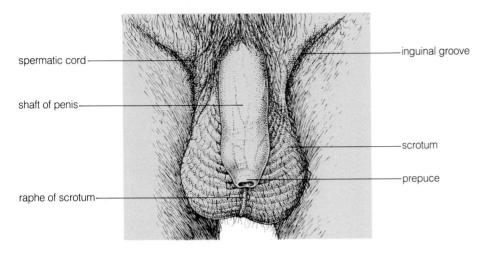

spermatic cord

shaft of penis

raphe of scrotum

inguinal groove

scrotum

prepuce

Figure 24-6   The male external genitalia, anterior view.

24-1). The secretions are rich in *fructose,* a simple sugar serving as an energy source for the sperm, and prostaglandins that stimulate uterine contractions to facilitate sperm movement through the female reproductive tract.

Prostate Gland   The prostate gland is located just inferior to the neck of the urinary bladder and surrounds the urethra (Figs. 24-1 and 24-5). It is a compound tubuloalveolar gland with sixteen to thirty-two small ducts (prostatic ductules) opening separately into the *urethra.* The cytology of the secretory elements varies according to their physiological state from low cuboidal to pseudostratified epithelium whose cells contain numerous secretory granules. The secretion of the prostate is a milky fluid containing several enzymes. The prostate glands contain smooth fibers that contract during ejaculation, forcing the secretions into the urethra.

> The posterior surface of the prostate gland can be palpated or massaged through the rectum. Palpation reveals information about the size of the prostate and its consistency. Massage is frequently performed to obtain fluid for microscopic identification of cells or bacteria that may be present.

Bulbourethral Glands   The paired bulbourethral glands (Figs. 24-1 and 24-5) are small, compound tubuloaveolar glands embedded within the urogenital diaphragm. Each gland has a single duct that courses distally to open into the *cavernous* (penile) *urethra.* The secretory cells are usually columnar in shape, resembling mucous gland cells, and elaborate a clear, viscous substance that lubricates the urethra, facilitating passage of sperm.

The Male External Genitalia

The male external genitalia, located between the anus and the pubic symphysis, consist of the *scrotum* and the *penis* (Figs. 24-1 and 24-6).

The Scrotum   The scrotum, located posterior to the penis and anterior to the anus, is a pouch of skin and fascia containing the testes, the efferent ducts and epididymis, and the nerves and blood vessels supplying these structures. The skin of the scrotum is covered with coarse hairs after puberty and is usually wrinkled owing to the unique structure of its superficial fascia, called the **dartos tunic.** This fascia contains no fat but has muscle fibers, collectively called the *dartos.* The dartos fibers contract under the stimulus of cold or sexual stimulation, causing the scrotum to wrinkle and shorten, which helps regulate

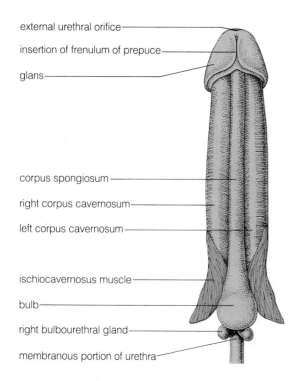

Figure 24-7 A penis, partially dissected to show major components, ventral aspect.

external urethral orifice
insertion of frenulum of prepuce
glans
corpus spongiosum
right corpus cavernosum
left corpus cavernosum
ischiocavernosus muscle
bulb
right bulbourethral gland
membranous portion of urethra

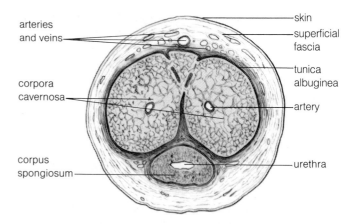

Figure 24-8 Cross section through the penis.

arteries and veins
skin
superficial fascia
tunica albuginea
corpora cavernosa
artery
corpus spongiosum
urethra

the temperature of the testes. High temperature affects spermatogenesis, so a man will become sterile if the scrotal temperature is elevated to body temperature for extended periods. In the cold the smooth muscle of the dartos tunic contracts, bringing the testis close to the body to reduce heat loss, but in hot weather the muscle relaxes, allowing the testes to move farther from the body to increase heat loss. The scrotum is divided internally by a septum of fascia and dartos, which is represented on the surface by the raphe, the line of fusion of the two embryonic structures (scrotal swellings) from which the scrotum develops. The raphe of the scrotum is continuous anteriorly with the raphe of the penis and posteriorly with the raphe of the perineum.

The Penis The penis (Figs. 24-1, 24-7, and 24-8) transmits the urethra to its external opening and is the organ of copulation. It consists of a *shaft,* or *body,* and a terminal *glans.* The shaft contains three cylinders of erectile tissue, each surrounded by a tough collagenous tunica albuginea. Two cylinders, the *corpora cavernosa,* lie adjacent to one another on the dorsum (upper side) of the penis; the *corpus spongiosum* is a single cylinder located ventral to the other two in a median position. Each corpus

cavernosum passes toward the perineum and diverges toward the ischium, forming the *crus penis* which is covered by the *ischiocavernosus muscle.* The corpus spongiosum, which contains the urethra, expands into the glans distally and ends proximally at the urogenital diaphragm as the *bulb* of the *penis,* which is covered by the *bulbospongiosus muscle.*

The penis is supplied with blood by a dorsal midline artery and an artery within each of the corpora cavernosa. The erectile tissues of the corpora consist of a network of connective tissue septa covered by endothelium. The cavernous spaces between the septa become filled with blood during periods of sexual excitement as stimulation by the parasympathetic nerve fibers causes the arteries to dilate. The penis swells and becomes rigid in the state of erection. Enlargement of the corpora exerts pressure on the veins, retarding venous return and maintaining erection during intercourse. The ejaculation of semen is largely a function of the sympathetic nervous system. Sympathetic impulses initiate peristaltic waves in the epididymis and ductus deferens, propelling sperm toward the urethra where they are mixed with seminal vesicle and prostatic secretions, producing semen. Rhythmic contractions of the perineal muscles expel the semen from the penile urethra; each ejaculate contains about 300 million sperm. After ejaculation, blood gradually drains from the erectile tissue, returning the penis to its flaccid state.

Functional Anatomy of the Male Reproductive System

The male reproductive system is responsible for the production of mature sperm as well as the substances providing an environment in which the sperm may survive while stored and transported to the female reproductive tract. The substances produced by the accessory glands of the male reproductive tract constitute the *seminal plasma* to which the *sperm* are added, forming *semen.*

When mature sperm (Fig. 24-4) are discharged from the seminiferous tubules, each is composed of a head which contains the nuclear material and has an *acrosome* containing proteolytic enzymes at the tip, a *middle piece* containing mitochondria with the oxidative enzymes for energy production, and the long *tail* that must be motile if the sperm is to attain the capacity to fertilize the ovum. Large numbers of sperm are stored in the tail of the epididymis. The total mass is stored in the epididymis and is discharged at ejaculation as a viscid material adequate for the survival of the sperm but inadequate for sperm motility. At ejaculation this thick mass is propelled down the ductus deferens by peristalsis and is mixed with secretions of the *seminal vesicles,* a thin fluid serving to dilute the sperm so that they may become motile. The products of the seminal vesicles include fructose, used by the mitochondrial enzymes of the sperm middle piece to provide the energy for swimming. The seminal vesicle fluid also contains large quantities of potassium, essential for sperm motility, proteins that act as buffers to prevent rapid alteration in pH (to which sperm are sensitive), and prostaglandins that stimulate the uterine contractions thought to be important in transporting sperm through the female reproductive tract.

The *prostatic secretions,* discharged along with the ejaculation of the seminal fluid and sperm, mix with them in the urethra. Prostatic fluid, containing several enzymes, is slightly alkaline and helps neutralize the acid environment of the vagina. The mucus produced by the *bulbourethral glands* serves as a lubricant in the urethra and is not an important factor in the reactions within the vagina. Initially the semen is somewhat coagulated within the vagina, but it undergoes liquefaction within a minute or two. The sperm undergo changes, known as the *acrosomal reaction,* in which the plasma membrane covering the acrosome becomes permeable to the proteolytic enzymes within it. The release of these enzymes is necessary to enable the sperm to penetrate the ovum. Many of the sperm die within the vagina and never enter the uterus, but some become motile and enter the cervical canal by swimming in the cervical mucus, to be transported to the uterine tube where fertilization occurs.

The average volume of semen in each ejaculate is about 3.5 ml, with sperm accounting for only about 10% of the volume. The sperm count varies from 50 to 150 million per ml, with each ejaculate containing 200 million or more. Although only 200 to 500 sperm are found in the uterine tube in the vicinity of the ovum at the time it is fertilized, many more sperm are transported, with many passing through the uterine tubes into the peritoneal cavity.

THE FEMALE REPRODUCTIVE SYSTEM

The principal internal organs of the female reproductive system (Figs. 24-9 and 24-10) are a pair of *ovaries* in which ova are produced, a pair of **uterine tubes** for the transport of ova and sperm, a median unpaired **uterus** in which the embryo develops, and the *vagina,* which opens to the exterior, serving as a receptacle for the penis and ejaculate and as a birth canal for exit of the new individual at birth.

The Ovaries

The *ovaries* (Figs. 24-9 and 24-10) produce *ova* and also act as *endocrine glands,* producing the sex hormones *estrogen* and *progesterone,* which are important in the maintenance of sex characteristics and in the preparation for and maintenance of pregnancy. In women who have not borne children the ovary is intraperitoneal; it is located in the pelvic cavity against the lateral wall at the level of the anterior superior iliac spine and may be palpated just medial to the lateral plane. In women who have borne children the position of the ovary is variable, because the ovary is moved superiorly during pregnancy and may not return to its former position. The surface of the ovary is covered with a single layer of cuboidal epithelium, termed *germinal epithelium* (Fig. 24-11b), derived as modified serous membrane. In spite of its name the germinal epithelium does not give rise to germ cells.

The ovary is an oval structure with its long axis oriented transversely. Its upper pole is attached to the wall of the pelvic cavity by the *suspensory ligament* (Fig. 24-10) containing the ovarian blood vessels; the lower pole is attached to the lateral wall of the uterus by a fibromuscular cord, the *proper ligament* of the *ovary,* which lies within the **broad ligament** of the *uterus.* The anterior or mesovarian border of the ovary is attached to the broad ligament by the *mesovarium,* a peritoneal fold continuous with the posterior layer of the broad ligament. The upper pole of the ovary lies close to the fimbriated (irregular or fringed) end of the uterine tube and is sometimes attached to one of the fimbriae (fingerlike projections) by means of fibrous tissue within the mesovarium.

Internally (Fig. 24-11a, b) the ovary has a thick *cortex,* in which the ova are produced, surrounding the *medulla,* in which the larger blood vessels are found. Beneath the surface epithelium a layer of dense connective tissue, the *tunica albuginea,* is superficial to the connective tissue (stroma) in which the follicles are embedded. At puberty the two ovaries contain about 400,000 *primordial follicles,* each consisting of a large, round oocyte surrounded by a single layer of flattened follicular cells. During the reproductive span, from puberty to menopause, the ovaries release only 400 to 500 ova. The remaining follicles and

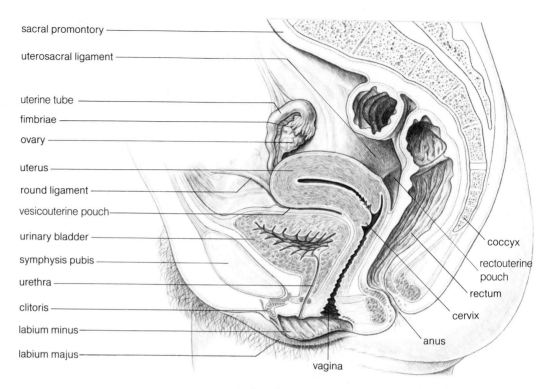

sacral promontory

uterosacral ligament

uterine tube

fimbriae

ovary

uterus

round ligament

vesicouterine pouch

urinary bladder

symphysis pubis

urethra

clitoris

labium minus

labium majus

coccyx

rectouterine pouch

rectum

cervix

anus

vagina

Figure 24-9 The female reproductive system, a medial view of the right half of the female pelvis.

ova may begin to mature but degenerate, becoming *atretic* (collapsed, without follicular fluid) before ovulation can occur. The ovary of a young adult always contains a large number of follicles in various stages of maturation; as the follicles enlarge and increase in thickness, fluid begins to accumulate within the follicle, pushing the ovum to one side (Fig. 24-11c). The *mature (Graafian) follicle* reaches the surface of the ovary (Fig. 24-11b), where it may cause a visible bulge before it ruptures, discharging the ovum near the opening (ostium) of the uterine tube. The ovum is discharged into the serous fluid of the peritoneal cavity but is usually carried to the uterine tube by the current created by contractions of the fimbriae around the expanded end (infundibulum) of the uterine tube. The ovum undergoes the *first meiotic division* prior to ovulation, but it does not complete the *second division* unless fertilized.

The Uterine Tubes

The *uterine tubes* (oviducts, or Fallopian tubes) are the ducts that receive the ovum and the sperm, provide a proper environment for fertilization, and convey the zygote or (if fertilization does not occur) the ovum to the uterus. They are muscular

tubes about 10 cm long; each begins laterally as a funnel-shaped *infundibulum* (Fig. 24-10) with fingerlike extensions called *fimbriae* projecting from its margin. The infundibulum leads into the *ampulla,* the longest and largest portion of the uterine tube, which is the usual site of fertilization. The uterine tubes narrow into an *isthmus,* continuous with the *uterine portion* of the tube that passes through the uterine wall to open into the uterine cavity.

Since the uterine tubes open into the peritoneal cavity, they provide direct communication into the cavity from the exterior and are a potential pathway for infection of the peritoneum, which may occur in the case of gonorrhea.

The wall of the uterine tube consists of mucosa, a layer of muscle, and a covering of serous membrane. The epithelium of the mucosa is simple columnar, containing secretory cells mixed with ciliated cells. The secretions nourish the ovum, and

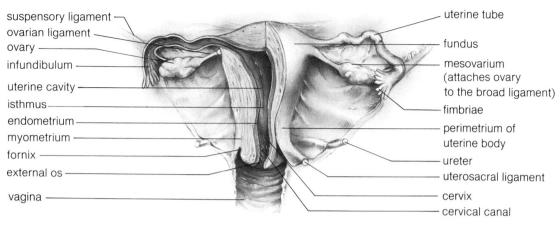

suspensory ligament
ovarian ligament
ovary
infundibulum
uterine cavity
isthmus
endometrium
myometrium
fornix
external os
vagina

uterine tube
fundus
mesovarium
(attaches ovary
to the broad ligament)
fimbriae
perimetrium of
uterine body
ureter
uterosacral ligament
cervix
cervical canal

a

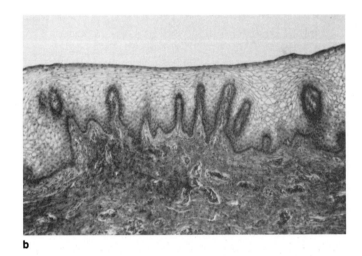

b

Figure 24-10 The organs of the female reproductive system and their supporting structures. (**a**) Anterior view of the uterine tube, uterus, and vagina have been opened by removing portions of their anterior walls. (**b**) Photomicrograph of the epithelial lining of the vagina (courtesy of W. K. Elwood).

the cilia help propel the ovum toward the uterus. The lamina propria is a relatively thin, loose, areolar connective tissue bounded externally by two layers of smooth muscle, an inner circular layer and an outer longitudinal layer covered by serous membrane. Rhythmic contractions of the muscle layers are important in the transport of the ovum to the uterus.

The Uterus

The *uterus* (Fig. 24-9) is a thick-walled muscular structure located posterior to the urinary bladder. When the bladder is empty, the uterus tilts forward, extending at a right angle to the vagina across the superior margin of the urinary bladder. The non-pregnant uterus has a narrow lumen with lateral openings superiorly for the uterine tubes and inferiorly for communication

with the vagina. The part of the uterus superior to the openings of the uterine tubes (Fig. 24-10) is called the *fundus*. The expanded *body* of the uterus extends from the openings of the uterine tubes to a central, narrow region termed the *isthmus* that is continuous with the *cervix,* the inferior part of the uterus. The cervix projects into the lumen of the vagina, creating a circular pouch, the *vaginal fornix,* surrounding the inferior rim of the cervix. The opening from the fornix into the cervical canal is called the external os of the cervix.

In the female the ureter passes close to the lateral fornix. At this site a kidney stone in the ureter can be palpated on vaginal examination.

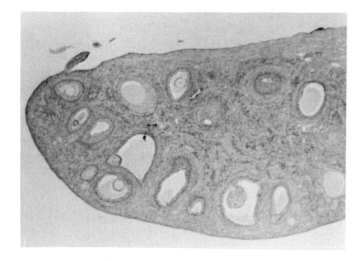

a

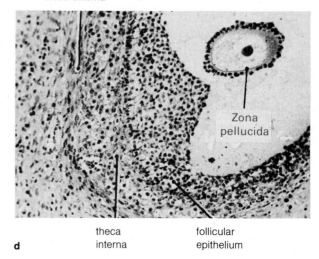

primary follicles

granulosa (follicle cells)

growing follicles

theca

medulla

cortex

capsule

germinal epithelium

tunica albuginea

blood vessels

corpus albicans (residue of a corpus luteum)

corpus luteum

stroma (connective tissue fibers)

zona pellucida

antrum

mature Graafian follicle

b

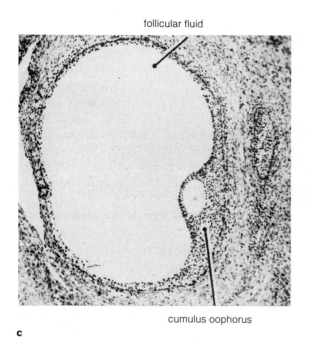

follicular fluid

cumulus oophorus

c

theca externa

Zona pellucida

theca interna

follicular epithelium

d

Figure 24-11 (**a**) Photomicrograph of a cat ovary, ×25. (Photograph by Joan Creager.) (**b**) Diagram of a cross section through an ovary, showing maturation of follicles. (**c**) Very low-power photomicrograph of a developing follicle distended with follicular fluid. The primary oocyte is surrounded by follicular cells collectively called the cumulus oophorus. (**d**) Low-power photomicrograph of a more mature follicle containing a secondary oocyte, surrounded by a thick membrane called the zona pellucida. The oocyte and its associated cells are still attached to the wall of the follicle, even though the attachment is not apparent in this particular microscopic section through the follicle. [(**c**) and (**d**) reproduced with permission from A. W. Ham, *Histology*, J. P. Lippincott, 1979.]

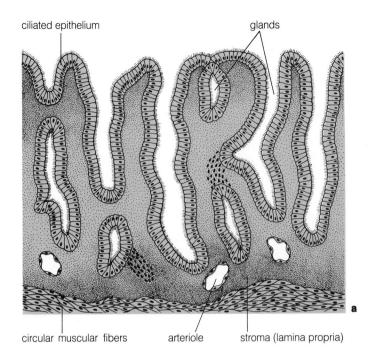

ciliated epithelium

glands

circular muscular fibers

arteriole

stroma (lamina propria)

a

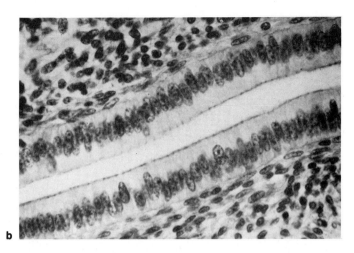

b

Figure 24-12 (**a**) Diagram of a section of the endometrium to show the endometrial glands and stroma. (**b**) Photomicrograph of the simple columnar epithelium of the endometrial glands (courtesy of W. K. Elwood).

The uterus is almost entirely covered by peritoneum, which dips inferiorly between the anterior uterine wall and the urinary bladder, forming a pocket in the peritoneal cavity called the *vesicouterine pouch* (Fig. 24-9). Posteriorly the peritoneum covers the body of the uterus and the anterior surface of the rectum, forming a peritoneal pouch, the *rectouterine pouch*. Along the lateral borders of the uterus the peritoneum is continuous with the two layers of the *broad ligament* that attaches the uterus to the lateral walls of the pelvis (Fig. 24-10). The broad ligaments contain the uterine tubes in their upper free borders, with the ligaments of the ovaries extending through them horizontally. Fibrous strands called the *round ligaments* of the *uterus,* attached to the upper end of the uterine body, pass through the broad ligaments and the inguinal canals, terminating in the skin of the **labia majora.** The uterosacral ligament is the fibrous tissue in the root of the broad ligament.

The wall of the uterus consists of three distinct layers (Fig. 24-10): the *endometrium* or mucosa of the uterus, which undergoes cyclic changes under the influence of sex hormones during the menstrual cycle; the *myometrium,* or muscle layer of the uterus, capable of remarkable expansion during pregnancy; and the *perimetrium* at the surface.

The endometrium (Fig. 24-12) is lined at its luminal surface by simple columnar epithelium composed of a mixture of ciliated and secretory cells. From the surface epithelium tubular glands extend into the thick lamina propria, termed *endome-*

trial stroma. The uterine glands are mostly simple tubular glands whose secretory cells are similar to those of the surface epithelium. The stroma contains many stellate cells that resemble embryonic mesenchymal cells, but its most striking morphological feature is in its blood supply. In the basal part of the stroma the highly coiled final branches of the larger arteries passing into the stroma between the glands, the "coiled arterioles," provide a rich blood supply and have an important role in menstruation.

The myometrium consists of several layers of smooth muscle oriented circularly, obliquely, and longitudinally, arranged in bundles separated from one another by connective tissue. During pregnancy individual smooth muscle cells may enlarge as much as ten times, accounting for part of the expansion of the uterus. In addition to the hypertrophy of muscle cells there is evidence that the numbers of cells increase by division of smooth muscle cells and by differentiation of new smooth muscle from connective tissue cells. During birth, waves of contraction of the myometrium are an important factor in the expulsion of the fetus.

The Vagina

The *vagina* (Figs. 24-9 and 24-10) is a tubular organ extending from the uterine cervix to its external opening at the vestibule. It is posterior to the urinary bladder and urethra and anterior

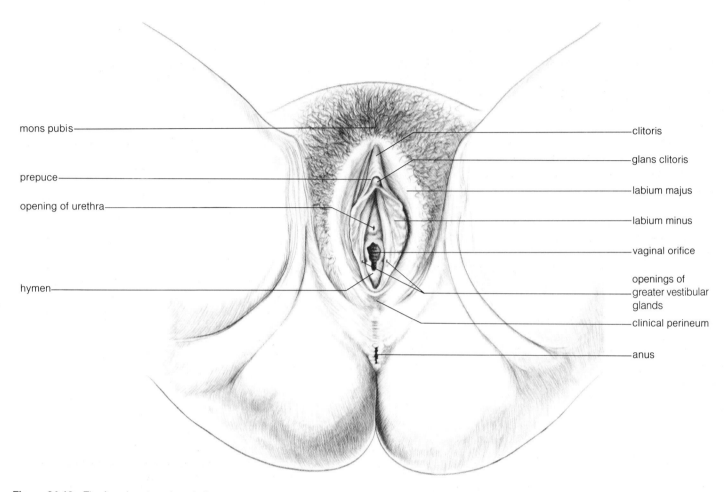

mons pubis

prepuce

opening of urethra

hymen

clitoris

glans clitoris

labium majus

labium minus

vaginal orifice

openings of greater vestibular glands

clinical perineum

anus

Figure 24-13 The female external genitalia.

to the rectum. Its superior end contains the uterine cervix, with the lumen of the vagina forming an *anterior fornix* anterior to the cervix, a *posterior fornix* posterior to the cervical lip, and a *lateral fornix* on either side. The lumen is slitlike, with a transverse diameter of about 3.5 cm. In a virgin, the lumen is often closed at the lower end of the vagina by a *hymen,* a thin, fenestrated membrane (a membrane with one or more openings).

The wall of the vagina is composed of three layers: *mucosa, muscular layer,* and a *connective tissue coat* (adventitia). The surface epithelium of the mucosa is stratified squamous that usually does not become keratinized but is moistened by mucus from the cervix. The lamina propria is a dense connective tissue containing a plexus of veins near the muscle layer. The muscle layer consists of circular and longitudinal bundles of smooth muscle that form an interlacing network. Around the opening of the vagina into the vestibule skeletal muscle fibers of the bulbocavernosus muscle form a sphincterlike arrangement. The

muscle coat is covered externally by loose connective tissue, binding the vagina to adjacent structures including the muscles and fascia of the pelvic floor.

The Female External Genitalia

The *female external genitalia* (Fig. 24-13) consist of the **mons pubis,** the **clitoris,** and the paired **labia majora** and **minora.** The *pudendal cleft* is bounded anteriorly by the mons pubis and laterally by the labia majora; within it are the labia minora, the clitoris, and the urethral and vaginal orifices.

The Mons Pubis The mons pubis (Fig. 24-13) consists of an elevated area of fatty tissue covered with skin containing coarse hairs in the adult. It is located anterior to the pubic symphysis and acts as a cushion during sexual intercourse.

The Clitoris The clitoris (Fig. 24-13) is homologous with part of the male penis in that it is derived from the same embryonic tissue, but it is shorter and does not contain a corpus spongiosum or the urethra. The *corpora cavernosa* end anteriorly in a rudimentary *glans clitoridis*. Posteriorly the *crura* of the *corpora cavernosa* lie near the *vaginal bulbs,* two elongated masses of erectile tissue surrounding the vaginal orifice. The *crura* of the *clitoris* are covered by the *ischiocavernosus muscles;* the *vaginal bulbs* are covered by the bulbospongiosus muscle, which functions as a sphincter, tightening the vagina during intercourse.

The Labia Majora The labia majora (Fig. 24-13) are paired, fat-filled folds of skin covered with hair; they are located on either side of the vaginal orifice and are continuous anteriorly with the mons pubis. The labia majora form the lateral boundaries of the pudendal cleft and are homologous with the scrotum.

The Labia Minora The labia minora (Fig. 24-13) are skin folds without hairs. They are located medial to the labia majora and are continuous posteriorly by a transverse skin fold posterior to the vaginal orifice. Anteriorly the folds continue onto the clitoris, forming the *prepuce* of the *clitoris*. The labia minora form the boundaries of the *vaginal vestibule,* a shallow sinus homologous with the male penile urethra.

The Reproductive Cycle

Beginning at puberty the human female undergoes a reproductive cycle of approximately 28 days that is normally repeated continuously until menopause unless interrupted by pregnancy. This cycle involves the brain (hypothalamus), hypophysis (pituitary gland), and the reproductive organs in the preparation of these organs for pregnancy.

The Ovarian Cycle Changes in the ovary before and after ovulation are regulated by hormones called *gonadotropins* (follicle-stimulating hormone, FSH, and luteinizing hormone, LH) produced in the *anterior lobe* of the *pituitary gland*. The release of these hormones into the blood stream is controlled by substances called *releasing hormones* produced in the *hypothalamus* (see Chapter 18).

At the beginning of the reproductive cycle, the onset of *menstruation,* there is an increase in the FSH released into the blood stream, which stimulates growth of the primary follicles in the ovary (Fig. 24-14). The follicular cells surrounding the ovum divide rapidly, producing a stratified layer of secretory cells next to the ovum known as the **granulosa cells.** These cells produce the hormone *estrogen,* a female sex hormone that

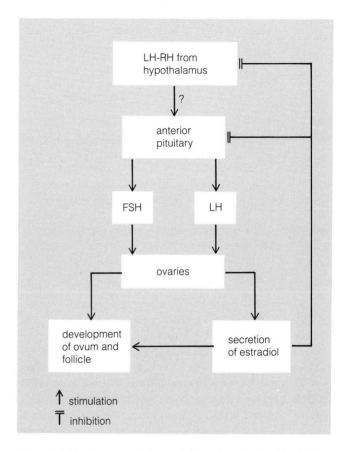

Figure 24-14 Summary of the regulation of ovarian function in the preovulatory portion of the menstrual cycle.

affects all the reproductive organs, with a specific stimulating effect upon growth of the endometrium.

As the growing follicle matures, LH is released from the pituitary gland and, along with the high level of FSH present, causes a sudden spurt of growth of the follicle, leading to *ovulation* at about the fourteenth day of the cycle. When the ovum is discharged, the follicle collapses and its walls are thrown into the folds. Under the influence of LH the follicle cells differentiate into a secretory structure known as the **corpus luteum** (yellow body), which produces the hormone *progesterone* as well as estrogen. These hormones stimulate additional changes in the endometrium, designed to prepare the uterus for *implantation*. If pregnancy occurs, the corpus luteum is maintained through most of the gestational period. Failure of an embryo to implant results in a drop in LH and a degeneration of the corpus luteum into a white scar called the *corpus albicans* (white body). The drop in ovarian hormone secretion brings on the onset of menstruation, concluding the cycle.

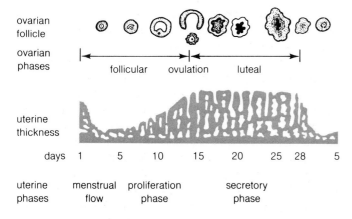

ovarian follicle

ovarian phases

follicular   ovulation   luteal

uterine thickness

days   1   5   10   15   20   25   28   5

uterine phases

menstrual flow   proliferation phase   secretory phase

Figure 24-15 Diagrams of the changes occurring during ovarian and uterine phases of the female reproductive cycle.

The Uterine (Endometrial) Cycle The endometrium undergoes morphological changes during the 28 day reproductive cycle that are correlated with the events of the *ovarian cycle* (Fig. 24-15). Ovulation usually occurs at about 14 days before the onset of menstruation and can be used as an event dividing the reproductive cycle into two phases: a *preovulatory phase,* or *follicular phase,* during which at least one follicle is growing rapidly and secreting *estrogen,* and a *postovulatory phase,* or *luteal phase,* during which the corpus luteum is producing estrogen and *progesterone.*

The endometrium (Fig. 24-12) is a mucous membrane composed of a lining of simple columnar epithelium and simple straight tubular glands extending into the lamina propria to the myometrium with no intervening submucosa. The basal portions of the tubular glands receive their blood supply from separate branches of the arteries that enter the endometrium to supply the superficial part of the glands. This separate blood supply can be considered to divide the endometrium into a *basal layer,* which does not slough during menstruation, and a *superficial layer* (functional layer), which undergoes morphological changes during the uterine cycle and is discharged during menstruation.

At the onset of menstruation, brought on by the drop in hormonal level when the corpus luteum degenerates, the coiled arterioles supplying the superficial layer of the endometrium constrict, reducing the blood flow and causing the death of the endometrium. The blood vessels are also damaged and subsequently rupture as blood flows into the vessels under arterial pressure. Blood flows into the tissue spaces, disrupting the endometrium and causing it to disintegrate. This process occurs in patches over a period of two to seven days as the superficial layer of the endometrium is sloughed, leaving only the basal portions of the tubular glands as the source of epithelium for replacement of the surface epithelium and superficial portion of the glands. Blood flow is stopped by spasm of the blood vessels in the basal layer of the endometrium.

Regeneration of the endometrium begins even before menstrual loss is complete, by proliferation of the epithelium of the basal parts of the glands and migration of cells to cover the surface. This proliferation, which occurs in the follicular phase, is under the influence of estrogen produced in the ovary and affects not only the epithelium but also the connective tissue stroma and the arterioles. As the glands lengthen and the surface is again covered with epithelium, the stroma increases in cellularity and the coiled arterioles increase in length, becoming more tortuous. By the time of ovulation the glands have lengthened to such an extent that the endometrium has attained almost maximum thickness (Fig. 24-15).

Following ovulation, during the luteal phase, the endometrium comes under the influence of progesterone in addition to estrogen. This phase is characterized by a decline in proliferation concurrent with an increase in the secretory activity of the epithelium. The gland cells actively elaborate glycogen and mucopolysaccharides, discharging these substances into the lumen of the glands. The simple straight tubular glands increase in complexity by becoming *sacculated* as the metabolically active cells increase in size. The stroma also undergoes changes, with the cells increasing in size and with fluid accumulating in the deeper layer of the endometrium. The coiled arteries increase in length to a greater extent than the endometrial glands and thus become more highly coiled. If implantation does not occur, these arteries begin a period of alternately constricting and relaxing that soon results in the onset of menstruation, completing the uterine cycle.

Sequence of Events Leading to Pregnancy

Ovulation usually occurs at about the middle of the menstrual cycle by rupture of the ovarian follicle, which discharges the ovum into the peritoneal cavity near the infundibulum of the uterine tube. Oscillations of the fimbriae and ciliary action of the mucosa, assisted by muscular contractions of the tubular wall, move the ovum into the ampulla. At the time of ovulation the ovum has completed the first meiotic division by giving off the first polar body and is in metaphase of the second meiotic division. It remains in this phase until a sperm penetrates.

About 300 to 500 million sperm are deposited near the external os (opening) of the cervix during intercourse. The sperm, now motile, move into the *cervical canal* by action of their tails, but most of them are killed by the cervical mucus, with relatively few surviving to enter the lumen of the body of the uterus. From this point the passage of the sperm through the female reproductive tract is accomplished largely by muscular contractions of the uterus and uterine tubes. It probably takes less than an hour for the sperm to reach the *ampulla,* in the vicinity of the ovum. If the timing is right, a viable sperm makes contact with a viable ovum and *fertilization* occurs. The ovum is capable of being fertilized for only about 24 hours after ovulation, but sperm may remain in the reproductive tract of the female for as long as four days, retaining the capacity for fertilization.

The sperm penetrates the ovum by releasing enzymes from the tip of the head (acrosome) that digest a pathway through the follicular cells (corona radiata) adhering to the ovum and the zona pellucida surrounding the ovum. The plasma membrane of the sperm head fuses with that of the ovum, releasing the nucleus of the sperm into the cytoplasm of the ovum. After the sperm has penetrated the ovum, the second polar body is given off (second meiotic division); the pronuclei fuse, and cell division (cleavage) by mitosis begins. The developing embryo moves slowly toward the uterus, taking about six days to reach the lumen of the body of the uterus where the process of implantation begins. The most common site of implantation is the posterior wall of the body of the uterus; implantation can occur at any point at which the embryo happens to adhere to the endometrium. By about eleven days following fertilization (usually about day 26 of the menstrual cycle) implantation has proceeded to the point that production of the hormone *chorionic gonadotropin* by the fetal membranes can begin, preventing degeneration of the corpus luteum in the ovary. The continued production of hormones by the corpus luteum maintains the endometrium for the first few months of pregnancy, after which the placenta begins to produce enormous quantities of estrogen and progesterone.

> During pregnancy, ligaments between the sacrum and os coxae relax, and movements at the sacroiliac joints increase. During labor the symphysis pubis relaxes, and the distance between the pubic bones increases, allowing the birth canal to enlarge.

Methods of Preventing Pregnancy

Various contraceptive measures use widely different mechanisms to prevent pregnancy, with varying degrees of effectiveness (Fig. 24-16). The discharge of sperm from the male repro-ductive tract may be prevented by **vasectomy,** the cutting of the vas deferens to prevent sperm from entering the semen, a safe and effective method. The passage of sperm into the vagina may be prevented by the use of a *condom,* which has the advantage of also preventing the transfer of venereal disease. The capacity of the sperm to fertilize the ovum may be destroyed while the sperm are in the vagina by the use of *acid foam* or *jelly* which immobilizes the sperm. The entrance of the sperm into the cervical canal may be blocked by a *diaphragm,* with spermicidal jelly applied to the rim as a seal; this is inserted into the vaginal fornix and covers the cervical os. Fertilization may be effectively prevented by *ligation* of the *uterine tubes,* preventing passage of ova or sperm, so that no contact between the two is possible. Alternatively fertilization may be permitted to take place but pregnancy prevented by an **intrauterine device,** a ring, coil, or loop that alters the endometrium, causing a failure of the implantation process. A physician inserts the device into the lumen of the body of the uterus; it is normally removed only when the woman wants to become pregnant. Finally, the ovarian cycle may be altered by ingestion of hormones ("the pill"), so that ovulation does not occur. The most commonly used *oral contraceptive* includes both estrogen and progesterone at a daily dose level that will suppress ovulation but will allow menstruation to occur upon interruption of the pill-taking. This is one of the most reliable of the contraceptive methods, but some serious side effects (increase in thrombosis in some women) suggest the exercise of caution in long-term use.

Functional Anatomy of the Female Reproductive System

The female produces a single ovum, discharged near the midpoint of each ovarian cycle by the process of ovulation. The ovum contains the maternal contribution of genetic material, but the specific constitution of the chromosomes is not identical with that of the chromosomes of the mother. During maturation of the ovum (and sperm), the chromosomes undergo structural alterations, so that the offspring is more genetically variant than would be expected from merely pooling one-half of the maternal and paternal chromosomes. This uniqueness of chromosomal structure is accomplished by the exchange of genetic material (crossing over) between chromosomes during meiosis as the chromosomal complement is reduced to the haploid number. Maturation of the ovum also includes the accumulation within its cytoplasm of proteins essential for the metabolic processes of the early phases of development prior to implantation.

Implantation Sites *Implantation* usually occurs in the *body* of the *uterus* (Fig. 24-10), most frequently on its posterior wall.

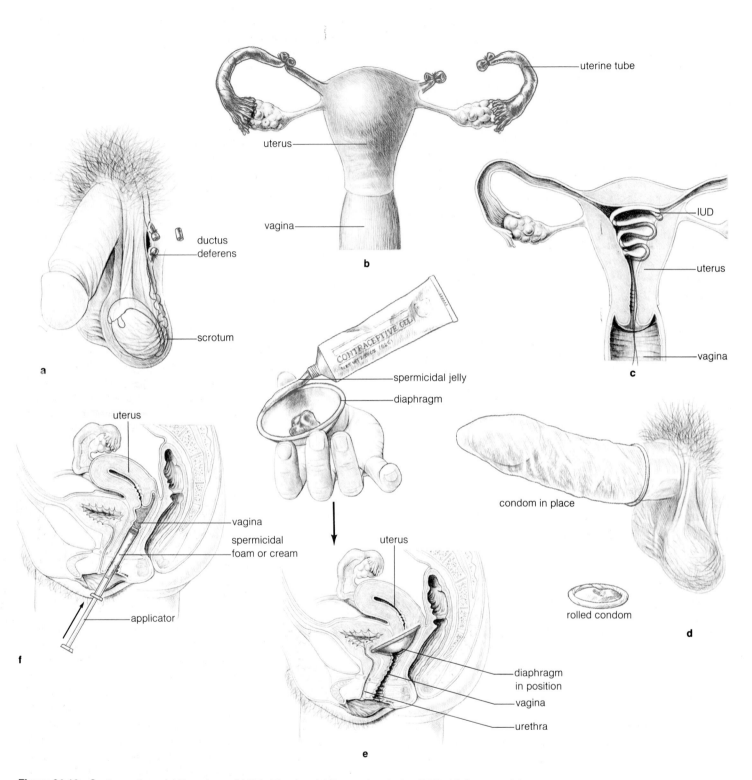

Figure 24-16 Contraceptives. (**a**) Vasectomy. (**b**) Tubal ligation. (**c**) Intrauterine device (IUD). (**d**) Condom. (**e**) Diaphragm. (**f**) Application of spermicidal substance.

During early implantation phases the endometrium becomes modified in a response known as the *decidual reaction* (the decidua are the parts of the endometrium that are shed after childbirth), during which there is a rapid accumulation of glycogen and mucopolysaccharides that are presumably nutrients for the embryo. Implantation of the embryo is followed by development of the placenta, from maternal and fetal tissues, bringing fetal blood vessels into close relationship with the maternal blood supply. The placenta can develop in any location in which maternal blood vessels are available, but there will be problems if the implantation site does not allow for expansion as the fetus enlarges. For example, implantation low in the body of the uterus may result in the cervical canal being overlapped by the fetal placenta, a situation called *placenta praevia*. Later as the growth of the fetus causes the upper cervical canal to dilate, the fetal blood vessels may be torn, causing bleeding that may result in death of the fetus.

Implantation sometimes ocurs within the uterine tubes with serious consequences for the mother. The uterine tube is incapable of stretching to accommodate the rapid expansion of the fetus; instead it usually ruptures, most commonly during the fourth month of pregnancy, resulting in severe internal bleeding, with the possibility of death of the woman. On the other hand the fetus may herniate through the uterine tube with little maternal bleeding and continue development within the coelom, keeping its blood supply in the uterine tube. Such a fetus occasionally develops to term, but it must be removed surgically by incision through the body wall. A similar procedure is used to remove the fetus in cases of coelomic pregnancies, in which fertilization and implantation occur in the peritoneal cavity, with development entirely outside the reproductive tract. Any implantation site other than the normal place within the uterus is considered an *ectopic site* and increases the hazards to mother and fetus.

Changes in the Uterus During Pregnancy   The uterus not only allows implantation of the embryo and provides for nutrition of the rapidly growing conceptus but also is instrumental in expulsion of the fetus at birth. During the growth of the fetus and fetal membranes the uterus must expand from an organ with a cavity that is only a potential space, to an organ capable of holding a volume of about four quarts by the end of the pregnancy. Expansion of the uterus involves both passive stretching and active growth. Individual smooth muscle cells of the myometrium enlarge during pregnancy to about ten times their previous length. Additional smooth muscle cells are added by mitosis of smooth muscle and differentiation of mesenchymal cells during the first half of pregnancy. During the last half of pregnancy the uterus expands largely by reduction in thickness of its wall and by dilation of the superior half of the cervical canal. The inferior half of the cervix does not dilate due to its

collagenous fiber content, but the muscular wall of the upper half expands along with the body of the uterus, providing for growth of the fetus.

Parturition

The specific causes of the onset of *parturition* (childbirth) are unknown, but the events of the birth process occur in a specific sequence that usually follows a predictable pattern. Myometrial contractions occur throughout pregnancy, but such contractions have little force. At term the contractions increase in force and frequency, applying pressure on the fibrous wall of the inferior half of the cervix. The increased muscular contraction may be due to an increase in oxytocin, a hormone released in the posterior lobe of the pituitary gland, which has a generalized stimulatory effect upon smooth muscle. As pressure within the uterus increases, the cervix dilates, enlarging the birth canal, the amniotic membrane ruptures, releasing the amniotic fluid, and the fetus is expelled through the vagina. The fibrous symphysis pubis undergoes dissolution under the influence of relaxin, a hormone produced by the corpus luteum during the last three months of pregnancy, allowing the bony pelvis to spread as the fetus descends in the birth canal and the diameter of the canal to increase. The muscles and fascia of the perineum stretch or are torn as the fetus descends. Uterine contractions continue after the fetus emerges, expelling the fetal membranes, commonly called the *afterbirth*.

THE DEVELOPMENT OF THE REPRODUCTIVE SYSTEM

The sex of each individual is determined at fertilization by the chromosomal content of the sperm (see Chapter 3), but development of the reproductive system is morphologically indistinguishable in the two sexes during an early "indifferent" stage (Fig. 24-17a). The *gonads,* which become testes or ovaries, begin as thickenings of the *coelomic epithelium* on the medial aspect of the mesonephros. The primordia of both male and female reproductive ducts are present in both sexes during early development, with the fate of the ducts differing in the course of differentiation. Development of the reproductive systems is closely related to that of the urinary system, with the relationship being most obvious in the male, who utilizes the embryonic excretory (mesonephric) duct as a major part of the adult reproductive duct system: the mesonephros gives rise to the efferent ductules, epididymis, and ductus deferens.

The thickening of the coelomic epithelium, called the **genital ridge,** begins on the medial aspect of the middle portion

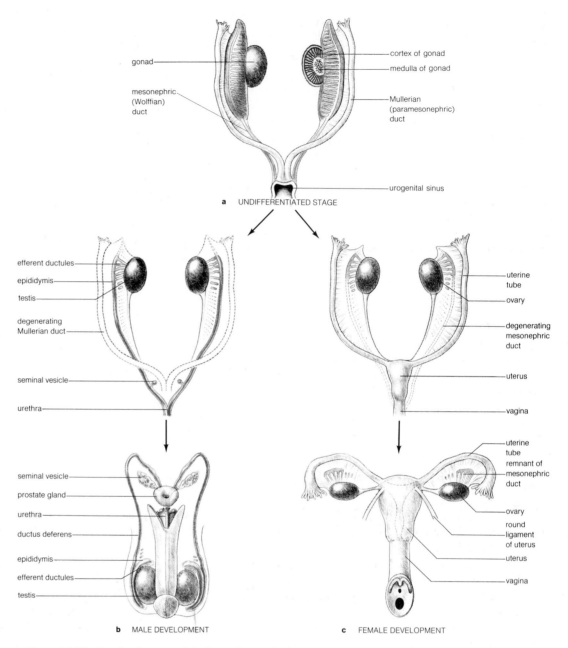

Figure 24-17 The development of the internal reproductive systems.

of the mesonephros. *Primordial germ cells* (the future sperm or ova) originate from special cells on the caudal aspect of the yolk sac and migrate into the genital ridge as it thickens, creating an indifferent (sexually undifferentiated) gonad. In both sexes fingerlike extensions of the proliferating coelomic epithelium, called **sex cords,** grow into the genital ridge, where they are separated by mesenchymal septa. In the male the sex cords canalize to become the *seminiferous tubules* and *rete testis*. In the female the sex cords break up into numerous oval masses, becoming the primary follicle cells that surround the germ cells. This differentiation of the gonads begins at about five weeks of intrauterine age.

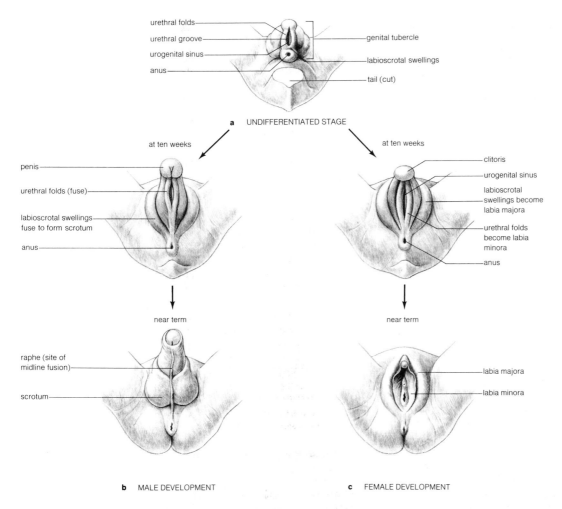

Figure 24-18 The development of the external reproductive structures.

In both sexes two sets of ducts are formed: the *mesonephric ducts* (Wolfian ducts), which receive urine produced by the mesonephric tubules and empty into the urogenital sinus, and the *paramesonephric ducts* (Mullerian ducts), which parallel the mesonephric ducts but end blindly posterior to the urogenital sinus (Fig. 24-17a). In the male the paramesonephric ducts degenerate; the mesonephros begins to degenerate, but in the vicinity of the genital ridge the rete testis joins some of the mesonephric tubules that persist to form the *efferent ductules* (Fig. 24-17b). The mesonephric duct adjacent to the genital ridge elongates and becomes the convoluted *epididymis,* with the remaining mesonephric duct persisting as the *ductus deferens* and *ejaculatory duct.*

In the female the mesonephros and most of the mesonephric duct degenerate, leaving the ovary with a mesentery attachment to the body wall but no direct communication with

a duct (Fig. 24-17c). The caudal portions of the *paramesonephric ducts* fuse to form the single midline *uterus,* and the cranial portions of the ducts remain separate as the laterally placed *uterine tubes.* The caudal end of the fused paramesonephric ducts pushes against the back wall of the urogenital sinus but does not open into it. A solid cord of cells grows caudally from the blind end of the fused paramesonephric duct (uterus) and canalizes, giving rise to the *vagina.*

During the third week of development mesodermal cells form elevated *urogenital folds* around the plate (cloacal membrane) that closes the hindgut. The *genital tubercle,* or phallus, forms anterior to the cloacal membrane, which soon becomes divided into an *anal membrane* and a *urogenital membrane.* These changes are similar in both sexes, and this stage (the "indifferent stage") lasts until about six weeks of development (Fig. 24-18a).

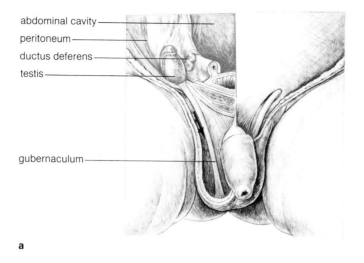

abdominal cavity
peritoneum
ductus deferens
testis

gubernaculum

a

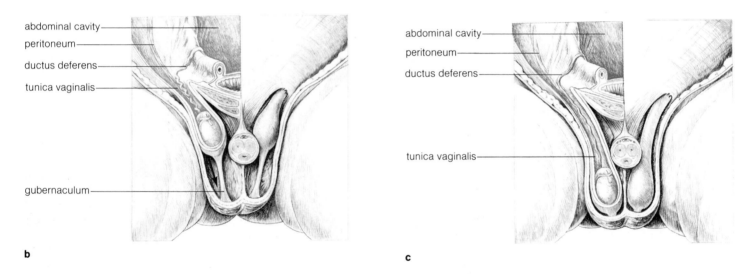

abdominal cavity
peritoneum
ductus deferens
tunica vaginalis

gubernaculum

b

abdominal cavity
peritoneum
ductus deferens

tunica vaginalis

c

Figure 24-19   The descent of the testes into the scrotum. (**a**) Testis beginning descent. (**b**) Tunica vaginalis being pushed ahead of testes as gubernaculum guides movement. (**c**) Testes arriving in scrotum.

Another pair of swellings, the *scrotal swellings,* appear lateral to the urogenital folds and differentiate into the *scrotum* in the male (Fig. 24-18b) and the *labia majora* in the female (Fig. 24-18c). In the male the genital tubercle elongates rapidly, pulling the urogenital folds forward to form the edge of the urethral groove (Fig. 24-18b). The urogenital folds fuse beneath the urethral groove, giving rise to the *penile urethra.* The line of fusion of the urethral folds is continuous with the line of fusion of the scrotal swellings and remains visible in the adult as the *raphe* of the *penis* and *scrotum.* The testes descend from the abdominal cavity into the scrotum toward the end of fetal development (Fig. 24-19).

WORDS IN REVIEW

Broad ligament	Mediastinum testis
Bulbourethral glands	Mons pubis
Clitoris	Processus vaginalis
Corpus luteum	Prostate gland
Dartos tunic	Rete testis
Ductus deferens	Semen
Epididymis	Seminal vesicles
External spermatic fascia	Sex cord
Gamete	Tunica albuginea
Genital ridge	Tunica vasculosa
Granulosa cells	Uterine tube
Internal spermatic fascia	Uterus
Intrauterine device	Vasectomy
Labia majora	Zygote
Labia minora	

SUMMARY OUTLINE

I. The Male Reproductive System
   A. The testis contains seminiferous tubules which produce sperm; the sperm leave via the rete testis which opens into ductuli efferentes which convey the sperm from the testis to the epididymis.
   B. The epididymis is the highly coiled tube within the scrotum that receives sperm from ductuli efferentes and stores them until ejaculation.
   C. The ductus (vas) deferens extends from the epididymis to the ejaculatory duct; it is a component of the spermatic cord. It passes through the inguinal canal, turns medially inside the abdominal wall to pass posterior to the neck of the bladder and is joined by the seminal vesicles to form the ejaculatory duct, which extends from the seminal vessals to the urethra.
   D. Accessory glands add secretions to sperm, forming semen.
      1. The seminal vesicles are elongated saccular glands that produce a fluid rich in fructose which provides energy for sperm.
      2. The prostate gland is a compound gland located below the neck of the bladder; it has multiple ducts that convey secretions containing enzymes to the urethra.
      3. The bulbourethral glands are paired compound glands embedded in the urogenital diaphragm; their secretions lubricate the urethra, facilitating the passage of sperm.

   E. The male external genitalia consist of organs located outside the body wall.
      1. The scrotum is a skin pouch that contains the testes.
      2. The penis is the copulatory organ; it contains the urethra and erectile tissue.

II. The Female Reproductive System
   A. The ovaries produce ova and the hormone estrogen, from growing follicles, and progesterone, from the corpus luteum.
   B. The uterine tubes receive ova; they are the site of fertilization and convey the zygote to the uterus.
   C. The uterus, a thick, muscular organ, is the site of implantation and of the development of the embryo.
   D. The vagina is a muscular tube that receives sperm at sexual intercourse and provides a passageway for the fetus at birth.
   E. The female external genitalia are located in the perineal area.
      1. The mons pubis is a fatty mound of tissue located anterior to the pubic symphysis.
      2. The clitoris is the homologue of part of the penis; it contains erectile tissue but does not contain the urethra.
      3. The labia majora are hair-covered skin folds lateral to the pudendal cleft; they are homologous with the scrotum.
      4. The labia minora are skin folds lateral to the vaginal orifice; they form the boundaries of the vaginal vestibule, which is homologous with the penile urethra.

III. The Reproductive Cycle
   A. The ovarian cycle is regulated by the releasing hormones produced by the hypothalamus. These hormones cause the anterior pituitary to release FSH which stimulates the ovarian follicle to grow and produce estrogen. Then LH and FSH stimulate ovulation, and LH causes formation of the corpus luteum. The corpus luteum produces estrogen and progesterone, which prepare the uterus for implantation of the embryo. If pregnancy does not occur, the corpus luteum degenerates, and maturation of another follicle begins.
   B. In the uterine cycle the endometrium sloughs at the onset of the cycle (menstruation) due to the drop in hormone level as the corpus luteum degenerates. Estrogen from the growing follicle stimulates regeneration of the endometrium; progesterone stimulates sacculation of the glands; and the endometrium sloughs again unless implantation of an embryo occurs.

IV. Events Leading to Pregnancy
   The sequence of events leading to pregnancy includes ovulation, intercourse, fertilization, and implantation.

V. Methods of Preventing Pregnancy

Methods of preventing pregnancy include vasectomy and use of condom for the male. The female may use acid foam or jellies, alone or with a diaphragm, intrauterine devices, ligation of uterine tubes, or oral contraceptives composed of estrogen and/or progesterone.

VI. Development of the Reproductive System

   A. The ovaries or testes originate from thickenings of coelomic epithelium. The testes descend from the abdominal cavity to the scrotum late in fetal life.

   B. Male ducts (including the epididymis and vas deferens) arise from mesonephric ducts initially present in embryos of both sexes.

   C. Female ducts (the uterine tubes and uterus) arise from paramesonephric ducts, initially present in both sexes; the vagina arises from a solid cord of cells extending below the fused paramesonephric ducts (the future uterus).

   D. Two genital swellings give rise to the scrotum in the male and the labia majora in the female; the two halves of the scrotum fuse, whereas the labia majora retain a cleft between them.

   E. Both the penis and clitoris arise from the genital tubercle.

REVIEW QUESTIONS

1. Sperm differentiate in what specific part of the testis?

2. Mature sperm move from the rete testis into what structures as they exit from the testis?

3. Where are mature sperm stored prior to ejaculation?

4. What structure is formed by the junction of the vas deferens with the seminal vesicles?

5. By what ovarian attachment do the blood vessels pass from the trunk wall to the ovary?

6. Fertilization usually occurs in what structure?

7. What attaches the uterus to the wall of the pelvis?

8. What is the relationship of the vagina to the rectum?

9. What morphological events occur in the uterus during the period of follicular growth in the ovary?

10. Implantation occurs most frequently at what site?

SELECTED REFERENCES

Finn, C. A., and Porter, D. G. *The Uterus. Handbooks of Reproductive Biology.* London: Paul Elek, Ltd., 1974.

Guyton, A. C. *Textbook of Medical Physiology.* 4th ed. Philadelphia: W. B. Saunders, 1971.

Mann, T. *Biochemistry of Semen and of the Male Reproductive Tract.* London: Methuen and Co., 1964.

Moore, K. L. *The Developing Human, Clinically Oriented Embryology.* Philadelphia: W. B. Saunders, 1979.

Moore, K. L. *Clinically Oriented Anatomy.* Baltimore: Williams and Wilkins, 1980.

Smith, D. E., ed. *The Ovary.* Baltimore. Williams and Wilkins, 1962.

Velardo, J. T., and Kasprow, B. T., eds. *Biology of Reproduction.* Pan American Congress of Anatomy, 1972.

DEFENSE MECHANISMS

After completing this chapter you should be able to:

1. Describe the morphological features that enable the skin to prevent loss of water and repel bacterial invasion.

2. Describe the features of mucous membranes important in protecting against bacterial invasion.

3. Name the lymphoid organs and give the relation of each to the lymphatic vessels and/or to the blood vessels.

4. Describe the structure of a lymph node and give the major functions of each component.

5. Locate the tonsils and describe their structure. Contrast the structure of the tonsil with that of a lymph node.

6. Locate the spleen and describe its structure. Relate the structure of the red and white pulp of the spleen to the special function of each of these components.

7. Locate the thymus and describe its structure. Demonstrate an understanding of the role of the thymus in cell-mediated immunity.

8. Name and describe the cells present in peripheral blood, giving the general functions of each cell type.

9. Describe the distribution and functions of macrophages.

10. Describe the sequence of events in the primary and secondary immune responses. Compare and contrast the roles of B- and T-lymphocytes in these responses.

The environment in which we live contains all the elements essential for the maintenance of the body's cells. Individually these cells have a very limited capacity to adjust to changes in their environment and can maintain a healthy state only when bathed in tissue fluid of a constant acidity, salt concentration, and temperature. Because fluctuations of the external environment exceed those compatible with the life of the cells, protective mechanisms exist that isolate the cells from the harmful elements of the external environment while allowing them to obtain needed materials. The external environment also contains numerous substances, including bacteria and viruses, that are potentially harmful; the outer covering of the body is designed to discourage them from gaining entrance. However, because we must take in useful substances and discharge wastes, we cannot have a completely impermeable barrier isolating us from the environment. Thus harmful agents may penetrate the outer defenses, and we must have mechanisms for detecting and eliminating them.

The entry of unwanted foreign material is restricted by the relatively impermeable outer covering of the body, the skin, and by the mucous membranes that line the tubes of the digestive and respiratory tracts. These tubes are open to the external environment and are not sterile, that is, they always contain bacteria and other foreign particulate matter, but they are constructed to reduce penetration. The other tubes open to the outside, the excretory and reproductive ducts, are normally sterile but may be a pathway for penetration by bacteria under some circumstances. If bacteria do succeed in penetrating any of these membranes, the body recognizes them as foreign and initiates an *immune response* to combat the invasion. The response includes the activation of **macrophages** which cooperate with **lymphocytes** in eliminating the foreign elements. The lymphocytes are components of lymphoid tissue; macrophages may also be described as components of lymphoid tissue, but they have a broader distribution in the body as components of the reticuloendothelial system.

The major components of the defense system are the *skin, mucous membranes, lymphoid tissue,* and the *reticuloendothelial system* (the system of macrophages). The first two serve primarily to keep unwanted matter out of the body; the second two constitute the *immune system,* which combats invaders that penetrate the first line of defense. We will first consider the anatomy of each component and then the role of each in defending against disease.

THE SKIN

The *skin* (Fig. 5-1) may be considered the first line of defense, providing a barrier that is impermeable even to water. The *pigment* of the skin shields the deeper structures from the harmful effects of the ultraviolet rays of sunlight yet allows adequate penetration for the synthesis of vitamin D. The outer layers of the epidermis are *keratinized,* so that they resist abrasion and prevent the passage of water, enabling us to live in a relatively dry environment without becoming dehydrated. The skin eliminates water and other waste products by means of *sweat glands,* derived from the epidermis. These glands, along with the blood vessels of the dermis, also have an important role in temperature regulation. However, the most important function of the skin in its role in the defense against disease is to act as a barrier that cannot be penetrated by bacteria or other infectious agents.

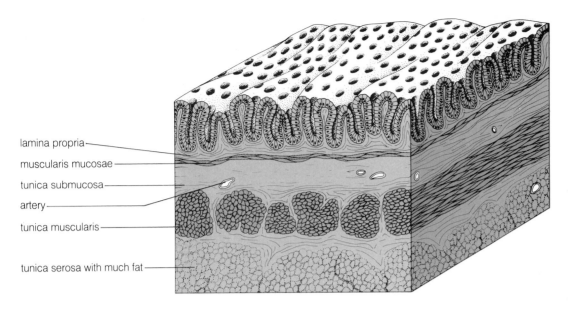

lamina propria
muscularis mucosae
tunica submucosa
artery
tunica muscularis
tunica serosa with much fat

Figure 25-1 A block of the wall of the intestine showing the structure of a mucous membrane.

THE MUCOUS MEMBRANES

All the tubes that open externally, such as the digestive and respiratory tracts, are lined with moist epithelium that, with its associated connective tissue, constitutes a *mucous membrane* (Fig. 25-1). Because each membrane is continuous with the skin at the external opening of the tube (muco-cutaneous junction), there is continuity of the dry epithelium of the epidermis with the moist epithelium of the mucous membrane at the edge of the opening. Although the epithelium of the mucous membrane is more permeable than that of the epidermis, it does present a continuous sheet of cells that is not readily penetrated by foreign agents. In the respiratory tract the epithelium is *ciliated pseudostratified columnar* epithelium, and bacteria or other particulate matter trapped in its mucous layer are moved toward the mouth by ciliary action. The urinary tract is normally sterile, due to the presence of sphincters that can close the tubes and by the passage of urine toward the exterior, which tends to wash substances from the tract. Secretions of the reproductive tract of the female contain antibacterial substances. The lower digestive tract is not normally sterile, but the upper digestive tract contains enzymes and hydrochloric acid that reduce the bacterial population.

The connective tissue of the mucous membranes contains cells, such as macrophages and lymphocytes (*antibody-producing cells*), that are part of the defenses against foreign substances. Some macrophages and lymphoid cells are always present as scattered elements in the connective tissue beneath the epithelium of the mucous membrane; they respond to bacterial invasion of the mucous membrane by accumulating at the invasion site in large numbers. The presence of large numbers of these cells in a localized area, termed *inflammation,* indicates that the mucous membrane has been penetrated by foreign substances, which have triggered a local response by specialized blood and connective tissue cells. In the mucous membranes these specialized cells may show little anatomical organization, but in specific regions of the body they are present in distinct structures termed *lymphoid organs.*

THE LYMPHOID ORGANS

The principal lymphoid organs (Table 25-1) are the **lymph nodes, tonsils, spleen** and **thymus.** These are encapsulated masses of lymphoid (lymphatic) tissues consisting of an aggregation of unattached *lymphocytes* and *free* or *fixed macrophages* held in a meshwork of *reticular fibers. Lymph nodes* are interposed in a channel of lymph drainage, *tonsils* are located at the beginning of lymph vessels, and the *spleen* is lymphoid tissue interposed in a blood vascular channel. The *thymus* has a typical relationship to both the blood vascular system and the lymph

Table 25-1   The Lymphoid Organs

Organ	Location	Relation to Vascular Channels
Lymph nodes	Throughout the body, in aggregations	Placed within lymph channels
Tonsils	Near the entrances of food and air passages	Placed at the beginning of lymph channels
Spleen	In the abdominal cavity to the left of the stomach	Placed within blood channels
Thymus	In the anterior mediastinum of the thorax	Has a blood supply and lymph drainage with no special vascular pattern.

vessels; that is, it is supplied by blood vessels and has lymph vessels draining the lymph produced within the organ.

Lymph Nodes

Lymph nodes are a major part of the *immune system,* serving as filters of the lymph and providing resistance against disease. They are located in series in lymph channels and usually occur in aggregates at specific points in the pathway of lymph drainage of the major parts of the body (Fig. 25-2). For example, the lymph from the lower extremity passes through the inguinal nodes located in the groin region. Because these nodes are in chains interrupting the lymph vessels, lymph must pass through several nodes prior to entering the main lymph vessel of the trunk, the *thoracic duct.*

Each lymph node (Fig. 25-3) is an oval mass of lymphoid tissue enclosed within a capsule of collagenous connective tissue. The capsule is penetrated by several *afferent lymph vessels* bringing lymph into the node, and there are usually two *efferent lymph vessels* that drain the lymph from the node, conveying it to the next lymph node in the chain. The connective tissue of the capsule is continuous with fibrous strands, called *trabeculae,* that extend into the substance of the node, providing a pathway for small blood vessels and a framework supporting the meshwork of *reticular fibers* that serves as a filter, through which the lymph percolates slowly as it passes through the node. The reticular fibers are strands to which the fixed macrophages are attached; they retard the flow of lymph and tend to retain the unattached lymphocytes within the node. The macrophages remove particulate matter from the lymph, cleansing it of potentially harmful material.

The predominant cell type in all lymphoid tissue is the *lymphocyte* (Fig. 25-4), a mononuclear (agranular) leukocyte which is a normal component of blood. The lymphocyte of peripheral blood is a small cell (about 8 to 10 μm in diameter) with a round nucleus surrounded by a thin rim of basophilic cytoplasm. Within the lymph node lymphocytes vary in size from the small lymphocyte, about 9 μm in diameter similar to that of peripheral blood, to a cell approximately 16 to 18 μm in diameter. The larger lymphocytes do not normally leave the lymph nodes but may differentiate into small lymphocytes that work their way slowly from the lymph node into efferent lymph vessels, for passage into the blood. The lymph nodes are a major source of peripheral blood lymphocytes, following stimulation by an **antigen** (the term for any substance that provokes an immune response). The lymph nodes actively engaged in the production of lymphocytes always contain several **lymph nodules,** dense aggregations of free lymphocytes, located chiefly beneath the capsule. These nodules have a definite pattern of organization, with each nodule characteristically consisting of a light central portion composed mostly of large lymphocytes and a darker rim of closely packed small lymphocytes. The light areas within the lymph node between the nodules are the *lymph sinuses,* which provide the pathway of least resistance for lymph passage through the lymph node. Scattered along the lymph sinuses are many macrophages, cells actively involved in the removal of particulate matter from the lymph. Thus the lymph node functions as a filter cleansing the lymph, in addition to producing lymphocytes.

Some of the lymphocytes, the **T-lymphocytes,** leave the lymph node and enter the blood stream, where they are available to react with and destroy invading antigenic substances;

this defense mechanism is called *cell-mediated immunity.* Others, the **B-lymphocytes,** remain in the lymph nodes and respond to antigenic stimulation by differentiating into *plasma cells,* which produce proteins called **antibodies** that enter the circulation to participate in defensive mechanisms. Defense by antibodies is called *humoral immunity,* because the antibodies are a component of blood and lymph, "humor" being an old term for the body fluids.

The Tonsils

The *tonsils* are aggregations of lymphoid tissue beneath the epithelium covering the *tongue* (lingual tonsils), in the *tonsillar fossa* between the anterior and posterior pillars of the fauces (palatine tonsils), and in the posterior wall of the *nasopharynx* (pharyngeal tonsil, or adenoids). Each tonsil (Fig. 25-5) consists of several lymph nodules collected within the lamina propria of the mucous membrane. The epithelium, serving as the capsule on one side of the tonsil, is usually folded into the lymphoid tissue, forming *crypts* between the lymph nodules. The fibrous connective tissue at the deep surface serves as the underlying part of the capsule and contains the blind-ending lymph capillaries that drain the lymph from the tonsil. Tonsils remove particulate matter and produce lymphocytes in much the same manner as the lymph nodes.

The Spleen

The *spleen* (Fig. 25-6) is a lymphoid organ located in the upper left quadrant of the abdominal cavity to the left of the greater curvature of the stomach, to which it is attached by means of a mesentery, the *gastrolienal ligament.* It is also attached to the posterior abdominal wall by the *lienorenal ligament,* a mesentery continuous with the parietal peritoneum covering the left kidney. The spleen is an oval, flattened structure encased by a connective tissue capsule indented on one side, at the *hilum,* where the blood vessels enter. The splenic artery, a branch of the celiac artery, enters at the hilum, giving rise to a number of small arteries that course in the connective tissue trabeculae that extend into the substance of the organ from the hilum. The trabeculae consist of collagenous fibrous connective tissue, providing a highly branching supporting framework for the lymphoid tissue filling the spaces between the trabeculae.

The lymphoid tissue of the spleen (Fig. 25-7) is organized around the blood vessels. The small arteries of the trabeculae give rise to branches that leave the trabeculae to enter the lymphoid tissue. When these arteries enter the lymphoid tissue, they are surrounded by a dense aggregation of lymphocytes held in a loose framework of reticular fibers. This morphologic

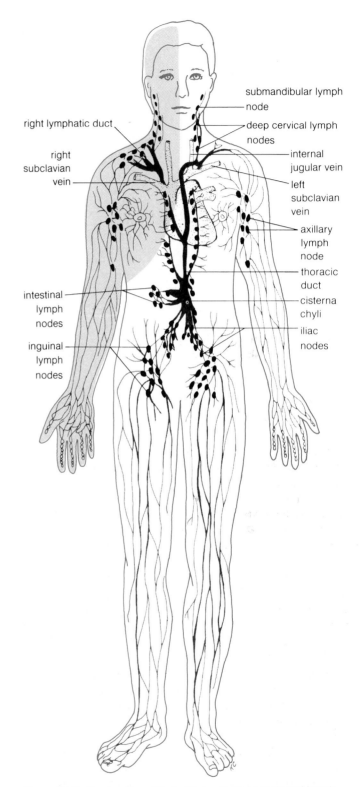

Figure 25-2 The location of the major lymphatic vessels and lymph nodes. The shaded area in the figure shows the portion of the body drained by the right lymphatic duct.

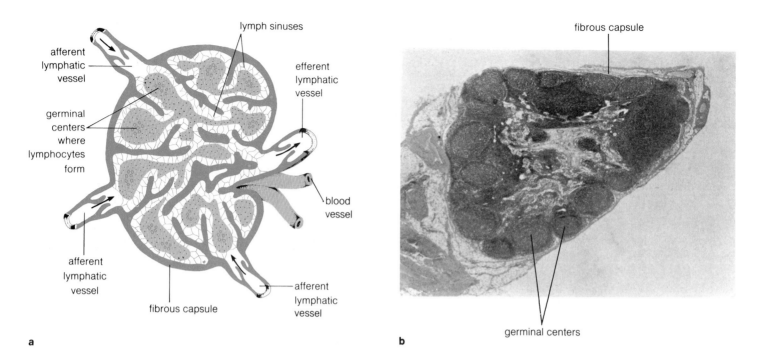

Figure 25-3 The structure of a lymph node, shown in cross section. (**a**) Diagram. (**b**) Photomicrograph, × 19 (photograph by Joan Creager).

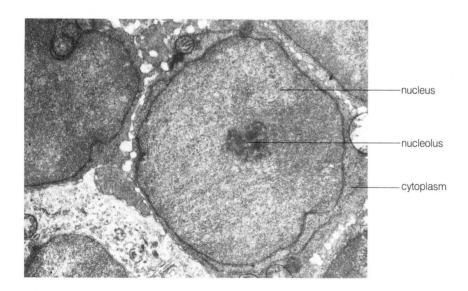

Figure 25-4 The lymphocyte. Electron photomicrograph of a lymphocyte of the rat spleen, × 11,200. (Courtesy of D. H. Matulionis.)

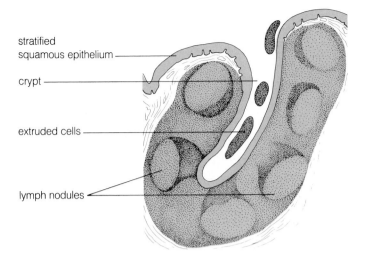

stratified
squamous epithelium

crypt

extruded cells

lymph nodules

Figure 25-5 Section through one of the crypts of the tonsil. See text for description.

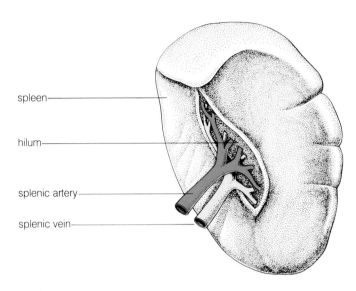

spleen

hilum

splenic artery

splenic vein

Figure 25-6 The spleen, medial aspect showing the area of the hilum.

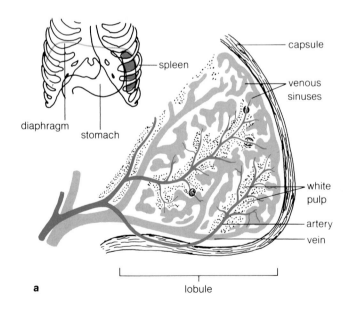

spleen

diaphragm

stomach

capsule

venous sinuses

white pulp

artery

vein

a

lobule

b

Figure 25-7 The structure of the spleen, shown in cross-section. (**a**) Diagram. (**b**) Photomicrograph, × 175 (courtesy of W. K. Elwood).

arrangement is similar to the arrangement of the lymphocytes in the nodules of a lymph node. Because lymphocytes are white blood cells and this unit does not contain visible blood, this tissue is designated the *white pulp* of the spleen.

The artery of the white pulp gives rise to branches that exit from the dense aggregation of lymphocytes to open directly into a network of *sinusoids* between the scattered masses of white pulp. The sinusoids are lined with macrophages, with definite gaps between the cells through which blood may pass freely. The area between adjacent sinusoids, called *pulp cords (of Bilroth)*, is occupied by a meshwork of reticular fibers supporting scattered macrophages. Blood cells that leave the sinusoids are trapped within the meshwork, imparting a red color to the tissue. The sinusoids and pulp cords constitute the *red pulp* of the spleen, the area in which all components of blood are brought into close contact with macrophages, either in the sinusoids or in the cords. The blood from the spleen drains into the splenic vein which empties into the hepatic portal vein.

The functions of the spleen are suggested by the morphology of the white and red pulp. The white pulp, like the lymph nodes, is important in the production of lymphocytes and in the immune response, and the lymph nodules of the white pulp of the spleen closely resemble the nodules of lymph nodes. The macrophages lining the sinusoids and cords of the red pulp are important in the removal of foreign particles, including bacteria, from blood. The most important function of the macrophages, however, is the removal of dead erythrocytes (red blood cells), which live only about 120 days. Breakdown components of hemoglobin, the major protein of erythrocytes, are discharged by the macrophages into the venous drainage of the spleen and transported via the hepatic portal vein to the liver. Within the liver some of the breakdown products of hemoglobin become major components of *bile* and are excreted through the digestive tract with the feces. Other components of hemoglobin, such as iron, are conserved by the liver and returned to bone marrow for use in synthesis of hemoglobin for new erythrocytes.

> The spleen retains its hematopoietic potential throughout life even though it does not normally produce blood cells other than lymphocytes. During certain pathological states it may become active in the production of all types of blood cells.

The slow passage of the blood through the spleen facilitates the removal of worn out erythrocytes by the macrophages lining the extensive network of sinusoids. The sinusoids of the red pulp are readily distensible; so that as the blood flows slowly through them, a considerable quantity of blood may be stored in the spleen, to be discharged when needed in other parts of the body. The storage function of the spleen is suggested by the fact that its size is directly related to the amount of blood it contains at any particular time.

During embryonic life the spleen functions as a blood-forming (hematopoietic) organ, giving rise to all types of blood cells. This hematopoietic function is gradually reduced in the later stages of prenatal development, so that by birth lymphocytes and perhaps monocytes are the only blood cells produced by the spleen. In the adult the spleen can be removed with no harmful effects; the spleen functions are taken over by the liver and lymphoid organs.

The Thymus

The *thymus gland* (Fig. 25-8a) is a lymphoid organ located in the thorax, where it lies just deep to the sternum. It undergoes fatty degeneration with age and is present only in young individuals as an active lymphoid organ. The thymus is enclosed in a connective tissue capsule from which strands, or septa, extend into the gland, separating its lymphoid tissue into lobules (Fig. 25-8b). Each lobule consists of a *cortex,* a dense outer rim of lymphocytes, and a *medulla,* a less dense central aggregation. The *medulla* usually contains one or more degenerating epithelial masses, called *Hassel's bodies,* which are the remains of the embryonic epithelium from which the thymus is derived. The lymphocytes of the thymus (called thymocytes) are free cells, held in place only by the meshwork of reticular fibers that support the gland and serve for the attachment of fixed macrophages.

During prenatal and early postnatal development the thymus is essential in establishing a population of lymphocytes for cell-mediated immunity. These cells, the *T-lymphocytes,* have the capacity to destroy foreign cells. For example, skin transplanted from one individual to another is likely to die within a period of three to four weeks and slough due to the reaction of the T-lymphocytes of the recipient against the foreign cells of the transplanted skin. The T-lymphocytes eventually migrate from the thymus to the lymph nodes and spleen.

THE BLOOD

The blood also contains substances that have a major role in defense against disease: the antibodies and several types of leukocytes (white blood cells, Fig. 25-9). These components are

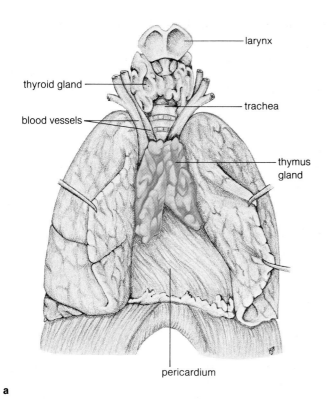

a

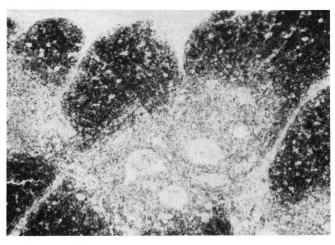

b

Figure 25-8 (**a**) The location of the thymus gland. (**b**) Photomicrograph of the human thymus (courtesy of W. K. Elwood).

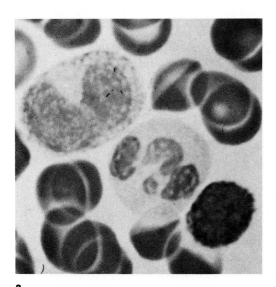

a

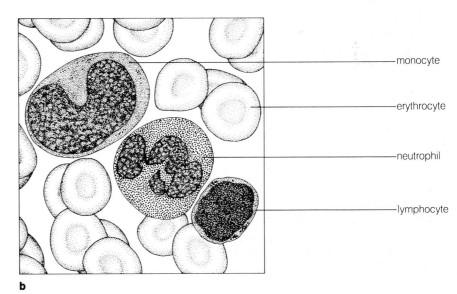

b

Figure 25-9 Human leukocytes. (**a**) Monocyte, neutrophil, and lymphocyte, × 3500 (courtesy of W. K. Elwood). (**b**) Sketch of the photomicrograph.

Table 25-2 Leukocytes

Cell Type	Site of Formation	Function
Lymphocyte	Bone marrow, lymphoid organs	Humoral and cell-mediated immunity
Monocyte	Bone marrow	Blood cell that moves into tissues and becomes tissue macrophage
Neutrophil	Bone marrow	Phagocytic cell that leaves blood to accumulate at sites of infection
Eosinophil	Bone marrow	Phagocytic cell involved in allergic reactions
Basophil	Bone marrow	Contains heparin (prevents blood from clotting) and histamine (increases capillary permeability during allergic reactions)

always present in blood and circulate continuously until a foreign substance penetrates the protective membranes. Penetration of the outer defenses by foreign substances elicits a response by the walls of nearby blood vessels, with the result that antibodies, leukocytes, or both, leave the blood at the point of attack and accumulate at the invasion site. From their sites of formation in bone marrow or lymphoid tissue the leukocytes (Table 25-2) are transported rapidly to the parts of the body where they are needed, providing an effective defense against infectious agents.

THE SYSTEM OF MACROPHAGES

Most of the organs and tissues of the body contain phagocytic cells, which arise from undifferentiated stem cells of the bone marrow and are collectively referred to as the *reticuloendothelial system of macrophages*. In loose connective tissue these cells are known as *histiocytes* and may be free or attached to reticular fibers. In some organs, such as the liver, phagocytic cells contribute to the lining of vascular channels, called *sinusoids,* by intermingling with nonphagocytic endothelial cells. In lymph nodes and tonsils the macrophages may be free or fixed and are scattered throughout the reticular fiber framework, where they filter the lymph. In the brain the macrophages are special *neuroglia cells,* called *microglia,* which aid in preventing potentially harmful substances from gaining access to the neurons. In the blood the leukocytes, including the *neu-*

trophils, eosinophils, and *monocytes* (see Chapter 4), are the circulating phagocytic cells capable of migrating from the blood stream to meet the challenge at a specific site.

Leukocytes can squeeze through pores in the capillary walls even when the pores are much smaller than the cell. The part of the cell that is passing through the pore is temporarily constricted to the size of the pore.

Macrophages in different locations in the body differ in their morphological appearance, but all *tissue macrophages* are apparently derived from *monocytes*. All macrophages contain *hydrolytic enzymes* in membrane-enclosed *lysosomes*. Ingested foreign particles such as bacteria or cell debris usually fuse with the lysosome and are broken down into their constituent amino acids by action of the hydrolytic enzymes. These small molecules are then used in the metabolism of the cell or thrown into the general circulation for use elsewhere or for elimination from the body. Macrophages ingest and continue to digest foreign material such as bacteria until the accumulation of toxic substances kills the phagocytes themselves. The accumulated toxic substances are then released into the intercellular fluid to be removed by entering the circulation or by being ingested by other phagocytes.

THE IMMUNE SYSTEM

The immune system includes the lymphoid organs and their products (lymphocytes and antibodies) as well as the reticular system of macrophages. When foreign agents penetrate the skin and mucous membranes, a complex sequence of events called the immune response is set into motion. To illustrate this response, which can involve all of the cells of the lymphoid system, let us consider the events following a cut on the finger that allows bacteria to penetrate the skin. The reaction to such a local infection occurs in two stages. The first, a local reaction at the invasion site, involves macrophages; it is not a specific reaction, that is, the defending cells do not distinguish one antigen from another. The second stage of the response is a systemic reaction involving the proliferation of lymphocytes and production of antibodies; these defend against the specific invading antigen and provide future immunity against it. Essentially the same responses are involved in the defense against viruses and other potentially harmful antigens such as plant toxins. Similar mechanisms are also involved in the rejection of tissue transplants and in the defense against cancerous cells of one's own body.

Local Response (Macrophages)

The first thing that happens as bacteria enter the connective tissue at the site of an injury is the migration of granulocytes, especially neutrophils, from the capillaries in the vicinity; upon reaching the infection site these macrophages begin to ingest the bacteria and to cause the local heat and reddening called inflammation (Fig. 25-10). Shortly after inflammation begins, the area of injury is walled off from the surrounding tissue by clots of a plasma protein (fibrinogen) which inhibit the passage of fluid from the area, delaying the spread of the bacteria or their toxic products. The bacteria may multiply at the site, but even so, they are often eliminated by phagocytosis without spreading to other locations. The site of the infection soon contains only neutrophils, which eventually begin to die and disintegrate, sometimes accumulating in such large numbers that they form the white substance called pus. As the neutrophils die, tissue macrophages (histiocytes) move in to clear away the dead cells and cell fragments by phagocytosis and subsequent digestion. Tissue macrophages are larger than neutrophils and can ingest more foreign matter, including dead neutrophils; thus they are responsible for the final cleanup as inflammation subsides and the wound heals.

Systemic Response (Lymphocytes)

When bacteria enter the body in sufficient numbers to overcome the system of macrophages at the site of local infection,

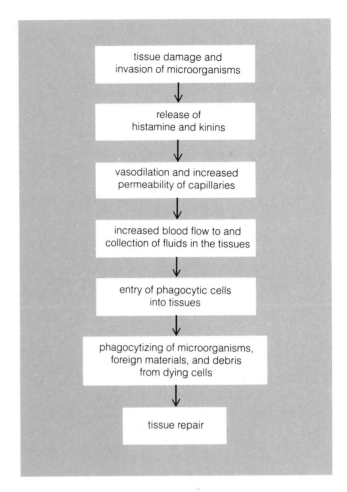

Figure 25-10 A summary of the inflammatory response in nonspecific defense mechanisms.

the next line of defense consists of the lymphocytes and antibodies that mediate the immune response to specific antigens. Antigens are large molecules, usually proteins, that elicit specific immune response; small molecules, such as amino acids, do not act as antigens. In the case of a bacterial invasion the antigenic substance may be a component of the bacterial wall, or it may be a toxin exuded by the bacteria. In a cut causing even a minor infection the bacteria or their products may spread beyond the infection site and initiate an immune response in the lymph nodes draining the area. If the body has never been exposed to the particular kind of bacteria before, the response is a *primary immune reaction,* in which specific lymphocytes of the lymph nodes are stimulated to divide repeatedly.

Every kind of antigen has a distinguishing portion of its molecule, called the *antigenic determinant,* which can bind to

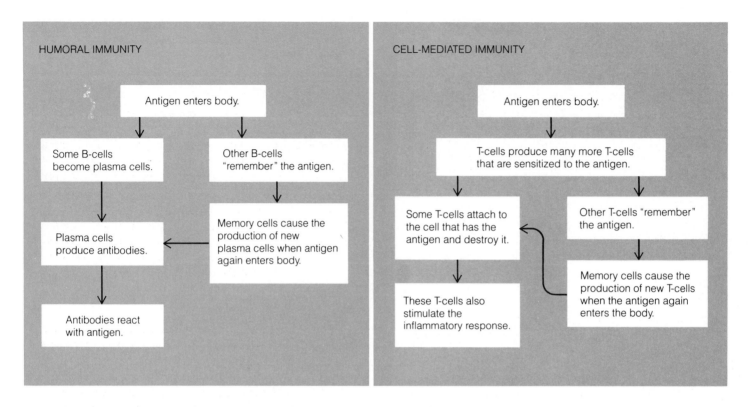

Figure 25-11 A summary of humoral and cell-mediated immunity, the two kinds of specific immunity.

molecules called *receptor sites* on a lymphocyte membrane. A given lymphocyte has receptors for only one antigenic determinant and thus can react with only one kind of antigen. According to the widely accepted clonal selection theory each lymphocyte is genetically programmed to have receptors that bind only to an antigenic determinant with a particular structure. The body initially contains only a few lymphocytes programmed for each of the many possible proteins and other antigenic molecules, but the binding of a lymphocyte to its specific antigen stimulates the lymphocyte to proliferate, giving rise to a large clone of cells capable of defending against that antigen. (A clone is a group of genetically identical cells.)

Two kinds of lymphocytes are present in lymph nodes and other lymphoid tissue: B-lymphocytes, responsible for the humoral (antibody) reaction, and T-lymphocytes, responsible for the cell-mediated reaction (Fig. 25-11). Whether an immune response is humoral or cellular is determined by whether a specific antigen reacts with B-lymphocyte or T-lymphocyte.

Humoral Immunity   The B-lymphocytes, or small lymphocytes, do not circulate but remain within the lymph node or other lymphoid tissue. When a B-lymphocyte is stimulated by its specific antigen, it proliferates, and some of the progeny cells undergo morphological transformation into larger *plasma cells* that secrete antibodies into the lymph. Antibodies, or **immunoglobulins,** are large molecules, most of them belonging to a class of plasma proteins called *gamma globulins.* The antibodies are conveyed by the lymph to the blood stream and are carried to all parts of the body, including the original site of infection, where they react with and neutralize any remaining antigens. If the antigen is a protein in the bacterial wall, antibodies bind to the antigen molecules and may initiate a reaction that causes the bacterium to lyse (rupture). In other cases the coating of the bacteria by the antibodies facilitates the ingestion of the bacteria by macrophages. If the antigen is a toxin, the antibodies bind to the toxin molecules, neutralizing them and causing them to clump into large aggregates that are easily engulfed by macrophages.

The proliferation of the lymphocytes, their differentiation into plasma cells, and the defensive action of the antibodies are all part of the primary immune response. In addition some of the progeny of the stimulated lymphocytes do not become plasma cells; they divide to become long-lived small lymphocytes called *memory cells,* available for a *secondary immune response.* Following the primary response the memory cells work their way from the regional lymph nodes into the general circulation for distribution throughout the body, ending up in all the lymphoid organs and in the loose connective tissue of all regions and organs of the body. Thus there are far more lymphocytes capable of responding to the specific antigen than there were before the initial exposure to the antigen; in the event of a second invasion by the same kind of bacteria, antibodies are produced much more quickly and in much larger quantities, and the infection is more likely to be suppressed before causing disease. This secondary response by the memory cells, also called the *anamnestic response* (Gr., not forgetting, that is, remembering), is the basis of immunity to specific diseases.

Cell-Mediated Immunity When antigens that are capable of eliciting a cellular immune response reach the regional lymph nodes, they stimulate T-lymphocytes so that they become "sensitized" to the specific antigen involved. The process of sensitization is poorly understood and is curently the subject of intensive investigation. It is known that, once sensitized, T-lymphocytes can destroy cells, such as cancer cells, by a cytolytic effect upon direct contact between the lymphocytes and the cancer cells. This interaction initiates the release of a variety of factors, called **lymphokines,** that have a toxic effect upon the cells or bacteria to which the T-lymphocytes were sensitized. T-lymphocytes are the major factor in the rejection of transplants such as kidney transplants; their activity must be suppressed if the donor organ is to survive in a genetically foreign host.

Unlike B-lymphocytes which change into plasma cells when specifically stimulated by antigens, T-lymphocytes do not undergo structural alteration; but they do multiply rapidly, adding to the numbers of sensitized lymphocytes that are available to combat the invading foreign substances or cells. Some of these progeny are long-lived memory cells which are distributed to the lymphoid and connective tissues of the body; like the B-lymphocyte memory cells they are available for the more rapid secondary (anamnestic) response when exposed again to the same antigen. A major difference between humoral immunity and cell-mediated immunity is the longer persistence of the cell-mediated type. Humoral antibodies rarely remain effective for more than a few months, whereas sensitized T-lymphocytes may resist infection for as long as ten years.

Human T-lymphocytes are a heterogeneous population of cells, which are separated into subpopulations by cell surface antigenic markers and cell surface receptors. They also have a variety of different roles in the immune response. Some T-lymphocytes are *helper* cells in antibody production; others regulate the amount of antibody produced by acting as *suppressor* cells that inhibit the activity of B-lymphocytes.

Artificial Immunization

Because of the slowness of the primary immune response, the natural immunization caused by invading organisms may fail to prevent disease on first exposure, though it often prevents it on subsequent exposures to the same organism. However, people can be immunized artificially by vaccines that contain specific antigens but not viable organisms, which stimulate the proliferation of lymphocytes and memory cells without producing the disease. The vaccines may contain fragments of bacterial cell walls, bacterial toxins, or attenuated bacteria or viruses. An attenuated organism is one that has been weakened so that it is no longer virulent (disease-causing). Vaccines are usually injected, but a few kinds are administered orally.

> In the case of exposure to a disease such as hepatitis, for which no specific immunization technique has been developed, a person may be given an injection of gamma globulins pooled from many donors, on the assumption that some of the donors have antibodies to the disease in question. The transfer of antibodies from one person to another is called passive immunization, in contrast to the active immunization by one's own immune system resulting from natural or artificial exposure to an antigen. Passive immunization has only a brief protective effect.

The Origin of Lymphocytes

The lymphocytes arise from bone marrow stem cells that migrate to organs of the lymphoid system (Fig. 25-6). Those that migrate to the thymus are acted upon, or processed (presumably by thymic hormones), to become T-lymphocytes, which then migrate from the thymus to peripheral lymphoid organs. It is not certain where the B-lymphocytes acquire their characteristics before migrating to their final locations, but there is evidence that the processing occurs in lymphoid tissue of the digestive tract.

ANTIBODY PRODUCTION AND FUNCTION

The body's ability to recognize foreign antigens and to respond by producing antibodies is an essential factor in combating disease. Immunological memory (the anamnestic response)

Clinical Applications: **Immunological Disorders**

The immunological mechanisms that protect us from foreign agents, such as bacteria, viruses, and parasites, can lead to abnormal situations, producing symptoms of clinical disease. For example, the formation of antigen–antibody complexes, useful in enhancing phagocytosis, can be excessive in local areas, blocking capillaries and depriving local tissues of nutrients. This response, known as *Arthus reaction,* can produce necrosis (cell death) and discomfort. Antigens may react with antibodies in the respiratory tract, producing bronchiolar constriction (asthma) and difficulty in breathing, or the same antigen–antibody reaction may stimulate excessive production of mucous by the mucous membrane of the upper respiratory tract (hay fever). Thus immunological reactions that normally serve to protect us from disease can, under certain conditions, lead to symptoms interpreted as disease.

Allergy

The term allergy was first used by von Pirquet in 1906 to indicate the altered reaction to repeated injections of gamma globulin to diphtheria toxin. At present the term allergy usually implies disease characterized by abnormal sensitivity (hypersensitivity), reactions such as hay fever and asthma, which are mediated by IgE antibodies. Hypersensitivity has been used to describe non-immune reactions to drugs as well as to describe immune reactions. The tendency to be allergic may be inherited, but the genetic mechanism is not known.

The antigens that cause allergic reactions are called allergens, substances that stimulate the production of IgE immunoglobulins in the tissues of susceptible individuals. The IgE anti-bodies cause the release of histamine from mast cells. The histamine alters the permeability of capillary endothelium which leads to an increase in tissue fluid, especially in the nasal epithelium. The resultant runny nose is a major symptom of an allergic reaction. Other symptoms include local edema (swelling), constriction of bronchioles of the lungs in asthma, and skin rash.

Contact Dermatitis

Contact dermatitis occurs in individuals after exposure to soluble substances that penetrate the skin; these substances combine with skin proteins, forming antigenic complexes that cause the sensitization of T-lymphocytes. The reaction (rash) takes up to 48 hours to develop and is an example of delayed hypersensitivity. A common example of this type of reaction is the rash resulting from contact with poison ivy. The spread of the rash in cases of poison ivy is due not to spread of the inducing substances but to the distribution via the lymph vessels and blood stream of the sensitized T-lymphocytes which have been exposed to the antigens.

Autoimmune Disease

The term *autoimmunity* refers to destruction of one's own tissues by the body's immune system, in most cases by reactions of the *delayed hypersensitivity* type, that is, caused by cellular immune responses. Autoimmune diseases are characterized by the accumulation of lymphocytes and macrophages within an organ, creating a chronic inflammation that results in tissue

enhances the immune reaction upon re-exposure to the same antigens, resulting in the development of specific resistance or immunity to the particular disease-producing bacteria or virus. The development of immunity may be considered a "learning" experience, during which lymphocytes programmed to respond to a specific antigen become more numerous and more efficient in antibody production.

At present there is no full explanation of the mechanism by which an antigen stimulates B-lymphocytes to differentiate into plasma cells and begin antibody production. The best known of the current theories is the clonal selection theory of Sir Macfarlane Burnet, which proposes that the antigen stimulates only those few cells that have the genetic ability to make antibody of the proper configuration. The antigen triggers the antibody-producing process within the cells and also stimulates their multiplication, increasing the numbers of the clone. This theory also postulates that during fetal life an individual's own proteins come into contact with the immature lymphoid cells programmed to make antibodies against those particular proteins, blocking their antibody-producing system permanently; thus one's own proteins are not destroyed when the lymphoid cells mature.

Antibodies, or immunoglobins (Ig), are released into the blood stream, where they become a significant component of the plasma proteins. The great majority of immunoglobulins

destruction with reduction in the functional capacity of the organ involved. Organs and tissues that may be subject to these diseases are the thyroid gland, suprarenal cortex, stomach and colon mucosa, and the connective tissue of joints.

Immunological Deficiency Disorders

Deficiencies of the immune system are seen most frequently in young people, because such conditions do not permit survival for a normal lifespan. The deficiencies may be in humoral or cellular immunity, or both. The humoral deficiency state, known as *agammaglobulinemia* (lack of gamma globulin, the type of plasma protein that includes most antibodies), is inherited as a sex-linked characteristic. In cases of cellular immune deficiencies the individuals are unable to reject tissue grafts and do not respond in a normal way to some antigens. They are susceptible to fungal and viral infections and usually die in childhood. Deficiencies in both humoral and cellular immunity sometimes occur, usually in association with other anomalies; survival in these cases is usually brief.

Abnormal Antibody Molecules

In some diseases abnormal immunoglobulins may be produced in large quantities. The best known of these diseases is *multiple myeloma,* a cancer of the lymphoid organs in which an abnormal protein, called *Bence Jones protein,* is produced in such large quantities that it is excreted in the urine.

Table 25-3 Immunoglobulins

Class	Location	Function
IgG	Serum	Neutralizes foreign proteins
IgM	Serum	Lyses cells in the presence of complement
IgA	Serum and secretions	Destroys bacteria and viruses on mucous membranes
IgD	Surface of lymphocytes	Serves as the surface receptor in the immune response
IgE	Epithelial membranes	Plays a role in allergic reactions and tissue sensitization

belong to a structurally defined group of proteins called *gamma globulins* (which also include proteins that have no antibody activity.) Immunoglobulins of the gamma globulin group are subdivided into five main classes (IgG, IgM, IgA, IgD, and IgE) on the basis of antigenic differences between the molecules (Table 25-3).

Classes of Immunoglobulins

The IgG is present in serum in a concentration of 12 mg/ml, and each IgG molecule has a molecular weight of 150,000. IgG has a major role in neutralizing foreign proteins such as bacterial toxins. Immunoglobulins of the second class are known as IgM and have a molecular weight of about 900,000. They are present in serum in a concentration of 1 mg/ml, and each antibody molecule is more effective than IgG in causing the lysis of cells in the presence of **complement** (a mixture of nine proteins in blood serum, which reacts with antigen–antibody complexes to destroy cells).

Immunoglobulins of the third main class, IgA, have a molecular weight of about 150,000 and are present in serum at about 4 mg/ml. IgA is present in body secretions at a considerably higher level than it is in blood, and it may destroy bacteria and viruses on the surfaces of mucous membranes. It is part of the defense against infections in the respiratory tract, for example.

The fourth type of immunoglobulin, IgD, is found on the surface of lymphocytes. It may have a role as a specific surface receptor in the initiation of the immune response.

The final type of immunoglobulin, IgE, is less well known, but it appears that IgE molecules attach to epithelial cell membranes, where they take part in allergic reactions such as hay fever and asthma. The reaction between the antigen and IgE on the cell surface releases *histamine,* which increases capillary permeability as the initial step in the local allergic response. As fluid leaves the circulation and enters the tissues beneath the epithelium, it causes a local swelling called a *rash.* Susceptibility to allergies appears to have a genetic basis and can be detected by the presence of IgE antibodies. Reaction of IgE antibodies (called reagin antibodies) with antigens (called allergens) causes damage to cells in various parts of the body such as the nasal mucosa.

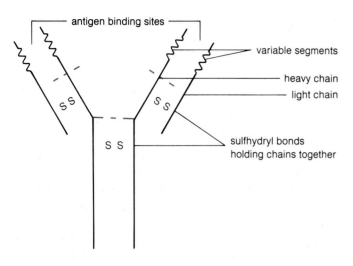

Figure 25-12 The general structure of an immunoglobulin.

Antibody Structure and Reactions

The antibody molecule (Fig. 25-12) consists of two pairs of polypeptide chains, called light and heavy chains, held together by molecular bridges called disulfide bonds. Individual light chains have a molecular weight of 25,000, and the heavy chains have a molecular weight of 50,000. Each kind of chain has a variable region, and it is the variations in molecular structure from one antibody to another that give antibodies their specificity for particular antigens. Each antibody has two identical combining sites, each composed of the variable region of one light and one heavy chain. Thus an antibody can combine with two identical antigen molecules at the same time. Because invaders such as bacteria have many antigenic molecules on their surfaces, antigens and antibodies can form large, easily phagocytized aggregations.

WORDS IN REVIEW

Antibodies	Lymphokines
Antigen	Macrophages
B-lymphocytes	Reticuloendothelial system
Complement	Spleen
Immunoglobulins	Thymus
Lymph nodes	T-lymphocytes
Lymph nodule	Tonsils
Lymphocytes	

SUMMARY OUTLINE

I. The Epidermis
The epidermis of the skin is keratinized stratified epithelium that acts as a barrier to harmful bacteria and other agents.

II. Mucous Membranes
The mucous membranes elaborate secretions on the surface that aid in the removal of harmful agents.

III. Lymphoid Organs
Lymphoid organs are encapsulated aggregates of lymphocytes and macrophages held in a reticular fiber framework.
   A. Lymph nodes are located in a pathway of lymph drainage.

1. Lymphocytes of lymph nodes respond in immune reactions.
2. Macrophages of lymph nodes ingest particulate matter in lymph.
3. The reticular fiber network traps cells and retards their passage through the lymph channels.
   B. Tonsils are lymphoid tissue located at the beginning of lymph channels and function similarly to lymph nodes.
1. The lingual tonsils are beneath the epithelium of the tongue.
2. The palatine tonsils lie on the edge of the soft palate.
3. The pharyngeal tonsils (adenoids) are located in the posterior wall of the nasopharynx.
   C. The spleen is located within the pathway of the blood flow.
1. The white pulp of the spleen has a structure and functions similar to those of lymph nodes.
2. The red pulp of the spleen filters the blood and is especially important in removing dead erythrocytes.
   D. The thymus is located in the anterior mediastinum and is essential in processing lymphocytes that will function in cell-mediated immunity. The thymus functions only in prenatal and early postnatal life.

IV. Blood
Blood functions as the distribution system of the body. It conveys leukocytes (chiefly lymphocytes and granulocytes) to sites of invasion of harmful agents.

V. Macrophages

The system of macrophages includes all phagocytic cells regardless of their location in the body. Cells of this system may be located in blood, loose connective tissue, lymphoid organs, or in special locations in organs such as the brain, liver, and suprarenal glands.

VI. The Immune Response

The immune response includes the local response of macrophages to infection, as well as humoral response and cell-mediated response of lymphocytes.

A. Granulocytes and tissue macrophages destroy foreign organisms at the invasion site.

B. In the primary humoral response specific antigens stimulate B-lymphocytes to differentiate into plasma cells, which produce antibodies directed against those antigens. The B-cells and plasma cells remain within the lymph organs and the antibodies circulate in the blood.

C. In the primary cell-mediated response specific antigens stimulate T-lymphocytes to proliferate, enter the circulation, and react with those antigens.

D. Both B-lymphocytes and T-lymphocytes give rise to large numbers of memory cells as part of the primary response. These undergo a rapid secondary response on subsequent exposure to the sensitising antigen.

REVIEW QUESTIONS

1. What are the morphological characteristics of the epidermis that enable the skin to resist penetration by bacteria?

2. What is the functional role of the epithelium of a mucous membrane in preventing bacterial invasion?

3. What type of lymphoid organ has both afferent and efferent lymph vessels associated with it?

4. What type of lymphoid organ has only efferent lymph vessels associated with it?

5. How does the structure of the tonsil differ from that of the lymph node?

6. Compare the white pulp of the spleen with a lymph nodule in structure and in function.

7. What structural features of the red pulp of the spleen facilitates its role in removal of dead red blood cells?

8. What is the principal role of the thymus in cell-mediated immunity?

9. Define and describe the reticuloendothelial system of macrophages.

10. What are the roles of B- and T-lymphocytes in the primary immune response? In the secondary immune response?

SELECTED REFERENCES

Bach, J. F. *Immunology*. 2nd ed. New York: John Wiley & Sons, 1982.

Barrett, J. T. *Basic Immunology and its Medical Application*. St. Louis: C. V. Mosby, 1980.

Golub, E. *The Cellular Basis of the Immune Response*. 2nd ed. Sunderland, Mass.: Sinauer Associates, 1981.

Nisonoff, A. *Introduction to Molecular Immunology*. Sunderland, Mass.: Sinauer Associates, 1982.

This glossary includes the most commonly used anatomical terms that many students have difficulty defining precisely. Some of the words that are frequently mispronounced are followed by a phonetic guide to pronunciation. In the usual pronunciation, the sound is long if the vowel occurs at the end of a syllable and is short if it is followed by a consonant. The macron (-) indicates a long sound and the breve (ˇ) a short sound when the general rule does not apply. Strongly accented syllables are marked by an accent (ʹ). Phonetic spellings are given for words that seem to be most frequently mispronounced by beginning students of anatomy.

The glossary does not include muscles, nerves, or blood vessels.

abdomen (ab-doʹmen) the part of the trunk that lies between the thorax and the pelvis.

abduction (ab-dukʹshun) movement away from the midline of the body.

abnormality an anomaly, deformity, or congenital malformation.

abortion (ă-borʹshun) the process of giving birth to an embryo or fetus prior to the stage of viability. May be either spontaneous or induced.

absorption the taking in of substances, such as gases, liquids, heat, or light.

accommodation of the eye the increase or decrease of the convexity of the lens in order to focus upon a near or distant external object.

accretion (ă-kreʹshun) increase in size by the addition of material at the periphery or surface of a structure.

acellular without cells.

acetabulum (as-ĕ-tabʹu-lum) the cup-shaped socket on the external surface of the hip bone, in which the head of the femur fits.

acetylcholine (as-ĕ-til-kōʹ-lēn) the neurotransmitter substance released by pre- and postganglionic parasympathetic nerve fibers and preganglionic sympathetics.

ascinus (as-i-nus) grape-shaped secretory portions of certain glands.

acromioclavicular joint the articulation between the acromion process of the scapula and the clavicle.

acromion the lateral end of the spine of the scapula. It projects superior to the glenoid fossa.

acrosome (akʹro-sōm) the thin covering over the tip of the nucleus of the spermatozoon. It contains hydrolytic enzymes.

actin (akʹtin) one of the protein components of myofibrils.

adduction movement toward the midline of the body.

adenoids hypertrophy of the pharyngeal tonsil.

adhesion (ad-heʹzhun) the union of two surfaces, such as the opposing surfaces of a wound.

adipose denoting fat.

aditus an entrance to a cavity, such as the superior opening into the larynx.

adrenal (ă-dreʹnal) near the kidney; denoting the suprarenal glands.

adrenaline denoting the hormone epinephrine that is produced by the suprarenal medulla.

adrenergic (ă-drĕ-nerʹjik) relating to neurons of the autonomic nervous system that release norepinephrine as their neurotransmitter.

adrenocorticotropic hormone a hormone of the anterior pituitary gland that stimulates the suprarenal cortex.

afferent denoting movement toward a center, such as flow of a nerve impulse from the periphery toward the central nervous system.

afterbirth the placenta and membranes that are discharged from the uterus after birth.

agenesis the failure of development of a structure or part of the body.

agranulocyte a white blood cell that has no granules in the cytoplasm.

ala any winglike structure, such as the flared wall of the nose.

aldosterone a steroid hormone produced by the suprarenal cortex. It regulates reabsorption of sodium and potassium ions in the kidney.

alimentary (al-ĭ-menʹtă-rĭ) relating to food or nutrition.

allantois (al-anʹto-is) a fetal membrane developing from the hindgut; it contributes to the formation of the urinary bladder.

allergy a condition in which an individual is abnormally sensitive to an antigen.

alveolar duct the smallest of the air passages in the lungs into which the alveoli open.

alveolus one of the terminal sac-like air spaces that open into the alveolar ducts in the lungs.

amniocentesis (amʹnĭ-o-sen-teʹsis) aspiration of fluid from the amniotic sac through the abdominal wall of the mother.

amnion (amʹni-on) the innermost fetal membrane, which is filled with the amniotic fluid that protects the fetus.

amorphous (ă-morʹfus) without definite form or structure.

ampulla a dilated portion of a duct or canal, such as the ampulla of the uterine tube.

anastomose to open one structure, such as a blood vessel, into another, either directly or by connecting channels.

anatomical position the position in which the body is upright, facing forward with the feet together, the hands down at the sides with the palms oriented anteriorly.

androgen a general term for a hormone that stimulates the accessory sex organs of the male.

GLOSSARY

anemia any condition in which the red blood cells or the amount of hemoglobin in the blood is below normal.

anesthesia (an′es-the′zĭ-ah) a condition characterized by the loss of sensation.

aneurysm (an′u-rizm) a condition in which the wall of an artery balloons outward.

anular ligament a circular ligament encircling the radius at its articulation with the ulna.

anulus a circular or ring-shaped structure.

anomaly a structure that deviates from the usual or normal.

antagonist something that opposes or resists the action of another; usually refers to a muscle action.

antebrachium (an-te-bra′kĭ-um) the forearm.

anterior in human anatomy, this denotes the front or ventral surface of the body.

anti- a prefix denoting against or opposing.

antibody an immune or protective protein that is produced in response to exposure to an antigen.

antidiuretic hormone an agent released in the posterior lobe of the pituitary gland that reduces the output of urine.

antigen (an′tĭ-jen) any substance that, as a result of coming in contact with appropriate cells, induces a state of sensitivity.

antigenic determinant the portion of a molecule that confers antigenic specificity.

antihistamine any drug that has an action antagonistic to that of histamine, a natural substance that enhances allergic reactions.

antrum any nearly closed cavity; usually used when the cavity has bony walls.

apatite (ap′ă-tīt) general term for substances that are similar to the salts of bones and teeth.

aperture (ap′er-tŭr) an opening or orifice.

apex the tip of a tapered structure such as the heart or lung.

aponeurosis (ap-o′nu-ro′sis) a fibrous sheet; a flattened tendon.

appendicular skeleton a collective term for the bones of the pectoral and pelvic girdles and the attached limbs.

appositional growth increase in size by means of addition of new material at the surface of a structure.

aqueduct (ak′we-dukt) a conduit or canal.

aqueous humor the watery material in the anterior chamber of the eyeball.

arachnoid resembling a cobweb, the filmy material forming the middle layer of the brain and spinal cord.

arcus any structure resembling a bent bow or an arch.

areola one of the spaces in areolar (loose) connective tissue.

arrector pili muscle a bundle of smooth muscle between the hair follicle and the skin; contraction of the muscle erects the hair.

arteriole (ar-tēr′ĭ-ōl) a minute artery with a muscular wall.

arterio-venous shunt a connection between an artery and vein so that blood may pass from the artery to the vein without passing through capillaries.

articulation a connection between two structures that allows movement between the parts.

atresia absence of a normal opening or lumen in a tubular structure.

atrioventricular node specialized conductile tissue in the atrioventricular septum of the heart. Fibers from the node extend through the ventricular septum into the cardiac muscle of the ventricles, regulating their contraction.

atrium a chamber or cavity that communicates with several cavities.

atrophy a decrease in size or wasting of tissues or organs.

axial skeleton the central portion of the skeleton, consisting of the skull, vertebrae, ribs, sternum, and hyoid bone.

axon the nerve cell process that conducts nerve impulses away from the cell body.

basal ganglia the large masses of gray matter at the base of the cerebral hemisphere, consisting of caudate, putamen, globus palidus, and claustrum as the nuclei.

basal lamina the filamentous ultramicroscopic layer at the base of epithelium and some other tissues. It is part of the basement membrane.

basement membrane a thin layer found between epithelium and the underlying connective tissue.

basophil a white blood cell containing granules that stain specifically with basic dyes. The granules contain heparin and histamine.

biceps a structure, such as the biceps brachii muscle, that has two heads.

bicipital groove the sulcus (intertubercular sulcus) between the greater and lesser tubercles of the humerus. The sulcus contains the tendon of the long head of the biceps brachii muscle.

bifurcation a division into two branches, such as occurs when an artery branches.

bile the yellowish brown or green fluid produced in the liver and discharged into the duodenum, where it emulsifies fats.

bipennate having a structure that resembles a double feather.

blastocoel (blas′to-sel) the cavity in the blastula of a developing embryo.

blastocyst the inner cell mass stage of mammalian embryos.

blood the fluid and its formed elements (cells) that circulates through the heart and blood vessels and is the means by which substances are distributed throughout the body or are removed from the body.

blood island an aggregation of mesodermal cells in the embryo that differentiate into endothelium and primitive blood cells.

blood platelet an irregularly shaped disk found in blood that is essential in the formation of blood clots.

B-lymphocyte the white blood cell that when properly stimulated by an antigen will differentiate into a plasma cell and and produce antibody specifically directed against the antigen.

bolus a masticated morsel of food ready to be swallowed.

bone a hard tissue consisting of cells in a matrix of ground substance and collagen fibers that is impregnated with minerals, chiefly calcium, phosphate, and carbonate.

bony labyrinth a series of cavities in the petrous portion of the temporal bone which contain the membranous labyrinth of the cochlea duct and semicircular canals.

Bowman's capsule the expanded beginning of a renal tubule that surrounds a tuft of capillaries, the glomerulus.

brachium the arm, the segment of the upper extremity between the shoulder and the elbow.

brain the part of the central nervous system contained within the cranium.

braincase the part of the skull that encloses the brain.

brainstem the portion of the central nervous system immediately rostral to the spinal cord, including the medulla, pons, and midbrain.

brain ventricle the cavity within the brain that contains the cerebrospinal fluid.

breathing the inhalation and exhalation of air.

bregma the point on the skull corresponding to the junction of the coronal and sagittal sutures.

broad ligament the mesentery attached to the lateral edges of the uterus.

bronchiole (brong′kĭ-ōl) the air passages that are 1 mm or less in diameter and without cartilage in their walls.

bronchitis (brong-ki′tis) inflammation of the mucous membrane of the bronchial tubes.

bronchopulmonary segment that portion of the lungs supplied by the tertiary branches of the bronchi.

buccal pertaining to the cheek.

bulbar relating to the pons and medulla (hindbrain); resembling a bulb.

bulbourethral glands glands located within the urogenital diaphragm that empty their secretions into the cavernous urethra.

bundle of His a band of specialized conductile fibers that arise in the atrioventricular node of the heart and extend toward the interventricular septum, consisting of fibers that stimulate the heart muscle to contract.

bursa a fluid-filled sac lined with synovial membrane.

bursitis inflammation of a bursa.

calcification a process by which a tissue becomes hardened as the result of deposits of insoluble salts of calcium.

calcitonin a hypocalcemic hormone secreted by the thyroid gland.

calyx (ka′liks) one of the subdivisions of the renal pelvis into which the orifices of the renal pyramids project.

canal (kă-nal′) a duct or channel; a tubular structure.

canaliculus (kan-ă-lik′u-lus) a small canal.

cancer (kan′ser) a general term to indicate any of various types of malignant tumors.

capillary a small blood vessel with a thin wall that consists only of a single layer of flattened endothelium.

capitulum (kă-pit′u-lum) a small head or rounded articular surface of a bone.

capsule (kap′sūl) a fibrous tissue layer enveloping an organ or joint.

capsular ligament a thickened portion of a joint capsule that connects two bones that articulate at a joint.

cardiac skeleton the dense supporting connective tissue of the heart; it surrounds the valves of the heart and great vessels.

cardiovascular system the heart and blood vessels, considered as a whole.

carotene a general term for yellow-red pigments found in plants and animals that include precursors of vitamin A.

carotid body a small epitheloid structure located near the origin of the internal carotid artery; it serves as a chemoreceptor that helps regulate the rate of heartbeat.

carotid sinus a slight dilation near the bifurcation of the common carotid artery that helps regulate blood pressure by causing the heart rate to slow when pressure rises in the carotid artery.

carpus (kar′pus) the wrist.

cartilage a type of connective tissue characterized by its nonvascularity and firm consistency.

caudal (kaw′dal) pertaining to the tail.

cavernous relating to a cavern or a cavity.

cavity a hollow space.

cecum (se′kum) the first part of the large intestine; a blind-ending sac lying below the opening of the ileum into the large intestine.

celiac plexus the largest autonomic plexus in the abdominal region; it lies in front of the aorta at the level of the origin of the celiac artery.

cell a minute structure composed of protoplasm; it contains a nucleus and is enclosed by a delicate plasma membrane.

cell cycle the cyclic biochemical and structural events occurring during proliferation of cells.

celom (se′lom) the general body cavity in the adult.

cementum (se-men′tum) a layer of modified bone on the external surface of the root of a tooth.

central nervous system the part of the nervous system that is housed within the skull and vertebral canal.

centriole (sen′tri-ōl) usually paired organelles composed of microtubules; they are involved in production of the mitotic spindle during cell division.

centromere (sen′tro-mēr) the nonstaining point of attachment of the spindle fiber to the chromosome during cell division.

centrum a center of any kind such as the body of the vertebra.

cephalic (se-fal′ik) cranial; toward the head.

cerebellum the large posterior brain mass lying dorsal to the pons and medulla; it consists of two lateral hemispheres connected by a middle portion, the vermis.

cerebral aqueduct the narrow canal in the midbrain, containing cerebrospinal fluid; it connects the third and fourth brain ventricles.

cerebral cortex the layer of gray matter at the surface of the cerebral hemisphere.

cerebrospinal fluid a fluid secreted by the choroid plexuses of the brain ventricles, filling the ventricles and the subarachnoid space around the brain and spinal cord.

cerebrum the largest part of the central nervous system; the part derived from the embryonic telencephalon.

cerumen (se-roo′men) the waxy secretion of the ceruminous glands of the external auditory meatus.

cervical (ser′vĭ-kal) referring to a neck in any structure.

cervix any necklike structure, such as the uterine cervix.

chamber a compartment or enclosed space, such as the anterior chamber of the eye.

cheek the side of the face forming the lateral wall of the oral cavity.

chiasma (ki-az′mah) a decussation or crossing of two tracts or nerves, such as the optic nerves.

chin mentum, the prominence formed by the anterior projection of the lower jaw.

choana (ko′an-ah) posterior nares; the opening into the nasopharynx of the nasal cavity.

chondrocranium the cartilaginous parts of the developing skull.

chondrocyte (kon′dro-sit) a cell that occupies a lacuna within the cartilage matrix.

chondroitin (kon-dro-ī-tin) a mucopolysaccharide, generally present as chondroitin sulfate.

chondromucoid a mucoprotein found in cartilage; it contains chondroitin sulfate.

chorda (kor′dah) a tendinous or cordlike structure.

chorda tendineae tendinous strands between the papillary muscles of the heart and the atrioventricular valves.

chorion (ko′rĭ-on) the outermost fetal membrane.

chorionic villi fingerlike projections of the embryo that enter into the formation of the placenta.

choroid (ko′royd) resembling the chorion or any membrane.

choroid coat the vascular layer of the eyeball between the retina and the sclera.

choroid plexus tufts of capillaries in the brain ventricles that elaborate the cerebrospinal fluid.

chromatid each of two strands of a chromosome that becomes visible during the prophase stage of mitosis.

chromatin the genetic material of the nucleus seen as granules; during mitosis this material condenses into chromosomes.

ciliary (sil-ĭ-ă-rī) relating to hairlike processes such as the eyelashes.

cilium a motile extension of a cell surface, composed of microtubules.

circle of Willis an arterial circle at the base of the brain from which branches arise that supply the brain.

circulation movement through a course that leads back to the starting point, such as occurs in the circulation of blood.

circumduction movement of a part, such as a limb, in a circular direction.

cirrhosis (sĭr-ro′sis) disease of the liver characterized by increase in the connective tissue that interferes with blood flow through the liver.

cisterna (sis-ter′nah) any enclosed space or cavity serving as a reservoir for fluid.

cisterna chyli a dilated sac at the lower end of the thoracic duct into which the intestinal lamphatics and lymph from the lower extremities drain.

cleavage a series of cell divisions occurring in the ovum immediately following fertilization.

cleft a fissure.

clitoris a small erectile body in the female that is homologous with the male penis.

clivus (kli′vus) the sloping surface in the floor of the skull from the dorsum sellae to the foramen magnum.

cloaca the common chamber in birds and monotremes into which the gut, urinary bladder, and genital ducts open.

clone a group of cells derived from a single cell, all having identical genetic characteristics.

coagulation (ko-ag′u-la′shun) the process of changing a liquid, such as blood, to a solid.

coccyx (kok′siks) three or four small bones at the inferior end of the vertebral column.

cochlear duct a spiral membranous tube within the petrous portion of the temporal bone containing the organ for hearing, the organ of Corti.

codon a sequence of three nucleotides in a strand of DNA or RNA that is the genetic code for an amino acid.

collagen (kol′lă-jen) the major protein in the white fibers of connective tissue.

collateral a side branch of a nerve or blood vessel.

colliculus (kol-lik′u-lus) a small elevation.

collateral ganglia sympathetic ganglia that are in addition to those in the sympathetic trunks.

colon (ko′lon) the division of the large intestine extending from the cecum to the rectum.

column an anatomical part in the form of an elongated rod or pillar.

commissure a bundle of nerve fibers crossing from one side to the other in the brain or spinal cord.

compact bone a bone that consists of solid lamellae (layers) of bone matrix with no marrow spaces.

complement a serum protein complex that is essential in certain immunological reactions and is destructive to some bacteria.

component a general term for any element forming a part of the whole.

concave (kon′kāv) a depression with more or less evenly curved sides.

conception the fertilization of the ovum by a spermatozoon.

concha (kon′kah) a structure comparable to a shell, such as a turbinate bone on the nose.

concretion an aggregation or formation of solid material.

conductivity the property of living protoplasm of transmitting a state of excitation, such as a nerve impulse.

condylar joint a joint, such as the knee, formed by articulation of rounded surfaces called condyles.

condyle rounded articular surface of a bone.

cones of the retina one of the two types of photoreceptor cells of the retina, essential for sharp vision.

confluens a meeting place; a joining of dural sinuses at the internal occipital protuberance.

congenital a condition, such as an abnormality, that exists at birth.

conjunctiva the mucous membrane lining the eyelids and covering the eyeball lateral to the cornea.

connective tissue one of the four fundamental tissues of the body; forms the supporting framework for the entire body.

contra a prefix indicating opposed or against.

contraceptive any agent for the prevention of pregnancy.

contractility the ability or property of cells, such as muscle cells, to shorten.

conus elasticus the thicker lower portion of the elastic membrane whose upper portion is the vocal fold of the larynx.

conus medullaris the tapered lower end of the spinal cord.

convex a surface that is curved outwards.

convolution a coiling or elevation of a part of an organ, such as a gyrus of the cerebrum.

coracoclavicular ligament a band of collagenous fibers connecting the clavicle and the coracoid process of the scapula.

coracoid shaped like a crow's beak; a process of the scapula.

corium the dermal layer of the integument.

cornea the transparent part of the anterior portion of the outer wall of the eyeball.

cornified keratinized.

cornu any structure resembling a horn in shape, such as the greater cornu of the hyoid bone.

coronal plane a frontal section, at right angle to the sagittal plane.

coronary circuit the blood supply of the heart, consisting of the right and left coronary arteries and the veins that return the blood to the right atrium via the coronary sinus.

coronary sinus the venous channel by which blood from the heart wall is returned to the right atrium.

coronary sulcus a groove on the outer surface of the heart marking the division between the atria and the ventricles.

coronoid shaped like a crow's beak, such as the coronoid process of the mandible.

corpus any body or mass, denoting the main part of any anatomical structure.

corpus albicans an atretic ovarian corpus luteum.

corpus callosum commissure of the cerebral hemispheres.

corpus cavernosum penis one of two parallel columns of erectile tissue forming the dorsal part of the body of the penis.

corpus luteum yellow body; the glandular structure formed on the ruptured ovarian follicle immediately after ovulation.

corpus spongiosum the median column of erectile tissue in the penis, containing the urethra.

corpuscle (kor′pus-el) a blood cell.

cortex the outer portion of an organ, as distinguished from the inner, or medullary, portion.

costa a rib.

coxa hip or hip joint.

cranial (kra′nĭ-al) relating to the head.

cranium (kra′nĭ-um) the bones of the skull or head.

crest a ridge, usually a bony ridge.

cricoid ring-shaped; the inferior cartilage ring of the larynx.

cricothyroid membrane the fascial sheet between the cricoid and thyroid cartilages of the larynx.

crista (kris′tah) a ridge projecting from a rounded surface.

cubital (ku′bi-tal) relating to the elbow.

cuboid (ku′boyd) cube-shaped.

cupula (ku′pu-lah) a dome-shaped structure.

cutaneous (ku-ta′ne-us) pertaining to the skin.

cyclic Amp (cAMP) adenosine 3′:5′—cyclic phosphate.

cytology (si-tol′o-ji) study of the cell.

cytoplasm the substance of a cell exclusive of the nucleus.

decidua (de-sid′u-ah) the maternal part of the placenta.

decussate to cross from one side of a structure to the other side.

deep away from the surface.

deformity any deviation from the normal.

degeneration (de-jen-er-a′shun) a retrogressive change in cells or tissues, usually with loss of function.

dendrite a nerve process that transmits a nerve impulse toward the cell body.

dentition a collective term for the natural teeth.

dermatitis inflammation of the skin.

dermatome the area of skin that is supplied by cutaneous branches of a single spinal nerve.

dermis the connective tissue layer of the skin.

desmosome (dez′mo-sōm) a site of adhesion between two cells.

diaphragm a thin partition separating adjacent regions.

diaphysis (di-af′i-sis) the shaft of a long bone.

diastole (de-as′to-le) dilation or relaxation of the heart, during which it fills with blood.

diencephalon (di-en-sef′a-lon) that part of the embryonic brain that becomes the thalamus and hypothalamus.

differentiation (dif′er-en′shi-a′shun) the production of new structural proteins within a cell, producing a new cell type.

digastric (di·gas′trik) having two bellies.

digestion the process of converting food into material suitable for absorption into the blood stream.

diploe (dip′lo-e) the central layer of spongy bone between the two layers of compact bone of the flat cranial bones.

diploid (dip′loyd) the state of a cell containing twice the gametic number of chromosomes.

dissect to cut apart or separate the structures of the body for study.

distal away from the center of the body.

diverticulum (di·ver·tik′u-lum) an outpocketing from a tubular structure.

dorsal in human anatomy, synonymous with posterior.

dorsal primary ramus one of the two first branches of the spinal nerve; it innervates the skin and muscles near the midline of the back.

Down's syndrome condition of mental retardation due to the presence of chromosome 21 three times instead of twice.

duct a tubular structure providing exit to glandular secretions.

ductus deferens spermatic duct, beginning at the epididymis and terminating as the ejaculatory duct that empties into the prostatic urethra.

duodenum (du-o-de′num) the first part of the small intestine, extending from the pylorus to the jejunum.

dural sinus a blood vascular channel principally between the meningeal and periosteal layers of the dura mater.

ectoderm the outer layer of cells of the embryo, after the primary germ layers are formed.

edema (e-de′mah) accumulation of excess fluid in tissues.

effector a peripheral tissue that reacts to a nerve impulse by contraction (muscle) or secretion (gland).

ejaculatory duct the continuation of the ductus deferens after it is joined by the seminal vesicles.

elastic fibers connective tissue fibers that return to the original length after being stretched.

ellipsoid joint a joint in which the articular surfaces of the bones have an oval shape.

embolism (em′bo-lizm) occlusion of a blood vessel by a clot.

embryo (em′bri-o) the developing organism prior to the development of body form.

emphysema (em-fi-se′mah) a condition of the lung characterized by destructive changes in the walls of the alveoli.

enamel the hard substance covering the exposed portion of the tooth.

endocardium the inner layer of the heart wall, including the endothelium and underlying connective tissue.

endochondral ossification the process of bone formation in which cartilage is replaced by bone.

endocrine (en′do-krin) glands that have no ducts and whose secretions are delivered by way of blood vessels.

endocytosis (en′do-si-to′sis) the process by which materials are taken into a cell by the invagination of the plasma membrane, which is then pinched off as a vesicle.

endoderm the inner primary germ layer of the embryo; forms the lining of the digestive tract.

endolymph (en′do-limf) the fluid contained within the membranous labyrinth of the inner ear.

endometrium (en′do-me′trĭ-um) the mucous membrane comprising the inner layer of the uterine wall.

endomysium (en′do-miz-ĭ-um) the fine connective tissue covering of a muscle fiber.

endoneurium (en-do-nu′ry-um) the fine connective tissue sheath of individual nerve fibers of peripheral nerves.

endoplasm the inner part of the cytoplasm of a cell.

endosteum (en-dos′te-um) a thin membrane lining the inner surface of bone in the marrow cavity.

endothelium (en′do-the′lĭ-um) a layer of flat cells lining the heart, blood, and lymph vessels.

enzyme (en′zīm) a protein that acts as a catalyst in chemical reactions within cells.

eosinophil a cell that stains readily with eosin dyes.

epaxial dorsal (or above) the spinal axis of the body.

ependyma (ep-en′dĭ-mah) the epithelial layer lining the cavity within the central nervous system.

epicardium the outer covering of the heart, consisting of connective tissue and serous membrane.

epicondyle a projection from a long bone near the articular surface above or upon the condyle.

epidermis the outer epithelial layer of the skin.

epididymis (ep-ĭ-did′ĭ-mis) the convoluted sperm duct on the posterior surface of the testis.

epidural outside the dura mater.

epiglottis a leaf-shaped projection above the opening into the larynx.

epimere (ep′ĭ-mēr) the dorsal portion of the myotome.

epimysium (ep-ĭ-miz′ĭ-um) the connective tissue sheath surrounding a skeletal muscle.

epinephrine a hormone produced by the adrenal medulla.

epineurium (ep-ĭ-nu′rĭ-um) the connective tissue sheath enclosing a peripheral nerve.

epiphyseal line the last point of a long bone to ossify; between the epiphysis and the diaphysis or shaft.

epiphysis (e-pif′ĭ-sis) the end of a long bone that develops from a secondary center of ossification.

epithelium the layer of cells that covers all free surfaces of the body.

erythrocyte (ĕ-rith′ro-sit) a mature red blood cell.

erythropoietin (ĕ-rith′ro-poy′ĕ-tin) a substance produced by the kidney that enhances production of red blood cells.

estrogen (es′tro-jen) a general term for a female sex hormone produced by the growing follicles of the ovary.

ethmoid air cells paranasal sinuses of the ethmoid bone.

eversion (e-ver-shun) a turning outward.

evagination (e-vaj-ĭ-na′shun) the protrusion of a part from its original position.

excretion the elimination of metabolic wastes from the body.

exhale to breathe out.

exocytosis the process by which secretory granules are released from a cell.

expiration (eks′pĭ-ra′shun) to breathe out.

extension movement that increases the angle at a joint.

extensor retinaculum a fibrous band across the back of the wrist that holds the extensor tendons in place.

extraembryonic outside the embryonic body; the fetal membranes.

extrapyramidal system motor neural pathways other than the pyramidal system.

extrinsic (eks-trin′sik) located partially outside of the part upon which it acts.

face the front portion of the head.

falciform (fal′sĭ-form) crescentic or sickle-shaped.

fascial cleft area of loose connective tissue (fascia) between two denser areas.

fascial compartment an area enclosed by a connective tissue membrane.

fascia sheets of fibrous tissue that encase all the structures of the body.

fascicle a bundle or band of fibers, usually of muscle or nerve fibers.

fauces the space between the oral cavity and the pharynx.

fenestrated having window-like openings.

fertilization the fusion of the spermatozoon with the ovum.

fetus the product of conception from the end of the eighth week until birth.

fiber extra cellular thread-like structures or elongated cellular processes.

fibroblast a connective tissue cell capable of producing collagenous fibers.

fibrocartilage a type of cartilage that contains visible fibers.

fibrous pericardium a heavy sheet of fascia that is external to the parietal layer of the serous pericardium.

fibula (fib′u-lah) the lateral and smaller of the two bones of the leg.

fissure deep sulcus or slit.

flexion a movement that reduces the angle at a joint.

flexor retinaculum a fibrous band across the anterior aspect of the wrist.

folium a broad, thin, leaf-like structure, such as a gyrus of the cerebellum.

follicle (fol′i-kle) a small sac-like structure.

fontanelle a membranous interval or gap between cranial bones in the infant.

foramen an aperture or perforation through a bone.

foramen cecum a median pit on the dorsum of the posterior part of the tongue.

foramen magnum the large opening in the basal part of the occipital bone through which the spinal cord becomes continuous with the medulla.

fossa a depression below the level of the surface.

fovea a cup-shaped depression or pit.

fovea centralis a depression in the center of the macula of the retina, containing cones only.

fracture a break, usually of a bone.

frenulum a small fold of mucous membrane passing from a movable part to a more fixed part, such as the fold underneath the tongue.

frontal sinus the paranasal sinus in the frontal bone.

fundus the part of a hollow organ farthermost from the opening or exit.

fusiform (fu′zĭ-form) spindle-shaped; tapering at both ends.

galea aponeurotica the aponeurosis connecting the frontalis and occipitalis muscles across the superior aspect of the skull.

gamete (gam′ēt) a germ cell; an ovum or spermatozoon.

ganglion (gan′glĭ-on) an aggregation of nerve cell bodies located in the peripheral nervous system.

gastric relating to the stomach.

genial indicating the chin.

geniculate (jĕ-nik′u-lāt) bent like a knee.

genu any structure resembling a flexed knee.

genital relating to reproduction.

gestation (jes-ta′shun) pregnancy.

gingiva the gum; the mucous membrane surrounding the necks of the teeth.

gland a secreting organ.

glenoid fossa articular surface of the scapula that articulates with the head of the humerus.

globulin a family of proteins of plasma that includes the antibodies (immunoglobulins).

glomerulus (glo-mĕr′u-lus) a tuft of capillary loops at the beginning of each nephron in the kidney.

glucocorticoid any steroid-like compound capable of influencing intermediary metabolism.

glycogen (gli′ko-jen) the animal storage form of glucose.

goiter (goy′ter) a chronic enlargement of the thyroid gland.

goblet cell a columnar glandular cell that secretes mucus.

Golgi complex a complex of parallel, flattened saccules, vesicles, and vacuoles; it is important in packaging secretory products for discharge from the cell.

gonad an organ that produces gametes.

gonadotropin (gon′ă-do-tro′pin) a hormone capable of promoting growth and function of the gonad.

gonion (go′nĭ-on) the lowest posterior point of the angle of the mandible.

granule (gran′ul) in the cell, a small discrete mass.

granulocyte leukocytes that contain cytoplasmic granules (neutrophils, eosinophils, basophils).

granulosa cells ovarian follicle cells.

gray matter areas of the central nervous system composed chiefly of neuron cell bodies.

groove a narrow, elongated depression or furrow.

gross anatomy structure of the human body that can be observed without magnification.

growth the increase in size of a structure by cell division, by cellular hypertrophy, or by the elaboration of intercellular material.

growth hormone (somatotrophin) a hormone elaborated by the anterior pituitary that affects growth of tissues, especially the epiphyseal plates of long bones.

gyrus convolution; a rounded elevation on the surface of the cerebral hemispheres.

hair a keratinized outgrowth of the epidermis.

hair follicle the cup-shaped sac of the epidermis that produces hair.

hamstring muscles the flexors of the knee (biceps femoris, semitendinosis, semimembranous).

hamulus any hook-like structure.

haploid denoting the number of chromosomes in sperm or ova; one-half the number of chromosomes in somatic cells.

hare lip cleft lip.

Hassall's bodies concentric epithelial structures in the thymus.

haustra coli the sacculations of the colon, caused by the teniae coli which are shorter than the gut.

Haversian system a column of bone consisting of concentric lamellae of bone surrounding a blood vessel.

hematoma (he′mă-to′mah) a localized mass of blood confined within an organ or tissue.

hemi a prefix signifying one-half.

hemoglobin the red respiratory protein of erythrocytes.

hemopoiesis (he′mo-poy-e′sis) the process of formation of the various types of blood cells.

hemorrhage (hem′ŏ-rij) bleeding.

heparin an anticoagulant produced by mast cells.

Hering-Breuer reflex the effects of afferent impulses of the vagus whereby stretching the lungs stops inspiration, while deflation initiates inspiration.

hepatic relating to the liver.

hiatus (hi-a′tus) an aperture.

hilum the part of an organ where the nerves and blood vessels enter and leave.

hinge joint a joint in which movement is largely restricted to flexion and extension.

histamine a substance that dilates capillaries during allergic reactions.

histology the study of tissues; microscopic anatomy.

holo prefix denoting the whole.

holocrine gland a gland whose secretion consists of disintegrated cells.

homeostasis (ho′me-o-sta′sis) the state of equilibrium of chemical compositions of the fluids and tissues of the body.

homologous (ho-mol′o-gus) similar in origin and structure but not necessarily in function.

hormone a substance produced in one organ or part and carried via the blood to its target organ or part.

humerus the bone of the arm.

hyaline (hi′ă-lin) a glassy, translucent appearance.

hydrocephalus (hi′dro-sef′ă-lus) a condition characterized by excessive accumulation of fluid in the brain ventricles, causing a separation of the cranial bones.

hydroxyapatite a natural mineral forming the crystal lattice of bone matrix.

hymen a thin membranous fold partly occluding the vagina in the virgin.

hyoid bone a U-shaped bone in the upper neck.

hyper prefix denoting excessive or above the normal.

hyperplasia (hi′per-pla′zĭ-ah) an increase in number of cells in an organ or tissue.

hypertrophy (**hi-per′tro-fĭ**) increase in size but no increase in number of cells.

hypo prefix denoting a position underneath something else or an amount less than normal.

hypodermis the subcutaneous layer of connective tissue; superficial fascia.

hypomere the part of the myotome that gives rise to the muscles innervated by the ventral primary rami of the spinal nerves.

hypophyseal portal system the vascular pattern consisting of capillaries in the hypothalamus that drain via veins into the capillary bed in the anterior lobe of the pituitary gland.

hypothalamus the ventral portion of the diencephalon involved principally in autonomic and endocrine functions.

hypothenar eminence the fleshy mass at the medial side of the palm.

idiogram karyotype; a diagrammatic representation of the chromosomes of a cell or an individual.

ileum (**il′e-um**) the part of the small intestine between the jejunum and the colon.

iliac (**il′ĭ-ac**) relating to the ilium.

iliotibial tract a fibrous band extending from the ilium to the tibia.

ilium the broad portion of the hip bone.

immune resistant to an infectious disease.

immunization (**im′u-nĭ-za′shun**) the process by which one is rendered immune.

immunoglobulin proteins that function as antibodies.

immunology (**im′u-nol′o-jĭ**) the science of the study of immunity.

implantation the process of embedding of the embryo in the endometrium.

incisura (**in′si-su′rah**) a notch; an indentation at the edge of any structure.

incus the middle of the three ossicles in the middle ear.

infarct (**in′farkt**) an area of necrosis resulting from lack of adequate blood supply.

inferior situated nearer the soles of the feet.

inferior nasal concha a separate bone in the inferior aspect of the lateral wall of the nasal cavity.

inferior salivatory nucleus the autonomic nucleus of the glossopharyngeal nerve.

inflammation a complex of cytologic and histologic reactions in response to an injury.

infrahyoid (**in′fra-hi′oyd**) below the hyoid bone.

infundibulum (**in-fun-dib′u-lum**) a funnel-shaped structure or passage.

inguinal canal the pathway through the inferior part of the anterior body wall which the ductus deferens traverses in reaching the urethra.

inguinal hernia a protrusion of a viscus, usually intestine, through the body wall in the inguinal region.

inguinal ligament a fibrous band between the anterior superior iliac spine and the pubic tubercle.

inion (**in-ĭ-on**) a point located at the external occipital protuberance.

inner cell mass the cells of the blastocyst stage that will become the embryo proper.

inorganic matrix the portion of the intercellular material of bone or cartilage that is composed of minerals rather than organic molecules.

insertion the point of muscle attachment that is most movable.

inspiration inhalation.

insulin a polypeptide hormone secreted by the pancreatic islets of Langerhans that promotes glucose utilization.

integument (**in-teg′u-ment**) the outer covering of the body.

inter a prefix denoting in between.

internal auditory meatus the bony canal leading from the inner ear to the cranial cavity.

interosseous membrane a fascial sheet between two bones.

interphase (**in′ter-fāz**) the stage between two successive divisions of a cell nucleus.

intersection (**in′ter-sek′shun**) the site of crossing of two structures.

interstitial (**in-ter-stish′al**) relating to spaces in any structure.

interstitial growth increase in mass within a structure as opposed to being added at the surface.

interstitial substance material located between cells or other structures.

intervertebral disc the fibrocartilage and fibrous tissue forming the joint between adjacent vertebrae.

intestine the digestive tube between the stomach and the anus.

intra-articular ligament a fibrous cord between the articular surfaces of adjacent bones; within a synovial joint.

intracellular within a cell or cells.

intrafusal applied to structures within the muscle spindle.

intramembranous ossification bone formation in connective tissue rather than within cartilage.

intrauterine device a contraceptive device placed within the lumen of the uterus.

intrinsic in anatomy, refers to muscles whose origins and insertions are entirely within the structure, such as a limb.

inversion (**in-ver′shun**) a turning inward or upside down.

iris the vascular tunic of the eyeball that surrounds the pupil.

irritability the property inherent in protoplasm of reacting to a stimulus.

ischial tuberosity a large projection on the ischium to which the hamstring muscles are attached.

ischium the posterior-inferior part of the hip bone.

isometric contraction the condition when muscle contraction produces increased tension without movement.

isotonic contraction a condition when a muscle shortens against a constant load.

islets of Langerhans cellular masses within the pancreas that produce the hormones insulin and glucagon.

isthmus a narrow part connecting two larger parts.

joint the point of articulation of two adjacent bones.

jugular (**jug-u-lar**) relating to the throat or neck.

juxtaglomerular complex a group of cells around the afferent arteriole near the glomerulus, produces renin.

karyotype the chromosomes characteristic of an individual.

keratin a hard protein found in structures such as hair and nails.

keratohyalin (**ker′ă-to-hi′ă-lin**) the substance in the granules of the stratum granulosum of the epidermis.

kyphos (**ki′fos**) a hump.

labium majus one of two rounded folds of skin forming the lateral boundaries of the pudental cleft.

labium minus one of two narrow folds of mucous membrane lateral to the orifice of the vagina.

labrum glenoidale a rim of hyaline cartilage around the glenoid fossa.

lacrimal relating to tears.

lacrimal gland the gland in the superior-lateral aspect of the orbit that produces tears.

lactiferous duct a duct of the mammary gland that delivers milk to the surface of the nipple.

lacuna (**lă-ku′nah**) a small space, cavity, or depression.

lamellar bone bone in which the matrix is in the form of layers with cells in rows between the layers.

lamina (**lam′ĭ-nah**) a thin plate or flat layer.

lateral farther from the midline of the body.

lateral cerebral fissure the groove between the temporal lobe and the frontal and parietal lobes.

lateral corticospinal tract the spinal cord tract that controls voluntary movements initiated in the motor area of the cerebral cortex.

lateral geniculate body the nuclear mass of the thalamus in which the optic tracts terminate.

lateral spinothalamic tract the spinal cord tract that conveys impulses for pain sensation to the thalamus.

lens of the eyeball a transparent biconvex disc in the anterior portion of the eyeball that functions in focusing light rays on the retina.

lesion (le'zhun) a wound or injury.

lesser trochanter a bony projection on the femur near its head for muscle attachments.

leukocyte a general term for a white blood cell.

lien spleen.

ligament (lig'a-ment) a fibrous band connecting two bones.

linea aspera a rough ridge running down the posterior surface of the shaft of the femur.

lingual (ling'gwal) relating to the tongue.

lipid substance that can be extracted by fat solvents.

lobe one of the subdivisions of an organ or other part bounded by fissures, septa, or other structural demarcations.

lobule (lob'ul) a small lobe or subdivision of a lobe.

longitudinal running parallel to the long axis of the body.

lordosis anteroposterior curvature of the spine with the convexity towards the anterior.

lumbar relating to the loins; the part of the body between the ribs and the pelvis.

lumen the cavity in a tubular structure.

luteinization (lu'te-in-ĭ-za'shun) the transformation of the ovarian follicle into a corpus luteum.

lymph a relatively clear fluid that enters lymph vessels from the tissues of the body.

lymph node an encapsulated mass of lymphoid tissue.

lymph nodule a small mass of lymphoid tissue organized as a light central area surrounded by a darker rim.

lymphocyte a small mononuclear leukocyte formed in lymphoid tissue.

lymphokines (lim'fo-kīnz) soluble substances released by sensitized lymphocytes when stimulated with specific antigens; these substances take part in the cellular immune response.

lysosome (li'so-sōm) a membrane-bound cytoplasmic particle containing hydrolyzing enzymes.

macrophage (mak'ro-fāj) any large ameboid, mononuclear, phagocytic cell.

macula lutea a yellow spot in the sensory retina that contains at its center the fovea centralis, which contains only retinal cones.

major calyx one of the major subdivisions of the renal pelvis.

magnus denoting a structure of large size.

malformation abnormal development.

malleolus a rounded, bony prominence such as those on either side of the ankle joint.

malleus (mal'e-us) the largest of the three bones of the middle ear cavity.

mammary gland the breast; milk-producing gland of the anterior thoracic wall.

mandible (man'di-bl) the bone of the lower jaw.

manus the hand.

mast cell a connective tissue cell that produces heparin.

maternal (mă-ter'nal) denoting the mother.

matrix (ma'trĭks) the intercellular substance of a tissue.

maturation (mat'u-ra'shun) attaining full development or growth.

maxilla (mak-sil'ah) bone of the upper jaw.

maxillary sinus a paranasal air space in the maxilla.

mastication (mas'tĭ-ka'shun) the process of chewing food.

meatus (me-a'tus) the external opening of a canal or passage.

medial nearer the midline of the body.

medial geniculate nucleus the nucleus of the thalamus that receives auditory impulses.

medial lemniscus a tract in the brainstem that conveys sensory impulses from the spinal cord to the thalamus.

mediastinum (me'dĭ-as-tĭ-num) the median part of the thorax containing all the thoracic viscera except the lungs.

mediastinum testis fibrous tissue projecting into the posterior aspect of the testis.

medulla oblongata the lowest division of the brainstem, continuous with the spinal cord.

megakaryocyte a large cell with a multilobed nucleus found in bone marrow; gives rise to blood platelets.

meiosis (mi-o'sis) cell division in which the number of chromosomes is reduced to one-half.

melanin dark brown pigment normally occurring in skin.

melanocyte (mel'ă-no-sīt) pigment cell of the skin.

melatonin a hormone produced by the pineal gland that appears to depress gonadal function in mammals.

membranous epithelium epithelial cells that form a continuous sheet.

membranous labyrinth the membrane-enclosed space within the bony labyrinth of the inner ear.

meninges (mĕ-nin'jes) connective tissue protective coverings of the central nervous system.

menstruation (men-stru-a'shun) cyclic shedding of the endometrium.

mental in anatomy, refers to the chin.

merocrine a gland that elaborates secretions without loss or destruction of cells.

mesencephalon (mes'en-sef'ă-lon) the primary brain vesicle of the embryo that becomes the midbrain.

mesentery (mes'in-tĕr-ĭ) a double layer of serous membrane that suspends an abdominal viscus from the abdominal wall.

mesoderm the middle of the three primary germ layers of the embryo.

mesonephros (mes'o-nef'ros) the embryonic kidney that functions in the mammalian embryo.

mesothelium (mes'o-the'lĭ-um) the epithelium of the serous membranes, originates in the embryo from mesoderm.

metanephros the mammalian adult kidney.

metaphase the stage of cell division in which the chromosomes become aligned on the equatorial plate of the cell.

metencephalon (met'en-sef'ă-lon) the anterior of the two secondary brain vesicles derived from the rhombencephalon, gives rise to the pons and cerebellum.

micro a prefix denoting small size.

microfilament (mi-kro-fil'ă-ment) the finest of the fibrous elements of a cell or tissue.

microfibril (mi-kro-fi'bril) a very small fibrous strand, having an average diameter of 130 Å.

microtubule a cylindrical cytoplasmic element 200 to 270 Å in diameter and variable in length.

microvillus a minute fingerlike projection of the cell membrane.

micturition (mik-tu-rish'un) urination.

midsagittal plane a longitudinal section through the middle of the body or organ, dividing it into two equal halves.

mineralocorticoids the steroid hormones of the adrenal cortex that influence salt metabolism.

minor calyx a secondary division of the renal cortex into which a renal pyramid empties.

mitochondrion a cytoplasmic organelle that is the principal energy source of the cell.

mitosis (mi-to'sis) cell division producing two daughter cells with the same chromosome content as that of the original cell.

mitotic spindle the fusiform figure of dividing cells consisting of microtubules, some of which become attached to each chromosome.

mitral valve the valve between the left atrium and left ventricle of the heart; resembles the shape of a headband or turban.

monocyte a relatively large mononuclear leukocyte that is phagocytic.

mononuclear leukocyte a white blood cell that has an oval or slightly indented nucleus.

monosomy a cell or individual that has lost one member of a pair of homologous chromosomes.

mons pubis the prominence caused by a pad of fatty tissue over the symphysis pubis in the female.

morphology (mor-fol′o-jĭ) the study of structure of animals and plants.

morula (mor′u-lah) the solid ball stage of cleavage of the embryo.

motor cortex the area of the cerebral cortex that controls voluntary movements.

motor unit a single neuron and all of the skeletal muscle fibers that it innervates.

mucin a secretion containing mucopolysaccharides.

mucosa (mu-ko′sah) a sheet of epithelium and underlying connective tissue that is kept moist by a secretion of mucus.

multipolar neuron a nerve cell in which processes project from several points.

muscle an organ consisting principally of contractile cells or fibers.

muscles of facial expression skeletal muscles that originate or insert in the skin of the face and neck.

muscles of mastication the four muscles innervated by the trigeminal nerve that are the chief muscles involved in chewing.

myelencephalon the embryonic brain vesicle that becomes the medulla oblongata.

myelin the lipoproteinaceous material that envelops nerve fibers in concentric layers.

myeloid (mi′e-loyd) pertaining to the bone marrow.

myoblast a cell that will develop into a muscle fiber.

myocardium (mi′o-kar′dĭ-um) the middle layer of the heart, consisting of cardiac muscle.

myosin a globulin protein in muscle that, in combination with actin, forms actomyosin, the contractile unit of muscle.

myotome that part of a somite that gives rise to skeletal muscle.

myotube a skeletal muscle fiber during a developmental stage.

nail a thin, horny, translucent plate covering the dorsal surface of the distal end of the fingers and toes.

nasal bones the pair of bones forming the bridge of the nose.

nasolacrimal canal the bony canal leading from the orbit to the inferior meatus of the nasal cavity.

nasion (na′zĭ-on) a point on the skull corresponding to the middle of the nasofrontal suture.

nasopharynx the part of the pharynx superior to the soft palate.

necrosis (nĕ-kro′sis) cell death.

nephron the functional unit of the kidney, consisting of Bowman's capsule, proximal convoluted tubule, loop of Henle, and distal convoluted tubule.

nerve a bundle of nerve fibers outside the central nervous system.

nerve fiber a process extending from the cell body of a neuron.

nerve plexus exchange of nerve fibers between peripheral nerves by means of communicating branches.

nerve tract a group of functionally related nerve fibers within the central nervous system.

neural crest a band of neural cells along either side of the embryonic neural tube that develop into dorsal root ganglia, autonomic ganglia, adrenal medulla, neurilemma cells, and pigment cells.

neural tube the embryonic spinal cord.

neuraxis the unpaired central nervous system.

neurilemma (nu′rĭ-lĕm′ah) the sheath of Schwann, protective cellular covering of peripheral nerve fibers.

neurohypophysis the posterior lobe of the pituitary gland, a derivative of the diencephalon.

neuromuscular spindle a fusiform end organ in skeletal muscle that is particularly sensitive to stretch of the muscle.

neurotransmitter a specific chemical agent released by a neuron that crosses the synapse to stimulate or inhibit the postsynaptic neuron.

neutrophil (nu′tro-fil) a granulocytic leukocyte with a nucleus of three to five lobes and cytoplasm that contains fine granules that are neither distinctly acidophilic nor basophilic.

node an enclosed mass of differentiated tissue.

node of Ranvier a short interval in the myelin sheath of a nerve fiber, occurring between each two successive segments of the myelin sheath.

nodule (nod′ūl) a small node.

norepinephrine the neurotransmitter released by most postganglionic sympathetic fibers; one of the hormones produced by the adrenal medulla.

nostril the anterior opening of the nasal cavity.

nuchal (nu′kal) relating to the back of the neck.

nuclear envelope the membrane enclosing the nucleus.

nucleic acid (nu-kle′ik) a family of substances, found in chromosomes and other parts of all cells; in combinations with proteins they are called nucleoproteins.

nucleolus (nu-kle′o-lus) a small rounded mass in the nucleus, composed largely of RNA.

nucleoplasm the protoplasm of the nucleus of a cell.

nucleus (nu′kle-us) "a little nut"; an oval mass within the cell, surrounded by a nuclear envelope and containing the chromosomes.

nucleus of CNS an aggregation of cell bodies within the central nervous system that have a similar function.

nucleus pulposus the soft central portion of the intervertebral disk.

obturator foramen an opening in the lower part of the hip bone, occluded by a membrane.

occipital (ok-sip′ĭ-tal) the back of the head.

occipital lobe the posterior lobe of the cerebral hemisphere.

odontoblast a cell that produces dentin.

olecranon (o-lek′ră-non) the proximal process of the ulna at the elbow.

olfaction (ol-fak′shun) the sense of smell.

oligodendroglia (ol′ĭ-go-den-drog′lĭ-ah) the type of neuroglia that forms the myelin sheath around nerve fibers.

omentum a fold of peritoneum that covers the anterior surface of the intestine.

oocyte (o′o-sit) an immature ovum.

oogenesis (o-o-jen′e-sis) the process of formation of the ovum.

oogonium the primitive cell from which the oocytes are developed.

ootid the nearly mature ovum after the first maturation division has been completed.

opsonins a substance that enhances phagocytosis.

optic chiasma the point at which one-half of the fibers of the right and left optic nerves cross to the opposite side.

optic disc an area of the retina that is devoid of rods and cones.

orbit the bony cavity containing the eyeball.

organ two or more tissues organized to perform one or more specific functions.

organ of Corti a specialized structure within the cochlear duct for the perception of sound waves.

organelle a small structure within a cell that performs a specific function.

orifice any aperture or opening.

origin the less movable of two points of muscle attachment; the early beginning of an embryonic structure.

os coxae the hip bone.

osseous bony.

ossification (os′ĭ-fĭ-ka′shun) the formation of bone.

osteoblast a bone-forming cell.

osteocyte a bone cell that occupies a lacuna within bone matrix.

ostium a small opening, usually leading into a hollow organ or canal.

otic relating to the ear.

otoliths particles of calcium carbonate on a gelatinous membrane in the organ of equilibrium of the inner ear.

oval window the membrane-closed window that transmits sound waves to the perilymph around the cochlear duct in the inner ear.

ovarian cycle the normal sex cycle, including development of the ovarian follicle, ovulation, and formation and regression of a corpus luteum.

ovary one of the paired reproductive organs of the female.

ovulation (o′vu-la′shun) the release of an ovum from the ovarian follicle.

ovum the female germ cell.

oxytocin (ox-sĭ-to′sin) a hormone released in the neurohypophysis that causes contraction of the uterine muscle.

palate the partition between the oral and nasal cavities.

palmar aponeurosis the thickened anterior fascia ensheathing the hand.

paramesonephric ducts the embryonic ducts that develop into the uterine tubes cranially and fuse caudally to form the uterus.

paranasal sinuses the air sinuses that communicate with the nasal cavity.

palpebra the eyelid.

pancreas (pan′kre-as) a digestive gland in the abdomen that empties into the duodenum.

papilla (pă-pil′ah) any small nipple-like process.

para a prefix meaning alongside of.

parasympathetic nervous system the division of the autonomic nervous system that arises from the brainstem and sacral spinal cord; craniosacral outflow.

parathyroid glands four endocrine glands embedded in the posterior aspect of the thyroid gland.

parietal (paŕ-ri′ĕ-tal) relating to the wall of any cavity.

parietal lobe of cerebral hemisphere the portion of the cerebral hemisphere that lies adjacent to the parietal bone, posterior to the central sulcus.

parotid (pă-rot′id) the salivary gland near the ear.

pars a part, a portion.

patella the kneecap, the large sesamoid bone in the tendon of the extensor of the knee.

patent open, exposed.

pathology (pă-thol′o-jĭ) the medical science concerned with all aspects of disease.

pectoral relating to the chest.

pectoral girdle the skeleton of the shoulder which supports the upper limb; consists of the scapula and clavicle.

pectineal line a ridge on the anterior aspect of the pubis.

peduncle (pĕ-dung′kl) a stalk or stem; in neuroanatomy it is used for a variety of stalk-like structures.

pelvic brim the superior aperture of the pelvis, above the arcuate line and the sacral promontory.

pelvic girdle the hip bones that connect the lower limb to the sacrum.

penis the organ of copulation in the male.

pericardial cavity the potential space between the visceral and parietal layers of the serous membrane (pericardium) that surrounds the heart.

pericardium the serous membrane around the heart.

perichondrium (pĕr-ĭ-kon′drĭ-um) the dense, irregular connective tissue sheet around cartilage.

perilymph the fluid contained within the bony labyrinth and protecting the membranous labyrinth of the inner ear.

perimysium (pĕr-ĭ-mis′i-um) the connective tissue sheath of muscle fascicles.

perineurium (pĕr-ĭ-nu′rĭ-um) the connective tissue sheath of nerve fascicles.

periosteum (pĕr-ĭ-os′te-um) the dense, irregular connective tissue covering of bone.

peripheral nerve a bundle of nerve fibers outside the central nervous system.

peristalsis (pĕr′ĭ-stal′sis) waves of alternate contraction and relaxation of the smooth muscle of the gut or other tubular structure.

peritoneum the serous membranes of the abdominal cavity.

pes the foot; any foot-like structure.

phagocyte (fag′o-sit) a cell capable of ingesting bacteria, foreign particles, or other cells.

phagocytosis (fag′o-si-to′sis) the process of ingestion and digestion of solid substances by cells.

pharyngeal tonsil aggregation of lymphoid tissue in the posterior wall of the nasopharynx.

phalanx a bone of the fingers or toes.

pia mater the layer of meninges of the central nervous system that is in direct contact with the nervous system.

pillar a structure resembling a column or pillar.

pineal gland a small, unpaired, flattened body, shaped like a pine cone, lying on the posterior aspect of the brain below the corpus callosum.

pinocytosis (pin′o-si-to′sis) the cellular process of actively engulfing liquid.

pituitary relating to the pituitary gland, located below the hypothalamus in the sella turcica of the sphenoid bone.

placenta the organ of metabolic exchange between the mother and fetus.

placentation (plas′en-ta′shun) the process of formation of the placenta; the establishment of the attachment of fetal to maternal tissues.

plane a flat surface.

plasma (plaz′mah) the liquid portion of blood.

plasma cell a cell derived from a lymphocyte that produces antibodies.

plasma membrane the outer-limiting membrane of the cell.

platelet an irregularly shaped disk found in blood.

pleura the serous membrane enveloping the lungs and lining the wall of the pleural cavities.

pleural cavity the potential space between the parietal and visceral layers of pleura.

pleural recess a potential space between two layers of parietal pleura into which the lung expands during inspiration.

plexus a network or interlacing of nerves or blood vessels.

plica an anatomical structure in which there is a folding over of the parts.

polar body one of two small cells formed by the ovum during its maturation.

polarity the property of having two opposite poles.

polymorphonuclear leukocyte white blood cells of more than one kind or form of nucleus.

polypeptide a molecule formed by the union of an indefinite number of amino acids.

polyribosomes (pol-ĭ-ri′bo-sŏmz) two or more ribosomes connected by a molecule of RNA, active in protein synthesis.

polysaccharide (pol-ĭ-sak′ă-rid) a carbohydrate containing a large number of starch groups.

pons that part of the brainstem that lies between the midbrain and the medulla.

popliteal space the area posterior to the knee joint.

porta (por′tah) gate; the point of entrance to a structure.

portal system a vascular pattern that includes two capillary beds connected by a vein.

posterior toward the back surface of the body.

postganglionic fiber a nerve fiber that passes into the periphery from its cell body in a ganglion.

preganglionic fiber a nerve process that is situated proximal to or preceding a ganglion.

prenatal preceding birth.

prevertebral muscles muscles that originate and insert anterior to the transverse processes of the vertebral column.

prime mover a muscle whose contraction produces a movement.

primitive reticular cell a relatively undifferentiated connective tissue cell that can become phagocytic, associated with reticular fibers.

primitive streak an ectodermal ridge in the midline at the caudal end of the embryonic disc that gives rise to the mesoderm.

process (pros′es) in anatomy, usually means a projection or outgrowth.

processus vaginalis a peritoneal diverticulum in the fetus that traverses the inguinal canal, preceding descent of the testis.

progesterone (pro-jes′ter-ōn) a hormone produced by the corpus luteum and placenta, prepares the endometrium for implantation.

prolactin a hormone of the anterior pituitary that stimulates milk production.

pronation turning the forearm so that the palm is turned backwards.

pronephros (pro-nef'ros) head kidney, the functional kidney of primitive fishes; a vestigial kidney in the embryo of higher vertebrates.

prophase the first stage of mitosis or meiosis.

proprioception (pro'pri-o-sep'shun) the sensation that arises when a muscle is moved.

prosencephalon the rostral primitive cerebral vesicle, which gives rise to the cerebral hemispheres and diencephalic structures.

prostate gland a chestnut-shaped body that surrounds the urethra in the male; it secretes a milky fluid that becomes a component of semen.

proximal nearest the midline of the body or nearest the point of attachment of a structure such as a limb.

pseudostratified a simple columnar epithelium that gives the appearance of having more than one layer of cells.

pterygomandibular raphe a tendinous band extending from the pterygoid hamulus to the mandible; it separates the buccinator muscle and the superior pharyngeal constrictor.

pterygoid (těr'ĭ-goyd) wing-shaped; a process of the sphenoid bone.

pterygopalatine ganglion a parasympathetic ganglion on which preganglionic fibers of the facial nerve synapse; located in the pterygopalatine fossa.

ptosis (to'sis) a drooping of the upper eyelid.

pubic symphysis the midline joint between the right and left pubic bones; a joint in which the bones are joined by fibrocartilage.

pubic tubercle the small elevation on the pubis to which the inguinal ligament attaches.

pubis the bone that forms the anterior portion of the hip bone (os coxa).

pulmonary (pul'mo-něr-ĭ) relating to the lungs.

pupil the round opening in the center of the iris, through which light rays enter the eyeball.

Purkinje fibers modified heart muscle fibers that conduct electrical impulses to heart muscle fibers.

pyramidal system the nerve fibers that control voluntary contraction of skeletal muscle fibers; these fibers pass through the medulla in the pyramids.

quadriceps (kwah'drĭ-seps) having four heads.

radius in anatomy, the lateral and shorter of the two bones of the forearm.

ramus a branch; one of two divisions of a nerve or blood vessel.

raphe (ra'fe) the line of union of two bilaterally symmetrical structures.

Rathke's pouch the evagination from the ectodermal lining of the roof of the oral cavity in the embryo that develops into the anterior lobe of the pituitary gland.

receptor one of the various sensory nerve endings in the skin or other tissues; a point on a cell surface that will react with a molecule or substance.

recess a small depression or indentation.

rectum (rek'tum) straight; the terminal portion of the digestive tube.

rectus sheath the aponeuroses of the lateral abdominal wall muscles that enclose the rectus abdominus muscles.

reflex (re'fleks) an involuntary response to a stimulus applied in the periphery and transmitted to the central nervous system.

region (re'jun) a portion of the body having a special nervous or vascular supply.

releasing factors substances produced in the hypothalamus that stimulate release of hormones by the anterior lobe of the pituitary.

renal capsule the connective tissue covering of the kidney.

renal column kidney cortical type of tissue that extends into the renal medulla between the renal pyramids.

renal cortex kidney tissue near the surface; consists largely of nephrons.

renal fascia the thickened transversalis fascia that splits to enclose the kidneys.

renal medulla kidney tissue in the central portion of the kidney; consists largely of collecting ducts.

renal pelvis the dilated portion of the ureter within the renal sinus.

renal pyramid the aggregation of renal collecting ducts that empty into a minor calyx.

renal sinus the area of the kidney at which the blood vessels enter and which is occupied by the renal pelvis and calyces.

renin an enzyme that converts angiotensinogen to angiotensin.

reproduction the process by which organisms produce offspring.

resorption the removal of a substance, such as a blood clot, by absorption.

respiration the use of oxygen to oxidize molecules, providing a source of energy.

respiratory bronchioles small air passageways that have alveoli at irregular intervals in their walls.

rete testis a network of tubules in the mediastinum of the testis.

reticular fibers fine collagenous connective tissue fibers that usually occur as an irregular meshwork.

reticular lamina a major component of the basement membrane; consists largely of reticular fibers and ground substance.

reticuloendothelial system the cells in different organs that function as macrophages.

reticulum (re-tik'u-lum) a fine network formed by cells or fibers.

retina the sensory layer of the eyeball; contains receptor cells, relay neurons, and pigment cells.

retinaculum a fibrous band that functions to hold structures in their proper place.

rhodopsin (ro-dop'sin) visual purple, a protein found on the rods of the retina.

rhombencephalon the primary brain vesicle of the embryo that gives rise to the cerebellum, pons, and medulla.

rhomboid (rom'boyd) having unequal but parallel sides.

ribs bones supporting the body wall of the thorax.

ribonucleic acid RNA, a macromolecule found in all cells.

ribosomes a cytoplasmic granule of ribonucleoprotein that is the site of protein synthesis.

ridge a rough elongated elevation.

rigor mortis the stiffness or inflexibility that develops after death.

rods of the retina the photosensitive cells of the eye that contain rhodopsin.

root of the lung the area of the lung at which the bronchi and blood vessels enter.

rostral toward the beak or any structure resembling a beak.

rotation a movement around the axis of a part.

rotator cuff the short muscles around the shoulder joint that rotate the arm and stabilize the shoulder joint.

sacral (sa'kral) relating to the sacrum of the vertebral column.

sacrospinous ligament the fibrous band between the sacrum and the ischial spine.

sacrum that part of the lower vertebral column composed of five vertebrae fused into a single unit.

sagittal plane any longitudinal plane that is parallel to the long axis of the body.

saliva (sǎ-li'va) the viscous fluid elaborated by the salivary glands of the oral mucosa.

salpingo prefix denoting a tube.

sarcolemma (sar'ko-lim'ah) the plasma membrane of a muscle cell or fiber.

sarcomere (sar'ko-mēr) the portion of a skeletal muscle fiber between two adjacent Z-lines.

sarcoplasm the nonfibrillar part of the cytoplasm of a muscle fiber.

scala tympani the lower part of the perilymph-containing canal of the cochlea.

scala vestibuli the upper part of the perilymph-containing canal of the cochlea.

scapula the shoulder blade; the posterior bone of the pectoral girdle.

sclera a fibrous tunic forming the outer covering of the eyeball.

scrotum (skro'tum) the sac of skin containing the testes.

sebaceous gland a gland adjacent to the hair follicle that produces an oily secretion.

segment a division of a structure into approximately equal parts.

semen the penile ejaculate, containing spermatozoa and the secretions of seminal vesicles, the prostate gland, and bulbourethral glands.

semicircular canals the membranous labyrinth of the inner ear containing endolymph.

semilunar valve the half-moon shaped valves of the aorta and the pulmonary trunk that prevent the backflow of blood into the heart.

seminal vesicles a gland that is a diverticulum of the ductus deferens and secretes a component of semen.

seminiferous tubules the tubules of the testis in which the spermatozoa are produced.

sensation the conscious feeling that results from a stimulus exciting any of the sense organs or receptors.

septum a thin wall forming a partition between two cavities or soft structures.

serosa a membrane composed of epithelium and its underlying connective tissue that is moistened by serous fluid.

serous (sēr'us) a thin, watery fluid containing some protein in solution.

serum the fluid portion of blood with the fibrin removed.

sinus a channel for the passage of blood or lymph that has a wall that is less complete than ordinary blood vessels.

sinoatrial node the modified cardiac muscle fibers in the right atrium that serves as the pacemaker of the heart.

skull the bones that together form the cranium.

somatic (so-mat'ik) relating to the wall of the body cavity, or body in general.

somatostatin a factor that inhibits the release of growth hormone by the anterior lobe of the pituitary gland.

somite one of the paired segmentally arranged masses of embryonic mesoderm.

spermatic cord the vas deferens and nerves and blood vessels that accompany the vas deferens between the body wall and the testis; these structures are wrapped in fascia, forming a cord-like structure.

spermatid (sper'ma-tid) a cell in a late stage of the development of the spermatozoon.

spermatocyte (sper'mă-to-sit) a cell derived from a spermatogonium and destined to give rise to spermatozoa.

spermatogenesis the process of formation of spermatozoa.

spermatogonium (sper'mă-to-go'nĭ-um) the primitive germ cell derived from the germ cell by mitosis.

spermatozoon the male gamete or sex cell.

sphenoid bone a bone of the skull forming part of the nasal septum and lateral walls of the nasal cavity.

sphenoid sinuses the paranasal air cells in the sphenoid bone.

sphincter a circular muscle that constricts an opening.

spinal cord segment that portion of the spinal cord that gives rise to a single pair of spinal nerves.

spinal nerve the bundle of peripheral nerve fibers formed by the junction of dorsal and ventral roots of the spinal nerve.

spinal nerve root the motor (ventral) or sensory (dorsal) fibers found within the vertebral canal that unite at the intervertebral foramen to form a spinal nerve.

spine the vertebral column; a sharp bony process.

splanchnic nerve a nerve supplying sensory and autonomic fibers to the viscera.

spleen a large lymphoid organ to the left of the stomach in the upper abdomen.

spongy bone bone that contains numerous irregular marrow spaces.

squamous (skwa'mus) relating to or covered with scales.

stapes the smallest of the three auditory ossicles.

stenosis (stĕ-no'sis) a narrowing of any tubular structure.

stereo prefix denoting a solid condition.

stereocilia nonmotile cilia.

sternoclavicular joint the articulation of the clavicle with the sternal manubrium; this joint has a complete articular disc.

sternum the breastbone; the bone in the anterior midline that articulates with the first seven ribs and the clavicle.

stimulus (stim'u-lus) any change in the internal or external environment that can evoke a response in a muscle, nerve, or gland.

stomodeum the membrane of ectoderm and endoderm that closes the entrance into the foregut of the embryo.

strap muscles the muscles of the anterior neck that insert on the hyoid bone; infrahyoid muscles.

stratum a layer of differentiated tissue.

stratum corneum the outer, keratinized layer of the epidermis.

stratum germinativum the layer of the epidermis in which mitosis occurs.

stratum granulosum the layer of the epidermis that contains granules of keratohyalin and eleidin.

stratum lucidum clear layer of the epidermis; lies just deep to the stratum corneum.

stratum spinosum layer of the epidermis in which the cells are connected by desmosomes (intercellular bridges), giving the layer a spiny appearance.

stroma the framework, usually connective tissue, of an organ.

styloid process the slender bony process on the temporal bone.

subarachnoid space the space between the arachnoid and the pia mater containing the cerebrospinal fluid.

submandibular ganglion the parasympathetic ganglion on which preganglionic fibers of the facial nerve synapse; postganglionics distributed to the submandibular and sublingual glands.

sulcus a groove or furrow on the surface of the brain.

superficial located nearer the surface of the body in relation to a specific reference point.

superior situated nearer the head in relation to a specific reference point.

superior cervical ganglion the sympathetic ganglion located in the upper neck at the cranial end of the sympathetic trunk.

superior colliculus masses of gray matter in the midbrain that function in visual pathways.

superior salivatory nucleus the nucleus of origin of the preganglionic parasympathetic fibers of the facial nerve.

supination (su'pĭ-na'shun) rotation of the forearm so that the palm faces forward.

supraglenoid tubercle a small bony prominence superior to the glenoid fossa from which the long head of the triceps originates.

suprahyoid muscles muscles that attach to the hyoid bone and pass to their other attachment superior to the hyoid bone.

suprapatellar bursa a bursa that communicates with the synovial cavity of the knee joint and is located beneath the quadriceps tendon above the patella.

suprarenal cortex the outer portion of the suprarenal gland that produces steroid hormones.

suprarenal gland the endocrine gland lying on the superior pole of the kidney; produces steroid hormones in the cortex and epinephrine and norepinephrine in the medulla.

suprarenal medulla the central portion of the suprarenal gland; produces epinephrine and norepinephrine.

suprascapular notch the gap in the superior border of the scapula through which the suprascapular nerve and blood vessels pass to the posterior aspect of the scapula.

superficial fascia the layer of fascia just beneath the skin, usually containing a considerable amount of fat.

surfactant (sur-fak'tant) a lipoprotein that stabilizes alveolar volume by reducing surface tension.

suspensory ligament of the lens fibrous connective tissue connecting the lateral edge of the lens to the ciliary body of the eye; maintains tension on the lens that is lessened by contraction of the ciliaris muscle.

suture a fibrous joint in which two bones are joined by a fibrous membrane that is continuous with the periosteum.

sweat gland sudoriferous glands of the skin that produce sweat (perspiration).

sympathetic nervous system that portion of the autonomic nervous system in which the preganglionic fibers arise in spinal cord segments T1 to L2.

sympathetic trunk the ganglionated chain of sympathetic fibers that lies on each side of the bodies of the vertebrae, extending the entire length of the vertebral column.

symphysis a joint on which the bones are connected by fibrocartilage.

synapse (sin′aps) the membrane-to-membrane contact point at which a nerve impulse is transmitted from one neuron to another.

synarthrosis (sin′ar-thro′sis) a general category of joints that do not contain a synovial membrane.

synchondrosis (sin′kin-dro′sis) a joint in which the union between two bones is formed by hyaline cartilage.

syncytium (sin-sish′ĭ-um) a multinucleated protoplasmic mass formed by the fusion of originally separate cells.

syndesmosis (sin′diz-mo′sis) a fibrous joint in which the bones are relatively far apart and are held together by ligaments.

syndrome (sin-drōm) the signs and symptoms of a disease or abnormal condition.

synergist (sin′er-jist) in anatomy, a muscle that assists another in its action.

synostosis bony union between two bones.

synovial membrane a connective tissue sheet that lines the inner surface of the capsule of synovial joints.

synovial fluid a clear fluid that serves as a lubricant in synovial joints.

systole (sis′to-le) the contraction of the ventricles of the heart.

tactile (tac′til) the sense of touch.

target organ an organ that is specifically affected by a hormone.

tarsus the seven bones of the ankle.

taste bud the special receptor on the tongue and at the base of the epiglottis for the sense of taste.

tegmentum a covering structure; in the midbrain is the central portion dorsal to the cerebral peduncles.

telencephalon (tel-en-sef′a-lon) the anterior division of the prosencephalon of the embryo, gives rise to the cerebrum.

telophase the final stage of mitosis or meiosis, the stage in which the cytoplasm is divided to produce two cells.

temporal lobe the part of the cerebral hemisphere below the lateral fissure.

tendon a fibrous cord or band that connects a muscle to a bone.

tendon sheath a double layer of synovial membrane enclosing a tendon.

testis one of the two male reproductive organs, located in the scrotum.

testosterone the male sex hormone, the most potent androgen.

tetrad a chromosome that divides into four during meiosis.

thalamus a large gray mass in the diencephalon that functions as relay nuclei for sensory impulses destined for the cerebral cortex.

theca interna the inner layer of the sheath of the ovarian follicle.

thenar eminence the fleshy mass on the thumb side of the palm.

thigh the part of the lower limb between the hip and the knee.

thoracic duct the main lymph vessel that begins at the cisterna chyli in the abdomen and empties into the blood stream at the junction of the left subclavian and left internal jugular veins.

thoracic vertebra the twelve vertebrae of the posterior wall of the thorax; each articulates with a pair of ribs.

thoracolumbar outflow the preganglionic sympathetic fibers that originate in spinal cord segments T1 to L2.

thymus a lymphoid organ located in the superior mediastinum and lower part of the neck, essential in early life for development of cellular immunity.

thyrocalcitonin a hormone secreted by parafollicular cells of the thyroid gland; inhibits bone resorption.

thyroglobulin a protein that contains the thyroid hormone.

thyroglossal duct a remnant of the embryonic connection of the thyroid gland to its point of origin near the base of the tongue.

thyrohyoid membrane a fascial sheet between the superior aspect of the thyroid gland and the hyoid bone.

thyroid gland an endocrine gland located in the anterior neck, closely related to the trachea and larynx.

thyroid stimulating hormone a hormone secreted by the anterior lobe of the pituitary gland that increases the secretory activity of the thyroid gland.

tissue a collection of cells of similar structure or function and the intercellular substance surrounding them.

T-lymphocyte the category of lymphocyte that mediates the cellular immune response.

tonsil a collection of lymphoid tissue found beneath epithelium.

tracheoesophageal fistula an opening that connects the cavities of the trachea and esophagus.

trabecula (trǎ-bek′u-lah) one of the supporting bundles of fibers within the substance of an organ; a spicule of spongy bone.

trachea (tra′ke-ah) the air tube between the larynx and bronchi.

tract a bundle of functionally related nerve fibers within the central nervous system.

transcription the process by which messenger RNA is synthesized on a template of complementary DNA.

transitional epithelium a stratified sheet of cells, consisting of cuboidal cells that can slide on one another when stretched.

translation the process by which messenger RNA, transfer RNA, and ribosomes produce protein from amino acids.

transversalis fascia the thin fascia of the abdominal wall between the muscles of the body wall and the parietal peritoneum.

transverse plane a plane or section at right angles to the long axis of the body.

trapezoid line the ridge on the inferior surface of the clavicle on which the lateral part of the coracoclavicular ligament attaches.

trauma an injury or wound.

tricuspid valve the valve between the right atrium and right ventricle having three points or parts called cusps.

trigone (tri′gon) the area of the urinary bladder between the orifices of the two ureters and the urethra.

tri-iodothyronine a thyroid hormone.

trisomy instead of a normal pair of homologous chromosomes, there are three of a particular chromosome, resulting in a total of 47 chromosomes.

trochanter (tro-kan′ter) a bony prominence near the head of the femur.

trochlea (trok′le-ah) a structure serving as a pulley.

trophic (trof′ik) relating to nutrition.

trophoblast (trof′o-blast) the outer layer of cells covering the blastocyst of the early embryo.

tropomyosin a fibrous protein of muscle.

troponin a globular protein that binds to tropomyosin.

tubercle (tu′ber-kl) a small elevation on a bone or tooth; a lesion due to an inflammatory reaction to infection.

tuberosity a bony prominence or elevation.

tubule a small tube or duct.

tubulin protein subunit of microtubules.

tumor a general term for an abnormal growth or swelling.

tunica a coat or covering of a structure.

tunica albuginea (testis) a thick white fibrous connective tissue membrane forming the outer covering of the testis.

tunica vasculosa (testis) the vascular layer enclosing the testis beneath the tunica albuginea.

tympanic membrane the eardrum; the epithelial and connective tissue sheet separating the external and middle ear cavities.

ulna the medial and larger of the two bones of the forearm.

ultrastructure cells and tissues as seen with the electron microscope.

umbilicus naval; the pit in the anterior abdominal wall marking the point where the umbilical cord entered the fetus.

undifferentiated primitive embryonic, immature cells not having specialized structure or function.

unicellular composed of a single cell.

unipennate muscle a muscle having a featherlike arrangement on one side.

unipolar neuron a nerve cell from which processes project on one side only.

ureter (u-re′ter) the tube conducting urine from the kidney to the urinary bladder.

urethra (u-re′thrah) a tube leading from the urinary bladder to the outside.

urogenital sinus the embryonic ventral part of the cloaca after its separation from the rectum; gives rise to the urinary bladder.

urine the fluid excreted by the kidney.

urogenital a general term that includes both the urinary system and the reproductive system.

uterine cycle the cyclic changes in the uterine endometrium due to the hormones produced in the ovary.

uterine tube the oviduct; the tube by which the fertilized ovum reaches the uterus.

uterus the muscular organ in which the fertilized ovum is implanted.

uvula a conical projection from the posterior aspect of the soft palate.

vaccine (vak′sen) a preparation that causes active immunity against a disease.

vacuole (vak′u-ōl) a minute space in any tissue; a clear space in a cell.

vagina (vă-ji′nah) the genital canal of the female, between the uterus and the vulva.

vascular relating to blood vessels.

vasectomy (vă-sek′to-mī) excision of a segment of the vas deferens, to produce sterility.

vasopressin antidiuretic hormone.

vein a blood vessel that conveys blood toward the heart.

ventilation movement of air into and out of the lungs.

ventral primary ramus the branch of the spinal nerve that continues into the anterior body wall after the dorsal ramus is given off.

ventricle a normal cavity, as of the brain or heart.

venule a small vein.

vermis the narrow middle zone between the two hemispheres of the cerebellum.

vertebral canal the space in the vertebral column occupied by the spinal cord.

vertebral column the longitudinal skeletal structure composed of vertebrae arranged on top of one another.

vertebral foramen the opening in a vertebra bounded by the neural arch and the vertebral body.

vesicle a small sac-like structure.

vestibular fold the false vocal fold of the larynx.

vestibular relating to the structures of the inner ear that are concerned with equilibrium.

villus a minute, finger-like projection of the cell surface.

visceral splanchnic, referring to the soft organs of the body.

visceral arch the embryonic mesodermal arches of the lateral walls of the pharynx.

visceral pouch the endodermal evagination of the lateral wall of the pharynx between adjacent visceral arches.

vitreous body the glassy gel in the eyeball posterior to the lens and iris.

vocal fold the true vocal cord; the elastic membrane between the arytenoid cartilage and thyroid cartilage that is essential in sound production.

vomer a flat bone forming the inferior and posterior part of the nasal septum.

woven bone immature bone in which the bone matrix is not arranged in layers.

yolk sac the dilated ventral extension of the embryonic midgut.

zygomatic bone the bone of the cheek, forming the lateral and inferior one third of the rim of the orbit.

zygote the fertilized ovum.

Note: When an item appears only in a table or an illustration, the page number is italicized, but when an item appears in the text or in both the text and a table or illustration, the page number is not italicized.

A-band, *82*, 186, *187*, *188*
abdominal regions, 516
abdominal viscera
  surface landmarks, 516
  topographical anatomy, 515
abduction, 12, 202
abortion, spontaneous, 54
absorption, *20*, 23
accommodation, mechanism, 365
acetabulum, 157, 178, 266
acetylcholine, 282, 302, 349
achondroplasia, 110
acidophil, 385
acini, 506
acromegaly, 400
acromion process, 149
acrosomal reaction, 549
acrosome, 45, *48*
actin, 80, 186
active transport, 23
Adam's apple, 467, 469
Addison's disease, 402
adduction, 12, 202
adenohypophysis, 382, *383*, 384
adenoids (see tonsil, pharyngeal)
adenosine diphosphate (ADP), 21
  breakdown, 22
  synthesis, 22
adenosine monophosphate, cyclic, 380
adenosine triphosphate (ATP), 21, 380
  breakdown, 22
  muscle contraction and relaxation, 191
  synthesis, 22
adenyl cyclase, 380
adhesion
  cellular, 35
  desmosome, 35
  intermediate junction, 36
  synovial membrane, 172
  tight junction, 35
adipose tissue, 34, *70*
aditus, 467, 478
adrenocorticotropic hormone (ACTH), 385
adrenocorticotropic-hormone-releasing-hormone, *387*
afterbirth, 53, 559
agammaglobulinemia, 579
aging theories
  molecular damage, 57
  programmed senescence, 57
  somatic mutation, 57
ala, nasal, 463

orbitalis, *129*
temporalis, *129*
albumin, 538
aldosterone, 391, *393*, 529
allantois, 53, 534
allergen, 579
allergy, 578
alveolar duct, 473, *474*
alveolar sac, 436, 473
alveolus, 434, *436*, 473
amino acids, 17, 538
amniocentesis, 53
amnion, 54
amniotic cavity, 53
amorphous ground substance, 69, 74
  interstitial, 69
  matrix, 69
amygdala, 356, 375
amyotrophic lateral sclerosis, 342
anabolism (see metabolism)
anal columns, 509
anamnestic response, 577
anaphase, *41*, 42, *44*
anatomical position, *10*, 11
anatomists, renaissance, 2
anatomy, comparative, 2
  definition, 2
  gross, 2, 4
  radiological, 5
  regional, 5, 6
  microscopic, 2
  surface, 5
  systemic, 5
androgen, *393*, 543
anemia, 457
  sickle cell, 18
aneurysm, 120, 457
angiotensin, 529
ankle (see joint, general)
ankylosis, 182
anopsia, 369
anosmia, 362
antagonist, 201, 338
antebrachium, 149
anterior pillar of the fauces (see fauces)
antibody, function, 577
  production, 577
  structure, 580
antidiuretic hormone, 382, 383, *384*, 529
antigen, 568, 575
antigenic determinant, 575
anus, 486
aorta (see arteries, specific)
aortic arches, 457
apatite crystals, 489
aponeurosis, 195, 196
  bicipital, 254
  palmar, 255
  plantar, 275

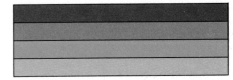

INDEX

hyperparathyroid bone disease, 110
hyperpituitarism, 400
hypersensitivity, 578
    delayed, 578
hypertension, essential, 457
    renal, 457
hyperthyroidism, 31
hypertrophy, heart, 411
    muscle, 81
    bone, 109
hypesthesia, 300
hypodermis, 88, *89*
hypoglycemia, 402
hypomere, 237, *239*
hypophyseal portal system, 382, 386, 392
hypopituitarism, 400
hypothalamus, 326, 327, 356, 375
hypothalamic-adenohypophyseal system (see
    neurovascular pathway)
hypothalamic-neurohypophyseal system (see
    neurosecretory pathway)
hypothenar eminence, 259
hypothyroidism, 31
H-zone, 186

I-band, *82*, 186, *187*, *188*
IgE antibodies, 578
ileum, 499
iliac crest, 247, 515
iliotibial tract, 263
ilium, 159
immune response, 566
immune system, 575
    local response, 575
    primary response, 575
    secondary response, 577
immunity, cell-mediated, 569, *576*, 577
    humoral, 569, 576
immunization, active, 577
    passive, 577
immunoglobulin
    IgA, 579
    IgD, 579
    IgE, 578, 579
    IgG, 579
    IgM, 579
implantation, 50, 557
inclusions, cytoplasmic, 17, 33, *34*
    lipid droplets, 34
    secretory granules, 33
incus, 370, *372*
infectious mononucleosis, 457
inferior nasal concha, 122, 126
inflammation, 567
infundibular stem, 382
infundibulum, uterine tube, 550
incisor, 490
ingestion, 486

inguinal ring, superficial, 545
inner cell mass, 50, *52*
inspiration, 476
inspiratory capacity, 476
insulin, 381, 394, *395*, 504
integument, 88
interneuron, *285*
internuncial neuron (see neuron, association)
interoceptors, 297
interphase, 41
interstitial cells (of Leydig), 397, *398*, 543, *544*
intestine, large, anatomy, 506
    development, 514
    function, 508
intestine, small, 499
    anatomy, 499
    development, 514
    function, 500
intrauterine device (IUD), 557, 558
intraperitoneal position, 499, 518, 549
iodine, 400
ion, 23
iris, 364, 365
iron, 572
irritability, cellular, *20*, 24
    nerve cell, 282
ischemia, 422, 537
ischium, 160
islets of Langerhans, 393, 506
isthmus, uterus, 551
itch, 91, 300

jejunum, 499
jogging, 191
joint, general
    ankle, 275
    diseases of, 182
    elbow, *252*
    hip, 266
    knee, 268, *269*
    stability
        bony configuration, 178
        ligament, 179
        muscles, 179
    synovial, *174*, 176, *177*, *181*
        classification, 180
    type
        ball and socket, 180, *181*
        condylar, 182
        ellipsoid, *180*, *181*, 182
        gliding, *181*
        hinge, 180, *181*
        pivotal, 180, *181*
        plane, 180
        saddle, *180*, *181*, 182
joint, specific
    acromioclavicular, 245
    carpometacarpal, 261

glenohumeral, 249
humeroulnar, 251
interphalangeal, 261
intertarsal, 278
metacarpophalangeal, 261
metatarsophalangeal, 278
midcarpal, 261
radiocarpal, 260
radioulnar, 260
    proximal, 251, *252*
sternoclavicular, 245
subtalar, 278
talocalcaneonavicular, 278
temperomandibular, 178, 215
tibiofibular
    distal, 175
    inferior, 276
    superior, 270
junction, intermediate, 35
    tight, 35
juxtaglomerular apparatus, 529

karyoplasm (see nucleoplasm)
karyotype, 47, 48
keratin, 66, 88, 90
keratohyaline, 88
kidney
    artificial, 537
    gross anatomy, *525*
    internal organization, 523
    location, 523
    regulation, 529
    vascular pattern, 524
kinetochore, *44*
knee, muscles
    adductor compartment, 267
    extensor compartment, 267
    posterior compartment, 268
knuckle, 155
kyphosis, 140

labium majus, *550*, 553, *554*
labium minus, *554*
labyrinth
    bony, 370
    membranous, 370
lacrimal bones, *116*, 122, 124, *126*
lacrimal sac, 364
lacteals, 499
lacuna, 74, 109
lamellae, interstitial, 104
    circumferential, 104
lamina, basal, 61
    propria, 487, 551
    reticular, 61
Langer's lines, 91

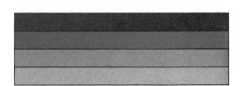

ILLUSTRATORS